The Complexity of Zadeh's Pivot Rule

Vom Fachbereich Mathematik der Technischen Universität Darmstadt
Zur Erlangung des Grades eines Doktors der Naturwissenschaften (Dr. rer. nat.)
Genehmigte Dissertation von Alexander Vincent Hopp aus München
Tag der Einreichung: 8. Juli 2020, Tag der Prüfung: 7. Oktober 2020

Referent: Yann Disser
Korreferent: Bernd Gärtner
Korreferent: Martin Skutella
Darmstadt 2020

Hopp, Alexander Vincent:
The Complexity of Zadeh's Pivot Rule
Tag der mündlichen Prüfung: 7. Oktober 2020
Logos Verlag Berlin, 2020
ISBN: 978-3-8325-5206-0
Zugl.: Darmstadt, Technische Universität Darmstadt, Dissertation

Bibliografische Information der Deutschen Nationalbibliothek

Die Deutsche Nationalbibliothek verzeichnet diese Publikation in der Deutschen Nationalbiografie; detaillierte bibliografische Daten sind im Internet über http://dnb.d-nb.de abrufbar

ISBN 978-3-8325-5206-0

Logos Verlag Berlin GmbH
Georg-Knorr-Str. 4, Gebäude 10
D-12681 Berlin
Tel.: +49 (0)30 42 85 10 90
Fax: +49 (0)30 42 85 10 92
https://www.logos-verlag.de

Abstract

The question whether linear programs can be solved in strongly polynomial time is a major open problem in the field of optimization. One promising candidate for an algorithm that potentially guarantees to solve any linear program in such time is the simplex algorithm of George Dantzig. This algorithm can be parameterized by a pivot rule, and providing a pivot rule guaranteeing a polynomial number of iterations in the worst case would resolve this open problem.

For all known classical natural pivot rules, superpolynomial lower bounds have been developed. Starting with the famous Klee-Minty cube, a series of exponential lower bound constructions have been developed for a majority of pivot rules. There were, however, two classes of pivot rules whose worst-case behavior remained unclear for a long time – randomized and memorizing rules.

Only in the 2010s, the works of Fearnley, Friedmann, Hansen and their colleagues provided superpolynomial bounds for those rules, starting a second series of lower bounds. The arguably most remarkable of these bounds was Friedmann's construction for which Zadeh's LeastEntered pivot rule requires at least a subexponential number of iterations.

This pivot rule is the main focus of this thesis. Following the work of Friedmann, we introduce parity games, Markov decision processes and linear programs and investigate certain subclasses of the first two structures. We discuss connections between these three frameworks, generalize previous definitions and provide a clean framework for working with so-called sink games and weakly unichain Markov decision processes.

We then revisit Friedmann's subexponential lower bound and discuss several of its technical aspects in full detail and exhibit several flaws in his analysis. The most severe is that the sequence of steps performed by Friedmann does not consistently obey Zadeh's pivot rule. We resolve this issue by providing a more sophisticated sequence of steps, which is in accordance with the pivot rule, without changing the macroscopic structure of Friedmann's construction.

The main contribution of this thesis is the newest member of the second wave of lower bound examples – the first exponential lower bound for Zadeh's pivot rule. This closes a long-standing open problem by ruling out this pivot rule as a candidate for a deterministic, subexponential pivot rule in several areas of linear optimization and game theory.

Zusammenfassung

Bis heute ist die Frage, ob lineare Programme in stark polynomieller Zeit gelöst werden können, eines der größten ungeklärten Probleme der mathematischen Optimierung. Ein Kandidat für einen Algorithmus, der solch eine Laufzeit garantiert ist der Simplex Algorithmus von George Dantzig. Dieser Algorithmus kann durch eine Pivotregel parametrisiert werden, und eine Pivotregel die eine polynomielle Anzahl von Iteration garantieren würde, wäre eine mögliche Lösung dieses offenen Problems.

Im Laufe der Zeit wurden für alle klassischen natürlichen Pivotregeln superpolynomielle untere Schranken entwickelt. Beginnend mit dem berühmten Klee-Minty Würfel gab es geradezu eine Welle von unteren Schranken, die für eine Vielzahl von Pivotregeln entwickelt wurden. Es gab jedoch zwei Klassen von Pivotregeln deren schlechtestmögliche Laufzeit für lange Zeit unklar blieb – randomisierte Regeln und Regeln, die sich frühere Entscheidungen merken und spätere Entscheidungen von diesen abhängig machen.

Erst mit Beginn der 2010er Jahre haben die Arbeiten von Fearnley, Friedmann, Hansen und deren Kollegen untere Schranken für diese Regeln geliefert und somit eine zweite Welle eingeleitet. Die vielleicht bemerkenswerteste Schranke war Friedmanns Konstruktion, für die Zadehs LeastEntered Pivotregel stets eine subexponentielle Anzahl von Iterationen benötigt.

Diese Pivotregel ist das Hauptthema dieser Arbeit. Friedmanns Ansatz folgend führen wir Paritätsspiele, Markoventscheidungsprobleme und lineare Programme ein und untersuchen zwei Unterklassen der beiden zuerst genannten Strukturen. Wir diskutieren verschiedene Zusammenhänge zwischen diesen drei Bereichen, schärfen einige Definitionen und Aussagen und entwickeln ein klares Gerüst für das Arbeiten mit sogenannten Senkenparitätsspielen und schwach einkettigen Markoventscheidungsproblemen.

Im Anschluss untersuchen wir Friedmanns subexponentielle untere Schranke, diskutieren einige technischen Aspekte dieser Konstruktion im Detail und zeigen drei Makel in Friedmanns Analyse auf. Der gravierendste ist, dass die durch Friedmann durchgeführte Folge an Operationen nicht durchgängig Zadehs Pivotregel befolgt. Wir beheben diesen Makel durch Angabe einer komplexeren Folge von Operationen die Zadehs Pivotregel befolgt und die makroskopische Struktur von Friedmanns Konstruktion nicht ändert.

Der Hauptbeitrag dieser Arbeit ist ein neues Mitglied der zweiten Welle von unteren Schranken – die erste exponentielle untere Schranke für Zadeh's Pivotregel. In Folge dieses Ergebnisses ist diese Regel nicht länger ein Kandidat für die erste deterministische, subexponentielle Pivotregel in verschiedenen Bereichen der linearen Optimierung oder Spieltheorie.

Acknowledgments

First of all, I want to thank Prof. Dr. Yann Disser for supervising me during my studies and giving me the opportunity to obtain a doctoral degree. We are working together for nearly 8 years now, and I benefited very much from his knowledge during these years. He convinced me to move to Darmstadt, and is thus partly responsible for one of the best decisions of my life.

Furthermore, I want to thank Prof. Bernd Gärtner from ETH Zürich and Prof. Dr. Martin Skutella from TU Berlin for acting as co-reviewers of my thesis. I also want to thank Prof. Dr. Marc Pfetsch and Prof. Dr. Volker Betz from TU Darmstadt for participating in my examination committee, and Prof. Große-Brauckmann for TU Darmstadt for his role as chairman during my examination.

Many thanks also go to Dr. Oliver Friedmann for sharing his ideas, thoughts and insights on both his subexponential construction and the exponential construction. I benefited extremely from his knowledge and intuition regarding the design of lower bound examples and am very thankful for being able to cooperate with him.

Further thanks go to all of my colleagues in the research group optimization and all members of the Graduate School of Computational Engineering at TU Darmstadt. Working with you was always a pleasure, and I very much enjoyed the atmosphere and togetherness of the group.

I am also very grateful to Alexander Birx, Mirco Kraenz, Daniel Nowak and Sylvain Spitz for proof-reading parts of my thesis and providing their thoughts and comments. I also want to thank Nils Mosis for his comments of my thesis, and especially for his insights on the linear programming formulation of Section 3.3.

Special thanks go out to all my friends that supported me during the past years, especially to my band colleagues Christian Altenhofen, Ralf Gutbell and Michel Krämer. Without their constant support, friendship and our regular band rehearsals, I would not have managed to finish my thesis.

Last, but definitely not least, I want to thank Saskia Hollweg for her support, her love and for enduring all my moods and quirks. She endured and accepted it when I was physically present but only had mathematical problems on my mind and was always able to motivate me and cheer me up when I was frustrated with my work.

This work is supported by the 'Excellence Initiative' of the German Federal and State Government and the Graduate School of Computational Engineering at Technische Universität Darmstadt, and I gratefully acknowledge their support.

Contents

List of Figures

List of Tables

Notation and general assumptions

Throughout the thesis, we use the following notation and assumptions. Note that some of the notation introduced here uses terms that are defined in the thesis. This is done intentionally in order to have one summary for all of the notation that is introduced and used in this thesis.

Let $A \in \mathbb{R}^{m \times n}, x \in \mathbb{R}^n$ and $i \in \{1, \ldots, m\}, j \in \{1, \ldots, n\}$.

Let $G = (V, E)$ be a directed graph.

- For $n \in \mathbb{N}$, we define $[n] := \{1, \ldots, n\}$.
- The symmetrical difference of two sets A, B is denoted by $A \Delta B$.
- For a countable set S and an index $k \leq |S|$, we denote the k-th entry of S by $S[k]$.
- We denote the i-th row of A by $A_{i,\bullet}$ and the j-th column A by $A_{\bullet,j}$.
- For a set $I \subseteq [n]$ of column indices with $|I| = k$, we denote the matrix induced by the corresponding columns of A by A_I. That is, $A_I := (A_{\bullet,I[1]} \cdots A_{\bullet,I[k]}) \in \mathbb{R}^{m \times k}$.
- We analogously define $x_I := (x_{I[1]}, \ldots, x_{I[k]}) \in \mathbb{R}^k$ for $x \in \mathbb{R}^n$.
- We define $x \geq 0$ if $x_i \geq 0$ for all $i \in [n]$ and define $x > 0, x \leq 0$ and $x < 0$ analogously. For two vectors $x, y \in \mathbb{R}^n$, we write $x \geq y$ if $x - y \geq 0$ and define other relations analogously.
- The support of x is defined as $\operatorname{supp}(x) := \{i : x_i \neq 0\}$.
- The indicator function is denoted by $\mathbf{1}_{x=y}$, so $\mathbf{1}_{x=y} = 1$ if $x = y$ and $\mathbf{1}_{x=y} = 0$ if $x \neq y$.
- The i-th unit vector is denoted by e_i (where the dimension should be clear from the context).
- The n-dimensional vector that has a 1 in every entry is denoted by $\mathbb{1}_n$.
- For a vertex $v \in V$, we denote the set of vertices such that v has an outgoing edge to them by $\Gamma^+(v) := \{u \in V : (v, u) \in E\}$. The set $\Gamma^-(v) := \{u \in V : (u, v) \in E\}$ is defined analogously.
- The set of all n-digit binary numbers is denoted by $\mathfrak{B}_n$, i.e., $\mathfrak{B}_n := \{0, \ldots, 2^n - 1\}$.
- For $\mathfrak{b} = (\mathfrak{b}_n, \ldots, \mathfrak{b}_1) \in \mathfrak{B}_n \setminus \{0\}$, we denote the least significant set bit of $\mathfrak{b}$ by $\ell(\mathfrak{b})$, so $\ell(\mathfrak{b}) := \min\{i \in [n] : \mathfrak{b}_i \neq 0\}$. For $i \in [n]$, we define $\ell_i(\mathfrak{b}) := \min\{i' \geq i : \mathfrak{b}_{i'} \neq 0\}$ analogously.
- If the number $\mathfrak{b}$ is fixed or clear from the context, we typically write ℓ instead of $\ell(\mathfrak{b})$ and ν as abbreviation for $\ell(\mathfrak{b} + 1)$.
- For $\mathfrak{b} \in \mathfrak{B}_n$ and $i \in [n]$, we define $\sum(\mathfrak{b}, i) := \sum_{l<i} \mathfrak{b}_l \cdot 2^{l-1}$.
- We use the symbol $*$ as a general wildcard. More precisely, when using the symbol $*$, this means that any suitable index, vertex, object and so on can be inserted such that the corresponding statement, definition and so on is valid.

- The transition from a strategy or policy $\sigma_{\mathfrak{b}}$ to the strategy $\sigma_{\mathfrak{b}+1}$ is abbreviated by $\sigma_{\mathfrak{b}} \to \sigma_{\mathfrak{b}+1}$.
- When not stated otherwise, we assume sets to be ordered.
- For an n-digit binary number, we interpret bit $n+1$ as being equal to 0.
- If we consider a parity game, the term "strategy" always refers to a player 0 strategy.
- When considering boolean expressions, the precedence level of "=" and "≠" is higher than the precedence level of $\wedge$ and $\vee$. That is, an expression $x \wedge y = z$ is interpreted as $x \wedge (y = z)$.
- For a strategy σ and an edge (v, w), we say that v *points to* or *moves to* w if $\sigma(v) = w$.

Henceforth, let σ be a strategy for the exponential construction introduced in Chapter 5.

- We define the following function $\bar{\sigma}$ (see Table 5.4). We interpret the values of this functions as boolean values.

Symbol	Encoded expression
$\bar{\sigma}(b_i)$	$\sigma(b_i) = g_i$
$\bar{\sigma}(s_{i,j})$	$\sigma(s_{i,j}) = h_{i,j}$
$\bar{\sigma}(g_i)$	$\sigma(g_i) = F_{i,1}$
$\bar{\sigma}(d_{i,j,k})$	$\sigma(d_{i,j,k}) = F_{i,j}$
$\bar{\sigma}(e_{i,j,k})$	$\sigma(e_{i,j,k}) = b_2$

Symbol	Encoded expression
$\bar{\sigma}(s_i)$	$\bar{\sigma}(s_{i\bar{\sigma}(g_i)})$
$\bar{\sigma}(d_{i,j})$	$\bar{\sigma}(d_{i,j,0}) \wedge \bar{\sigma}(d_{i,j,1})$
$\bar{\sigma}(d_i)$	$\bar{\sigma}(d_{i,\bar{\sigma}(g_i)})$
$\bar{\sigma}(eg_{i,j})$	$\bigvee_{k\in\{0,1\}}[\neg\bar{\sigma}(d_{i,j,k}) \wedge \neg\bar{\sigma}(e_{i,j,k})]$
$\bar{\sigma}(eb_{i,j})$	$\bigvee_{k\in\{0,1\}}[\neg\bar{\sigma}(d_{i,j,k}) \wedge \bar{\sigma}(e_{i,j,k})]$
$\bar{\sigma}(eg_i)$	$\bar{\sigma}(eg_{i,\bar{\sigma}(g_i)})$
$\bar{\sigma}(eb_i)$	$\bar{\sigma}(eb_{i,\bar{\sigma}(g_i)})$

- The set of incorrect levels for σ is $\mathcal{I}^\sigma := \{i \in [n] : \bar{\sigma}(b_i) \wedge \bar{\sigma}(g_i) \neq \bar{\sigma}(b_{i+1})\}$.
- We define the next relevant bit of σ as

$$\mu^\sigma := \begin{cases} \min\{i > \max\{i' \in \mathcal{I}^\sigma\} : \bar{\sigma}(b_i) \wedge \bar{\sigma}(g_i) = \bar{\sigma}(b_{i+1})\} \cup \{n\}, & \text{if } \mathcal{I}^\sigma \neq \emptyset, \\ \min\{i \in [n+1] : \sigma(b_i) = b_{i+1}\}, & \text{if } \mathcal{I}^\sigma = \emptyset. \end{cases}$$

- For $x \in \{b, s, g\}$, we define $m_x^\sigma := \min(\{i \in [n] : \bar{\sigma}(x_i)\} \cup \{n+1\})$ as well as $\overline{m}_x^\sigma := \min(\{i \in [n] : \neg\bar{\sigma}(x_i)\} \cup \{n+1\})$.
- We let $\mathfrak{D}^\sigma := \{(d_{i,j,k}, F_{i,j}) : \sigma(d_{i,j,k}) \neq F_{i,j}\}$.
- Let $\mathfrak{b} \in \mathfrak{B}_n$ and $\nu := \ell(\mathfrak{b}+1)$. We typically define $m := \max\{i \in [n] : \sigma(b_i) = g_i\}$ and $\mathfrak{E}^\sigma := \{(d_{i,j,k}, F_{i,j}), (e_{i,j,k}, b_2) : \sigma(e_{i,j,k}) = g_1\}$ if $\nu > 1$. Analogously, we let $\mathfrak{E}^\sigma := \{(d_{i,j,k}, F_{i,j}), (e_{i,j,k}, g_1) : \sigma(e_{i,j,k}) = b_2\}$ if $\nu = 1$.
- For $\mathfrak{b} \in \mathfrak{B}_n$ we define $\mathfrak{m} := \lfloor(\mathfrak{b}+1)/2\rfloor$ as this quantity describes the maximum occurrence records that edges have with respect to a canonical strategy $\sigma_{\mathfrak{b}}$.
- For $\mathfrak{b} \in \mathfrak{B}_n, i \in [n]$ and $j \in \{0, 1\}$, we let

$$\ell^{\mathfrak{b}}(i,j,k) := \left\lceil \frac{\operatorname{lfn}(\mathfrak{b}, i, \{(i+1, j)\} + 1 - k)}{2} \right\rceil + \mathfrak{b} - \mathbf{1}_{j=0}\operatorname{lfn}(\mathfrak{b}, i+1) - \mathbf{1}_{j=1}\operatorname{lufn}(\mathfrak{b}, i+1).$$

- Additional notation regarding binary counting can be found at the beginning of Chapter 4.

1. Introduction

In this thesis, we prove that a range of algorithms applicable for several problems of discrete and combinatorial optimization can require an exponential number of iterations when using Zadeh's pivot rule. We consider *parity games, Markov decision processes* and *linear programs*, which are important fields of mathematical optimization and game theory. For each of these fields, we discuss one of the most important algorithms, namely the *strategy improvement algorithm* for parity games, the *policy iteration algorithm* for Markov decision processes and the *simplex algorithm* for linear programs. These algorithms are closely connected to each other and can all be parameterized by specifying a *pivot rule*.

We investigate the worst-case running times of these algorithms when they are parameterized with Zadeh's pivot rule and prove that they require exponentially many iterations in the worst case.

1.1. Parity Games and Markov Decision Processes

Parity games and Markov decision processes can be interpreted as infinite duration perfect information games played on a directed graph. Parity games are played by two deterministic players, whereas Markov decision processes are played by one deterministic and one randomized player.

In a parity game, each vertex of the underlying graph is owned by one of the two players, called Even and Odd, and each vertex is assigned an integer priority. At the beginning of a play, a pebble is placed on one of the vertices. Then, the player that owns the current vertex chooses an adjacent vertex and moves the pebble along the corresponding edge. This process is then iterated, and the two players construct an infinite walk. The play is won by Even if the largest priority occurring infinitely often is even, and won by Odd if that priority is odd. *Solving* a parity game corresponds to finding *winning strategies* for the two players. An example of a parity game is given in Figure 1.1.

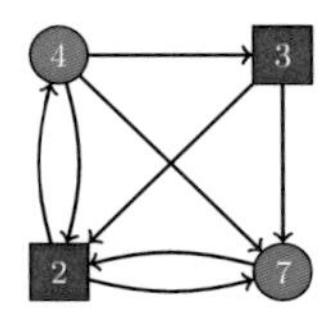

Figure 1.1.: Example of a parity game. Blue circular vertices are owned by Even, red rectangular vertices by Odd. Vertex labels show the priority of the vertex.

Parity games arise in many fields of mathematics. They are closely related to other and more general classes of games [Pur95, Sti95, Jur98], the problem of μ-calculus model checking [EJ91, EJS93, GTW02] and are also central for several problems regarding computer-aided verification [AVW03, FLL10]. Parity games are also very interesting from a complexity theoretical point of view. The natural decision problems corresponding to parity games belong to NP∩coNP[EJS93] and even UP∩coUP [Jur98], while computing winning strategies is known to be in CLS [DP11]. Quite recently, a breakthrough result of Calude et al. showed that parity games can be solved in quasi-polynomial time [CJK+17]. The results and techniques were then extended to prove that quasi-linear space is sufficient [FJdK+19] and were applied to improve the running time of classical algorithms [Par19]. However, it is a major open question whether they can be solved in polynomial time.

In Markov decision processes, each vertex belongs to either the deterministic or the randomization player. As in a parity game, a play in a Markov decision process begins by placing a pebble in the underlying graph. If the pebble is placed on a vertex of the deterministic player, then the player chooses an edge, moves the pebble along this edge and collects a *reward* that depends on the chosen edge. If the pebble is placed on a vertex of the randomized player, then one of the outgoing edges is chosen at random according to a given probability distribution and the pebble is moved along the chosen edge. A play is infinite, and the objective is to maximize a given function of the expected reward collected by the deterministic player, for example the average reward obtained per movement of the pebble. An example of a Markov decision process is given in Figure 1.2.

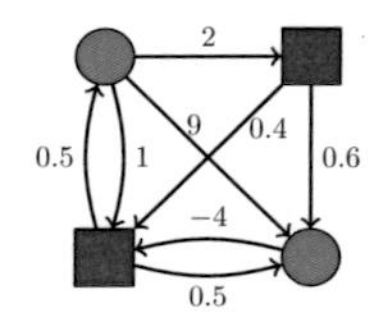

Figure 1.2.: Example of a Markov decision process. Blue circular vertices are owned by the deterministic player, red rectangular vertices by the randomization player. Labels on edges denote the reward collected by traversing it resp. the probability of choosing the edge.

Markov decision processes were introduced and studied independently by several authors [Sha53, Bel57, How60]. They are typically used to model long-term decision making under uncertainty. One famous example of a problem that is typically modeled in this fashion is to manage the inventory of a store that sells a single good and orders its stock on a monthly basis while not knowing exactly how much the customers will buy in the upcoming month [Put05]. They are used in a variety of applications like reinforcement learning [SB18], finance, communication networks and several more [FS02, BvD17]. Markov decision processes can be formulated as linear programs [Man60, d'E63, Put05] and can thus be solved in weakly polynomial time [Kha80, Kar84]. Although there are variants that can be solved in strongly polynomial time [Ye11, PY15], it is unknown whether this is true for general Markov decision processes.

1.2. Strategy Improvement

Strategies, also called *policies* in the Markov decision process community, are rules used by the deterministic player(s) to describe how they move the pebble if it is placed on one of their vertices. The goal of the deterministic player(s) in parity games or Markov decision processes is to find *optimal* strategies. Intuitively, a strategy is optimal for a player in a parity game if using this strategy maximizes the number of plays they win regardless of the choices of the other player. In a Markov decision process, a strategy is optimal if using this strategy maximizes the predefined function of the expected collected reward.

In theory, strategies might depend on the history of a play, like the previous movement of the pebble, or the decisions of the other player. It is a major result in the theory of parity games and Markov decision processes that such strategies do not need to be considered for finding optimal strategies. More precisely, it is sufficient to consider *memoryless, deterministic* strategies, so strategies that do not depend on the history of the play and always choose an outgoing edge deterministically [How60, Zie98].

One of the key algorithmic frameworks to find optimal strategies is based on the following idea. If every strategy is assigned a *valuation*, this defines a pre-order on the set of all strategies. Now, if this valuation is defined in such a way that a strategy is optimal if and only if it maximizes the valuation among all strategies, then the problem of finding an optimal strategy can be solved by improving strategies with respect to the valuation until this is no longer possible. If the valuations are defined in such a way that it is easy to calculate them and to improve a non-optimal strategy, then this framework yields a viable algorithm for finding optimal strategies. This framework is called *strategy improvement* or *policy iteration* and is a standard technique for both parity games and Markov decision processes [How60, VJ00], although it can also be applied for more general classes of games [HK66, Con92]. As there is only a finite number of strategies, strategy improvement always terminates and guarantees to find an optimal strategy in finite time. The exact number of iterations of strategy improvement highly depends on the implementation. In particular, the chosen *improvement rule*, that is, the procedure deciding how to change the current strategy, highly influences the behavior of the algorithm. For both parity games and Markov decision processes, superpolynomial lower bounds were established for the most important and natural improvement rules [Fri09, Fea10a, Fri11a, Fri11c, FHZ11b, FHZ11a, AF17, DH19]. It is an open question whether there is an efficiently computable improvement rule guaranteeing a polynomial number of iterations in the worst case.

Of course, there are many more algorithms that can be used to calculate optimal strategies. For parity games, there are, for example, the recursive algorithm of Zielonka [Zie98], the small progress measure algorithm [Jur00] and the subexponential deterministic algorithm of Jurdziński, Paterson and Zwick [JPZ08] and its big-step variant [Sch17]. Rather recently, Calude et al. provided the first quasipolynomial algorithm [CJK$^+$17] which is considered a major breakthrough and allowed to improve several other algorithms [FJdK$^+$19, Par19]. Besides using techniques of linear programming, some of the most notable and important algorithms used for Markov decision processes are value iteration and modified policy iteration, and we refer to [Put05] for a discussion of these algorithms.

Compared to most of these algorithms, the beauty of strategy improvement lies in its simplicity and that it can be applied to a variety of problems. In particular, there is a strong and natural connection between the strategy improvement algorithm and the famous simplex algorithm used for solving linear programs which we analyze in detail in Chapter 3.

1.3. Linear Programming and the Simplex Algorithm

The field of linear programming was developed in the 1940's during World War II, and its original purpose was to assist with logistics and the planning of military operations. After the war, linear programming was developed further and further, leading to what is known today as operations research.

The goal of linear programming is the maximization or minimization of a linear objective function under linear constraints. One of the first and most important contributions to this area of optimization is the *simplex algorithm* of George Dantzig [Dan51, Dan63]. Given a feasible system of linear inequalities and equations and a linear objective function, this algorithm operates as follows. It calculates a vertex of the polyhedron defined by the given system of inequalities and equations and checks whether this vertex is optimal with respect to the objective function. This is done by checking whether the vertex is locally optimal, which is sufficient as polyhedra are convex sets. If this is the case, then the algorithm has found an optimal solution and terminates. Otherwise, it calculates an improving direction which corresponds to an edge of the polyhedron. If this edge does not end in another vertex but is infinite, then the value of the objective function is unbounded and the algorithm terminates. If this is not the case, then the algorithm proceeds along the edge until it reaches the next vertex and iterates. The algorithm thus traverses the vertices and edges of the polyhedron until it either finds an optimal vertex or confirms that the value of the objective function is unbounded. A visualization of this algorithm is given in Figure 1.3.

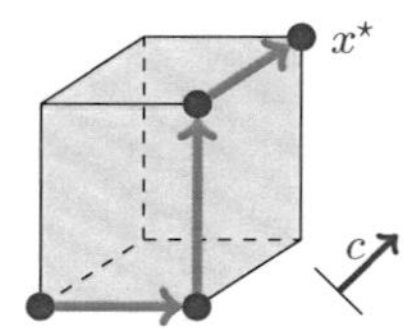

Figure 1.3.: An example of a possible execution of the simplex algorithm on a three-dimensional cube. The vertex x^* visualizes the optimal solution with respect to the objective function c. Blue vertices are the vertices visited by the algorithm and the green edges mark the path the algorithm takes.

Until today, the simplex algorithm is one of the most important algorithms in both theory and practice. One of its key features is that it is highly flexible, as the algorithm does not dictate exactly which improving direction to choose in each step. It can thus be parameterized by a *pivot rule* that determines which improving direction the algorithm

Name	Lower Bound	Proven via
Dantzig's rule [Dan51]	**exponential** [KM72]	Klee-Minty cube
Shadow vertex rule [GS55, Bor87]	**exponential** [Mur80]	Klee-Minty cube
Lexicographic rule [DOW55]	**exponential** [DS15]	Klee-Minty cube
LARGESTINCREASE rule[Jer73]	**exponential** [Jer73]	Klee-Minty cube
Bland's rule [Bla77]	**exponential** [AC78]	Klee-Minty cube
STEEPESTEDGE rule [FG92]	**exponential** [GS79]	Klee-Minty cube
RANDOMEDGE rule	subexponential [FHZ11b]	Markov decision process
Cunningham's rule [Cun79]	**exponential** [AF17]	Markov decision process
RAISINGTHEBAR [Kal91]	subexponential [FHZ11b]	Markov decision process
Zadeh's LEASTENTERED rule [Zad80]	**exponential** [DFH19]	Markov decision process
RANDOMFACET [Kal92, SW92, Kal97]	**subexponential** [FHZ11b]	Markov decision process
Randomized Bland [Mat94]	subexponential [Han12]	Markov decision process

Table 1.1.: An overview over important pivot rules that were developed for the simplex algorithm. Bounds of randomized pivot rules hold in expectation. **Bold** lower bounds are tight in the sense that there is an asymptotically matching upper bound, provided that cycling of the algorithm is prevented. The given sources do not necessarily refer to the first mention of a pivot rule or a lower bound.

should take. In the past 70 years, many pivot rules were invented and investigated, and an overview over some of the most important pivot rules is given in Table 1.1. For a long time, there was the hope that the simplex algorithm using Dantzig's original pivot rule might be a polynomial algorithm for solving linear programs. However, Klee and Minty showed in 1972 that the simplex algorithm requires an exponential number of iterations [KM72] in the worst case. Following their line of work, it was proven that several of the most important and most natural pivot rules require a superpolynomial number of iterations in the worst case. An overview over some results for classical and natural pivot rules is given in Table 1.1. Of course, there are many more pivot rules, and there are variants of the simplex algorithm that implement similar ideas but might for example consider points outside of the polyhedron instead of vertices. We refer to [TZ93, Han12, APR14] for further details.

Most of these first worst-case examples were adjusted versions of the *Klee-Minty cube* first used in [KM72]. In fact, several of these constructions were proven to be special cases of a general class of polyhedra, called *deformed products* [AZ98]. There were however still pivot rules whose worst-case running times were not proven to be superpolynomial. These pivot rules were randomized pivot rules in which the choice of the next improving direction is not deterministic, and memorizing pivot rules in which this choice depends on previously chosen directions. While it is known since the 1980s that linear programs can in general be solved in weakly polynomial time via interior point or ellipsoid methods [Kha80, Kar84], the search for a polynomial time pivot rule continued as such a pivot rule would yield the first strongly polynomial algorithm for linear programming.

Moreover, such a pivot rule would have immediate consequences for the famous Hirsch conjecture. This conjecture was stated by Warren M. Hirsch in 1957 and was first published by George Dantzig in 1963 [Dan63]. It states that the (combinatorial) diameter of a polytope in dimension d with n facets is at most $n - d$. This conjecture was open for over 50 years until it was proven to be incorrect by Francisco Santos [San12]. Nevertheless, a weaker variant, the so-called polynomial Hirsch conjecture, is still open. This conjecture claims that there is a polynomial function $p\colon \mathbb{R} \to \mathbb{R}$ such that the diameter of every polytope with n facets is bounded from above by $p(n)$. The connection between the simplex algorithm and this conjecture is very natural and strong: The diameter of a polytope is a lower bound for the number of pivot steps the simplex algorithm has to perform on the polytope. Moreover, any pivot rule guaranteeing a finite number of pivot steps directly implies an upper bound for the diameter of any polytope. In particular, the simplex algorithm is a potential tool for solving the polynomial Hirsch conjecture.

After the Klee-Minty cube was adjusted successfully for a variety of pivot rules, it took nearly 40 years until a new technique was introduced that allowed for new lower bound constructions. In 2011, starting with the work of Friedmann [Fri11b], a new class of worst-case instances based on the connection between linear programs and Markov decision process was established. This new class then allowed to prove that all of the remaining candidates for natural and potentially polynomial pivot rules are in fact superpolynomial. There was however one pivot rule whose *exact* complexity status remained unclear – Zadeh's pivot rule [Zad80].

1.4. Zadeh's Pivot Rule

Zadeh's pivot rule was invented in 1980 by Norman Zadeh [Zad80]. The motivation was the pathological behavior of most pivot rules when applied to deformed cubes. Zadeh observed that the examples based on the Klee-Minty cube all behaved as follows: There are directions that would lead the simplex algorithm quickly to the optimal vertex. By carefully designing the system, these directions *appear* to only slightly increase the objective function value, tricking the simplex algorithm into performing a lot of unnecessary steps. More precisely, the algorithm typically visits all vertices of a facet before moving to the next facet, although switching to the other facet was a valid choice. Then, after performing all of these unnecessary steps, the algorithm performs one good pivot step. This idea is then iterated, forcing the algorithm to perform an exponential number of steps in total. Geometrically, a good pivot step corresponds to moving to a new facet of the polytope, while the unnecessary steps correspond to staying in one facet of the polytope. An algorithm that behaves like this visits all vertices of the cube and thus requires an exponential number of operations. A visualization of this behavior is given in Figure 1.4.

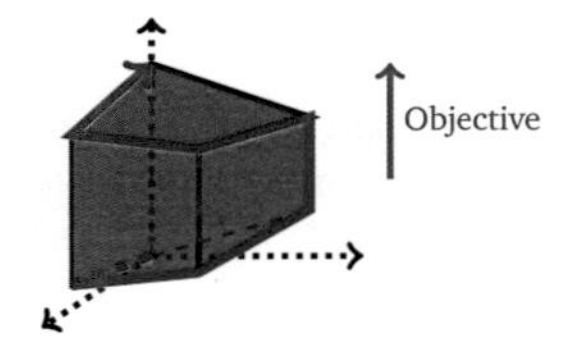

Figure 1.4.: Sketch of the worst-case behavior of the simplex algorithm on a three-dimensional Klee-Minty cube.

Zadeh's idea to prevent this somewhat unbalanced behavior was to enforce balance artificially. He thus proposed his famous LEASTENTERED pivot rule that is specifically designed to avoid this behavior: Whenever the algorithm performs a pivot step, it chooses an improving direction that was chosen least often before. This can be achieved by maintaining an *occurrence record* that counts for every direction how often it was chosen. Then, whenever the algorithm has to choose the next direction, a direction minimizing the occurrence record is chosen. This however might not be sufficient to fully specify the choices of the algorithm as there might be several eligible directions that can be chosen and minimize the occurrence record. Zadeh's pivot rule thus needs an additional tie-breaking rule that decides which direction to choose in such a case.

As Zadeh's pivot rule depends on previous iterations, it is a *memorizing* pivot rule. Next to Cunningham's pivot rule [Cun79], it was one of the first and most important memorizing pivot rules. Although a naive implementation of Zadeh's pivot rule might lead to cycling, it is very unlikely to do so [Avi09]. Using standard anti-cycling procedures like the lexicographic rule for choosing the leaving variable, this can be prevented.

This pivot rule defeated all previously known lower bound examples as it only required a polynomial number of iterations on those. For over 30 years, it was unclear whether this pivot rule might guarantee a polynomial worst-case running time and interest in either a proof confirming this conjecture or a counterexample was exceptionally high. In particular, the price of 1000$ promised by Zadeh for either of the two results promoted the interest even further and quickly became a part of the folklore of linear optimization. This offer was made by Norman Zadeh in a letter to Victor Klee, and the letter itself is one of the most famous notes of linear optimization. The letter first appeared in Günter Ziegler's paper [Zie04] and is included here with his kind permission.

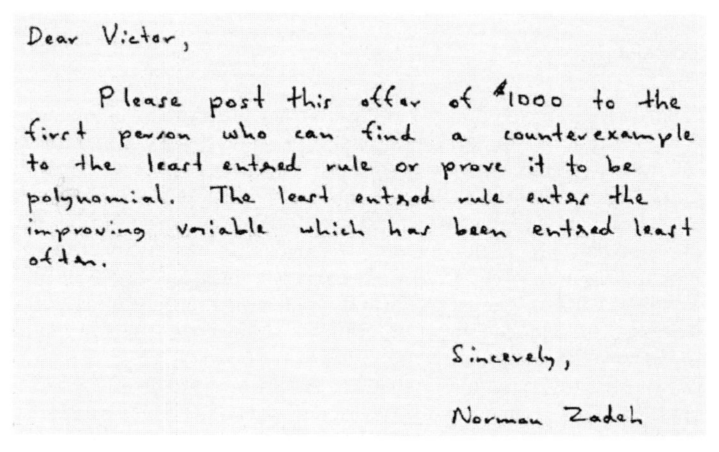

Dear Victor,

Please post this offer of $1000 to the first person who can find a counterexample to the least entered rule or prove it to be polynomial. The least entered rule enters the improving variable which has been entered least often.

Sincerely,

Norman Zadeh

Figure 1.5.: The famous letter promising 1000$ to the first person to prove or disprove that Zadeh's pivot rule is polynomial.

For over 30 years, it was unclear whether this rule might guarantee a polynomial number of iterations, and hopes were high that this pivot rule might provide a strongly polynomial algorithm for linear programming. In 2011, Oliver Friedmann was able to solve this problem by proving that Zadeh's pivot rule might require a subexponential number of iterations in the worst case [Fri11c]. Although his original proof contained some flaws that were corrected later (see Chapter 4 and [DH19]), his result ruled out Zadeh's LeastEntered pivot rule as one of the final remaining promising candidates for a polynomial time pivot rule. There was, however, still a gap as his construction did not yield an *exponential* lower bound. Since a polyhedron has at most exponentially many vertices [McM70, AZ98], any deterministic pivot rule that visits each vertex at most once requires at most an exponential number of iterations. It was thus unclear what the exact worst-case behavior of Zadeh's pivot rule was, and although Friedmann's result proved that it is worse than polynomial, Zadeh's pivot rule remained as the last natural deterministic pivot rule whose exact worst-case behavior was not established. This is of particular interest as there is a *randomized* pivot rule that guarantees a subexponential number of iterations – the RandomFacet pivot rule [Kal92, SW92, Kal97]. Intuitively, this pivot rule chooses a facet containing the vertex that is currently considered uniformly at random, finds the optimal vertex contained in that facet, moves to this vertex, and iterates. This raises the question whether the use of randomization might in general yield better pivot rules or if there exists a natural deterministic pivot rule that always terminates after a subexponential number of iterations. Even after Friedmann's result, this question was not answered. Although exponential lower bounds for Zadeh's pivot rule were developed in more abstract frameworks like Acyclic Unique Sink Orientations [Tho17], the constructions was applicable for linear programs. Thus, hopes were high that Zadeh's pivot rule might be the first natural pivot rule guaranteeing a subexponential number of iterations.

1.5. Our contribution

The main contribution of this thesis is the proof that the simplex algorithm for linear programs using Zadeh's LeastEntered pivot rule requires an exponential number of iterations in the worst case. Forty years after its invention, this settles the worst-case complexity for Zadeh's pivot rule. In particular, it remains unclear whether there is a deterministic pivot rule that can compete asymptotically with the RandomFacet rule. This result is not only proven for the simplex algorithm but for the general strategy improvement algorithm that can be applied to parity games, Markov decision processes and several other classes of games. We also formalize the relationship between Markov decision processes and induced linear programs, and make some terms and definitions originally introduced by Friedmann more precise.

Moreover, we discuss several flaws in the original proof of Friedmann's subexponential lower bound [Fri11c]. We show that the description given in [Fri11c] contradicts Zadeh's pivot rule, and prove that the proof needs to be significantly changed in order to retain the subexponential lower bound. We then provide these changes and prove that Friedmann's historical result is correct.

1.6. Outline

In Chapter 2, we introduce the mathematical background. More precisely, we introduce and discuss parity games, Markov decision processes and linear programs and establish the notation used in this thesis.

In Chapter 3, we discuss the strategy improvement algorithm for parity games and Markov decision processes. We generalize the definitions of the terms "sink game" and "weakly unichain Markov decision process", and introduce the strategy improvement algorithm for these frameworks. We furthermore prove under which conditions a lower bound obtained for the strategy improvement algorithm applied to a Markov decision process implies the same bound for the simplex algorithm when applied to the induced linear program in Theorem 3.3.4 and Corollary 3.3.5.

In Chapter 4, we discuss Friedmann's subexponential lower bound construction. We discuss the key ideas and point out one major flaw in Issue 4.3.12 and several minor flaws in Issues 4.3.1, 4.3.3 and 4.3.4. These flaws are then corrected in Section 4.4, allowing us to retain Friedmann's historic result (Theorem 4.4.15). The results of this chapter were previously published at IPCO 2019 [DH19] and an extended version is available online [DH18].

Using the key ideas introduced in this chapter, we introduce the exponential lower bound construction in Chapter 5. We discuss the main ideas and explain how the strategy improvement algorithm behaves when applied to this construction. We give a first informal idea of the proof of the correctness of our main statements in this chapter, since the formal proof is quite complicated and involved. The formal proofs are then given in Chapter 6, proving that Zadeh's pivot rule requires an exponential number of iterations in the worst case (Theorem 5.3.20). The results of these chapters are available online in a preliminary version [DFH19].

Finally, we conclude our findings in Chapter 7.

2. Preliminaries

In this chapter, we introduce the main mathematical frameworks and objects of this thesis. We begin by discussing *parity games*, which are two-player games played on a directed graph, in Section 2.1. In Section 2.2, we introduce *Markov decision processes*. These provide a model for making decisions under uncertainty such that a certain objective function is maximized. We introduce the computational tasks that are associated with both frameworks. As especially Markov decision processes and linear programs have a close connection, we give an introduction into linear programming in Section 2.3. We furthermore discuss the simplex algorithm for solving linear programs and especially Zadeh's LeastEntered pivot rule.

2.1. Parity Games

Parity games are a class of games that are played by two players, called player 0 (or Even) and player 1 (or Odd), on a directed graph. Every vertex of the graph has a natural number assigned to it, called *priority*. Priorities are unique, so no two vertices have the same priority. In addition, every vertex either belongs to player 0 or to player 1. A *play* in a parity game begins by choosing a starting vertex and placing a pebble on this vertex. If the starting vertex belongs to player $p \in \{0,1\}$, then player p chooses an edge adjacent to the current vertex and moves the pebble to the endpoint of this edge. The pebble is thus placed on another vertex, and the player who owns this vertex then chooses an adjacent edge again. This procedure is now iterated ad infinitum, and we identify the play with the sequence of vertices visited by the pebble. Since the play is infinite, some of the vertices are visited infinitely often. Among all vertices that are visited infinitely often, consider the vertex with the highest priority. If its priority is even, then player 0 wins the play. If its priority is odd, then player 1 wins the play. Thus, the *parity* of the highest priority seen infinitely often during a play determines the winner of the play, giving parity games their name.

This intuitive description is formalized by the following definition.

Definition 2.1.1 (Parity game). A *parity game* is a tuple $G = (V_0, V_1, E, \Omega)$. We set $V := V_0 \cup V_1$ and require (V, E) to be a directed graph with $|\Gamma^+(v)| \geq 1$ for every $v \in V$. The function $\Omega\colon V \to \mathbb{N}$ is the *priority function*. For $p \in \{0,1\}$, the set V_p is the set of vertices of player p and the set $E_p := \{(v,w) \in E : v \in V_p\}$ is the set of edges of player p. We assume all sets to be finite.

Note that we do not require the priority function Ω to be injective. In fact, this condition is typically relaxed when considering parity games as there is a weaker condition that can be imposed on the priority function that intuitively ensures that it behaves as if it was injective when injectivity is actually needed. This issue is discussed in more detail in Chapter 3.

We now formalize the term play and what winning a parity game means formally. Our notation and presentation is based on the description given in [Fea10b]. For the remainder of this section, let $G = (V_0, V_1, E, \Omega)$ denote a parity game.

Definition 2.1.2 (Play). Let $v_0 \in V$. A *play starting at* v_0 is an infinite sequence $\pi = v_0, v_1, \dots$ of vertices such that $(v_i, v_{i+1}) \in E$ for all $i \geq 0$.

Let π be a play starting in some vertex v. As π is infinite, it has to contain a cycle since G only contains a finite number of vertices. This implies that every vertex contained in π is either contained exactly once or infinitely often. The play π can thus be partitioned uniquely into a *path component* and a *cycle component*, and these components are used to define the winner of the play.

Definition 2.1.3 (Components, winning a play). Let π be a play. The set of all vertices occurring exactly once resp. infinitely often in π is called the *path component* resp. *cycle component* of π and is denoted by $P(\pi)$ resp. $C(\pi)$. We also write $\pi = P(\pi), C(\pi)^\infty$ as a representation of π. Player p *wins the play* π if $\max\{\Omega(w) : w \in C(\pi)\} \bmod 2 = p$.

Consider a partial play $\pi = v_0, \dots, v_k$ and let $p \in \{0, 1\}$ such that $v_k \in V_p$. Then, player p has to choose the next vertex v_{k+1} such that $(v_k, v_{k+1}) \in E$. This decision could possibly depend on the previously encountered vertices. For example, player p might choose differently if vertex v_k was already encountered previously. It is not immediately clear if it is beneficial for the player to base their decision on the history of the play or to even randomize their choices. A central result in the theory of parity games states that neither of the aforementioned are necessary and that it suffices to consider *deterministic, memoryless* strategies for the players (see e.g. [EJ91, Zie98]). Such strategies also induce a play in the parity game in a natural way.

Definition 2.1.4 (Strategy (PG), induced play). Let $p \in \{0, 1\}$. A function $\sigma_p : V_p \to V$ with $(v, \sigma_p(v)) \in E$ for all $v \in V_p$ is a (deterministic, memoryless) *strategy for player* p.

Let σ_0, σ_1 be strategies for player $0, 1$, respectively, and $v_0 \in V$. The *play induced by* σ_0 *and* σ_1 *starting at* v_0 is the play $\pi_{v_0,\sigma_0,\sigma_1} := v_0, v_1, \dots$ where $v_i \in V_p$ implies $v_{i+1} = \sigma_p(v_i)$ for all $i \geq 0$.

Example 1 (A small parity game). *Consider the parity game given in Figure 2.1. It contains 3 player* 0 *vertices, marked in blue, and two player* 1 *vertices, marked in red. Edges are colored accordingly. To further distinguish the two types of vertices, player* 0 *vertices are circular while player* 1 *vertices are rectangular. The label of a vertex shows its priority and is also used when referring to the vertex. The example also visualizes two strategies* σ_0, σ_1 *by marking the edges of the strategies in bold. The strategies are defined via* $\sigma_0(6) = 14, \sigma_0(14) = 7$ *and* $\sigma_0(11) = 11$ *respectively* $\sigma_1(7) = 6$ *and* $\sigma_1(14) = 11$. *The play* $\pi_{19,\sigma_0,\sigma_1}$ *thus has the path component* 19 *and the cycle component* 11, *so* $\pi_{19,\sigma_0,\sigma_1} = 19, (11)^\infty$.

Strategies can also be used to extend the notion of winning a play to winning a vertex. We only define the corresponding terms for player 0, the definitions in terms of player 1 are completely analogous.

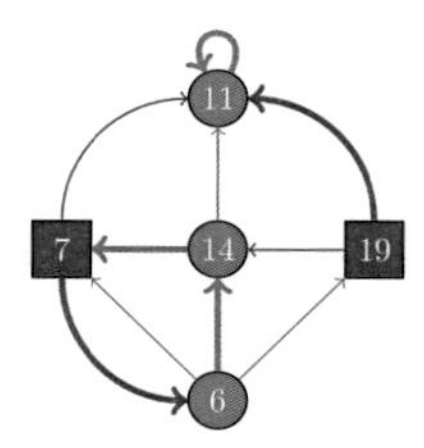

Figure 2.1.: A small parity game with 5 vertices. The blue circular vertices belong to player 0, the red rectangular vertices belong to player 1. Vertex labels denote priorities.

Definition 2.1.5 (Winning a vertex). Player 0 *wins vertex* $v \in V$ if there is a strategy σ for player 0 such that player 0 wins the play π_{v,σ,σ_1} for every strategy σ_1 for player 1. Player 0 *wins the set* $W \subseteq V$ if player 0 wins every vertex $v \in W$.

The following theorem shows that restricting the definition of strategies to deterministic and memoryless strategies is not a real restriction. It states that every vertex is either won by player 0 or by player 1. In particular, this means that parity games are *determined.* Moreover, it shows that winning strategies for the players do not depend on the starting vertices. This means that if player p wins a set $W \subseteq V$ of vertices, then they can win all vertices of this set by using the same strategy.

Theorem 2.1.6 (See e.g. [EJ91]). *There is a partition* $W_0 \cup W_1$ *of* V *such that player* p *has a single strategy* σ_p *in the sense of Definition 2.1.4 winning for* W_p*. The set* W_p *is the* winning set *of player* p*.*

The problem of *solving* a parity game is to find the winning sets W_0, W_1 alongside the corresponding strategies σ_0, σ_1. There are several algorithms for solving this problem, and the complexity status of finding and calculating is also very interesting.

Solving a parity game can be phrased as a decision problem by asking the question which of the two players has a winning strategy for a given starting vertex. This problem is one of the few problems contained in NP∩coNP and even UP∩coUP [Jur98] for which no polynomial time algorithm has been found yet. Most people, however, believe that solving parity games is possible in polynomial time, and this belief was strengthened by the breakthrough result of Calude et al. who provided a quasi-polynomial algorithm for solving parity games [CJK+17]. This algorithm was then improved and investigated further by other researchers, and the techniques of Calude et al. also allowed to improve previously developed algorithms [FJdK+19, Par19]. It was however also shown that the techniques used in the design of this algorithm do not allow for algorithms with strictly better running time [CDF+19], so new approaches are necessary. For a more general discussion on algorithms for solving parity games, we refer to [Fea10a, Fri11b].

The problem of actually *calculating* winning strategies is also interesting from a complexity theoretic point of view. It is known to be in the class PLS containing problems for which it is possible to verify local optimality of a solution in polynomial time. Moreover, it is also contained in the complexity class PPAD that was introduced in [Pap94]. Informally, this class is defined as the class of problems for which existence of solutions can be proven by Polynomial Parity Arguments on Directed graphs (hence the abbreviation PPAD). This complexity class was originally defined by specifying one of its complete problems, the so-called *end-of-the-line* problem. The class gained significant attention in the field of algorithmic game theory when it was proven that the problem of computing a Nash equilibrium is complete for this class [DGP09]. It was then even proven that the problem is in the class CLS which is a specific subclass of PPAD∩PLS, capturing problems of local optimization in which the domain and the functions involved are continuous [DP11].

In this thesis, we focus on the *discrete strategy improvement algorithm* developed in [VJ00]. This algorithm can be interpreted as an specification of a general algorithmic scheme applied to parity games. It is in particular deeply connected to the *policy iteration algorithm* for solving Markov decision processes and the *simplex algorithm* for solving linear programs. We thus postpone the discussion of this algorithm and the general algorithmic scheme to Section 3.3 and introduce the next central mathematical framework of this thesis.

2.2. Markov Decision Processes

Markov decision processes provide a mathematical framework for making decisions under uncertainty to maximize some accumulated reward. Typically, a Markov decision process consists of a set of states a system can be in. In each state, a rational decision maker, the so-called *player*, has a set of actions available from which they can choose. Depending on the chosen action, the player receives a reward (or has to pay costs), and the system transitions into another state. Typically, the player tries to maximize a certain function of the reward that is accumulated over time when starting in a certain state. This function might, for example, be the total reward or the average reward per action. If choosing an action, the new state of the system might be determined completely by the current state and the chosen action, but it might also be drawn out of an probability distribution which depends on the chosen action. The tuple consisting of all states, actions, rewards and transition probabilities (of which some might be deterministic) then constitutes a *Markov decision process*.

Markov decision processes were first introduced in the late 1950s and early 60s and it is hard to determine who was the first to investigate them formally. Among the first and most influential works on Markov decision processes are [Sha53, Bel57, How60]. They are typically used to model decision making under uncertainty, and there is a rich theory on different types of Markov decision processes. We refer here to [Put05] for a modern and in-depth discussion of Markov decision processes, and for proofs of all statements given here.

There is, however, another interpretation of Markov decision processes that does not use the notions of states and actions, and this interpretation is used in this thesis. It is important to mention that both formulations are interchangeable, and that they are both used in the literature. We refer to [Han12] for a formalization of how to transfer one formulation into the other.

The idea of the alternative formulation used here is to model a Markov decision process as a bipartite graph. To implement this idea, there is one vertex per state, and the set of all these vertices is the set of *player vertices*. Similarly, there is one *randomization vertex* per action, and the underlying graph contains an edge from player vertex u to randomization vertex v if and only if v is an action that can be chosen at u. These edges are assigned the reward of choosing the corresponding actions. Each randomization vertex then has edges to all player vertices representing states the system might transition to when this action is chosen. These edges are assigned the corresponding probabilities. Consequently, a Markov decision process is a bipartite graph with two types of vertices and two types of edges. This is formalized in the following definition. Note however that we explicitly allow edges between player vertices as deterministic actions (i.e., actions that lead to one state with probability 1) can be modeled without using randomization vertices.

Definition 2.2.1 (Markov decision process). A *Markov decision process* (or *MDP*) is a tuple $G = (V_0, V_R, E_0, E_R, r, p)$. The set V_0 is the set of *player vertices*, V_R is the set of *randomization vertices*, and we set $V := V_0 \cup V_R$. Similarly, $E_0 \subseteq V_0 \times V$ is the set of *player edges*, $E_R \subseteq V_R \times V_0$ is the set of *randomization edges*, and we set $E := E_0 \cup E_R$. In particular, (V, E) forms a directed graph.

The function $r\colon E_0 \to \mathbb{R}$ is the *reward function*, and $p\colon E_R \to (0, 1]$ is the *probabilistic transition function* fulfilling $\sum_{v \in \Gamma^+(u)} p(u, v) = 1$ for all $u \in V_R$.

In all of the upcoming definitions and statements, we let $G = (V_0, V_R, E_0, E_R, r, p)$ be a Markov decision process.

As mentioned earlier, Markov decision processes model decision making under uncertainty. The player begins in a specific state and makes sequential decisions that maximize a function of the accumulated reward. In our framework, this can be interpreted as the player moving a pebble along the edges of the Markov decision process. At player vertices, the player can choose which edge they want to take. At randomization vertices, the player has no control which edge will be chosen as each edge is chosen according to the corresponding probability distribution.

Depending on the function that is maximized, different strategies for moving the pebble might be optimal for the player. The player might for example not always choose the same edge when they visits a player vertex more than once, or their decisions might be based on the previously encountered vertices. However, for most relevant objective functions, it is known that strategies maximizing the objective functions do not need to have these properties. In fact, the best strategies are typically *memoryless* and *deterministic*. That is, they do not depend on previous choices and the player always chooses the same edge when visiting a vertex more than once. The strategy of the player can thus be described as a memoryless, deterministic *strategy* which is defined as follows.

Definition 2.2.2 (Strategy (MDP)). A (deterministic, memoryless) *strategy* for G is a function $\sigma\colon V_0 \to V$ such that $(v, \sigma(v)) \in E_0$ for all $v \in V_0$.

Strategies in Markov decision processes are also often called *policies* in the literature. We deliberately use the same term that we introduced for parity games, since later statements will be proven for parity games and Markov decision processes simultaneously, and overloading notation then streamlines these statements.

Observation 2.2.3. Let σ be a strategy for G. Then, the pair (G, σ) induces a Markov chain, and this Markov chain is denoted by $\mathrm{MC}(G, \sigma)$.

For this thesis, we assume that the reader is familiar with the basics of Markov chains as we do not discuss them here and refer to [Put05, Appendix A] instead.

Example 2. *Consider the Markov decision process given in Figure 2.2. It contains 3 player vertices, marked in blue, and two randomization vertices, marked in red. Edges are colored accordingly. To further distinguish the vertices, player vertices are circular while randomization vertices are rectangular. Vertex labels show the names of the corresponding vertex. Labels on player edges show the reward of these edges, and labels on randomization edges show the transition probabilities. Note that we do not have rewards on randomization edges, although Definition 2.2.1 allows for them. The example also visualizes the strategy σ which is defined via $\sigma(a) = b, \sigma(b) = c$ and $\sigma(c) = c$ by bold edges.*

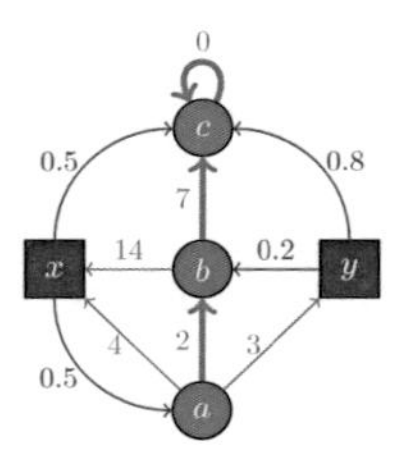

Figure 2.2.: A small Markov decision process with 3 player and 2 randomization vertices. The strategy σ which is defined via $\sigma(a) := b, \sigma(b) := c$ and $\sigma(c) := c$ is visualized by bold edges.

As discussed previously, the player collects rewards when transitioning between states, and typically aims to maximize some objective function of the collected rewards. We consider two of the most important objective functions in this thesis, the *expected total reward* criterion and the *expected average reward* criterion. As the names suggest, the player aims to either maximize the expected total reward resp. expected average reward per turn. Although the first objective is not well-defined for general Markov decision processes, this will not be an issue for the processes discussed in this thesis. In fact, the total expected reward of all "relevant" strategies will be finite. Moreover, the two criteria will actually turn out to be equivalent for the Markov decision processes considered in this thesis.

There are, however, also other objective functions. The most important one of those is the *discounted reward criterion*. The intuition between this criterion is that collecting a reward earlier is more beneficial than collecting the same reward later. Formally, this is modeled by introducing a *discount factor* $\gamma \in (0, 1)$, and the i-th reward collected by the player is discounted by the factor γ^i. This guarantees that the expected total discounted reward is always finite, so this objective function is well-defined for all Markov decision processes. For this reason and since several algorithms behave particular nicely for the discounted reward criterion [Ye11], this criterion is widely used. However, most of the results obtained for the expected total reward criterion also apply to the discounted reward criterion if the discount factor is sufficiently close to 1.

The choices of the player are modeled by strategies. The computational task associated with solving a Markov decision process is thus typically the following: *Find a strategy maximizing the given objective function.* Such a strategy is called *optimal* for the respective objective. Although there are also other computational tasks associated with Markov decision processes, like finding a strategy maximizing or minimizing the probability for reaching a certain state [HM18], we only consider the objective of maximizing a given function of the collected reward.

Of course, there are several ways of finding an optimal strategy. The majority of them is based on techniques that were originally developed for linear programming, as Markov decision processes and linear programs are closely related. Algorithms that are designed specifically for Markov decision processes typically depend on the *optimality equations*, which were proposed by Bellman [Bel57]. In principle, they state that the optimality of a strategy can be characterized and verified by solving a system of equations. More precisely, given a strategy σ, we can assign values to the vertices representing the expected collected reward when starting at the vertex and following σ forever. If the strategy is optimal, then these values are the solution of a specific set of equations. If the strategy is not optimal, then these values can be used to measure "how far" the current solution is from an optimal solution. This insight can then be used to improve the current strategy, yielding an iterative procedure to find an optimal strategy. We now formalize this idea and base our explanation on [Put05].

We begin by introducing the expected total reward criterion. As mentioned earlier, this is not well-defined for arbitrary Markov decision processes unless we allow an infinite total reward. We later define a special class of Markov decision processes guaranteeing finiteness of the expected total reward.

Definition 2.2.4 (Expected total reward criterion (cf. Theorem 7.1.3. in [Put05])). For a strategy σ, the *values* $\mathrm{Val}_\sigma(u)$ of the vertices $u \in V$ are defined as the unique solution (if existing) of the system

$$\mathrm{Val}_\sigma(u) = \begin{cases} r(u,\sigma(u)) + \mathrm{Val}_\sigma(\sigma(u)), & u \in V_0 \\ \sum_{v\in\Gamma^+(u)} p(u,v)\,\mathrm{Val}_\sigma(v), & u \in V_R \end{cases} \tag{2.1}$$

together with the condition that the value of every vertex contained in an irreducible recurrent class of $\mathrm{MC}(G,\sigma)$ is 0 (making the solution unique). The *expected total reward criterion (ETRC)* asks for a strategy σ^* such that $\mathrm{Val}_{\sigma^*}(v) \geq \mathrm{Val}_\sigma(v)$ for all strategies σ and $v \in V$. Such a strategy is called *optimal* for the expected total reward criterion.

Example 3. *Consider the Markov decision process discussed in Example 2 alongside the strategy σ given in that example. Then,* $\mathrm{Val}_\sigma(c) = 0 + \mathrm{Val}_\sigma(c)$. *The value of the vertex c is thus not determined by the Equation* (2.1). *Since this vertex is an irreducible recurrent class* *of* $\mathrm{MC}(G, \sigma)$*, its value is set to 0, which is in fact the expected total reward of this vertex. As* $\sigma(b) = c$ *and* $r(b, c) = 7$*, this implies* $\mathrm{Val}_\sigma(b) = 7$ *and, analogously,* $\mathrm{Val}_\sigma(a) = 9$*. For the two randomization vertices, we obtain* $\mathrm{Val}_\sigma(x) = 0.5 \cdot \mathrm{Val}_\sigma(c) + 0.5 \cdot \mathrm{Val}_\sigma(a) = 4.5$ *and* $\mathrm{Val}_\sigma(y) = 0.8 \cdot \mathrm{Val}_\sigma(c) + 0.2 \cdot \mathrm{Val}_\sigma(b) = 1.4$.

We mention here that optimal strategies indeed exist if the values of all vertices are finite for all strategies [Put05, Chapter 7]. We discuss optimal strategies in more detail in Chapter 3.

We now consider the expected average reward criterion. As for the ETRC, there is a set of optimality equations that can be used to define the expected average reward criterion. We now introduce this system and base our description on the explanation given in [Fea10b]. To simplify the equations, we assume that the Markov decision process is bipartite with respect to player and randomization vertices. However, the criterion is also applicable to arbitrary Markov decision processes when using a more complicated notation or the "action-state" formulation.

For general Markov decision processes, it is not sufficient to assign a single value to every vertex representing the collected average reward per turn. The optimality equations thus consist of two interlaced systems of equations that need to be solved simultaneously. These are the *gain* and the *bias* equations. The gain equation for vertex $u \in V_0$ is defined via

$$G(u) := \max_{x \in \Gamma^+(u)} \sum_{v \in \Gamma^+(x)} p(x, v) \cdot G(v). \tag{2.2}$$

The bias equation is based on the gain equation. For $u \in V_0$, let M_u denote the set of vertices in $\Gamma^+(u)$ that achieve the maximum in the gain equation at vertex u, so

$$M_u := \left\{ x \in \Gamma^+(u) : G(u) = \sum_{v \in \Gamma^+(x)} p(x, v) \cdot G(v) \right\}. \tag{2.3}$$

Then, the bias equation at vertex $u \in V_0$ is defined as

$$B(u) := \max_{x \in M_u} \left(r(u, x) - G(u) + \sum_{v \in \Gamma^+(x)} p(x, v) \cdot B(v) \right). \tag{2.4}$$

It is well-known that a solution to Equations (2.2) and (2.4) yields the expected average reward as follows [Put05, Theorem 9.1.3].

Theorem 2.2.5. *Let* G^*, B^* *be solutions to Equations* (2.2) *and* (2.4). *Then, for every* $u \in V_0$*, the gain* $G^*(u)$ *is the maximal expected average reward obtainable when starting in* u.

It is however unclear how to find solutions to the optimality equations. One approach is to generalize the equations and introduce the gain and bias of a vertex with respect to a given strategy. Then, driven by the optimality equations, the strategy can be changed until a strategy is found such that the gain and bias of this strategy solves the optimality conditions. This algorithmic scheme is known as *strategy improvement* and is discussed in Section 3.3.

Definition 2.2.6 (Expected average reward criterion). Let G be bipartite, $u \in V_0$, and let σ be a strategy for G. The *gain* $G_\sigma\colon V_0 \to \mathbb{R}$ and the *bias* $B_\sigma\colon V_0 \to \mathbb{R}$ (with respect to σ) are the solution of the system

$$\begin{aligned} G_\sigma(u) &= \sum_{v \in \Gamma^+(\sigma(u))} p(\sigma(u), v) \cdot G_\sigma(v) \\ B_\sigma(u) &= r(u, \sigma(u)) - G_\sigma(u) + \sum_{v \in \Gamma^+(\sigma(u))} p(\sigma(u), v) \cdot B_\sigma(v). \end{aligned} \tag{2.5}$$

The *expected average reward criterion (EARC)* requires us to find a strategy σ^* such that G_{σ^*} and B_{σ^*} fulfill Equations (2.2) and (2.4). Such a strategy is called *optimal* for the EARC.

The solution of the system (2.5) might however not be unique. Analogously to Theorem 2.2.5, $G_\sigma(v)$ is the expected average reward collected when starting in vertex v and choosing the edge $(u, \sigma(u))$ when encountering $u \in V_0$. This implies that the gains are unique. This is not the case for the biases. There are several ways of making the biases unique, for example by including more equations or restricting the Markov decision process. We will not discuss this aspect in more detail, as the biases will be unique for the special class of Markov decision processes considered in this thesis. This class of Markov decision processes implements the same idea that is used to simplify the treatment of the parity games. It is thus introduced alongside the corresponding class of parity games in Section 3.2.

2.3. Linear Programming

Linear programming is a discipline of mathematics that was developed during World War II to aid in logistics and planning of military operations. Quickly after the end of the war, companies and industries realized that the techniques and algorithms of linear programming can be applied to reduce costs and increase profits. As a result, the field of linear programming and all related aspects of mathematics, like operations research and integer programming, grew exceptionally fast and became one of the most important areas in discrete mathematics. Until today, techniques, results and insights of linear programming are at the core of many algorithms which are used in a variety of software, making it a very important topic for practical applications. In addition, there are several theoretical questions that are related to linear programming, with the P-NP-problem being the most famous of them, emphasizing the importance of linear programming for purely theoretical research.

In this section, we give a brief introduction to the theory of linear programming. We focus on the theory necessary to develop Dantzig's famous *simplex algorithm* [Dan51] and omit other aspects like duality, interior point methods or the ellipsoid algorithm. As the theory of linear programming is a widely discussed topic and content of many books about optimization, it cannot be attributed to a single publication. This overview is thus based on many different publications, the main ones being [BT97, Chapters 2-4],[Fri11b, Chapter 2.2] and [Han12, Chapter 1].

2.3.1. The Basics of Linear Programming

The main task of linear programming is to answer the following question: Given a linear objective function and a set of linear constraints, how can a solution maximizing (or minimizing) the objective function subject to the constraints be calculated? More formally, given a vector $c \in \mathbb{R}^n$ and a system

$$\begin{array}{ccccccccc} a_{1,1} \cdot x_1 & + & a_{1,2} \cdot x_2 & + & \dots & + & a_{1,n} \cdot x_n & = & b_1 \\ a_{2,1} \cdot x_1 & + & a_{2,2} \cdot x_2 & + & \dots & + & a_{2,n} \cdot x_n & = & b_2 \\ \vdots & & \vdots & & & & \vdots & & \vdots \\ a_{m,1} \cdot x_1 & + & a_{m,2} \cdot x_2 & + & \dots & + & a_{m,n} \cdot x_n & = & b_m \end{array} \tag{2.6}$$

of linear equalities, the goal is to find a vector $x \in \mathbb{R}^n$ maximizing $c^T x$ subject to (2.6) and $x \geq 0$. If we define A as the matrix of constraints, this yields the following definition.

Definition 2.3.1 (Linear program). Let $m, n \in \mathbb{R}, c \in \mathbb{R}^n, b \in \mathbb{R}^m$ and $A \in \mathbb{R}^{m \times n}$. Then, the *linear program* in standard form induced by A, b and c is the optimization problem

$$\begin{array}{lll} \max & c^T x & \\ \text{subject to} & Ax & = b \\ & x & \geq 0 \end{array} \tag{2.7}$$

and is denoted by LP(A, b, c). The function $c^T x$ is the *objective function*, the matrix A is the *constraint matrix* and the vector b is the *right-hand side*.

Linear programs can also be defined in a more general way, for example, by also considering inequalities instead of or in addition to equalities, considering minimization instead of maximization, or restricting the variable values in a different way. By introducing artificial variables, it is possible to transform any linear program into the standard form presented here. One such transformation is performed in Example 4. We do not discuss this transformation here in general and refer to any book on linear optimization for details.

For the remainder of this section, let $m, n \in \mathbb{R}$ and consider some fixed matrix $A \in \mathbb{R}^{m \times n}$, vectors $c \in \mathbb{R}^n, b \in \mathbb{R}^m$ and the linear program LP(A, b, c). The task to find a vector $x \in \mathbb{R}^n$ maximizing $c^T x$ subject to $Ax = b, x \geq 0$ requires the analysis of LP(A, b, c) and the introduction of additional terms and notation.

Definition 2.3.2 (Properties of linear programs). Let LP $=$ LP(A, b, c) be a linear program.

1. A vector $x \in \mathbb{R}^n$ with $Ax = b$ and $x \geq 0$ is *feasible* for LP or a *feasible solution*. The set of all feasible solutions of LP is denoted by $P_{\text{LP}} := \{x \in \mathbb{R}^n : Ax = b, x \geq 0\}$. If $P_{\text{LP}} = \emptyset$, then LP is called *infeasible*, otherwise it is called *feasible*.
2. A vector $x^* \in \mathbb{R}^n$ is *optimal* for LP if $x^* \in \arg\max_{x \in P_{\text{LP}}} c^T x$.
3. LP is *unbounded* if for every $\lambda \in \mathbb{R}$ there exists a $x \in P_{\text{LP}}$ with $c^T x \geq \lambda$. If LP is not unbounded, it is called *bounded*.

The following theorem is fundamental. It states that a linear program either admits at least one optimal solution, can have arbitrary good solutions or does not allow for any solution at all.

Theorem 2.3.3. *A linear program is either (i) bounded and feasible, (ii) unbounded or (iii) infeasible and it has exactly one of these properties.*

Example 4 (Basic terms of linear programming)**.** *Consider the following linear program.*

$$\begin{aligned} \max \quad & x_2 \\ s.t. \quad & x_1 - x_2 \geq 0 \\ & x_1 + x_2 \leq 4 \\ & x_1, x_2 \;\; \geq 0 \end{aligned} \tag{2.8}$$

This linear program is not in standard form. However, in the given form, the set of feasible solutions of this linear program can be visualized in two dimensions. As each of the four inequalities (if we interpret the sign restrictions as inequalities) defines a half-space, the set of feasible solutions is exactly the intersection of these half-spaces.

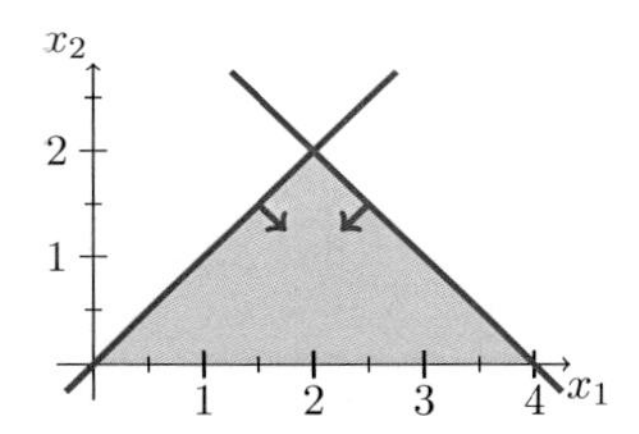

Figure 2.3.: Representation of the set of feasible solutions of the linear program discussed in Example 4.

In this example, every point (x_1, x_2) *contained in the triangle with the three corners* $(0,0), (4,0)$ *and* $(2,2)$ *is a feasible solution. The point* $(x_1^*, x_2^*) = (2,0)$ *is the unique optimal solution. If the inequality* $x_1 + x_2 \leq 4$ *is removed, then the linear program becomes unbounded. If the inequality* $x_1 - x_2 \geq 5$ *is added, then the linear program becomes infeasible.*

By introducing two artificial slack variables s_1 *and* s_2*, the linear program can be transformed into standard form:*

$$\begin{aligned} \max \quad & x_2 \\ s.t. \quad & x_1 - x_2 - s_1 = 0 \\ & x_1 + x_2 + s_2 = 4 \\ & x_1, x_2 \qquad \geq 0 \\ & s_1, s_2 \qquad \geq 0 \end{aligned} \tag{2.9}$$

In Example 4, the two-dimensional object that is defined by the constraints of LP(A, b, c) has a very special and distinct geometry. In fact, it is a *polytope*, and we now introduce the terms and notation necessary to formally describe the geometry of linear programs.

Intuitively, every constraint of a linear program separates the n-dimensional space into one "inner" and one "outer" part. The set of feasible solutions of a linear program is then the intersection of the "inner" parts. Formally, each constraints defines a *hyperplane* and a corresponding *halfspace*, and intersecting finitely many halfspaces yields a polyhedron.

Definition 2.3.4 (Hyperplane and halfspace). Let $a \in \mathbb{R}^n \setminus \{(0, \ldots, 0)\}$ and $b \in \mathbb{R}$. Then, $\{x \in \mathbb{R}^n : a^T x = b\}$ is the *hyperplane* defined by a and b. Similarly, $\{x \in \mathbb{R}^n : a^T x \leq b\}$ is the *halfspace* defined by a and b.

Definition 2.3.5 (Polyhedron and polytope). A *polyhedron* $P \subseteq \mathbb{R}^n$ is the intersection of finitely many halfspaces in $\mathbb{R}^n$. A polyhedron is called *bounded* if there exists a constant $M \in \mathbb{R}$ such that $|x_i| \leq M$ for every $x \in P$ and $i \in [n]$. A bounded polyhedron is called *polytope*.

Observation 2.3.6. Let LP(A, b, c) be a linear program. Then, P_{LP} is a polyhedron.

Polyhedra, and consequently also polytopes, have a very special property. Let $P \subseteq \mathbb{R}^n$ be a non-empty polytope and let $x, y \in P$. Then, every $z \in \mathbb{R}^n$ that lies on the line between x and y is also contained in P. Formally, this means that for every $\lambda \in [0, 1]$, the point $\lambda x + (1 - \lambda) y$ is contained in the polyhedron P, provided $x, y \in P$. This property is called *convexity*, and it is a central term in the theory of optimization.

Definition 2.3.7 (Convex set). A set $S \subseteq \mathbb{R}^n$ is *convex* if for every $x, y \in S$ and $\lambda \in [0, 1]$, it holds that $\lambda x + (1 - \lambda) y \in S$.

As the following lemma suggests, the geometric objects introduced previously are in fact all convex sets.

Lemma 2.3.8. *Halfspaces in $\mathbb{R}^n$ are convex sets. The intersection of a finite number of convex sets is a convex set. In particular, hyperplanes and polyhedra are convex.*

Another important concept are *vertices* of polyhedra. There are several equivalent ways of defining what exactly vertices are. For optimization, the most intuitive definition is that every vertex is the unique maximizer of *some* linear function on P.

Definition 2.3.9 (Vertex). Let $P \subseteq \mathbb{R}^n$ be a polyhedron. Then, $x \in P$ is a *vertex* of P if there exists a $c \in \mathbb{R}^n$ such that $c^T x > c^T y$ for all $y \in P \setminus \{x\}$.

Now consider LP(A, b, c) and assume that P_{LP} has an optimal solution with respect to c. The following central theorem highlights the importance of vertices for linear optimization, as it implies that it is sufficient to focus on vertices of P_{LP} when searching for optimal solutions.

Theorem 2.3.10 (Cf. Theorem 2.7 in [BT97]). *Assume that P_{LP} has at least one vertex and at least one optimal solution with respect to c. Then, there is a vertex $x^* \in P_{\text{LP}}$ optimal for LP(A, b, c), and such a vertex is then called* optimal.

We now develop the theory related to vertices that will then lead to the description of the simplex algorithm. For simplicity, let $P = P_{\text{LP}} = \{x \in \mathbb{R}^n : Ax = b, x \geq 0\}$ and assume $P \neq \emptyset$. Henceforth, we further assume that the rows of A are linearly independent. Note that this implies $n \leq m$ as there is at most one solution of the system $Ax = b, x \geq 0$ otherwise.

We begin by investigating the connection between vertices and solutions of the system $\{Ax = b, x \geq 0\}$ defining P. By Definition 2.3.9, every vertex can be interpreted as the optimum solution of the problem $c^T x$ subject to $x \in P$ for some suitable $c \in \mathbb{R}^n$. The following theorem gives an equivalent characterization. More precisely, it says that a solution x of $\{Ax = b, x \geq 0\}$ is a vertex of P if and only if the columns corresponding to non-zero entries are linearly independent.

For the remainder of this thesis, we fix the following notation. The set of the first m integers is denoted by $[m]$, so $[m] := \{1, \dots, m\}$. For a countable set S and an index $k \leq |S|$, we denote the k-th entry of S by $S[k]$. Let $A \in \mathbb{R}^{m \times n}$ and $i \in [m], j \in [n]$. We denote the i-th row by of A by $A_{i,\bullet}$ and the j-th column of A by $A_{\bullet,j}$. For a set $I \subseteq [n]$ of column indices with $|I| = k$, we denote the matrix induced by the corresponding columns of A by A_I,so $A_I := (A_{\bullet,I[1]} \cdots A_{\bullet,I[k]}) \in \mathbb{R}^{m \times k}$.

Theorem 2.3.11 (Cf. Theorem 2.4 in [BT97]). *Let P be a polyhedron and $x \in P$. Then, x is a vertex of P if and only if the columns of the matrix $A_{\mathrm{supp}(x)}$ are linear independent.*

Under these assumptions, there is a strong connection between vertices of P, feasible solutions of $\{Ax = b, x \geq 0\}$ and linear independent columns of A, or, more precisely, *bases* of A.

Definition 2.3.12 (Basis). A *basis* $B \subseteq [n]$ of a polyhedron $P \subseteq \mathbb{R}^n$ is an ordered set $B = (B_1, B_2, \dots, B_m)$ of column indices such that the *basis matrix* A_B is non-singular.

Given a basis B, we define the vector $\bar{x}_B$ by setting

$$(\bar{x}_B)_j = \begin{cases} ((A_B)^{-1}b)_j, & j \in B \\ 0, & \text{otherwise.} \end{cases} \tag{2.10}$$

This vector is called the *basis vector* of B. Note that $\bar{x}_B \neq x_B$ in our notation as x_B is the vector containing the entries of x corresponding to the elements of B. By the definition of this vector, $A\bar{x}_B = b$. It is, however, not guaranteed that $\bar{x}_B \in P$ since it might contain negative entries. In the case that all of the entries are positive, we call the basis B *feasible*.

We now define another type of possible solutions of $\{x \in \mathbb{R}^n : Ax = b, x \geq 0\}$ that uses the notion of bases. It will turn out that these so-called *basic solutions* are exactly the vertices of P if they are feasible.

Definition 2.3.13 (Basic (feasible) solution). Let $x \in \{x \in \mathbb{R}^n : Ax = b\}$. Then, x is a *basic solution* if there is a basis B such that $\mathrm{supp}(x) \subseteq B$. This basis is then called the *corresponding basis*. The indices $j \in B$ are called *basic* and the indices $j \notin B$ are called *non-basic*. A basic solution x is called *basic feasible solution* if $x \geq 0$.

It is possible that more than one basis corresponds to a basic solution x. This can happen if $|\mathrm{supp}(x)| < m$. In this case, the vertex x is called *degenerate*.

By Theorem 2.3.11, every vertex of P is a basic feasible solution. If a vertex x is a basic feasible solution for P, then there is at least one basis B such that $\mathrm{supp}(x) \subseteq B$. It is easy to verify that $x = \bar{x}_B$, i.e., the vertex x is the basis vector for the basis B. In particular, by the definition of a basis, the matrix A_B is non-singular, implying that the columns are linearly independent. This implies that every basic feasible solution is a vertex. In particular, by Theorems 2.3.10 and 2.3.11, this yields the following central characterization of vertices of a linear program.

Theorem 2.3.14 (Cf. Theorems 2.3 and 2.4 in [BT97]). *Let $LP(A, b, c)$ be a linear program and $x \in P = P_{LP}$. The following statements are equivalent:*

1. *x is a vertex of P.*
2. *x is a basic feasible solution for P.*
3. *$x = \bar{x}_B$ for some basis B.*
4. *There is a $z \in \mathbb{R}^n$ such that x is the unique optimal solution of $\max z^T x$ s.t. $x \in P$.*

These terms and theorems build the foundation of one of the most important algorithms in optimization, the simplex algorithm, which is the topic of the next subsection.

2.3.2. The Simplex Algorithm

The simplex algorithm was one of the first algorithms able to solve general linear programs. It was developed by George Dantzig in 1947 (see e.g. [Dan63]), and several improved and modern variants are still used to solve linear programs today. It is an iterative procedure inspired by Theorem 2.3.10. This theorem states that, provided there is at least one optimal solution, there is a vertex of the polyhedron P_{LP} that is optimal. The algorithm thus searches for this optimal vertex by using ideas of local search. It begins in a vertex and checks whether this vertex is optimal. It terminates in that case, and calculates an improving direction otherwise. This direction typically corresponds to an edge of the polyhedron P_{LP}. If the algorithm does not terminate, it then walks along this edge until it reaches another vertex. This procedure is then iterated until either an optimal vertex is found or until the algorithm verified that it can improve infinitely in a direction. Since a polyhedron has finitely many vertices, this procedure terminates.

A description of the simplex algorithm is given in Algorithm 1 and an exemplary run is shown in Figure 2.4. The goal of this section is to develop all the theoretical background needed for understanding and applying the simplex algorithm in its most basic version. This description is heavily inspired by [BT97], and we refer to their book for proofs and further details.

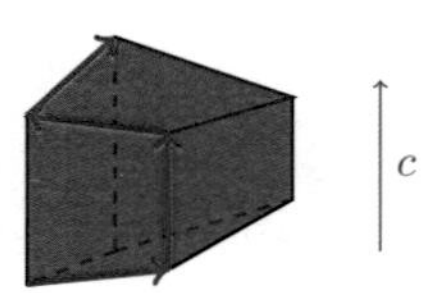

Figure 2.4.: Exemplary run of the simplex algorithm.

The simplex algorithm needs to be able to (i) find an initial vertex, (ii) identify improving directions and corresponding adjacent vertices and (iii) identify optimal solutions. As it is easier to present how an initial vertex can be found when it is clear how the simplex algorithm finds adjacent vertices, we postpone (i) for now.

Henceforth, consider a feasible linear program LP(A, b, c), let x be a basic solution and B be a corresponding basis. The first observation is that the value of the basis variables x_B

```
Input: Feasible linear program LP(A, b, c)
Output: Optimal vertex x* or confirmation that LP(A, b, c) is unbounded
k := 0
Let x^(k) be a vertex of P_LP
while x^(k) is not optimal do
    Let d^(k) be an improving direction with respect to c at vertex x^(k)
    if x^(k) + λd^(k) ∈ P_LP for all λ > 0 then
        return "unbounded"
    else
        Choose maximal λ > 0 such that x^(k) + λd^(k) ∈ P_LP
        x^(k+1) := x^(k) + λd^(k)
        k := k + 1
return x^(k)
```

Algorithm 1: The simplex algorithm

is fully determined by the values of the non-basic variables as $Ax = b$ is equivalent to

$$A_B x_B + \sum_{j \notin B} A_{\bullet,j} x_j = b \qquad \text{resp.} \qquad x_B = A_B^{-1} b - \sum_{j \notin B} A_B^{-1} A_{\bullet,j} x_j. \tag{2.11}$$

In particular, Equation (2.11) describes exactly how the values of the basis variables change if the values of non-basis variables change. This observation is formalized by the notion of a *basic direction*.

Definition 2.3.15 (Basic direction). Let $\mathrm{LP}(A, b, c)$ be a linear program, B be a basis and $j \notin B$. Then, the *j-th basic direction* $d \in \mathbb{R}^n$ is defined via

$$d_i := \begin{cases} 1, & i \notin B, i = j, \\ 0, & i \notin B, i \neq j, \\ [-(A_B)^{-1} A_{\bullet,j}]_i, & i \in B. \end{cases}$$

Basic directions are of particular importance since $A(x+\lambda d) = b$ for any $\lambda \in \mathbb{R}$. However, even if the basic solution x is feasible, not all points on the line $x+\lambda d$ are necessarily feasible for $\mathrm{LP}(A, b, c)$. The reason is that large values of λ might cause negative components. Nevertheless, this property shows that basic directions are reasonable directions for searching new vertices.

Lemma 2.3.16. *Let B be a basis and d be the j-th basic direction for some $j \notin B$. Then $A(x + \lambda d) = b$ for all $\lambda \in \mathbb{R}$.*

For a vertex x of P_{LP} and a corresponding basis B, there might be several basic directions d such that $x + \lambda d \in P_{\mathrm{LP}}$ for small values of λ. The simplex algorithm thus needs to decide which direction to choose. There are several possible ways of choosing directions, and we discuss this point in more detail when discussing pivot rules in Section 2.3.3. To evaluate the directions and verify which of them are actually improving the objective function value, the simplex algorithm calculates *reduced costs*. Intuitively, the reduced cost of a variable x_j indicates by how much the objective function is improved when moving in the j-th basic direction.

Definition 2.3.17 (Reduced costs). Let B be a basis. The *reduced cost* of variable $j \in [n]$ with respect to B is defined as $\bar{c}_j^B := c_j - c_B^T(A_B)^{-1}A_{\bullet,j}$. The vector of reduced cost with respect to B is $\bar{c}^B = (\bar{c}_1^B, \ldots, \bar{c}_n^B)$.

It is easy to calculate that the reduced cost of a basis variable is equal to 0 and that the objective function value with respect to any $x \in P_{\text{LP}}$ can be expressed using $\bar{c}_B$ and $\bar{x}_B$.

Lemma 2.3.18. *Let B be a basis. Then $(\bar{c}_B)_j = 0$ for all $j \in B$. If x satisfies $Ax = b$, then $c^T x = c^T \bar{x}_B + (\bar{c}^B)^T x$.*

As claimed previously, the reduced costs can be used to evaluate the current basic feasible solution. This is formalized by the following lemma.

Lemma 2.3.19 (Cf. Theorem 3.1 in [BT97]). *Let x be a basic feasible solution for LP(A, b, c) with corresponding basis B.*

1. *If $\bar{c}^B \leq 0$, then x is an optimal solution.*
2. *If x is an optimal solution and non-degenerate, then $\bar{c} \leq 0$.*

This lemma justifies the extension of the notion of optimality to bases as these are sufficient to describe reduced costs. A basis B is thus called *optimal* if (i) $(A_B)^{-1}b \geq 0$ and (ii) $\bar{c}^B \leq 0$.

We are now able to describe the behavior of the simplex method in more detail. Let x be a basic feasible solution with corresponding basis B. As explained earlier and as motivated by Lemma 2.3.19, it is desirable to move in the j-th basic direction for some $j \notin B$ with $\bar{c}_j^B > 0$. If there is no such direction, then the basis B is optimal and the algorithm terminates. Thus, assume that this is not the case and fix some $j \notin B$ with $\bar{c}_j^B > 0$. The algorithm now calculates the maximum $\lambda > 0$ such that $x + \lambda d \in P_{\text{LP}}$. This can be achieved by performing the so-called *minimum ratio test*. This test guarantees (i) that the value of the j-th variable that enters the basis is positive and (ii) that the values of all other basis variables remain positive. It can also be used to verify that the linear program is unbounded.

The minimum ratio test is performed by calculating

$$\lambda^* := \min_{i: u_i > 0} \frac{x_{B[i]}}{u_i}, \tag{2.12}$$

where $u := (A_B)^{-1}A_{\bullet,j}$. If this minimum does not exist, then any $x + \lambda d$ is feasible for P_{LP} and the problem is unbounded. Thus, consider the case that this minimum is attained for some $\ell \in [m]$. Then, the algorithm replaces the basic variable ℓ by the non-basic variable j and iterates with the new basis $B' := B \setminus \{\ell\} \cup \{j\}$.

Input: Vertex x and corresponding basis B of LP(A, b, c)
Output: Vertex x' and basis B' with higher objective function value or confirmation that LP(A, b, c) is unbounded or confirmation that x is optimal

---

Calculate reduced costs $\bar{c}_j^B := c_j - c_B^T u_j$ for all $j \notin B$
if $\bar{c}_j^B \leq 0$ *for all* $j \notin B$ **then**
  return x is optimal
Let $j \in \{j \notin B : \bar{c}_j^B > 0\}$ and $u := (A_B)^{-1} A_{\bullet,j}$
if $u \leq 0$ **then**
  return LP(A, b, c) is unbounded
Let $\lambda^* := \min_{i:u_i>0} \frac{x_{B[i]}}{u_i} = \frac{x_{B[\ell]}}{u_\ell}$ for some $\ell \in [m]$
Let $x' \in \mathbb{R}^n$ with $x'_j := \lambda^*, x'_{B[i]} := x_{B[i]} - \lambda^* u_i$ and $x'_i := 0$ else
Let $B' := B \setminus \{\ell\} \cup \{j\}$
return x', B'

Algorithm 2: An iteration of the simplex algorithm

If every basic feasible solution of LP(A, b, c) is non-degenerate, i.e., if every basic feasible solution has only one corresponding basis, then this procedure terminates after a finite number of steps. This is formalized by the following two theorems. The first theorem shows that a single iteration of the simplex algorithm is correct, while the second shows that the iterative application of Algorithm 2 is correct.

Theorem 2.3.20 (Cf. Theorem 3.2 in [BT97])**.** *Let x be a non-degenerate basic feasible solution, $j \notin B$ with $\bar{c}_j^B > 0$, d the j-th basic direction and $\lambda^* = \max\{\lambda : x + \lambda d \in P_{LP}\} < \infty$.*

1. *There exists an index $\ell \in [m]$ such that $\lambda^* = \min_{i \in B: d_i < 0} \frac{-x_i}{d_i} = \frac{-x_{B[\ell]}}{d_{B[\ell]}}$.*
2. *Let $B'[i] := B[i]$ for $i \neq \ell$ and $B'[\ell] = j$. Then, $B' = (B'[1], \ldots, B'[m]')$ is a basis.*
3. *$x' := x + \lambda^* d$ is a basic feasible solution with corresponding basis B and $c^T x' > c^T x$.*

Theorem 2.3.21. *If every basic feasible solution of LP(A, b, c) is non-degenerate, then the simplex algorithm terminates after a finite number of iterations. It either returns an optimal solution with corresponding optimal basis or confirms that LP(A, b, c) is unbounded.*

We now discuss the simplex algorithm if basic feasible solutions can be degenerate. In this case, it might happen that $\lambda^* = 0$. Then, the algorithm also calculates another basis, but the basic feasible solution and the corresponding vertex remains identical. If the variables for entering and leaving the basis are chosen badly, it might then happen that the algorithm *cycles*. This means that it calculates different bases for the same basic feasible solution without ever finding a new basic feasible solution. There are, however, pivot rules guaranteeing that the algorithm does not cycle, and Theorems 2.3.20 and 2.3.21 are also applicable for these pivot rules.

To conclude our discussion of the simplex algorithm, we next explain how an initial basic feasible solution is found. Since we will provide our algorithms with initial solutions, we only briefly discuss this topic here.

The idea presented here is the so-called *two-phase simplex algorithm*. The first phase is the calculation of an initial solution, the second phase is the iterative application of Algorithm 2. This variant of the algorithm uses the capability of the second phase of finding an optimal solution if it is provided an initial solution. More precisely, the first phase considers a slightly changed problem with the following two properties:

1. It is trivial to find an initial basic feasible solution.
2. Optimal solutions of the changed problem yield initial solutions of the original problem.

The initial problem LP(A, b, c) is altered in the following way. First, all rows $i \in [m]$ with $b_i < 0$ are multiplied by -1. This does not change the sets of feasible or optimal solutions, but enables us to assume $b \geq 0$ without loss of generality. Then, the matrix A is extended by adding the m-dimensional identity matrix I_m. We thus set $A' := (A, I_m) \in \mathbb{R}^{m \times (n+m)}$. This creates m new variables $x_{n+1}, \ldots, x_{n+m}$. Since the right-hand side b is non-negative, setting $x_{n+j} := b_j$ for $j \in [m]$ and $B := \{n+1, \ldots, n+m\}$ yields a basic feasible solution for this problem. Now, to obtain an initial solution for the original problem, phase two of the simplex algorithm is used in such a way that all the variables $x_{n+1}, \ldots, x_{n+m}$ leave the basis. This can be achieved by using the objective function vector $c' \in \mathbb{R}^{n+m}$ where $c'_i = 0$ for $i \in [n]$ and $c'_i = -1$ for $i \geq n+1$. Thus, an initial basic feasible solution for LP(A, b, c) can be found by applying the second phase of the simplex algorithm to LP(A', b, c') using the initial basis $B = \{n+1, \ldots, n+m\}$. More precisely, the first phase is able to find a basic feasible solution if and only if LP(A, b, c) is feasible. and can thus also be used to detect whether LP(A, b, c) is feasible.

We described the two-phase simplex algorithm using Dantzig's pivot rule here. Intuitively, a pivot rule is a routine determining the leaving and entering variable with respect to a given basis. Different pivot rules as well as the computational complexity are the focus of the next section, in which also includes a detailed discussion of Zadeh's pivot rule.

2.3.3. Pivot Rules and the Complexity of Linear Programming

Pivot rules are used in a variety of algorithms. Consider an arbitrary optimization algorithm that works by first calculating an initial solution and then produces a sequence of intermediate solutions until either finding an optimal solution or confirming that the given instance is unbounded. Such an algorithm can be interpreted as a local search algorithm that, given a solution, always looks for an adjacent solution that is in some sense better than the current solution. However, depending on the algorithm and the optimization problem, there might be several such solutions, and the algorithm can choose any of these solutions. A *pivot rule* is now a subroutine that specifies exactly which adjacent solution the algorithm should choose.

If interpreted like this, pivot rules can be stated very generally and can be applied for a range of algorithms. Although there are attempts of formally defining pivot rules, at least for the simplex algorithm [APR14], there is no general formal definition for what exactly is considered a pivot rule. We thus briefly discuss some of the most important pivot rules in terms of the simplex algorithm. An overview over these pivot rules can be found in Table 1.1. We also discuss the running time of the simplex algorithm in general and if these pivot rules are used and conclude this by discussing the complexity of solving linear programs. For discussions on even more pivot rules and some more “exotic” or general pivot rules (e.g. pivot rules that allow for intermediate solutions to be infeasible) we refer to [TZ93, Ter01a, Ter01b, APR14].

The running time of the simplex algorithm highly depends on the chosen pivot rule. Assume that the algorithm uses a pivot rule that calculates an initial basic feasible solution as well as an adjacent basic feasible solution to a given solution in strongly polynomial time. Then, the overall running time of the simplex algorithm is strongly polynomial if and only if the pivot rule guarantees that only a polynomial number of vertices is calculated. Since all polynomial algorithms for solving linear programs are only *weakly* polynomial [Kha80, Kar84], finding such a pivot rule would immediately answer the still open question whether linear programming can be done in strongly polynomial time. This motivates the quest for developing new pivot rules and for developing worst-case instances for known pivot rules.

An overview over classical and common pivot rules

We now introduce and discuss the most common pivot rules. An overview over these pivot rules as well as a selection of corresponding literature is given in Table 1.1. We also refer to [TZ93, Han12] for further details.

1. **Dantzig's pivot rule:** Dantzig's pivot rule chooses a non-basic variable $j \notin B$ maximizing the reduced cost. This pivot rule was originally used in Dantzig's development of the simplex algorithm [Dan51].

   It was proven that the worst-case running time using this pivot rule can be exponential by Klee and Minty in 1972 [KM72], providing the first super-polynomial lower bound for the simplex algorithm.

2. **The shadow vertex rule:** Intuitively, given a basic feasible solution x, this rule first finds a new cost function c' and right-hand side b' such that x is an optimal solution for $\mathrm{LP}(A, b', c')$. It then considers the “path” between the two linear programs $\mathrm{LP}(A, b, c)$ and $\mathrm{LP}(A, b'c')$ by considering $\mathrm{LP}(A, \lambda b + (1 - \lambda b'), \lambda c + (1 - \lambda)c')$ for increasing $\lambda \in [0, 1]$. If the reduced cost of some variable becomes positive before reaching c, then the corresponding variable is chosen as the entering variable. It is known that the basic feasible solution x is optimal if this does not happen before reaching c. This rule was first developed in [GS55], and we refer to [Bor87] for a more clear and modern presentation.

   It was shown by Murty in 1980 that this pivot rule has exponential running time in the worst-case [Mur80].

3. **The lexicographic pivot rule:** This pivot rule was proposed in [DOW55] and is rather a specification of Dantzig's original pivot rule as it decides which variable should *leave* the basis. As choosing a leaving variable corresponds to choosing a row

of the tableau that is typically maintained when actually performing the simplex algorithm, this pivot rule decides which variable leaves the basis by choosing the lexicographically smallest row that may be chosen. This pivot rule was developed to prevent *cycling*, a phenomenon that might occur when the polyhedron defining the linear program is degenerate. In this case, it can happen that the simplex algorithm reaches a vertex that can be represented by several bases and alternates between those bases without terminating. The lexicographic pivot rule ensures that this cannot happen and that the algorithm terminates, even if the polyhedron is degenerate.

We are not aware of an explicit proof regarding the worst-case running time of this pivot rule. However, as the example used by Disser and Skutella in [DS15] is non-degenerate, it follows that the lexicographic pivot rule can be exponential in the worst case.

4. **The LargestIncrease pivot rule:** It is not clear when exactly this pivot rule was proposed, and we refer to [Jer73] for a detailed discussion. The LargestIncrease pivot rule always chooses the entering and leaving variable in such a way that the objective function value increases by the maximum amount possible.

   Adapting the example of Klee and Minty, Jeroslow proved that this pivot rule as well as a generalization of this pivot rule visits exponentially many vertices in the worst case [Jer73].

5. **Bland's pivot rule:** This pivot rule was developed by Bland in 1977 [Bla77] to prevent the simplex algorithm from cycling. Among all variables that can enter the basis, this pivot rule chooses the variable with the smallest index and chooses the corresponding leaving variable analogously.

   Only one year after this pivot rule was proposed, Avis and Chvátal proved that this pivot rule may require exponentially many iterations in the worst case [AC78].

6. **The SteepestEdge pivot rule:** This pivot rule always chooses the steepest edge that is incident to the current vertex of the polyhedron. This corresponds to choosing the variable $j \notin B$ maximizing $\bar{c}_j^B/\|[A_B]^{-1}A_{\bullet,j}\|$. It is not clear where this pivot rule was proposed originally and we refer to [FG92] for a discussion of this rule.

   The exponential lower bound for this pivot rule was proven in [GS79].

7. **The RandomEdge pivot rule:** As the name suggests, this pivot rule is a randomized pivot rule and it chooses an edge incident to the current vertex uniformly at random. Again, it is unclear where this pivot rule was proposed first as it is the most natural use of randomization applicable to the simplex algorithm. As the number of iterations is not deterministic when using this pivot rule, one investigates the *expected* number of iterations.

   Although it was hoped that randomized pivot rules maybe have a polynomial worst-case running time in expectation, Friedmann, Hansen and Zwick proved that the expected worst-case running time can be subexponential [FHZ11b].

8. **Cunningham's pivot rule:** This pivot rule is a *memorizing* pivot rule as it depends on previous iterations of the algorithm. Before the algorithm is executed, a cyclic

order of the variables is fixed. In addition, the pivot rule remembers the last variable to enter the basis. When a new variable should enter the basis, the pivot rule then chooses the first variable that is allowed to enter the basis with respect to the fixed order, starting from the last chosen variable.

Rather recently, Avis and Friedmann proved that the worst-case running time using this pivot rule is also exponential [AF17].

9. **The RaisingTheBar pivot rule:** This pivot rule is a generalization of the RandomEdge rule and was introduced in [Kal91]. It can be described as follows. A parameter M is chosen and starting at the current vertex x, the algorithm takes a random walk along the vertices x' of the polyhedron with value at least $c^T x$. That is, $c^T x$ is a "bar", and the random walk is forced to stay above this bar. After performing M steps, the process is repeated, so the bar is raised. For $M = 1$, this behavior is identical to the RandomEdge pivot rule.

10. **Zadeh's LeastEntered pivot rule:** This pivot rule was developed by Zadeh in 1980 [Zad80] as a pivot rule that behaves well on the worst-case examples for other pivot rules. It is a memorizing pivot rule as it remembers for each variable how often it entered the basis. When determining which variable should enter the basis, this pivot rule always chooses a variable that was chosen least often until now.

    Until Friedmann's breakthrough result in 2011 [Fri11c], the worst-case complexity of Zadeh's pivot rule was unclear. The historic subexponential lower bound of Friedmann is the central topic of Chapter 4. The first exponential lower bound is the main topic of this thesis and presented in Chapters 5 and 6.

11. **The RandomFacet pivot rule:** This pivot rule is again a randomized rule, though it is more involved than the RandomEdge rule. It was independently introduced by Kalai [Kal92, Kal97] and Sharir and Welzl [SW92] and can be described geometrically as follows. Given a vertex x, pick a random facet containing this vertex. Then, solve the problem of finding an optimal vertex restricted to this facet first. If it is possible, move towards this vertex and iterate.

    As for the RandomEdge pivot rule, it was proven that this pivot rule requires an expected subexponential number of iterations in the worst case [FHZ11a]. However, it was also proven that the RandomFacet rule requires no more than that many iterations in expectation [MSW96]. It is thus the only pivot rule guaranteeing a better than exponential number of iterations, at least in expectation.

12. **The randomized Bland's rule:** This pivot rule is a natural randomization of Bland's pivot rule. It first orders the variables uniformly at random and then uses Bland's rule with respect to this ordering.

    It can also be interpreted as a variant of the RandomFacet pivot rule, and we refer to [Mat94] for more details on this.

For all of these pivot rules, the worst case running time is at least subexponential. Interestingly, each of the original worst-case constructions belongs to one of two frameworks. The constructions used for the first six listed pivot rules are all deformed hypercubes that are based on the first lower bound construction of Klee and Minty [KM72]. Amenta and

Ziegler also formalized this observation by introducing *deformed products* and proving that all of these lower bound constructions are in fact deformed products [AZ98].

Although deformed products were applied successfully to many different pivot rules, there were two classes of pivot rules for which they did not provide superpolynomial lower bounds, namely memorizing and randomized pivot rules. Only in 2011, Friedmann et al. were able to devise meaningful lower bounds for those pivot rules (see e.g. [Fri11b, Han12] and the sources mentioned in Table 1.1). Interestingly, all of these lower bounds share many similarities as they are all obtained by applying the strategy improvement algorithm to a Markov decision process that models a binary counter. However, until today, there is no general construction that is comparable to the notion of deformed products generalizing the lower bound examples based on binary counting Markov decision processes. We discuss similarities between these lower bound constructions when discussing the subexponential lower bound of Zadeh's pivot rule in Chapter 4 and develop a new member of this family when proving that Zadeh's pivot rule requires an exponential number of iterations in the worst case in Chapter 5.

Since it seems to be very challenging to find a pivot rule for the simplex algorithm guaranteeing a polynomial number of iterations, several authors began investigating the simplex algorithm from another perspective. Disser and Skutella introduced a new complexity theoretic notion describing the capability of an optimization algorithm [DS15]. They define an algorithm to be *NP-mighty* if it can solve any problem in NP. If P$\neq$NP, then an NP-mighty algorithm cannot be polynomial. They proved that the original version of the simplex algorithm is NP-mighty. This result can be interpreted as the simplex algorithm, which was designed to solve a problem solvable in polynomial time, being "too mighty" for the problem it was originally designed for. Their work thus gave a first hint explaining why it is hard to find a polynomial pivot rule for the simplex algorithm. Since then, similar results were obtained, proving that the simplex algorithm is able to solve hard problems and that several decision problems that are related directly to the simplex algorithm are hard to decide [FS15]. One such result is, for example, that it is PSPACE-hard to decide whether the simplex algorithm ever visits a certain basis.

3. Strategy Improvement and Policy Iteration

Strategy improvement, also often referred to as policy iteration, is an algorithmic framework that is applied to a variety of games. Its main idea is that there are players that use strategies to model their choices. These strategies are pre-ordered in such a way that the computational task associated with the corresponding game is solved if and only if optimal strategies for the players have been found. Then, the process of iteratively improving the players' strategies until optimal strategies have been found is referred to as *strategy improvement* or *policy iteration*, depending on the underlying game.

In this thesis, we focus on the classical policy iteration algorithm for Markov decision processes developed by Howard in 1960 [How60] and the discrete strategy improvement algorithm of Vöge and Jurdziński, developed in 2000 [VJ00]. To simplify our presentation, we refer to both algorithms as "strategy improvement".

3.1. A General Framework for Strategy Improvement

In this chapter, we develop a general algorithmic framework for calculating optimal strategies in Markov decision processes and winning sets in parity games. This framework can be interpreted as a generalization of the discrete strategy improvement algorithm for parity games [VJ00], the policy iteration algorithm for Markov decision processes [How60] and many more. The approach and description given here is mainly based on [Fri11b] and an earlier version of this introduction can be found in [DFH19].

We begin by describing the framework in terms of parity games. Afterwards, we discuss in what way this framework is also applicable in the context of Markov decision processes. Consequently, let G be a parity game. We denote the underlying graph by (V, E) and fix one of the two players, say player 0.

The key idea is to take the perspective of player 0 and develop a notion of optimality for this player by assigning a meaningful *valuation* to every vertex $v \in V$. This valuation encodes how "profitable" the vertex is for player 0. By defining a suitable pre-order on these valuations, this enables us to compare vertices by comparing their valuations. In particular, for a fixed vertex v and a strategy σ, we can compare the valuation of $\sigma(v)$ and other vertices $w \in \Gamma^+(v)$. If there is a vertex $w \in \Gamma^+(v)$ with a strictly better valuation than $\sigma(v)$, we can "improve" the strategy σ by re-defining $\sigma(v) := w$. Since there are only finitely many vertices and strategies, this iterative procedure terminates at some point. The final strategy σ^* is then "optimal" with respect to the previously defined pre-order. The key idea is thus to define a pre-order and vertex valuations in such a way that σ^* is a winning strategy in the sense of Theorem 2.1.6 for player 0. Note that the idea and motivation of such valuations share many similarities with the optimality equations for Markov decision processes discussed in Section 2.2.

We now formalize this idea. The *vertex valuations* are given as a totally ordered set $(U, \preceq)$. For every pair of strategies σ, τ for player $0, 1$, we are given a function $\mathrm{Val}_{\sigma,\tau}\colon V \to U$ assigning valuations to vertices. Note that we intentionally use the same notation that was used when discussing values of vertices in Markov decision processes in Section 2.2. The next step is to eliminate the dependency on the behavior of player 1. As we take the perspective of player 0, we thus assume that player 1 is an adversary working against us. Consequently, as we try to maximize the valuations of the vertices, we assume that player 1 tries to minimize the valuations of the vertices. This allows us to eliminate the dependency on player 1 by setting $\mathrm{Val}_\sigma(v) := \min_\prec \mathrm{Val}_{\sigma,\tau}(v)$ where the minimum is taken over all player 1 strategies τ. Formally, a player 1 strategy is called *counterstrategy* for σ if $\mathrm{Val}_{\sigma,\tau}(v) \preceq \mathrm{Val}_\sigma(v)$ for all $v \in V$. An arbitrary but fixed counterstrategy for σ is denoted by τ^σ. Although it is not obvious, counterstrategies exist and can be computed efficiently [VJ00]. This ordering can now be extended to a partial ordering of strategies. For two player 0 strategies σ, σ', we define $\sigma \trianglelefteq \sigma'$ if and only if $\mathrm{Val}_\sigma(v) \preceq \mathrm{Val}_{\sigma'}(v)$ for all $v \in V$. We write $\sigma \lhd \sigma'$ if $\sigma \trianglelefteq \sigma'$ and there exists a vertex $v \in V$ with $\mathrm{Val}_\sigma(v) \prec \mathrm{Val}_{\sigma'}(v)$, and define $\trianglerighteq, \rhd$ analogously.

The idea is to find a strategy that is maximal with respect to the partial ordering $\trianglelefteq$. In particular, given a player 0 strategy σ, we need to be able to find a strategy σ' with $\sigma \lhd \sigma'$, i.e., a strategy that is strictly better with respect to this partial ordering. This can be done by applying an *improving switch*. Intuitively, an improving switch is an edge such that including e in σ improves the strategy with respect to $\trianglelefteq$.

Definition 3.1.1 (Improving switch). Let G be a parity game and let $\trianglelefteq$ be a partial ordering of strategies induced by vertex valuations as described previously. Let $e = (u, v) \in E_0$ and $\sigma(u) \neq v$, and define the strategy σe via $\sigma e(u) := v$ and $\sigma e(u') := \sigma(u')$ for $u' \in V_0 \setminus \{u\}$. Then, e is *improving* or an *improving switch* for σ if $\sigma \lhd \sigma e$. The set of improving switches for a strategy σ is denoted by I_σ.

We will typically use σe to denote the strategy that is obtained by applying the improving switch e in the strategy σ. This now enables us to formulate the *strategy improvement algorithm* [VJ00]. It operates as follows. Given an initial strategy σ_0, apply improving switches until a strategy σ^* with $I_{\sigma^*} = \emptyset$ is reached. Such a strategy is called *optimal* and the following theorem justifies this name.

Theorem 3.1.2 ([VJ00]). *Let G be a parity game and let $\trianglelefteq$ denote a partial ordering of strategies induced by vertex valuations as described previously. Let σ denote a player 0 strategy for G. Then, $I_\sigma = \emptyset$ if an only if there is no player 0 strategy σ' with $\sigma \lhd \sigma'$.*

The ideas introduced here can be directly applied to Markov decision processes with the expected total reward criterion as follows. In the context of Markov decision processes, the set of vertex valuations is the set of real numbers $\mathbb{R}$ together with the natural ordering. The function assigning valuations is given by the values of the vertices defined by the system (2.1). The intermediate construction of counterstrategies is not necessary for

Markov decision processes as there is only one player. In particular, we can also define $\sigma \trianglelefteq \sigma'$ for policies σ, σ' by setting $\sigma \trianglelefteq \sigma'$ if and only if $\mathrm{Val}_\sigma(v) \leq \mathrm{Val}_{\sigma'}(v)$ for all $v \in V$. We thus define improving switches for the expected total reward criterion analogously to Definition 3.1.1. Since a strategy σ^* is optimal with respect to the expected total reward criterion if and only if $I_{\sigma^*} = \emptyset$, this enables us to describe the algorithms of [VJ00] and [How60] by one algorithmic scheme.

Input: Either parity game or Markov decision process G
Partial ordering $\trianglelefteq$ induced by vertex valuations
Output: Strategy σ^* optimal with respect to $\trianglelefteq$

Let σ be a strategy for G
while $I_\sigma \neq \emptyset$ **do**
  Let $e \in I_\sigma$
  Set $\sigma := \sigma e$
return σ

Algorithm 3: The strategy improvement algorithm.

It is not immediate how this scheme can be applied to the expected average reward criterion. We thus introduce a special subclass of Markov decision processes in the next section for which optimizing with respect to the expected average reward criterion is equivalent to optimizing with respect to the expected total reward criterion. Also, it is not obvious how the vertex valuations for parity games should be defined. Although strategy improvement is applicable to all parity games, we introduce a subclass of parity games that shares many similarities with the subclass considered for Markov decision processes. These valuations will then turn out to have the desired property that a strategy σ is optimal with respect to the induced partial ordering of strategies if and only if σ is a winning strategy for player 0.

3.2. Sink Parity Games and the (Weak) Unichain Condition

In this section, we introduce subclasses of parity games and Markov decision processes that allow us to simplify proofs and statements significantly. Both subclasses are already known and were investigated in the past. Nevertheless, we give a full introduction here, stressing in particular their similarities and providing a new perspective on them.

The subclass of parity games is the class of *sink games*. These were introduced and studied in [Fri11b]. For Markov decision processes, the subclass we consider is closely related to the *unichain condition*. This condition is a well-known and studied property of Markov decision processes and we refer to [Put05] for an in-depth discussion of them. The processes considered in this thesis will however not have the traditional unichain condition. We instead introduce and discuss the so-called *weak unichain condition*, and refer to Markov decision processes having the weak unichain condition as weakly unichain.

Although the weak unichain condition is also already known, it is rarely considered as there are only few cases where the weak unichain condition is fulfilled although the traditional unichain condition is not fulfilled.

The design principle of both sink games and weakly unichain Markov decision processes is very similar. The key idea is that the underlying graph contains a *sink*, which is a vertex looping to itself, and that the whole graph "leads" to this sink. That is, independent of the choices of the player(s), every play resp. every walk reaches the sink. For a Markov decision process, this guarantees that the values with respect to the expected total reward criterion are finite as follows. If the reward of the edge building the loop is zero and if the Markov Chain $\mathrm{MC}(G, \sigma)$ reaches the sink for every strategy σ with probability 1, then the expected total reward is the expected reward obtained until reaching the sink. Since the rewards are finite, this guarantees finiteness of the values. In particular, the expected average reward criterion then reduces to the expected total reward criterion as the gain of every vertex is equal to zero and the bias corresponds to the expected reward obtained until reaching the sink. For a parity game, we can use such a sink to ensure that player 1 wins every play by setting the priority of the sink to an odd number. This allows for a significantly easier definition of vertex valuations as general vertex valuations have to evaluate the vertices that are visited infinitely often. Our presentation here also gives a new perspective and generalizes the original definition of sink games as this relies on an initially chosen strategy. In contrast to this definition, we define sink games independent of strategies and introduce the new notion of *sink strategies* instead.

Sink games

We begin with the discussion and introduction of sink games. Our approach differs from the original approach of [Fri11b, Chapter 4.2] and we define some terms differently. The reason is that sink games were originally introduced as a subclass of parity games that simplified the vertex valuations. However, this approach can only be used if vertex valuations were introduced for general parity games before, as the original definition given in [Fri11b] uses vertex valuations. As we only consider sink games in this thesis and never use the general vertex valuations, we present a different approach for defining and introducing sink games that does not require introducing vertex valuations for general parity games.

We begin by defining the term *sink* for parity games and the corresponding term *sink game* (cf. the sink existence property in [Fri11b]). Note that [Fri11b] defines the term "sink game" differently as it depends on a given strategy. As we however want to provide a definition and framework that clearly distinguishes between games and strategies, we consider the following definition here.

Definition 3.2.1 (Sink (parity game)). Let G be a parity game. A vertex $t \in V$ is called *sink* of G if $\Gamma^+(t) = \{t\}, \Omega(t) = 1 < \Omega(v)$ for all $v \in V \setminus \{t\}$ and if it is reachable from all vertices. A parity game that contains a sink is a *sink game* if player 1 wins every vertex.

This allows us to define the term of a *sink strategy*. A sink strategy is a player 0 strategy in a sink game such that player 1 can force the corresponding play to end in the sink.

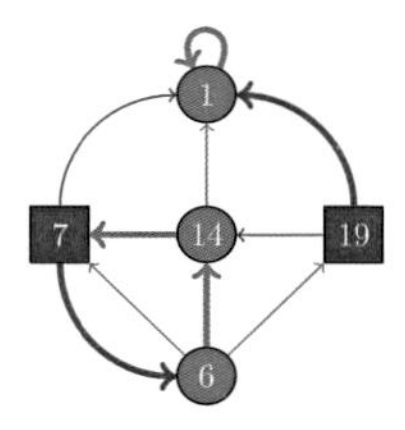

Figure 3.1.: An example of a sink game.

Definition 3.2.2 (Sink strategy). Let G be a sink game and let t denote the sink. Then, a strategy σ is called *sink strategy* if there is a player 1 strategy τ such that for every vertex $v \in V$, it holds that $\pi_{v,\sigma,\tau} = v, v_1, v_2, \dots, v_k, (t)^\infty$.

Example 5 (Sink game). *We consider a slightly altered version of the parity game introduced in Example 1 where the priority of the top vertex is changed to* 1*, see Figure 3.1. We in particular consider the same player* 0 *and player* 1 *strategies* σ_0, σ_1*. The top vertex of this parity game is a sink. Player* 1 *wins the plays starting in the two vertices* 19 *and* 1*. When these two strategies are considered, player* 1 *does not win the play starting in either of the vertices* 6, 14 *or* 7*. However, player* 1 *can win any play starting in these vertices by setting* $\sigma_1(7) := 1$*. Consequently, player* 1 *wins every vertex of this parity game and the game is thus a sink game. In particular,* σ_0 *is a sink strategy and, in fact, every strategy for player* 0 *is a sink strategy in this example.*

The next goal is to show the following. If G is a sink game and if σ is a sink strategy, then applying an improving switch to σ yields another sink strategy. The problem is that we still did not introduce vertex valuations for sink games and hence, the term "improving switch" is not yet well-defined. We thus investigate what happens when an *arbitrary* edge $e = (u, v) \in E_0$ with $\sigma(u) \neq v$ is switched. As it turns out, there are two possibilities. Either σe is a sink strategy, or there is a player 1 strategy τ such that the cycle component of $\pi_{u,\sigma e,\tau}$ contains a vertex with high odd priority. Arguably, the latter case is bad for player 0 and in particular worse than any play reaching the sink, motivating that such an edge e should never be applied. We then use this insight to define vertex valuations for sink games.

Lemma 3.2.3. *Let G be a sink game, σ be a sink strategy and $e = (u, v) \in E_0$ such that $\sigma(u) \neq v$. If σe is not a sink strategy, then there is a player* 1 *strategy τ such that* $\max\{\Omega(w) : w \in C(\pi_{u,\sigma e,\tau})\} \bmod 2 = 1$ *and* $\max\{\Omega(w) : w \in C(\pi_{u,\sigma e,\tau})\} > 1$.

Proof. Assume that σe is not a sink strategy. Then, there is a vertex $u \in V$ such that for every player 1 strategy τ, the play $\pi_{u,\sigma e,\tau}$ does not reach the sink t, so $t \notin C(\pi_{u,\sigma e,\tau})$. Now, for the sake of contradiction, assume that player 0 wins $\pi_{u,\sigma e,\tau}$ for every player 1 strategy τ. Then, for every player 1 strategy τ and every induced play $\pi_{u,\sigma e,\tau}$, it holds that $\max\{\Omega(w) : w \in C(\pi_{u,\sigma e,\tau})\}$ is even. But this implies that player 0 wins vertex u, contradicting the definition of a sink game, as player 1 wins every vertex in a sink game.

Consequently, there is at least one player 1 strategy τ' such that player 1 wins $\pi_{u,\sigma e,\tau'}$. Since σe is not a sink strategy, $t \notin C(\pi_{u,\sigma e,\tau'})$. Since player 1 wins $\pi_{u,\sigma e,\tau'}$, this implies $\max\{\Omega(w) : w \in C(\pi_{u,\sigma e,\tau'})\} \bmod 2 = 1$. In addition, the definition of a sink game implies that $\max\{\Omega(w) : w \in C(\pi_{u,\sigma e,\tau'})\} > 1$ as the sink has the lowest priority. □

Lemma 3.2.3 shows that including an edge e into a sink strategy σ has one of two effects. Either σe is again a sink strategy or σe is a strategy that is arguably more profitable for player 1 and, consequently, less profitable for player 0. The following definition of vertex valuations for sink games implements this insight. Assume that the initial strategy provided to the strategy improvement algorithm is a sink strategy. Then, any meaningful definition of vertex valuations should prohibit the application of an improving switch such that the obtained strategy is worse for player 0. Consequently, such a definition would imply that applying an improving switch to a sink strategy yields another sink strategy. By Lemma 3.2.3, this implies that all strategies obtained by applying improving switches are sink strategies. In order to compare strategies, it thus suffices to be able to define vertex valuations for comparing paths to the sink of the sink game. It is thus sufficient to define the valuation of a vertex v under the strategies σ, τ as the *path component* of $\pi_{v,\sigma,\tau}$. In fact, this is a well-studied choice of vertex valuations and is exactly the same choice used by [VJ00] and [Fri11b] when applied to sink games. To give a total ordering of the vertex valuations, it is thus sufficient to give an ordering of all subsets of V.

Let $M, N \subseteq V$ and $M \neq N$. Intuitively, N is better than M for player 0 if it contains a vertex with a high even priority not contained in M. Analogously, M is worse than N for player 0 if it contains a vertex with a high odd priority not contained in N. We thus need to analyze the symmetrical difference of M and N and thus introduce the following term.

Definition 3.2.4 (Most significant difference). Let G be a sink game and $M, N \subseteq V$ with $M \neq N$. The vertex $v \in M\Delta N$ is the *most significant difference of M and N* if $\Omega(v) > \Omega(w)$ for all $w \in M\Delta N, w \neq v$ and is denoted by $\Delta(M, N)$.

We now define an ordering "$\lhd$" on the set of subsets of V. For $M, N \subset V, M \neq N$ let

$$\begin{aligned} M \lhd N :\Longleftrightarrow \quad & [\Delta(M,N) \in N \wedge \Omega(\Delta(M,N)) \bmod 2 = 0] \\ & \vee[\Delta(M,N) \in M \wedge \Omega(\Delta(M,N)) \bmod 2 = 1]. \end{aligned}$$

Note that $\lhd$ only defines a pre-order on the subsets of V if the priority function is not injective. Although injectivity of Ω implies that $\lhd$ is a proper ordering of the subsets of V, it is sufficient if the most significant difference between any two vertex valuations is unique. This will be the case in our constructions.

The framework now justifies to define vertex valuations for sink games as follows.

Definition 3.2.5 (Vertex valuations (sink game)). Let G be a sink game and let $v \in V$. Let σ be a sink strategy and let τ denote a counterstrategy for σ. Then, the *valuation* of v with respect to σ and τ is the path component of the play $\pi_{v,\sigma,\tau}$.

The following theorem summarizes the most important aspects related to parity and sink games and vertex valuations and improving switches for these. As mentioned earlier, the vertex valuations constructed here are a simplified version of the general concept and construction of vertex valuations. They are, however, in accordance with the general construction for parity games and we refer to [Fri11b] for a detailed discussion.

Theorem 3.2.6 ([VJ00]). *Let G be a sink game and let σ be a sink strategy.*

1. *The vertex valuations of a player 0 strategy can be computed in polynomial time.*
2. *There is a sink strategy σ^* that is optimal with respect to the ordering $\lhd$.*
3. *It holds that $I_\sigma = \emptyset$ if and only if there is no strategy σ' with $\sigma \lhd \sigma'$.*
4. *It holds that $I_\sigma = \{(u,v) \in E_0 : \mathrm{Val}_\sigma(\sigma(u)) \lhd \mathrm{Val}_\sigma(v)\}$ and $\sigma \lhd \sigma e$ for all $e \in I_\sigma$.*

Example 6. *We again consider the sink game introduced in Example 5. As argued previously, player 1 can choose his strategy in such a way that every play has cycle component $(1)^\infty$. An example of a player 0 strategy σ together with the corresponding counterstrategy of player 1 is given in Figure 3.2*

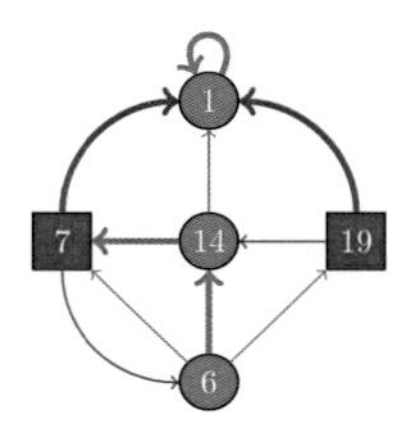

Figure 3.2.: A sink game in which player 1 chooses the strategy such that the cycle component of every play is $(1)^\infty$.

The valuation of the vertices is the path leading to the sink. Thus, $\mathrm{Val}_\sigma(19) = \{19\}$ *resp.* $\mathrm{Val}_\sigma(7) = \{7\}$ *for the vertices of player* 1 *and* $\mathrm{Val}_\sigma(1) = \emptyset, \mathrm{Val}_\sigma(14) = \{14, 7\}$ *resp.* $\mathrm{Val}_\sigma(6) = \{6, 14, 7\}$. *In particular, this implies that* $\Delta(\mathrm{Val}_\sigma(7), \mathrm{Val}_\sigma(1)) = \{7\}$. *Since* 7 *is odd, this implies* $\{7\} \prec \emptyset$ *and thus* $\mathrm{Val}_\sigma(7) \lhd \mathrm{Val}_\sigma(1)$. *Consequently, the edge* $(14, 1)$ *is an improving edge in this example.*

This concludes our discussion regarding sink games. We now discuss how the concept of a sink vertex can be transferred to Markov decision processes.

Markov decision processes and sinks

We now consider Markov decision processes that contain a *sink* $t \in V_0$. This vertex has a single outgoing edge (t,t) with reward 0. As for sink games, the idea is that *every* infinite walk reaches this vertex at some point - regardless the strategy. If this is the case, then the expected total reward is well defined and equal to the expected reward collected until reaching t. This reduces the complexity of the vertex valuations by considering a subclass such that the valuation is equal to the valuation of the path leading to the sink. In particular, this then implies the expected average reward is equal to 0 for every strategy.

More precisely, for the expected average reward criterion, the system of gain and bias equations reduces to a significantly simpler system as only biases need to be considered. As it turns out, this "reduced" system of biases is equivalent to the system (2.1), and the two criteria are therefore identical.

We now formalize this intuition. The property that we just described is known as the *unichain condition* and Markov decision processes that have this condition are called *unichain*. Unichain Markov decision processes are well-studied and understood objects, and we refer to [Put05] for more details. For our purposes, the unichain condition is however too strong. As it turns out, it is not necessary that the vertex t is reached by every strategy, it is only required for *certain* strategies. More precisely, when applying the strategy iteration algorithm, it turns out that it suffices if the optimal strategy reaches the sink with probability 1 and if a suitable initial strategy is chosen. This yields the *weak unichain condition*. In all of the following definitions and statements, we let G be a Markov decision process and use the notation of Definition 2.2.1.

Definition 3.2.7 (Sink (Markov decision process)). A vertex $t \in V_0$ is called *sink* of G if $\Gamma^+(t) = \{t\}, r(t,t) = 0$ and if it is reachable from all vertices.

Definition 3.2.8 (Weak unichain condition). Let σ be a strategy for G. Then, σ is a *weak unichain strategy for G* if G has a sink t that is the single irreducible recurrent class of $\mathrm{MC}(G,\sigma)$. If there is at least one weak unichain strategy for G, then G is *weakly unichain*.

We now investigate weakly unichain Markov decision processes. We first prove that the values of the vertices defined in Definition 2.2.4 are finite for weak unichain strategies.

Lemma 3.2.9. *Consider the total expected reward criterion. Let σ be a weak unichain strategy for G. Then, the values of the vertices are finite.*

Proof. Since G is weakly unichain, it has a sink t. By definition, t is the single irreducible recurrent class of $\mathrm{MC}(G,\sigma)$. As we impose the condition that the value of each vertex contained in such a class is zero, this implies $\mathrm{Val}_\sigma(t) = 0$. In addition, $\mathrm{MC}(G,\sigma)$ reaches t with probability 1 after finitely many steps. This implies that the value of any vertex is equal to the expected sum of rewards obtained before reaching t. Since rewards are finite, this implies that all values are finite. □

We investigate the expected average reward criterion next. We prove that the gain of every vertex is 0 with respect to a weak unichain strategy. We furthermore prove that setting the bias of the sink t to 0 implies that the optimality criteria introduced in Definitions 2.2.4 and 2.2.6 are equivalent for weak unichain Markov decision processes. As we introduced gain and bias only for bipartite Markov decision processes, we again assume here that G is bipartite. However, by using a more complicated definition of gains and biases, the following theorem can be generalized for general Markov decision processes.

Lemma 3.2.10. *Let G be bipartite and σ be a weak unichain strategy for G. Let $t \in V_0$ be the sink. Then, $G_\sigma(v) = 0$ for all $v \in V_0$. Furthermore, when setting $B_\sigma(t) = 0$ and $B_\sigma(v) = \sum_{w\in\Gamma^+(v)} p(v,w)B_\sigma(w)$ for all $v \in V_R$, then the values B_σ are a solution of the system defining the total expected reward criterion.*

Proof. By [Put05, Theorem 8.2.6], the gain $G_\sigma(u)$ of a state $u \in V_0$ is the expected average reward obtained by starting in u and following σ. Since σ is a weak unichain strategy for G, the graph contains a sink $t \in V_0$ such that $\mathrm{MC}(G,\sigma)$ reaches t after a finite time. By construction, the only outgoing edge of this vertex is (t,t) with $r(t,t) = 0$. But this implies that the expected average reward obtained for every state $u \in V_0$ is equal to 0.

We next show that it is feasible to set $B_\sigma(t) = 0$. By Definition 2.2.6 and since $G_\sigma(t) = 0$,

$$B_\sigma(t) = r(t,\sigma(t)) - G_\sigma(t) + B_\sigma(t) = r(t,t) + B_\sigma(t) = B_\sigma(t).$$

We can hence set $B_\sigma(t) := 0$.

We extend the notion of the bias to randomization vertices. For $u \in V_R$, we define $B_\sigma(u) := \sum_{v\in\Gamma^+(u)} p(u,v)\cdot B_\sigma(v)$. Then, since $G_\sigma(u) = 0$ for all $u \in V_0$, the system (2.5) can be rewritten equivalently as

$$B_\sigma(u) = \begin{cases} r(u,\sigma(u)) + B_\sigma(\sigma(u)), & u \in V_0 \\ \sum_{v\in\Gamma^+(u)} p(u,v)\cdot B_\sigma(v), & u \in V_R \end{cases}.$$

This is exactly the system (2.1) defining the total expected reward criterion. In addition, since t is the single irreducible recurrent class of $\mathrm{MC}(G,\sigma)$, the value of any such vertex is equal to zero. □

Corollary 3.2.11. *Let σ be a weak unichain strategy for G. Then, σ is optimal for the expected total reward criterion if and only if it is optimal for the expected average reward criterion.*

These results show that it is sufficient to consider weakly unichain Markov decision processes with the expected total reward criterion. It remains to discuss what happens if an improving switch is applied to a strategy that has the weak unichain condition. In theory, it might happen that applying an improving switch either creates a cycle such that the reward collected in this cycle is not finite or that another irreducible recurrent class is created. We address the first issue by simply assuming that the expected total reward criterion is well-defined for all strategies. A Markov decision process with this property is called *finite*, and although it might seem like a very strong condition, it can be verified rather easily.

Definition 3.2.12 (Finite Markov decision process). G is called *finite* if the expected total reward criterion introduced in Definition 2.2.4 is well-defined for all strategies.

To verify whether a Markov decision process is finite, it suffices to provide a strategy that solves the optimality equations. Such a strategy is known to maximize the expected total rewards and hence the values of the vertices. It is thus a witness that the values of all vertices are finite. Although this is not a feasible method to detect finiteness in practice, it is sufficient for the construction of specific examples for which optimal policies are known.

Observation 3.2.13. G is finite if and only if there is a strategy fulfilling the optimality equations.

To conclude this section, we now prove that applying an improving switch to a weak unichain strategy yields another weak unichain strategy. To avoid confusion, we restate the definition of an improving switch for finite Markov decision processes. Note however that this definition is just a reformulation of Definition 3.1.1.

Definition 3.2.14 (Improving Switch (MDP)). Let G be finite and let σ be a strategy for G. Consider the expected total reward criterion. An edge $e = (u, v) \in E_0$ with $\sigma(u) \neq v$ is called *improving switch*, if $r(u, v) + \mathrm{Val}_\sigma(v) > r(u, \sigma(u)) + \mathrm{Val}_\sigma(\sigma(u))$.

Lemma 3.2.15. *Let G be finite and consider the total expected reward criterion. Let σ be a weak unichain strategy for G. Assume $I_\sigma \neq \emptyset$ and let $e \in I_\sigma$. Then σe is a weak unichain strategy for G.*

Proof. Let $e = (u, v)$. Since σ is a weak unichain strategy for G, there is a sink t which is the single irreducible recurrent class of $\mathrm{MC}(G, \sigma)$. Consider the Markov chain $\mathrm{MC}(G, \sigma e)$. Then, t is still an irreducible recurrent class of this Markov chain as $\sigma(t) = \sigma e(t) = t$ due to $\Gamma^+(t) = \{t\}$. It thus suffices to prove that it is the only such class.

For the sake of a contradiction, assume that there is another irreducible recurrent class in $\mathrm{MC}(G, \sigma e)$. We denote the states of this class by C and observe that $t \notin C$ since t is an absorbing state and C is irreducible. Since every state in C is recurrent, every state in C is encountered infinitely many times. In particular, when the system is in some state $c \in C$, then it will only visit states contained in C in the future.

Fix some state $c \in C$ and consider the Markov chain $\mathrm{MC}(G, \sigma)$ with initial state c. Since σ is a weak unichain strategy for G, this chain reaches state t with probability 1. Now, consider $\mathrm{MC}(G, \sigma e)$ with initial state c. If the probability of reaching either u or v is 0 in $\mathrm{MC}(G, \sigma)$ with initial state c, then the same is true for $\mathrm{MC}(G, \sigma e)$. But this directly implies that $\mathrm{MC}(G, \sigma e)$ reaches t with probability 1 as $\mathrm{MC}(G, \sigma)$ reaches t with probability 1, contradicting $t \notin C$. We hence need to have either $u \in C$ or $v \in C$. We now show that both statements are true. If $u \in C$, then this also implies $v \in C$ due to $\sigma e(u) = v$. Assume $v \in C$ but $u \notin C$. Then, the Markov chain $\mathrm{MC}(G, \sigma e)$ never reaches the sink t as $t \notin C$. However, since $u \notin C$, the choice of e implies $\sigma(w) = \sigma e(w)$ for all $w \in C$. But this implies that the set of states reachable in $\mathrm{MC}(G, \sigma)$ and $\mathrm{MC}(G, \sigma e)$ is the same, contradicting that t is reached with probability 1 in $\mathrm{MC}(G, \sigma)$ when starting in v. Consequently, $u, v \in C$.

Since G is finite, the expected total reward is well-defined for all policies, so in particular for σe. Since C is an irreducible recurrent class, this implies $\mathrm{Val}_{\sigma e}(w) = 0$ for all $w \in C$. Now consider $w \in C \cap V_0$. Then $\sigma e(w) \in C$, so $\mathrm{Val}_{\sigma e}(w) = r(w, \sigma e(w)) + \mathrm{Val}_{\sigma e}(\sigma e(w)) = 0$. But then, $\mathrm{Val}_{\sigma e}(\sigma e(w)) = 0$ implies that $r(w, \sigma e(w)) = 0$. This in particular implies $r(w, \sigma(w)) = 0$ for all $w \in C \cap V_0, w \neq u$ and $r(u, v) = 0$.

Now consider the Markov chain $\mathrm{MC}(G, \sigma)$ and let v be the initial state. Then, since $\sigma(w) = \sigma e(w)$ for all $w \neq u$ and since C is an irreducible recurrent class of $\mathrm{MC}(G, \sigma e)$, this implies that $\mathrm{MC}(G, \sigma)$ will only visit states contained in C until reaching state u for the first time. Since $r(w, \sigma(w)) = 0$ for all $w \in C \cap V_0, w \neq u$, this implies $\mathrm{Val}_\sigma(v) = \mathrm{Val}_\sigma(u)$. But this is a contradiction as $e = (u, v)$ being improving for σ yields

$$\mathrm{Val}_\sigma(u) = \mathrm{Val}_\sigma(v) = r(u, v) + \mathrm{Val}_\sigma(v) > r(u, \sigma(u)) + \mathrm{Val}_\sigma(\sigma(u)) = \mathrm{Val}_\sigma(u). \qquad \square$$

Corollary 3.2.16. *Let G be finite and let σ be a strategy for G. Let $u, v \in V_0$ be two vertices that are contained in an irreducible recurrence class of $\mathrm{MC}(G, \sigma)$. Then $r(u, v) = 0$.*

This implies that the strategy improvement algorithm only produces weak unichain strategies if the Markov decision process is finite and the initial strategy is weakly unichain.

Corollary 3.2.17. *Let G be finite and let σ be a weak unichain strategy. If the strategy improvement algorithm uses σ as initial strategy, then every strategy calculated by the algorithm is weakly unichain. In particular, the optimal strategy is weakly unichain.*

Corollary 3.2.18. *If G is a finite Markov decision process that admits a weak unichain strategy, then there is an weakly unichain optimal strategy σ^* for G.*

3.3. Strategy Improvement and the Simplex Algorithm

We now discuss the connection between the strategy improvement algorithm for Markov decision processes and the simplex algorithm for linear programs. Already in the 1960s, Markov decision processes were formulated as linear programs and techniques developed for linear programs were used for solving them [Man60, d'E63]. In particular, all complexity theoretic results applicable to linear programs imply the corresponding statement for Markov decision processes. Most notably, this connection proves that Markov decision processes can be solved in weakly polynomial time [Kha80, Kar84]. We provide and discuss a linear program such that basic feasible solutions of this program are in bijection with strategies of a given weakly unichain Markov decision process. Although this linear program is well-known and discussed in the literature, it is typically only considered for unichain Markov decision processes, i.e., for processes for which every strategy reaches the sink with probability 1. Also, linear programs are usually described for Markov decision processes that are defined in the terms of actions and states instead of a graph containing player and randomization vertices. In this thesis, we discuss this connection for weak unichain Markov decision processes in the "player-random" setting. The results and the discussion here are adapted from [Han12, Chapter 2.4], and we refer to [Put05, Han12] for proofs and further details.

The idea of the linear programming formulation for the expected total reward criterion is the following. Let G be a finite Markov decision process, let $|V_0| = n_0, |V_R| = n_R$ and assume without loss of generality that G is bipartite. Since we only have rewards on the player edges, we want to maximize a linear objective function of the type $\sum_{e \in E_0} r(e)x(e)$, where $x \in \mathbb{R}^{|E_0|}$. Intuitively, $x(e)$ should be the expected number of times the edge e is traversed by a strategy. Note however that this interpretation is not valid for edges between vertices in an irreducible recurrent class of a strategy since these edges are taken an infinite number of times. For convenience, the value of the variables of such edges is set to 1. This does also not interfere with the objective function as the reward between two such vertices is always 0 by Corollary 3.2.16.

As we only consider Markov decision processes that have a sink t in this thesis, we only give the linear programming formulation for such Markov decision processes. Further note

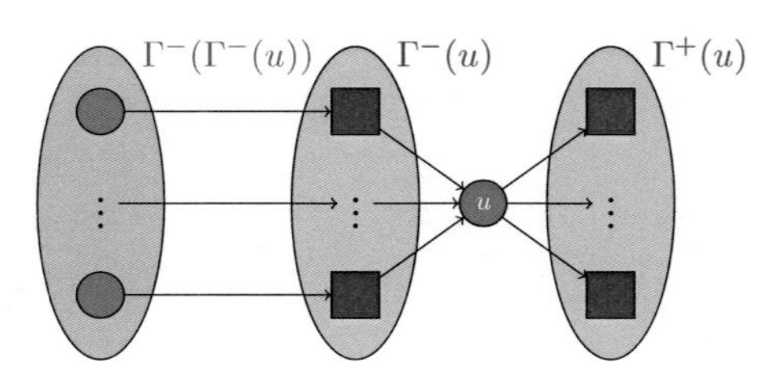

Figure 3.3.: Visualization of the constraints of the linear program (3.1) for solving finite bipartite Markov decision processes.

that the given formulation only applies to bipartite Markov decision processes. For non-bipartite Markov decision processes, a similar but more involved linear program can be formulated. We also assume for simplicity that the given Markov decision process is finite.

Using the previous given interpretation, the problem of finding an optimal strategy in a finite bipartite Markov decision process can be solved by the following linear program.

$$\begin{aligned} \max \quad & \sum_{e \in E_0} r(e)x(e) \\ \text{subject to} \quad & \sum_{(u,v)\in E_0} x(u,v) - \sum_{\substack{(v',w)\in E_0 \\ (w,u)\in E_R}} x(v',w)p(w,u) = 1 \quad \forall u \in V_0 \setminus \{t\} \\ & x(t,t) - \sum_{\substack{(v',w)\in E_0 : v' \neq t \\ (w,u)\in E_R}} x(v',w)p(w,t) = 1 \\ & x(u,v) \geq 0 \quad \forall (u,v) \in E_0 \end{aligned} \tag{3.1}$$

The constraints are visualized in Figure 3.3. For a vertex $u \in V_0 \setminus \{t\}$, the constraint can be interpreted as a combination of flow conservation and choosing an outgoing edge. The flow leaving u is described by the first sum, the flow entering u is described by the second sum. Classical flow conservation would then demand that these two quantities are the same, hence the right-hand side would be equal to zero. However, in this setting, every player vertex should also select an outgoing edge. Consequently, the right-hand side in this linear program is set to 1. A slightly adapted version of this constraint can also be used to describe the flow on the edge (t,t), although it is also possible to just set the value of the corresponding variable to 1. Of course, the values of the variables are bounded from below by 0.

We now show that this linear program can be used to find an optimal strategy if G is weakly unichain. We begin by describing how a weak unichain strategy for G defines a basic feasible solution of the linear program (3.1). Hence, let σ be a weak unichain strategy for G and let t denote the sink of G. Then, the values of the vertices are the unique solution of the system

$$\begin{aligned} \mathrm{Val}_\sigma(t) &= 0, \\ \mathrm{Val}_\sigma(u) &= \begin{cases} r(u,\sigma(u)) + \mathrm{Val}_\sigma(\sigma(u)), & u \in V_0 \setminus \{t\}, \\ \sum_{v \in \Gamma^+(u)} p(u,v)\,\mathrm{Val}_\sigma(v), & u \in V_R \end{cases}. \end{aligned} \tag{3.2}$$

In this system, the values of the randomization vertices can be calculated if the values of the player vertices are known. In particular, since G is bipartite, the system can be written equivalently as a system that only considers player vertices. We thus obtain the system

$$\begin{aligned} \mathrm{Val}_\sigma(t) = r(t,\sigma(t)) = 0, \\ \mathrm{Val}_\sigma(u) - \sum_{v \in \Gamma^+(\sigma(u))} p(\sigma(u),v) \cdot \mathrm{Val}_\sigma(v) = r(u,\sigma(u)) \qquad \forall u \in V_0 \setminus \{t\}. \end{aligned} \tag{3.3}$$

For the remainder of this section, fix an ordering S of V_0 and identify vertices with their position in that ordering. We denote the reduced vector of rewards corresponding to the edges used in σ with respect to the ordering by $r^\sigma := (r_{(1,\sigma(1))}, \dots, r_{(n_0,\sigma(n_0))}) \in \mathbb{R}^{n_0}$. We can now analogously define a matrix $P^\sigma \in \mathbb{R}^{n_0 \times n_0}$ where row and column i correspond to vertex i. More precisely, we define the matrix elementwise via

$$P^\sigma_{i,j} := \begin{cases} 1, & i = j = t, \\ 1 - p(i,i), & i = j \neq t, \\ -p(\sigma(i),j), & j \in \Gamma^+(\sigma(i)), \\ 0, & \text{else.} \end{cases}$$

Using this notation, the system (3.3) can then be rewritten as $P^\sigma \cdot v = r^\sigma$. In particular, since this system has a unique solution as it describes the values of the vertices, the matrix P^σ is non-singular and it holds that $v = (P^\sigma)^{-1} r_\sigma$. The matrix P^σ can thus be used to calculate the values of the vertices. It can, in addition, also be used to find feasible solutions of the linear program (3.1) as follows. If the reward of every player edge was equal to 1, then $v = (P^\sigma)^{-1} \mathbb{1}_{n_0}$, where $\mathbb{1}_{n_0}$ denotes the n_0-dimensional vector containing 1 in every element. In particular, the values of the vertices are given by the row sums of $(P^\sigma)^{-1}$. As it turns out, the column sums have a similar interpretation. Consider the vector $x^\sigma := (\mathbb{1}_{n_0}^T (P^\sigma)^{-1})^T \in \mathbb{R}^{n_0}$ of column sums of $(P^\sigma)^{-1}$. This vector is sometimes also called the *flux vector* as it describes the amount of "flow" that is sent along the edges $(u, \sigma(u))$. In particular, this corresponds directly to feasible solutions of the linear program (3.1), as the variables of this program exactly describe this quantity. This solution x is called the solution *induced* by the strategy σ. The flux vector can thus naturally be extended to a solution of the linear program (3.1) by setting $x(u,\sigma(u)) := x^\sigma_u$ and $x(u,v) := 0$ if $v \neq \sigma(u)$. By interpreting the matrix P^σ as the sum of the n_0-dimensional identity and a matrix of transition probabilities, it can be shown that the vector x is in fact a feasible solution if σ is a weak unichain strategy.

Lemma 3.3.1 (cf. Lemma 2.4.8 of [Han12]). *Let σ be a weak unichain strategy for a finite bipartite Markov decision process G. Then, $x^\sigma \geq \mathbb{1}_{n_0}$ and the induced solution x is a feasible solution for the linear program (3.1).*

It is thus possible to identify weak unichain strategies with feasible solutions of the linear program (3.1). There is a similar lemma that indicates that a feasible solution corresponds to at least one possible choice of a strategy. Given a feasible solution x, there is at least one edge (u, v) for every $u \in V_0$ such that $x(u, v) > 0$ (see Lemma 2.4.9 in [Han12]). In particular, it can be shown that feasible solutions obtained via the calculation described above are indeed basic feasible solutions and that there are no other basic feasible solutions.

Lemma 3.3.2 (cf. Lemma 2.4.10 in [Han12]). *Let G be a finite bipartite Markov decision process. For every weak unichain strategy σ, the induced vector x is a basic feasible solution of the linear program (3.1) with basis $\{(u, \sigma(u)) : u \in V_0\}$. Every basic feasible solution is induced by some weak unichain strategy σ' for G. In particular, there is a bijection between basic feasible solutions of the linear program (3.1) and policies of G.*

These are not all connections between the Markov decision process and the given linear program. Most importantly for this work, applying the simplex algorithm to the linear program (3.1) and applying the strategy improvement algorithm to G is practically the same. First, it is possible to define reduced costs for Markov decision processes as follows.

Definition 3.3.3 (Reduced costs). Let G be a finite bipartite Markov decision process and let σ be a weak unichain strategy for G. Then, $r(u, v) + \mathrm{Val}_\sigma(v) - \mathrm{Val}_\sigma(u)$ is the *reduced cost* of the player edge $(u, v) \in E_0$.

This definition of reduced costs is the same as the definition of reduced costs in the sense of Definition 2.3.17 for the linear program (3.1). Furthermore, it is possible to define the term "improving switch" in terms of Definition 3.3.3. More precisely, a player edge is an improving switch if and only if its reduced cost is strictly larger than 0. In addition, given a weak unichain strategy σ, the reduced cost of an edge in the sense of Definition 3.3.3 is the same as the reduced cost of the corresponding variable with respect to the basic feasible solution induced by σ. This implies that applying an improving switch $(u, v) \in E_0$ in the Markov decision process corresponds to changing the corresponding basis by exchanging $(u, \sigma(u))$ with (u, v). In particular, each such step is non-degenerate and we obtain the following result.

Theorem 3.3.4. *Let G be a finite bipartite Markov decision process and let σ be a weak unichain strategy for G. Then, applying the strategy improvement algorithm to G is equivalent to applying the simplex algorithm to the linear program (3.1) if the same rule for choosing the entering variable is used and only one improving switch is applied per iteration.*

Since a non-bipartite Markov decision process can be transformed into bipartite Markov decision process by only including a linear number of additional vertices and edges, we obtain the following result.

Corollary 3.3.5. *Let G be a finite Markov decision process and let σ be a weak unichain strategy for G. Assume that applying the strategy improvement algorithm with initial strategy σ and a fixed rule of choosing improving switches requires N iterations. Further assume that the algorithm only performs one improving switch per iteration. Then, there is a linear program of the same asymptotic size as G such that applying the simplex algorithm with the corresponding fixed rule of choosing entering variables requires N iterations.*

4. On Friedmann's Subexponential Lower Bound for Zadeh's Pivot Rule

In this chapter we describe and discuss the subexponential lower bound for Zadeh's pivot rule for the strategy improvement algorithm applied to parity games and Markov decision processes and the simplex algorithm for linear programs. This lower bound was originally proven by Oliver Friedmann in 2011 [Fri11c] and answered the question whether the simplex algorithm has polynomial running time when using Zadeh's pivot rule which was unclear for more than 30 years. The original proof presented in [Fri11c] however contains several minor and one major flaw. The minor flaws only require small changes to the specifications of the initial policy and the occurrence records, respectively, and can be resolved rather easily. The major flaw is more severe, as we prove that the sequence of improving switches applied in [Fri11c] does not consistently follow Zadeh's pivot rule (Issues 4.3.5 and 4.3.7). We furthermore prove that the way improving switches are applied is a special case of a general class of possible applications of improving switches, and that no application that is a member of this class follows Zadeh's pivot rule (Issue 4.3.12). We resolve this issue by providing a significantly more sophisticated ordering and associated tie-breaking rule that are in accordance with Zadeh's pivot rule (Theorems 4.4.14 and 4.4.15). As our result does not require us to change the structure of Friedmann's construction, we are able to retain is original result.

All of the results presented here were previously published in [DH18, DH19]. The author of this thesis contributed all of the results and proofs of these works.

4.1. Description of the Construction

We begin by describing the idea and conceptual structure of the construction developed. The key observation that is implemented by the construction is that an n-digit binary counter enumerates 2^n numbers when counting from 0 to $2^n - 1$. The idea is thus to design a weakly unichain Markov decision process that implements such a binary counter. In this Markov decision process, certain strategies can be interpreted as binary numbers. If the strategy iteration algorithm enumerates the strategies such that each number is represented at least once, this yields a (sub-)exponential lower bound for the strategy iteration algorithm and the simplex algorithm if the size of the Markov decision process is polynomial in n.

The general design principle and binary counting

This general idea of constructing lower bounds was used by several authors in the last years. In particular, it was not only applied for Markov decision processes but also parity games and other classes of games. Among other, all of the constructions presented in [Fea10a, Fri11c, FHZ11b, FHZ11a, AF17] implement this idea and even share the same design principle. Before discussing Friedmann's subexponential lower bound construction in detail, we thus briefly discuss this general design principle.

The idea is that every bit of the n-bit counter is represented by a *level* of the structure, and these levels are connected with each other. To implement this idea, (most of) the constructions additionally require a set of *global* vertices that are not associated with a single level. One such vertex that is present in all constructions is, for example, a *sink* ensuring that the construction is (at least weakly) unichain resp. a sink game if the construction is a parity game. These global vertices are then also connected to some or all of the levels. A visualization of this general design principle is shown in Figure 4.1.

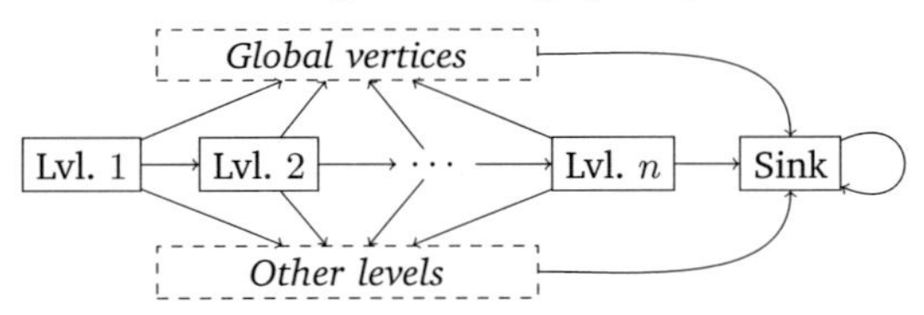

Figure 4.1.: A general framework for implementing n-bit binary counters.

To mimic a binary counter, the constructions are typically designed in such a way that there is only a single path leading to the sink. Each level is either traversed or ignored by the path. A bit of the counter is then interpreted as being equal to 1 if and only if the corresponding level is traversed. The idea is then to make it in general beneficial to traverse a level and to ensure that traversing level i is more beneficial than traversing *all* levels 1 to $i-1$. To prevent levels from being traversed too early, *entering* a level without traversing it is very expensive. This is the core idea of all of the previously mentioned lower bound constructions.

As both the description and the analysis of such constructions require some notation regarding binary numbers and binary counting, we first introduce these terms here before investigating Friedmann's construction in detail. Henceforth, let $n \in \mathbb{N}$ be fixed. We define $\mathfrak{B}_n := \{0, 1, \dots, 2^n - 1\}$ as the set of all numbers that can be represented by using n bits. A number $\mathfrak{b} \in \mathfrak{B}_n$ is typically represented as $\mathfrak{b} = (\mathfrak{b}_n, \dots, \mathfrak{b}_1)$. Here, $\mathfrak{b}_1$ denotes the least significant bit of $\mathfrak{b}$, so $\mathfrak{b} = \sum_{i \in [n]} \mathfrak{b}_i 2^{i-1}$. For $\mathfrak{b} \in \mathfrak{B}_n \setminus \{0\}$, we define the least significant set bit of $\mathfrak{b}$ by $\ell(\mathfrak{b})$, so $\ell(\mathfrak{b}) = \min\{i \in [n] : \mathfrak{b}_i \neq 0\}$. We will typically abbreviate $\ell := \ell(\mathfrak{b})$ and $\nu := \ell(\mathfrak{b}+1)$. If we consider the least significant set bit with index at least i, we add a lower index i, so formally, $\ell_i(\mathfrak{b}) := \min\{i' \geq i : \mathfrak{b}_{i'} \neq 0\}$.

By binary counting, we refer to the process of enumerating binary numbers with n digits in increasing order, beginning with 0. The analysis of the construction requires us to determine how often specific edges were applied as improving switches. As it turns out in the analysis, the following terms (introduced in [Fri11c]) are central for determining these quantities.

Let $\mathfrak{b} \in \mathfrak{B}_n$. Intuitively, we are interested in *schemes* that we observe when counting from 0 to $\mathfrak{b}$ in binary, or, more formally, in the set of numbers that *match a scheme* with respect to the following definition.

Definition 4.1.1 (Scheme, match set). A *scheme* is a set $S \subseteq \mathbb{N} \times \{0, 1\}$. A number $\mathfrak{b} \in \mathfrak{B}_n$ *matches* S if $\mathfrak{b}_i = q$ for all $(i, q) \in S$. We define the *match set*

$$M(\mathfrak{b}, S) := \{\mathfrak{b}' \in \{0, \dots, \mathfrak{b}\} : \mathfrak{b}'_i = q \quad \forall (i, q) \in S\}$$

as the set of all numbers between 0 and $\mathfrak{b}$ that match S.

The next definition introduces the *flip set* with respect to a number $\mathfrak{b}$, an index i and a scheme S. This is a subset of $M(\mathfrak{b}, S)$ that fixes the first $i-1$ bits as 0 and bit i as 1.

Definition 4.1.2 (Flip set, flip number). Let $\mathfrak{b} \in \mathfrak{B}_n, i \in [n]$ and S be a scheme. We define the *flip set* corresponding to $\mathfrak{b}, i$ and S as

$$F(\mathfrak{b}, i, S) := M(\mathfrak{b}, S \cup \{(i, 1)\} \cup \{(j, 0); j \in \{1, \dots, i-1\}\}).$$

The *flip number* is defined as $f(\mathfrak{b}, i, S) := |F(\mathfrak{b}, i, S)|$. We set $F(\mathfrak{b}, i) := F(\mathfrak{b}, i, \emptyset)$ and $f(b, i) := f(\mathfrak{b}, i, \emptyset)$ for convenience.

Finally, we define the *maximal flip number* with respect to a number $\mathfrak{b}$, an index i and a scheme S. It is the largest number contained in $F(\mathfrak{b}, i, S)$ smaller than $\mathfrak{b}$ or 0 if $F(\mathfrak{b}, i, S) = \emptyset$.

Definition 4.1.3 (Maximal flip number). Let $\mathfrak{b} \in \mathfrak{B}_n, i \in [n]$ and S be a scheme. The *maximal flip number* is $\mathfrak{g}(\mathfrak{b}, i, S) := \max(\{0\} \cup \{\mathfrak{b}' : \mathfrak{b}' \in F(\mathfrak{b}, i, S)\})$.

The following lemma summarizes several properties of flip numbers. Its proof is provided in Appendix A.

Lemma 4.1.4. *Let $\mathfrak{b} \in \mathfrak{B}_n$ and $i, j \in [n]$. Then the following hold:*

1. *Let S, S' be schemes and $S \subseteq S'$. Then $M(\mathfrak{b}, S') \subseteq M(\mathfrak{b}, S)$.*
2. *Let S, S' be schemes and $S \subseteq S'$. Then $f(\mathfrak{b}, i, S') \leq f(\mathfrak{b}, i, S)$.*
3. *It holds that $f(\mathfrak{b}, j) = f(\mathfrak{b}, j, \{(i, 0)\}) + f(\mathfrak{b}, j, \{(i, 1)\})$ and $f(\mathfrak{b}, j) = \lfloor (\mathfrak{b} + 2^{j-1})/2^j \rfloor$.*
4. *Let $i \leq j$ and S be a scheme. Then $f(\mathfrak{b}, j, S) \leq f(\mathfrak{b}, i, S)$ and thus $f(\mathfrak{b}, j) \leq f(\mathfrak{b}, i)$.*
5. *Let $i < j$. Then $F(\mathfrak{b}, j) = F(\mathfrak{b}, j, \{(i, 0)\})$ and thus $f(\mathfrak{b}, j, \{(i, 0)\}) = f(\mathfrak{b}, j)$.*

Friedmann's construction

In [Fri11c], these ideas are implemented as follows. Friedmann describes how to construct a weakly unichain Markov decision process implementing an n-bit binary counter such that applying the strategy improvement algorithm with Zadeh's pivot rule requires $\Omega(2^n)$ iterations. Since the size of the construction is quadratic in n, this yields a subexponential lower bound. Using the same arguments we presented in Section 3.3, he then argues that the same lower bound is valid for the simplex algorithm. He then briefly discusses how the Markov decision process can be used to construct a sink game such that the strategy improvement algorithm behaves identical in the Markov decision process and the sink game. Since the main focus of [Fri11c] is the Markov decision process and since we discussed the connection to the simplex algorithm in detail previously, we only describe the construction of the Markov decision process here.

Friedmann constructs a Markov decision process G_n such that there is a strategy $\sigma_{\mathfrak{b}}$ for every $\mathfrak{b} \in \mathfrak{B}_n$ and that applying the strategy improvement algorithm enumerates the strategies $\sigma_0, \sigma_1, \ldots, \sigma_{2^n-1}$ when using Zadeh's pivot rule. As described earlier, G_n implements the idea of connecting n levels, where level i of G_n represents the i-th bit of the counter. The Markov decision process also contains a *source* s with $\Gamma^-(s) = \emptyset$ and a *sink* t as defined in Definition 3.2.7. The idea is that a strategy $\sigma_{\mathfrak{b}}$ defines a path starting in s and ending in t such that all levels i with $\mathfrak{b}_i = 1$ are traversed and all levels with $\mathfrak{b}_i = 0$ are "skipped". This behavior is achieved by making levels i with $\mathfrak{b}_i = 1$ profitable for the player while making levels i with $\mathfrak{b}_i = 0$ expensive.

The main challenge in implementing this idea is Zadeh's pivot rule. Intuitively, this pivot rule forces the algorithm to apply improving switches approximately equally often. However, a classical binary counter does not switch individual bits equally often. For instance, the least significant bit of a binary counter switches whenever the counter is incremented, whereas the most significant bit only switches once. Such a counter is thus highly unbalanced, so the construction needs to implement a counter that works correctly when bits are switched in a "balanced" fashion.

The key idea to overcome this obstacle is to use not one, but *two* gadgets per level for representing bits. These gadgets are called *bicycles*, and the bicycles of level i are denoted by A_i^0 and A_i^1. For every strategy σ encountered during the application of the algorithm, only one of these two gadgets is interpreted as encoding the bit of level i. This allows for manipulating the other gadget of level i without losing the interpretation of bit i. If the construction ensures that the two gadgets alternate in representing bit i, this idea yields a balanced binary counter.

Which of the two bicycles encodes $\mathfrak{b}_i$ depends on the setting of the next bit $\mathfrak{b}_{i+1}$. More precisely, A_i^j with $j = \mathfrak{b}_{i+1}$ represents bit i and this bicycle is called *active*. The other bicycle is called *inactive*. A bicycle can be in one of two possible *configurations*, and these configurations are used to determine the setting of bit i. These two configurations are called *open* and *closed*, and bit i is interpreted as 1 if and only if the active bicycle of level i is closed. Using this interpretation when counting from 0 to $2^n - 1$ then results in the desired alternating and balanced usage of both bicycles for representing bit i as bit $i + 1$ switches every second time bit i switches. A visualization of this idea is given in Figure 4.2.

We now describe the construction in full detail and provide a precise definition of the Markov decision process G_n. As all levels are constructed identically, it suffices to describe an arbitrary level i. The vertices are defined via

$$V_0 := \bigcup_{i\in[n]} \{k_i, b_{i,0}^0, b_{i,0}^1, b_{i,1}^0, b_{i,1}^1, d_i^0, d_i^1, h_i^0, h_i^1, c_i^0, c_i^1\} \cup \{k_{n+1}, s, t\}, \quad V_R := \bigcup_{i\in[n]} \{A_i^0, A_i^1\}.$$

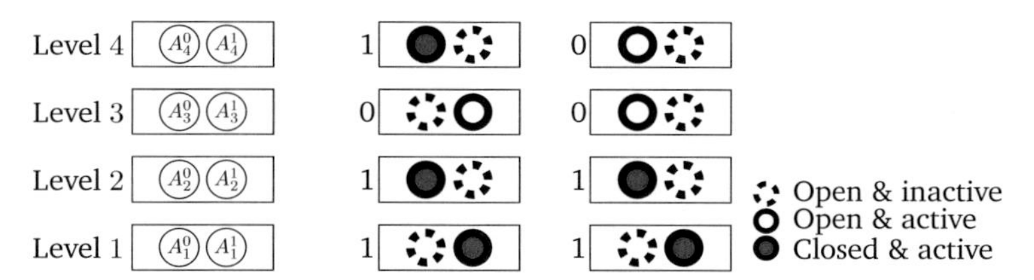

Figure 4.2.: Visualization of the intuitive idea of the binary counter of [Fri11c] for $n = 4$. The left picture shows the bicycles and their positioning within the levels. The two pictures on the right give examples for settings of the cycles representing the numbers 11 and 3, respectively.

A visualization of level i is given in Figure 4.3. We define the edges, rewards and probabilities after discussing the design principles of the construction.

As mentioned previously, the goal is that applying the strategy improvement algorithm enumerates the strategies $\sigma_0, \sigma_1, \ldots, \sigma_{2^n-1}$ representing the corresponding numbers in binary. The Markov decision process is constructed in such a way that a strategy $\sigma_{\mathfrak{b}}$ induces a path starting at the source s and leading to the sink t with probability 1. This path then traverses all levels i with $\mathfrak{b}_i = 1$ while "ignoring" all levels i with $\mathfrak{b}_i = 0$. This is achieved by making *entering* a level very expensive and *traversing* a level very profitable. It is however only possible to traverse a level if the active bicycle of this level is closed.

Ignoring and including levels into the path is controlled by the *entry vertex* k_i of level i. Every edge leaving the vertex k_i has a reward of $(-N)^{2i+7}$, where $N > 0$ is a very large parameter that is specified later. It is thus very expensive for the player to enter a level. As discussed before, the entry vertex should direct the path towards the active bicycle $A_i^j, j = \mathfrak{b}_{i+1}$, of level i if $\mathfrak{b}_i = 1$. If $\mathfrak{b}_i = 0$, then the entry vertex moves to level $\ell(\mathfrak{b})$. We now focus on the case $\mathfrak{b}_i = 1$ and formalize the idea of a bicycle. A bicycle is a gadget that ensures that the player can collect a large positive reward if the bicycle is closed while "hiding" this reward if the bicycle is open. To provide this functionality, the bicycle A_i^j contains one randomization vertex (to which we also refer as A_i^j) and two player vertices $b_{i,0}^j, b_{i,1}^j$, called *bicycle vertices*. A visualization of a single bicycle is given in Figure 4.4. The two vertices $b_{i,0}^j, b_{i,1}^j$ each constitute a cycle with A_i^j via the edges $(b_{i,*}^j, A_i^j), (A_i^j, b_{i,*}^j)$, giving the gadget its name. The vertex A_i^j also has one additional outgoing edge to a vertex d_i^j. This vertex is used to connect level i with higher levels and make the level very profitable if $\mathfrak{b}_i = 1$ and the sink s without making the level profitable if $\mathfrak{b}_i = 0$.

We now formalize the terms "open" and "closed".

Definition 4.1.5 (Configurations of bicycles in [Fri11c]). The bicycle A_i^j is *closed* for a strategy σ if $\sigma(b_{i,0}^j) = \sigma(b_{i,1}^j) = A_i^j$. The bicycle is *open* for σ if it is not closed.

We consider the vertex d_i^j next. It is connected to a vertex h_i^j which leads to k_{i+1} if $j = 1$ and to $k_{i+2}, \ldots, k_n$ if $j = 0$. Every edge leaving this vertex has a reward of $(-N)^{2i+8}$. In particular, since $N^{2i+8} - N^{2i+7} > 0$, it is profitable for the player to traverse the level completely if he can get access to this vertex. As the player should however

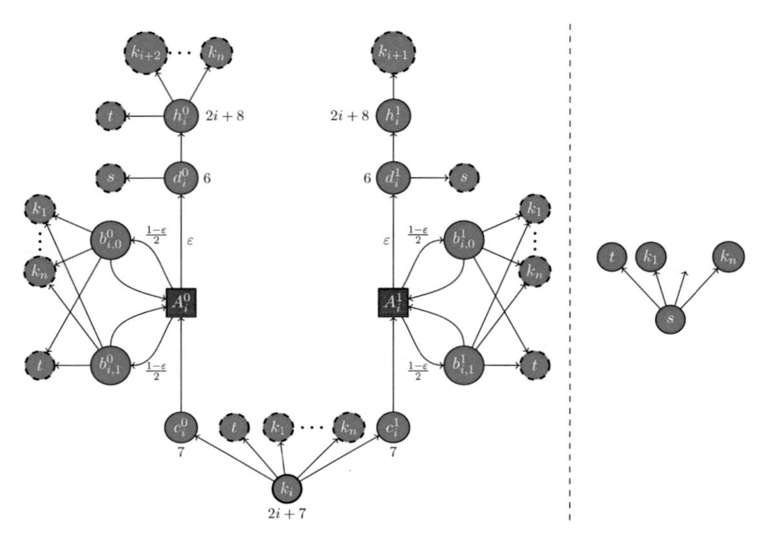

Figure 4.3.: Level i of G_n. Circular vertices are player vertices, rectangular vertices are randomization vertices. Vertex labels show the names of the vertex, edge labels show their probability. A number $\Omega(v)$ below or next to a vertex depicts that every edge leaving this vertex has a reward of $(-N)^{\Omega(v)}$. Dotted vertices do (typically) not belong to level i.
The right part shows the global vertices s and t and how s is connected to the entry levels of the vertices.

only traverse the level if the active bicycle is closed, the edge (A_i^j, d_i^j) is assigned a very small probability $\varepsilon > 0$. The edges $(A_i^j, b_{i,*}^j)$ are each assigned the probability $(1-\varepsilon)/2$. This has the effect that the valuation of d_i^j and thus the very profitable edges of h_i^j has nearly no impact on the valuation of A_i^j if the bicycle is open. However, if it is closed, then its valuation is equal to the valuation of d_i^j as the Markov decision process will move to the state d_i^j with probability 1 after a finite number of steps. Since the rewards that can be collected in the levels are increasing with the levels, closing a bicycle significantly improves its valuation and makes it very profitable. This is the reason why the rewards are defined as powers of a very large number. The idea is that closing a bicycle A_i^j increases its valuation in such a way that it becomes profitable for all closed bicycles $A_{i'}^j$ with $i' < i$ to open and try to get access to A_i^j. This is achieved by defining the rewards that can be collected in level i as powers of a sufficiently large parameter N.

Consequently, the entry vertex of level i needs to have access to both bicycles of level i. For technical reasons, there are no direct edges (k_i, A_i^*) but there is an intermediate vertex c_i^j between k_i and A_i^j.

We now provide the exact set of edges alongside their probabilities and rewards. To

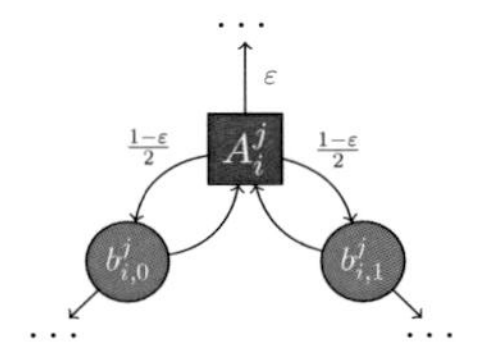

Figure 4.4.: A bicycle gadget, consisting of the randomization vertex A_i^j and the two player vertices $b_{i,0}^j, b_{i,1}^j$. Edge labels show the respective probabilities. We identify a bicycle gadget with its randomization vertex.

define the probabilities and rewards, we need to define the two parameters ε and N. In principle, traversing a level that is representing a bit which is equal to 1 yields a reward of (approximately) $N^{2i+8} - N^{2i+7}$. Since setting bit i to 1 implies that all bits $i' < i$ need to be switched from 1 to 0, the parameter N needs to be chosen such that traversing level i is more profitable than traversing all levels $i' < i$. This can be achieved by setting $N \geq 7n+1$. To ensure that the bicycles can in fact hide the very profitable edges if a bicycle is open, it is required that $\varepsilon \leq N^{-(2n+11)}$, and we define $\varepsilon := N^{-(2n+11)}$ henceforth. Note that both of these parameters can be encoded with a polynomial number of bits.

The construction has the property that all edges leaving a fixed vertex v have the same reward assigned to them, and this reward is some power of N. To define the rewards of the edges, it is thus sufficient to assign a priority $\Omega(v)$ to the vertices and setting $r(v,w) := (-N)^{\Omega(v)}$ for all $w \in \Gamma^+(v)$. This enables us to define the set of edges via the following table where $r(v,w) := 0$ for all $w \in \Gamma^+(v)$ if the vertex v is not assigned a priority. Table 4.1 thus fully describes the edges, rewards and probabilities, concluding the description of the Markov decision process G_n.

Vertex v	$\Gamma^+(v)$	Probability
A_i^j	d_i^j	ε
	$b_{i,*}^j$	$\frac{1-\varepsilon}{2}$

Vertex v	$\Gamma^+(v)$	Priority
t	t	–
s	$t, k_1, \ldots, k_n$	–
$b_{i,*}^j$	$t, A_i^j, k_1, \ldots, k_n$	–

Vertex v	$\Gamma^+(v)$	$\Omega(v)$
k_{n+1}	t	$2n+9$
k_i	$c_i^*, t, k_a, \ldots, k_n$	$2i+7$
h_i^0	$t, k_{i+2}, \ldots, k_n$	$2i+8$
h_i^1	k_{i+1}	$2i+8$
c_i^j	A_i^j	7
d_i^j	h_i^j, s	6

Table 4.1.: The edges of the subexponential construction [Fri11c] alongside their rewards and probabilities, see Table 1 therein.

This concludes the formal description of the subexponential lower bound construction of [Fri11c]. We now discuss the application of the strategy improvement algorithm to this Markov decision process. As this application depends on the initial strategy, we also discuss this and two issues related to the initial strategy.

4.2. Application of the Strategy Improvement Algorithm

In this section, we discuss the application of the strategy improvement algorithm using Zadeh's pivot rule and an (implicit) tie-breaking rule to the Markov decision process defined in Section 4.1. The algorithm is provided an initial strategy σ_0 representing the number 0 and then calculates the strategies $\sigma_1, \dots, \sigma_{2^n-1}$ representing the respective numbers. We thus begin by formally defining a strategy $\sigma_{\mathfrak{b}}$ representing a number $\mathfrak{b} \in \mathfrak{B}_n$. All explanations here are extracted from [Fri11c]. To simplify explanation, we say that vertex v *points to* w if $\sigma(v) = w$.

Definition 4.2.1 (Representing a number). The strategy $\sigma_{\mathfrak{b}}$ *represents the number* $\mathfrak{b} \in \mathfrak{B}_n$ if it has the following properties:

1. The bicycle $A_i^{\mathfrak{b}_{i+1}}$ is closed if and only if $\mathfrak{b}_i = 1$.
2. If $\mathfrak{b}_i = 1$, then $\sigma_{\mathfrak{b}}(k_i) = c_i^j$ where $j = \mathfrak{b}_{i+1}$. If $\mathfrak{b}_i = 0$, then $\sigma_{\mathfrak{b}}(k_i) = k_{\ell(\mathfrak{b})}$.
3. The source s points to the level of the least significant set bit, i.e., $\sigma_{\mathfrak{b}}(s) = k_{\ell(\mathfrak{b})}$.
4. All vertices h_i^0 point to the entry vertex of the first level after level $i+1$ corresponding to a bit equal to 1, so $\sigma_{\mathfrak{b}}(h_i^0) = k_{\ell_{i+2}(\mathfrak{b})}$. If no such index exists, then $\sigma_{\mathfrak{b}}(h_i^0) = t$.
5. The vertex d_i^j points to h_i^j if and only if $\mathfrak{b}_{i+1} = j$.

This definition also applies to $\mathfrak{b} = 0$ by substituting $k_{\ell(\mathfrak{b})}$ with t.

It is clear that several strategies can represent the same binary number with respect to Definition 4.2.1. We will later fix a specific strategy $\sigma_{\mathfrak{b}}$ for every $\mathfrak{b} \in \mathfrak{B}_n$ which will be the interpreted as the "canonical" strategy representing the number $\mathfrak{b}$. As the definition of this specific strategy requires more knowledge regarding the construction, we postpone it for now. For the moment, it is sufficient to just interpret a strategy $\sigma_{\mathfrak{b}}$ as *some* strategy representing $\mathfrak{b}$.

Since the application of the improving switches depends on the initially chosen strategy, we begin by discussing the initial strategy. In [Fri11c], the initial strategy σ^* is defined as follows: "*As designated initial strategy σ^*, we use $\sigma^*(d_i^j) = h_i^j$ and $\sigma^*(\_) = t$ for all other player 0 nodes with non-singular out-degree.*" This initial strategy is however inconsistent with two other aspects of [Fri11c]. Since we did not introduce these aspects yet, we do not discuss these issues here. For the sake of completeness, we already introduce the alternative initial strategy $\sigma^\star$ that avoids these issues. We discuss the issues related to the original initial strategy later.

Definition 4.2.2 (Alternative initial strategy $\sigma^\star$). We define the strategy $\sigma^\star$ by setting $\sigma^\star(d_i^0) := h_i^0$ and $\sigma^\star(d_i^1) := s$ for all $i \in [n]$ and $\sigma^\star(v) := t$ for all other player vertices v with non-singular out-degree.

This strategy is visualized in Figure 4.5. As every vertex but the vertices d_1^0, d_i^1 directly point to the vertex t which is obviously a sink, the following statement is immediate.

Lemma 4.2.3. *The alternative initial strategy $\sigma^\star$ is weakly unichain.*

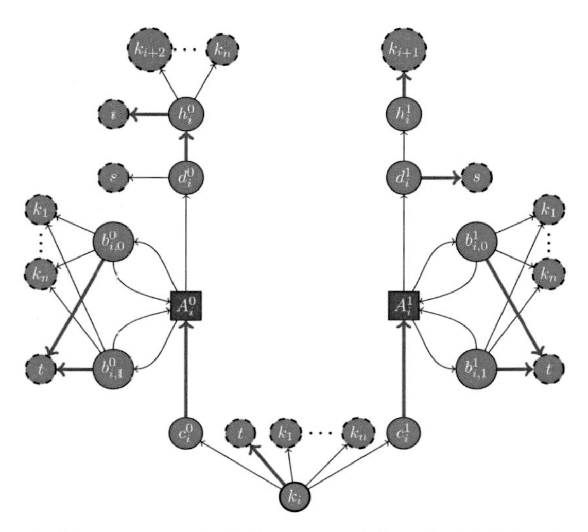

Figure 4.5.: A level i of the alternative initial strategy σ^* described in Definition 4.2.2. The edges of the strategy are marked by thick red edges.

It is clear that the strategy $\sigma_{\mathfrak{b}+1}$ cannot be reached by applying a single improving switch to $\sigma_{\mathfrak{b}}$. Thus, intermediate strategies need to be introduced for the transition from $\sigma_{\mathfrak{b}}$ to $\sigma_{\mathfrak{b}+1}$. These intermediate strategies are divided into six *phases*. The idea is that only one or two "tasks" are performed within the counter during each phase. Examples of such tasks are, for example, the opening and closing of bicycles, adjusting the targets of the vertices d_i^j and so on. This allows for simplifying the majority of the proofs and arguments as they can be based on phases instead of transitions. We mention here that our description partly differs from the original description given in [Fri11c, Pages 8,9], and explain the reason in the following two sections.

Before discussing the phases, it is important to mention how the application of improving switches is handled in [Fri11c]. Instead of proving that applying improving switches following Zadeh's pivot rule proceeds along the phases, the order in which improving switches is applied is described explicitly and it is later argued that this order of application is in accordance with Zadeh's pivot rule. We thus also explain the application of the improving switches in this fashion as well. Consider the strategy $\sigma_{\mathfrak{b}}$ representing $\mathfrak{b} \in \mathfrak{B}_n$ and let $\nu := \ell(\mathfrak{b}+1)$.

1. In phase 1, the algorithm applies improving switches within the bicycles. Some of these switches have to be applied as they minimize the occurrence record, while some are applied trying to keep the occurrence records as balanced as possible. For every open bicycle, at least on of its bicycle edges is switched during phase 1. Some inactive bicycles switch both of their edges which can be interpreted as "catching up" with the other edges. The algorithm also switches both bicycle edges of the active bicycle in level ν as this bicycle needs to be closed with respect to $\sigma_{\mathfrak{b}+1}$.

2. In phase 2, the new least significant set bit $(\mathfrak{b}+1)_\nu$ is made accessible by the rest of the Markov decision process. Consequently, the target of k_ν is switched to c_ν^j, where $j := (\mathfrak{b}+1)_\nu$.
3. Phase 3 is responsible for resetting the counter, i.e., opening the bicycles in levels below ν. The entry vertices of all levels i with $(\mathfrak{b}+1)_i = 0$ are switched to k_ν. The same is done for all vertices $b_{*,*}^*$ contained in inactive cycle centers and all vertices $b_{i,*}^*$ with $(\mathfrak{b}+1)_i = 0$. As there is a major flaw in this phase, we discuss it in more detail in Section 4.3.
4. In phase 4, the vertices h_i^0 are updated for all $i < \nu$. This is necessary as bits 1, to ν switch when transitioning from $\mathfrak{b}$ to $\mathfrak{b}+1$.
5. In phase 5, the target of the source vertex is switched to the entry vertex of the level corresponding to the new least significant set bit.
6. In phase 6, the targets of the vertices d_i^j is changed such that h_i^j is the target of d_i^j if and only if $(\mathfrak{b}+1)_{i+1} = j$.

These phases and the improving switches that are applicable in each phase are formally described by three tables in [Fri11c]. In the remainder of this section, we introduce and briefly discuss these tables.

The tables use an alternative notation for describing strategies in G_n. This notation uses integers to describe the targets of the vertices with respect to a strategy σ, allowing for a simpler description of strategies. We also use this notation henceforth and thus define the function $\bar{\sigma}$ as specified by Table 4.2.

$\sigma(v)$	t	k_i	h_*^*	s	A_*^*	c_i^j
$\bar{\sigma}(v)$	$n+1$	i	1	0	0	j
In addition	$\bar{\sigma}(A_i^j) := 1$ if A_i^j is closed and $\bar{\sigma}(A_i^j) := 0$ else					

Table 4.2.: The function $\bar{\sigma}$ in the subexponential construction.

The first table is Table 4.3 which is a slightly adapted variant of [Fri11c, Table 2]. This table formally defines when a strategy σ belongs to phase p. More precisely, σ is a phase p strategy (with respect to a number $\mathfrak{b}$) if every vertex is mapped by σ to a choice included in the respective cell of the table and if the strategy fulfills the side conditions of the phase (if there are any). Cells that contain more than one choice indicate that strategies of the respective phase are allowed to match any of the choices. As this table is only used for formalizing the description and intuition given earlier, we do not explain it in detail.

Since Table 4.3 formalizes the term of a phase p strategy, we can now formally define the "canonical" strategy $\sigma_\mathfrak{b}$ representing the number $\mathfrak{b} \in \mathfrak{B}_n$. This strategy is called *initial phase* 1 *strategy* in [Fri11c] as it is the first phase 1 strategy for $\mathfrak{b}$ calculated by the strategy improvement algorithm. For $\mathfrak{b} \in \mathfrak{B}_n$, we henceforth denote the initial phase 1 strategy for $\mathfrak{b}$ with respect to the following definition by $\sigma_\mathfrak{b}$.

Definition 4.2.4 (Initial phase 1 strategy)**.** The unique phase 1 strategy $\sigma_\mathfrak{b}$ for $\mathfrak{b}$ with $\sigma_\mathfrak{b}(b_{i,0}^j) = \sigma_\mathfrak{b}(b_{i,1}^j) = A_i^j$ if and only if $\mathfrak{b}_i = 1$ and $\mathfrak{b}_{i+1} = j$ is called *initial phase* 1 *strategy for* $\mathfrak{b}$.

Phase	1	2	3	4	5	6
$\bar{\sigma}(s)$	$\ell(\mathfrak{b})$	$\ell(\mathfrak{b})$	$\ell(\mathfrak{b})$	$\ell(\mathfrak{b})$	$\ell(\mathfrak{b})$	$\ell(\mathfrak{b}')$
$\bar{\sigma}(d_i^0)$	$1-\mathfrak{b}_{i+1}$	$1-\mathfrak{b}_{i+1}$	$1-\mathfrak{b}_{i+1}$	$1-\mathfrak{b}_{i+1}$	$1-\mathfrak{b}_{i+1}$	$1-\mathfrak{b}_{i+1}, 1-\mathfrak{b}'_{i+1}$
$\bar{\sigma}(d_i^1)$	$\mathfrak{b}_{i+1}$	$\mathfrak{b}_{i+1}$	$\mathfrak{b}_{i+1}$	$\mathfrak{b}_{i+1}$	$\mathfrak{b}_{i+1}$	$\mathfrak{b}_{i+1}, \mathfrak{b}'_{i+1}$
$\bar{\sigma}(h_i^0)$	$\ell_{i+2}(\mathfrak{b})$	$\ell_{i+2}(\mathfrak{b})$	$\ell_{i+2}(\mathfrak{b})$	$\ell_{i+2}(\mathfrak{b}), \ell_{i+2}(\mathfrak{b}')$	$\ell_{i+2}(\mathfrak{b}')$	$\ell_{i+2}(\mathfrak{b}')$
$\bar{\sigma}(b_{*,*}^*)$	$0, \ell(\mathfrak{b})$	$0, \ell(\mathfrak{b})$	$0, \ell(\mathfrak{b}), \ell(\mathfrak{b}')$	$0, \ell(\mathfrak{b}')$	$0, \ell(\mathfrak{b}')$	$0, \ell(\mathfrak{b}')$
$\bar{\sigma}(A_i^{\mathfrak{b}_{i+1}})$	$\mathfrak{b}_i$	$*$	$*$	$*$	$*$	$*$
$\bar{\sigma}(A_i^{\mathfrak{b}'_{i+1}})$	$*$	$\mathfrak{b}'_i$	$\mathfrak{b}'_i$	$\mathfrak{b}'_i$	$\mathfrak{b}'_i$	$\mathfrak{b}'_i$

Phase	1-2	3-4	5-6
$\bar{\sigma}(k_i)$	$\begin{cases} \ell(\mathfrak{b}) & \text{if } \mathfrak{b}_i = 0 \\ -\mathfrak{b}_{i+1} & \text{if } \mathfrak{b}_i = 1 \end{cases}$	$\begin{cases} \ell(\mathfrak{b}), \ell(\mathfrak{b}') & \text{if } \mathfrak{b}'_i = 0 \wedge \mathfrak{b}_i = 0 \\ -\mathfrak{b}_{i+1}, \ell(\mathfrak{b}') & \text{if } \mathfrak{b}'_i = 0 \wedge \mathfrak{b}_i = 1 \\ -\mathfrak{b}'_{i+1} & \text{if } \mathfrak{b}'_i = 1 \end{cases}$	$\begin{cases} \ell(\mathfrak{b}') & \text{if } \mathfrak{b}'_i = 0 \\ -\mathfrak{b}'_{i+1} & \text{if } \mathfrak{b}'_i = 1 \end{cases}$

Phase 3	Side Conditions
(a)	$\forall i\colon ([\mathfrak{b}'_i = 0 \text{ and } (\exists j, l\colon \bar{\sigma}(b_{i,l}^j) = \ell(\mathfrak{b}'))] \text{ implies } \bar{\sigma}(k_i) = \ell(\mathfrak{b}'))$
(b)	$\forall i, j\colon ([\mathfrak{b}'_i = 0, \mathfrak{b}'_j = 0, \bar{\sigma}(k_i) = \ell(\mathfrak{b}')' \text{ and } \bar{\sigma}(k_j) \neq \ell(\mathfrak{b}')] \text{ implies } i > j)$

Table 4.3.: Definition of the strategy phases in [Fri11c]. We let $\mathfrak{b}' = \mathfrak{b} + 1$. This table is an adapted version of [Fri11c, Table 2] such that it is in line with our notation.

Table 4.4 (which is an adapted version of [Fri11c, Table 3]) is related to the improving switches that are applied by the strategy improvement algorithm. It however does not contain the exact set of improving switches. For a given phase p strategy σ, it contains a subset L_σ and a superset U_σ of the set of improving switches I_σ. Given two such sets, it is then sufficient to prove that some switch $e \in L_\sigma$ minimizes the occurrence record among all edges contained in U_σ. This guarantees that the switch e can be applied next according to Zadeh's pivot rule, even if the exact set of improving switches is not known. It is then in particular not necessary to determine the exact set of improving switches for every strategy. We already mention here that there is an issue with the set L_σ^6 defining the subset for phase 6 strategies. We discuss this issue later in more detail and provide the original table as it is contained in [Fri11c].

The final table is Table 4.5, an adapted version of [Fri11c, Table 4]. For $\mathfrak{b} \in \mathfrak{B}_n$, this table contains the occurrence records of the edges with respect to the initial phase 1strategy $\sigma_\mathfrak{b}$ representing $\mathfrak{b}$. Here, the occurrence records with respect to a strategy σ are described by a function $\phi^\sigma\colon E \to \mathbb{N} \cup \{0\}$. More precisely, $\phi^\sigma(e)$ is the number of times the edge e was applied as improving switch during the execution of the algorithm until reaching the strategy σ. For most of the edges, the table gives the exact occurrence record with respect to $\sigma_\mathfrak{b}$. For the bicycle edges, the occurrence records are not given exactly. Instead, the table shows that the occurrence records of two bicycle edges differs by at most one and shows that the sum of their occurrence records can be described exactly. The entries of the table use the notation introduced when discussing binary numbers in the beginning of Section 4.1. There is an issue related to the so-called "complicated conditions" which we discuss in Section 4.3.

Ph. p	Improving switches subset L_σ^p	Improving switches superset U_σ^p
1	$\{(b_{i,l}^j, A_i^j) : \sigma(b_{i,l}^j) \neq A_i^j\}$	L_σ^1
2	$\{(k_\nu, c_\nu^j)\}$ where $j = \mathfrak{b}'_{\nu+1}$	$L_\sigma^1 \cup L_\sigma^2$
3	$\{(k_i, k_\nu) : \bar{\sigma}(k_i) \neq \nu \wedge \mathfrak{b}'_i = 0\} \cup$ $\{(b_{i,l}^j, k_\nu) : \bar{\sigma}(b_{i,l}^j) \neq \nu \wedge \mathfrak{b}'_i = 0\} \cup$ $\{(b_{i,l}^j, k_\nu) : \bar{\sigma}(b_{i,l}^j) \neq \nu \wedge \mathfrak{b}'_{i+1} \neq j\}$	$U_\sigma^4 \cup \{(k_i, k_z) : \bar{\sigma}(k_i) \notin \{z, \nu\}, z \leq \nu \wedge \mathfrak{b}'_i = 0\} \cup$ $\{(b_{i,l}^j, k_z) : \bar{\sigma}(b_{i,l}^j) \notin \{z, \nu\}, z \leq \nu \wedge \mathfrak{b}'_i = 0\} \cup$ $\{(b_{i,l}^j, k_z) : \bar{\sigma}(b_{i,l}^j) \notin \{z, \nu\}, z \leq \nu \wedge \mathfrak{b}'_{i+1} \neq j\}$
4	$\{(h_i^0, k_{\ell_{i+2}(\mathfrak{b}')}) : \bar{\sigma}(h_i^0) \neq \ell_{i+2}(\mathfrak{b}')\}$	$U_\sigma^5 \cup \{(h_i^0, k_l) \mid l \leq \ell_{i+2}(\mathfrak{b}')\}$
5	$\{(s, k_\nu)\}$	$U_\sigma^6 \cup \{(s, k_i) : \bar{\sigma}(s) \neq i \wedge i < \nu\} \cup$ $\{(d_i^j, x) : \sigma(d_i^j) \neq x \wedge i < \nu\}$
6	$\{(d_i^0, v) : \sigma(d_i^0) \neq v \wedge \bar{\sigma}(d_i^0) \neq \mathfrak{b}'_{i+1}\} \cup$ $\{(d_i^1, v) : \sigma(d_i^1) \neq v \wedge \bar{\sigma}(d_i^1) = \mathfrak{b}'_{i+1})\}$	$L_\sigma^1 \cup L_\sigma^6$

Table 4.4.: Sub- and supersets of the improving switches of phase p strategies. We let $\mathfrak{b}' := \mathfrak{b} + 1$ and $\nu := \ell(\mathfrak{b}')$. This table is an adapted version of [Fri11c, Table 3] such that it is in line with our notation.

Other than correcting the issues in the next section, we rely on these tables.

4.3. Flaws in the Original Proofs

There are three flaws in the construction of [Fri11c]. Two of these flaws are rather minor and can be repaired relatively easily. These flaws are related to the initial strategy (Issues 4.3.1 and 4.3.3) and the description of the occurrence records given in Table 4.5 (Issue 4.3.4). These also do not have a huge impact on the correctness of the results of [Fri11c]. There is, however, one major flaw that needs to be corrected. This flaw is related to the application of the improving switches during phase 3. We prove that a general framework of applying improving switches in phase 3 does not obey Zadeh's pivot rule (Issue 4.3.12) and argue that the application described in [Fri11c] is one special case of this framework.

The initial strategy

We begin by discussing two issues related with the original initial strategy σ^* described in [Fri11c]. As a reminder, the initial strategy is described as follows: *"As designated initial strategy σ^*, we use $\sigma^*(d_i^j) = h_i^j$ and $\sigma^*(\_) = t$ for all other player 0 nodes with non-singular out-degree."* This initial strategy is however inconsistent with the sub- and supersets of improving switches given in Table 4.4 and by [Fri11c, Lemma 4].

Issue 4.3.1 (Initial strategy I). *The initial strategy σ^* for G_n as described in [Fri11c, Page 10] contradicts Table 4.4 since $I_{\sigma^*} \neq \{(b_{i,k}^j, A_i^j) : \sigma^*(b_{i,k}^j) \neq A_i^j\}$.*

Edge e	$(*, t)$	(s, k_ℓ)	(h_*^0, k_ℓ)	(k_i, k_ℓ)
$\phi^{\sigma_\mathfrak{b}}(e)$	0	$f(\mathfrak{b}, \ell)$	$f(\mathfrak{b}, \ell)$	$f(\mathfrak{b}, \ell, \{(i, 0)\})$

Edge e	$(b_{i,*}^j, k_\ell)$
$\phi^{\sigma_\mathfrak{b}}(e)$	$f(\mathfrak{b}, \ell, \{(i, 0)\}) + f(\mathfrak{b}, \ell, \{(i, 1), (i+1, 1-j)\})$

Edge e	(d_i^j, s)	(d_i^j, h_i^j)	(k_i, c_i^j)
$\phi^{\sigma_\mathfrak{b}}(e)$	$f(\mathfrak{b}, i+1) - j \cdot \mathfrak{b}_{i+1}$	$f(\mathfrak{b}, i+1) - (1-j) \cdot \mathfrak{b}_{i+1}$	$f(\mathfrak{b}, i, \{(i+1, j)\})$

Complicated Conditions
$\lvert \phi^{\sigma_\mathfrak{b}}(b_{i,0}^j, A_i^j) - \phi^{\sigma_\mathfrak{b}}(b_{i,1}^j, A_i^j) \rvert \leq 1$
$\phi^{\sigma_\mathfrak{b}}(b_{i,0}^j, A_i^j) + \phi^{\sigma_\mathfrak{b}}(b_{i,1}^j, A_i^j) = \begin{cases} \mathfrak{g}^* + 1 & \text{if } \mathfrak{b}_i = 1 \text{ and } \mathfrak{b}_{i+1} = j \\ \mathfrak{g}^* + 1 + 2 \cdot z & \text{if } \mathfrak{b}_{i+1} \neq j \text{ and } z := \mathfrak{b} - \mathfrak{g}^* - 2^{i-1} < \frac{1}{2}(\mathfrak{b} - 1 - \mathfrak{g}^*) \\ \mathfrak{b} & \text{otherwise} \end{cases}$

Table 4.5.: Occurrence records for the initial phase 1 strategy $\sigma_\mathfrak{b}$ for $\mathfrak{b}$ calculated by the strategy improvement algorithm. We let $\ell := \ell(\mathfrak{b})$ and $\mathfrak{g}^* = \mathfrak{g}(\mathfrak{b}, i, \{(i+1, j)\})$.

This is proven using the following lemma which is proven in Appendix A.

Lemma 4.3.2. *None of the edges $(b_{i,k}^1, A_i^1)$ for $i \in [n]$ and $k \in \{0, 1\}$ is an improving switch with respect to σ^*.*

Proof of Issue 4.3.1. By [Fri11c], σ^* is the initial phase 1 strategy representing 0. In particular, it is a phase 1 strategy. Thus, according to Table 4.4 and since $L_\sigma^1 = U_\sigma^1 = I_\sigma$ for all phase 1 strategies σ, it holds that $I_{\sigma*} = \{(b_{i,k}^j, A_i^j) : \sigma^*(b_{i,k}^j) \neq A_i^j\}$.

Let $i \in [n], k \in \{0, 1\}$. By definition of σ^*, it holds that $\sigma^*(b_{i,k}^1) = t$ and in particular $\sigma^*(b_{i,k}^1) \neq A_i^1$. Therefore, $(b_{i,r}^1, A_i^1) \in I_{\sigma^*}$ should hold according to Table 4.4. But, by Lemma 4.3.2, $(b_{i,k}^1, A_i^1)$ is not an improving switch. Therefore, σ^* contradicts Table 4.4. □

This is not the only issue related to the initial strategy. If the strategy improvement algorithm is applied with σ^* as initial strategy, then at least one of the tables describing the strategies calculated by the algorithm is incorrect.

Issue 4.3.3 (Initial strategy II). *When the strategy iteration algorithm is started using σ^* as initial strategy, then either Table 4.4 or Table 4.5 is incorrect for σ_1.*

Proof. Let σ denote the first phase 6 strategy calculated by the strategy improvement algorithm, let $i \in [n]$ and $\text{Val} := \text{Val}_\sigma$. Then, since the initial strategy represents the number 0, Table 4.3 implies $\sigma(s) = k_1$ and $\sigma(k_{i'}) = k_1$ for all $i' \in \{2, \dots, n\}$. Since $r(s, k_1) = 0$, this implies $\text{Val}(s) = \text{Val}(k_1)$ and consequently

$$\text{Val}(\sigma(d_i^1)) = \text{Val}(h_i^1) = N^{2i+8} + \text{Val}(k_{i+1}) = N^{2i+8} - N^{2i+9} + \text{Val}(k_1) < \text{Val}(s).$$

This implies that (d_i^1, s) is an improving switch for every $i \in [n]$. Now, the algorithm can either (i) apply (some or all of) these improving switches or (ii) apply none of these improving switches now. Consider the case that the algorithm applies all improving switches (d_*^1, s) next (that is, before applying any improving switch that is not of this type). The application of these switches obeys Zadeh's pivot rule as it is easy to verify that none of them was applied earlier when transitioning from σ^* to σ. By definition, phase 6 ends after all these switches are applied and the initial phase 1 strategy σ_1 for 1 is obtained. But, according to Table 4.5, it should now hold that $\phi^{\sigma_1}(d_i^1, s) = f(1, i+1) = 0$ since the first bit is the only bit that is not equal to zero. But this is a contradiction to the fact that the edges (d_i^1, s) were already switched once. Note that this argument also applies if only a subset of (d_*^1, s) is applied.

Thus consider the case that none of these switches is applied and let σ_1 again denote the initial phase 1 strategy for 1. It is easy to verify that that the edges (d_*^1, s) are still improving switches for σ_1. This however contradicts Table 4.4. □

As mentioned previously, the alternative initial strategy $\sigma^\star$ defined in Definition 4.2.2 avoids both of these issues. We prove this in Section 4.4 and discuss another issue next.

The occurrence record of bicycle edges

We now discuss an issue related to the occurrence records of the bicycle edges as specified in Table 4.5. Let $\mathfrak{b} \in \mathfrak{B}_n$ and consider a fixed bicycle A_i^j. We define $\mathfrak{g} := \mathfrak{g}(\mathfrak{b}, i, \{(i+1, j)\})$, $z := \mathfrak{b} - \mathfrak{g} - 2^{i-1}$ and $\phi^{\sigma_\mathfrak{b}}(A_i^j) := \phi^{\sigma_\mathfrak{b}}(b_{i,0}^j, A_i^j) + \phi^{\sigma_\mathfrak{b}}(b_{i,1}^j, A_i^j)$. Using this notation, Table 4.5 states the following regarding the occurrence records of the bicycle edges:

$$\left|\phi^{\sigma_\mathfrak{b}}(b_{i,0}^j, A_i^j) - \phi^{\sigma_\mathfrak{b}}(b_{i,1}^j, A_i^j)\right| \leq 1, \tag{4.1}$$

$$\phi^{\sigma_\mathfrak{b}}(A_i^j) = \begin{cases} \mathfrak{g}+1 & \text{if } \mathfrak{b}_i = 1 \wedge \mathfrak{b}_{i+1} = j, \\ \mathfrak{g}+1+2z & \text{if } \mathfrak{b}_{i+1} \neq j \wedge z < \frac{1}{2}(\mathfrak{b}-1-\mathfrak{g}), \\ \mathfrak{b}, & \text{else.} \end{cases} \tag{4.2}$$

We now prove that there is an inconsistency regarding Equation (4.2) as follows. Assuming that the occurrence records of the bicycle edges are described by Equations (4.1) and (4.2) and that the other entries of Table 4.5 are correct implies that some edges have a negative occurrence record. More formally, the following issue arises.

Issue 4.3.4 (Occurrence record of bicycle edges). *Let $\mathfrak{b} < 2^{n-k-1} - 1$ for some $k \in \mathbb{N}$. Then, there is at least one edge $(b_{i,*}^j, A_i^j)$ that has a negative occurrence record.*

Proof. Let $i \in \{n-k, \dots, n-1\}$ and $j = 1$. Since $\mathfrak{b} \leq 2^{n-k-1} - 1$ and $i \geq n-k$, it follows that $\mathfrak{b} < 2^i - 1$. This in particular implies $\mathfrak{b}_i = 0$ and $\mathfrak{b}_{i+1} = 0 \neq 1 = j$ as well as $\mathfrak{b}'_{i+1} = 0$ for all $\mathfrak{b}' \leq \mathfrak{b}$. By definition, this implies $\mathfrak{g} = \mathfrak{g}(\mathfrak{b}, i, \{(i+1, j)\}) = 0$. Since $\mathfrak{b} < 2^i - 1$ is equivalent to $2^i > \mathfrak{b} + 1$, we obtain

$$2z = 2(\mathfrak{b} - 2^{i-1}) = 2\mathfrak{b} - 2^i < 2\mathfrak{b} - (\mathfrak{b}+1) = \mathfrak{b} - 1,$$

or, equivalently, $z < \frac{1}{2}(\mathfrak{b}-1) = \frac{1}{2}(\mathfrak{b}-1-\mathfrak{g})$. Consequently, all conditions for the second case of Equation (4.2) are fulfilled, implying

$$\begin{aligned}\phi^{\sigma_{\mathfrak{b}}}(A_i^j) &= \mathfrak{g}+1+2z = 2z+1 = 2(\mathfrak{b}-2^{i-1})+1 < 2(2^{n-k-1}-1-2^{i-1})+1 \\ &\leq 2(2^{n-k-1}-1-2^{n-k-1})+1 = -1 < 0.\end{aligned}$$

Hence, at least one edges has a negative occurrence record. □

We resolve this issue in the next section by providing a system similar to the one described by Equations (4.1) and (4.2) that avoids this issue.

Before doing so, we discuss the main flaw of [Fri11c], the application of improving switches during phase 3.

Application of improving switches during phase 3

Fix some $\mathfrak{b} \in \mathfrak{B}_n$ and let $\ell := \ell(\mathfrak{b}), \nu := \ell(\mathfrak{b}+1)$. In Section 4.2, we stated that during phase 3, improving switches need to be applied for every entry vertex k_i belonging to a level with $(\mathfrak{b}+1)_i = 0$. In addition, several bicycles are opened during this phase, for example, some inactive bicycles. According to the informal description given in [Fri11c, Pages 9,10], these updates are only performed in those levels with an index smaller than ν. To be precise the following is stated (where $r \in \{0,1\}$ is arbitrary and the notation was adapted to be in line with our paper): *"In the third phase, we perform the major part of the resetting process. By resetting, we mean to unset lower bits again, which corresponds to reopening the respective bicycles. Also, we want to update all other inactive or active but not set bicycles again to move to the entry point k_ν. In other words, we need to update the lower entry points k_z with $z < \nu$ to move to k_ν, and the bicycle nodes $b_{z,r}^j$ to move to k_i. We apply these switches by first switching the entry node k_z for some $z < \nu$ and then the respective bicycle nodes $b_{z,r}^j$."* We prove that applying improving switches in this way violates Zadeh's pivot rule and is inconsistent with the tables used in [Fri11c]. As we do not consider the occurrence records of bicycle edges, all results therefore hold independently of Issue 4.3.4.

Issue 4.3.5 (Informal description of phase 3). *For every $\mathfrak{b} \in [2^{n-2}-1]$, the informal description of phase 3 given in [Fri11c, Pages 9,10] contradicts Tables 4.3 and 4.5. It additionally violates Zadeh's pivot rule during the transition from $\sigma_{\mathfrak{b}}$ to $\sigma_{\mathfrak{b}+1}$ for every $\mathfrak{b} \in \{3,\dots,2^{n-2}-2\}$.*

Many of the following proofs require us to discuss the application of improving switches when transitioning from $\sigma_{\mathfrak{b}}$ to $\sigma_{\mathfrak{b}+1}$. We thus abbreviate this transition by $\sigma_{\mathfrak{b}} \to \sigma_{\mathfrak{b}+1}$.

Since we need to analyze the values of the vertices in detail we need an additional lemma. It is an extraction of some estimations contained in the proof of [Fri11c, Lemma 3].

Lemma 4.3.6. *Let σ be a strategy calculated by the strategy improvement algorithm during $\sigma_{\mathfrak{b}} \to \sigma_{\mathfrak{b}+1}$. Denote the reward of each edge emanating from vertex v by$\langle v\rangle$. Let*

$$S_i := \sum_{\substack{j\geq i:\\ \mathfrak{b}_j=1}} \left(\langle k_j\rangle + \langle c_j^0\rangle + \langle d_j^0\rangle + \langle h_j^0\rangle\right) \quad \text{and} \quad T_i := \sum_{\substack{j\geq i:\\ (\mathfrak{b}+1)_j=1}} \left(\langle k_j\rangle + \langle c_j^0\rangle + \langle d_j^0\rangle + \langle h_j^0\rangle\right).$$

Then $\operatorname{Val}_\sigma(k_i) \in [\langle k_i\rangle + S_1, T_i]$.

Proof of Issue 4.3.5. Let $\mathfrak{b} \in [2^{n-2} - 1]$ and consider the transition from $\sigma_\mathfrak{b}$ to $\sigma_{\mathfrak{b}+1}$. According to Table 4.3, for each phase 1 or phase 2 strategy σ, it should hold that $\sigma(k_i) = k_\ell$ if $\mathfrak{b}_i = 0$ and $\sigma(k_i) = c_i^j, j = \mathfrak{b}_{i+1}$ if $\mathfrak{b}_i = 1$. But, since $\mathfrak{b} < 2^{n-2}$, we have $\mathfrak{b}'_n = 0$ for all $\mathfrak{b}' \leq \mathfrak{b}$. In particular, $n > \ell(\mathfrak{b}')$ for all of those $\mathfrak{b}'$. Since phase 3 is the only phase in which the target of k_n can be changed, the target of k_n has thus never been changed. But for every strategy σ considered so far, $\sigma(k_n) = t$ held due to $\sigma^\star(k_n) = \sigma^*(k_n) = t$. Since $\sigma_\mathfrak{b}$ is a phase 1 strategy by definition, this contradicts Table 4.3, even if the alternative initial strategy described in Definition 4.2.2 is used. These arguments furthermore imply $\mathrm{Val}_{\sigma_\mathfrak{b}}(\sigma_\mathfrak{b}(k_n)) = 0$ for all $\mathfrak{b} \in [2^{n-2} - 1]$.

As a consequence, the occurrence records of all edges (k_n, k_i) for $i \in [n-1]$ are zero. We now discuss how this violates Table 4.5. Consider some $i \in \mathbb{N}$ such that $\mathfrak{b} \geq 2^{i-1}$. Then, by Table 4.5, $\phi^{\sigma_\mathfrak{b}}(k_n, k_i) = f(\mathfrak{b}, i, \{(n,0)\})$. But, due to $\mathfrak{b}'_n = 0$ for all $\mathfrak{b}' \leq \mathfrak{b}$, we have $f(\mathfrak{b}, i, \{(n,0)\}) = f(\mathfrak{b}, i)$. Thus, by Lemma 4.1.4 (3) and since $\mathfrak{b} \geq 2^{i-1}$,

$$f(\mathfrak{b}, i, \{(n,0)\}) = f(\mathfrak{b}, i) = \left\lfloor \frac{\mathfrak{b} + 2^{i-1}}{2^i} \right\rfloor \geq \left\lfloor \frac{2^{i-1} + 2^{i-1}}{2^i} \right\rfloor = 1.$$

This contradicts the occurrence records of all edges (k_n, k_i) for $i \in [n-1]$ being zero.

It remains to show that applying the improving switches as described before contradicts the LeastEntered pivot rule. This is achieved by proving that (k_n, k_1) is improving during $\sigma_2 \to \sigma_3$. We discuss the case of $\mathfrak{b} \in \{3, \dots, 2^{n-2} - 2\}$ afterwards. By Table 4.4, $L_\sigma^5 = \{(s, k_\nu)\}$ for any phase 5 strategy σ. Since only switches contained in the subsets L_σ^p are chosen as improving switches, this implies that (s, k_1) is chosen in phase 5 of $\sigma_2 \to \sigma_3$. But, since $\ell(1) = \ell(3) = 1$, this edge has already been chosen in phase 5 of $\sigma_0 \to \sigma_1$. Therefore, the edge has a non-zero occurrence record throughout $\sigma_2 \to \sigma_3$. Thus, the result follows once we showed that (k_n, k_1) is an improving switch, since we already observed that it has an occurrence record of zero but is not switched.

Consider $\sigma_\mathfrak{b}$ for $\mathfrak{b} = 2$. The only set bit in the binary representation of $\mathfrak{b}$ is $\mathfrak{b}_2$. As observed before, $\sigma_2(k_n) = t$, implying $\mathrm{Val}_{\sigma_2}(\sigma_2(k_n)) = 0$. In addition, by Lemma 4.3.6, for every strategy σ calculated during the transition from σ_2 to σ_3, it holds that

$$\begin{aligned}
\mathrm{Val}_{\sigma_2}(k_1) &\geq \langle k_1 \rangle + S_1 = (-N)^{2\cdot 1+7} + S_1 \\
&\geq \sum_{j \in [n]:\mathfrak{b}_j = 1} \left[(-N)^{2j+7} + (-N)^{2j+8} + (-N)^7 + (-N)^6\right] - N^9 \\
&= N^{12} - N^{11} - N^9 - N^7 - N^6 > 0.
\end{aligned}$$

Thus, (k_n, k_1) is an improving switch during the whole transition from σ_2 to σ_3.

Since $\mathrm{Val}_{\sigma_\mathfrak{b}}(k_n) = 0$ for all $\mathfrak{b} \in \{3, \dots, 2^{n-2} - 2\}$, since $\ell(\mathfrak{b}) \neq n$ for those $\mathfrak{b}$, and since the values are non-decreasing, (k_n, k_1) remains improving for all $\mathfrak{b} \in \{3, \dots, 2^{n-2} - 2\}$. In addition, due to $\mathfrak{b} \geq 3$, both bicycles of level 1 have been closed at least once. This implies that the edges of these bicycles have an occurrence of at least 1. Also, at least one of the edges of the inactive bicycle of level 1 is switched when transitioning from $\sigma_\mathfrak{b}$ to $\sigma_{\mathfrak{b}+1}$ for any $\mathfrak{b} \in \mathfrak{B}_n$. Because this edge has a non-zero occurrence record whereas the edge (k_n, k_1) has an occurrence record of zero and is an improving switch, this shows that following the informal description contradicts the Least-Entered pivot rule at least once during the transition from $\sigma_\mathfrak{b}$ to $\sigma_{\mathfrak{b}+1}$ for every $\mathfrak{b} \in \{3, \dots, 2^{n-1} - 2\}$. □

There is however another description of the application of improving switches during phase 3. According to this description, the switches are not only be applied in levels with a lower index than the least significant set bit but for all levels. Especially, the side conditions specified in Table 4.3 rely on the fact that these switches are applied for all levels i with $(\mathfrak{b}+1)_i = 0$. According to the proof of [Fri11c, Lemma 5], the switches need to be applied as follows (where the notation is again adapted): "*In order to fulfill all side conditions for phase 3, we need to perform all switches from higher indices to smaller indices, and k_i to k_ν before $b^j_{i,k}$ with $(\mathfrak{b}+1)_i \neq j$ or $(\mathfrak{b}+1)_i = 0$ to k_ν.*" However, applying improving switches in this fashion violates Zadeh's pivot rule.

Issue 4.3.7 (Alternative description of phase 3). *Applying the improving switches as described in [Fri11c, Lemma 5] does not obey Zadeh's* LEASTENTERED *pivot rule.*

Proving this issue requires a more involved analysis of the subset of improving switches applied during phase 3 and the occurrence records of these switches. As the proofs to the following statements are rather technical, we do not include them here. They can however be found in Appendix A.

Let σ be a fixed phase 3 strategy. We first partition the subset L^3_σ of the set of improving switches for a phase 3 strategy into three sets $L^{3,1}_\sigma$, $L^{3,2}_\sigma$ and $L^{3,3}_\sigma$ (cf. Table 4.4):

- $L^{3,1}_\sigma := \{(k_i, k_\nu) : \sigma(k_i) = k_{\ell'} \land (\mathfrak{b}+1)_i = 0\}$
- $L^{3,2}_\sigma := \{(b^j_{i,k}, k_\nu) : \sigma(b^j_{i,k}) \neq k_{\ell'} \land (\mathfrak{b}+1)_i = 0\}$
- $L^{3,3}_\sigma := \{(b^j_{i,k}, k_\nu) : \sigma(b^j_{i,k}) \neq k_{\ell'} \land (\mathfrak{b}+1)_{i+1} \neq j\}$

The next lemma shows that the improving switches that are applied during phase 3 are fully characterized by the strategy $\sigma_\mathfrak{b}$. This lemma is just a summary and reformulation of the description of the application of improving switches during phase 3, verified by the definition of the strategy phases provided by Table 4.3. We thus do not prove it here.

Lemma 4.3.8. *Let $\mathfrak{b} \in \mathfrak{B}_n$ and let σ denote the first phase 3 strategy calculated after $\sigma_\mathfrak{b}$. Then, $L^3_\sigma = L^3_{\sigma_\mathfrak{b}}$, and $L^3_{\sigma_\mathfrak{b}}$ is the set of improving switches that should be applied during phase 3 according to Table 4.3.*

The following lemma now shows that applying a single improving switch in phase 3 reduces the size of L^3_σ by 1. As a reminder, for a strategy σ and an improving switch $e \in I_\sigma$, the strategy σe is the strategy obtained from σ by applying the switch e.

Lemma 4.3.9. *Let σ be a phase 3 strategy and let $e \in L^3_\sigma$. Then $L^3_{\sigma e} = L_\sigma \setminus \{e\}$.*

This lemma immediately implies that improving switches stay improving during phase 3 in the following sense.

Corollary 4.3.10. *Let σ be a phase 3 strategy and $e \in I_\sigma$. Let σ' be a phase 3 calculated by the strategy iteration algorithm during the same transition. If e was not applied when transitioning from σ to σ', then $e \in I_{\sigma'}$.*

The next lemma now relates the occurrence records of edges of the type (k_*, k_ν).

Lemma 4.3.11. *Let $i \in \{2, \ldots, n-2\}$ and $l < i$. Then, there is a number $\mathfrak{b} \in \mathfrak{B}_n$ with $\ell(\mathfrak{b}+1) = \nu = l$ such that for all $j \in \{i+2, \ldots, n\}$, it holds that $\phi^{\sigma_\mathfrak{b}}(k_i, k_\nu) < \phi^{\sigma_\mathfrak{b}}(k_j, k_\nu)$ and $(k_i, k_\nu), (k_j, k_\nu) \in L^3_{\sigma_\mathfrak{b}}$.*

These results enable us to prove that the application as described in [Fri11c, Lemma 5] does not obey Zadeh's rule.

Proof of Issue 4.3.7. According to [Fri11c, Lemma 5], the improving switches of phase 3 should be applied as follows: "*[...] we need to perform all switches from higher indices to smaller indices, and k_i to k_ν before $b^j_{i,k}$ with $(\mathfrak{b}+1)_{i+1} \neq j$ or $(\mathfrak{b}+1)_i = 0$ to k_ν*". This description is also further formalized in the side conditions of Table 4.3.

Let $i \in \{2, \ldots, n-2\}, l < i$ and $j \in \{i+2, \ldots, n-2\}$. By Lemma 4.3.11, there is a number $\mathfrak{b} \in \mathfrak{B}_n$ such that $l = \nu = \ell(\mathfrak{b}+1)$ and $\phi^{\sigma_\mathfrak{b}}(k_i, k_\nu) < \phi^{\sigma_\mathfrak{b}}(k_j, k_\nu)$. In addition, $(k_i, k_\nu), (k_j, k_\nu) \in L^3_{\sigma_\mathfrak{b}}$. Therefore, by Lemma 4.3.8, (k_j, k_ν) should be applied before (k_i, k_ν) during $\sigma_\mathfrak{b} \to \sigma_{\mathfrak{b}+1}$ when following the description of [Fri11c].

Consider the phase 3 strategy σ of this transition in which the switch (k_j, k_ν) should be applied. Then, since $j > i$ and since we "*perform all switches from higher indices to smaller indices*", the switch (k_i, k_ν) was not applied yet. But, by Corollary 4.3.10, it is an improving switch for the current strategy σ. This implies $\phi^{\sigma_\mathfrak{b}}(k_j, k_\nu) = \phi^{\sigma}(k_j, k_\nu)$ and $\phi^{\sigma_\mathfrak{b}}(k_i, k_\nu) = \phi^{\sigma}(k_i, k_\nu)$. Consequently, $\phi^{\sigma}(k_i, k_\nu) < \phi^{\sigma}(k_j, k_\nu)$. Thus, since (k_i, k_ν) is improving for σ and has a lower occurrence record than (k_j, k_ν) and σ was chosen as the strategy in which (k_j, k_ν) should be applied, the LeastEntered rule is violated. □

The application described in [Fri11c, Lemma 5] can be interpreted as a special case of a general framework for applying improving switches during phase 3. This framework is that improving switches are applied "one level after another". That is, during the transition from $\sigma_\mathfrak{b}$ to $\sigma_{\mathfrak{b}+1}$, a fixed ordering $S^{\ell(\mathfrak{b}+1)}$ depending on $\ell(\mathfrak{b}+1)$ of the levels 1 to n is considered. When level i_1 now precedes level i_2 within $S^{\ell(\mathfrak{b}+1)}$, all improving switches that correspond to edges (u, v) with u being part of level i_1 need to be applied before any such switch of level i_2 is applied. This ordering $S^{\ell(\mathfrak{b}+1)}$ only depends on $\ell(\mathfrak{b}+1)$, so during transitions $\sigma_\mathfrak{b} \to \sigma_{\mathfrak{b}+1}$ and $\sigma_{\mathfrak{b}'} \to \sigma_{\mathfrak{b}'+1}$ with $\ell(\mathfrak{b}+1) = \ell(\mathfrak{b}'+1)$, the same ordering is used. It is clear that the description described in [Fri11c, Lemma 5] is of this kind.

Our goal is now to prove the following: Let $l \in [n-4]$. If the improving switches of phase 3 are applied level by level according to a fixed ordering S^l during all transitions from $\sigma_\mathfrak{b}$ to $\sigma_{\mathfrak{b}+1}$ for which $\ell(\mathfrak{b}+1) = l$, then the application violates Zadeh's pivot rule at least once. This shows that an entire class of orderings of the improving switches of phase 3 violates the LeastEntered pivot rule, including the ordering used in [Fri11c]. In some sense, this proves that the ordering used in [Fri11c] needs to be changed fundamentally and cannot be fixed by slight adaption. We therefore interpret this issue as a major issue as it might have a significant impact on the results of [Fri11c].

Issue 4.3.12. *Consider an arbitrary tie-breaking rule for the LeastEntered pivot rule such that the improving switches of phase 3 are applied one level after another as described previously. That is, let the ordering of the levels in the transition from $\sigma_\mathfrak{b}$ to $\sigma_{\mathfrak{b}+1}$ only depend on $\ell(\mathfrak{b}+1)$ for all $\mathfrak{b} \in \mathfrak{B}_n$. Then, the LeastEntered pivot rule is violated.*

Proving this issue requires another lemma similar to Lemma 4.3.11. This level relates the occurrence records of bicycle edges of level i and (k_{i+1}, k_ν). It in particular implies that level $i+1$ has to be applied before level i if improving should be applied one level after another.

Lemma 4.3.13. *Assume that all edges of $L^3_{\sigma_\mathfrak{b}}$ are applied during phase 3 of the transition from $\sigma_\mathfrak{b}$ to $\sigma_{\mathfrak{b}+1}$ for all $\mathfrak{b} \in \mathfrak{B}_n$. Let $i \in \{2, \dots, n-2\}$ and $l < i$ be fixed. Then, there is a $\mathfrak{b} \in \mathfrak{B}_n$ with $\ell(\mathfrak{b}+1) = l$ such that $\phi^{\sigma_\mathfrak{b}}(k_{i+1}, k_\nu) < \phi^{\sigma_\mathfrak{b}}(b^1_{i,k}, k_\nu)$ for some $k \in \{0,1\}$ and $(k_{i+1}, k_\nu), (b^1_{i,k}, k_\nu) \in L^3_{\sigma_\mathfrak{b}}$.*

Using Lemmas 4.3.11 and 4.3.13 now allows us to prove Issue 4.3.12.

Proof of Issue 4.3.12. To prove that applying improving switches level by level cannot obey Zadeh's pivot rule, we show the following statement. For every $i \in [n]$, we let S^i denote a fixed ordering of $[n]$. Suppose that the improving switches of phase 3 are applied level by level according to the ordering $S^{\ell(\mathfrak{b}+1)}$ when transitioning from $\sigma_\mathfrak{b}$ to $\sigma_{\mathfrak{b}+1}$ for all $\mathfrak{b} \in \mathfrak{B}_n$. Then, for every $l \in [n-4]$, assuming that applying the improving switches according to S^l obeys the LEASTENTERED pivot rule yields a contradiction.

Let $l \in [n-4]$ and consider the ordering $S^l = (s_1, \dots, s_n)$ of $[n]$. For $k \in [n]$, we denote the position of k within S^l by k^*. That is, k^* is the unique number in $[n]$ such that $s_{k^*} = k$.

Assume that applying the improving switches level by level according to S^l obeys Zadeh's pivot rule. We prove that this yields $(l+1)^* < (n-1)^*$ as well as $(n-1)^* < (l+1)^*$.

Let $i \in \{l+1, \dots, n-2\}$. Then, by Lemma 4.3.13, there exists at least one $\mathfrak{b} \in \mathfrak{B}_n$ with $\ell(\mathfrak{b}+1) = \nu = l$ and $\phi^{\sigma_\mathfrak{b}}(k_{i+1}, k_\nu) < \phi^{\sigma_\mathfrak{b}}(b^1_{i,k}, k_\nu)$ as well as $(k_{i+1}, k_\nu), (b^1_{i,k}, k_\nu) \in L^3_{\sigma_\mathfrak{b}}$. By Lemma 4.3.8, both switches are applied during $\sigma_\mathfrak{b} \to \sigma_{\mathfrak{b}+1}$. As the application of the improving switches obeys Zadeh's pivot rule, this implies $(i+1)^* < i^*$. This argument can be applied for all $i \in \{l+1, \dots, n-2\}$, hence $(n-1, n-2, \dots, l+1)$ is a (not necessarily consecutive) subsequence of S^l. In particular, $(n-1)^* < (l+1)^*$ as $l+1 \neq n-1$ if we choose n sufficiently large.

Now, let $i = l+1$ and $j \in \{i+2, \dots, n\}$. By Lemma 4.3.11, there is some $\mathfrak{b} \in \mathfrak{B}_n$ with $\ell(\mathfrak{b}+1) = l$ such that $\phi^{\sigma_\mathfrak{b}}(k_i, k_\nu) < \phi^{\sigma_\mathfrak{b}}(k_{i+2}, k_\nu)$. Lemma 4.3.8 implies that these switches are applied during $\sigma_\mathfrak{b} \to \sigma_{\mathfrak{b}+1}$. This implies that any level $i \in \{l+1, \dots, n-2\}$ has to precede any level $j \in \{i+2, \dots, n\}$ within the ordering S^l. Consequently, the sequence $(l+1, \dots, n-1, n)$ is a (not necessarily consecutive) subsequence of S^l. This in particular implies $(l+1)^* < (n-1)^*$ since $n-1 \geq l+3$ as we have $l \leq n-4$ by assumption. But this contradicts $(n-1)^* < (l+1)^*$.

Therefore, applying the improving switches level by level according to the ordering S^l does not obey Zadeh's LEASTENTERED pivot rule. □

This concludes our discussion of the issues. In the next section, all of these issues are resolved by (i) analyzing the alternative initial strategy given in Definition 4.2.2, (ii) giving a more sophisticated description of the occurrence records of the bicycle edges and (iii) proving that there is *some* way of applying the improving switches of phase 3 during phase 3 without violating Zadeh's pivot rule.

4.4. Correction of the Flaws

In this section, we resolve each issue that was discussed in Section 4.3. We first prove that the alternative initial strategy given in 4.2.2 avoids Issues 4.3.1 and 4.3.3 (Theorems 4.4.1 and 4.4.2). We then provide another system for describing the occurrence records of the bicycle edges and prove that this system correctly specifies the occurrence records (Theorem 4.4.3). Finally, we prove that it is possible to apply the improving switches of phase 3 while obeying Zadeh's pivot rule (Theorem 4.4.14) and that this does not affect the overall correctness of Friedmann's original result (Theorem 4.4.15). We begin by discussing the initial strategy.

The initial strategy

We have proven that there are two issues regarding the initial strategy σ^*. In Issue 4.3.1, we showed that the set of improving switches I_{σ^*} does not conform to the sub- and supersets of Table 4.4. Furthermore, we proved that the strategy contradicts either Table 4.4 or Table 4.5 with respect to $\mathfrak{b} = 1$. We gave an alternative initial strategy $\sigma^\star$ in Definition 4.2.2, claiming that this strategy does not have the issues the original initial strategy has. This claim is now proven.

As a remainder, $\sigma^\star$ is defined via $\sigma^\star(d_i^0) := h_i^0$ and $\sigma^\star(d_i^1) := s$ for all $i \in [n]$ and $\sigma^\star(v) := t$ for all other player vertices v with non-singular out-degree. We now prove that $\sigma^\star$ avoids Issues 4.3.1 and 4.3.3.

Theorem 4.4.1. *The set of improving switches for $\sigma^\star$ is $I_{\sigma^\star} = \{(b_{i,k}^j, A_i^j) : \sigma^\star(b_{i,k}^j) \neq A_i^j\}$.*

Proof. In comparison with the original initial strategy, the changes can only have an effect on bicycle edges $(b_{*,*}^1, A_*^1)$ and and on edges of the type (d_*^1, h_*^1). It thus suffices to prove that none of the edges (d_*^1, h_*^1) is an improving switch whereas all edges $(b_{*,*}^1, A_*^1)$ are improving for $\sigma^\star$.

Fix some $i \in [n]$ and let $\mathrm{Val} := \mathrm{Val}_{\sigma^\star}$. By the definition of $\sigma^\star$, it holds that $\sigma^\star(d_i^1) = s$ and $\sigma^\star(s) = t$, so $\mathrm{Val}(\sigma^\star(d_i^1)) = 0$. Thus, $(d_i^1, h_i^1) \notin I_{\sigma^\star}$ follows from

$$\mathrm{Val}(h_i^1) = (-N)^{2i+8} + \mathrm{Val}(k_{i+1}) = N^{2i+8} + (-N)^{2i+9} + \mathrm{Val}(t) = N^{2i+8} - N^{2i+9} < 0.$$

Let $k \in \{0,1\}$. Since $\mathrm{Val}(\sigma^\star(b_{i,k}^1)) = \mathrm{Val}(\sigma^\star(b_{i,1-k}^1)) = 0$, it suffices to prove $\mathrm{Val}(A_i^1) > 0$. As $\sigma^\star(d_i^1) = s$, this follows from $\mathrm{Val}(A_i^1) = \varepsilon\,\mathrm{Val}(d_1^1) = \varepsilon[N^6 + \mathrm{Val}(s)] = \varepsilon N^6 > 0$. Consequently, $(b_{i,k}^1, A_i^1)$ is an improving switch for $\sigma^\star$. □

Theorem 4.4.2. *If the strategy iteration algorithm uses $\sigma^\star$ as initial strategy, then both Tables 4.4 and 4.5 are correct for the initial phase 1 strategy σ_1 for $\mathfrak{b} = 1$.*

Proof. Let σ denote the first phase 6 strategy calculated during the transition from $\sigma^\star$ to σ_1. As no improving switch $(d_*^1, *)$ is applied when transitioning from $\sigma^\star$ to σ, the definition of $\sigma^\star$ implies that $\sigma(d_i^1) = s$ holds for all $i \in [n]$. This in particular implies that none of the edges (d_i^1, s) is an improving switch for σ, and none of these edges can be switched. Thus, once σ_1 is reached, the occurrence record of all these edges is 0 which is in accordance with Table 4.5. This furthermore implies that none of the edges (d_i^1, s) is improving during phase 1 of the transition from σ_1 to σ_2, resolving the contradiction regarding Table 4.4. □

This concludes our discussion of the corrections of the issues related to the initial strategy. We now discuss the next issue, the occurrence records of the bicycle edges.

The occurrence records of bicycle edges

As proven in Issue 4.3.4, the description of the bicycle edges given by Table 4.5 is not entirely accurate. The problem is that the system describing these occurrence records does not properly distinguish between bicycles that were already closed at least once during some previous transition and bicycles that have never been closed until now. We now provide a system of equations properly distinguishes between these types of bicycles and thus describes the occurrence records of the bicycle edges properly. Let $\mathfrak{b} \in \mathfrak{B}_n$ be fixed, let A_i^j be a fixed bicycle and define $\mathfrak{g} := \mathfrak{g}(\mathfrak{b}, i, \{(i+1, j)\}), z := \mathfrak{b} - \mathfrak{g} - 2^{i-1}$ and $\phi^{\sigma_\mathfrak{b}}(A_i^j) := \phi^{\sigma_\mathfrak{b}}(b_{i,0}^j, A_i^j) + \phi^{\sigma_\mathfrak{b}}(b_{i,1}^j, A_i^j)$. We consider the following system:

$$|\phi^{\sigma_\mathfrak{b}}(b_{i,0}^j, A_i^j) - \phi^{\sigma_\mathfrak{b}}(b_{i,1}^j, A_i^j)| \leq 1 \tag{4.3}$$

$$\phi^{\sigma_\mathfrak{b}}(A_i^j) = \begin{cases} \mathfrak{g}+1, & A_i^j \text{ is closed and active} \\ \mathfrak{b}, & A_i^j \text{ is open and active} \\ \mathfrak{b}, & A_i^j \text{ is inactive and } \mathfrak{b} < 2^{i-1} + j \cdot 2^i \\ \mathfrak{g}+1+2z, & A_i^j \text{ is inactive and } \mathfrak{b} \geq 2^{i-1} + j \cdot 2^i \end{cases} \tag{4.4}$$

Before proving that this system in fact correctly describes the occurrence record of the bicycle edges, we describe these occurrence records and the conditions given here informally. Ideally, the occurrence records of bicycle edges in a bicycle A_i^j with respect to $\sigma_\mathfrak{b}$ should be as follows:

- If the bicycle is closed and active, then $\phi^{\sigma_\mathfrak{b}}(A_i^j)$ should correspond to the last time the bicycle closed. This is the last time a number $\mathfrak{b}'$ with $\ell(\mathfrak{b}') = i$ and $\mathfrak{b}'_{i+1} = j$ was calculated, which is exactly given by $\mathfrak{g}$.
- If it is open and active, then its occurrence record should be equal to the currently represented number $\mathfrak{b}$.
- If it is inactive, then its occurrence record either has to "catch up" to $\mathfrak{b}$ since it was closed for a very long time or it already successfully caught up.

To give more intuition why the system given by Equation (4.4) correctly formalizes this behavior, we compare this to the system given by Equations (4.1) and (4.2).

Both systems contain an inequality that encodes that the difference of the occurrence records of the two bicycle edges may differ by at most 1. We thus focus on Equations (4.2) and (4.4). Consider the second condition of Equation (4.2). This models the case that A_i^j is inactive and does not have an occurrence record of $\mathfrak{b}$. This is handled by the condition $z < \frac{1}{2}(\mathfrak{b} - 1 - \mathfrak{g})$ which is equivalent to $\mathfrak{g} + 1 + 2z < \mathfrak{b}$. As shown in Issue 4.3.4, this condition does not described inactive bicycles properly. This distinction can be included by an additional condition regarding the relation between $\mathfrak{b}$ and $2^{i-1}+j\cdot 2^i$. More precisely, since $2^{i-1}+j\cdot 2^i$ is the smallest number for which the cycle center A_i^j needs to be closed, this condition is used to distinguish inactive bicycles as follows. If $\mathfrak{b}\geq 2^{i-1}+j\cdot 2^i$, then the bicycle has already been active and closed once and might need to catch up as the occurrence record of the bicycle edges might be very low. If $\mathfrak{b}<2^{i+1}+j\cdot 2^{i-1}$, then the bicycle does not not need to catch up because it has not been active yet.

To prove that the system correctly describes the occurrence records, we need to explain how improving switches within the bicycles are applied according to [Fri11c]. Our description is a reformulation of the description given in the proof of [Fri11c, Lemma 5]. The following rules summarize the application of the improving switches within a bicycle A_i^j during phase 1 of the transition from $\sigma_{\mathfrak{b}}$ to $\sigma_{\mathfrak{b}+1}$ (rules are not stated in the order of their application):

1. If A_i^j is open and active, one of the two switches of the bicycle A_i^j is switched.
2. Let $j:=\mathfrak{b}_{\ell(\mathfrak{b}+1)+1}$. In addition to the first rule, the second edge of $A_{\ell(\mathfrak{b}+1)}^j$ is switched.
3. If A_i^j is inactive and $\mathfrak{b}<2^{i-1}+j\cdot 2^i$, one of the two edges of the bicycle is switched.
4. If A_i^j is inactive, $\mathfrak{b}\geq 2^{i-1}+j\cdot 2^i$ and $z<\frac{1}{2}(\mathfrak{b}-1-\mathfrak{g})$, both edges of A_i^j are switched.
5. If A_i^j is inactive, $\mathfrak{b}\geq 2^{i-1}+j\cdot 2^i$ and $z\geq\frac{1}{2}(\mathfrak{b}-1-\mathfrak{g})$, only one edge is switched.

Applying the improving switches according to these 5 rules yields the occurrence records as described by Equation (4.4).

Theorem 4.4.3. *Let $\mathfrak{b}\in\mathfrak{B}_n$ and A_i^j be a bicycle. If the improving switches within A_i^j are applied by as described by rules 1 to 5, then Equations (4.3), (4.4) correctly specify the occurrence records $\phi^{\sigma_{\mathfrak{b}}}(A_i^j)$.*

To simplify the proof, we introduce the following notion. Fix some $\mathfrak{b}\in\mathbb{B}_n$ and a bicycle A_i^j. We say that A_i^j is (a bicycle) *of type k for $\sigma_{\mathfrak{b}}$* when it fulfills the k-th condition of Equation (4.4) for $\sigma_{\mathfrak{b}}$. We additionally establish the following abbreviations and state a lemma that is implicitly contained in the proof of [Fri11c, Lemma 5].

- Similarly to defining $\mathfrak{g}:=\mathfrak{g}(\mathfrak{b},i,\{(i+1,j)\}$, we let $\mathfrak{g}':=\mathfrak{g}(\mathfrak{b}+1,i,\{(i+1,j)\})$.
- We define $z:=\mathfrak{b}-\mathfrak{g}-2^{i-1}$ and $z':=\mathfrak{b}+1-\mathfrak{g}'-2^{i-1}$ analogously.
- We define $\ell:=\ell(\mathfrak{b})$ and $\nu:=\ell(\mathfrak{b}+1)$.

Lemma 4.4.4 ([Fri11c]). *For every $\mathfrak{b}\in\mathfrak{B}_n, i\in[n]$ with $i\neq\ell(\mathfrak{b}+1)$ and $j\in\{0,1\}$, $\mathfrak{g}=\mathfrak{g}'$.*

We also make use of the following lemma, formalizing the intuition we gave previously on the further distinction regarding the occurrence records of the bicycles.

Lemma 4.4.5. *Let $\mathfrak{b}\in\mathfrak{B}_n$ and A_i^j be a bicycle. Then, A_i^j was closed at least once during the application of the strategy iteration algorithm upto strategy $\sigma_{\mathfrak{b}}$ if and only if $\mathfrak{b}\geq 2^{i-1}+j\cdot 2^i$.*

Proof. The bicycle A_i^j is closed the first time when a number $\tilde{\mathfrak{b}}\leq\mathfrak{b}$ is reached such that $\tilde{\mathfrak{b}}_i=1,\tilde{\mathfrak{b}}_{i+1}=j$ and $\tilde{\mathfrak{b}}_l=0$ is calculated by the strategy number algorithm. As this number is exactly $2^{i-1}+j\cdot 2^i$, the statement follows. □

This enables us to prove Theorem 4.4.3 Whenever we discuss how a bicycle should look like, we implicitly refer to the invariants introduced in Section 4.2.

Proof of Theorem 4.4.3. Alongside the main statements of the theorem, we also prove

$$\phi^{\sigma_{\mathfrak{b}}}(A_i^j) \leq \mathfrak{b} + 1 \tag{4.5}$$

where equality holds if and only if $i = \ell$ and $j = \mathfrak{b}_{\ell+1}$. The reason is that this statement is needed in some cases and simplifies the proof.

We show the statement of the theorem and Equation (4.5) via induction on $\mathfrak{b}$. Let $\mathfrak{b} = 0$. By the definition of both the original and the alternative initial strategy, the target of $b_{*,*}^*$ under the corresponding strategy is t. Therefore, all bicycles are open, regardless which of the two initial strategies is considered. As $\mathfrak{b} = 0$, we have $0 = \mathfrak{b} < 2^{i-1} + j \cdot 2^i$ for all $i \in [n]$ and $j \in \{0, 1\}$. This implies that every bicycle is either of type 2 or of type 3. Therefore, the occurrence record of every bicycle needs to be equal to $\mathfrak{b} = 0$. Since we consider the initial strategies, no improving switch was applied yet, implying the statement. Therefore, $\phi^{\sigma_0}(A_i^j) = 0$ for all bicycles A_i^j. Consequently, Equation (4.4) holds. In particular, Equation (4.3) holds as well. Furthermore, there is no least significant set bit ℓ by the choice of $\mathfrak{b}$. Hence, since $\phi^{\sigma_0}(A_i^j) = 0 < \mathfrak{b} + 1$ for all A_i^j, and no bicycle is closed, Equation (4.5) holds as well.

Suppose that the statements holds for all $\mathfrak{b}' \in \mathfrak{B}_n$ with $\mathfrak{b}' \leq \mathfrak{b}$ for some fixed $\mathfrak{b} \in \mathfrak{B}_n$. We show that the two statements also hold for $\mathfrak{b} + 1$. We distinguish between the induction hypotheses with respect to Equation (4.4) and Equation (4.5) and always state to which we refer. We discuss Equation (4.3) at the end of the proof.

Let $\in [n], j \in \{0, 1\}$ and fix a bicycle A_i^j. The proof is organized as follows. We distinguish all "states" the bicycle could be in for $\sigma_{\mathfrak{b}}$. We investigate of which type the bicycle is for $\sigma_{\mathfrak{b}}$ and if this type changes when transitioning to $\sigma_{\mathfrak{b}+1}$. We state how many improving switches are applied according to the rules and why Equation (4.4) remains valid for $\sigma_{\mathfrak{b}+1}$.

1. **A_i^j is open, active and $i = \nu$.** Then A_i^j is the active bicycle corresponding to the least significant set bit of $\mathfrak{b} + 1$. By construction, it is open for $\sigma_{\mathfrak{b}}$ but needs to be closed for $\sigma_{\mathfrak{b}+1}$. The bicycle remains active as $\mathfrak{b}_{\nu+1} = (\mathfrak{b} + 1)_{\nu+1}$, so A_i^j is of type 1 for $\sigma_{\mathfrak{b}+1}$. As both bicycle edges are switched, we prove $\phi^{\sigma_{\mathfrak{b}}}(A_i^j) + 2 = \mathfrak{g}' + 1$.

   By the induction hypothesis (4.4), $\phi^{\sigma_{\mathfrak{b}}}(A_i^j) = \mathfrak{b}$ since A_i^j is a type 2 bicycle for $\sigma_{\mathfrak{b}}$. To show Equation (4.4), it therefore suffices to show $\mathfrak{g}' = \mathfrak{b} + 1$. This however follows since both $\mathfrak{g}'$ and $\mathfrak{b} + 1$ end on the subsequence $(\mathfrak{b}_{\nu+1}, 1, 0, \dots, 0)$ of length $\nu + 1$.

   In addition, $\phi^{\sigma_{\mathfrak{b}+1}}(A_i^j) = (\mathfrak{b} + 1) + 1$, hence Equation (4.5) remains valid.

2. **A_i^j is open and active, but $i \neq \nu$.** We prove that A_i^j remains open and active. By the definition of open and active, $\mathfrak{b}_i = 0$ and $j = \mathfrak{b}_{i+1}$. In addition, $\nu = \ell(\mathfrak{b} + 1)$ implies $\mathfrak{b}_{i'} = 1$ for all $i' \in [\nu - 1]$. As all active bicycles in the levels 1 to $\nu - 1$ are closed for $\sigma_{\mathfrak{b}}$ and $i \neq \nu$, this implies $i > \nu$. As only the bits $\mathfrak{b}_1$ to $\mathfrak{b}_\nu$ are switched, the bicycle A_i^j remains active. Since the active bicycle of level ν is the only bicycle that is open for $\sigma_{\mathfrak{b}}$ but closed for $\sigma_{\mathfrak{b}+1}$, A_i^j remains open. Hence, A_i^j is of type 2 for $\sigma_{\mathfrak{b}+1}$. As only one improving switch is applied in A_i^j (rule 1), we therefore need

to show that $\phi^{\sigma_{\mathfrak{b}}}(A_i^j)+1=\mathfrak{b}+1$. By the induction hypothesis (4.4), $\phi^{\sigma_{\mathfrak{b}}}(A_i^j)=\mathfrak{b}$, so $\phi^{\sigma_{\mathfrak{b}}}(A_i^j)+1=\mathfrak{b}+1$. Therefore, Equations (4.4) and (4.5) hold.

3. **A_i^j is closed, active and $i>\nu$.** We prove that A_i^j is of type 1 for $\sigma_{\mathfrak{b}+1}$. By the definition of closed and active, $\mathfrak{b}_i=1$ and $\mathfrak{b}_{i+1}=j$. As only bits in levels below ν switch, A_i^j remains active and closed since $i>\nu$. Thus, A_i^j is of type 1 for $\sigma_{\mathfrak{b}+1}$ and none of its bicycle edges are switched. We thus need need to show $\phi^{\sigma_{\mathfrak{b}}}(A_i^j)=\mathfrak{g}'+1$.

   By the induction hypothesis (4.4), $\phi^{\sigma_{\mathfrak{b}}}(A_i^j)=\mathfrak{g}+1$, so it suffices to show $\mathfrak{g}+1=\mathfrak{g}'+1$. Since $i\neq\nu$, this follows from Lemma 4.4.4. In addition, Equation (4.5) remains valid since $\phi^{\sigma_{\mathfrak{b}}}(A_i^j)\leq\mathfrak{b}$ by the induction hypothesis (4.5). Since $\phi^{\sigma_{\mathfrak{b}+1}}(A_i^j)=\phi^{\sigma_{\mathfrak{b}}}(A_i^j)$ as argued before, this implies $\phi^{\sigma_{\mathfrak{b}+1}}(A_i^j)<\mathfrak{b}+1$.

4. **A_i^j is closed, active and $i<\nu$.** We show that A_i^j is of type 4 for $\sigma_{\mathfrak{b}+1}$. Since $i<\nu$, the bits $\mathfrak{b}_i$ and $\mathfrak{b}_{i+1}$ both switch. Thus, $(\mathfrak{b}+1)_i=0$ as $i<\nu$. Hence A_i^j is open for $\sigma_{\mathfrak{b}+1}$. Since A_i^j is active for $\sigma_{\mathfrak{b}}$, the choice of i yields $\mathfrak{b}_{i+1}=j$ and $(\mathfrak{b}+1)_{i+1}\neq j$. The bicycle is thus inactive for $\sigma_{\mathfrak{b}+1}$. Since A_i^j is closed, Lemma 4.4.5 implies $\mathfrak{b}\geq 2^{i-1}+j\cdot 2^i$. Therefore, A_i^j is a bicycle of type 4 for $\sigma_{\mathfrak{b}+1}$. As A_i^j is closed, the bicycle edges are not switched. We thus prove $\phi^{\sigma_{\mathfrak{b}}}(A_i^j)=\mathfrak{g}'+1+2z'$.

   By the induction hypothesis (4.4), it follows that $\phi^{\sigma_{\mathfrak{b}}}(A_i^j)=\mathfrak{g}+1$. It thus suffices to prove $\mathfrak{g}+1=\mathfrak{g}'+1+2z'$. Since $i\neq\nu$, Lemma 4.4.4 implies $\mathfrak{g}=\mathfrak{g}'$, so it suffices to prove $z'=\mathfrak{b}+1-\mathfrak{g}'-2^{i-1}=0$. As $i<\nu$ and A_i^j is closed and active, $\mathfrak{b}_i=1$ and $j=\mathfrak{b}_{i+1}$ follow. This implies $\mathfrak{g}=(\mathfrak{b}_n,\dots,\mathfrak{b}_{i+1},1,0,\dots,0)$. Therefore, since $i<\nu$ implies $\mathfrak{b}=(\mathfrak{b}_n,\dots,\mathfrak{b}_{i+1},1,1,\dots,1)$, we obtain $\mathfrak{b}-\mathfrak{g}=2^{i-1}-1$. As $\mathfrak{g}=\mathfrak{g}'$ by Lemma 4.4.4, this yields $z'=\mathfrak{b}+1-\mathfrak{g}'-2^{i-1}=0$, so Equation (4.4) remains valid.

   As in Case 2, $\phi^{\sigma_{\mathfrak{b}+1}}(A_i^j)=\phi^{\sigma_{\mathfrak{b}}}(A_i^j)$ and since $\phi^{\sigma_{\mathfrak{b}}}(A_i^j)\leq\mathfrak{b}$ by the induction hypothesis (4.5), also Equation (4.5) follows.

5. **A_i^j is closed, active and $i=\nu$.** This cannot happen as both bicycles of level ν are open with respect to $\sigma_{\mathfrak{b}}$ since it is the initial phase 1 strategy.

6. **A_i^j is closed and inactive.** This cannot happen since closed bicycles are always active for the initial phase 1 strategy $\sigma_{\mathfrak{b}}$ of a transition.

7. **A_i^j is inactive and $\mathfrak{b}<2^{i-1}+j\cdot 2^i$.** Then, A_i^j is of type 3, and A_i^j being inactive implies that A_i^j is open. We consider the possible types of A_i^j for $\sigma_{\mathfrak{b}+1}$.

   It is impossible that A_i^j is closed for $\sigma_{\mathfrak{b}+1}$ as the active bicycle of level ν is the only bicycle which is open for $\sigma_{\mathfrak{b}}$ and closed for $\sigma_{\mathfrak{b}+1}$, and A_i^j is inactive.

   Suppose that A_i^j is of type 3 for $\sigma_{\mathfrak{b}+1}$. As only one improving switch is applied (rule 3), we thus need to prove $\phi^{\sigma_{\mathfrak{b}}}(A_i^j)+1=\mathfrak{b}+1$. But this follows immediately as $\phi^{\sigma_{\mathfrak{b}}}(A_i^j)=\mathfrak{b}$ by the induction hypothesis (4.4).

   Suppose that A_i^j is of type 2 for $\sigma_{\mathfrak{b}+1}$. We then need to prove $\phi^{\sigma_{\mathfrak{b}}}(A_i^j)+1=\mathfrak{b}+1$, which follows from the induction hypotheses (4.4).

   Suppose that A_i^j is of type 4 for $\sigma_{\mathfrak{b}+1}$. Then, since $\mathfrak{b}<2^{i-1}+j\cdot 2^i$, it follows that $\mathfrak{b}+1=2^{i-1}+j\cdot 2^i$. But, by Lemma 4.4.5, this is only possible if A_i^j is closed during the transition from $\sigma_{\mathfrak{b}}$ to $\sigma_{\mathfrak{b}+1}$, contradicting the inactivity of A_i^j for $\sigma_{\mathfrak{b}}$.

Therefore, $\phi^{\sigma_{\mathfrak{b}}}(A_i^j) + 1 = \mathfrak{b} + 1$ in all possible cases, and both Equation (4.4) and Equation (4.5) stay valid.

8. $\boldsymbol{A_i^j}$ **is inactive,** $\boldsymbol{\mathfrak{b} \geq 2^{i-1} + j \cdot 2^i}$ **and** $\boldsymbol{z < \frac{1}{2}(\mathfrak{b} - 1 - \mathfrak{g})}$**.** Then, A_i^j is a bicycle of type 4 for $\sigma_{\mathfrak{b}}$. We prove that it is also of type 4 for $\sigma_{\mathfrak{b}+1}$. It then remains to prove $\phi^{\sigma_{\mathfrak{b}}}(A_i^j) + 2 = \mathfrak{g}' + 1 + 2z'$, or, since $\phi^{\sigma_{\mathfrak{b}}}(A_i^j) = \mathfrak{g} + 1 + 2z$ by the induction hypothesis (4.4), $\mathfrak{g} + 1 + 2z + 2 = \mathfrak{g}' + 1 + 2z'$.

   First, $\mathfrak{b} + 1 \geq 2^{i-1} + j \cdot 2^i$ follows from $\mathfrak{b} \geq 2^{i-1} + j \cdot 2^i$. Assume that A_i^j was active for $\sigma_{\mathfrak{b}+1}$. Since only bits with an index smaller or equal to ν are switched, only inactive bicycles in levels 1 to $\nu - 1$ can become active. As a consequence, $i < \nu$.

   We next show that $\mathfrak{b} - \mathfrak{g} = 2^i + 2^{i-1} - 1$. First assume $i \neq \nu - 1$. Then, since $i < \nu - 1$ and $\mathfrak{b} = (\mathfrak{b}_n, \dots, \mathfrak{b}_{\nu+1}, 0, 1, \dots, 1)$, it follows that $\mathfrak{b}_{i+1} = 1$. Hence, by the inactivity of A_i^j with respect to $\sigma_{\mathfrak{b}}$, we obtain $j = 0$. Therefore,

$$\mathfrak{g} = (\mathfrak{b}_n, \dots, \mathfrak{b}_{\nu+1}, 0, 1, \dots, 1, \underbrace{0}_{\mathfrak{g}_{i+1}}, \underbrace{1}_{\mathfrak{g}_i}, 0, \dots, 0),$$

   since $\mathfrak{g}_i = 1$ and $\mathfrak{g}_{i+1} = j = 0$ by definition. Consequently, $\mathfrak{b} - \mathfrak{g} = 2^i + 2^{i-1} - 1$.

   Now let $i = \nu - 1$. Then $\mathfrak{b}_{i+1} = \mathfrak{b}_\nu = 0$ and hence $j = 1$ as A_i^j is inactive. Therefore,

$$\mathfrak{g} = (\tilde{\mathfrak{b}}_n, \dots, \tilde{\mathfrak{b}}_{\nu+1}, 1, \underbrace{1}_{\mathfrak{g}_i = \mathfrak{g}_{\nu-1}}, 0, \dots, 0)$$

   where $(\tilde{\mathfrak{b}}_n, \dots, \tilde{\mathfrak{b}}_{\nu+1}) = (\mathfrak{b}_n, \dots, \mathfrak{b}_{\nu+1}) - 1$. This implies $\mathfrak{g} + 2^i + 2^{i-1} = \mathfrak{b} + 1$ which is equivalent to $\mathfrak{b} - \mathfrak{g} = 2^i + 2^{i-1} - 1$.

   Using the identities $\mathfrak{b} - \mathfrak{g} = 2^i + 2^{i-1} - 1$ and $\phi^{\sigma_{\mathfrak{b}}}(A_i^j) = \mathfrak{b} + 1 + 2z$ which follows from the induction hypothesis (4.4), we obtain

$$\phi^{\sigma_{\mathfrak{b}}}(A_i^j) = \mathfrak{b} + 2^i + 2^{i-1} - 1 - 2^i + 1 = \mathfrak{b} + 2^{i-1} > \mathfrak{b}. \tag{4.6}$$

   Additionally, by assumption, $z < \frac{1}{2}(\mathfrak{b} - 1 - \mathfrak{g})$, which implies

$$\phi^{\sigma_{\mathfrak{b}}}(A_i^j) = \mathfrak{g} + 1 + 2z < \mathfrak{g} + 1 + \mathfrak{b} - 1 - \mathfrak{g} = \mathfrak{b}. \tag{4.7}$$

   But this is a contradiction to Equation (4.6). Therefore, A_i^j cannot be active for $\sigma_{\mathfrak{b}+1}$, hence it must be inactive for $\sigma_{\mathfrak{b}+1}$ and thus be of type 4.

   It remains to prove $\phi^{\sigma_{\mathfrak{b}}}(A_i^j) + 2 = \mathfrak{g} + 1 + 2z + 2 = \mathfrak{g}' + 2 + 2z'$. As A_i^j is inactive for $\sigma_{\mathfrak{b}+1}$, it follows that $i \neq \nu$ and thus, by Lemma 4.4.4, also $\mathfrak{g} = \mathfrak{g}'$. Therefore,

$$\mathfrak{g} + 1 + 2z + 2 = \mathfrak{g} + 1 + 2\mathfrak{b} - 2\mathfrak{g} - 2^i + 2 = \mathfrak{g}' + 1 + 2z',$$

   hence Equation (4.4) still holds.

   It remains to show Equation (4.5). By Equation (4.7), we have $\phi^{\sigma_{\mathfrak{b}}}(A_i^j) < \mathfrak{b}$, and thus, by integrality, $\phi^{\sigma_{\mathfrak{b}}}(A_i^j) \leq \mathfrak{b} - 1$. Thus, $\phi^{\sigma_{\mathfrak{b}+1}}(A_i^j) = \phi^{\sigma_{\mathfrak{b}}}(A_i^j) + 2 \leq \mathfrak{b} - 1 + 2 = \mathfrak{b} + 1$ follows since both bicycle edges of A_i^j are switched.

9. **A_i^j is inactive, $\mathfrak{b} \geq 2^{i-1} + j \cdot 2^i$ and $z \geq \frac{1}{2}(\mathfrak{b} - 1 - \mathfrak{g})$.** In this case, we do not distinguish the type of A_i^j for $\sigma_{\mathfrak{b}+1}$ and prove $\mathfrak{g} + 1 + 2z = \mathfrak{b}$ instead. This suffices as A_i^j cannot become closed and active for $\sigma_{\mathfrak{b}+1}$ and, by rule 5, the occurrence record of A_i^j increases by 1. Therefore, we do not need to specify the type of A_i^j if we prove that its occurrence record before applying the switch is equal to $\mathfrak{b}$.

   We prove $z = \frac{1}{2}(\mathfrak{b} - 1 - \mathfrak{g})$. Assume $z > \frac{1}{2}(\mathfrak{b} - 1 - \mathfrak{g})$. Then, since A_i^j is of type 4, the induction hypothesis (4.4) implies $\phi^{\sigma_\mathfrak{b}}(A_i^j) = \mathfrak{g} + 1 + 2z$. Thus

$$\phi^{\sigma_\mathfrak{b}}(A_i^j) = \mathfrak{g} + 1 + 2z > \mathfrak{g} + 1 + \mathfrak{b} - 1 - \mathfrak{g} = \mathfrak{b},$$

   contradicting the induction hypothesis (4.5) requiring $\phi^{\sigma_\mathfrak{b}}(A_i^j) \leq \mathfrak{b}$. Therefore, equality holds, implying $\phi^{\sigma_\mathfrak{b}}(A_i^j) = \mathfrak{g} + 1 + (\mathfrak{b} - 1 - \mathfrak{g}) = \mathfrak{b}$. As a single switch is applied, we obtain $\phi^{\sigma_\mathfrak{b}}(A_i^j) + 1 = \mathfrak{b} + 1$ as claimed.

Thus, the occurrence records given in Equation (4.4) and the estimation given in Equation (4.5) hold. Since the switches can be applied alternatingly within a single bicycle, Equation (4.3) holds at all times during the application of the improving switches. □

The improving switches of phase 3

We now prove that the improving switches of phase 3 can be applied without violating Zadeh's pivot rule and that this application can be extended in such a way that the improving switches can be applied in all phases without violating the pivot rule.

Let σ be a phase 3 strategy. The set L_σ^3 contains all edges that should be applied as improving switches since $L_\sigma^3 \subseteq I_\sigma$ by [Fri11c, Lemma 4]. Similarly, U_σ^3 contains the edges that might be applied as improving switches since $I_\sigma \subseteq U_\sigma^3$. We thus compare and analyze these sets in detail. This comparison enables us to prove that there is always a switch contained in L_σ^3 minimizing the occurrence record. This justifies that "*we will only use switches from L_σ^p*" [Fri11c, page 12] for phase $p = 3$. We then prove our main statement and main contribution regarding the subexponential lower bound of [Fri11c]: All improving switches that should be applied during phase 3 according to [Fri11c] can be applied during phase 3 while obeying Zadeh's LeastEntered pivot rule.

As discussed in Section 4.1, a transition between two consecutive initial phase 1 strategies is partitioned into 6 phases. In each phase, a different "task" is performed within the construction, and the task of phase 3 is to reset the Markov decision process. More precisely, some bicycles are opened and the targets of some of the entry vertices are adjusted according to the new least significant set bit. In particular, a phase 3 strategy is always associated with such a transition and we implicitly consider the underlying transition from $\sigma_\mathfrak{b}$ to $\sigma_{\mathfrak{b}+1}$ whenever discussing a phase 3 strategy and use the typical abbreviations $\ell := \ell(\mathfrak{b})$ and $\nu := \ell(\mathfrak{b}+1)$.

By Lemma 4.3.8, $L_{\sigma_\mathfrak{b}}^3$ is the set of all improving switches that should be applied during phase 3. We begin by providing an upper bound on the occurrence record of these edges.

Lemma 4.4.6. *Let σ be a phase 3 strategy. Then $\max_{e \in L_\sigma^3} \phi^\sigma(e) \leq f(\mathfrak{b}, \nu)$.*

We now focus on the set U_σ^3. This set contains L_σ^6, hence this set needs to be analyzed as well. There is, however, a small error in the definition of this set that needs to be corrected. As we believe that this error is just a typo in [Fri11c], we do not discuss it in detail here.

Issue 4.4.7. *For every $\mathfrak{b} \in \mathfrak{B}_n$ with $\ell(\mathfrak{b}+1) > 1$, there is an improving switch that should be applied during phase 6 of $\sigma_\mathfrak{b} \to \sigma_{\mathfrak{b}+1}$ but is not contained in L_σ^6 for any phase 6 strategy σ of this transition.*

Proof. Let $\mathfrak{b} \in \mathfrak{B}_n$ with $\nu = \ell(\mathfrak{b}+1) > 1$ and consider the vertex $d_{\nu-1}^0$. We prove that $(d_{\nu-1}^0, s)$ needs to be applied during phase 6 of $\sigma_\mathfrak{b} \to \sigma_{\mathfrak{b}+1}$ but is not contained in L_σ^6 for any phase 6 strategy σ. By analyzing Table 4.3, it is easy to verify that $\mathfrak{b}_\ell = 0$ implies $\sigma_\mathfrak{b}(d_{\nu-1}^0) = h_i^0$. Since bit ℓ switches during $\sigma_\mathfrak{b} \to \sigma_{\mathfrak{b}+1}$, Table 4.3 implies that $\sigma_{\mathfrak{b}+1}(d_{\nu-1}^0) = s$ needs to hold. Therefore, $(d_{\nu-1}^0, s)$ needs to be an improving switch for some strategy σ calculated during $\sigma_\mathfrak{b} \to \sigma_{\mathfrak{b}+1}$.

For the sake of a contradiction, assume that there was a strategy σ in which $(d_{\nu-1}^0, s)$ is applied. Since only the subsets of phase 6 strategies can contain this edge, σ is a phase 6 strategy. By [Fri11c, Lemma 4], $(d_{\nu-1}^0, s) \in L_\sigma^6$ for such a strategy σ. Analyzing Table 4.4, it is easy to verify that $(d_{\nu-1}^0, s) \in L_\sigma^6$ then implies $\sigma(d_{\nu-1}^0) \neq s$ and $\sigma(d_{\nu-1}^0) = s$. This is a contradiction, so there is no strategy σ in which $(d_{\nu-1}^0, s)$ is applied. □

We believe that the set was intended to be defined as follows and use this definition henceforth.

Theorem 4.4.8 ("Correction" of L_σ^6). *Let σ be a phase 6 strategy for $\mathfrak{b} \in \mathfrak{B}_n$ and $\mathfrak{b}' := \mathfrak{b}+1$. Then, the subset L_σ^6 of I_σ should be defined as*

$$L_\sigma^6 := \{(d_i^0, v) : \sigma(d_i^0) \neq v \wedge \bar{\sigma}(d_i^0) = \mathfrak{b}'_{i+1}\} \cup \{(d_i^1, v) : \sigma(d_i^1) \neq v \wedge \bar{\sigma}(d_i^1) \neq \mathfrak{b}'_{i+1}\}.$$

It is easy to verify that this definition of L_σ^6 resolves Issue 4.4.7.

We return to the discussion of the set U_σ^6 and partition this set into 9 subsets as follows:

$$\begin{aligned}
U_\sigma^{3,1} &:= \{(k_i, k_l) : \sigma(k_i) \notin \{k_l, k_\nu\} \wedge l \leq \nu \wedge (\mathfrak{b}+1)_i = 0\} \\
U_\sigma^{3,2} &:= \{(b_{i,k}^j, k_l) : \sigma(b_{i,k}^j) \notin \{k_l, k_\nu\} \wedge l \leq \nu \wedge (\mathfrak{b}+1)_i = 0\} \\
U_\sigma^{3,3} &:= \{(b_{i,k}^j, k_l) : \sigma(b_{i,k}^j) \notin \{k_l, k_\nu\} \wedge l \leq \nu \wedge (\mathfrak{b}+1)_{i+1} \neq j\} \\
U_\sigma^{3,4} &:= \{(h_i^0, k_l) : l \leq \min(\{n+1\} \cup \{j \geq i+2 : \mathfrak{b}_j = 1\})\} \\
U_\sigma^{3,5} &:= \{(s, k_i) : \sigma(s) \neq k_i \wedge i < \nu\} \\
U_\sigma^{3,6} &:= \{(d_i^j, v) : \sigma(d_i^j) \neq v \wedge i < \nu\} \\
U_\sigma^{3,7} &:= \{(d_i^0, v) : \sigma(d_i^0) \neq v \wedge \bar{\sigma}(d_i^0) = (\mathfrak{b}+1)_{i+1}\} \\
U_\sigma^{3,8} &:= \{(d_i^1, v) : \sigma(d_i^1) \neq v \wedge \bar{\sigma}(d_i^1) \neq (\mathfrak{b}+1)_{i+1}\} \\
U_\sigma^{3,9} &:= \{(b_{i,l}^j, A_i^j) . \sigma(b_{i,l}^j) \neq A_i^j\}
\end{aligned}$$

By Lemma 4.4.6, the occurrence records of the edges that should be applied during phase 3 are bounded by $f(\mathfrak{b}, \nu)$. We now provide a matching lower bound regarding the switches that should be applied *after* phase 3. This bound will also be used to estimate the occurrence records of all edges contained in U_σ^3.

Lemma 4.4.9. *Let σ be a phase 3 strategy. Assume that the strategy iteration algorithm is started with the initial strategy $\sigma^\star$. Then $\min_{e \in L_\sigma^4 \cup L_\sigma^5 \cup L_\sigma^5} \phi^{\sigma_\mathfrak{b}}(e) \geq f(\mathfrak{b}, \nu)$.*

This lemma can also be used to prove that none of the edges contained in any of the sets $U_\sigma^{3,3}, \ldots, U_\sigma^{3,9}$ is applied during phase 3. The reason is that the occurrence record of these edges is too large, so the LeastEntered pivot rule will not choose to apply them.

Lemma 4.4.10. *Let σ be a phase 3 strategy. Let $e_1 \in L_\sigma^3$ and $e_2 \in I_\sigma \cap (U_\sigma^{3,4} \cup \cdots \cup U_\sigma^{3,9})$. Then $\phi^\sigma(e_1) \leq \phi^\sigma(e_2)$.*

It remains to analyze $U_\sigma^{3,1}, U_\sigma^{3,2}$ and $U_\sigma^{3,3}$. These sets do not interfere with the application of the improving switches for another reason: Applying specific switches of L_σ^3 prevents certain subsets of these sets from being applied as they are no longer improving. To prove this, we introduce subsets of these sets, called *slices*.

Definition 4.4.11 (Slice). Let σ be a phase 3 strategy, $i \in [n], j, l \in \{0, 1\}$. Then

- $S_{i,\sigma}^{3,1} := \{(k_i, k_z) : \sigma(k_i) \notin \{k_z, k_\nu\} \wedge z \leq \nu \wedge (\mathfrak{b}+1)_i = 0\}$ is called *slice* of $U_\sigma^{3,1}$,
- $S_{i,j,l,\sigma}^{3,2} := \{(b_{i,l}^j, k_z) : \sigma(k_i) \notin \{k_z, k_\nu\} \wedge z \leq \nu \wedge (\mathfrak{b}+1)_i = 0\}$ is called *slice* of $U_\sigma^{3,2}$,
- $S_{i,j,l,\sigma}^{3,3} := \{(b_{i,l}^j, k_z) : \sigma(k_i) \notin \{k_z, k_\nu\} \wedge z \leq \nu \wedge (\mathfrak{b}+1)_{i+1} \neq j\}$ is called *slice* of $U_\sigma^{3,3}$.

It is easy to see that the set of all slices of a specific set is a partition of that set. We now formalize the idea that applying specific improving switches prevents whole slices from being applied later on.

Lemma 4.4.12. *Let σ be a phase 3 strategy and let e denote the switch that is applied in σ. Let σ' denote an arbitrary phase 3 strategy of $\sigma_\mathfrak{b} \to \sigma_{\mathfrak{b}+1}$ calculated after the strategy σ.*

1. *If $e = (k_i, k_\nu)$, then $I_{\sigma'} \cap S_{i,\sigma'}^{3,1} = \emptyset$.*
2. *If $e = (b_{i,l}^j, k_\nu)$ with $\sigma(b_{i,l}^j) \neq k_\nu$ and $(\mathfrak{b}+1)_i = 0$, then $I_{\sigma'} \cap S_{i,j,l,\sigma'}^{3,2} = \emptyset$.*
3. *If $e = (b_{i,l}^j, k_\nu)$ with $\sigma(b_{i,l}^j) \neq k_\nu$ and $(\mathfrak{b}+1)_{i+1} \neq j$, then $I_{\sigma'} \cap S_{i,j,l,\sigma'}^{3,3} = \emptyset$.*

All of these lemmas now enable us to prove that it is possible to always apply some improving switch contained in L_σ^3 without violating Zadeh's pivot rule.

Lemma 4.4.13. *Let σ be a phase 3 strategy. Then $L_\sigma^3 \cap \arg\min_{e' \in I_\sigma} \phi^\sigma(e') \neq \emptyset$.*

This is not yet sufficient for proving that the improving switches of phase 3 can be applied as it is intended in [Fri11c] as it is not clear why it does not happen that a phase 4 strategy is calculated before all switches of phase 3 are applied. We thus prove that the improving switches of phase 3 can be applied ensuring that this does not happen.

Theorem 4.4.14. *There is an ordering of the improving switches of phase 3 and an associated tie-breaking rule compatible with the LeastEntered pivot rule such that*

1. *all improving switches contained in $L_{\sigma_\mathfrak{b}}^3$ are applied and*
2. *the LeastEntered pivot rule is obeyed during phase 3.*

Proof. Let σ denote the first phase 3 strategy of $\sigma_{\mathfrak{b}} \to \sigma_{\mathfrak{b}+1}$. Then, $L_{\sigma}^{3} = L_{\sigma_{\mathfrak{b}}}^{3}$ by Lemma 4.3.8. By Lemma 4.4.13, there is an edge $e_1 \in L_{\sigma}^{3}$ minimizing the occurrence record among all improving switches. By Lemma 4.3.9, applying e_1 yields a new phase 3 strategy $\sigma' := \sigma e_1$ such that $L_{\sigma'}^{3} = L_{\sigma}^{3} \setminus \{e_1\}$. Now, again by Lemma 4.4.13, there is an edge $e_2 \in L_{\sigma'}^{3}$ minimizing the occurrence record $I_{\sigma'}$ among all improving switches.

This argument can be applied until we reach a phase 3 strategy $\hat{\sigma}$ such that $|L_{\hat{\sigma}}^{3}| = 1$ while only switches contained in $L_{\sigma_{\mathfrak{b}}}^{3}$ are applied. Then, by construction and by Lemma 4.4.13, $(e_1, e_2, \dots)$ defines an ordering of the edges of $L_{\sigma_{\mathfrak{b}}}^{3}$ and an associated tie-breaking rule that obeys the LEASTENTERED pivot rule. When the strategy $\hat{\sigma}$ with $|L_{\hat{\sigma}}^{3}| = 1$ is reached, applying the remaining improving switch results in a phase 4 strategy. Then, all improving switches contained in $L_{\sigma_{\mathfrak{b}}}^{3}$ were applied and the LEASTENTERED pivot rule was obeyed. □

The ordering that is implicitly given in the proof of Theorem 4.4.14 avoids Issue 4.3.12. This issue showed that it is not possible to apply the improving switches of phase 3 of $\sigma_{\mathfrak{b}} \to \sigma_{\mathfrak{b}}$ "level by level" where the ordering of the levels depends only on $\ell(\mathfrak{b}+1)$ without violating Zadeh's pivot rule. The ordering given in Theorem 4.4.14 always chooses an improving switch minimizing the occurrence record among all improving switches. This choice is made regardless of the level of the switch. Consequently, the application of the improving switches is not performed "level by level" in an order that only depends on the least significant set bit.

Theorem 4.4.14 proves that the improving switches of phase 3 can be applied while obeying Zadeh's pivot rule. It does however not imply that the transition from $\sigma_{\mathfrak{b}}$ to $\sigma_{\mathfrak{b}+1}$ can be executed as intended in [Fri11c]. More precisely, Theorem 4.4.14 does not imply that the application of the improving switches in phase 3 is compatible with the application of the switches during the other phases. Analyzing the remaining phases, it can however be proven that this transition can in fact be executed as intended.

Theorem 4.4.15. *Fix some $\mathfrak{b} \in \mathbb{B}_n$ and consider the transition from $\sigma_{\mathfrak{b}}$ to $\sigma_{\mathfrak{b}+1}$. There is an order in which to apply the improving switches of this transition such that*

1. *the application obeys Zadeh's pivot rule and*
2. *for every $p \in [5]$, all switches of phase p are applied before any switch of phase $p+1$.*

This theorem proves that the result of [Fri11c] remains correct despite the flaws contained in the original proofs. Proving it however requires to analyze the remaining phases. We do not analyze these phases in all detail, and refer to [Fri11c] for results and descriptions related to these phases. We however prove that there is always an improving switch $e \in L_{\sigma}^{p}$ that can be applied without violating the LEASTENTERED pivot rule if σ is a phase p strategy and use this to prove Theorem 4.4.15.

Lemma 4.4.16. *Let $p \in \{1, 2, 4, 5, 6\}$ and let σ be a phase p strategy. Then, there is an improving switch $e \in L_{\sigma}^{p}$ such that $\phi^{\sigma}(e) \leq \min_{e' \in U_{\sigma}^{p} \cap I_{\sigma}} \phi^{\sigma}(e')$.*

Proof of Theorem 4.4.15. Consider the initial phase 1 strategy $\sigma_{\mathfrak{b}}$ for $\mathfrak{b}$. By Lemma 4.4.16, there is an improving switch contained in $L_{\sigma_{\mathfrak{b}}}^{1}$ minimizing the occurrence record among all improving edges. Thus, this switch can be applied without violating Zadeh's pivot rule.

By [Fri11c, Lemma 5], the resulting strategy is either a phase 2 strategy or a phase 1 strategy for $\mathfrak{b}$. In the second case, the same argument can be applied iteratively until a strategy is reached such that applying the next improving switch yields a phase 2 strategy. After applying a finite number of improving switches we thus obtain a phase 2 strategy σ^2. By Lemma 4.4.16 we can apply the single improving switch of $L^2_{\sigma^2}$ without violating Zadeh's pivot rule. By [Fri11c, Lemma 5], the resulting strategy is a phase 3 strategy. As proven in Theorem 4.4.14, all improving switches that should be applied during phase 3 can now be applied in some order. This then yields a phase 4 strategy σ^4.

By Lemma 4.4.16, there is a switch contained in $L^4_{\sigma^4}$ minimizing the occurrence record among all improving edges since $I_{\sigma^4} \subseteq U^4_{\sigma^4}$ by [Fri11c, Lemma 4]. The resulting strategy is either another phase 4 strategy or a phase 5 strategy. In the first case, the same argument can be applied iteratively until a phase 5 strategy is reached. Thus, after applying a finite number of improving switches, we obtain a phase 5 strategy.

By applying Lemma 4.4.16 for $p = 5$ and $p = 6$, the same arguments used for phase 4 can now be used for phase 5 and 6, concluding the proof. □

Conclusion

In this chapter, we discussed Friedmann's subexponential lower bound construction for Zadeh's pivot rule [Fri11c]. We described the construction in detail and explained its design concept as well as the application of the strategy improvement algorithm. We highlighted several issues present in the original analysis and discussed why one of the issues can be considered a major issue that needs to be resolved. We then clarified and proposed alterations regarding the application of the strategy iteration algorithm to resolve all of these issues. More precisely, we proved that the initial strategy needs to be changed (Issues 4.3.1 and 4.3.3) and provided an alternative initial strategy (Definition 4.2.2 and Theorems 4.4.1 and 4.4.2). Furthermore, we showed that the description of the occurrence records of the bicycle edges is not entirely accurate (Issue 4.3.4) and corrected the inaccuracy by giving an alternative description of the occurrence records of the bicycle edges (Theorem 4.4.3).

Most importantly, we discussed a major issue regarding phase 3 and investigated the application of the improving switches in this phase. We argued why the informal description given in [Fri11c] of this phase cannot be correct (Issue 4.3.5). We then proved that the more formal description does not obey Zadeh's pivot rule (Issue 4.3.7). More severely, we proved that the application of the improving switches during phase 3 as described by [Fri11c] can be interpreted as a realization of a whole framework, and that applying improving switches according to this framework cannot obey Zadeh's pivot rule (Issue 4.3.12). This issue was then resolved by implicitly providing a more involved ordering and associated tie-breaking rule that overcome this issue (Theorem 4.4.14). Finally, we showed that this ordering is compatible with the application of the improving switches during the other phases (Theorem 4.4.15).

Crucially, our changes do not alter the macroscopic structure of the original construction. Consequently, we are able to recover Friedmann's subexponential lower bound. However, we are not able to explicitly give the ordering in which improving switches should be applied during phase 3, and the tie-breaking rule remains highly artificial.

In the next chapter, we provide a new construction that implements similar ideas to obtain a Markov decision process of size $\mathcal{O}(n)$ implementing an n-bit binary counter. This construction thus significantly improves the lower bound discussed in this chapter, providing the first truly exponential lower bound for the LeastEntered pivot rule for Markov decision processes, parity games, linear programs and other stochastic games.

5. An Exponential Lower Bound for Zadeh's Pivot Rule

Since Zadeh's pivot rule was developed in 1980 [Zad80], it remained a promising candidate for a potential polynomial time pivot rule for over 30 years. In 2011, Oliver Friedmann proved that the running time can be subexponential in the worst case, using the construction we discussed in Chapter 4. Although Zadeh's pivot rule was no longer a candidate for a polynomial time pivot rule, it remained the most prominent candidate for the first deterministic pivot rule with guaranteed subexponential running time. As the RandomFacet pivot rule has an expected subexponential worst case running time [MSW96], hopes were high that Zadeh's pivot rule matches this worst case running time. In this chapter, we present a construction proving this pivot rule is in fact exponential in the worst case.

This chapter is organized as follows. We begin by describing a sink game such that applying the strategy improvement algorithm using Zadeh's pivot rule requires an exponential number of iterations in Section 5.1. As all of the modern lower bound constructions, the sink game is based on a binary counter. Then, in Section 5.2, we discuss how this sink game can be transformed into a weakly unichain Markov decision process. The idea is that the strategy improvement algorithm using Zadeh's pivot rule behaves nearly identical in the Markov decision process. This is however not always the case as there is no known transformation from sink games to Markov decision processes ensuring that the algorithms behave the same. Hence, we discuss the main differences between the sink game and the Markov decision process. Finally, we give the proof of the exponential lower bound in Section 5.3. Since the Markov decision process is weakly unichain, this then implies an exponential lower bound for the simplex algorithm using Zadeh's pivot rule by Corollary 3.3.5. In particular, it is sufficient to consider the sink game and the Markov decision process and not necessary to explicitly investigate the linear program induced by the Markov decision process.

The results of this chapter were partly verified using the PGSolver library [FL17]. Oliver Friedmann provided an implementation of the sink game construction, and applied the strategy improvement algorithm using this library. We then compared our results and the behavior of the algorithm manually for small examples representing binary counters with up to 10 bits, verifying our results for at least the sink game construction and small examples. Visualizations of full executions of the algorithm for 3 and 4 levels are available online [Fri19]. A preliminary version of the results presented here were previously published and are available online [DFH19].

5.1. The Basic Sink Game Construction

In this section, we describe a sink game $S_n = (V_0, V_1, E, \Omega)$ of size $\mathcal{O}(n)$ such that the strategy improvement algorithm performs at least 2^n iterations when using Zadeh's pivot rule and a specific tie-breaking rule. The key idea of the sink game is again the implementation of a binary counter and is very similar to the general design principle introduced in the beginning of Section 4.1. We thus do not discuss the ideas and the notation related to binary counting in all detail here and refer to Section 4.1 for a more detailed explanation, and compare our construction with similar constructions later.

The intuitive idea

The sink game S_n consists of n (nearly identical) levels, each representing one bit of the counter, and one sink vertex t. The i-th level is shown in Figure 5.1, the full graph S_3 in Figure 5.2.

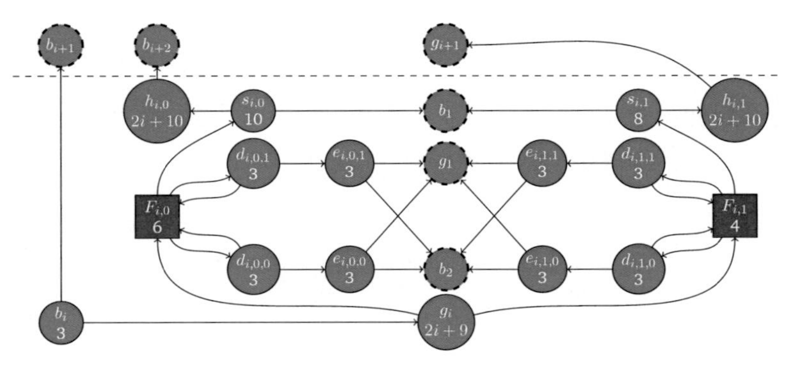

Figure 5.1.: Level i of S_n for $i \in [n-2]$. Circular vertices are player 0 vertices, rectangular vertices are player 1 vertices. Labels below vertex names denote their priorities. Dashed vertices do not (necessarily) belong to level i.

As for the subexponential construction, the main challenge in designing S_n is that a classical binary counter is highly unbalanced, whereas Zadeh's pivot rule enforces the algorithm to use improving switches applied least often during the execution. Again, the key idea to overcome this obstacle is to have two gadgets per level representing the bit. At any time, exactly one of these gadgets of level i is interpreted as encoding bit i. In order to not confuse them with the bicycle gadgets of the subexponential construction, these gadgets are called *cycle centers* and the cycle centers of level i are denoted by $F_{i,0}, F_{i,1}$. Given a number $\mathfrak{b} \in \mathfrak{B}_n$, the idea again is that the cycle center $F_{i,\mathfrak{b}_{i+1}}$ encodes bit i, and we call this cycle center the *active cycle center* of level i with respect to $\mathfrak{b}$. Consequently, $F_{i,1-\mathfrak{b}_{i+1}}$ is the *inactive cycle center* of level i. Cycle centers can again be either closed or open, and the idea is that bit i is equal to 1 if and only if the active cycle center of level i

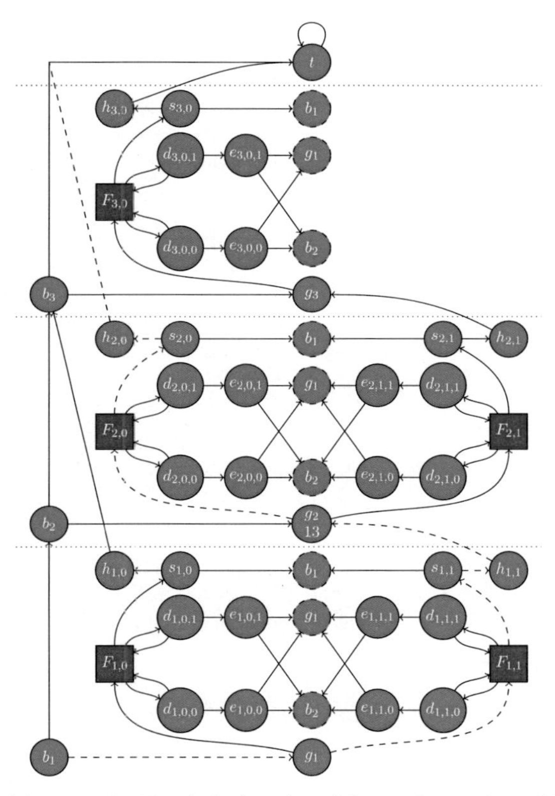

Figure 5.2.: The sink game S_3. The dashed copies of the vertices g_1, b_1 and b_2 all refer to the corresponding vertices of levels 1 and 2. The vertex priorities are not shown here.

is closed. In particular, Figure 4.2 still represents the intuitive idea for the exponential lower bound.

These ideas are the same used by Friedmann in his subexponential construction [Fri11c]. Our construction is designed in such a way that the basic ideas of the subexponential construction still apply, as they are strong enough to provide a lower bound for Zadeh's pivot rule while significantly reducing the size of the construction. Intuitively, the biggest change in design in the exponential construction is the connection of the levels with each other. In the subexponential construction, all levels are connected to each other, since the entry vertices have to get access to the level of the least significant set bit. In the exponential construction, the levels are only connected to the next two levels and to the first two levels. It is then sufficient for all levels representing bits that are equal to 0 to have access to the first level if the represented number is odd and to the second if it is even. More precisely, when representing a number $\mathfrak{b}$, the levels of the subexponential construction require access to level $\ell(\mathfrak{b})$ while the levels of the exponential construction only require access to level $\ell(\mathfrak{b}) \bmod 2$. This significantly reduces the size of the construction, yielding an improved bound.

However, there are more changes that need to be performed, and the details of the constructions are fundamentally different. We discuss these differences in more detail at the end of Section 5.2 after having introduced the Markov decision process.

The full construction

We now describe the construction of the sink game implementing these ideas. Henceforth, let $n \in \mathbb{N}$ be fixed and let $S_n = (V_0, V_1, E, \Omega)$ denote the sink game that is to be constructed here. The vertex sets V_0 and V_1 of player 0 resp. player 1 are defined via

$$\begin{aligned} V_0 :=& \{b_i, g_i : i \in [n]\} \cup \{d_{i,j,k}, e_{i,j,k}, s_{i,j} : i \in [n-j], j, k \in \{0,1\}\} \\ &\cup \{h_{i,j} : i \in [n-j], j \in \{0,1\}\} \cup \{t\}, \\ V_1 :=& \{F_{i,j} : i \in [n-j], j \in \{0,1\}\}. \end{aligned}$$

For convenience of notation, we identify the vertex names b_i, g_i for $i > n$ with t. The priorities of the vertices as well as the edges of the construction are given by Table 5.1. The priorities are not unique, although this is technically required for the definition of the vertex valuations. It is however sufficient to demand that the most significant difference as defined in Definition 3.2.4 be unique whenever comparing valuations, which will turn out to be the case for our construction.

Vertex	Successors	Priority
t	t	1
b_i	g_i, b_{i+1}	3
$e_{i,j,k}$	b_2, g_1	3
$d_{i,j,k}$	$F_{i,j}$, $e_{i,j,k}$	3

Vertex	Successors	Priority
g_i	$F_{i,0}(, F_{i,1})$	$2i+9$
$h_{i,0}$	b_{i+2}	$2i+10$
$h_{i,1}$	g_{i+1}	$2i+10$
$s_{i,j}$	$h_{i,j}$, b_1	$10-2j$
$F_{i,j}$	$d_{i,j,0}$, $d_{i,j,1}$, $s_{i,j}$	$6-2j$

Table 5.1.: Edges and vertex priorities of the sink game S_n.

By construction, every vertex $v \in V_0$ has at most two outgoing edges. The construction implements the general idea of a binary counter, and it can be separated into n different levels. The first $n-2$ levels are structurally identical, and the levels $n-1$ and n only differ slightly from the other levels.

The idea of the construction is that there are player 0 strategies σe and player 1 counterstrategies τ^σ such that σ and τ^σ together represent a number $\mathfrak{b} \in \mathfrak{B}_n$. Such a pair of strategies induces a path in S_n. This path starts in b_1, ends in t and traverses exactly the levels $i \in [n]$ with $\mathfrak{b}_i = 1$ while ignoring levels with $\mathfrak{b}_i = 0$. This path is called the *spinal path* with respect to $\mathfrak{b}$. The idea is that it is not profitable for player 0 to *enter* a level, but very profitable to *traverse* a level. As player 1 tries to minimize the valuation of the vertices, they will try to prevent player 0 from traversing levels. However, the sink game is constructed in such a way that player 0 can always traverse levels representing bits which are equal to 1 while player 1 is able to prevent this for levels representing bits equal to 0. Also, traversing a level $i > 1$ is better than traversing *all* levels $1, 2, \ldots, i-1$. This already implies that strategies representing higher numbers are better for player 0 than strategies representing smaller numbers.

Consider some fixed level $i \in [n]$. Then, whether level i is traversed or ignored is controlled by the *entry vertex* $b_i \in V_0$ of level i. More precisely, the entry vertex of level i is intended to point to the *selector vertex* $g_i \in V_0$ if and only if $\mathfrak{b}_i = 1$. If $\mathfrak{b}_i = 0$, then b_i instead points to the entry vertex of the next level, i.e., to b_{i+1}. The vertex g_i has a high odd priority, making it in general unprofitable for player 0 to choose the edge (b_i, g_i).

Assume $i \leq n-2$ for the moment and consider some $\mathfrak{b} \in \mathfrak{B}_n$. Attached to the selector vertex g_i are the *cycle centers* $F_{i,0}, F_{i,1} \in V_1$ of level i. As explained previously, these cycle centers are used for determining whether bit i is equal to 1 and they function similar to the bicycles in the subexponential construction of [Fri11c]. That is, the cycle centers alternate in encoding bit i since we interpret the *active cycle center* $F_{i,\mathfrak{b}_{i+1}}$ as encoding bit i. Consequently, the *inactive cycle center* $F_{i,1-\mathfrak{b}_{i+1}}$ does not interfere with the encoding, enabling us to manipulate the "inactive" part of level i without loosing the encoded value of $\mathfrak{b}_i$. Therefore, the selector vertex is used to ensure that the active cycle center is contained in the spinal path. More precisely, if $\mathfrak{b}_i = 1$, then g_i should select $F_{i,\mathfrak{b}_{i+1}}$, while its selection is not specified if $\mathfrak{b}_i = 0$ for technical reasons. Also, since the cycle center is a player 1 vertex, it can be used to prevent player 0 from traversing a level and reaching vertices with high even priorities unless player 1 is forced to grant access.

A cycle center $F_{i,j}$ can have several different configurations, and we refer to Figure 5.3 for an overview. The configuration of the cycle center $F_{i,j}$ is defined via its *cycle vertices* $d_{i,j,0}, d_{i,j,1}$ and the two *cycle edges* $(d_{i,j,0}, F_{i,j}), (d_{i,j,1}, F_{i,j})$. Most importantly, a cycle center can be *closed*. Intuitively, a cycle center is closed if it either represents a bit being equal to 1 or if the occurrence record of the cycle edges is too low and has to "catch up". In all other cases, it is in one of three possible different states. We introduce the different states formally after having described the full construction.

The mechanism of closing a cycle center works as follows. If a cycle center is closed, then player 1 cannot choose one of the cycle edges $(F_{i,j}, d_{i,j,*})$. The reason is that this would close a cycle, contradicting that S_n is a sink game (see Lemma 5.3.3). Thus, by closing a cycle center, player 0 can force player 1 to grant access to the "higher" and better vertices of this level. Consequently, $F_{i,j}$ is connected to one further vertex, called the *upper selection vertex* $s_{i,j} \in V_0$. This vertex has the purpose of connecting the cycle center $F_{i,j}$ with the other levels of the graph and granting player 0 access to a vertex with high even priority. More precisely, it connects the cycle center with the first level via the edge $(s_{i,j}, b_1)$, and, depending on whether $j = 0$ or $j = 1$, with either level $i+1$ or $i+2$. The connection to level $i+1$ resp. $i+2$ uses an intermediate vertex $h_{i,j} \in V_0$, the edge $(s_{i,j}, h_{i,j})$ and the edge $(h_{i,0}, b_{i+2})$ resp. $(h_{i,1}, g_{i+1})$. The priority of $h_{i,j}$ is large and even and is chosen in such a way that it compensates for the odd priority of g_i. It is thus very desirable for player 0 to get access to this vertex. As they can enforce this by closing cycle centers, this implies that closing cycle centers is always desirable from the perspective of player 0. The upper selection vertex is thus central in granting access to either the beginning of the spinal path or the next level contained in the spinal path.

We discuss the cycle vertices $d_{i,j,k} \in V_0$ next. If the cycle center $F_{i,j}$ is not closed, then these vertices need to be able to access the spinal path as this will typically be very profitable for player 0 and is also used for "resetting" the bits of the counter. As accessing the spinal path via the cycle center $F_{i,j}$ would close this cycle center by definition, they need to be able to "escape" the level in another way. This is handled by the *escape vertex* $e_{i,j,k} \in V_0$ of $d_{i,j,k}$. The escape vertices $e_{i,j,0}, e_{i,j,1}$ are used to connect the cycle vertices of $F_{i,j}$ to the first two levels, thus granting the cycle vertices access to the spinal path. More precisely, the escape vertices are connected with the entry vertex b_2 of level 2 and the selection vertex g_1 of level 1. In principle, the escape vertices will point to g_1 if and only if the currently represented number is odd.

Having introduced all details of the construction, we now formalize the different *states* a cycle center can be in. Note that we henceforth typically use the term "strategy" instead of "player 0 strategy".

Definition 5.1.1 (States of cycle centers). Let σ be a strategy. The cycle center $F_{i,j}$ is *closed* for σ if $\sigma(d_{i,j,*}) = F_{i,j}$. It is g_1*-halfopen* for σ if $\sigma(d_{i,j,k}) = F_{i,j}, \sigma(d_{i,j,1-k}) = e_{i,j,1-k}$ and $\sigma(e_{i,j,1-k}) = g_1$ for some $k \in \{0,1\}$. It is g_1*-open* for σ if $\sigma(d_{i,j,k}) = e_{i,j,k}$ and $\sigma(e_{i,j,k}) = g_1$ for both $k \in \{0,1\}$. The terms b_2-halfopen and b_2-open are defined analogously. It is *mixed* if $\sigma(d_{i,j,*}) = e_{i,j,*}$ and $\sigma(e_{i,j,k}) = g_1, \sigma(e_{i,j,1-k}) = b_2$ for some $k \in \{0,1\}$.

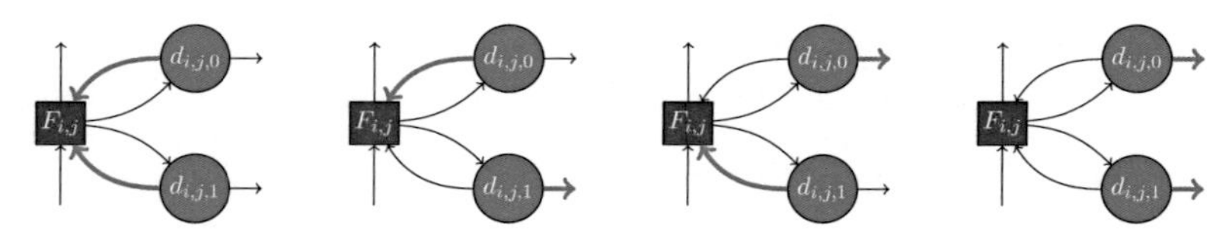

Figure 5.3.: A closed, two halfopen and an open or mixed cycle center, depending on the choices of the escape vertices. Thick blue edges indicate the corresponding choices of player 0.

This concludes our description of the sink game. We now formalize the idea of a strategy encoding a binary number by introducing the term *canonical strategy*. A canonical strategy is the analogue of the term "initial phase 1 policy" that was used when discussing the subexponential lower bound in Chapter 4. We consequently use the same symbol. This definition includes some aspects that are purely technical and are needed for some proofs and uniqueness and do not have an immediate intuitive explanation.

Definition 5.1.2 (Canonical strategy for S_n). Let $\mathfrak{b} \in \mathfrak{B}_n$. A strategy $\sigma_{\mathfrak{b}}$ for S_n is called *canonical strategy for* $\mathfrak{b}$ if it has the following properties.

1. All escape vertices point to g_1 if $\mathfrak{b}_1 = 1$ and to b_2 if $\mathfrak{b}_1 = 0$.
2. The following hold for all levels $i \in [n]$ with $\mathfrak{b}_i = 1$:
   a) Level i needs to be accessible, i.e., $\sigma_{\mathfrak{b}}(b_i) = g_i$.
   b) The cycle center $F_{i,\mathfrak{b}_{i+1}}$ is closed while $F_{i,1-\mathfrak{b}_{i+1}}$ is not closed.
   c) The selector vertex selects the active cycle center, i.e., $\sigma_{\mathfrak{b}}(g_i) = F_{i,\mathfrak{b}_{i+1}}$.

3. The following hold for all levels $i \in [n]$ with $\mathfrak{b}_i = 0$:
   a) Level i is not accessible and needs to be "avoided", i.e., $\sigma_{\mathfrak{b}}(b_i) = b_{i+1}$.
   b) The cycle center $F_{i,\mathfrak{b}_{i+1}}$ is not closed.
   c) If the cycle center $F_{i,1-\mathfrak{b}_{i+1}}$ is closed, then $\sigma_{\mathfrak{b}}(g_i) = F_{i,1-\mathfrak{b}_{i+1}}$.
   d) If none of the cycle centers $F_{i,0}, F_{i,1}$ is closed, then $\sigma_{\mathfrak{b}}(g_i) = F_{i,0}$.
4. Let $\mathfrak{b}_{i+1} = 0$. Then, $\sigma_{\mathfrak{b}}(s_{i,0}) = h_{i,0}$ and $\sigma_{\mathfrak{b}}(s_{i,1}) = b_1$.
5. Let $\mathfrak{b}_{i+1} = 1$. Then, $\sigma_{\mathfrak{b}}(s_{i,0}) = b_1$ and $\sigma_{\mathfrak{b}}(s_{i,1}) = h_{i,1}$.
6. Both cycle centers of level $\ell(\mathfrak{b}+1)$ are open.

An example of the canonical strategy σ_3 representing the number 3 in the sink game S_3 is depicted in Figure 5.4.

The main structure that is used for the encoding of binary numbers are the cycle centers. In particular, every possible configuration of the cycle centers, and thus every strategy, induces *some* binary number. Introducing this so-called *induced bit state* will turn out to be helpful, as it allows us to identify the currently represented number for non-canonical strategies.

Definition 5.1.3 (Induced bit state). Let σ be a player 0 strategy for S_n. Then, the *induced bit state* $\beta^\sigma = (\beta^\sigma_n, \dots, \beta^\sigma_1)$ is defined as follows: We define $\beta^\sigma_n := 1$ if and only if $F_{n,0}$ is closed. For $i < n$, we define $\beta^\sigma_i = 1$ if and only if $F_{i,\beta^\sigma_{i+1}}$ is closed.

When the strategy is clear from the context or the induced bit state is identical for all currently considered strategies, we often skip the upper index and just write β instead of β^σ. This definition is in accordance with the interpretation of encoding a number via canonical strategies. In fact, it is easy to verify that the definition of a canonical strategy immediately implies $\beta^{\sigma_{\mathfrak{b}}} = \mathfrak{b}$. Furthermore, this enables us to give a definition of active and inactive cycle centers independent of a given binary number. Consequently, we call the cycle center $F_{i,\beta^\sigma_{i+1}}$ the *active* cycle center of level i while $F_{i,1-\beta^\sigma_{i+1}}$ is the *inactive* cycle center of level i.

This concludes our definition of the sink game. Before proving that S_n is in fact a sink game (Lemma 5.3.3) and discussing the application of the strategy improvement algorithm to it, we discuss how S_n can be transformed into a Markov Decision Process implementing the same ideas.

5.2. Transforming the Sink Game into a Markov Decision Process

We now discuss how the sink game S_n constructed in Section 5.1 is altered to obtain a Markov decision process M_n. The idea is that the strategy improvement algorithm should behave nearly identical on M_n and S_n when using Zadeh's pivot rule and similar tie-breaking. We first discuss the changes performed to the sink game S_n intuitively and define the Markov decision process M_n formally at the end of this section. Since sink games are 2-player games but there is only a single player in a Markov decision process, we need to change the sink game such that only one player remains. This will be achieved

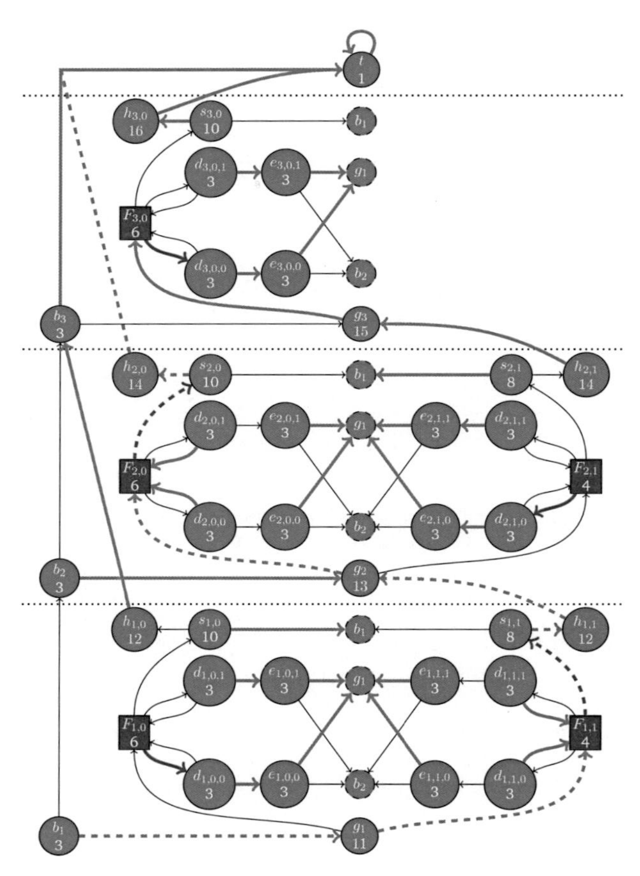

Figure 5.4.: The sink game S_3 together with a canonical strategy representing the number 3 in S_3. The dashed copies of the vertices g_1, b_1 and b_2 all refer to the corresponding vertices of levels 1 and 2. Blue edges belong to the strategy of player 0, red edges belong to the counterstrategy of player 1. The dashed edges indicate the spinal path, the dotted lines separate the levels.

by replacing player 1 by randomization. This is a common technique used for obtaining Markov decision processes that behave similarly to given parity games and was used by several authors before (for example [Fea10a, Fri11c, AF17] among others). Although the ideas used in their transformations are all quite similar, there is no standard reduction from parity games to Markov decision processes preserving all properties. Before discussing how to replace player 1 in our construction, we first discuss how the vertices and other aspects related to player 0 change.

In the sink game S_n, every vertex is assigned an integer priority. Priorities have the effect that once a vertex with a very high priority is reached, the priorities of all vertices with smaller priorities become irrelevant. This is, for example, used to make higher levels more profitable than lower levels and to make it unprofitable to enter a level without fully traversing it. Ideally, the Markov decision process should also have this property. This can be achieved by introducing a sufficiently large natural number $N \in \mathbb{N}$ and defining the reward obtained by traversing vertex v as $(-N)^{\Omega(v)}$, where $\Omega(v)$ denotes the priority of v. Note that it is easily possible to introduce and interpret the reward obtained by traversing a vertex v by assigning the same reward to all edges $(v,*) \in E$. This has the effect that it is still profitable to traverse vertices with high priority while it is expensive to traverse vertices with odd priority. It turns out that it is not required to assign a non-zero reward to every edge of the graph, and that it is sufficient to choose N as a natural number at least equal to the number of vertices in M_n with a priority assigned to them. We thus define $N := 7n$. We state which vertices are assigned a priority and which edges have a reward of 0 precisely when formalizing the Markov decision process.

We next discuss how to replace player 1. The only vertices of player 1 are the cycle centers $F_{i,j}$ for $i \in [n], j \in \{0,1\}$. They are designed in such a way that player 1 only chooses the edge $(F_{i,j}, s_{i,j})$ if the cycle center $F_{i,j}$ is closed. Although this behavior cannot be modeled exactly by randomization, it can be modeled approximately by defining $F_{i,j}$ as randomization vertex and assigning suitable probabilities to its edges. Here, suitable means that the probability of $(F_{i,j}, s_{i,j})$ is extremely small. This has the effect that the very profitable vertex $h_{i,j}$ is "hidden". As mentioned previously, this use of randomization is very similar to the use in the subexponential construction of Friedmann, but was also used by several other authors. We thus define $\varepsilon := N^{-(2n+11)}$ and set $p(F_{i,j}, s_{i,j}) := \varepsilon$ and $p(F_{i,j}, d_{i,j,*}) := (1-\varepsilon)/2$.

We now give the formal definition of $M_n = (V_0, V_R, E_0, E_R, r, p)$ and refer to Figure 5.5 for a visualization of level i of M_n for $i < n$. The player vertices V_0 and the randomization vertices V_R are defined analogously to the definition of V_0 and V_1 in S_n via

$$\begin{aligned}
V_0 :=& \{b_i, g_i : i \in [n]\} \cup \{d_{i,j,k}, e_{i,j,k}, s_{i,j} : i \in [n-j], j,k \in \{0,1\}\} \\
& \cup \{h_{i,j} : i \in [n-j], j \in \{0,1\}\} \cup \{t\}, \\
V_R :=& \{F_{i,j} : i \in [n-j], j \in \{0,1\}\}.
\end{aligned}$$

The edges of M_n are defined by Table 5.2. The first table shows player vertices v and corresponding successors $w \in \delta^+(v)$ with $r(v,w) := 0$. Consequently, no priority is assigned to these vertices. The second table analogously shows player vertices v with an assigned priority $\Omega(v)$ and successors $w \in \delta^+(v)$ such that $r(v,w) := (-N)^{\Omega(v)}$. The third

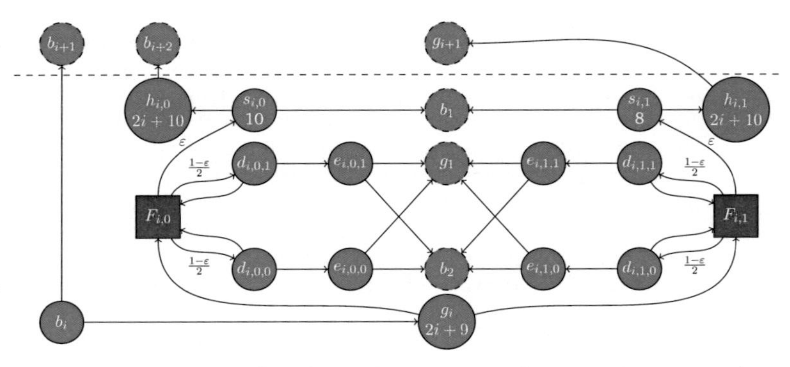

Figure 5.5.: Level i of M_n for $i \in [n-2]$. Circular vertices are player 0 vertices, rectangular vertices are randomization vertices. Labels on edges denote their probability, labels on vertices denote their priority (if they have a priority assigned to them). Dashed vertices do not (necessarily) belong to level i.

table contains the randomization vertices and their successors. The last column of that table shows the probability of each of the corresponding edges.

Vertex	Successors
b_i	g_i, b_{i+1}
$e_{i,j,k}$	b_2, g_1
$d_{i,j,k}$	$F_{i,j}, e_{i,j,*}$
t	t

Vertex	Successors	Priority
$s_{i,j}$	$h_{i,j}, b_1$	$10-2j$
g_i	$F_{i,0}(, F_{i,1})$	$2i+9$
$h_{i,0}$	b_{i+2}	$2i+10$
$h_{i,1}$	g_{i+1}	$2i+10$

Vertex	Successors	Probability
$F_{i,j}$	$s_{i,j}$	ε
$F_{i,j}$	$d_{i,j,*}$	$\frac{1-\varepsilon}{2}$

Table 5.2.: Edges and vertex priorities of the Markov decision process M_n.

We now discuss the main differences between the sink game S_n and the Markov decision process M_n. One of the main differences is the definition of canonical strategies. Consider a strategy $\sigma_{\mathfrak{b}}$ representing $\mathfrak{b}$, some fixed level $i \in [n-2]$ and two cycle centers $F_{i,0}, F_{i,1}$. In S_n, both cycle centers have an even priority and the priority of $F_{i,0}$ is larger than the priority of $F_{i,1}$. Thus, if none of the cycle centers is closed and both cycle centers escape, the valuation of $F_{i,0}$ is better than the valuation of $F_{i,1}$. In this case, this implies that $(g_i, F_{i,0})$ is improving if $\sigma_{\mathfrak{b}}(g_i) \neq F_{i,0}$. In some sense, this can be interpreted as the sink game "preferring" $F_{i,0}$ over $F_{i,1}$. A similar, but not quite identical phenomenon occurs in M_n. If both cycle centers $F_{i,0}, F_{i,1}$ are in the same "state", then the valuation of the upper selection vertices $s_{i,0}, s_{i,1}$ determines which cycle center has the better valuation. More precisely, it can happen that $\mathrm{Val}_{\sigma_{\mathfrak{b}}}(F_{i,0}) - \mathrm{Val}_{\sigma_{\mathfrak{b}}}(F_{i,1}) = \varepsilon[\mathrm{Val}_{\sigma_{\mathfrak{b}}}(s_{i,0}) - \mathrm{Val}_{\sigma_{\mathfrak{b}}}(s_{i,1})]$. It turns out that the valuation of $s_{i,\mathfrak{b}_{i+1}}$ is typically better than the valuation of $s_{i,1-\mathfrak{b}_{i+1}}$. Most importantly, in contrast to S_n, the valuation of $s_{i,0}$ is *not* typically better than the valuation of $s_{i,1}$. Hence, M_n "prefers" cycle centers $F_{i,\mathfrak{b}_{i+1}}$ over cycle center $F_{i,1-\mathfrak{b}_{i+1}}$. This has some serious consequences for the exact application of the improving switches, and implies that S_n and M_n do not behave exactly identically, even if the same pivot and tie-breaking rule are used. Although we are not yet able to fully explain all of these differences, we already adjust the definition of a canonical strategy for M_n.

Definition 5.2.1 (Canonical strategy for M_n)**.** Let $\mathfrak{b} \in \mathfrak{B}_n$. A strategy $\sigma_{\mathfrak{b}}$ for the Markov decision process M_n is called *canonical strategy for* $\mathfrak{b}$ if it has the properties given in Definition 5.1.2 where Property 3.(d) is replaced by the following: If neither of the cycle centers $F_{i,0}, F_{i,1}$ is closed, then $\sigma_{\mathfrak{b}}(g_i) = F_{i,\mathfrak{b}_{i+1}}$.

To conclude the description of the counter, we briefly compare our construction to similar lower bound examples. More precisely, we consider the exponential lower bound for Cunningham's pivot rule [AF17] and the subexponential lower bounds for the RandomEdge and the RandomFacet pivot rule [FHZ11b]. The corresponding constructions are shown in Figure 5.6 alongside a sketch of our construction, each for the parameter $n = 3$.

The four examples are constructed similarly and implement similar ideas. Most importantly, each of the constructions implements a binary counter on n bits. They all use the idea of having one level per bit, and each construction contains a gadget that looks similar to the cycle gadgets that we introduced. Similar to the use of cycle gadgets in our construction, these gadgets are the main tool used for interpreting certain bits as being equal to 1 and thus for representing binary numbers through strategies. A further examination also shows that each construction uses a "barrier vertex" that makes it unprofitable to enter a level and a "reward vertex" that compensates for entering a level. As for our construction, the reward vertex can only be reached if the gadgets are in a certain state. In particular, these constructions can be interpreted as a family of lower bound constructions, and it might be possible to generalize this framework in a similar way Amenta and Ziegler generalized the constructions based on the Klee-Minty cube [AZ98].

5.3. Proof of the Lower Bound

In this section, we prove the exponential lower bound for the strategy improvement and simplex algorithm when using Zadeh's pivot rule. We however do not provide all details and formal aspects here, these can be found in Chapter 6. Instead, this chapter presents the core ideas and arguments, and the next chapter proves that the statements presented here are correct.

As several of the statements and arguments are applicable for both the sink game S_n and the Markov decision process M_n we introduce notation that allows us to discuss both constructions simultaneously. We use the symbol G_n to signify that a statement holds for both S_n and M_n. All of the following arguments and explanations hold for both S_n and M_n unless explicitly stated otherwise.

Although there are attempts of unifying notation, algorithms and research in the field of stochastic optimization which includes Markov decision processes [Pow19], we are not aware of a unified treatment of parity games and Markov decision processes. As we pointed out in Chapter 3, there are deep connections between sink games and weakly unichain Markov decision processes, and we believe that there are classes of sink games and weakly unichain Markov decision processes that can be treated in a unified manner.

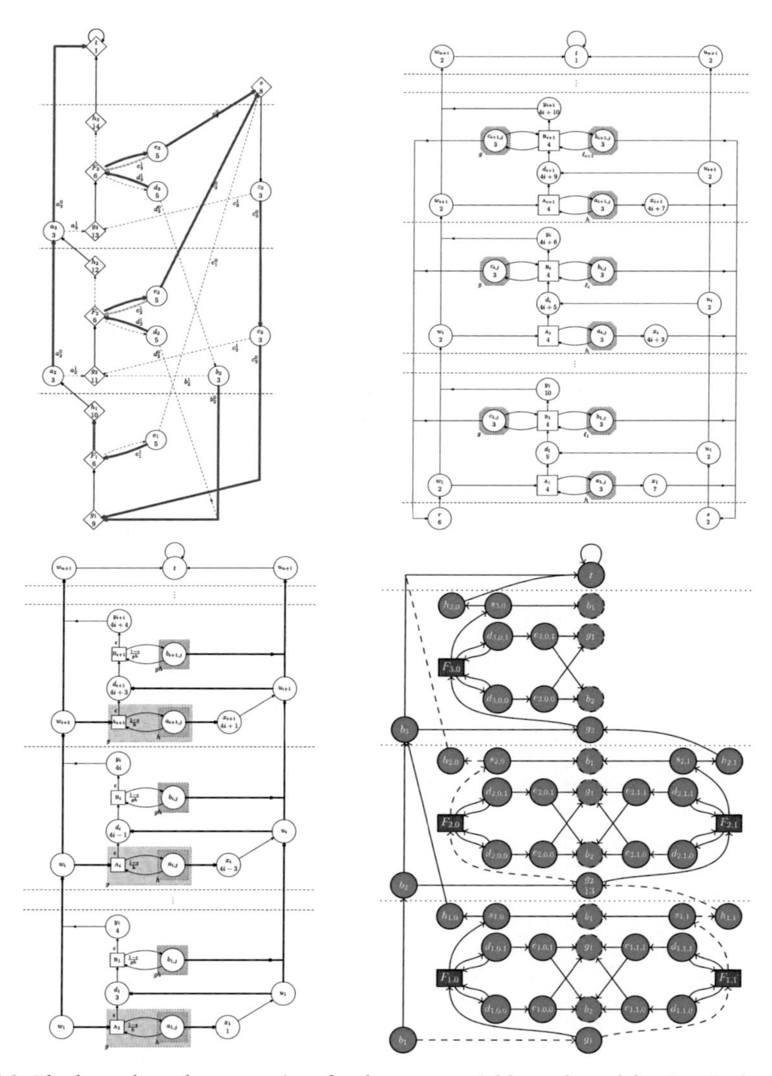

Figure 5.6.: The lower bound constructions for the exponential lower bound for Cunningham's pivot rule [AF17], the RANDOMEDGE and the RANDOMFACET pivot rule [FHZ11b] and our construction for 3 levels. All figures but the sketch of the exponential construction are taken from the given sources.

We now discuss the following key components of our proof separately:

1. In Section 5.3.1, we provide an initial strategy σ_0 that is weakly unichain for M_n and a sink strategy for S_n. Moreover, we provide the optimal strategy σ^* and prove that it is a sink strategy resp. weakly unichain as well. This proves that S_n is a sink game resp. that M_n is weakly unichain as claimed in Sections 5.1 and 5.2. We furthermore discuss the concept of occurrence records and revisit Zadeh's pivot rule.
2. In Section 5.3.2, we discuss and formally state the tie-breaking rule. The tie-breaking rule is implemented as an ordering of the player edges, and we need to distinguish between S_n and M_n for its definition. This completely describes the exact application of the improving switches performed by the strategy improvement algorithm in G_n. It is then the main challenge to prove that this application yields the desired behavior and, in particular, exponential lower bound which is shown in the following steps.

   In addition, we discuss the topic of tie-breaking rules and their importance in general.
3. We then focus on a single transition $\sigma_{\mathfrak{b}} \to \sigma_{\mathfrak{b}+1}$ between two consecutive canonical strategies. Such a transition requires the application of many improving switches, and many intermediate strategies need to be considered. As with the approaches of similar lower bounds (e.g. [Fea10a, Fri11c, AF17] and others), this application is then divided into disjoint *phases*. We give both intuitive and formal descriptions and definition of these phases in Section 5.3.3.
4. To prove that the pivot and tie-breaking rules proceed along the previously described phases, we specify how often edges are applied as improving switches in Section 5.3.4. This is formalized by the occurrence record, and we provide the occurrence records of the edges for canonical strategies. We also briefly explain how the provided occurrence records are related to the previously given description of the application of the improving switches.
5. Finally, in Section 5.3.5, we combine the previous aspects to prove that applying improving switches using Zadeh's pivot rule and our tie-breaking rule yields an exponential number of iterations. Since the size of G_n is linear in n, this yields an exponential lower bound with respect to the input size.

5.3.1. The Initial and Optimal Strategies

We begin by providing an initial strategy σ_0 for G_n. This strategy is (i) a canonical strategy for 0 in the sense of Definitions 5.1.2 and 5.2.1, (ii) a sink strategy for S_n and (iii) a weak unichain strategy for M_n.

Definition 5.3.1 (Initial strategy for the exponential construction). The initial strategy $\sigma_0 \colon V_0 \to V$ is defined as follows:

v	b_i for $i < n$	t	g_i	$d_{i,j,k}$	$e_{i,j,k}$	$s_{i,0}$	$s_{i,1}$	$h_{i,0}$	$h_{i,1}$
$\sigma_0(v)$	b_{i+1}	t	$F_{i,0}$	$e_{i,j,k}$	b_2	$h_{i,0}$	b_1	b_{i+2}	g_{i+1}

The following easy observation justifies that we use the symbol σ_0 for the initial strategy.

Observation 5.3.2. The initial strategy σ_0 is a canonical strategy for 0 in G_n.

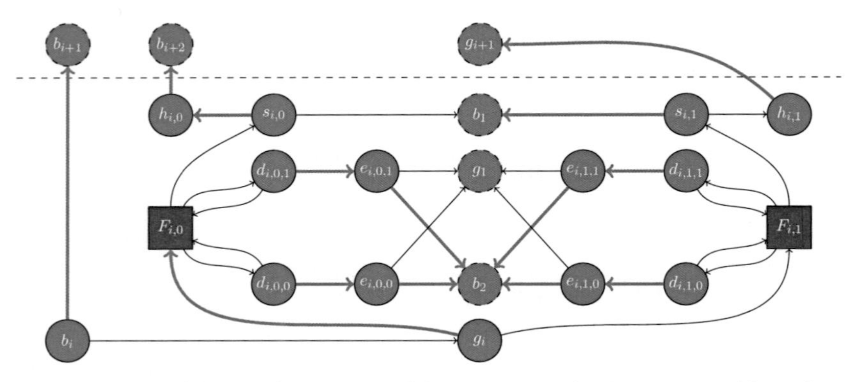

Figure 5.7.: Level i of the initial strategy σ_0 of the construction for the exponential lower bound for $i \in [n-2]$. For simplicity, we do not provide rewards, probabilities or priorities here. Thick blue edges show the choices made by the strategy.

A visualization of the initial strategy is given in Figure 5.7.

We now prove that S_n is a sink game and that M_n is weakly unichain.

Lemma 5.3.3. *The strategy σ_0 is a sink strategy for S_n and a weak unichain strategy for M_n. In particular, S_n is a sink game and M_n is a finite weakly unichain Markov decision process.*

Proof. We first prove that S_n is a sink game. By Definition 3.2.1, it suffices to prove that (i) S_n has a sink and (ii) that player 1 wins every vertex in S_n. It is easy to verify that t is a sink as it has $\Gamma^+(t) = \{t\}, \Omega(t) = 1 < \Omega(v)$ for all $v \in V \setminus \{t\}$ and since it is reachable from all vertices. It remains to prove that player 1 wins every vertex of S_n.

Consider the player 1 strategy τ defined via $\tau(F_{*,*}) := s_{*,*}$. Let σ denote an arbitrary player 0 strategy. Player 1 wins every vertex v for which the play $\pi_{\sigma,\tau,v}$ reaches the sink t. It thus suffices to investigate plays that do not end in t and prove that player 1 wins these.

Since $\tau(F_{i,j}) = s_{i,j}$ for all suitable indices i, j, it is impossible that any play has the cycle component $\{F_{i,j}, d_{i,j,k}\}$. By construction, this implies that cycle components can only be formed by higher levels escaping to one of the first two levels via some upper selection vertex $s_{*,*}$ or escape vertex $e_{*,*,*}$. In particular, any cycle that is not formed by a cycle center and one of its cycle vertices needs to use an edge $(s_{*,*}, b_1), (e_{*,*,*}, b_2)$ or $(e_{*,*,*}, g_1)$. By the choice of τ, each possible cycle component thus contains a unique edge $(s_{i,j}, b_1)$. But this implies that the highest priority occurring infinitely often is the priority of g_i which is odd, so player 1 wins the cycle. Since this argument holds for all cycles, player 1 wins all vertices of all possible cycle components and thus all vertices. Hence, S_n is a sink game.

Consider the strategy σ_0 in S_n. We prove that there is a player 1 strategy τ such that every play $\pi_{\sigma_0,\tau,v}$ ends in t. Since $\sigma_0(b_i) = b_{i+1}$ for all $i \in [n-1]$ and $\sigma(b_n) = t$, this is true for all entry vertices b_i. In addition, as all vertices $d_{*,*,*}, e_{*,*,*}, s_{*,*}$ and $h_{*,0}$ point towards some vertex b_*, all of the corresponding plays end in t. Since $\sigma(h_{i,1}) = g_{i+1}$ for $i \in [n-1]$, it suffices to consider the vertices g_i. But then, by choosing $\tau(F_{i,j}) = d_{i,j,k}$ for

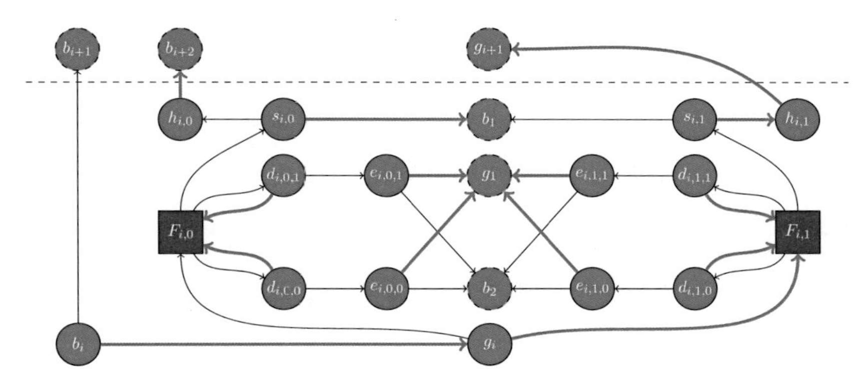

Figure 5.8.: Level i of the optimal strategy given in the proof of Lemma 5.3.3 for levels $i \in [n-2]$. For simplicity, we do not provide rewards, probabilities or priorities here. Thick blue edges show the edges of the strategy.

some $k \in \{0, 1\}$, player 1 can enforce that the plays starting in these vertices also reach t. Hence, σ_0 is a sink strategy for S_n.

Now consider the Markov decision process M_n. It suffices to prove that every vertex reaches t with probability 1. By the same arguments used previously, it immediately follows that every vertex reaches some vertex b_i after a finite number of steps with probability 1. Since $\sigma_0(b_i) = b_{i+1}$ for all $i \in [n-1]$ and $\sigma(b_n) = t$, this implies that every vertex reaches the sink t after a finite number of steps. Hence, σ_0 is a weak unichain strategy for M_n.

It remains to give an optimal sink strategy for S_n and an optimal weak unichain strategy for M_n and to prove that M_n is finite. Consider the following strategy σ^* and its visualization given in Figure 5.8. Note that we do not need to specify the targets of vertices $h_{i,j}$ as these have only one outgoing edge.

v	b_i	$d_{i,j,k}$	$e_{i,j,k}$	g_i for $i < n$	g_n	$s_{i,0}$ for $i < n$	$s_{n,0}$	$s_{i,1}$
$\sigma^*(v)$	g_i	$F_{i,j}$	g_1	$F_{i,1}$	$F_{n,0}$	b_1	$h_{n,0}$	$h_{i,1}$

Consider S_n first. If player 1 selects the counterstrategy τ by setting $\tau(F_{i,j}) := s_{i,j}$ for all suitable indices i, j, then for every vertex $v \in V$, the play $\pi_{\sigma^*,\tau,v}$ ends in t. The reason is that every level i (except n) is traversed using the vertices $b_i, g_i, F_{i,1}, s_{i,1}$ and $h_{i,1}$, leading to the vertex g_{i+1}. Then, level $i+1$ is traversed similarly and the final level is traversed in such a way that the sink t is reached. As every vertex that is not part of this spinal path beginning in b_1 reaches either a selector or entry vertex after a finite number of steps, this implies the statement. Thus, σ^* is a sink strategy. This argument also implies that the sink t is reached with probability 1 in M_n, implying that σ^* is a weak unichain strategy for M_n. In particular, by Lemma 3.2.9, the values of the vertices are finite.

It remains to show that σ^* is optimal. We do so by proving that $I_{\sigma^*} = \emptyset$. Note that this proves that σ^* solves the optimality equations for the expected total reward criterion, implying that M_n is finite. We however refrain from explicitly calculating all vertex valuations but argue why there are no improving switches with respect to σ^*.

By the definition of improving switches, it suffices to show that no edge $(u, v) \in E_0$ with $\sigma^*(u) \neq v$ is improving. First of all, traversing a level is profitable for the player. This implies that it is better to enter a level than to skip a level since either outgoing edges of entry vertices do not yield any reward, or, these vertices have a very low odd priority. Consequently, no edge (b_i, b_{i+1}) for $i \in [n-1]$ or (b_n, t) is improving. For the same reason, no edge $(e_{*,*,*}, b_2)$ is improving, since taking this edge would decrease the valuation of the corresponding escape vertex. Consider some vertex $s_{i,0}$. If $\sigma^*(s_{i,0}) = h_{i,0}$ held, then the reward associated with the edge $(h_{i,0}, b_{i+2})$ but not the reward of level $i+1$ would be collected. It is easy to verify that traversing level $i+1$ completely is more beneficial than taking the edge $(s_{i,0}, h_{i,0})$ and skipping level $i+1$. Consequently, it is better for $s_{i,0}$ to move to b_1, implying that no edge $(s_{i,0}, h_{i,0})$ is improving.

Since $\sigma^*(g_i) = F_{i,1}$ for all $i \in [n-1]$ and since all cycle centers are closed, any edge $(s_{i,1}, b_1)$ would create a cycle. But this would contradict that G_n is a sink game resp. weakly unichain. Consequently, no edge $(s_{i,1}, b_1)$ is improving.

Consider some cycle vertex $d_{i,j,k}$. If this vertex escaped level i via its escape vertex $e_{i,j,k}$, this again created a cycle. Thus, no edge of $(d_{i,j,k}, e_{i,j,k})$ is improving. Since $\sigma^*(s_{i,0}) = b_1$ for all $i \in [n-1]$, this also implies that none of the edges $(g_i, F_{i,0})$ is improving.

Consequently, there are no improving switches with respect to σ^*, hence σ^* is optimal. As σ^* is weakly unichain for M_n, the Markov decision process is thus finite. □

We now introduce further notation and begin with the term *reachable strategy*. Intuitively, a strategy σ' is reachable from a strategy σ if there is a sequence of improving switches such that applying the sequence to σ yields σ'. In particular, the notion of reachability does not depend on the pivot or the tie-breaking rule, and every strategy calculated by the strategy improvement algorithm is reachable from the initial strategy by definition.

Definition 5.3.4 (Reachable strategy). Let σ be a strategy for G_n. The set of all strategies that can be obtained from σ by applying an arbitrary sequence of improving switches is denoted by $\rho(\sigma)$. A strategy σ' is *reachable* from σ if $\sigma' \in \rho(\sigma)$.

Reachability is a transitive property and we let $\sigma \in \rho(\sigma)$ since the empty sequence is technically a sequence of improving switches.

5.3.2. The Tie-Breaking Rule

Before discussing the tie-breaking rule in detail, we revisit Zadeh's LEASTENTERED pivot rule and introduce related notation. In each step, Zadeh's pivot rule chooses an improving switch that was chosen least often until now. It is thus a memorizing pivot rule, and it needs to keep track of how often an improving switch was applied. Formally, this is handled by a function $\phi^\sigma\colon E_0 \to \mathbb{N}$ which is called the *occurrence record* of the strategy σ. We define the occurrence record with respect to the initial strategy by setting $\phi^{\sigma_0}(e) := 0$ for all $e \in E_0$. Now, let $\sigma \in \rho(\sigma_0)$ and $e \in I_\sigma$ and let ϕ^σ denote the occurrence record with respect to σ. Then, the occurrence record with respect to the strategy σe that is obtained by applying e in σ is given by $\phi^{\sigma e}(e') := \phi^{\sigma}(e')$ if $e' \neq e$ and $\phi^{\sigma e}(e) = \phi^{\sigma}(e) + 1$. Using this framework, Zadeh's pivot rule can be phrased as follows: Given a strategy $\sigma \in \rho(\sigma_0)$ and an associated occurrence record ϕ^{σ}, apply an improving switch $e \in \arg\min_{e \in I_\sigma} \phi^{\sigma}(e)$ next.

In general, this pivot rule does not uniquely determine which improving switch is applied next as there might be more than one switch minimizing the occurrence record. For example, if there is more than one improving switch for the initial strategy, then Zadeh's pivot rule is already ambiguous. The algorithm thus needs an additional rule deciding which switch to apply in such a case. Such a rule is called *tie-breaking rule*. Before introducing the tie-breaking rule that used for our construction, we briefly discuss the importance of breaking ties when applying Zadeh's LeastEntered pivot rule.

A general discussion of tie-breaking rules

For an arbitrary pivot rule, a tie-breaking rule decides which switch to apply if there are multiple candidates, all of which are valid choices with respect to the considered pivot rule. There are pivot rules for which a tie-breaking rule might not be needed. Consider, for example, Dantzig's original pivot rule which chooses a non-basic variable with maximal reduced cost to enter the current basis. If this variable is unique in every step, then the pivot rule does not require an additional tie-breaking rule as there is never more than one eligible candidate for the pivot rule. Another example of a pivot rule that does not require a tie-breaking rule is Cunningham's pivot rule in which a cyclic order of the variables is fixed in the beginning. The pivot rule the chooses the first improving variable with respect to this order, starting from the last chosen variable.

For Zadeh's pivot rule, an additional tie-breaking rule is necessary, however. Consider, for example, the very first iteration of the strategy improvement algorithm using Zadeh's pivot rule. If there is more than one improving switch for the initial strategy, then the algorithm already requires tie-breaking as the occurrence records of all edges is 0 with respect to the initial strategy. The same also holds for later iterations whenever there are multiple improving switches minimizing the occurrence records. Consequently, the choice of a tie-breaking rule is unavoidable when using Zadeh's pivot rule.

It is of course just natural to then ask for a tie-breaking rule that guarantees a small number of iterations. From an extreme point of view, one might argue that a worst-case example for Zadeh's pivot rule only yields a "valid" lower bound if it applies to all possible tie-breaking rules. At least in the context of Markov decision processes, asking for such a lower bound construction is not realistic, however. The reason is that there is always a way to break ties such that the strategy improvement algorithm using Zadeh's pivot rule requires at most n iterations in a Markov decision process with n vertices [Fri11b, Corollary 4.79]. In particular, it is thus not possible to find a Markov decision process that provides a lower bound independent of the chosen tie-breaking rule. As formulated by Oliver Friedmann in his theses, "*the question whether Zadeh's pivoting rule solves MDPs (and LPs) in polynomial time should therefore be phrased independently of the heuristic of breaking ties. In other words, we as "lower bound designers" are the ones that choose a particular tie breaking rule*" [Fri11b, page 191].

It is not clear how a "good" tie-breaking rule looks like. Of course, a tie-breaking rule that leads to less iterations is more desirable than a tie-breaking rule that results in a lot of iterations. Another important measurement for the quality of a tie-breaking rule is how natural it is. If a tie-breaking rule heavily depends on the execution of the algorithm, specific configurations or similar aspects, it is hard to understand and thus potentially hard to use. Thus, the question arises if there is a tie-breaking rule that is natural and results in as few iterations as possible. Candidates for such natural tie-breaking rules might, for example, be other pivot rules. One could, for example, use the tie-breaking rule "among all improving switches minimizing the occurrence record, choose the one with largest reduced cost" and thus combine Zadeh's and Dantzig's pivot rules.

The tie-breaking rule that we use in our construction is of a different type, but we still consider it a natural rule. Our tie-breaking rule is an ordering of the edges, and the algorithm always chooses the improving switch minimizing the occurrence records which appears first in this ordering. This rule is arguably natural and easy to implement. However, our tie-breaking rule depends on the current strategy, so it is not as natural as a tie-breaking rule that is based on ordering the edges of the instance could be.

The tie-breaking rule used for our construction

In our case, the tie-breaking rule is implemented as an ordering of all edges that depends on the current strategy. Then, whenever there are multiple improving switches minimizing the occurrence record, the algorithm chooses the first edge that is an improving switch with respect to this ordering. Although our ordering depends on the current strategy, a tie-breaking rule that is implemented by an ordered list of edges is one of the most natural ways to define such a rule.

For the remainder of this section, let σ be a strategy for G_n. It turns out that it is not necessary to give a full ordering of E_0, significantly simplifying the presentation of the tie-breaking rule. In fact, it is sufficient to describe a pre-order of E_0, and any linear extension of this pre-order can then be used. As $\Gamma^+(t) = \{t\}$, we do not include the edge (t,t) here. We thus define the following subsets of $E_0 \setminus \{(t,t)\}$:

- $\mathbb{G} := \{(g_*, F_{*,*})\}$ is the set of all edges leaving selector vertices.
- $\mathbb{E}^0 := \{(e_{i,j,k}, *) \in E_0 \colon \sigma(d_{i,j,k}) \neq F_{i,j}\}$ is the set of edges leaving escape vertices whose cycle vertices do not point towards their cycle center. Similarly, the set of edges leaving escape vertices whose cycle vertices point towards their cycle center is defined as $\mathbb{E}^1 := \{(e_{i,j,k}, *) \colon \sigma(d_{i,j,k}) = F_{i,j}\}$.
- $\mathbb{D}^1 := \{(d_{*,*,*}, F_{*,*})\}$ is the set of cycle edges an $\mathbb{D}^0 := \{(d_{*,*,*}, e_{*,*,*})\}$ is the set of the other edges leaving cycle vertices.
- $\mathbb{B}^0 := \{(b_i, b_{i+1}) : i \in [n-1]\} \cup \{(b_n, t)\}$ is the set of all edges between entry vertices. The set $\mathbb{B}^1 := \{(b_*, g_*)\}$ is defined analogously and we let $\mathbb{B} := \mathbb{B}^0 \cup \mathbb{B}^1$.
- $\mathbb{S} := \{(s_{*,*}, *)\}$ is the set of all edges leaving upper selection vertices.

We next define a pre-order $\prec_\sigma$ based on σ and these sets. However, we need to give a finer pre-order for $\mathbb{E}^0, \mathbb{E}^1, \mathbb{S}$ and $\mathbb{D}^1$ first.

Finer pre-order for certain sets

The finer pre-order of certain sets simplifies formal proofs and arguments. Intuitively, it forces the algorithm to behave in a more controlled fashion as we can then ensure that certain improving switches are only applied after a certain "setup" was performed. More precisely, the pre-order on $\mathbb{E}^0$ forces the algorithm to (i) favor edges contained in higher levels, (ii) favor $(e_{i,0,*},*)$ over $(e_{i,1,*},*)$ in S_n and (iii) favor $(e_{i,\beta_{i+1},*},*)$ over $(e_{i,1-\beta_{i+1},*},*)$ in M_n. For S_n, we define $(e_{i,j,x},*) \prec_\sigma (e_{k,l,y},*)$ if (i) $i > k$ or (ii) $i = k$ and $j < l$. For M_n, we define $(e_{i,j,x},*) \prec_\sigma (e_{k,l,y},*)$ if (i) $i > k$ or (ii) $i = k$ and $j = \beta_{i+1}$.

Similarly, the pre-order on $\mathbb{S}$ forces the algorithm to favor edges contained in higher levels as well. We thus define $(s_{i,j},*) \prec_\sigma (s_{k,l},*)$ if $i > k$.

We next describe the pre-order on $\mathbb{E}^1$. Let $(e_{i,j,x},*), (e_{k,l,y},*) \in \mathbb{E}^1$.

1. The first criterion encodes that switches contained in higher levels are applied first. We thus define $(e_{i,j,x},*) \prec_\sigma (e_{k,l,y},*)$ if $i > k$.
2. If $i = k$, then we consider the states of the cycle centers $F_{i,j}$ and $F_{k,l} = F_{i,1-j}$. If exactly one of them is closed, say $F_{i,j}$, then the improving switches within this cycle center are applied first. We thus define $(e_{i,j,x},*) \prec_\sigma (e_{k,l,y},*)$ if (i) $i = k$, (ii) $F_{i,j}$ is closed and (iii) $F_{i,1-j}$ is not closed.
3. Consider the case where $i = k$ but no cycle center of level i is closed. Let $t^\rightarrow := b_2$ if $\ell(\beta) > 1$ and $t^\rightarrow := g_1$ if $\ell(\beta) = 1$. If there is exactly one $t^\rightarrow$-halfopen cycle center in level i, then switches within this cycle center have to be applied first. Formally, we thus define $(e_{i,j,x},*) \prec_\sigma (e_{k,l,y},*)$ if (i) $i = k$, (ii) $F_{i,j}$ is $t^\rightarrow$-halfopen and (iii) $F_{i,1-j}$ is neither closed nor $t^\rightarrow$-halfopen.
4. Assume that none of the prior criteria applied. This includes the case where both cycle centers are in the same state and implies $i = k$. Then, the order of application differs for S_n and M_n. In S_n, switches within $F_{i,0}$ are applied first. In M_n, switches within $F_{i,\beta_{i+1}}$ are applied first. We thus define $(e_{i,0,x},*) \prec_\sigma (e_{i,1,y},*)$ if (i) $i = k$, (ii) we consider S_n and (iii) none of the previous criteria applied. Analogously, we define $(e_{i,\beta_{i+1},x},*) \prec_\sigma (e_{i,1-\beta_{i+1},y},*)$ for M_n.

We next give a pre-order for $\mathbb{D}^1$. The main purpose of this pre-order is that the cycle center $F_{i,j}$ with $i = \ell(\mathfrak{b}+1)$ and $j = (\mathfrak{b}+1)_{i+1}$ is the only active cycle center that is closed when transitioning from $\sigma_\mathfrak{b}$ to $\sigma_{\mathfrak{b}+1}$. Let $(d_{i,j,x}, F_{i,j}), (d_{k,l,y}, F_{k,l}) \in \mathbb{D}^1$.

1. Improving switches contained in open cycles are applied first. We thus define $(d_{i,j,x}, F_{i,j}) \prec_\sigma (d_{k,l,y}, F_{k,l})$ if $\sigma(d_{k,l,1-y}) = F_{k,l}$ but $\sigma(d_{i,j,1-x}) \neq F_{i,j}$.
2. The second criterion states that among all halfopen cycle centers, those contained in levels i with $\beta_i^\sigma = 0$ are applied first. If the first criterion does not apply, we thus define $(d_{i,j,x}, F_{i,j}) \prec_\sigma (d_{k,l,y}, F_{k,l})$ if $\beta_k > \beta_i$.
3. The third criterion states that among all halfopen cycle centers, improving switches of lower levels are applied first. If none of the first two criteria apply, we thus define $(d_{i,j,x}, F_{i,j}) \prec_\sigma (d_{k,l,y}, F_{k,l})$ if $k > i$.
4. The fourth criterion states that, within one level, improving switches of the active cycle center are applied first. If none of the previous criteria apply, we thus define $(d_{i,j,x}, F_{i,j}) \prec_\sigma (d_{k,l,y}, F_{k,l})$ if $\beta_{k+1} \neq l$ and $\beta_{i+1} = j$.

5. The last criterion states that, within one cycle center, edges with last index equal to zero are preferred. That is, if none of the previous criteria apply, we define $(d_{i,j,x}, F_{i,j}) \prec_\sigma (d_{k,l,y}, F_{k,l})$ if $x < y$. If this criterion does not apply either, the edges are incomparable.

Definition of the tie-breaking rule

Definition 5.3.5 (Tie-breaking rule). Let σ be a strategy for G_n and $\phi^\sigma\colon E_0 \to \mathbb{N}$ be an occurrence record for σ. We define the pre-order $\prec_\sigma$ on E_0 via

$$\mathbb{G}_\sigma \prec_\sigma \mathbb{D}^0 \prec_\sigma \mathbb{E}^1 \prec_\sigma \mathbb{B} \prec_\sigma \mathbb{S} \prec_\sigma \mathbb{E}^0 \prec_\sigma \mathbb{D}^1$$

where the sets $\mathbb{E}^0, \mathbb{E}^1, \mathbb{S}$ and $\mathbb{D}^1$ are additionally pre-ordered as described before.

We extend the pre-order $\prec_\sigma$ to an arbitrary but fixed total ordering on E_0, also denoted by $\prec_\sigma$. We define the following tie-breaking rule: *Let $I_\sigma^{\min} := \arg\min_{e \in I_\sigma} \phi^\sigma(e)$ denote the set of improving switches minimizing the occurrence record. Apply the first improving switch contained in $I_\sigma^{\min}$ with respect to the ordering $\prec_\sigma$ with the following exception: If $\phi^\sigma(b_1, b_2) = \phi^\sigma(s_{1,1}, h_{1,1}) = 0$, then apply $(s_{1,1}, h_{1,1})$ instead of (b_1, b_2).*

We briefly discuss the exception and explain why it is needed. During the execution of the algorithm, it will typically be the case that the occurrence record of $(s_{1,1}, h_{1,1})$ is lower than the occurrence record of (b_1, b_2). In particular, when both of these edges are improving, the edge (b_1, b_2) does not minimize the occurrence record then. Hence, it is not even a candidate for being applied, and even though (b_1, b_2) precedes $(s_{1,1}, h_{1,1})$ in the tie-breaking rule, (b_1, b_2) will not be applied. This, however, is not true in the very beginning, that is, when the occurrence record of both edges is equal to zero. Then, both minimize the occurrence record, so the tie-breaking rule decides which switch to apply. As $\mathbb{B} \prec_\sigma \mathbb{S}$ is required for other applications of improving switches during the execution of the algorithm, we have to include this exception. To prove that the tie-breaking rule defined in Definition 5.3.5 is computationally tractable, it remains to prove that it can be evaluated efficiently.

Lemma 5.3.6. *Given a strategy $\sigma \in \rho(\sigma_0)$ and an occurrence record $\phi^\sigma\colon E_0 \mapsto \mathbb{N}$, the tie-breaking rule can be evaluated in polynomial time.*

Proof. Let $\sigma \in \rho(\sigma_0)$. Identifying the subsets of E_0 can be done by iterating over E_0 and checking $\sigma(v)$ for all $v \in E_0$. Therefore, the pre-order of the sets can be calculated in polynomial time. Since expending the chosen pre-order to a total order is possible in polynomial time [Szp30], the tie-breaking rule can be computed in polynomial time. Whenever the tie-breaking rule needs to be considered, the algorithm needs to iterate over the chosen ordering. Since this can also be done in time polynomial in the input, the tie-breaking rule can be applied in polynomial time. Also, handling the exception described in Definition 5.3.5 can be done in polynomial time. □

5.3.3. The Five Phases and the Application of Improving Switches

Our goal is to prove that applying the strategy improvement algorithm to G_n using Zadeh's pivot and our tie-breaking rule enumerates one canonical strategy $\sigma_{\mathfrak{b}}$ per $\mathfrak{b} \in \mathfrak{B}_n$. This will be proven in an inductive fashion as follows by proving the following statement: Given a canonical strategy $\sigma_{\mathfrak{b}}$ for $\mathfrak{b} \in \mathfrak{B}_n$, the algorithm calculates a canonical strategy $\sigma_{\mathfrak{b}+1}$ for $\mathfrak{b}+1$. This process is called *transition* from $\sigma_{\mathfrak{b}}$ to $\sigma_{\mathfrak{b}+1}$ and is usually abbreviated by $\sigma_{\mathfrak{b}} \to \sigma_{\mathfrak{b}+1}$. To analyze a single transition, we divide it into four to five *phases* which are inspired by the macroscopic tasks performed by the algorithm to transform $\sigma_{\mathfrak{b}}$ into $\sigma_{\mathfrak{b}+1}$. These tasks are, for example, the opening and closing of cycle centers, updating the escape vertices or adjusting some of the selection vertices.

The exact number of phases depends on whether we consider S_n or M_n and on $\ell(\mathfrak{b}+1)$. Phases $1, 3$ and 5 always take place, while phase 2 is only present if $\ell(\mathfrak{b}+1) > 1$ since the targets of several vertices in levels $i < \ell(\mathfrak{b}+1)$ are updated in this phase. The same is true for phase 4, although this phase only exists for S_n. If we consider M_n, then the switches that are applied during phase 4 in S_n are already applied during phase 3 of M_n and there is no separate phase 4.

Informal description

We begin by giving an intuitive description and explanation of the individual phases. A very simplified and schematic sketch of the different phases and their interaction in the Markov decision process M_n is given in Figure 5.9. Consider the canonical strategy $\sigma_{\mathfrak{b}}$ for some $\mathfrak{b} \in \mathfrak{B}_n$ and let $\nu := \ell(\mathfrak{b}+1)$.

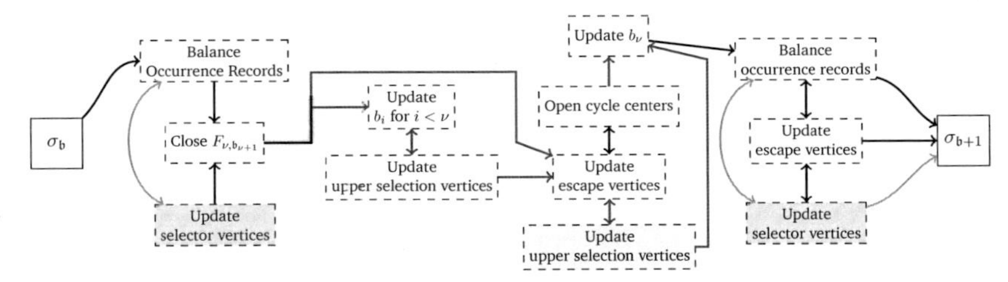

Figure 5.9.: Sketch of some of the tasks performed during the different phases by the strategy improvement algorithm when applied to M_n. Each box marks one task that has to be performed, and each vertical set of boxes corresponds to one of the phases $1, 2, 3$ and 5 (from left to right). Gray boxes and edges represent tasks that are not performed during all transitions. Red/blue edges represent that the corresponding boxes are only relevant if $\mathfrak{b}+1$ is odd/even. The yellow box represents the task that constitutes phase 4 in S_n but part of phase 3 in M_n.

1. During phase 1, several cycle vertices switch towards their cycle centers. The primary purpose of this phase is that the strategy σ obtained after the application of the final switch represents $\mathfrak{b}+1$. A secondary purpose is to balance the occurrence records of the cycle edges by applying additional improving switches. This application

might close inactive cycle centers $F_{i,1-\mathfrak{b}_{i+1}}$ and can thus make edges $(g_i, F_{i,1-\mathfrak{b}_{i+1}})$ improving. As balancing occurrence records might close additional cycle centers, more edges of the type $(g_*, F_{*,*})$ can become improving and are then applied. The final improving switch applied during phase 1 closes the cycle center $F_{\nu,(\mathfrak{b}+1)_{\nu+1}}$. Depending on the parity of $\mathfrak{b}+1$, either phase 2 (if $\mathfrak{b}+1 \bmod 2 = 0$) or 3 (if $\mathfrak{b}+1 \bmod 2 = 1$) begins.

2. In phase 2, the upper selection vertices $s_{i,j}$ for $i \in [\nu-1]$ and $j = (\mathfrak{b}+1)_{i+1}$ change their targets from b_1 to $h_{i,j}$. This is necessary as the induced bit state is now representing $\mathfrak{b}+1$, so the ν least significant bits changed. Furthermore, the entry vertices b_i of these levels switch towards b_{i+1} (with the exception of b_1). Since $\mathfrak{b}_i = \mathfrak{b}_{i+1}$ for all $i \neq 1$ if $\nu = 1$, these operations only need to be performed if $\nu > 1$.
3. Phase 3 is partly responsible for applying improving switches involving escape vertices. Since the parities of $\mathfrak{b}$ and $\mathfrak{b}+1$ are not the same, all escape vertices have to change their targets. During phase 3, exactly the escape vertices $e_{i,j,k}$ with $i \in [n], j, k \in \{0,1\}$ whose cycle vertex $d_{i,j,k}$ points to the cycle center $F_{i,j}$ change their targets. In addition, exactly these cycle vertices then also change their targets to $e_{i,j,k}$ unless the cycle center $F_{i,j}$ is closed and active. This enables the application of $(d_{i,j,k}, F_{i,j})$ which is necessary to balance the occurrence records of the cycle edges.

   At the end of this phase either (b_1, g_1) (if $\nu = 1$) or (b_1, b_2) (if $\nu > 1$) is applied. In M_n, the improving switches of phase 4 are also applied during phase 3.
4. During phase 4, the remaining upper selection vertices $s_{i,j}$ with $i \in [\nu-1]$ and $j = 1-(\mathfrak{b}+1)_{i+1}$ are updated by changing their targets to b_1. These updates are necessary to allow the cycle centers and cycle vertices to access the spinal path. Similarly to phase 2, these switches are only applied if $\nu > 1$.
5. During phase 5, the remaining improving switches involving escape vertices are applied. Moreover, some of the edges $(d_{*,*,*}, F_{*,*})$ that have a very low occurrence record are also applied in order to increase their occurrence records. In some sense, the switches "catch up" to the other edges that have been applied more often. This application might close some inactive cycle centers $F_{*,*}$ and consequently make the corresponding edge $(g_*, F_{*,*})$ improving. This switch is then also applied. Phase 5 ends once the set of improving switches only contains edges of the type $(d_{*,*,*}, F_{*,*})$.

Before giving the formal definition of the phases, we want to briefly discuss Oliver Friedmann's implementation of our construction. He used the PGSolver library to implement the general sink game S_n for arbitrary values of n. We then applied the strategy improvement algorithm using Zadeh's pivot rule and the tie-breaking rule provided in Definition 5.3.5 to examples with 3 up to 10 levels. To verify our findings, we manually validated the full sequence of strategies produced by the algorithm for $n = 3$ and $n = 4$ and specific subsequences of produced strategies for $n \in \{5, \dots, 10\}$. In addition, we encoded some of the formal assumptions that are imposed on the strategies in Chapter 6 to validate these as well. As the implementation proceeded exactly along the phases we just described and additionally checks whether the produced strategies fulfill the formal and technical assumptions that are required for the rigorous proof, this validates our results for at least the sink game construction. We also want to mention here that an implementation of the Markov decision process was not feasible as the probabilities are extremely small, making a numerically stable implementation extremely challenging.

Formal definition of the phases

To give the formal definition of the phases, we require additional notation for describing strategies. In particular, we encode the choices of σ by using integers. For this purpose, we introduce a function $\bar{\sigma}$ that will be extended later to describe more complex configurations of G_n. At this point, we only provide the first layer of complexity by defining $\bar{\sigma}(v)$ for all $v \in V_0$ using Table 5.3. In principle, $\bar{\sigma}$ is used to abbreviate boolean expressions. These expressions are either true (i.e., equal to 1) or false (i.e., equal to 0). For example, $\bar{\sigma}(b_i)$ denotes the boolean expression $\sigma(b_i) = g_i$, so $\bar{\sigma}(b_i) = 1$ if and only if $\sigma(b_i) = g_i$.

Symbol	$\bar{\sigma}(b_i)$	$\bar{\sigma}(s_{i,j})$	$\bar{\sigma}(g_i)$	$\bar{\sigma}(d_{i,j,k})$	$\bar{\sigma}(e_{i,j,k})$
Encoded expression	$\sigma(b_i) = g_i$	$\sigma(s_{i,j}) = h_{i,j}$	$\sigma(g_i) = F_{i,1}$	$\sigma(d_{i,j,k}) = F_{i,j}$	$\sigma(e_{i,j,k}) = b_2$

Table 5.3.: Definition of the function $\bar{\sigma}$ for the player vertices and a strategy σ in G_n.

For convenience, we define $\bar{\sigma}(t) := 0$. Since every player vertex has an outdegree of at most two, the value of $\bar{\sigma}(v)$ is in bijection to $\sigma(v)$. We can thus use $\neg\bar{\sigma}(v)$ to denote $\bar{\sigma}(v) = 0$. For convenience of notation, the precedence level of "$=$" and "$\neq$" is higher than the precedence level of $\wedge$ and $\vee$. That is, $x \wedge y = z$ is interpreted as $x \wedge (y = z)$.

Using this notation, we now introduce a strategy-based parameter $\mu^\sigma \in [n+1]$. This parameter is called *the next relevant bit* of the strategy σ. Before defining this parameter formally, we briefly explain its importance and how it can be interpreted.

One of the central concepts of G_n is that the two cycle centers of a fixed level alternate in representing bit i. Consequently, the selector vertex g_i of level i needs to select the correct cycle center. Moreover, b_i should point to g_i if and only if bit i is equal to 1 (see Definition 5.1.2 resp. 5.2.1). This in particular implies that the selector vertex g_{i-1} of level $i-1$ needs to be in accordance with the entry vertex of level b_i if bit $i-1$ is equal to 1. More precisely, it should not happen that $\sigma(b_i) = g_i, \sigma(b_{i+1}) = g_{i+1}$ and $\sigma(g_i) = F_{i,0}$. However, it cannot be guaranteed that this does not happen for some intermediate strategies encountered during $\sigma_{\mathfrak{b}} \to \sigma_{\mathfrak{b}+1}$. Such a configuration is then an indicator that *some* operations have to be performed in the levels i and $i+1$. This is captured by the parameter μ^σ as it is defined as the lowest level higher than any level that is set "incorrectly" in that sense. If there are no such levels, then μ^σ is the lowest level i with $\sigma(b_i) = b_{i+1}$. The parameter can thus be interpreted as an indicator encoding where "work needs to be done next". Formally, it is defined as follows.

Definition 5.3.7 (Next relevant bit). Let $\sigma \in \rho(\sigma_0)$. The *set of incorrect levels* is defined as $\mathcal{I}^\sigma := \{i \in [n] : \bar{\sigma}(b_i) \wedge \bar{\sigma}(g_i) \neq \bar{\sigma}(b_{i+1})\}$. The *next relevant bit* μ^σ of the strategy σ is

$$\mu^\sigma := \begin{cases} \min\{i > \max\{i' \in \mathcal{I}^\sigma\} : \bar{\sigma}(b_i) \wedge \bar{\sigma}(g_i) = \bar{\sigma}(b_{i+1})\} \cup \{n\}, & \text{if } \mathcal{I}^\sigma \neq \emptyset, \\ \min\{i \in [n+1] : \sigma(b_i) = b_{i+1}\}, & \text{if } \mathcal{I}^\sigma = \emptyset. \end{cases}$$

The next relevant bit now enables us to give a formal definition of the phases. These phases are described by a set of *properties* that the strategies of the corresponding phase have to fulfill. Before we list and explain the properties that are used for defining the phases, we extend the function $\bar{\sigma}$. The definition given in Table 5.3 allows us to use the function $\bar{\sigma}$ to describe the state of individual vertices. It is however convenient to also describe more complex configuration by encoding them as boolean expressions. An example for such a configurations is the setting of the cycle centers. We thus extend the notation of $\bar{\sigma}$ and refer to Table 5.4 for an overview over the complete definition of the function.

Symbol	Encoded expression
$\bar{\sigma}(b_i)$	$\sigma(b_i) = g_i$
$\bar{\sigma}(s_{i,j})$	$\sigma(s_{i,j}) = h_{i,j}$
$\bar{\sigma}(g_i)$	$\sigma(g_i) = F_{i,1}$
$\bar{\sigma}(d_{i,j,k})$	$\sigma(d_{i,j,k}) = F_{i,j}$
$\bar{\sigma}(e_{i,j,k})$	$\sigma(e_{i,j,k}) = b_2$

Symbol	Encoded expression
$\bar{\sigma}(s_i)$	$\bar{\sigma}(s_{i\bar{\sigma}(g_i)})$
$\bar{\sigma}(d_{i,j})$	$\bar{\sigma}(d_{i,j,0}) \wedge \bar{\sigma}(d_{i,j,1})$
$\bar{\sigma}(d_i)$	$\bar{\sigma}(d_{i,\bar{\sigma}(g_i)})$
$\bar{\sigma}(eg_{i,j})$	$\bigvee_{k\in\{0,1\}}[\neg\bar{\sigma}(d_{i,j,k}) \wedge \neg\bar{\sigma}(e_{i,j,k})]$
$\bar{\sigma}(eb_{i,j})$	$\bigvee_{k\in\{0,1\}}[\neg\bar{\sigma}(d_{i,j,k}) \wedge \bar{\sigma}(e_{i,j,k})]$
$\bar{\sigma}(eg_i)$	$\bar{\sigma}(eg_{i,\bar{\sigma}(g_i)})$
$\bar{\sigma}(eb_i)$	$\bar{\sigma}(eb_{i,\bar{\sigma}(g_i)})$

Table 5.4.: Extension and full definition of the function $\bar{\sigma}$ given in Table 5.3 to describe more complex configurations. Here, $\neg\bar{\sigma}(v)$ is the logical negation of $\bar{\sigma}(v)$.

Formally, a strategy belongs to one of the five phases if it has a certain set of properties. These properties can be partitioned into several categories depending on the vertices or terms that are involved. The properties might also depend on one or more parameters like a level or a cycle center.

Consider some fixed strategy σ, $\mathfrak{b} \in \mathfrak{B}_n$ and let $\nu := \ell(\mathfrak{b}+1)$ denote the least significant set bit of $\mathfrak{b}+1$. The first three properties are related to the **E**ntry **V**ertices. Property $(\text{EV1})_i$ states that the entry vertex of level i should point to g_i if and only if the the active (with respect to the induced bit state) cycle center $F_{i,,\beta^\sigma_{i+1}}$ is closed, so

$$\bar{\sigma}(b_i) = \bar{\sigma}(d_{i,\beta^\sigma_{i+1}}). \tag{EV1}$$

Similarly, Property $(\text{EV2})_i$ states that $\sigma(b_i) = g_i$ implies that the selector vertex g_i of level i should point to the corresponding active cycle center, so

$$\bar{\sigma}(b_i) \implies \bar{\sigma}(g_i) = \beta^\sigma_{i+1}. \tag{EV2}$$

Property $(\text{EV3})_i$ states that $\sigma(b_i) = g_i$ implies that the inactive cycle center is not closed, so

$$\bar{\sigma}(b_i) \implies \neg\bar{\sigma}(d_{i,1-\beta^\sigma_{i+1}}). \tag{EV3}$$

This property is a good example for a property that will be violated during specific phases as several inactive cycle centers will be closed when the induced bit state increases.

The next five properties are all related to the **ESC**ape vertices $e_{*,*,*}$. Property (ESC1) states that the escape vertices are set "correctly", that is, as they should be set for a canonical strategy representing β, so

$$[\beta_1^\sigma = 0 \implies \sigma(e_{*,*,*}) = b_2] \wedge [\beta_1^\sigma = 1 \implies \sigma(e_{*,*,*}) = g_1]. \tag{ESC1}$$

Property (ESC2) states that all escape vertices point to g_1, so

$$\sigma(e_{*,*,*}) = g_1. \tag{ESC2}$$

Although this property seems redundant due to Property (ESC1), it is crucial for properly defining the second phase.

The next three properties are used to describe the access of $F_{i,j}$ to the vertices g_1 and b_2 via the escape vertices. More precisely, they state whether $F_{i,j}$ has access to only g_1 (Property (ESC3)$_{i,j}$), only b_2 (Property (ESC4)$_{i,j}$) or to both of these vertices (Property (ESC5)$_{i,j}$). We mention here that Property (ESC3) is technically not used for the definition of the phases, but as it will be used within several proofs and fits the other properties related to the escape vertices, we already provide it here. Formally, these properties are given via

$$\bar{\sigma}(eg_{i,j}) \wedge \neg\bar{\sigma}(eb_{i,j}), \tag{ESC3}$$

$$\bar{\sigma}(eb_{i,j}) \wedge \neg\bar{\sigma}(eg_{i,j}), \tag{ESC4}$$

$$\bar{\sigma}(eb_{i,j}) \wedge \bar{\sigma}(eg_{i,j}). \tag{ESC5}$$

The next three properties are concerned with the **U**pper **S**election **V**ertices $s_{*,*}$. Property (USV1)$_i$ states that both upper selection vertices of level i are set "correctly" with respect to the induced bit state, while Property (USV3)$_i$ states that both of these vertices are set incorrectly. Property (USV2)$_{i,j}$ simply states $\sigma(s_{i,j}) = h_{i,j}$ and will be used to identify strategies for which the upper selection vertices of lower levels need to be updated since the induced bit state changed.

$$\sigma(s_{i,\beta_{i+1}^\sigma}) = h_{i,\beta_{i+1}^\sigma} \wedge \sigma(s_{i,1-\beta_{i+1}^\sigma}) = b_1 \tag{USV1}$$

$$\sigma(s_{i,j}) = h_{i,j} \tag{USV2}$$

$$\sigma(s_{i,\beta_{i+1}^\sigma}) = b_1 \wedge \sigma(s_{i,1-\beta_{i+1}^\sigma}) = h_{i,1-\beta_{i+1}^\sigma} \tag{USV3}$$

The next two properties are related to the **C**ycle **C**enters. Property (CC1)$_i$ states that at least one cycle center of level i has to be open or halfopen if $i < \mu^\sigma$. Property (CC2) states that the active cycle center of level $\nu = \ell(\mathfrak{b}+1)$ is closed and that the selector vertex of level ν chooses the correct cycle center with respect to $\mathfrak{b}+1$, so

$$i < \mu^\sigma \implies \neg\bar{\sigma}(d_{i,0}) \vee \neg\bar{\sigma}(d_{i,1}), \tag{CC1}$$

$$\bar{\sigma}(d_\nu) \wedge \bar{\sigma}(g_\nu) = (\mathfrak{b}+1)_{\nu+1} \tag{CC2}$$

The following properties are related to the **S**elector **V**ertices and are unique for either the **M**arkov decision process M_n (Property (SVM)) or the sink **G**ame S_n (Property (SVG)). They are related to the setting of selector vertices if the represented bit is equal to 0. According to Definition 5.1.2 resp. 5.2.1, the cycle center chosen by g_i is fixed in this case depends on whether we consider M_n or S_n, see condition 3.(d). Properties (SVM) and (SVG) now state that the selector vertex can only choose the other cycle center $F_{i,1}$ resp. $F_{i,1-\beta^\sigma_{i+1}}$ if this cycle center is closed, so

$$\bar{\sigma}(g_i) = 1 - \beta^\sigma_{i+1} \implies \bar{\sigma}(d_{i,1-\beta^\sigma_{i+1}}), \tag{SVM}$$

$$\bar{\sigma}(g_i) = 1 \implies \bar{\sigma}(d_{i,1}). \tag{SVG}$$

The final two properties are related to the next **REL**evant bit μ^σ defined in Definition 5.3.7. Property (REL1) states that the set of incorrect levels is empty. This in particular implies that $\mu^\sigma = \min\{i \in [n+1] : \sigma(b_i) = b_{i+1}\}$. Property (REL2) states that this parameter is equal to the least significant set bit of the bit state induced by σ, so

$$\nexists i : \sigma(b_{i-1}) = g_{i-1} \wedge \bar{\sigma}(b_i) \neq \bar{\sigma}(g_{i-1}), \tag{REL1}$$

$$\mu^\sigma = \ell(\beta^\sigma). \tag{REL2}$$

Together with the induced bit state, these properties are now used to formally define the phases. This is done by providing a table where each row corresponds to one of the properties and each column to one phase. In addition, there are some special conditions that have to be fulfilled during some phases that cannot be phrased as a simple property.

Definition 5.3.8 (Phase-k-strategy)**.** Let σ be a strategy for G_n and $k \in [5]$. Then, σ is a *phase-k-strategy* if it has the properties of the corresponding column of Table 5.5 and fulfills the corresponding special conditions.

This concludes the formal definition of the phases and the intuitive description regarding the application of the improving switches. In the next section, we discuss the occurrence records that emerge when applying improving switches as described here.

5.3.4. The Occurrence Records

The occurrence records are described using the terms related to binary counting introduced in the beginning of Section 4.1. We introduce two additional terms. For a number $\mathfrak{b} \in \mathfrak{B}_n$, an index $i \in [n]$ and some $j \in \{0,1\}$, these terms describe the last time bit i was switched to 1 resp. 0 while bit $i+1$ was equal to j.

Definition 5.3.9 (Last (un-)flip number)**.** Let $\mathfrak{b} \in \mathfrak{B}_n, i \in [n]$ and $j \in \{0,1\}$. Then, the *last flip number* $\operatorname{lfn}(\mathfrak{b}, i, \{(i+1, j)\})$ is the largest $\mathfrak{b}' \leq \mathfrak{b}$ with $\ell(\mathfrak{b}') = i$ and $\mathfrak{b}'_{i+1} = j$. Similarly, the *last unflip number* $\operatorname{lufn}(\mathfrak{b}, i, \{(i+1, j)\})$ is the largest $\mathfrak{b}' \leq \mathfrak{b}$ with $\mathfrak{b}'_{i'} = 0$ for all $i' \leq i$ and $\mathfrak{b}'_{i+1} = j$. If there are no such numbers, then both quantities are defined as 0. If the setting of bit $i+1$ should not be enforced we use the terms $\operatorname{lfn}(\mathfrak{b}, i)$ and $\operatorname{lufn}(\mathfrak{b}, i)$ which are defined analogously.

Property	Phase 1	Phase 2	Phase 3	Phase 4	Phase 5
$(\text{EV1})_i$	$i \in [n]$	$i > \mu^\sigma$	$i > 1$	$i \in [n]$	$i \in [n]$
$(\text{EV2})_i$	$i \in [n]$	$i \geq \mu^\sigma$	$i > 1$	$i \in [n]$	$i \in [n]$
$(\text{EV3})_i$	$i \in [n] \setminus \{\nu\}$	$i > \mu^\sigma$	$i > 1, i \neq \mu^\sigma$	$i \in [n]$	$i \in [n]$
$(\text{USV1})_i$	$i \in [n]$	$i \geq \mu^\sigma$	$i \geq \mu^\sigma$	$i \geq \nu$	$i \in [n]$
$(\text{USV2})_{i,j}$	-	$(i, 1-\beta_{i+1})\colon i < \mu^\sigma$	$(i, *)\colon i < \mu^\sigma$	$(i, \beta_{i+1})\colon i < \nu$	-
(ESC1)	True	-	-	-	False*
(ESC2)	-	True	-	-	-
$(\text{ESC4})_{i,j}$	-	-	-	$(i,j) \in S_1$	-
$(\text{ESC5})_{i,j}$	-	-	-	$(i,j) \in S_2$	-
(REL1)	True	-	-	True	True
(REL2)	-	True	True	False	False
$(\text{CC1})_i$	$i \in [n]$	$i \in [n]$	$i \in [n]$	$i \in [n]$	$i \in [n]$
(CC2)	-	True†	True†	True	True
$(\text{SVM})_i/(\text{SVG})_i$	$i \in [n]$	$i \in [n]$	-	-	-*
$\beta =$	$\mathfrak{b}$	$\mathfrak{b}+1$	$\mathfrak{b}+1$	$\mathfrak{b}+1$	$\mathfrak{b}+1$
Special	Phase 2:	$\exists i < \mu^\sigma\colon (\text{USV3})_i \wedge \neg(\text{EV2})_i \wedge \neg(\text{EV3})_i$			
	Phase 2,3:	†A phase-2- resp. phase-3-strategy without Property (CC2) is called *pseudo* phase-2- resp. phase-3-strategy.			
	Phase 4:	$\exists i < \nu(\mathfrak{b}+1)\colon (\text{USV2})_{i,1-\beta_{i+1}}$			
	Phase 5:	*If σ has Property (ESC1) and there is an index i such that σ does not have Property $(\text{SVM})_i \setminus(\text{SVG})_i$, it is defined as a phase-5-strategy			

Set	Definition
$S_1 =$	$\{(i, 1-\beta_{i+1})\colon i \in [\nu-1]\} \cup \{(i, 1-\beta_{i+1})\colon i \in \{\nu, \dots, m-1\} \wedge \beta_i = 0\}$ $\cup \begin{cases} \emptyset, & \exists k \in \mathbb{N}\colon \mathfrak{b}+1 = 2^k \\ \{(\nu, 1-\beta_{\nu+1})\}, & \nexists k \in \mathbb{N}\colon \mathfrak{b}+1 = 2^k \end{cases}$
$S_2 =$	$\{(i, \beta^\sigma_{i+1})\colon i \in [\nu-1]\} \cup \{(i, 1-\beta_{i+1})\colon i \in \{\nu+1, \dots, m\} \wedge \beta_i = 1\}$ $\cup\{(i, \beta_{i+1})\colon i \in \{\nu, \dots, m-1\} \wedge \beta_i = 0\} \cup \{(i,k)\colon i > m, k \in \{0,1\}\}$ $\cup \begin{cases} \{(\nu, 1)\}, & \exists k \in \mathbb{N}\colon \mathfrak{b}+1 = 2^k \\ \emptyset, & \nexists k \in \mathbb{N}\colon \mathfrak{b}+1 = 2^k \end{cases}$
$S_3 =$	$\{(i, 1-\beta_{i+1})\colon i \in [u]\} \cup \{(i, 1-\beta_{i+1})\colon i \in \{u+1, \dots, m\} \wedge \beta_i = 1\} \cup$ $\cup\{(i, \beta_{i+1})\colon i \in \{u+1, \dots, m-1\} \wedge \beta_i = 0\}$ $\cup\{(i,k)\colon i > m, k \in \{0,1\}\} \cup \{(u, \beta_{u+1})\}$
$S_4 =$	$\{(i, 1-\beta_{i+1})\colon i \in \{u+1, \dots, m-1\} \wedge \beta_i = 0\}$

Table 5.5.: Definition of the phases for a strategy σ and a number $\mathfrak{b} \in \mathfrak{B}_n$. The entries show for which indices the strategy has the corresponding property resp. whether the strategy has the property at all. Expressions of the type "$i \in [n]$" or similar are meant as "$\forall i \in [n]$". A '-' signifies that it is not specified whether σ has the corresponding property. The last row contains further properties used for the definition of the phases. The lower table contains all sets used for the definition of the phases and two additional sets S_3, S_4 that are necessary for later proofs. We use the abbreviations $\nu := \ell(\mathfrak{b}+1), m := \max\{i\colon \beta_i = 1\}$ and $u := \min\{i\colon \beta_i = 0\}$.

Note that the third argument in these definitions is again a set, although it would be sufficient to use only the pair $(i+1, j)$ as an argument. This is done intentionally in order to have the same notation used for the previous terms related to binary counting. We do not explicitly determine the occurrence record of every edge for every strategy. Instead, we focus on canonical strategies and give a table describing the occurrence records for a canonical strategy. We then prove the following statement inductively: If the table correctly describes the occurrence records for $\sigma_{\mathfrak{b}}$ and Zadeh's pivot rule with our tie-breaking rule is applied, then the table correctly describes the occurrence records for $\sigma_{\mathfrak{b}+1}$.

Edge e	$\phi^{\sigma_{\mathfrak{b}}}(e)$	Edge e	$\phi^{\sigma_{\mathfrak{b}}}(e)$
$(e_{*,*,*}, g_1)$	$\lceil \frac{\mathfrak{b}}{2} \rceil$	(b_i, g_i)	$\mathrm{fl}(\mathfrak{b}, i)$
$(e_{*,*,*}, b_2)$	$\lfloor \frac{\mathfrak{b}}{2} \rfloor$	(b_i, b_{i+1})	$\mathrm{fl}(\mathfrak{b}, i) - \mathfrak{b}_i$
$(s_{i,j}, h_{i,j})$	$\mathrm{fl}(\mathfrak{b}, i+1) - (1-j) \cdot \mathfrak{b}_{i+1}$	$(d_{i,j,k}, e_{i,j,k})$	$\leq \begin{cases} \phi^{\sigma_{\mathfrak{b}}}(e_{i,j,k}, g_1), & \mathfrak{b}_1 = 0 \\ \phi^{\sigma_{\mathfrak{b}}}(e_{i,j,k}, b_2), & \mathfrak{b}_1 = 1 \end{cases}$
$(s_{i,j}, b_1)$	$\mathrm{fl}(\mathfrak{b}, i+1) - j \cdot \mathfrak{b}_{i+1}$	$(g_i, F_{i,j})$	$\leq \min\limits_{k \in \{0,1\}} \phi^{\sigma_{\mathfrak{b}}}(d_{i,j,k}, F_{i,j})$

Condition	$\phi^{\sigma_{\mathfrak{b}}}(d_{i,j,k}, F_{i,j})$	Tolerance
$\mathfrak{b}_i = 1 \wedge \mathfrak{b}_{i+1} = j$	$\left\lceil \frac{\mathrm{lfn}(\mathfrak{b}, i, \{(i+1, j)\}) + 1 - k}{2} \right\rceil$	0
$\mathfrak{b}_i = 0 \vee \mathfrak{b}_{i+1} \neq j$	$\min\left(\left\lfloor \frac{\mathfrak{b}+1-k}{2} \right\rfloor, \ell^{\mathfrak{b}}(i,j,k) + t_{\mathfrak{b}}\right)$	$t_{\mathfrak{b}} \in \begin{cases} \{0\}, & i = 1 \vee \mathfrak{b}_i = 1 \\ \{0,1\}, & i \neq 1 \wedge \mathfrak{b}_1 = 0 \\ \{-1, 0, 1\}, & i \neq 1 \wedge \mathfrak{b}_1 = 1 \end{cases}$

$$\ell^{\mathfrak{b}}(i,j,k) := \left\lceil \frac{\mathrm{lfn}(\mathfrak{b}, i, \{(i+1, j)\} + 1 - k)}{2} \right\rceil + \mathfrak{b} - \mathbf{1}_{j=0}\mathrm{lfn}(\mathfrak{b}, i+1) - \mathbf{1}_{j=1}\mathrm{lufn}(\mathfrak{b}, i+1)$$

Table 5.6.: Occurrence records for the canonical strategy $\sigma_{\mathfrak{b}}$. For each edge, we either give the exact occurrence record, an upper bound, or the occurrence record up to a certain tolerance. A parameter $t_{\mathfrak{b}}$ fulfilling the assumptions for the case $\mathfrak{b}_i = 0 \vee \mathfrak{b}_{i+1} \neq j$ is called feasible for $\mathfrak{b}$.

Let $\mathfrak{b} \in \mathfrak{B}_n$ and $\sigma_{\mathfrak{b}}$ be a canonical strategy for $\mathfrak{b}$. Table 5.6 gives an overview over the occurrence records of the edges. For each edge $e \in E_0$, the occurrence record $\phi^{\sigma_{\mathfrak{b}}}(e)$ is either given exactly, bounded by the occurrence record of another edge or given exactly with a certain tolerance. For all edges whose occurrence record is only bounded, it turns out that it is not necessary to provide an exact occurrence record. Note that we only use tolerances for the occurrence records of the cycle edges $(d_{i,j,k}, F_{i,j})$. The reason is that describing the occurrence records of these edges exactly is complicated, whereas we are able to state the occurrence record relatively easily if we allow a small error. This specification will turn out to be sufficiently good.

We now give an intuitive explanation for some of the entries of Table 5.6. In all of the following, let $i \in [n], j, k \in \{0, 1\}$ be arbitrary indices. Consider some edge (b_i, g_i). This edge is applied as an improving switch whenever bit i switches from 0 to 1. That is, it is applied exactly during transitions $\sigma_{\mathfrak{b}'} \to \sigma_{\mathfrak{b}'+1}$ with $\mathfrak{b}' \leq \mathfrak{b}$ and $\ell(\mathfrak{b}'+1) = i$. Therefore, by definition, $\phi^{\sigma_{\mathfrak{b}}}(b_i, g_i) = \mathrm{fl}(\mathfrak{b}, i)$. Now consider (b_i, b_{i+1}). This edge is only applied as an improving switch when bit i switches from 1 to 0. This can however only happen if bit i switched from 0 to 1 previously. That is, applying (b_i, b_{i+1}) can only happen when (b_i, g_i) was applied before. Also, (b_i, g_i) can only be applied again after bit i switched back to 0, i.e., after (b_i, b_{i+1}) was applied. Consequently, $\phi^{\sigma_{\mathfrak{b}}}(b_i, b_{i+1}) = \phi^{\sigma_{\mathfrak{b}}}(b_i, g_i) - \mathfrak{b}_i = \mathrm{fl}(\mathfrak{b}, i) - \mathfrak{b}_i$.

Next, consider some edge $(s_{i,1}, h_{i,1})$. This edge is applied as an improving switch if and only if bit $i+1$ switches from 0 to 1. Hence, $\phi^{\sigma_{\mathfrak{b}}}(s_{i,j}, h_{i,j}) = \mathrm{fl}(\mathfrak{b}, i+1)$. Consider $(s_{i,0}, h_{i,0})$ next. This switch is applied whenever bit $i+1$ switches from 1 to 0. This requires the bit to have switched from 0 to 1 before. Therefore, $\phi^{\sigma_{\mathfrak{b}}}(s_{i,0}, h_{i,0}) = \phi^{\sigma_{\mathfrak{b}}}(s_{i,1}, h_{i,1}) - \mathfrak{b}_{i+1}$. Further note that the switch $(s_{i,j}, b_1)$ is applied in the same transitions in which the switch $(s_{i,1-j}, h_{i,1-j})$ is applied. Hence, $\phi^{\sigma_{\mathfrak{b}}}(s_{i,j}, h_{i,j}) = \mathrm{fl}(\mathfrak{b}, i+1) - (1-j) \cdot \mathfrak{b}_{i+1}$ and $\phi^{\sigma_{\mathfrak{b}}}(s_{i,j}, b_1) = \mathrm{fl}(\mathfrak{b}, i+1) - j \cdot \mathfrak{b}_{i+1}$.

Next consider some edge $(e_{i,j,k}, g_1)$. This edge is applied as improving switch whenever the first bit switches from 0 to 1. Since 0 is even, this happens once for every odd number smaller than or equal to $\mathfrak{b}$, so $\lceil \mathfrak{b}/2 \rceil$ times. Since $(e_{i,j,k}, b_2)$ is applied during each transition in which the switch $(e_{i,j,k}, g_1)$ is not applied, we have $\phi(e_{i,j,k}, g_1) = \mathfrak{b} - \lceil \mathfrak{b}/2 \rceil = \lfloor \mathfrak{b}/2 \rfloor$.

Now consider some edge $(d_{i,j,k}, e_{i,j,k})$. This edge will only become improving after the application of either $(e_{i,j,k}, g_1)$ or $(e_{i,j,k}, b_2)$, depending on the parity of $\mathfrak{b}$ and is then applied immediately. As all edges $(e_{*,*,*}, g_1)$ resp. $(e_{*,*,*}, b_2)$ have the same occurrence record, providing the upper bound of $\phi^{\sigma_{\mathfrak{b}}}(e_{i,j,k}, g_1)$ resp. $\phi^{\sigma_{\mathfrak{b}}}(e_{i,j,k}, b_2)$ is thus sufficient.

By a similar argument, it is sufficient to upper bound the occurrence record of an edge $(g_i, F_{i,j})$. Such an edge only becomes improving after closing the cycle center $F_{i,j}$ and should then be applied immediately, so the proposed bound is sufficient.

Finally, consider some cycle edge $(d_{i,j,k}, F_{i,j})$. If $\mathfrak{b}_i = 1 \wedge \mathfrak{b}_{i+1} = j$, then its occurrence record has not changed since the last time it was closed. Since $F_{i,j}$ changes its state from open to closed whenever bit i becomes the least significant bit and bit $i+1$ is equal to j and cycle edges are typically applied in an alternating fashion, the occurrence record is approximately $\mathrm{lfn}(\mathfrak{b}, i, \{(i+1, j)\})/2$. The additional terms and the rounding operation are then necessary to give the exact description of the occurrence record. If $\mathfrak{b}_i \neq 1 \vee \mathfrak{b}_{i+1} \neq j$, then there are several possible cases. In the first case, the occurrence record of $(d_{i,j,k}, F_{i,j})$ is sufficiently large in comparison to the occurrence record of other cycle edges. It can be shown that the occurrence record of these edges that have been applied "sufficiently" often is around $\mathfrak{b}/2$. In the second case, the occurrence record of $(d_{i,j,k}, F_{i,j})$ is too low and needs to "catch up". This is encoded by the term $\ell^{\mathfrak{b}}(i,j,k)$ which is composed of two parts. The first part again corresponds to the last time the cycle center $F_{i,j}$ was closed, and we refer to our explanation for the case $\mathfrak{b}_i = 1 \wedge \mathfrak{b}_{i+1} = j$ here. The second part corresponds to the additional times the edge was applied as improving switch to catch up. It is thus equal to the currently represented number $\mathfrak{b}$ minus the last time the cycle center $F_{i,j}$ was opened, which is $\mathrm{lfn}(\mathfrak{b}, i+1)$ if $j = 0$ and $\mathrm{lufn}(\mathfrak{b}, i+1)$ if $j = 1$. However, this description is not entirely accurate as there are some special cases, making it necessary to include a tolerance term.

Although proving that Table 5.6 in fact specifies the occurrence records of the edges correctly for canonical strategies is technically involved, we already state the corresponding statement now. Its formal proof is deferred to Chapter 6 containing the technical details of the construction and all proofs.

Theorem 5.3.10. *Let $\sigma_{\mathfrak{b}}$ be a canonical strategy for $\mathfrak{b} \in \mathfrak{B}_n$ and assume that the improving switches are applied as described in Section 5.3.3. Then, Table 5.6 describes the occurrence records of all edges $e \in E_0$ with respect to $\sigma_{\mathfrak{b}}$.*

5.3.5. Proving the Lower Bound

We now describe how the exponential lower bound for Zadeh's pivot rule for the strategy improvement algorithm is proven, implying the same bound for the simplex algorithm applied to linear programs and similar algorithms. This section does however not contain all of the formal details. Instead, it presents the core concepts and aspects of the proofs as well as the most important ideas and arguments that are used to derive the lower bound. Of course, this approach does not suffice to give a formal proof. Incorporating both the intuitive ideas and the "core" of our proof as well as the technical details in the same chapter would make it extremely hard to understand the approach and statements. All of the necessary formalism is thus introduced, proven and discussed in detail in Chapter 6.

We prove that applying improving switches to G_n using Zadeh's pivot rule and our tie-breaking rule requires an exponential number of iterations using an inductive argument. Assume we are given a canonical strategy $\sigma_{\mathfrak{b}}$ for $\mathfrak{b} \in \mathfrak{B}_n$ that has some helpful additional properties as well as an occurrence record as described by Table 5.6. We prove that the application of improving switches eventually yields a canonical strategy $\sigma_{\mathfrak{b}+1}$ for $\mathfrak{b}+1$ that has the same additional properties and whose occurrence record is also described by Table 5.6 when interpreted for $\mathfrak{b}+1$. It is then sufficient to prove that the initial strategy σ_0 has these properties already and that ϕ^{σ_0} is described by Table 5.6, then the exponential lower bound on the number of iterations follows immediately. Since G_n has a linear number of vertices and edges and the priorities, rewards and probabilities can be encoded using a polynomial number of bits, this implies an exponential lower bound for the respective algorithms. We begin explaining the proof by introducing the mentioned set of properties.

The canonical properties and basic statements

The additional properties of the canonical strategies are called *canonical properties*. Two of these are straight-forward. They state that Table 5.6 correctly describes the occurrence records for $\sigma_{\mathfrak{b}}$ and that each improving switch was applied at most once per previous transition $\sigma_{\mathfrak{b}'} \to \sigma_{\mathfrak{b}'+1}$ with $\mathfrak{b}' < \mathfrak{b}$. The remaining properties are more involved and introduced in more detail. An overview over these properties can be found in Table 5.7.

Let $\mathfrak{b} \in \mathfrak{B}_n$. Consider some cycle center $F_{i,j}$ with $\mathfrak{b}_i = 0 \vee \mathfrak{b}_{i+1} \neq j$. Then, $F_{i,j}$ should not be closed for $\sigma_{\mathfrak{b}}$. It can however still happen that $\sigma_{\mathfrak{b}}(d_{i,j,k}) = F_{i,j}$ for some $k \in \{0,1\}$. Property (OR1) states that this can only happen if the occurrence record of $(d_{i,j,k}, F_{i,j})$ is sufficiently low. Formally, for a general strategy σ, the property is defined as

$$\sigma(d_{i,j,k}) = F_{i,j} \wedge (\mathfrak{b}_i = 0 \vee \mathfrak{b}_{i+1} \neq j) \implies \phi^{\sigma}(d_{i,j,k}, F_{i,j}) < \left\lfloor \frac{\mathfrak{b}+1}{2} \right\rfloor. \qquad \text{(OR1)}$$

Property (OR2) characterizes under which circumstances the parameter $t_{\mathfrak{b}}$ used for describing the occurrence records of cycle edges (see Table 5.6) is equal to 1. More precisely, it states that the parameter is equal to 1 if and only if the cycle vertex points towards the cycle center. Of course, this statement is only valid if either $\beta_i^\sigma = 0$ or $\beta_{i+1}^\sigma \neq j$ since the parameter is only then relevant for describing the occurrence record. Formally,

$$\beta_i^\sigma = 0 \vee \beta_{i+1}^\sigma \neq j \implies (\phi^\sigma(d_{i,j,k}, F_{i,j}) = \ell^{\mathfrak{b}}(i,j,k) + 1 \iff \sigma(d_{i,j,k}) = F_{i,j}). \qquad \text{(OR2)}$$

Analogously, Property (OR3) gives a characterization regarding the cases in which the parameter is equal to -1. This characterization is more involved as it also depends on the exact value of the occurrence record of a cycle edge. It states that the parameter can only be -1 without being equal to $\lfloor(\mathfrak{b}+1-k)/2\rfloor$ if and only if (i) $\mathfrak{b}$ is odd, (ii) $\mathfrak{b}+1$ is not a power of 2, (iii) $i = \ell(\mathfrak{b}+1)$, (iv) $j \neq \mathfrak{b}_{i+1}$ and (v) $k = 0$. Formally,

$$\begin{aligned} &\phi^\sigma(d_{i,j,k}, F_{i,j}) = \ell^{\mathfrak{b}}(i,j,k) - 1 \wedge \phi^\sigma(d_{i,j,k}, F_{i,j}) \neq \left\lfloor \frac{\mathfrak{b}+1-k}{2} \right\rfloor \\ \iff &\mathfrak{b} \bmod 2 = 1 \wedge \nexists l \in \mathbb{N} : \mathfrak{b}+1 = 2^l \wedge i = \ell(\mathfrak{b}+1) \wedge j \neq \mathfrak{b}_{i+1} \wedge k = 0. \end{aligned} \qquad \text{(OR3)}$$

The final property states that the occurrence record of any cycle edge with $\sigma(d_{i,j,k}) \neq F_{i,j}$ is relatively high. Formally,

$$\sigma(d_{i,j,k}) \neq F_{i,j} \implies \phi^\sigma(d_{i,j,k}, F_{i,j}) \in \left\{ \left\lfloor \frac{\mathfrak{b}+1}{2} \right\rfloor - 1, \left\lfloor \frac{\mathfrak{b}+1}{2} \right\rfloor \right\}. \qquad \text{(OR4)}$$

$(\text{OR1})_{i,j,k}$	$\sigma(d_{i,j,k}) = F_{i,j} \wedge (\mathfrak{b}_i = 0 \vee \mathfrak{b}_{i+1} \neq j) \implies \phi^\sigma(d_{i,j,k}, F_{i,j}) < \lfloor \frac{\mathfrak{b}+1}{2} \rfloor$
$(\text{OR2})_{i,j,k}$	$\beta_i^\sigma = 0 \vee \beta_{i+1}^\sigma \neq j \implies (\phi^\sigma(d_{i,j,k}, F_{i,j}) = \ell^{\mathfrak{b}}(i,j,k) + 1 \iff \sigma(d_{i,j,k}) = F_{i,j})$
$(\text{OR3})_{i,j,k}$	$\phi^\sigma(d_{i,j,k}, F_{i,j}) = \ell^{\mathfrak{b}}(i,j,k) - 1 \wedge \phi^\sigma(d_{i,j,k}, F_{i,j}) \neq \lfloor \frac{\mathfrak{b}+1-k}{2} \rfloor$ $\iff \mathfrak{b} \bmod 2 = 1 \wedge \nexists l \in \mathbb{N} : \mathfrak{b}+1 = 2^l \wedge i = \ell(\mathfrak{b}+1) \wedge j \neq \mathfrak{b}_{i+1} \wedge k = 0$
$(\text{OR4})_{i,j,k}$	$\sigma(d_{i,j,k}) \neq F_{i,j} \implies \phi^\sigma(d_{i,j,k}, F_{i,j}) \in \{\lfloor \frac{\mathfrak{b}+1}{2} \rfloor - 1, \lfloor \frac{\mathfrak{b}+1}{2} \rfloor\}$

Table 5.7.: The additional properties of canonical strategies.

This now allows to define the canonical properties formally.

Definition 5.3.11 (Canonical properties). Let $\sigma \in \rho(\sigma)$ be a strategy for G_n. Then, σ has the *canonical properties* if

1. the occurrence records ϕ^σ are described correctly by Table 5.6,
2. σ has Properties $(\text{OR1})_{*,*,*}$ to $(\text{OR4})_{*,*,*}$ and
3. every improving switch was applied at most once per previous transition between canonical strategies.

We begin our arguments by explicitly determining the set of improving switches for canonical strategies. We also extend our notation of transitions. Let $\sigma, \sigma' \in \rho(\sigma_0)$ denote two strategies and let σ' be reached after σ. We denote the sequence of strategies calculated by the algorithm while transitioning from σ to σ' by $\sigma \to \sigma'$. In addition, the sequence of actually applied improving switches is denoted by $\mathfrak{A}_{\sigma}^{\sigma'}$. Throughout this section, let $\mathfrak{b} \in \mathfrak{B}_n$ be fixed and $\nu := \ell(\mathfrak{b}+1)$. The following statement is proven in Chapter 6.

Lemma 5.3.12. *Let $\sigma_{\mathfrak{b}} \in \rho(\sigma_0)$ be a canonical strategy for $\mathfrak{b} \in \mathfrak{B}_n$. Then, $\sigma_{\mathfrak{b}}$ is a phase-1-strategy for $\mathfrak{b}$ and $I_{\sigma_{\mathfrak{b}}} = \{(d_{i,j,k}, F_{i,j}) : \sigma_{\mathfrak{b}}(d_{i,j,k}) \neq F_{i,j}\}$.*

As our proof is inductive, we need to give a basis for the induction.

Lemma 5.3.13. *The initial strategy σ_0 is a canonical strategy for $\mathfrak{b} = 0$ and has all canonical properties.*

Proof. As no improving switch was applied yet and it is easy to verify that σ_0 is a canonical strategy for 0, it suffices to prove that σ_0 has Properties (OR1)$_{*,*,*}$ to (OR4)$_{*,*,*}$.

Let $i \in [n]$ and $j,k \in \{0,1\}$. First, σ_0 has Property (OR1)$_{i,j,k}$ as $\sigma_0(d_{i,j,k}) = e_{i,j,k}$. In addition, $0 = \phi^{\sigma_0}(d_{i,j,k}, F_{i,j}) < 1 \leq \ell^{\mathfrak{b}}(i,j,k) + 1$, so σ_0 has Property (OR2)$_{i,j,k}$. Moreover, $\phi^{\sigma_0}(d_{i,j,k}, F_{i,j}) = 0 = \lfloor (1-k)/2 \rfloor = \lfloor (\mathfrak{b}+1-k)/2 \rfloor$, hence the premise of Property (OR3)$_{i,j,k}$ is incorrect. Thus, σ_0 has Property (OR3)$_{i,j,k}$. Since it is immediate that σ_0 has Property (OR4)$_{i,j,k}$, the statement follows. □

We now discuss how the main statement is proven in more detail. Consider a canonical strategy $\sigma_{\mathfrak{b}}$ for $\mathfrak{b}$. We prove that applying improving switches according to Zadeh's pivot rule and the tie-breaking rule given in Definition 5.3.5 produces a specific phase-k-strategy for $\mathfrak{b}$ for every $k \in [5]$. These strategies are the first phase-k-strategies for $\mathfrak{b}$ that the algorithm reaches and have several properties that allow us to simplify the proofs. These properties are summarized in Table 5.8. We furthermore explicitly state the improving switches with respect to these "initial phase-k-strategies" in Table 5.9. Both tables distinguish whether the number $\mathfrak{b}+1$ is even or odd and specify whether certain entries are only valid for S_n resp. M_n.

Detailed application of the improving switches

We now discuss the application of the improving switches during the individual phases more formally and link this description to the previously given tables. We illustrate this procedure by providing sketches of the first phase-k-strategies that are calculated by the algorithm in the sink game S_3 when transitioning from σ_3 to σ_4. As 4 is an even number, we have $\nu > 1$ in this example. The canonical strategy σ_3 in the sink game S_3 shown given in Figure 5.10.

By Table 5.9, the set of improving switches is given by all cycle edges $(d_{i,j,k}, F_{i,j})$ with $\sigma_{\mathfrak{b}}(d_{i,j,k}) \neq F_{i,j}$ at the beginning of phase 1. During phase 1, all improving switches $e = (d_{i,j,k}, F_{i,j}) \in I_{\sigma_{\mathfrak{b}}}$ with $\phi^{\sigma_{\mathfrak{b}}}(e) = \lfloor (\mathfrak{b}+1)/2 \rfloor - 1$ are applied first as these minimize the occurrence records, cf. Property (OR4). This might close an inactive cycle center $F_{i,1-\mathfrak{b}_{i+1}}$, making the edge $(g_i, F_{i,1-\mathfrak{b}_{i+1}})$ improving if $\mathfrak{b}_i = 0$ and $\sigma_{\mathfrak{b}}(g_i) \neq F_{i,1-\mathfrak{b}_{i+1}}$. Since

Phase	$\nu = 1$	$\nu > 1$	
		S_n	M_n
1	Canonical strategy for $\mathfrak{b}$ having the canonical properties		
2	-	$\sigma(d_{i,j,k}) \neq F_{i,j} \Rightarrow \phi^{\sigma}(d_{i,j,k}, F_{i,j}) = \lfloor \frac{\mathfrak{b}+1}{2} \rfloor$ $(g_i, F_{i,j}) \in \mathfrak{A}^{\sigma}_{\sigma_{\mathfrak{b}}} \Rightarrow [\mathfrak{b}_i = 0 \wedge \mathfrak{b}_{i+1} \neq j] \vee i = \nu$ and $F_{i,j}$ is closed $\mathfrak{A}^{\sigma}_{\sigma_{\mathfrak{b}}} \subseteq \mathbb{D}^1 \cup \mathbb{G}$ $\sigma(g_\nu) = F_{\nu,\beta^{\sigma}_{\nu+1}}$ and $\sigma(g_i) = F_{i,1-\beta^{\sigma}_{i+1}}$ for all $i < \nu$ $i < \nu \Rightarrow \bar{\sigma}(d_i)$ and Property (USV3)$_i$	
3	$\sigma(d_{i,j,k}) \neq F_{i,j} \Rightarrow \phi^{\sigma}(d_{i,j,k}, F_{i,j}) = \phi^{\sigma_{\mathfrak{b}}}(d_{i,j,k}, F_{i,j}) = \lfloor \frac{\mathfrak{b}+1}{2} \rfloor$ $\sigma(s_{i,*}) = h_{i,*}, \sigma(g_i) = F_{i,1-\beta^{\sigma}_{i+1}}$ and $\bar{\sigma}(d_i)$ for all $i < \nu$ as well as $\sigma(g_\nu) = F_{\nu,\mathfrak{b}_{\nu+1}}$ $\mathfrak{A}^{\sigma^3}_{\sigma_{\mathfrak{b}}} \subseteq \mathbb{D}^1 \cup \mathbb{G} \cup \mathbb{S} \cup \mathbb{B}$ and $(g_i, F_{i,j}) \in \mathfrak{A}^{\sigma}_{\sigma_{\mathfrak{b}}} \Rightarrow [\mathfrak{b}_i = 0 \wedge \mathfrak{b}_{i+1} \neq j] \vee i = \nu$ and $F_{i,j}$ is closed		
4	-	$\mu^{\sigma} = \min\{i : \beta^{\sigma}_i = 0\}$ $\sigma(d_{i,j,*}) = F_{i,j} \Leftrightarrow \beta^{\sigma}_i = 1 \wedge (\mathfrak{b}+1)_{i+1} = j$ $(i,j) \in S_1 \Rightarrow \bar{\sigma}(eb_{i,j}) \wedge \neg\bar{\sigma}(eg_{i,j})$ $(i,j) \in S_2 \Rightarrow \bar{\sigma}(eb_{i,j}) \wedge \bar{\sigma}(eg_{i,j})$ $(g_i, F_{i,j}) \in \mathfrak{A}^{\sigma}_{\sigma_{\mathfrak{b}}} \Rightarrow [\mathfrak{b}_i = 0 \wedge \mathfrak{b}_{i+1} \neq j] \vee i = \nu$ $(g_i, F_{i,j}) \in \mathfrak{A}^{\sigma}_{\sigma_{\mathfrak{b}}} \Rightarrow F_{i,j}$ is closed $\sigma(e_{i,j,k}) = b_2 \Rightarrow \phi^{\sigma}(d_{i,j,k}, F_{i,j}) = \phi^{\sigma_{\mathfrak{b}}}(d_{i,j,k}, F_{i,j})$ $\sigma(e_{i,j,k}) = b_2 \Rightarrow \phi^{\sigma}(d_{i,j,k}, F_{i,j}) = \lfloor \frac{\mathfrak{b}+1}{2} \rfloor$	-
5	$\mu^{\sigma} = \min\{i : \beta^{\sigma}_{i+1} = 0\}$ $\sigma(d_{i,j,k}) = F_{i,j} \Leftrightarrow \beta^{\sigma}_i = 1 \wedge \beta^{\sigma}_{i+1} = j$ $(g_i, F_{i,j}) \in \mathfrak{A}^{\sigma}_{\sigma_{\mathfrak{b}}} \Rightarrow [\mathfrak{b}_i = 0 \wedge \mathfrak{b}_{i+1} \neq j] \vee i = \nu$ $\sigma(e_{i,j,k}) = t^{\rightarrow} \Rightarrow \phi^{\sigma}(d_{i,j,k}, F_{i,j}) = \phi^{\sigma_{\mathfrak{b}}}(d_{i,j,k}, F_{i,j}) = \lfloor \frac{\mathfrak{b}+1}{2} \rfloor$		
	$(i,j) \in S_3 \Rightarrow \bar{\sigma}(eg_{i,j}) \wedge \bar{\sigma}(eb_{i,j})$ $(i,j) \in S_4 \Rightarrow \bar{\sigma}(eg_{i,j}) \wedge \neg\bar{\sigma}(eb_{i,j})$	$(i,j) \in S_1 \Rightarrow \bar{\sigma}(eb_{i,j}) \wedge \neg\bar{\sigma}(eg_{i,j})$ $(i,j) \in S_2 \Rightarrow \bar{\sigma}(eb_{i,j}) \wedge \bar{\sigma}(eg_{i,j})$ $i < \nu \Rightarrow \bar{\sigma}(g_i) = 1 - \beta^{\sigma}_{i+1}$	
1	Canonical strategy for $\mathfrak{b}+1$ having the canonical properties		

Table 5.8.: Properties of specific phase-k-strategies. To simplify notation, let $t^{\rightarrow} := g_1$ if $\nu = 1$ and $t^{\rightarrow} := b_2$ if $\nu > 1$. A '-' signifies that the corresponding combination does not occur during the execution of the algorithm.

Phase	$\nu = 1$	$\nu > 1$	
		S_n	M_n
1	$\mathfrak{D}^\sigma := \{(d_{i,j,k}, F_{i,j})\colon \sigma(d_{i,j,k}) \neq F_{i,j}\}$		
2	-	$\mathfrak{D}^\sigma \cup \{(b_\nu, g_\nu), (s_{\nu-1,1}, h_{\nu-1,1})\}$	
3	$\mathfrak{D}^\sigma \cup \{(b_1, g_1)\} \cup \{(e_{*,*,*}, g_1)\}$	$\mathfrak{D}^\sigma \cup \{(b_1, b_2)\} \cup \{(e_{*,*,*}, b_2)\}$	
4	-	$\mathfrak{E}^\sigma \cup X_0 \cup X_1 \cup \{(s_{\nu-1,0}, b_1)\} \cup \{(s_{i,1}, b_1)\colon i \leq \nu - 2\}$	-
5	$\mathfrak{E}^\sigma \cup \bigcup_{\substack{i=\mu^\sigma+1 \\ \beta_i^\sigma=0}}^{m-1} \{(d_{i,1-\beta_{i+1}^\sigma,*}, F_{i,1-\beta_{i+1}^\sigma})\}$	$\mathfrak{E}^\sigma \cup \bigcup_{i=1}^{\nu-1} \{d_{i,1-\beta_{i+1}^\sigma,*}, F_{i,-1-\beta_{i+1}^\sigma})\} \cup X_0 \cup X_1$	
1	$\mathfrak{D}^\sigma$		

$$X_k := \begin{cases} \emptyset, & \mathfrak{b}+1 \text{ is a power of two,} \\ \{(d_{\nu,1-\beta_\nu^\sigma,k}, F_{\nu,1-\beta_\nu^\sigma})\} \cup \bigcup_{\substack{i=\nu+1 \\ \beta_i=0}}^{m-1} \{(d_{i,1-\beta_{i+1}^\sigma,k}, F_{i,1-\beta_{i+1}^\sigma})\}, & \text{otherwise.} \end{cases}$$

Table 5.9.: The improving switches at the beginning of the different phases. We define $m := \max\{i\colon \sigma(b_i) = g_i\}$ and $\mathfrak{E}^\sigma := \{(d_{i,j,k}, F_{i,j}), (e_{i,j,k}, b_2)\colon \sigma(e_{i,j,k}) = g_1\}$ if $\nu > 1$. Analogously, we let $\mathfrak{E}^\sigma := \{(d_{i,j,k}, F_{i,j}), (e_{i,j,k}, g_1)\colon \sigma(e_{i,j,k}) = b_2\}$ if $\nu = 1$. We do not interpret 1 as a power of two.

the occurrence record of this edge is then smaller than the occurrence record of the corresponding cycle edges by Table 5.6, such a switch is then applied immediately. After all improving switches e with $\phi^{\sigma_\mathfrak{b}}(e) = \lfloor(\mathfrak{b}+1)/2\rfloor - 1$ are applied, the algorithm switches edges $e = (d_{i,j,k}, F_{i,j})$ with $\phi^{\sigma_\mathfrak{b}}(e) = \lfloor(\mathfrak{b}+1)/2\rfloor$ until there are no open cycle centers anymore. This behavior is enforced by the tie-breaking rule which then ensures that the cycle center $F_{\nu,\mathfrak{b}_{\nu+1}}$ is closed next. As before, this might "unlock" the improving switch $(g_\nu, \mathfrak{b}_{\nu+1})$. By Table 5.6, this switch minimizes the occurrence record among all improving switches and is thus applied which concludes phase 1. In any case, phase 2 then begins if $\nu > 1$, and phase 3 begins if $\nu = 1$.

Lemma 5.3.14. *Let $\sigma_\mathfrak{b} \in \rho(\sigma_0)$ be a canonical strategy for $\mathfrak{b} \in \mathfrak{B}_n$ with $\ell(\mathfrak{b}+1) > 1$ having the canonical properties. After applying finitely many improving switches, the strategy improvement algorithm produces a phase-2-strategy $\sigma^{(2)}$ for $\mathfrak{b}$ as described by the corresponding rows of Tables 5.8 and 5.9.*

A schematic example of the phase-2-strategy $\sigma^{(2)}$reached when transitioning from σ_3 to σ_4 in the sink game S_3 is given in Figure 5.11.

We now consider the case $\nu > 1$. Closing the cycle center $F_{\nu,\mathfrak{b}_{\nu+1}}$ at the end of phase 1 changes the induced bit state from $\mathfrak{b}$ to $\mathfrak{b}+1$. This implies that the targets of the entry vertices contained in levels $i \leq \nu$ and of all upper selection vertices contained in levels $i < \nu$ need to be changed accordingly. This is reflected by the set of improving switches containing the edges (b_ν, g_ν) and $(s_{\nu-1,1}, h_{\nu-1,1})$. Moreover, cycle edges that were improving for $\sigma_\mathfrak{b}$ but were not applied during phase 1 remain improving, see Table 5.9.

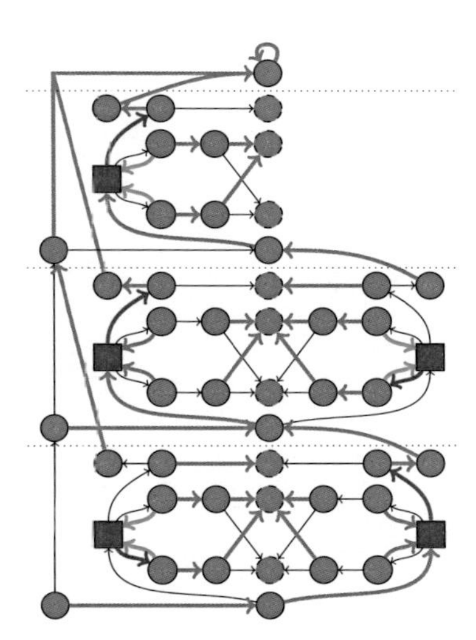

Figure 5.10.: The canonical strategy σ_3 in the sink game S_3. Blue edges represent choices of player 0, red choices represent choices of player 1, and green edges represent improving switches. For simplification, we omit the labels here and refer to Figure 5.8 instead.

These switches were not applied as they have an occurrence record of $\lfloor(\mathfrak{b}+1)/2\rfloor$, and this large occurrence record guarantees that the algorithm will not apply these switches. The algorithm thus applies the improving switches involving the entry vertices b_2 through b_ν as well as the improving switches $(s_{i,(\mathfrak{b}+1)_{i+1}}, h_{i,(\mathfrak{b}+1)_{i+1}})$ next. These switches are applied until (b_1, b_2) becomes improving. This switch is however not applied yet, and the algorithm reaches phase 3. Since none of these switches needs to be applied if $\nu = 1$, the algorithm directly produces a phase-3-strategy after phase 1 if $\nu = 1$.

Lemma 5.3.15. *Let $\sigma_{\mathfrak{b}} \in \rho(\sigma_0)$ be a canonical strategy for $\mathfrak{b} \in \mathfrak{B}_n$ having the canonical properties. After applying finitely many improving switches, the strategy improvement algorithm produces a phase-3-strategy $\sigma^{(3)}$ for $\mathfrak{b}$ as described by the corresponding rows of Tables 5.8 and 5.9.*

A schematic example of the phase-3-strategy $\sigma^{(3)}$ that is reached when transitioning from σ_3 to σ_4 in the sink game S_3 is given in Figure 5.12. Note that this is an example for the case $\nu > 1$ as 4 is an even number.

When phase 3 begins, all edges $(e_{*,*,*}, g_1)$ resp. $(e_{*,*,*}, b_2)$ become improving, depending on ν. The reason is that closing $F_{\nu,\mathfrak{b}_{\nu+1}}$ resp. closing this cycle center and updating the spinal path in the levels 1 to ν significantly increases the valuation of g_1 resp. b_2. Since the improving switches of the form $(d_{*,*,*}, F_{*,*})$ still maximize the occurrence records, the switches involving the escape vertices are applied next. As all of them have the same occurrence record, all improving switches $(e_{i,j,k}, *)$ with $\sigma^{(3)}(d_{i,j,k}) = F_{i,j}$ are applied due to the tie-breaking rule. If the corresponding cycle center is not closed and active, then

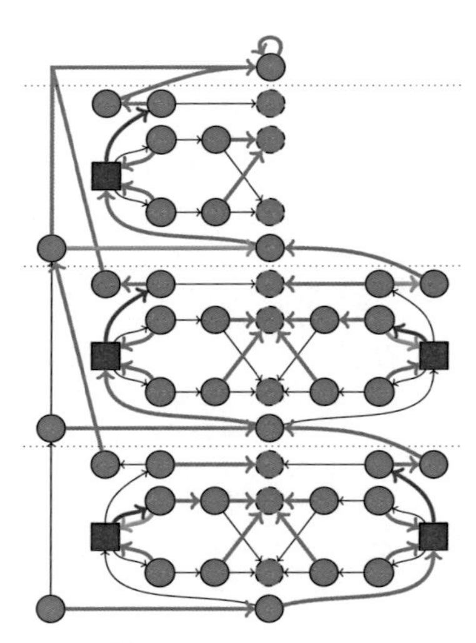

Figure 5.11.: The phase-2-strategy $\sigma^{(2)}$ calculated when transitioning from σ_3 to σ_4 in the sink game S_3. Blue edges represent choices of player 0, red choices represent choices of player 1, and green edges represent improving switches. For simplification, we omit the labels here and refer to Figure 5.8 instead.

this application unlocks the improving switch $(d_{i,j,k}, e_{i,j,k})$ as this edge allows the cycle vertex to gain access to the very profitable spinal path.

At this point, there is a major difference between the behavior of S_n and M_n. In S_n, the application of a switch $(d_{i,j,k}, e_{i,j,k})$ does not change the valuation of its cycle center $F_{i,j}$. The reason is that player 1 controls the cycle center and can then react by choosing vertex $d_{i,j,1-k}$, yielding the same valuation as before. This is also true if the cycle center was closed, since player 1 chooses the upper selection vertices in both cases. In M_n, the application of the switch $(d_{i,j,k}, e_{i,j,k})$ however has an immediate consequence regarding the valuation of $F_{i,j}$. As the valuation of the cycle center is (roughly) the arithmetic mean of the valuation of its cycle vertices, the increase of the valuations of $d_{i,j,k}$ also increases the valuation of $F_{i,j}$. This then makes the cycle center $F_{i,j}$ profitable since it grants access to the spinal path. Most importantly, it enables the upper selection vertex $s_{i,j}$ to use this access by switching to b_1 as the path starting in b_1 then leads to $F_{i,j}$. The reason is that the exact way that improving switches are applied in this phase since the tie-breaking rule dictates that switches of higher levels are applied prior to switches of lower levels.

In summary, all switches $(e_{i,j,k}, *)$ with $\sigma^{(3)}(d_{i,j,k}) = F_{i,j}$ are applied during phase 3. If $F_{i,j}$ is not closed and active, this makes $(d_{i,j,k}, e_{i,j,k})$ improving, and this switch is applied next. In M_n, this might also make switches $(s_{i,j}, b_1)$ improving if $i < \nu$ and $j = 1 - (\mathfrak{b} + 1)_{i+1}$, and this switch is then also applied immediately. Phase 3 ends with the application of (b_1, g_1) if $\nu = 1$ and (b_1, b_2) if $\nu > 1$. Depending on whether we consider M_n or S_n and depending on ν, we then either obtain a phase-4-strategy or a phase-5-strategy.

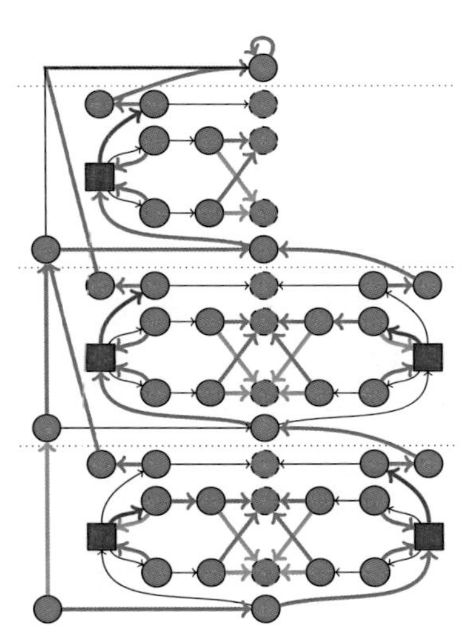

Figure 5.12.: The phase-3-strategy $\sigma^{(3)}$ calculated when transitioning from σ_3 to σ_4 in the sink game S_3. Blue edges represent choices of player 0, red choices represent choices of player 1, and green edges represent improving switches. For simplification, we omit the labels here and refer to Figure 5.8 instead.

Lemma 5.3.16. *Let $\sigma_{\mathfrak{b}} \in \rho(\sigma_0)$ be a canonical strategy for $\mathfrak{b} \in \mathfrak{B}_n$ having the canonical properties. After applying finitely many improving switches, the strategy improvement algorithm produces a strategy σ with the following properties. If $\nu > 1$, then σ is a phase-4-strategy for $\mathfrak{b}$ in S_n and a phase-5-strategy for $\mathfrak{b}$ in M_n. If $\nu = 1$, then σ is a phase-5-strategy for $\mathfrak{b}$. In any case, σ is described by the corresponding rows of Tables 5.8 and 5.9.*

In the sink game S_3 that we consider as an example in this section, the algorithm thus produces a phase-4-strategy when transitioning from σ_3 to σ_4 which is visualized in Figure 5.13.

Consider the case that there is a phase 4. This only happens in S_n and if $\nu > 1$. During this phase, the improving switches $(s_{i,j}, b_1)$ with $i < \nu$ and $j = 1 - (\mathfrak{b}+1)_{i+1}$, which were already applied during phase 3 in M_n, are applied. The reason that these switches only become improving now in S_n is that the application of (b_1, b_2) at the end of phase 3 significantly increases the valuation of b_1 as this vertex then enables accessing the spinal path. The switches are then applied from higher levels to lower levels, so the final improving switch applied in this phase is $(s_{1,1-(\mathfrak{b}+1)_2}, b_1)$, resulting in a phase-5-strategy. Most importantly, the algorithm produces a phase-5-strategy in any case.

Lemma 5.3.17. *Let $\sigma_{\mathfrak{b}} \in \rho(\sigma_0)$ be a canonical strategy for $\mathfrak{b} \in \mathfrak{B}_n$ having the canonical properties. After applying finitely many improving switches, the strategy improvement algorithm produces a phase-5-strategy $\sigma^{(5)}$ for $\mathfrak{b}$ as described by the corresponding rows of Tables 5.8 and 5.9.*

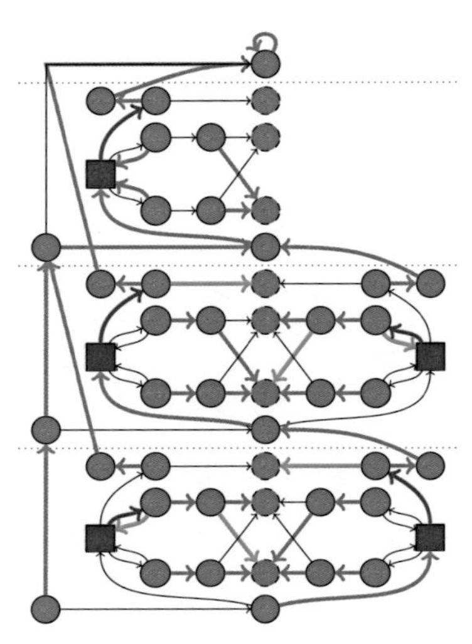

Figure 5.13.: The phase-4-strategy $\sigma^{(4)}$ calculated when transitioning from σ_3 to σ_4 in the sink game S_3. Blue edges represent choices of player 0, red choices represent choices of player 1, and green edges represent improving switches. For simplification, we omit the labels here and refer to Figure 5.8 instead.

A schematic example of the phase-5-strategy $\sigma^{(5)}$ that is reached when transitioning from σ_3 to σ_4 in the sink game S_3 is given in Figure 5.14.

We now discuss phase 5. During this phase, the remaining improving switches of the type $(e_{*,*,*}, *)$ are applied. Applying such a switch then forces every cycle center to point towards the spinal path. But this implies that the valuation of a cycle center$F_{i,j}$ with $i \in [n], j \in \{0,1\}$ increases significantly, making the corresponding cycle edges $(d_{i,j,*}, F_{i,j})$ improving again. Several of these cycle edges may now have very low occurrence records as their cycle center was closed for a large number of iterations, and are thus applied immediately after being unlocked. Similarly to phase 1, this can then make the edge $(g_i, F_{i,j})$ improving, and this edges is then applied immediately if it becomes improving. After all switches $(e_{*,*,*}, *)$ as well as possible switches $(d_{*,*,*}, F_{*,*})$ with low occurrence records and corresponding switches $(g_*, F_{*,*})$ are applied, this yields a canonical strategy for $\mathfrak{b}+1$. Then, phase 1 of the next transition begins.

All of this is formalized by the following two statements. Note that we use the expression of the "next feasible row" as certain phases may not be present in certain cases. Thus, "the next row" may not always be accurate.

Lemma 5.3.18. *Let $\sigma_{\mathfrak{b}} \in \rho(\sigma_0)$ be a canonical strategy for $\mathfrak{b} \in \mathfrak{B}_n$ having the canonical properties. Let σ be a strategy obtained by applying a sequence of improving switches to $\sigma_{\mathfrak{b}}$. Let σ and I_σ have the properties of row k of Table 5.8 and 5.9 for some $k \in [5]$. Then, applying improving switches according to Zadeh's pivot rule and the tie-breaking rule of*

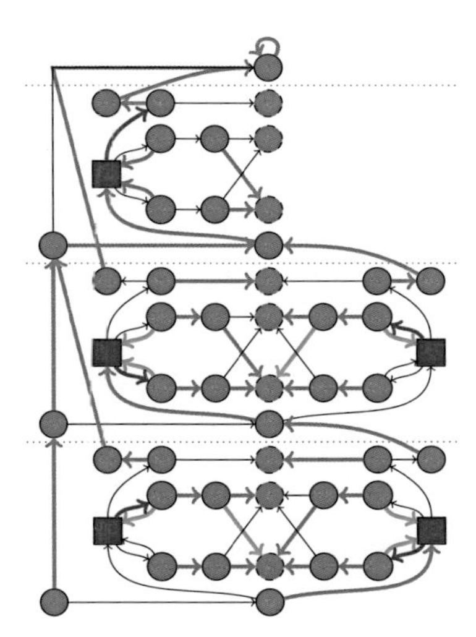

Figure 5.14.: The phase-5-strategy $\sigma^{(5)}$ calculated when transitioning from σ_3 to σ_4 in the sink game S_3. Blue edges represent choices of player 0, red choices represent choices of player 1, and green edges represent improving switches. For simplification, we omit the labels here and refer to Figure 5.8 instead.

Definition 5.3.5 produces a strategy σ' that is described by the next feasible rows of Tables 5.8 and 5.9.

This lemma then enables us to prove the two main theorems. The first theorem states that we reach the canonical strategy $\sigma_{\mathfrak{b}+1}$ having the canonical properties provided that we start with a canonical strategy $\sigma_{\mathfrak{b}}$ also having these properties.

Theorem 5.3.19. *Let $\sigma_{\mathfrak{b}} \in \rho(\sigma_0)$ be a canonical strategy for $\mathfrak{b} \in \mathfrak{B}_n$ having the canonical properties. After applying finitely many improving switches according to Zadeh's pivot rule and the tie-breaking rule of Definition 5.3.5, the strategy improvement algorithm calculates a strategy $\sigma_{\mathfrak{b}+1}$ with the following properties.*

1. $I_{\sigma_{\mathfrak{b}+1}} = \{(d_{i,j,k}, F_{i,j}) : \sigma_{\mathfrak{b}+1}(d_{i,j,k}) \neq F_{i,j}\}$.
2. *The occurrence records are described by Table 5.6 when interpreted for $\mathfrak{b}+1$.*
3. *$\sigma_{\mathfrak{b}+1}$ is a canonical strategy for $\mathfrak{b}+1$ and has Properties (OR1)$_{*,*,*}$ to (OR4)$_{*,*,*}$.*
4. *When transitioning from $\sigma_{\mathfrak{b}}$ to $\sigma_{\mathfrak{b}+1}$, every improving switch is applied at most once.*

In particular, $\sigma_{\mathfrak{b}+1}$ has the canonical properties.

A schematic example of the canonical strategy σ_4 that is reached when transitioning from σ_3 to σ_4 in the sink game S_3 is given in Figure 5.15.

The final theorem now states the core result of this thesis. It states that applying a variety of algorithms using Zadeh's pivot rule can require an exponential number of iterations in the worst case.

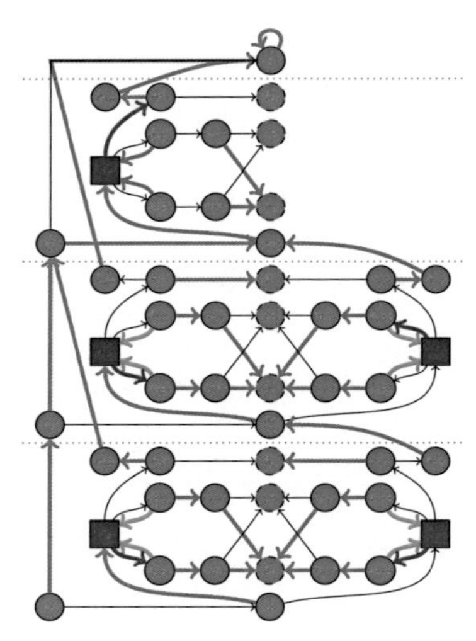

Figure 5.15.: The canonical strategy σ_4 calculated when transitioning from σ_3 to σ_4 in the sink game S_3. Blue edges represent choices of player 0, red choices represent choices of player 1, and green edges represent improving switches. For simplification, we omit the labels here and refer to Figure 5.8 instead.

Theorem 5.3.20. *Using Zadeh's pivot rule and the tie-breaking rule of Definition 5.3.5 when applying*

1. *the strategy improvement algorithm of [VJ00] to S_n,*
2. *the policy iteration algorithm of [How60] to M_n,*
3. *the simplex algorithm of [Dan51] to the linear program induced by M_n*

to the game G_n or the induced linear program requires at least 2^n iterations for finding the optimal strategy resp. solution when using σ_0 as initial strategy.

This concludes our informal proof of the exponential lower bound. The following chapter is now dedicated to give a rigorous formal treatment.

6. Technical Details of the Exponential Lower Bound Construction

This chapter contains all the technical and formal details required for properly proving the statements of Chapter 5. It is organized in three main parts. In Section 6.1 we characterize the valuations of the vertices in S_n resp. M_n as these are crucial for determining the improving switches. In Section 6.2, we then use these characterizations to prove how specific strategies behave and change when individual improving switches are applied. Finally, in Section 6.3 we then apply the corresponding statements to prove that the description we gave in Chapter 5 is correct. As most of the proofs are very technical, they are deferred to Appendix A.2.

6.1. Vertex Valuations and Well-Behaved Strategies

We begin by developing characterizations of the vertex valuations. This requires us to analyze the strategies calculated by the strategy improvement algorithm in more detail. As it will turn out, all strategies that the algorithm produces have a certain set of properties. The strategies are thus "well-behaved", and the properties represent the way the algorithm interacts with the instance. These properties drastically simplify the proofs but it is tedious to prove that every strategy is well-behaved. We thus prove that (i) the initial strategy is well-behaved and (ii) whenever the algorithm applies an improving switch to a well-behaved strategy, the resulting strategy is well-behaved. In particular, these two statements imply that any strategy calculated by the algorithm is well-behaved. This concept was not mentioned previously since it is solely used for proving our results and most of them have no clear intuitive explanation. We explicitly encoded the properties defining well-behaved properties in the implementation of the sink game provided by Oliver Friedmann, verifying that the produced strategies are indeed all well-behaved.

The properties are summarized in Table 6.1. We briefly discuss the properties next and introduce an additional set of parameters and additional notation. The parameters are abbreviations that denote the first level in which $\bar{\sigma}(x_*)$ is either true or false for $x_* \in \{b_*, s_*, g_*\}$. More precisely, we define $m_x^\sigma := \min(\{i \in [n] : \bar{\sigma}(x_i)\} \cup \{n+1\})$ as well as $\overline{m}_x^\sigma := \min(\{i \in [n] : \neg\bar{\sigma}(x_i)\} \cup \{n+1\})$ where $x \in \{b, s, g\}$. Furthermore, for a level $i \in [n]$ and a strategy σ for G_n, we refer to the cycle center $F_{i,\bar{\sigma}(g_i)}$, that is, the cycle center chosen by the selector vertex g_i, as the *chosen cycle center of level* i. Note that the chosen cycle center and the active cycle center of level i do not necessarily coincide.

Properties of well-behaved strategies

We now introduce the properties that all strategies produced by the algorithm have. The abbreviations used to refer to the properties are typically related to the vertices that they are related to, and similar or closely related properties have similar abbreviations. Note that several of the properties are in fact implications. For these properties, we thus demand that the full implication is true. In all of the following, let $i \in [n], j, k \in \{0,1\}$ be suitable indices.

Let σ be a strategy for G_n. Consider a level $i \geq \mu^\sigma$ such that $\sigma(b_i) = g_i$. Then, a well-behaved strategy has $\sigma(s_{i,j}) = h_{i,j}$ where $j = \bar{\sigma}(g_i)$. Intuitively, this states that in levels above μ^σ that represent a bit equal to one, the upper selection vertex is set correctly. Formally,

$$i \geq \mu^\sigma \wedge \sigma(b_i) = g_i \implies \sigma(s_{i,\bar{\sigma}(g_i)}) = h_{i,\bar{\sigma}(g_i)}. \tag{S1}$$

Let $i < \mu^\sigma$. Assume that either $\sigma(b_2) = g_2$ and $i > 1$ or that the chosen cycle center of level i is closed. Then, we demand that the upper selection vertex $s_{i,\bar{\sigma}(g_i)}$ points to $h_{i,\bar{\sigma}(g_i)}$. Formally,

$$i < \mu^\sigma \wedge ((\sigma(b_2) = g_2 \wedge i > 1) \vee \bar{\sigma}(d_i)) \implies \bar{\sigma}(s_i). \tag{S2}$$

Let $i < \mu^\sigma - 1$ and $\sigma(b_i) = \sigma(b_{i+1})$. Then, we demand that $\sigma(b_{i+1}) = b_{i+2}$. Intuitively, this encodes that entry vertices below level $\mu^\sigma - 1$ are reset "from top to bottom". Formally,

$$i < \mu^\sigma - 1 \wedge \sigma(b_i) = b_{i+1} \implies \sigma(b_{i+1}) = b_{i+2}. \tag{B1}$$

Assume that $\mu^\sigma \neq 1$ and that the entry vertex of level $\mu^\sigma - 1$ points towards the next entry vertex. Then, we demand that the entry vertex of level μ^σ points towards its selector vertex. Formally,

$$\mu^\sigma \neq 1 \wedge \sigma(b_{\mu^\sigma - 1}) = b_{\mu^\sigma} \implies \sigma(b_{\mu^\sigma}) = g_{\mu^\sigma}. \tag{B2}$$

Consider a level i such that $\sigma(s_{i,1}) = h_{i,1}$. Further assume $\sigma(b_{i+1}) = b_{i+2}$. Then the values of $\bar{\sigma}(g_{i+1})$ and $\bar{\sigma}(b_{i+2})$ do not coincide for well-behaved strategies. Formally,

$$\sigma(s_{i,1}) = h_{i,1} \wedge \sigma(b_{i+1}) = b_{i+2} \implies \bar{\sigma}(g_{i+1}) \neq \bar{\sigma}(b_{i+2}). \tag{B3}$$

Consider some level $i < \mu^\sigma$. For well-behaved strategies, $F_{i,1}$ is the chosen cycle center of level i if and only if $i \neq \mu^\sigma - 1$. This encodes the statuses of the selector vertices of levels below μ^σ. Formally,

$$i < \mu^\sigma \implies [\sigma(g_i) = F_{i,1} \iff i \neq \mu^\sigma - 1]. \tag{BR1}$$

We demand that the chosen cycle center of a level $i < \mu^\sigma$ does not escape towards g_1. Formally,

$$i < \mu^\sigma \implies \neg\bar{\sigma}(eg_{i,\bar{\sigma}(g_i)}). \tag{BR2}$$

Consider a level i such that $\sigma(b_i) = g_i$. Let either $i > 1, \mu^\sigma = 1$ or let $\sigma(b_2) = g_2$ be equivalent to $\mu^\sigma > 2$. In any of these cases, this implies that the cycle center $F_{i,\bar{\sigma}(g_i)}$ is closed for well-behaved strategies. Intuitively, this gives a list of situations in which the active cycle center of a level corresponding to a bit that should be equal to 1 is already closed. Formally,

$$\sigma(b_i) = g_i \wedge (i > 1 \vee \mu^\sigma = 1 \vee (\sigma(b_2) = g_2 \iff \mu^\sigma > 2)) \implies \bar{\sigma}(d_i). \tag{D1}$$

Let $\sigma(b_2) = g_2$ and $i \in \{2, \dots, \mu^\sigma - 1\}$. Then, we demand that the chosen cycle center of level i is closed. Formally,

$$\sigma(b_2) = g_2 \wedge (2 \leq i < \mu^\sigma) \implies \bar{\sigma}(d_i). \tag{D2}$$

Let $\mu^\sigma = 1, m_b^\sigma \leq \overline{m}_s^\sigma, \overline{m}_g^\sigma$ and $G_n = S_n$. Then, we demand that the chosen cycle center of level 1 does not escape towards b_2. Formally,

$$\mu^\sigma = 1 \wedge m_b^\sigma \leq \overline{m}_s^\sigma, \overline{m}_g^\sigma \wedge G_n = S_n \implies \neg\bar{\sigma}(eb_1). \tag{MNS1}$$

Let $\mu^\sigma = 1$ and consider a level i such that $i < \overline{m}_g^\sigma < \overline{m}_s^\sigma, m_b^\sigma$. Further assume that $G_n = S_n$ implies $\neg\bar{\sigma}(b_{\overline{m}_g^\sigma+1})$. For a well-behaved strategy, the chosen cycle center of level i does not escape towards b_2. Formally,

$$\mu^\sigma = 1 \wedge i < \overline{m}_g^\sigma < \overline{m}_s^\sigma, m_b^\sigma \wedge [G_n = S_n \implies \neg\bar{\sigma}(b_{\overline{m}_g^\sigma+1})] \implies \neg\bar{\sigma}(eb_i). \tag{MNS2}$$

Let $\mu^\sigma = 1$ as well as $i < \overline{m}_s^\sigma \leq \overline{m}_g^\sigma < m_b^\sigma$ and $G_n = M_n$. Then, we demand that the chosen cycle center of level i is closed. Formally,

$$\mu^\sigma = 1 \wedge i < \overline{m}_s^\sigma \leq \overline{m}_g^\sigma < m_b^\sigma \wedge G_n = M_n \implies \bar{\sigma}(d_i). \tag{MNS3}$$

Assume $\mu^\sigma = 1$ as well as $\overline{m}_s^\sigma \leq \overline{m}_g^\sigma < m_b^\sigma$. Then, we demand that the chosen cycle center of level $\overline{m}_g^\sigma$ escapes to b_2 but not to g_1. Formally,

$$\mu^\sigma = 1 \wedge \overline{m}_s^\sigma \leq \overline{m}_g^\sigma < m_b^\sigma \implies \bar{\sigma}(eb_{\overline{m}_s^\sigma}) \wedge \neg\bar{\sigma}(eg_{\overline{m}_s^\sigma}). \tag{MNS4}$$

Assume $\mu^\sigma = 1$ as well as $i < \overline{m}_s^\sigma < m_b^\sigma \leq \overline{m}_g^\sigma$ and $G_n = M_n$. Then, we demand that the chosen cycle center of level i is closed. Formally,

$$\mu^\sigma = 1 \wedge i < \overline{m}_s^\sigma < m_b^\sigma \leq \overline{m}_g^\sigma \wedge G_n = M_n \implies \bar{\sigma}(d_i). \tag{MNS5}$$

Assume $\mu^\sigma = 1$ as well as $\overline{m}_s^\sigma < m_b^\sigma \leq \overline{m}_g^\sigma$. Then, we demand that the chosen cycle center of level $\overline{m}_s^\sigma$ escapes to b_2 but not to g_1. Formally,

$$\mu^\sigma = 1 \wedge \overline{m}_s^\sigma < m_b^\sigma \leq \overline{m}_g^\sigma \implies \bar{\sigma}(eb_{\overline{m}_s^\sigma}) \wedge \neg\bar{\sigma}(eg_{\overline{m}_s^\sigma}). \tag{MNS6}$$

Consider a cycle center $F_{i,j}$ escaping only to g_1. Let $\mu^\sigma = 1$. Then, we demand that the upper selection vertex corresponding to the chosen cycle center of level i escapes to b_1. Formally,

$$\bar{\sigma}(eg_{i,j}) \wedge \neg\bar{\sigma}(eb_{i,j}) \wedge \mu^\sigma = 1 \implies \neg\bar{\sigma}(s_{i,j}). \tag{EG1}$$

Consider a cycle center $F_{i,j}$ escaping only to g_1. Let $\mu^\sigma = 1$. Then, we demand that the chosen cycle center of level 1 is closed. Formally,

$$\bar{\sigma}(eg_{i,j}) \wedge \neg\bar{\sigma}(eb_{i,j}) \wedge \mu^\sigma = 1 \implies \bar{\sigma}(d_1). \tag{EG2}$$

Consider a cycle center $F_{i,j}$ escaping only to g_1. Then, we demand that the upper selection vertex corresponding to the chosen cycle center of level 1 points towards the next level. Formally,

$$\bar{\sigma}(eg_{i,j}) \wedge \neg\bar{\sigma}(eb_{i,j}) \implies \bar{\sigma}(s_1). \tag{EG3}$$

Consider a cycle center $F_{i,j}$ escaping only to g_1. Let $\mu^\sigma = 1$. We demand that this implies $\bar{\sigma}(g_1) = \bar{\sigma}(b_2)$ for well-behaved strategies. Formally,

$$\bar{\sigma}(eg_{i,j}) \wedge \neg\bar{\sigma}(eb_{i,j}) \wedge \mu^\sigma = 1 \implies \bar{\sigma}(g_1) = \bar{\sigma}(b_2). \tag{EG4}$$

Consider a cycle center $F_{i,j}$ escaping only to g_1. Let $\mu^\sigma \neq 1$ and assume that the upper selection vertex of $F_{i,j}$ escapes towards b_1. Then, we demand that $\bar{\sigma}(b_{i+1}) = j$, i.e., the entry vertex of the next level is set correctly with respect to level i. Formally,

$$\bar{\sigma}(eg_{i,j}) \wedge \neg\bar{\sigma}(eb_{i,j}) \wedge \mu^\sigma \neq 1 \wedge \bar{\sigma}(s_{i,j}) \implies \bar{\sigma}(b_{i+1}) = j. \tag{EG5}$$

Consider a cycle center $F_{i,j}$ escaping only to b_2. Let $\sigma(b_1) = b_2$. Then $\bar{\sigma}(b_{i+1}) \neq j$. Formally,

$$\bar{\sigma}(eb_{i,j}) \wedge \neg\bar{\sigma}(eg_{i,j}) \wedge \sigma(b_1) = g_1 \implies \bar{\sigma}(b_{i+1}) \neq j. \tag{EB1}$$

Consider a cycle center $F_{i,0}$ escaping only to b_2. Let $\sigma(s_{i,0}) = h_{i,0}$ and $\sigma(b_1) = g_1$. Then, we demand that $\mu^\sigma = i + 1$. Formally,

$$\bar{\sigma}(eb_{i,0}) \wedge \neg\bar{\sigma}(eg_{i,0}) \wedge \sigma(b_1) = g_1 \wedge \sigma(s_{i,0}) = h_{i,0} \implies \mu^\sigma = i + 1. \tag{EB2}$$

Consider a cycle center $F_{i,j}$ escaping only to b_2. Let $\sigma(s_{i,j}) = h_{i,j}, \sigma(b_1) = g_1$ and $i > 1$. We then demand that the entry vertex of level 2 does not grant access to this level. Formally,

$$\bar{\sigma}(eb_{i,j}) \wedge \neg\bar{\sigma}(eg_{i,j}) \wedge \bar{\sigma}(s_{i,j}) = h_{i,j} \wedge i > 1 \wedge \sigma(b_1) = g_1 \implies \sigma(b_2) = b_3. \tag{EB3}$$

Consider a cycle center $F_{i,1}$ escaping only to b_2. Let $\sigma(s_{i,1}) = h_{i,1}$ and $\sigma(b_1) = g_1$. Then, we demand that $\mu^\sigma > i + 1$. Formally,

$$\bar{\sigma}(eb_{i,1}) \wedge \neg\bar{\sigma}(eg_{i,1}) \wedge \sigma(s_{i,1}) = h_{i,1} \wedge \sigma(b_1) = g_1 \implies \mu^\sigma > i + 1. \tag{EB4}$$

Consider a cycle center $F_{i,j}$ escaping only to b_2. Let $\sigma(b_1) = g_1$. Then, we demand that the entry vertex of level μ^σ grants access to this level. Formally,

$$\bar{\sigma}(eb_{i,j}) \wedge \neg\bar{\sigma}(eg_{i,j}) \wedge \sigma(b_1) = g_1 \implies \bar{\sigma}(b_{\mu^\sigma}) = g_{\mu^\sigma}. \tag{EB5}$$

Consider a cycle center $F_{i,j}$ escaping only to b_2. Let $\mu^\sigma > 2$. Then, we demand that the entry vertex of level 2 does not grant access to this level. Formally,

$$\bar{\sigma}(eb_{i,j}) \wedge \neg\bar{\sigma}(eg_{i,j}) \wedge \mu^\sigma > 2 \implies \sigma(b_2) = b_3. \tag{EB6}$$

Consider a cycle center $F_{i,j}$ that can escape towards both g_1 and b_2. Further assume that $\sigma(s_{i,j}) = h_{i,j}$. Then, we demand that $\sigma(b_{i+1}) = j$, so the entry vertex of level $i+1$ is set in accordance with the upper selection vertex of level i. Formally,

$$\bar{\sigma}(eb_{i,j}) \wedge \bar{\sigma}(eg_{i,j}) \wedge \sigma(s_{i,j}) = h_{i,j} \implies \bar{\sigma}(b_{i+1}) = j. \tag{EBG1}$$

Consider a cycle center $F_{i,j}$ that can escape towards both g_1 and b_2. Further assume that $\bar{\sigma}(g_1) = \bar{\sigma}(b_2)$. Then, we demand that the upper selection vertex corresponding to the chosen cycle center of level 1 points towards $h_{1,\bar{\sigma}(g_1)}$. Formally,

$$\bar{\sigma}(eb_{i,j}) \wedge \bar{\sigma}(eg_{i,j}) \wedge \bar{\sigma}(g_1) = \bar{\sigma}(b_2) \implies \bar{\sigma}(s_1). \tag{EBG2}$$

Consider a cycle center $F_{i,j}$ that can escape towards both g_1 and b_2. Further assume that $\bar{\sigma}(g_1) = \bar{\sigma}(b_2)$. Then the chosen cycle center of level 1 has to be closed for well-behaved strategies. Formally,

$$\bar{\sigma}(eb_{i,j}) \wedge \bar{\sigma}(eg_{i,j}) \wedge \bar{\sigma}(g_1) = \bar{\sigma}(b_2) \implies \bar{\sigma}(d_1). \tag{EBG3}$$

Consider a cycle center $F_{i,j}$ that can escape towards both b_2 and g_1. Let $F_{1,0}$ be the chosen cycle center of level 1 and $\sigma(b_2) = g_2$. Then, we demand that $\mu^\sigma \leq 2$. Formally,

$$\bar{\sigma}(eb_{i,j}) \wedge \bar{\sigma}(eg_{i,j}) \wedge \sigma(g_1) = F_{1,0} \wedge \sigma(b_2) = g_2 \implies \mu^\sigma \leq 2. \tag{EBG4}$$

Consider a cycle center $F_{i,j}$ that can escape towards both b_2 and g_1. Further assume that $\bar{\sigma}(g_1) \neq \bar{\sigma}(b_2)$. Then, we demand that $\mu^\sigma \neq 2$. Formally,

$$\bar{\sigma}(eb_{i,j}) \wedge \bar{\sigma}(eg_{i,j}) \wedge \bar{\sigma}(g_1) \wedge \neg\bar{\sigma}(b_2) \implies \mu^\sigma \neq 2. \tag{EBG5}$$

If the cycle center of level n is closed, then $\sigma(b_n) = g_n$ or $\sigma(b_1) = g_1$ has to hold for wee-behaved strategies. Formally,

$$\bar{\sigma}(d_n) \implies \bar{\sigma}(b_n) \vee \bar{\sigma}(b_1). \tag{DN1}$$

Let the cycle center of level n is closed or let $F_{i,1}$ be the chosen cycle center for all $i \in [n-1]$. Then, we demand that there is some level $i \in [n]$ such that $\sigma(b_i) = g_i$. Formally,

$$\bar{\sigma}(d_n) \vee \overline{m}_g^\sigma = n \implies \exists i \in [n] \colon \bar{\sigma}(b_i). \tag{DN2}$$

As mentioned previously, the abbreviations of the properties summarized in Table 6.1 are chosen according to configurations of G_n or individual vertices. An explanation of these names is given in Table 6.2.

Definition 6.1.1 (Well-behaved strategy). A strategy σ for G_n is *well-behaved* if it has all properties of Table 6.1.

	Premise	Conclusion
(S1)	$i \geq \mu^\sigma \wedge \sigma(b_i) = g_i$	$\bar{\sigma}(s_i)$
(S2)	$i < \mu^\sigma \wedge ((\sigma(b_2) = g_2 \wedge i > 1) \vee \bar{\sigma}(d_i) \vee \sigma(b_1) = b_2)$	$\bar{\sigma}(s_i)$
(B1)	$i < \mu^\sigma - 1 \wedge \sigma(b_i) = b_{i+1}$	$\sigma(b_{i+1}) = b_{i+2}$
(B2)	$\mu^\sigma \neq 1 \wedge \sigma(b_{\mu^\sigma - 1}) = b_{\mu^\sigma}$	$\sigma(b_{\mu^\sigma}) = g_{\mu^\sigma}$
(B3)	$\sigma(s_{i,1}) = h_{i,1} \wedge \sigma(b_{i+1}) = b_{i+2}$	$\bar{\sigma}(g_{i+1}) \neq \bar{\sigma}(b_{i+2})$
(BR1)	$i < \mu^\sigma$	$\sigma(g_i) = F_{i,1} \iff i \neq \mu^\sigma - 1$
(BR2)	$i < \mu^\sigma$	$\neg\bar{\sigma}(eg_{i,\bar{\sigma}(g_i)})$
(D1)	$\sigma(b_i) = g_i \wedge (i > 1 \vee \mu^\sigma = 1 \vee (\sigma(b_2) = g_2 \iff \mu^\sigma > 2))$	$\bar{\sigma}(d_i)$
(D2)	$\sigma(b_2) = g_2 \wedge (2 \leq i < \mu^\sigma)$	$\bar{\sigma}(d_i)$
(MNS1)	$\mu^\sigma = 1 \wedge m_b^\sigma \leq \overline{m}_s^\sigma, \overline{m}_g^\sigma \wedge G_n = S_n$	$\neg\bar{\sigma}(eb_1)$
(MNS2)	$\mu^\sigma = 1 \wedge i < \overline{m}_g^\sigma < \overline{m}_s^\sigma, m_b^\sigma \wedge [G_n = S_n \implies \neg\bar{\sigma}(b_{\overline{m}_g^\sigma + 1})]$	$\neg\bar{\sigma}(eb_i)$
(MNS3)	$\mu^\sigma = 1 \wedge i < \overline{m}_s^\sigma \leq \overline{m}_g^\sigma < m_b^\sigma \wedge G_n = M_n$	$\bar{\sigma}(d_i)$
(MNS4)	$\mu^\sigma = 1 \wedge \overline{m}_s^\sigma \leq \overline{m}_g^\sigma < m_b^\sigma$	$\bar{\sigma}(eb_{\overline{m}_s^\sigma}) \wedge \neg\bar{\sigma}(eg_{\overline{m}_s^\sigma})$
(MNS5)	$\mu^\sigma = 1 \wedge i < \overline{m}_s^\sigma < m_b^\sigma \leq \overline{m}_g^\sigma \wedge G_n = M_n$	$\bar{\sigma}(d_i)$
(MNS6)	$\mu^\sigma = 1 \wedge \overline{m}_s^\sigma < m_b^\sigma \leq \overline{m}_g^\sigma$	$\bar{\sigma}(eb_{\overline{m}_s^\sigma}) \wedge \neg\bar{\sigma}(eg_{\overline{m}_s^\sigma})$
(EG1)	$\bar{\sigma}(eg_{i,j}) \wedge \neg\bar{\sigma}(eb_{i,j}) \wedge \mu^\sigma = 1$	$\neg\bar{\sigma}(s_{i,j})$
(EG2)	$\bar{\sigma}(eg_{i,j}) \wedge \neg\bar{\sigma}(eb_{i,j}) \wedge \mu^\sigma = 1$	$\bar{\sigma}(d_1)$
(EG3)	$\bar{\sigma}(eg_{i,j}) \wedge \neg\bar{\sigma}(eb_{i,j})$	$\bar{\sigma}(s_1)$
(EG4)	$\bar{\sigma}(eg_{i,j}) \wedge \neg\bar{\sigma}(eb_{i,j}) \wedge \mu^\sigma = 1$	$\bar{\sigma}(g_1) = \bar{\sigma}(b_2)$
(EG5)	$\bar{\sigma}(eg_{i,j}) \wedge \neg\bar{\sigma}(eb_{i,j}) \wedge \mu^\sigma \neq 1 \wedge \bar{\sigma}(s_{i,j})$	$\bar{\sigma}(b_{i+1}) = j$
(EB1)	$\bar{\sigma}(eb_{i,j}) \wedge \neg\bar{\sigma}(eg_{i,j}) \wedge \sigma(b_1) = g_1$	$\bar{\sigma}(b_{i+1}) \neq j$
(EB2)	$\bar{\sigma}(eb_{i,0}) \wedge \neg\bar{\sigma}(eg_{i,0}) \wedge \sigma(b_1) = g_1 \wedge \sigma(s_{i,0}) = h_{i,0}$	$\mu^\sigma = i + 1$
(EB3)	$\bar{\sigma}(eb_{i,j}) \wedge \neg\bar{\sigma}(eg_{i,j}) \wedge \sigma(s_{i,j}) = h_{i,j} \wedge i > 1 \wedge \sigma(b_1) = g_1$	$\sigma(b_2) = b_3$
(EB4)	$\bar{\sigma}(eb_{i,1}) \wedge \neg\bar{\sigma}(eg_{i,1}) \wedge \sigma(s_{i,1}) = h_{i,1} \wedge \sigma(b_1) = g_1$	$\mu^\sigma > i + 1$
(EB5)	$\bar{\sigma}(eb_{i,j}) \wedge \neg\bar{\sigma}(eg_{i,j}) \wedge \sigma(b_1) = g_1$	$\sigma(b_{\mu^\sigma}) = g_{\mu^\sigma}$
(EB6)	$\bar{\sigma}(eb_{i,j}) \wedge \neg\bar{\sigma}(eg_{i,j}) \wedge \mu^\sigma > 2$	$\sigma(b_2) = b_3$
(EBG1)	$\bar{\sigma}(eb_{i,j}) \wedge \bar{\sigma}(eg_{i,j}) \wedge \sigma(s_{i,j}) = h_{i,j}$	$\bar{\sigma}(b_{i+1}) = j$
(EBG2)	$\bar{\sigma}(eb_{i,j}) \wedge \bar{\sigma}(eg_{i,j}) \wedge \bar{\sigma}(g_1) = \bar{\sigma}(b_2)$	$\bar{\sigma}(s_1)$
(EBG3)	$\bar{\sigma}(eb_{i,j}) \wedge \bar{\sigma}(eg_{i,j}) \wedge \bar{\sigma}(g_1) = \bar{\sigma}(b_2)$	$\bar{\sigma}(d_1)$
(EBG4)	$\bar{\sigma}(eb_{i,j}) \wedge \bar{\sigma}(eg_{i,j}) \wedge \sigma(g_1) = F_{1,0} \wedge \sigma(b_2) = g_2$	$\mu^\sigma \leq 2$
(EBG5)	$\bar{\sigma}(eb_{i,j}) \wedge \bar{\sigma}(eg_{i,j}) \wedge \bar{\sigma}(g_1) \wedge \sigma(b_2) = b_3$	$\mu^\sigma \neq 2$
(DN1)	$\bar{\sigma}(d_n)$	$\sigma(b_n) = g_n \vee \sigma(b_1) = g_1$
(DN2)	$\bar{\sigma}(d_n) \vee \overline{m}_g^\sigma = n$	$\exists i \in [n] : \sigma(b_i) = g_i$

Table 6.1.: Properties that all calculated strategies have. A strategy that has all of these properties is called well-behaved.

Abbreviation	Explanation: Property involves...
(S*)	upper **S**election vertex
(B*)	entry vertices $\boldsymbol{b_*}$
(BR*)	levels **B**elow the next **R**elevant bit μ^σ
(D*)	cycle vertices $\boldsymbol{d_{*,*,*}}$
(MNS*)	some parameter $\overline{\boldsymbol{m}}^\sigma$
(EG*)	cycle centers **E**scaping only to $\boldsymbol{g_1}$
(EB*)	cycle centers **E**scaping only to $\boldsymbol{b_2}$
(EBG*)	cycle centers **E**scaping only $\boldsymbol{b_2}$ and $\boldsymbol{g_1}$
(DN*)	the cycle vertices $\boldsymbol{d_{*,*,*}}$ of level $\boldsymbol{n}$.

Table 6.2.: Explanation of the abbreviations used for defining the properties of Table 6.1.

Derived properties of well-behaved strategies

We begin by giving results related to the next relevant bit μ^σ of a strategy σ. We first show that its definition can be simplified for well-behaved strategies. We recall here that the set of incorrect levels is defined as $\mathcal{I}^\sigma := \{i \in [n] : \bar{\sigma}(b_i) \wedge \bar{\sigma}(g_i) \neq \bar{\sigma}(b_{i+1})\}$. Note that we interpret expressions of the form $x \wedge x = y$ as $x \wedge (x = y)$, so the precedence level of "$=$" and "$\neq$" is higher than the precedence level of $\wedge$ and $\vee$. Also, we assume the parameter $n \in \mathbb{N}$ to be sufficiently large and in particular larger than 3.

Lemma 6.1.2. *Let $\sigma \in \rho(\sigma_0)$ have Properties (B2) and (BR1) and $\mathcal{I}^\sigma \neq \emptyset$. Then there is an index $i > \max\{i' \in \mathcal{I}^\sigma\}$ with $\bar{\sigma}(b_i)$ and $\bar{\sigma}(g_i) = \bar{\sigma}(b_{i+1})$. As a consequence, for arbitrary well-behaved strategies $\sigma \in \rho(\sigma_0)$ it holds that*

$$\mu^\sigma = \begin{cases} \min\{i > \max\{i' \in \mathcal{I}^\sigma\} : \bar{\sigma}(b_i) \wedge \bar{\sigma}(g_i) = \bar{\sigma}(b_{i+1})\}, & \text{if } \mathcal{I}^\sigma \neq \emptyset, \\ \min\{i \in [n+1] : \sigma(b_i) = b_{i+1}\}, & \text{if } \mathcal{I}^\sigma = \emptyset. \end{cases}$$

In particular, $\mu^\sigma = n+1$ implies $\mathcal{I}^\sigma = \emptyset$ for well-behaved strategies σ.

Proof. Let $\mathcal{I}^\sigma \neq \emptyset$. By construction and since we interpret t as b_{n+1}, it follows that $\bar{\sigma}(g_n) = 0 = \bar{\sigma}(b_{n+1})$. This implies $\max\{i' \in \mathcal{I}^\sigma\} \leq n-1$ and that indices larger than this maximum exist. For the sake of a contradiction, assume that there was no $i > \max\{i' \in \mathcal{I}^\sigma\}$ with $\bar{\sigma}(b_i)$ and $\bar{\sigma}(g_i) = \bar{\sigma}(b_{i+1})$. Then, $\mu^\sigma = n$ by definition and in particular $\mu^\sigma \neq 1$. By the definition of $\mathcal{I}^\sigma$ this implies $\sigma(b_n) = t$. Now let $\max\{i' \in \mathcal{I}^\sigma\} \neq n-1$. Then, Property (BR1) and $\mu^\sigma = n$ imply $\sigma(g_{n-1}) = F_{n-1,0}$. In particular, $\bar{\sigma}(g_{n-1}) = \bar{\sigma}(b_n)$, implying $\sigma(b_{n-1}) = b_n$ since we assume $\max\{i' \in \mathcal{I}^\sigma\} \neq n-1$. Consequently, $\sigma(b_{\mu^\sigma}) = g_{\mu^\sigma}$ by Property (B2) as $\mu^\sigma \neq 1$. But this is a contradiction to $\sigma(b_n) = t$. Now assume $\max\{i' \in \mathcal{I}^\sigma\} = n-1$. Then, $\bar{\sigma}(g_{n-1}) \neq \bar{\sigma}(b_n) = 0$ by the definition of $\mathcal{I}^\sigma$. Thus, $\sigma(g_{n-1}) = F_{n-1,1}$. But, since $n-1 = \mu^\sigma - 1$, we also have $\sigma(g_{n-1}) = F_{n-1,0}$ by Property (BR1) which is a contradiction.
Hence there is an index $i > \max\{i' \in \mathcal{I}^\sigma\}$ with $\bar{\sigma}(b_i) \wedge \bar{\sigma}(g_i) = \bar{\sigma}(b_{i+1})$. □

Whenever discussing μ^σ for a well-behaved strategy σ, we implicitly use Lemma 6.1.2 without explicitly mentioning it. We now prove that $\sigma(b_1) = b_2$ is equivalent to $\mu^\sigma = 1$ for well-behaved strategies and deduce similar helpful statements related to μ^σ.

Lemma 6.1.3. *Let $\sigma \in \rho(\sigma_0)$ have Property (B1) and let $\mu^\sigma \neq 1$. Then $\sigma(b_1) = g_1$. Consequently, if σ has Property (B1), then $\mu^\sigma = 1$ is equivalent to $\sigma(b_1) = b_2$.*

Proof. By the definition of μ^σ, it holds that $\mu^\sigma = 1$ implies $\sigma(b_1) = b_2$. It thus suffices to prove the first part of the statement, so let $\mu^\sigma \neq 1$, implying $\mu^\sigma > 1$.

Let $\mathcal{I}^\sigma = \emptyset$, implying $\mu^\sigma = \min\{i \in [n+1] : \sigma(b_i) = b_{i+1}\}$. Since $\mu^\sigma > 1$, it needs to hold that $\sigma(b_1) = g_1$ since the minimum would be attained for $i = 1$ otherwise.

Let $\mathcal{I}^\sigma \neq \emptyset$. Then $\mu^\sigma = \min\{i' > \max\{i \in \mathcal{I}^\sigma\} : \bar\sigma(b_{i'}) \wedge \bar\sigma(g_{i'}) = \bar\sigma(b_{i'+1})\}$. If $\mu^\sigma = 2$, then $\max\{i \in \mathcal{I}^\sigma\} = 1$, implying $\sigma(b_1) = g_1$ by the definition of $\mathcal{I}^\sigma$. If $\mu^\sigma > 2$, the contraposition of Property (B1) states

$$\sigma(b_{i+1}) = g_{i+1} \implies [i \geq \mu^\sigma - 1 \vee \sigma(b_i) = g_i].$$

Let $m := \max\{i \in \mathcal{I}^\sigma\}$. Then, by definition, $m < \mu^\sigma$ and $\sigma(b_m) = g_m$. We thus either have $m - 1 \geq \mu^\sigma - 1$ or $\sigma(b_{m-1}) = g_{m-1}$. Since $m - 1 \geq \mu^\sigma - 1$ contradicts $m < \mu^\sigma$, we have $\sigma(b_{m-1}) = g_{m-1}$. The argument can now be applied iteratively, implying $\sigma(b_1) = g_1$. □

The following statement now shows a deep connection between the choice of b_{μ^σ} and the set of incorrect levels if the strategy σ has certain properties.

Lemma 6.1.4. *Let $\sigma \in \rho(\sigma_0)$ have Properties (B1), (B2) and (BR1).*

1. *Let $\mathcal{I}^\sigma \neq \emptyset$. Then $\sigma(b_i) = g_i$ for all $i \leq \max\{i' \in \mathcal{I}^\sigma\}$ and $\sigma(b_{\mu^\sigma}) = g_{\mu^\sigma}$.*
2. *Let $\mathcal{I}^\sigma = \emptyset$. Then $\sigma(b_i) = g_i$ for all $i < \mu^\sigma$ and $\sigma(b_{\mu^\sigma}) = b_{\mu^\sigma+1}$. In addition, $\mu^\sigma > 1$ implies $\sigma(b_2) = g_2 \iff \mu^\sigma > 2$.*

Consequently, $\mathcal{I}^\sigma = \emptyset$ if and only if $\sigma(b_{\mu^\sigma}) = b_{\mu^\sigma+1}$.

Proof. The last statements follows directly from the first two, so only these are proven.

1. Since $\mathcal{I}^\sigma \neq \emptyset$ implies $\mu^\sigma \neq 1$, the first statement follows by the same arguments used in the proof of Lemma 6.1.3. The second statement follows directly from Lemma 6.1.2 as σ has Property (B2) and Property (BR1).
2. The first statement from $\mu^\sigma = \min\{i \in [n+1] \colon \sigma(b_i) = b_{i+1}\}$ in this case. The second statement follows directly since $\mu^\sigma > 1$ implies $\sigma(b_1) = g_1$ in this case. □

The second statement of Lemma 6.1.4 yields the following corollary for well-behaved strategies by Property (D1). This corollary allows us to simplify several proofs regarding valuations of vertices in M_n later.

Corollary 6.1.5. *Let σ be a well-behaved strategy with $\mathcal{I}^\sigma = \emptyset$. Then $i < \mu^\sigma$ implies $\bar\sigma(d_i)$.*

Before discussing canonical strategies in general, we provide one more lemma that significantly simplifies several proofs. It is closely related to properties of the type (MNS*) and proves that several of their assumptions imply useful statements.

Lemma 6.1.6. *Let $\sigma \in \rho(\sigma_0)$ have Properties (B1) and (B3) and let $\mu^\sigma = 1$.*

1. *If $m_b^\sigma \leq \overline{m}_s^\sigma, \overline{m}_g^\sigma$, then $m_b^\sigma = 2$.*
2. *If $\overline{m}_g^\sigma < \overline{m}_s^\sigma, m_b^\sigma$ and $\overline{m}_g^\sigma > 1$, then $\overline{m}_g^\sigma + 1 = m_b^\sigma$.*

Proof. Note that $\mu^\sigma = 1$ implies $\sigma(b_1) = b_2$.

1. By $\sigma(b_1) = b_2$, we have $m_b^\sigma \geq 2$. Assume $m_b^\sigma > 2$ and let $i := m_b^\sigma - 2$. Then $i < \overline{m}_g^\sigma$, implying $\sigma(g_i) = F_{i,1}$. In addition, $i < \overline{m}_s^\sigma$, implying $\sigma(s_{i,1}) = h_{i,1}$. Since $i+1 < m_b^\sigma$, also $\sigma(b_{i+1}) = b_{i+2}$. Consequently, by Property (B3), $\bar{\sigma}(g_{i+1}) \neq \bar{\sigma}(b_{i+2}) = 1$ since $i + 2 = m_b^\sigma$. But this implies $\sigma(g_{i+1}) = F_{i,0}$, contradicting $i + 1 \leq \overline{m}_g^\sigma$.
2. Let $i := \overline{m}_g^\sigma - 1$. Then $\sigma(g_i) = F_{i,1}$, implying $\sigma(s_{i,1}) = h_{i,1}$ by the choice of i. Furthermore, $\sigma(b_{i+1}) = b_{i+2}$ as $i + 1 = \overline{m}_g^\sigma < m_b^\sigma$. Consequently, by Property (B3), $0 = \bar{\sigma}(g_{i+1}) \neq \bar{\sigma}(b_{i+2})$, implying $\bar{\sigma}(b_{i+2}) = 1$ and thus $\overline{m}_g^\sigma + 1 = i + 2 = m_b^\sigma$. □

The framework for the vertex valuations

We now discuss the general framework used for describing and characterizing the vertex valuations. Whenever referring to valuations, we henceforth add one of three possible upper indices. If we consider the valuations exclusively in the sink game S_n resp. the Markov decision process M_n, we include an upper index S resp. M. If the arguments or statements apply to both S_n and M_n, then we use the general wildcard symbol $*$.

For most proofs, we do not consider the *real* valuations as described in Sections 2.2 and 3.2 but a "reduced" version, referred to as rVal. In S_n, the motivation for considering reduced valuations is that the game is constructed in such a way that the most significant difference between two vertex valuations will always be unique and typically have a priority larger than six. That is, vertices of priority three or four will rarely ever be relevant when comparing valuations. They can thus be ignored in most cases, simplifying the vertex valuations. In M_n, the reduced valuations are motivated differently. Consider some cycle center $F_{i,j}$. Its valuation is equal to

$$\varepsilon \operatorname{Val}_\sigma^{\mathrm{M}}(s_{i,j}) + \frac{1-\varepsilon}{2} \operatorname{Val}_\sigma^{\mathrm{M}}(d_{i,j,0}) + \frac{1-\varepsilon}{2} \operatorname{Val}_\sigma^{\mathrm{M}}(d_{i,j,1}).$$

Intuitively, if $F_{i,j}$ is not closed, then the contribution of $s_{i,j}$ to the valuation of $F_{i,j}$ is very likely to be negligible. However, if $F_{i,j}$ is closed, then $\varepsilon \operatorname{Val}_\sigma^{\mathrm{M}}(F_{i,j}) = \varepsilon \operatorname{Val}_\sigma^{\mathrm{M}}(s_{i,j})$, so $\operatorname{Val}_\sigma^{\mathrm{M}}(F_{i,j}) = \operatorname{Val}_\sigma^{\mathrm{M}}(s_{i,j})$ for every $\varepsilon > 0$. Thus, defining $\operatorname{rVal}_\sigma^{\mathrm{M}}$ as the limit of $\operatorname{Val}_\sigma^{\mathrm{M}}$ for $\varepsilon \to 0$ yields an easier way of calculating valuations as it eliminates terms of order $o(1)$. There are however several cases in which the real valuations $\operatorname{Val}_\sigma^{\mathrm{M}}$ need to be considered since $\varepsilon \operatorname{Val}_\sigma^{\mathrm{M}}(s_{i,j})$ is not always negligible. This motivation justifies the following definition.

Definition 6.1.7 (Reduced valuation). Let $v \in V$ and $\sigma \in \rho(\sigma_0)$. The *reduced valuation* of v with respect to σ in S_n is $\operatorname{rVal}_\sigma^{\mathrm{S}}(v) := \operatorname{Val}_\sigma^{\mathrm{S}}(v) \setminus \{v' \in \operatorname{Val}_\sigma^{\mathrm{S}}(v) \colon \Omega(v') \in \{3, 4, 6\}\}$. In M_n, the *reduced valuation* of $v \in V$ with respect to σ is $\operatorname{rVal}_\sigma^{\mathrm{M}}(v) := \lim_{\varepsilon \to 0} \operatorname{Val}_\sigma^{\mathrm{M}}(v)$.

We now introduce a unified notation for reduced valuations. This enables us to perform several calculations and arguments for S_n and M_n simultaneously. Since vertex valuations in S_n are sets of vertices, we begin by arguing that valuations in M_n can also be described as sets of vertices, although they are usually defined via edges.

Since M_n is weakly unichain, the reduced valuation of a vertex is typically a path ending in the vertex t with probability 1 after finitely many steps. The only exception are cycle centers escaping to both b_2 and g_1 which we discuss later. By construction, the reward of any edge leaving a vertex v is $\langle v \rangle := (-N)^{\Omega(v)}$, where $\Omega(v)$ denotes the priority of v. If the reduced valuation of a vertex v corresponds to a path P ending in t, then the total reward collected along the edges of P in M_n can thus be expressed as $\sum_{v \in P} \langle v \rangle$. This argument does however not apply to cycle centers. The reduced valuation of a cycle center $F_{i,j}$ might depend on both the reduced valuations of g_1 and b_2. This is the case if $F_{i,j}$ escapes towards both of these vertices using its cycle vertices and corresponding escape vertices. In this case, the reduced valuation of $F_{i,j}$ is the arithmetic mean of the reduced valuations of g_1 and b_2. Since M_n is weakly unichain and by the definition of the reduced valuations, the reduced valuations of these two vertices are disjoint paths ending in t. In particular, the previous interpretation can be applied to both of these vertices and it can thus be extended naturally to the cycle center $F_{i,j}$. In summary, the reduced valuation of any vertex can be interpreted as either a single path or a union of two disjoint paths leading to t.

Table 6.3 introduces a unified notation that can be used for discussing vertex valuations in both S_n and M_n simultaneously. In addition, it defines several subsets of vertices that turn out to be useful when describing vertex valuations. These sets are, for example, all vertices in a level i that contribute to the valuations of the vertices, or sets that will typically be part of several valuations. Although the sets contained in this table formally depend on the current strategy, we do not include an index denoting this strategy as it will always be clear from the context. To simplify this notation, we write $\langle v_1, v_2, \dots, v_k \rangle$ to denote $\sum_{i \in [k]} \langle v_i \rangle$ for arbitrary sets $\{v_1, \dots, v_k\}$ of vertices.

It is immediate that $\mathrm{rVal}_\sigma^{\mathrm{S}}(v) \lhd \mathrm{rVal}_\sigma^{\mathrm{S}}(w)$ implies $\mathrm{Val}_\sigma^{\mathrm{S}}(v) \lhd \mathrm{Val}_\sigma^{\mathrm{S}}(w)$. This is not however not completely obvious for M_n since it is not clear how much we "lose" by using the reduced instead of the real valuation. However, as shown by the following lemma, we only lose a negligible amount of $o(1)$. Hence, if $\mathrm{rVal}_\sigma^{\mathrm{M}}(v) > \mathrm{rVal}_\sigma^{\mathrm{M}}(w)$ and if the difference between the two terms is sufficiently large, then we can deduce $\mathrm{Val}_\sigma^{\mathrm{M}}(v) > \mathrm{Val}_\sigma^{\mathrm{M}}(w)$.

Lemma 6.1.8. *Let $P = \{g_*, s_{*,*}, h_{i*,*}\}$ be the set of vertices with priorities in M_n. Let $S, S', P \subseteq P$ be non-empty subsets, let $\sum(S) := \sum_{v \in S} \langle v \rangle$ and define $\sum(S')$ analogously.*

1. *$|\sum(S)| < N^{2n+11}$ and $\varepsilon \cdot |\sum(S)| < 1$ for every subset $S \subseteq P$, and*
2. *$|\max_{v \in S} \langle v \rangle| < |\max_{v \in S'} \langle v \rangle|$ if and only if $|\sum(S)| < |\sum(S')|$.*

Since $N = 7n$ is larger than the number of vertices with priorities, Lemma 6.1.8 implies that we can use reduced valuations in M_n in the following way.

Corollary 6.1.9. *Let $\sigma \in \rho(\sigma_0)$. Then $\mathrm{rVal}_\sigma^{\mathrm{M}}(w) > \mathrm{rVal}_\sigma^{\mathrm{M}}(v)$ implies $\mathrm{Val}_\sigma^{\mathrm{M}}(w) > \mathrm{Val}_\sigma^{\mathrm{M}}(v)$.*

Proof. Reduced valuations can be represented as sums of powers of N. If the reduced valuations of two vertices differ, then they thus differ by terms of order at least N. However, by Lemma 6.1.8, the reduced valuation of a vertex and its real valuation only differ by terms of order $o(1)$ since the difference between the real and the reduced valuation of a vertex is always an expression of the type $\varepsilon \cdot |\sum(S)|$ for some subset S. Consequently, if the reduced valuation of v is larger than the reduced valuation of w, the same is true for the real valuations. □

$W_i^{\mathrm{S}} := \{g_i, s_{i,\bar{\sigma}(g_i)}, h_{i,\bar{\sigma}(g_i)}\}$	$W_i^{\mathrm{M}} := \langle g_i, s_{i,\bar{\sigma}(g_i)}, h_{i,\bar{\sigma}(g_i)} \rangle$
$L_{i,\ell}^{\mathrm{S}} := \bigcup_{i'=i}^{\ell} \{W_{i'}^{P} : \sigma(b_{i'}) = g_{i'}\}$	$L_{i,\ell}^{\mathrm{M}} := \sum_{i'=i}^{\ell} \{W_{i'}^{\mathrm{M}} : \sigma(b_{i'}) = g_{i'}\}$
$R_{i,\ell}^{\mathrm{S}} := \bigcup_{i'=i}^{\mu^\sigma - 1} W_{i'}^{P} \cup \bigcup_{i'=\mu^\sigma+1}^{\ell} \{W_{i'}^{P} : \sigma(b_{i'}) = g_{i'}\}$	$R_{i,\ell}^{\mathrm{M}} := \sum_{i'=i}^{\mu^\sigma - 1} W_{i'}^{\mathrm{M}} + \sum_{i'=\mu^\sigma+1}^{\ell} \{W_{i'}^{\mathrm{M}} : \sigma(b_{i'}) = g_{i'}\}$
$B_{i,\ell}^{\mathrm{S}} := \begin{cases} R_{i,\ell}^{P} & \text{if } i < \mu^\sigma \text{ and } \sigma(b_i) = g_i \\ L_{i,\ell}^{P} & \text{otherwise} \end{cases}$	$B_{i,\ell}^{\mathrm{M}} := \begin{cases} R_{i,\ell}^{\mathrm{M}} & \text{if } i < \mu^\sigma \text{ and } \sigma(b_i) = g_i \\ L_{i,\ell}^{\mathrm{M}} & \text{otherwise} \end{cases}$

unified notation	$\oplus/\bigoplus_*^*$	$[\![\cdot]\!]$	$\prec / \succ$	0	$W \subset \mathrm{rVal}_\sigma^*(\cdot)$.
corr. notation for S_n	$\cup/\bigcup_*^*$	$\{\cdot\}$	$\lhd/\rhd$	$\emptyset$	All $w \in W$ are contained in $\mathrm{rVal}_\sigma^{\mathrm{S}}(\cdot)$
corr. notation for M_n	$+/\sum_*^*$	$\langle\cdot\rangle$	$< / >$	0	All $w \in W$ are summands of $\mathrm{rVal}_\sigma^{\mathrm{M}}(\cdot)$.

Table 6.3.: Abbreviations and notation used for unified arguments and vertex valuations. We also define $L_i^* := L_{i,n}^*, R_i^* := R_{i,n}^*, B_i^* := B_{i,n}^*$

It is possible that $\mathrm{rVal}_\sigma^{\mathrm{M}}(w) = \mathrm{rVal}_\sigma^{\mathrm{M}}(v)$ but $\mathrm{Val}_\sigma^{\mathrm{M}}(w) \neq \mathrm{Val}_\sigma^{\mathrm{M}}(v)$. This case can occur if there are two cycle centers $F_{i,0}$ and $F_{i,1}$ which are in the same state. Then, the valuations of the corresponding upper selection vertices decide which of these two vertices has the better valuation. Since the influence of these vertices is however neglected when considering the reduced valuation, we need to investigate the real valuations in such a case.

Before characterizing the vertex valuations, we state the following general statements regarding the terms of Table 6.3. We will not always refer to this lemma when we use it as it is used in nearly all calculations. However, we want to especially underline the last statement, as this formalizes the intuition that traversing a single level i completely is more beneficial then traversing all levels below level i.

Lemma 6.1.10. *Let $\sigma \in \rho(\sigma_0)$ be well-behaved.*

1. *Let $\sigma(b_{\mu^\sigma}) = b_{\mu^\sigma+1}$. Then $L_i^* \preceq R_i^*$ for all $i \in [n]$ and $L_i^* \prec R_j^*$ for $j < i \leq \mu^\sigma$.*
2. *Let $\sigma(b_{\mu^\sigma}) = g_{\mu^\sigma}$. Then $L_i^* \succeq R_i^*$ for all $i \in [n]$ and $L_i \succ R_j^*$ for $i \leq \mu^\sigma$ and $j \in [n]$ and $L_i \oplus [\![g_j]\!] \succ R_j$ for $i \leq \mu^\sigma$ and $j < \mu^\sigma$.*
3. *Let $i \geq \mu^\sigma > j$. Then $R_j^* \prec [\![s_{i,j}, h_{i,j}]\!] \oplus L_{i+1}^*$.*
4. *For all $i \in [n]$, it holds that $[\![g_i, s_{i,*}, h_{i,*}]\!] \succ \bigoplus_{i'<i} W_{i'}^*$ and $L_1^* \prec [\![s_{i,j}, h_{i,j}]\!] \oplus L_{i+1}^*$.*

Characterizing vertex valuations

The remainder of this section is dedicated to explicitly determine the vertex valuations for well-behaved strategies. Most of the proofs are very technical and are thus deferred to Appendix A.2. We however also provide some proofs here in the main part to show how these statements are proven. We begin by discussing the valuations of the entry vertices b_i for $i > 1$ and of selector vertices g_i when $i < \mu^\sigma$ and $\sigma(b_2) = g_2$.

Lemma 6.1.11. *Let $\sigma \in \rho(\sigma_0)$ be well-behaved and $i > 1$. Then $\mathrm{rVal}^*_\sigma(b_i) = B^*_i$ and $i < \mu^\sigma$ and $\sigma(b_2) = g_2$ imply $\mathrm{rVal}^*_\sigma(g_i) = R^*_i$.*

Proof. We prove both statements by backwards induction on i and begin with the first statement. Let $i = n$ and $\sigma(b_n) \neq g_n$. Then $\mathrm{rVal}^*_\sigma(b_n) = 0$. Since $B^*_n = L^*_n = 0$, the statement follows. Now let $\sigma(b_n) = g_n$. Then, by Property (D1), $\bar{\sigma}(d_n)$ and $\bar{\sigma}(s_i)$ by Property (S2) since $\mu^\sigma \leq n$ by Lemma 6.1.2. Hence $\mathrm{rVal}^*_\sigma(b_n) = W^*_n$. Since $W^*_n = B^*_n$ in this case, the statement follows for both S_n and M_n.

Now let $i < n, i > 1$, and assume that the statement holds for $i+1$. We show that it holds for i as well. This part of the proof uses the second statement of the lemma *directly*, i.e., in a non-inductive way. Since we use the first statement when proving the second *inductively*, the induction is correct. We distinguish several cases.

- Let $\sigma(b_i) = b_{i+1}$ and $\mu^\sigma = \min\{i' \colon \sigma(b_{i'}) = b_{i'+1}\}$. Then, $\mu^\sigma \leq i$ and we show $\mathrm{rVal}^*_\sigma(b_i) = L^*_i$. By the definition of μ^σ, there is no $i' \in [n]$ such that $\sigma(b_{i'}) = g_{i'}$ and $\bar{\sigma}(b_{i'+1}) \neq \bar{\sigma}(g_{i'})$. Since the statement holds if $\sigma(b_{i'}) = b_{i'+1}$ for all $i' > i$ consider the smallest $i' > i$ with $\sigma(b_{i'}) = g_{i'}$. Since $i' > i > 1$ we obtain $\mathrm{rVal}^*_\sigma(b_{i'}) = B^*_{i'}$ by the induction hypotheses. Furthermore, $B^*_{i'} = L^*_{i'}$ by $\mu^\sigma \leq i < i'$. By the choice of i' we have $\mathrm{rVal}^*_\sigma(b_i) = \mathrm{rVal}^*_\sigma(b_{i+1}) = \dots = \mathrm{rVal}^*_\sigma(b_{i'}) = L^*_{i'}$ as well as $L^*_{i'} = L^*_{i'-1} = \dots = L^*_i$. As $\mu^\sigma \leq i$ implies $L^*_i = B^*_i$, we thus have $\mathrm{rVal}^*_\sigma(b_i) = B^*_i$.
- Let $\sigma(b_i) = b_{i+1}$ and $\mu^\sigma = \min\{i' > \max\{i \in \mathcal{I}^\sigma\} \colon \bar{\sigma}(b_{i'}) \wedge \bar{\sigma}(g_{i'}) = \bar{\sigma}(b_{i'+1})\}$. Assume $i \geq \mu^\sigma$. Then $\mathrm{rVal}^*_\sigma(b_i) = \mathrm{rVal}^*_\sigma(b_{i+1}) = B^*_{i+1}$ by the induction hypotheses and $B^*_{i+1} = L^*_{i+1}$ by the choice of i. Since $B^*_i = L^*_i$ and $L^*_i = L^*_{i+1}$ as $\sigma(b_i) = b_{i+1}$, the statement follows. Hence assume $i < \mu^\sigma$. Since $\sigma(b_i) = b_{i+1}$ and since σ is well-behaved, Property (B1) yields $\sigma(b_{i'}) = b_{i'+1}$ for all $i' \in \{i, \dots, \mu^\sigma - 1\}$. By the induction hypothesis, we thus have $\mathrm{rVal}^*_\sigma(b_{i'}) = B^*_{i'} = L^*_{i'}$ for these indices. In particular, $\mathrm{rVal}^*_\sigma(b_{i+1}) = L^*_{i+1}$. Note that this also holds for the case $i+1 = \mu^\sigma$. Since $\sigma(b_i) = b_{i+1}$ implies $L^*_i = L^*_{i+1}$ and $\mathrm{rVal}^*_\sigma(b_i) = \mathrm{rVal}^*_\sigma(b_{i+1})$, this then yields $\mathrm{rVal}^*_\sigma(b_i) = L^*_{i+1} = L^*_i = B^*_i$.
- Let $\sigma(b_i) = g_i$ and $i \geq \mu^\sigma$. As before, the induction hypothesis yields $\mathrm{rVal}^*_\sigma(b_{i'}) = L^*_{i'}$ for all $i' > i$. Let $j := \bar{\sigma}(g_i)$. Then, $F_{i,j}$ is closed by Property (D1) since $i > 1$. Thus $\mathrm{rVal}^*_\sigma(F_{i,j}) = \mathrm{rVal}^*_\sigma(s_{i,j})$. In addition, $\sigma(s_{i,j}) = h_{i,j}$ by Property (S1). Since $i \geq \mu^\sigma$ and $\sigma(b_i) = g_i$ we have $j = \bar{\sigma}(g_i) = \bar{\sigma}(b_{i+1})$. By construction we thus have

$$\mathrm{rVal}^*_\sigma(b_i) = \mathrm{rVal}^*_\sigma(g_i) = W^*_i \oplus \mathrm{rVal}^*_\sigma(b_{i+1}) = W^*_i \oplus L^*_{i+1} = L^*_i = B^*_i.$$

- Finally, let $\sigma(b_i) = g_i$ and $i < \mu^\sigma$. Using the contraposition of Property (B1) we obtain that either $i - 1 \geq \mu^\sigma - 1$ or $\sigma(b_{i-1}) = g_{i-1}$. Since $i - 1 \geq \mu^\sigma - 1$ contradicts

$i < \mu^\sigma$, it follows that $\sigma(b_{i-1}) = g_{i-1}$. Applying this statement inductively then yields $\sigma(b_2) = g_2$. Using the second statement of this lemma *directly* we then obtain $\mathrm{rVal}^*_\sigma(b_i) = \mathrm{rVal}^*_\sigma(g_i) = R^*_i = B^*_i$.

We now show that $i < \mu^\sigma$ and $\sigma(b_2) = g_2$ imply $\mathrm{rVal}^*_\sigma(g_i) = R^*_i$. This proof uses the first statement in an inductive way. The statement is shown by backwards induction on i, so let $i = \mu^\sigma - 1$. Then $\sigma(g_i) = F_{i,0}$ by Property (BR1) and $\mathrm{rVal}^*_\sigma(F_{i,0}) = \mathrm{rVal}^*_\sigma(s_{i,0})$ since $F_{i,0}$ is closed by Property (D2). Also, $\sigma(s_{i,0}) = h_{i,0}$ by Property (S2). By construction, and using the first statement inductively we obtain

$$\mathrm{rVal}^*_\sigma(g_i) = W^*_i \oplus \mathrm{rVal}^*_\sigma(b_{i+2}) = W^*_i \oplus B^*_{i+2} = W^*_{\mu^\sigma-1} \oplus B^*_{\mu^\sigma+1} = W^*_{\mu^\sigma-1} \oplus L^*_{\mu^\sigma+1} = R^*_{\mu^\sigma-1}.$$

Let $i < \mu^\sigma - 1$. By Properties (BR1) and (D2), $\sigma(g_i) = F_{i,1}$ and $\mathrm{rVal}^*_\sigma(F_{i,1}) = \mathrm{rVal}^*_\sigma(s_{i,1})$. By Property (S2), also $\sigma(s_{i,1}) = h_{i,1}$. By construction, the induction hypotheses thus yields $\mathrm{rVal}^*_\sigma(g_i) = W^*_i \oplus \mathrm{rVal}^*_\sigma(g_{i+1}) = R^*_{i+1} \cup W^*_i = R^*_i = B^*_i$. □

The next lemma shows how the valuation of g_i might change if the additional requirements used in the second statement of Lemma 6.1.11 are not met resp. if $i = 1$. As its proof is rather involved, requires some case distinctions but uses again a backwards induction and 6.1.11, its proof is deferred to the appendix.

Lemma 6.1.12. *Let $\sigma \in \rho(\sigma_0)$ be well-behaved and $i < \mu^\sigma$. Then $\mathrm{rVal}^{\mathrm{S}}_\sigma(g_i) = R^{\mathrm{S}}_i$ and*

$$\mathrm{rVal}^{\mathrm{M}}_\sigma(g_i) = \begin{cases} B^{\mathrm{M}}_2 + \sum_{j=i}^{k-1} W^{\mathrm{M}}_j + \langle g_k \rangle, & \text{if } k := \min\{k \geq i \colon \neg\bar{\sigma}(d_k)\} < \mu^\sigma \\ \mathrm{rVal}^{\mathrm{M}}_\sigma(g_i) = R^{\mathrm{M}}_i, & \text{otherwise.} \end{cases}$$

This lemma can now be used to generalize the first statement of Lemma 6.1.11.

Lemma 6.1.13. *Let $\sigma \in \rho(\sigma_0)$ be well-behaved. Then $\mathrm{rVal}_\sigma(b_i)^{\mathrm{S}} = B^{\mathrm{S}}_i$ for all $i \in [n]$ and $\mathrm{rVal}^{\mathrm{M}}_\sigma(b_i) = B^{\mathrm{M}}_i$ for all $i \in \{2, \dots, n\}$. Furthermore,*

$$\mathrm{rVal}^{\mathrm{M}}_\sigma(b_1) = \begin{cases} B^{\mathrm{M}}_2 + \sum_{j=1}^{k-1} W^{\mathrm{M}}_j + \langle g_k \rangle, & \text{if } k := \min\{i \geq 1 \colon \neg\bar{\sigma}(d_i)\} < \mu^\sigma, \\ B^{\mathrm{M}}_1, & \text{otherwise.} \end{cases}$$

Proof. The case $i > 1$ follows by Lemma 6.1.11. It therefore suffices to consider the case $i = 1$. Let $\sigma(b_1) = b_2$. Then $\mu^\sigma = 1$ by Lemma 6.1.3. Therefore, by Lemma 6.1.11, we have $\mathrm{rVal}^*_\sigma(b_2) = B^*_2 = L^*_2$. Since $\sigma(b_1) = b_2$ implies $L^*_2 = L^*_1$ we thus obtain $\mathrm{rVal}^*_\sigma(b_1) = \mathrm{rVal}^*_\sigma(b_2) = B^*_2 = L^*_2 = L^*_1 = B^*_1$.

Assume $\sigma(b_1) = g_1$. Then $\mu^\sigma > 1$ by Lemma 6.1.3. Consider the case $G_n = S_n$ first. Then, $\mathrm{rVal}^{\mathrm{S}}_\sigma(g_1) = R^{\mathrm{S}}_1$ by Lemma 6.1.12. Hence, since $i = 1 < \mu^\sigma$ and $\sigma(b_1) = g_1$ it holds that $B^{\mathrm{S}}_1 = R^{\mathrm{S}}_1$. Thus $\mathrm{rVal}^{\mathrm{S}}_\sigma(b_1) = \mathrm{rVal}^{\mathrm{S}}_\sigma(g_1) = R^{\mathrm{S}}_1 = B^{\mathrm{S}}_1$.

Consider the case $G_n = M_n$ next. If $\mathrm{rVal}^{\mathrm{M}}_\sigma(g_1) = R^{\mathrm{M}}_1$, the statement follows by the same arguments used for the case $G_n = S_n$. Hence let $k := \min\{i \geq 1 \colon \neg\bar{\sigma}(d_i)\} < \mu^\sigma$. Then, since $\sigma(b_1) = g_1$ implies $\mathrm{rVal}^{\mathrm{M}}_\sigma(b_1) = \mathrm{rVal}^{\mathrm{M}}_\sigma(g_1)$, Lemma 6.1.12 implies the statement. □

We thus completely characterized the valuation of all vertices b_i. The next vertex valuation we discuss is the valuation of g_1 for the special case of $\mu^\sigma = 1$. As the vertex valuation of this vertex is rather complex and the proof requires several case distinctions, we defer it to Appendix A.2.

As always, we identify b_i for $i > n$ with t for convenience of notation.

Lemma 6.1.14. *Let $\mu^\sigma = 1$ and $m := \min\{\overline{m}_g^\sigma, \overline{m}_s^\sigma\}$. Then*

$$\operatorname{rVal}_\sigma^*(g_1) = \begin{cases} \langle g_1 \rangle + \operatorname{rVal}_\sigma^{\mathrm{M}}(b_2), & \text{if } m_b^\sigma \leq \overline{m}_s^\sigma, \overline{m}_g^\sigma \wedge G_n = M_n \wedge \neg\bar{\sigma}(d_1), \\ W_1^* \oplus \operatorname{rVal}_\sigma^*(b_2), & \text{if } m_b^\sigma \leq \overline{m}_s^\sigma, \overline{m}_g^\sigma, \\ & \quad \wedge (G_n = S_n \vee [G_n = M_n \wedge \bar{\sigma}(d_1)]), \\ \bigoplus_{i'=1}^{m} W_{i'}^* \oplus \operatorname{rVal}_\sigma^*(b_{\overline{m}_g^\sigma+2}) & \text{if } \overline{m}_g^\sigma < \overline{m}_s^\sigma, m_b^\sigma, \\ & \quad \wedge [(\bar{\sigma}(b_{\overline{m}_g^\sigma+1}) \wedge G_n = S_n) \vee \neg\bar{\sigma}(eb_{\overline{m}_g^\sigma})], \\ \bigoplus_{i'=1}^{m-1} W_{i'}^* \oplus [\![g_m]\!] \oplus \operatorname{rVal}_\sigma^*(b_2) & \text{otherwise.} \end{cases}$$

The next vertex valuation that we investigate in detail is the valuation of the vertices $F_{i,j}$, i.e., of the cycle centers. For these vertices we need to distinguish between the sink game S_n and the Markov decision process M_n. We begin with case $G_n = M_n$ as the corresponding statement follows directly from the definition of $\operatorname{rVal}_\sigma^{\mathrm{M}}$.

Lemma 6.1.15. *Let $G_n = M_n$. Let $\sigma \in \rho(\sigma_0)$ be well-behaved and $i \in [n], j \in \{0,1\}$. Then*

$$\operatorname{rVal}_\sigma^{\mathrm{M}}(F_{i,j}) = \begin{cases} \operatorname{rVal}_\sigma^{\mathrm{M}}(s_{i,j}), & \text{if } \bar{\sigma}(d_{i,j}), \\ \operatorname{rVal}_\sigma^{\mathrm{M}}(g_1), & \text{if } \bar{\sigma}(eg_{i,j}) \wedge \neg\bar{\sigma}(eb_{i,j}), \\ \operatorname{rVal}_\sigma^{\mathrm{M}}(b_2), & \text{if } \bar{\sigma}(eb_{i,j}) \wedge \neg\bar{\sigma}(eg_{i,j}), \\ \frac{1}{2}\operatorname{rVal}_\sigma^{\mathrm{M}}(g_1) + \frac{1}{2}\operatorname{rVal}_\sigma^{\mathrm{M}}(b_2), & \text{if } \bar{\sigma}(eg_{i,j}) \wedge \bar{\sigma}(eb_{i,j}). \end{cases}$$

The exact behavior of player 1 in the sink game S_n requires a more sophisticated analysis. The reason is that the behavior of player 1 very much depends on the configuration of the complete counter and the exact setting of several vertices in different levels. In particular, depending on the setting of the cycle vertices and the upper selection vertex of a cycle center, the valuations of $d_{i,j,0}, d_{i,j,1}$ $s_{i,j}$ can be completely different. Consequently, player 1 can theoretically choose from up to three different valuations. As the player always minimizes the valuation of the vertices, this requires us to analyze and compare a lot of valuations exactly. We thus do not provide its proof here but defer it to Appendix A.2.

Lemma 6.1.16. *Let $G_n = S_n$. Let $\sigma \in \rho(\sigma_0)$ be well-behaved and $i \in [n], j \in \{0,1\}$. Then*

$$\mathrm{rVal}_\sigma^{\mathrm{S}}(F_{i,j}) = \begin{cases} \mathrm{rVal}_\sigma^{\mathrm{S}}(s_{i,j}), & \text{if } \bar\sigma(d_{i,j}),\\ \{s_{i,j}\} \cup \mathrm{rVal}_\sigma^{\mathrm{S}}(b_2), & \text{if } \bar\sigma(eg_{i,j}) \wedge \neg\bar\sigma(eb_{i,j}) \wedge \mu^\sigma = 1,\\ \mathrm{rVal}_\sigma^{\mathrm{S}}(g_1), & \text{if } \bar\sigma(eg_{i,j}) \wedge \neg\bar\sigma(eb_{i,j}) \wedge \mu^\sigma \neq 1,\\ \mathrm{rVal}_\sigma^{\mathrm{S}}(b_2), & \text{if } \bar\sigma(eb_{i,j}) \wedge \neg\bar\sigma(eg_{i,j}) \wedge \mu^\sigma = 1\\ & \quad \wedge (\neg\bar\sigma(s_{i,j}) \vee \bar\sigma(b_{i+1}) = j),\\ \mathrm{rVal}_\sigma^{\mathrm{S}}(s_{i,j}), & \text{if } \bar\sigma(eb_{i,j}) \wedge \neg\bar\sigma(eg_{i,j})\\ & \quad \wedge (\mu^\sigma \neq 1 \vee (\bar\sigma(s_{i,j}) \wedge \bar\sigma(b_{i+1}) \neq j)),\\ \mathrm{rVal}_\sigma^{\mathrm{S}}(g_1), & \text{if } \bar\sigma(eb_{i,j}) \wedge \bar\sigma(eg_{i,j}) \wedge \bar\sigma(g_1) \neq \bar\sigma(b_2),\\ \mathrm{rVal}_\sigma^{\mathrm{S}}(b_2), & \text{if } \bar\sigma(eb_{i,j}) \wedge \bar\sigma(eg_{i,j}) \wedge \bar\sigma(g_1) = \bar\sigma(b_2). \end{cases}$$

This exact characterization of the valuations of the cycle centers can be used to determine the exact valuations of all selector vertices. We begin by considering the case $G_n = M_n$.

Corollary 6.1.17. *Let $G_n = M_n$. Let $\sigma \in \rho(\sigma_0)$ be well-behaved, $i \in [n]$ and define*

$$\lambda_i^{\mathrm{M}} := \min\{\ell \geq i \colon \sigma(b_\ell) = g_\ell \vee \sigma(g_\ell) = F_{\ell,0} \vee \sigma(s_{\ell,\bar\sigma(g_\ell)}) = b_1 \vee \neg\bar\sigma(d_\ell)\}.$$

Then $\mathrm{rVal}_\sigma^{\mathrm{M}}(g_i) = \sum_{\ell=i}^{\lambda-1} W_\ell^{\mathrm{M}} - \mathrm{rVal}_\sigma^{\mathrm{M}}(g_\lambda)$ *where* $\lambda := \lambda_i^{\mathrm{M}}$ *and*

$$\mathrm{rVal}_\sigma^{\mathrm{M}}(g_\lambda) = \begin{cases} \mathrm{rVal}_\sigma^{\mathrm{M}}(b_\lambda), & \text{if } \bar\sigma(b_\lambda),\\ \langle g_\lambda\rangle + \frac{1}{2}\mathrm{rVal}_\sigma^{\mathrm{M}}(g_1) + \frac{1}{2}\mathrm{rVal}_\sigma^{\mathrm{M}}(b_2), & \text{if } \neg\bar\sigma(b_\lambda) \wedge \bar\sigma(eg_\lambda) \wedge \bar\sigma(eb_\lambda),\\ \langle g_\lambda\rangle + \mathrm{rVal}_\sigma^{\mathrm{M}}(g_1), & \text{if } \neg\bar\sigma(b_\lambda) \wedge \bar\sigma(eg_\lambda) \wedge \neg\bar\sigma(eb_\lambda),\\ \langle g_\lambda\rangle + \mathrm{rVal}_\sigma^{\mathrm{M}}(b_2), & \text{if } \neg\bar\sigma(b_\lambda) \wedge \neg\bar\sigma(eg_\lambda) \wedge \bar\sigma(eb_\lambda),\\ \langle g_\lambda, s_{\lambda,\bar\sigma(g_\lambda)}\rangle + \mathrm{rVal}_\sigma^{\mathrm{M}}(b_1), & \text{if } \neg\bar\sigma(b_\lambda) \wedge \bar\sigma(d_\lambda) \wedge \neg\bar\sigma(s_\lambda),\\ W_\lambda^{\mathrm{M}} + \mathrm{rVal}_\sigma^{\mathrm{M}}(b_{\lambda+2}), & \text{otherwise.} \end{cases}$$

Proof. To simplify notation let $\lambda := \lambda_i^{\mathrm{M}}$. By the definition of λ, for all $\ell \in \{i, \dots, \lambda-1\}$, it holds that $\sigma(g_\ell) = F_{\ell,1}, \bar\sigma(d_{\ell,1})$ and $\sigma(s_{\ell,1}) = h_{\ell,1}$. This implies the first part of the statement as this yields $\mathrm{rVal}_\sigma^{\mathrm{M}}(g_\ell) = W_\ell^{\mathrm{M}} + \mathrm{rVal}_\sigma^{\mathrm{M}}(g_{\ell+1})$ for each such index.

Thus consider $\mathrm{rVal}_\sigma^{\mathrm{M}}(g_\lambda)$. The first four cases follow immediately resp. by Lemma 6.1.15. Consider the case $\neg\bar\sigma(b_\lambda) \wedge \bar\sigma(d_\lambda) \wedge \neg\bar\sigma(s_\lambda)$. Then $\sigma(s_{\lambda,\bar\sigma(g_\lambda)}) = b_1$ and the statement follows from $\mathrm{rVal}_\sigma^{\mathrm{M}}(F_{\lambda,\bar\sigma(g_\lambda)}) = \mathrm{rVal}_\sigma^{\mathrm{M}}(s_{\lambda,\bar\sigma(g_\lambda)})$. Hence consider the "otherwise" case, implying $\neg\bar\sigma(b_{\lambda_i}) \wedge \bar\sigma(d_\lambda) \wedge \bar\sigma(s_\lambda)$. By the definition of λ, this yields $\sigma(g_{\lambda_i}) = F_{\lambda,0}$. □

We now prove the corresponding statement for the case $G_n = S_n$.

Corollary 6.1.18. *Let $G_n = S_n$. Let $\sigma \in \rho(\sigma_0)$ be well-behaved, $i \in [n]$ and define*

$$\lambda_i^{\mathrm{S}} := \min\{\ell \geq i \colon \sigma(b_\ell) = g_\ell \vee \sigma(g_\ell) = F_{\ell,0} \vee \sigma(s_{\ell,\bar\sigma(g_\ell)}) = b_1 \vee \sigma(b_{\ell+1}) = g_{\ell+1}\}.$$

Then $\mathrm{rVal}_\sigma^{\mathrm{S}}(g_i) = \bigcup_{i'=i}^{\lambda_i-1} W_{i'}^{\mathrm{S}} \cup \mathrm{rVal}_\sigma^{\mathrm{S}}(g_{\lambda_i})$, *where* $\lambda := \lambda_i^{\mathrm{S}}$ *and*

$$\mathrm{rVal}_\sigma^{\mathrm{S}}(g_\lambda) = \begin{cases} \mathrm{rVal}_\sigma^{\mathrm{S}}(b_\lambda) & \text{if } \sigma(b_\lambda) = g_\lambda, \\ \{g_\lambda\} \cup \mathrm{rVal}_\sigma^{\mathrm{S}}(g_1) & \text{if } \neg\bar\sigma(b_\lambda) \wedge \bar\sigma(eg_\lambda) \wedge \neg\bar\sigma(eb_\lambda) \wedge \mu^\sigma \neq 1, \\ \{g_\lambda\} \cup \mathrm{rVal}_\sigma^{\mathrm{S}}(b_2) & \text{if } \neg\bar\sigma(b_\lambda) \wedge \bar\sigma(eb_\lambda) \wedge \neg\bar\sigma(eg_\lambda) \wedge \mu^\sigma = 1 \\ & \quad \wedge (\neg\bar\sigma(s_\lambda) \vee \bar\sigma(b_{\lambda+1}) = \bar\sigma(g_\lambda)), \\ \{g_\lambda\} \cup \mathrm{rVal}_\sigma^{\mathrm{S}}(g_1) & \text{if } \neg\bar\sigma(b_\lambda) \wedge \bar\sigma(eb_\lambda) \wedge \bar\sigma(eg_\lambda) \wedge \bar\sigma(g_1) \neq \bar\sigma(b_2), \\ \{g_\lambda\} \cup \mathrm{rVal}_\sigma^{\mathrm{S}}(b_2) & \text{if } \neg\bar\sigma(b_\lambda) \wedge \bar\sigma(eb_\lambda) \wedge \bar\sigma(eg_\lambda) \wedge \bar\sigma(g_1) = \bar\sigma(b_2), \\ \{g_\lambda, s_{\lambda,\bar\sigma(g_\lambda)}\} \cup \mathrm{rVal}_\sigma^{\mathrm{S}}(b_1) & \text{if none of the above and } \sigma(s_{\lambda,\bar\sigma(g_\lambda)}) = b_1, \\ W_\lambda^{\mathrm{S}} \cup \mathrm{rVal}_\sigma^{\mathrm{S}}(b_{\lambda+2}) & \text{if none of the above and } \sigma(g_\lambda) = F_{\lambda,0}, \\ W_\lambda^{\mathrm{S}} \cup \mathrm{rVal}_\sigma^{\mathrm{S}}(b_{\lambda+1}) & \text{otherwise.} \end{cases}$$

Proof. Let $\lambda := \lambda_i^{\mathrm{S}}$ and $\ell \in \{i, \dots, \lambda-1\}$. We prove $\mathrm{rVal}_\sigma^{\mathrm{S}}(F_{\ell,\bar\sigma(g_\ell)}) = \mathrm{rVal}_\sigma^{\mathrm{S}}(s_{\ell,\bar\sigma(g_\ell)})$. Since $\ell < \lambda$, it follows that $\sigma(b_\ell) = b_{\ell+1}, \sigma(g_\ell) = F_{\ell,1}$, $\sigma(s_{\ell,1}) = h_{\ell,1}$ and $\sigma(b_{\ell+1}) = b_{\ell+2}$. We show that this implies that none of the cases 3,4,6 and 7 of Lemma 6.1.16 can occur.

If the conditions of the third case were true, then $\sigma(s_{\ell,1}) = b_1$ by Property (EG1), contradicting $\sigma(s_{\ell,1}) = h_{\ell,1}$. If the conditions of the fourth case were true, then $\neg\bar\sigma(s_{\ell,1}) \vee \bar\sigma(b_{\ell+1}) = 1$. But, since $\sigma(s_{\ell,1}) = h_{\ell,1}$ and $\sigma(b_{\ell+1}) = b_{\ell+2}$, this cannot hold. If the conditions of the sixth or seventh case were true, then $\bar\sigma(eb_{\ell,1}) \wedge \bar\sigma(eg_{\ell,1})$. But then, since $\sigma(s_{\ell,1}) = h_{\ell,1}$, Property (EBG1) implies $\sigma(b_{\ell+1}) = g_{\ell+1}$, contradicting $\sigma(b_{\ell+1}) = b_{\ell+2}$.

Hence $\mathrm{rVal}_\sigma^{\mathrm{S}}(F_{\ell,\bar\sigma(g_\ell)}) = \mathrm{rVal}_\sigma^{\mathrm{S}}(s_{\ell,\bar\sigma(g_\ell)})$. Since also $\bar\sigma(g_\ell) = 1$ and $\sigma(s_{\ell,1}) = h_{\ell,1}$, this implies the first part of the statement. It thus remains to investigate $\mathrm{rVal}_\sigma^{\mathrm{S}}(g_\lambda)$.

The first five statements follow directly from Lemma 6.1.16. Thus consider the sixth. Since none of the five previous cases must hold, one of the following holds:

1. $\sigma(b_\lambda) = b_{\lambda+1} \wedge \neg\bar\sigma(eb_\lambda) \wedge [\neg\bar\sigma(eg_\lambda) \vee \mu^\sigma = 1]$
2. $\sigma(b_\lambda) = b_{\lambda+1} \wedge \neg\bar\sigma(eg_\lambda) \wedge [\neg\bar\sigma(eb_\lambda) \vee \mu^\sigma \neq 1 \vee (\bar\sigma(s_\lambda) \wedge \bar\sigma(b_{\lambda+1}) \neq \bar\sigma(g_\lambda))]$.

We now consider these two cases together with the assumption $\sigma(s_{\lambda,\bar\sigma(g_\lambda)}) = b_1$. It again suffices to show $\mathrm{rVal}_\sigma^{\mathrm{S}}(F_{\lambda,\bar\sigma(g_\lambda)}) = \mathrm{rVal}_\sigma^{\mathrm{S}}(s_{\lambda,\bar\sigma(g_\lambda)})$.

1. If $\neg\bar\sigma(eb_\lambda) \wedge \neg\bar\sigma(eg_\lambda)$, then $\bar\sigma(d_\lambda)$. Consequently, by Lemma 6.1.16, the statement follows. Otherwise we have $\neg\bar\sigma(eb_\lambda) \wedge \bar\sigma(eg_\lambda) \wedge \mu^\sigma = 1$. But then, the conditions of the second case of Lemma 6.1.16 hold and the statement follows again.
2. As previously, the statement follows if $\neg\bar\sigma(eg_\lambda) \wedge \neg\bar\sigma(eb_\lambda)$. Since $\sigma(s_{\lambda,\bar\sigma(g_i)}) = b_1$ by assumption, the conditions can thus only be fulfilled if $\neg\bar\sigma(eg_\lambda) \wedge \bar\sigma(eb_\lambda) \wedge \mu^\sigma \neq 1$. But then, the conditions of case 5 of Lemma 6.1.16 hold and the statement follows.

Next consider the seventh case. Then $\bar\sigma(g_\lambda) = 0$. Note that $\sigma(s_{\lambda,0}) = h_{\lambda,0}$ holds by assumption and that it again suffices to show $\mathrm{rVal}_\sigma^{\mathrm{S}}(F_{\lambda,0}) = \mathrm{rVal}_\sigma^{\mathrm{S}}(s_{\lambda,0})$. We thus again investigate the two cases mentioned before together with the assumption $\sigma(s_{\lambda,0}) = h_{\lambda,0}$ and $\bar\sigma(g_\lambda) = 0$.

1. Here, the same arguments used before can be applied again.
2. If either $\neg\bar\sigma(eg_\lambda) \wedge \neg\bar\sigma(eb_\lambda)$ or $\neg\bar\sigma(eg_\lambda) \wedge \bar\sigma(eb_\lambda) \wedge \mu^\sigma \neq 1$, then the statement follows by the previously given arguments. Hence consider the case $\neg\bar\sigma(eg_\lambda) \wedge \bar\sigma(eb_\lambda) \wedge$

$\mu^\sigma = 1 \wedge (\bar{\sigma}(s_\lambda) \wedge \bar{\sigma}(b_{\lambda+1}) \neq \bar{\sigma}(g_\lambda))$. But then, the conditions of the fifth case of Lemma 6.1.16 are fulfilled again, implying the statement.

Finally, consider the eighth case. We then have $\sigma(b_\lambda) = b_{\lambda+1}, \sigma(s_{\lambda,1}) = h_{\lambda,1}$ and $\bar{\sigma}(g_\lambda) = 1$. By the definition of λ, we thus need to have $\sigma(b_{\lambda+1}) = g_{\lambda+1}$. It hence suffices to prove that $\mathrm{rVal}_\sigma^{\mathrm{S}}(F_{\lambda,1}) = \mathrm{rVal}_\sigma^{\mathrm{S}}(s_{\lambda,1})$. This however follows by the same arguments used in the last case. □

To conclude the characterization of the vertex valuations, we state one additional lemma. This lemma allows us to simplify the evaluation of the valuation of the cycle centers under certain conditions without having to check the conditions of Lemma 6.1.15 resp. 6.1.16. It will in particular be used when analyzing cycle centers during phase 1.

Lemma 6.1.19. *Let $\sigma \in \rho(\sigma_0)$ be well-behaved, let $i \in [n], j \in \{0,1\}$ and consider the cycle center $F_{i,j}$. Assume that σ has Properties (ESC1), (EV1)$_1$ and (USV1)$_i$. It then holds that $\mathrm{rVal}_\sigma^*(F_{i,j}) = \mathrm{rVal}_\sigma^*(s_{i,j})$ if $\bar{\sigma}(d_{i,j})$ and $\mathrm{rVal}_\sigma^*(F_{i,j}) = \mathrm{rVal}_\sigma^*(b_1)$ otherwise.*

Proof. If $\bar{\sigma}(d_{i,j})$, then the statement follows from Lemma 6.1.15 resp. 6.1.16. Hence consider the case that $F_{i,j}$ is not closed and let $G_n = S_n$. We show that the conditions of either the first or the fourth case of Lemma 6.1.16 are fulfilled and that the corresponding valuations can be expressed as $\mathrm{rVal}_\sigma^{\mathrm{S}}(b_1)$.

By Property (ESC1), the last two cases of Lemma 6.1.16 cannot occur. Let, for the sake of contradiction, the conditions of the second case be fulfilled, i.e., $\bar{\sigma}(eg_{i,j}), \neg\bar{\sigma}(eb_{i,j})$ and $\mu^\sigma = 1$. Then $\sigma(b_1) = b_2$. By Property (EV1)$_1$ and the definition of the induced bit state, this implies $\beta_1 = 0$. Hence, by Property (ESC1), $\sigma(e_{*,*,*}) = b_2$. Since $F_{i,j}$ is not closed, this implies that there is at least one $k \in \{0,1\}$ such that $\sigma(d_{i,j,k}) = e_{i,j,k}$ and $\sigma(e_{i,j,k}) = b_2$. But then $\bar{\sigma}(eb_{i,j})$ contradicting $\neg\bar{\sigma}(eb_{i,j})$.

Now let, for the sake of contradiction, the conditions of the fifth case of Lemma 6.1.16 be fulfilled, i.e., $\bar{\sigma}(eb_{i,j}), \neg\bar{\sigma}(eg_{i,j})$ and either $\mu^\sigma \neq 1$ or $\bar{\sigma}(s_{i,j}) \wedge \bar{\sigma}(b_{i+1}) \neq j$. If $\mu^\sigma \neq 1$, we can deduce $\bar{\sigma}(eg_{i,j})$ by the same arguments used for the second case, again resulting in a contradiction. Thus let $\sigma(s_{i,j}) = h_{i,j}$ and $\bar{\sigma}(b_{i+1}) \neq j$. Then, by Property (USV1)$_i$, $j = \beta_{i+1}$. But then, the other condition states $\bar{\sigma}(b_{i+1}) \neq \beta_{i+1}$ which is a contradiction since $\beta_{i+1} = \bar{\sigma}(d_{i,\mathfrak{b}_{i+1}})$ by definition.

Consider the third case of Lemma 6.1.16. Then, $\mu^\sigma \neq 1$ implies $\sigma(b_1) = g_1$ by Lemma 6.1.3. Thus, $\mathrm{rVal}_\sigma^{\mathrm{S}}(b_1) = \mathrm{rVal}_\sigma^{\mathrm{S}}(g_1) = \mathrm{rVal}_\sigma^{\mathrm{S}}(F_{i,j})$. Consider the fourth case of Lemma 6.1.16. Then $\mu^\sigma = 1$, implying $\mathrm{rVal}_\sigma^{\mathrm{S}}(b_1) = \mathrm{rVal}_\sigma^{\mathrm{S}}(b_2) = \mathrm{rVal}_\sigma^{\mathrm{S}}(F_{i,j})$ as $\sigma(b_1) = b_2$.

Now let $G_n = M_n$ and let $F_{i,j}$ not be closed. By Property (ESC1), $F_{i,j}$ cannot escape towards both g_1 and b_2. By Property (EV1)$_1$, $\beta_1 = 1$ if and only if$\sigma(b_1) = g_1$. The statement thus follows since Property (ESC1) implies that $F_{i,j}$ escapes to g_1 if $\beta_1 = 1$ and to b_2 if $\beta_1 = 0$. □

This concludes our general results on the vertex valuations. As it will turn out that every strategy calculated by the strategy improvement resp. policy iteration algorithm is well-behaved. Consequently, these characterization can be applied to all strategies that are considered in the following sections.

6.2. The Application of Individual Improving Switches

This section contains technical details related to the application of individual improving switches and the different phases of a single transition. We consider a fixed number $\mathfrak{b} \in \mathfrak{B}_n$. Before analyzing the single phases, we develop some general statements that are either not related to a single phase or are used repeatedly in the upcoming proofs. Henceforth, $\mathfrak{b} \in \mathfrak{B}_n$ is a fixed number and $\nu := \ell(\mathfrak{b}+1)$ denotes the least significant set bit of $\mathfrak{b}+1$. We further define the abbreviation $\sum(\mathfrak{b}, i) := \sum_{l<i} \mathfrak{b}_l \cdot 2^{l-1}$.

Most of the proofs of this section are deferred to Appendix A.2.

Basic statements and statements independent of the phases

The first lemma enables us to compare the valuations of cycle centers in M_n for several well-behaved strategies calculated by the strategy improvement algorithm.

Lemma 6.2.1. *Let $G_n = M_n$. Let $\sigma \in \rho(\sigma_0)$ be a well-behaved phase-k-strategy for some $\mathfrak{b} \in \mathfrak{B}_n$ having Property (USV1)$_i$ and Property (EV1)$_{i+1}$ for some $i \in [n]$ where $k \in [5]$. If $F_{i,0}$ and $F_{i,1}$ are in the same state and if either $i \geq \nu$ or σ has Property (REL1), then $\mathrm{Val}_\sigma^{\mathrm{M}}(F_{i,\beta_{i+1}}) > \mathrm{Val}_\sigma^{\mathrm{M}}(F_{i,1-\beta_{i+1}})$.*

As mentioned at the beginning of Section 6.1, formal lemmas describing the applications of the improving switches are proven only for well-behaved strategies. We will always prove that the strategies obtained by the application of improving switches remain well-behaved and prove that canonical strategies are well-behaved. Consequently, *all* strategies calculated by the strategy improvement are well-behaved.

As a basis for these arguments, we prove that canonical strategies are well-behaved.

Lemma 6.2.2. *Let $\sigma_\mathfrak{b}$ be a canonical strategy for some $\mathfrak{b} \in \mathfrak{B}_n$. Then $\sigma_\mathfrak{b}$ is well-behaved.*

Proof. Let $\sigma := \sigma_\mathfrak{b}$ and let $i \in [n]$ such that $\sigma(b_i) = g_i$. Then, by the definition of a canonical strategy, we have $\mathfrak{b}_i = 1$, implying $\sigma(g_i) = F_{i,\mathfrak{b}_{i+1}}$. Hence, $\bar{\sigma}(g_i) = \mathfrak{b}_{i+1} = \bar{\sigma}(b_{i+1})$. Thus $\mathcal{I}^\sigma = \emptyset$, implying

$$\mu^\sigma = \min\{i \in [n+1] \colon \sigma(b_i) = b_{i+1}\}. \tag{6.1}$$

We prove that σ has all properties of Table 6.1. We investigate each property and show that either its premise is false or that both the premise and the conclusion are true.

(S1) Let $i \geq \mu^\sigma$ with $\sigma(b_i) = g_i$. Then, $\bar{\sigma}(g_i) = \mathfrak{b}_{i+1}$ and $\bar{\sigma}(s_{i,\mathfrak{b}_{i+1}}) = 1$, hence $\bar{\sigma}(s_i)$.

(S2) Let $i < \mu^\sigma$. Then, by (6.1), $\sigma(b_i) = g_i$. By Property (S1), this implies $\bar{\sigma}(s_i)$.

(B1) Let $i < \mu^\sigma - 1$. Then $\sigma(b_i) = g_i$ by (6.1), hence the premise is false.

(B2) By (6.1), we have $\sigma(b_{\mu^\sigma-1}) = g_{\mu^\sigma-1}$, so the premise is false.

(B3) Let $i \in [n]$ with $\sigma(b_{i+1}) = b_{i+2}$. This implies $\mathfrak{b}_{i+1} = 0$, so $\sigma(s_{i,1}) = b_1$.

(BR1) Let $i < \mu^\sigma$. This implies $\sigma(b_i) = g_i$ and $\sigma(g_i) = F_{i,\mathfrak{b}_{i+1}}$. Assume $\mathfrak{b}_{i+1} = 0$. We then have $\mu^\sigma \leq i+1$ by (6.1), implying $\mu^\sigma = i+1$. But then $\sigma(g_i) = F_{i,0}$ and $i = \mu^\sigma - 1$. Now assume $\mathfrak{b}_{i+1} = 1$. We then have $\mu^\sigma > i$ and $\mu^\sigma \neq i+1$, so $\mu^\sigma > i+1$. But then, $\sigma(g_i) = F_{i,1}$ and $i < \mu^\sigma - 1$ and in particular $i \neq \mu^\sigma - 1$.

(BR2) Since $\mathcal{I}^\sigma = \emptyset$, $i < \mu^\sigma$ implies $\bar{\sigma}(d_i)$ by Corollary 6.1.5, so $\neg\bar{\sigma}(eg_{i,\bar{\sigma}(eg_i)})$.

(D1) Since $\sigma(b_i) = g_i$ implies $\bar{\sigma}(g_i) = \mathfrak{b}_{i+1}$ and that $F_{i,\mathfrak{b}_{i+1}}$ is closed, both premise and conclusion are true.

(D2) This follows by the same arguments used in the last case since $\sigma(b_i) = g_i$ for any $i < \mu^\sigma$ by (6.1).

(MNS1) By Lemma 6.1.6, the premise implies $m_b^\sigma = 2$, hence $\sigma(b_2) = g_2$ and $\sigma(b_1) = b_2$. Consequently, $\sigma(g_1) = F_{i,1}$ as $m_b^\sigma \leq \overline{m}_g^\sigma, \overline{m}_s^\sigma$, contradicting the definition of a canonical strategy if $G_n = S_n$.

(MNS2) Assuming that there was some index $i < \overline{m}_g^\sigma < \overline{m}_s^\sigma, m_b^\sigma$ and $\mu^\sigma = 1$ implies $\sigma(b_1) = b_2$ as well as $\sigma(g_1) = F_{1,1}$, contradicting the definition of a canonical strategy for the case $G_n = S_n$. If $G_n = M_n$, then we need to have $\mathfrak{b}_2 = 1$, implying $m_b^\sigma = 2$. This is however a contradiction to $1 < \overline{m}_g^\sigma < m_b^\sigma$.

(MNS3) Let $\mu^\sigma = 1$ and assume there was some $i < \overline{m}_s^\sigma \leq \overline{m}_g^\sigma < m_b^\sigma$. As $G_n = M_n$ by assumption, $\sigma(s_{1,1}) = h_{1,1}$ then implies $\mathfrak{b}_2 = 1$ and thus $m_b^\sigma = 2$, contradicting $1 < \overline{m}_s^\sigma < m_b^\sigma$.

(MNS4) Let $\mu^\sigma = 1$, assume $\overline{m}_s^\sigma \leq \overline{m}_g^\sigma < m_b^\sigma$ and let $i := \overline{m}_s^\sigma$. We prove that the premise either yields a contradiction or implies $\bar{\sigma}(eb_i) \wedge \neg\bar{\sigma}(eg_i)$. Since $\mu^\sigma = 1$ implies $\sigma(b_1) = b_2$ and thus $\mathfrak{b}_1 = 0$, the definition of a canonical strategy implies that it suffices to prove that $F_{i,\bar{\sigma}(g_i)}$ is not closed. This follows from the definition of a canonical strategy if $i = 1$, so let $i > 1$. Then $1 < i = \overline{m}_s^\sigma \leq \overline{m}_g^\sigma$, hence $\sigma(g_1) = F_{1,1}$. This however contradicts the definition of a canonical strategy if $G_n = S_n$. If $G_n = M_n$, then this implies $\mathfrak{b}_2 = 1$ and hence $\sigma(b_2) = g_2$, thus $m_b^\sigma = 2$. But this contradicts the premise $\overline{m}_g^\sigma < m_b^\sigma$.

(MNS5) Let $\mu^\sigma = 1$. We show that there is no $i < \overline{m}_s^\sigma < m_b^\sigma \leq \overline{m}_g^\sigma$. Assume there was such an index i, implying that $1 < \overline{m}_s^\sigma < m_b^\sigma \leq \overline{m}_g^\sigma$. Thus, $\sigma(g_1) = F_{1,1}$ and $\sigma(s_{1,1}) = h_{1,1}$, implying $\mathfrak{b}_2 = 1$. But then $m_b^\sigma = 2$, contradicting $1 < \overline{m}_s^\sigma < m_b^\sigma$.

(MNS6) Let $\mu^\sigma = 1$. If $\overline{m}_s^\sigma \neq 1$, the same arguments used when discussing Property (MNS5) can be applied. However, for $\overline{m}_s^\sigma = 1$, the statement follows since both cycle centers of level 1 are open and since these cycle centers escape to b_2.

(EG1) By $\mu^\sigma = 1$, we have $\sigma(b_1) = b_2$, implying $\mathfrak{b}_1 = 0$. Thus, any cycle center which is not closed escapes towards b_2 by definition, hence the premise is incorrect.

(EG2) Follows by the same arguments used in the last case.

(EG3) Assume that there is some cycle center escaping towards g_1. Then $\mathfrak{b}_1 = 1$. This implies $\sigma(b_1) = g_1$ and by the same arguments used earlier in this proof, this implies $\bar{\sigma}(s_1)$.

(EG4) This follows by the same arguments used when discussing Properties (EG1) and (EG2).

(EG5) It is easy to see that $\bar{\sigma}(s_{i,j})$ implies $\bar{\sigma}(b_{i+1}) = j$.

(EB*) Every premise of any of the properties (EB*) contains $\sigma(b_1) = g_1$. Hence, we always have $\mathfrak{b} \bmod 2 = 1$, implying that no cycle center can escape towards b_2. But this implies that the premise any of these properties is false.

(EBG*) By the definition of a canonical strategy, no cycle center can escape towards both b_2 and g_1.

(DN1) By the definition of a canonical strategy, $\bar{\sigma}(d_n)$ holds if and only if $\sigma(b_n) = g_n$. Hence both the premise and the conclusion are true.

(DN2) If $\bar{\sigma}(d_n)$ the statement follows analogously as in the last case. Hence assume $\overline{m}_g^\sigma = n$. Then $\sigma(g_i) = F_{i,1}$ for all $i < n$, so in particular, $\sigma(g_1) = F_{1,1}$. But, by the definition of a canonical strategy, this immediately implies $\mathfrak{b}_1 = 1$ and $\sigma(b_1) = g_1$ if $G_n = S_n$ and $\mathfrak{b}_2$ and thus $\sigma(b_2) = g_2$ if $G_n = M_n$. □

Our goal is to prove Lemma 5.3.12 next as this statement describes the set of improving switches of canonical strategies. Before doing so, we analyze the terms used in Table 5.6 to describe the occurrence records of the cycle vertices in more detail. As the proofs of the following lemmas are rather technical, they are deferred to Appendix A.2.

Lemma 6.2.3. *Let $\mathfrak{b} \in \mathfrak{B}_n$. If $\mathbf{1}_{j=0}\mathrm{lfn}(\mathfrak{b}, i+1) + \mathbf{1}_{j=1}\mathrm{lufn}(\mathfrak{b}, i+1) = 0$ for $i \in [n], j \in \{0,1\}$, then $\ell^{\mathfrak{b}}(i,j,k) \geq \mathfrak{b}$ for , $k \in \{0,1\}$. Otherwise, the following hold:*

Setting of bits	$\mathfrak{b}_i = 1 \wedge \mathfrak{b}_{i+1} = 1-j$	$\mathfrak{b}_i = 0 \wedge \mathfrak{b}_{i+1} = j$	$\mathfrak{b}_i = 0 \wedge \mathfrak{b}_{i+1} = 1-j$
$\ell^{\mathfrak{b}}(i,j,k) =$	$\left\lceil \frac{\mathfrak{b}+\sum(\mathfrak{b},i)+1-k}{2} \right\rceil$	$\left\lceil \frac{\mathfrak{b}+2^{i-1}+\sum(\mathfrak{b},i)+1-k}{2} \right\rceil$	$\left\lceil \frac{\mathfrak{b}-2^{i-1}+\sum(\mathfrak{b},i)+1-k}{2} \right\rceil$

Lemma 6.2.4. *Let $\mathfrak{b} \in \mathfrak{B}_n$ and $i \in [n]$ and $j \in \{0,1\}$ such that $\mathfrak{b}_i = 0$ or $\mathfrak{b}_{i+1} \neq j$. Then,*

$$\mathbf{1}_{j=0}\mathrm{lfn}(\mathfrak{b}, i+1) - \mathbf{1}_{j=1}\mathrm{lufn}(\mathfrak{b}, i+1) = \mathbf{1}_{j=0}\mathrm{lfn}(\mathfrak{b}+1, i+1) - \mathbf{1}_{j=1}\mathrm{lufn}(\mathfrak{b}+1, i+1).$$

Moreover, if $i \neq \nu$, then $\ell^{\mathfrak{b}}(i,j,k) + 1 = \ell^{\mathfrak{b}+1}(i,j,k)$.

Lemma 6.2.5. *Let $\sigma_{\mathfrak{b}}$ be a canonical strategy for $\mathfrak{b}$ such that its occurrence records are described by Table 5.6. Assume that $\sigma_{\mathfrak{b}}$ has Properties (OR1)$_{*,*,*}$ to (OR4)$_{*,*,*}$. Then, the following hold.*

1. *Let $i \in [n]$ and $j \in \{0,1\}$ and assume that either $\mathfrak{b}_i = 0$ or $\mathfrak{b}_{i+1} \neq j$. Then, it holds that $\phi^{\sigma_{\mathfrak{b}}}(d_{i,j,*}, F_{i,j}) \leq \lfloor(\mathfrak{b}+1)/2\rfloor$.*
2. *Let $j := \mathfrak{b}_{\nu+1}$. Then, $\phi^{\sigma_{\mathfrak{b}}}(d_{\nu,j,0}, F_{\nu,j}) = \lfloor(\mathfrak{b}+1)/2\rfloor$. In addition, $\nu = 1$ implies $\phi^{\sigma_{\mathfrak{b}}}(d_{\nu,j,1}, F_{\nu,j}) = \lfloor(\mathfrak{b}+1)/2\rfloor$ and $\nu > 1$ implies $\phi^{\sigma}(d_{\nu,j,1}, F_{\nu,j}) = \lfloor(\mathfrak{b}+1)/2\rfloor - 1$.*
3. *If $i = 1$, then $\sigma_{\mathfrak{b}}(d_{1,1-\mathfrak{b}_2,*}) \neq F_{1,1-\mathfrak{b}_2}$ and $\phi^{\sigma_{\mathfrak{b}}}(d_{1,1-\mathfrak{b}_2,0}, F_{1,1-\mathfrak{b}_2}) = \lfloor(\mathfrak{b}+1)/2\rfloor$.*

Lemma 6.2.6. *Let $\mathfrak{b} \in \mathfrak{B}_n$ and $i \in [n]$. It holds that $\mathrm{fl}(\mathfrak{b}, i) = \lfloor(\mathfrak{b}+2^{i-1})/2^i\rfloor$ and $\mathrm{fl}(\mathfrak{b}+1, i) = \mathrm{fl}(\mathfrak{b}, i) + \mathbf{1}_{i=\nu}$. In addition, for indices $i_1, i_2 \in [n]$ with $i_1 < i_2$ and $\mathfrak{b} \geq 2^{i_1-1}$ imply $\mathrm{fl}(\mathfrak{b}, i_1) > \mathrm{fl}(\mathfrak{b}, i_2)$. Furthermore, if $k := \frac{\mathfrak{b}+1}{2^{\nu-1}}$ and $x \in [\nu-1]$, then $\mathrm{fl}(\mathfrak{b}, \nu-x) = k \cdot 2^{x-1}$.*

Now all general statements required for the upcoming proofs and statements are in place. We begin by analyzing phase 1, or, more precisely, the statements that prove the application of improving switches until reaching phase 2.

We first restate Lemma 5.3.12 and provide its formal proof.

Lemma 5.3.12. *Let $\sigma_{\mathfrak{b}} \in \rho(\sigma_0)$ be a canonical strategy for $\mathfrak{b} \in \mathfrak{B}_n$. Then, $\sigma_{\mathfrak{b}}$ is a phase-1-strategy for $\mathfrak{b}$ and $I_{\sigma_{\mathfrak{b}}} = \{(d_{i,j,k}, F_{i,j}) : \sigma_{\mathfrak{b}}(d_{i,j,k}) \neq F_{i,j}\}$.*

Proof. It is easy to verify that canonical strategies are phase-1-strategies. To simplify notation, let $\sigma := \sigma_{\mathfrak{b}}$ and $\mathfrak{D}^\sigma := \{(d_{i,j,k}, F_{i,j}) : \sigma(d_{i,j,k}) \neq F_{i,j}\}$. It then suffices to show $I_\sigma = \mathfrak{D}^\sigma$. We thus have to prove that $\sigma(d_{i,j,k}) \neq F_{i,j}$ implies $\mathrm{Val}^*_\sigma(F_{i,j}) > \mathrm{Val}^*_\sigma(e_{i,j,k})$ and that there are no other improving switches.

Let $e = (d_{i,j,k}, F_{i,j})$ with $\sigma(d_{i,j,k}) \neq F_{i,j}$. By Lemma 6.2.2, σ is well-behaved, and the results of Section 6.1 can be applied. By Lemma 6.1.19, $\mathrm{rVal}^*_\sigma(F_{i,j}) = \mathrm{rVal}^*_\sigma(b_1)$. Consider the case $G_n = S_n$. Then, $\mathrm{Val}^{\mathrm{S}}_\sigma(F_{i,j}) = \{F_{i,j}, d_{i,j,k'}, e_{i,j,k'}\} \cup \mathrm{Val}^{\mathrm{S}}_\sigma(\sigma(e_{i,j,k'}))$ for some $k' \in \{0,1\}$, implying $\mathrm{Val}^{\mathrm{S}}_\sigma(e_{i,j,*}) \lhd \mathrm{Val}^{\mathrm{S}}_\sigma(F_{i,j})$ since $\sigma(e_{i,j,0}) = \sigma(e_{i,j,1})$ by Property (ESC1). Now let $G_n = M_n$. By Property (ESC1) and Property $(\mathrm{EV1})_1$, $\mathrm{Val}^{\mathrm{M}}_\sigma(e_{*,*,*}) = \mathrm{Val}^{\mathrm{M}}_\sigma(b_1)$ for all escape vertices $e_{*,*,*}$. Thus, $\mathrm{Val}^{\mathrm{M}}_\sigma(F_{i,j}) = (1-\varepsilon)\mathrm{Val}_\sigma(b_1) + \varepsilon \mathrm{Val}^{\mathrm{M}}_\sigma(s_{i,j})$ and it suffices to prove $\mathrm{Val}^{\mathrm{M}}_\sigma(s_{i,j}) > \mathrm{Val}^{\mathrm{M}}_\sigma(b_1)$. If $\sigma(s_{i,j}) = b_1$, then this follows immediately from $\mathrm{Val}^{\mathrm{M}}_\sigma(s_{i,j}) = \langle s_{i,j}\rangle + \mathrm{Val}^{\mathrm{M}}_\sigma(b_1) > \mathrm{Val}^{\mathrm{M}}_\sigma(b_1)$. Thus assume $\sigma(s_{i,j}) = h_{i,j}$. Then, by Property $(\mathrm{USV1})_i$, $j = \beta_{i+1}$ and $\mathrm{rVal}^{\mathrm{M}}_\sigma(s_{i,j}) = \langle s_{i,j}, h_{i,j}\rangle + \mathrm{rVal}^{\mathrm{M}}_\sigma(b_{i+1})$ by Property $(\mathrm{EV1})_{i+1}$. Hence $F_{i,j}$ is the active cycle center of level i. Since it is not closed by assumption, we thus have $\mathfrak{b}_i = 0$ by Property $(\mathrm{EV1})_i$ and $i \geq \mu^\sigma$ by Property (REL1), implying $\mathrm{rVal}^{\mathrm{M}}_\sigma(b_{i+1}) = L^{\mathrm{M}}_{i+1}$. Furthermore, by Lemma 6.1.13 and Corollary 6.1.5, $\mathrm{rVal}^{\mathrm{M}}_\sigma(b_1) = B^{\mathrm{M}}_1$. If $B^{\mathrm{M}}_1 = L^{\mathrm{M}}_1$, then Lemma 6.1.10 (4.) implies $L^{\mathrm{M}}_1 < \langle s_{i,j}, h_{i,j}\rangle + L^{\mathrm{M}}_{i+1}$. If $B^{\mathrm{M}}_1 = R^{\mathrm{M}}_1$, then $i \geq \mu^\sigma$ and Lemma 6.1.10 (3.) yields $R^{\mathrm{M}}_1 < \langle s_{i,j}, h_{i,j}\rangle + L^{\mathrm{M}}_{i+1}$. Hence $\mathrm{rVal}^{\mathrm{M}}_\sigma(s_{i,j}) > \mathrm{rVal}^{\mathrm{M}}_\sigma(b_1)$, implying $\mathrm{rVal}^{\mathrm{M}}_\sigma(F_{i,j}) > \mathrm{rVal}^{\mathrm{M}}_\sigma(e_{i,j,*})$. Consequently, $\sigma(d_{i,j,k}) \neq F_{i,j}$ implies $\mathrm{Val}^*_\sigma(F_{i,j}) > \mathrm{Val}^*_\sigma(\sigma(d_{i,j,k}))$ in both S_n and M_n, so $(d_{i,j,k}, F_{i,j}) \in I_\sigma$. It remains to show that there are no other improving switches.

We first show that there is no improving switch $e = (b_i, *)$. Let $i \in [n]$ and $\sigma(b_i) = g_i$. We need to show $\mathrm{Val}^*_\sigma(b_{i+1}) \preceq \mathrm{Val}^*_\sigma(g_i)$. Since $\sigma(b_i) = g_i$ implies $\mathrm{rVal}^*_\sigma(b_i) = \mathrm{rVal}^*_\sigma(g_i)$, it suffices to show $\mathrm{rVal}^*_\sigma(b_{i+1}) \prec \mathrm{rVal}^*_\sigma(b_i)$. By Lemma 6.1.13 and Corollary 6.1.5, we have $\mathrm{rVal}^*_\sigma(b_{i+1}) = B^*_{i+1}$ and $\mathrm{rVal}^*_\sigma(b_i) = B^*_i$. Assume $i < \mu^\sigma$. Then $B^*_i = R^*_i$ and the statement follows directly if $B^*_{i+1} = R^*_{i+1}$. If $B^*_{i+1} = L^*_{i+1}$, we need to have $i+1 = \mu^\sigma$. But then $\sigma(b_{i+1}) = b_{i+2}$ and thus $L^*_{i+1} \prec R^*_i = W^*_i \oplus L^*_{i+1}$. Thus assume $i \geq \mu^\sigma$. Then $B^*_{i+1} = L^*_{i+1}$ and $B^*_i = L^*_i$ and the statement follows by $\sigma(b_i) = g_i$.

Now let $\sigma(b_i) = b_{i+1}$. We prove $\mathrm{rVal}^*_\sigma(g_i) \prec \mathrm{rVal}^*_\sigma(b_{i+1})$. Note that $\mathfrak{b}_i = 0$ implies $\mu^\sigma \leq i$, so $\mathrm{rVal}^*_\sigma(b_{i+1}) = L^*_{i+1} = L^*_i$. We use Corollary 6.1.17 resp. Corollary 6.1.18 to compute the valuation of g_i. We thus need to evaluate λ^{M}_i resp. λ^{S}_i. If $\sigma(g_i) = F_{i,0}$, we have $\lambda^*_i = i$. If $\sigma(g_i) = F_{i,1} \wedge \sigma(b_{i+1}) = b_{i+2}$, we have $\bar{\sigma}(g_i) = 1 \neq 0 = \mathfrak{b}_{i+1}$. Thus, by the definition of a canonical strategy, $\sigma(s_{i,\bar{\sigma}(g_i)}) = b_1$, implying $\lambda^*_i = i$. Hence assume $\sigma(g_i) = F_{i,1} \wedge \sigma(b_{i+1}) = g_{i+1}$. If $G_n = S_n$, then this implies $\lambda^{\mathrm{S}}_i = i$ by the definition of λ^{S}_i. If $G_n = M_n$, then, $\lambda^{\mathrm{M}}_i = i$ follows since we need to have $\neg\overline{\sigma_{\mathfrak{b}}}(d_i)$ due to $\sigma(b_i) = b_{i+1}, \sigma(g_i) = F_{i,1}$ and $\sigma(b_{i+1}) = g_{i+1}$. Hence, $\lambda^*_i = i$ in both cases.

Let $G_n = S_n$ and consider the different cases listed in Corollary 6.1.18 describing the vertex valuations for selection vertices in S_n. In order to show the statement we distinguish the cases listed in that corollary. Note that the first case cannot occur.

- Let $\bar{\sigma}(eg_i), \neg\bar{\sigma}(eb_i)$ and $\mu^\sigma \neq 1$. Then, $\mathrm{rVal}^{\mathrm{S}}_\sigma(g_i) = \{g_i\} \cup \mathrm{rVal}^{\mathrm{S}}_\sigma(g_1)$ by Corollary 6.1.18. Since $\mu^\sigma \neq 1$ implies $1 < \mu^\sigma$, $\mathrm{rVal}^{\mathrm{S}}_\sigma(g_1) = R^{\mathrm{S}}_1$ by Lemma 6.1.12.

Hence, since $i \geq \mu^\sigma$,

$$\mathrm{rVal}_\sigma^{\mathrm{S}}(g_i) = \{g_i\} \cup R_1^{\mathrm{S}} \lhd \bigcup_{i' \geq i} \{W_{i'}^{\mathrm{S}} \colon \sigma(b_{i'}) = g_{i'}\} = L_i^{\mathrm{S}}.$$

- Let $\bar{\sigma}(eb_i), \neg\bar{\sigma}(eg_i), \mu^\sigma = 1$ and $(\neg\bar{\sigma}(s_{i,j}) \vee \bar{\sigma}(b_{i+1}) = \bar{\sigma}(g_i))$. Then, by Corollary 6.1.18, $\mathrm{rVal}_\sigma^{\mathrm{S}}(g_i) = \{g_i\} \cup \mathrm{rVal}_\sigma^{\mathrm{S}}(b_2)$. Since $\mu^\sigma = 1$ we have $\mathrm{rVal}_\sigma^{\mathrm{S}}(b_2) = L_2^{\mathrm{S}}$. Thus, since $\sigma(b_i) = b_{i+1}$, we obtain $\mathrm{rVal}_\sigma^{\mathrm{S}}(g_i) = \{g_i\} \cup \mathrm{rVal}_\sigma^{\mathrm{S}}(b_2) = \{g_i\} \cup L_2^{\mathrm{S}} \lhd L_{i+1}^{\mathrm{S}}$.

This covers the first three cases. The fourth and fifth case cannot occur since they require a cycle center to escape towards both g_1 and b_2.

- Let the conditions of case six be fulfilled. Then $\mathrm{rVal}_\sigma^{\mathrm{S}}(g_i) = \{g_i, s_{i,j}\} \cup \mathrm{rVal}_\sigma^{\mathrm{S}}(b_1)$ where $j = \bar{\sigma}(g_i)$. If $\mathrm{rVal}_\sigma^{\mathrm{S}}(b_1) = L_1^{\mathrm{S}}$, then $\mathrm{rVal}_\sigma^{\mathrm{S}}(g_i) = \{g_i, s_{i,j}\} \cup L_1^{\mathrm{S}} \lhd L_{i+1}^{\mathrm{S}}$. If $\mathrm{rVal}_\sigma^{\mathrm{S}}(b_1) = R_1^{\mathrm{S}}$, then the statement follows by the same calculations used in the first case.
- Let the conditions of case seven be fulfilled. Then $\sigma(g_i) = F_{i,0}$. It is easy to verify that we then have $\sigma(b_i) = b_{i+1}$ and either $\neg\bar{\sigma}(eb_i) \wedge [\neg\bar{\sigma}(eg_i) \vee \mu^\sigma = 1]$ or $\neg\bar{\sigma}(eg_i) \wedge [\neg\bar{\sigma}(eb_i) \vee \mu^\sigma \neq 1 \vee \bar{\sigma}(b_{i+1})\neg\bar{\sigma}(g_i)]$. If $\neg\bar{\sigma}(eb_i) \wedge \neg\bar{\sigma}(eg_i)$, then $\bar{\sigma}(d_i)$. But then, $\sigma(b_i) = b_{i+1}$ implies $\bar{\sigma}(g_i) \neq \bar{\sigma}(b_{i+1})$. Hence, Property (USV1)$_i$ implies $\sigma(s_{i,0}) = b_1$, contradicting the currently considered case.

  Thus consider the case $\neg\bar{\sigma}(eb_i) \wedge \bar{\sigma}(eg_i) \wedge \mu^\sigma = 1$ next. Then, since $\mu^\sigma = 1$ and since σ is well-behaved, $\mathfrak{b}_1 = 0$. But $\bar{\sigma}(eg_i)$ implies $\mathfrak{b}_1 = 1$ which is a contradiction.

  Next, consider the case $\neg\bar{\sigma}(eg_i) \wedge \bar{\sigma}(eb_i) \wedge \mu^\sigma \neq 1$. As before, $\mu^\sigma \neq 1$ implies $\mathfrak{b}_1 = 1$ whereas $\sigma(eb_i)$ implies $\mathfrak{b}_1 = 0$, again resulting in a contradiction.

  Thus, consider the case $\neg\bar{\sigma}(eg_i) \wedge \bar{\sigma}(eb_i) \wedge \mu^\sigma = 1 \wedge \bar{\sigma}(b_{i+1}) \neq \bar{\sigma}(g_i)$. Then, since $\sigma(g_i) = F_{i,0}$, we have $\sigma(b_{i+1}) = g_{i+1}$. Since $\mu^\sigma = 1$ implies $\mathrm{rVal}_\sigma^{\mathrm{S}}(b_{i+1}) = L_{i+1}^{\mathrm{S}}$, we thus have $\mathrm{rVal}_\sigma^{\mathrm{S}}(g_i) = W_i^{\mathrm{S}} \cup \mathrm{rVal}_\sigma^{\mathrm{S}}(b_{i+2}) \lhd W_{i+1}^{\mathrm{S}} \cup \mathrm{rVal}_\sigma^{\mathrm{S}}(b_{i+2}) = \mathrm{rVal}_\sigma^{\mathrm{S}}(b_{i+1})$.
- Case eight can only occur if $F_{i,\mathfrak{b}_{i+1}}$ is closed, contradicting $\mathfrak{b}_i = 0$.

Let $G_n = M_n$ and consider Corollary 6.1.17. As before, we distinguish between the different cases listed in this corollary. The first two cases cannot occur due to $\sigma(b_i) = b_{i+1}$ resp. Property (ESC1). Consider the third case, implying $\sigma(b_1) = g_1$ and consequently $\mathrm{rVal}_\sigma^{\mathrm{M}}(g_1) = \mathrm{rVal}_\sigma^{\mathrm{M}}(b_1) = R_1^{\mathrm{M}}$. Using $i \geq \mu^\sigma$ and $\mathrm{rVal}_\sigma^{\mathrm{M}}(g_i) = \langle g_i \rangle + \mathrm{rVal}_\sigma^{\mathrm{M}}(g_1)$ we obtain

$$\mathrm{rVal}_\sigma^{\mathrm{M}}(g_i) = \langle g_1 \rangle + R_1^{\mathrm{M}} < \sum_{\ell \geq i} \{W_\ell^{\mathrm{M}} \colon \sigma(b_\ell) = g_\ell\} = L_i^{\mathrm{M}}.$$

Consider the fourth case. Then $\mathrm{rVal}_\sigma^{\mathrm{M}}(g_i) = \langle g_i \rangle + \mathrm{rVal}_\sigma^{\mathrm{M}}(b_2)$ and $\mathrm{rVal}_\sigma^{\mathrm{M}}(b_2) = L_2^{\mathrm{M}}$. Thus, by $\sigma(b_i) = b_{i+1}$, we obtain $\mathrm{rVal}_\sigma^{\mathrm{M}}(g_i) = \langle g_i \rangle + \mathrm{rVal}_\sigma^{\mathrm{M}}(b_2) = \langle g_i \rangle + L_2^{\mathrm{M}} < L_{i+1}^{\mathrm{M}}$. Consider the fifth case, implying $\mathrm{rVal}_\sigma^{\mathrm{M}}(g_i) = \langle g_i, s_{i,\bar{\sigma}(g_i)} \rangle + \mathrm{rVal}_\sigma^{\mathrm{M}}(b_1)$. Then, the statement follows analogously to the third case if $\sigma(b_1) = g_1$ and analogously to the fourth case if $\sigma(b_1) = b_2$. The sixth case requires that the active cycle center of level i is closed, contradicting $\mathfrak{b}_i = 0$ resp. Property (EV1)$_i$. Therefore there are no improving switches $e = (b_i, *)$.

Now consider some g_i with $i \in [n-1]$ since $\sigma(g_n) = F_{n,0}$ for every σ by construction. First assume $\mathfrak{b}_i = 0$. Then, by Definition 5.1.2 resp. 5.2.1, $F_{i,\mathfrak{b}_{i+1}}$ is not closed. Assume that $F_{i,1-\mathfrak{b}_{i+1}}$ is not closed either. Then, by Property (ESC1) and Property (REL1), $\mu^\sigma = 1$ implies $\bar{\sigma}(eb_{i,j}) \wedge \neg\bar{\sigma}(eg_{i,j})$ and $\mu^\sigma \neq 1$ implies $\bar{\sigma}(eg_{i,j}) \wedge \neg\bar{\sigma}(eb_{i,j})$ for both $j \in \{0,1\}$. Let $G_n = S_n$. Then, by Lemma 6.1.19, both cycle centers of level i escape towards the same vertex via some escape vertex. Since $\Omega(F_{i,0}) = 6$ and $\Omega(F_{i,1}) = 4$, this implies $\mathrm{Val}^{\mathrm{S}}_\sigma(F_{i,0}) \rhd \mathrm{Val}^{\mathrm{S}}_\sigma(F_{i,1})$. Thus, $(g_i, F_{i,1-\bar{\sigma}(g_i)}) \notin I_\sigma$ as $\sigma(g_i) = F_{i,0}$ by Definition 5.1.2. Let $G_n = M_n$. Then, $\mathrm{Val}^{\mathrm{M}}_\sigma(F_{i,\mathfrak{b}_{i+1}}) > \mathrm{Val}^{\mathrm{M}}_\sigma(F_{i,1-\mathfrak{b}_{i+1}})$ by Lemma 6.2.1, also implying the statement since $\sigma(g_i) = F_{i,\mathfrak{b}_{i+1}}$ by Definition 5.2.1. Thus consider the case that $F_{i,1-\mathfrak{b}_{i+1}}$ is closed. Then $\bar{\sigma}(g_i) = 1 - \mathfrak{b}_{i+1}$ by Definition 5.1.2 resp. 5.2.1. Since $F_{i,\mathfrak{b}_{i+1}}$ is not closed, Lemma 6.1.19 implies $\mathrm{rVal}^*_\sigma(F_{i,\mathfrak{b}_{i+1}}) = \mathrm{rVal}^*_\sigma(b_1)$. The statement thus follows since Property (USV1)$_i$ implies $\mathrm{rVal}^*_\sigma(F_{i,1-\mathfrak{b}_{i+1}}) = [\![s_{i,1-\mathfrak{b}_{i+1}}]\!] \oplus \mathrm{rVal}^*_\sigma(b_1)$.

Let $\mathfrak{b}_i = 1$, implying $\sigma(g_i) = F_{i,\mathfrak{b}_{i+1}}, \bar{\sigma}(d_{i,\mathfrak{b}_{i+1}})$ and $\mathrm{rVal}^*_\sigma(F_{i,\mathfrak{b}_{i+1}}) = \mathrm{rVal}^*_\sigma(s_{i,\mathfrak{b}_{i+1}})$. By the definition of a canonical strategy, $\mathrm{rVal}^*_\sigma(s_{i,\mathfrak{b}_{i+1}}) = [\![s_{i,\mathfrak{b}_{i+1}}, h_{i,\mathfrak{b}_{i+1}}]\!] \oplus \mathrm{rVal}^*_\sigma(b_{i+1})$ since $\sigma(b_{i+1}) = g_{i+1}$ if and only if $\mathfrak{b}_{i+1} = 1$, and $F_{i,1-\mathfrak{b}_{i+1}}$ is not closed. Hence, by Lemma 6.1.19, $\mathrm{rVal}^*_\sigma(F_{i,1-\mathfrak{b}_{i+1}}) = \mathrm{rVal}^*_\sigma(b_1)$. It thus suffices to show $\mathrm{rVal}^*_\sigma(b_1) \prec \mathrm{rVal}^*_\sigma(s_{i,\mathfrak{b}_{i+1}})$. This however follows immediately since $\sigma(b_i) = g_i$ implies $\mathrm{rVal}^*_\sigma(s_{i,\mathfrak{b}_{i+1}}) \subseteq \mathrm{rVal}^*_\sigma(b_1)$.

Next consider some escape vertex $e_{i,j,k}$ with $i \in [n], j, k \in \{0,1\}$ and let $\mathfrak{b}$ be even. Then $\sigma(e_{i,j,k}) = b_2$, so we prove $\mathrm{Val}^*_\sigma(g_1) \preceq \mathrm{Val}^*_\sigma(b_2)$. Since $\mathfrak{b}_1 = 0$, we have $\sigma(b_1) = b_2$ by Property (EV1)$_1$. Since we however already proved that $(b_1, g_1) \notin I_\sigma$, we need to have $\mathrm{Val}^*_\sigma(g_1) \preceq \mathrm{Val}^*_\sigma(b_2)$. Now let $\mathfrak{b} \bmod 2 = 1$, implying $\sigma(b_1) = g_1$. In this case, $\sigma(e_{i,j,k}) = g_1$, and $\mathrm{Val}^*_\sigma(b_2) \preceq \mathrm{Val}^*_\sigma(g_1)$ follows since $(b_1, b_2) \notin I_\sigma$.

Consider some upper selection vertex $s_{i,j}$ with $i \in [n]$ and $j = \mathfrak{b}_{i+1}$. Then $\sigma(s_{i,j}) = h_{i,j}$, so we prove $\mathrm{Val}^*_\sigma(b_1) \preceq \mathrm{Val}^*_\sigma(h_{i,j})$. By Property (EV1)$_{i+1}$, we have $\mathrm{rVal}^*_\sigma(h_{i,j}) = [\![h_{i,j}]\!] \oplus \mathrm{rVal}^*_\sigma(b_{i+1})$. There are two cases. If $\mathfrak{b}_i = 0$, then we have $h_{i,j} \notin \mathrm{rVal}^*_\sigma(b_1)$. If $\mathfrak{b}_i = 1$, then we have $g_i \in \mathrm{rVal}^*_\sigma(b_1)$. However, this implies $\mathrm{rVal}^*_\sigma(b_1) \prec \mathrm{rVal}^*_\sigma(h_{i,j})$ in either case since $\mathrm{rVal}^*_\sigma(b_{i+1}) \subseteq \mathrm{rVal}^*_\sigma(b_1)$. Now let $j \neq \mathfrak{b}_{i+1}$. In this case we prove $\mathrm{Val}^*_\sigma(h_{i,j}) \preceq \mathrm{Val}^*_\sigma(b_1)$. Consider the case $j = 0$ first. Then $\mathrm{rVal}^*_\sigma(h_{i,j}) = [\![h_{i,j}]\!] \oplus \mathrm{rVal}^*_\sigma(b_{i+2})$, so $W^*_{i+1} \notin \mathrm{rVal}^*_\sigma(h_{i,j})$. In particular we then have $\mathfrak{b}_{i+1} = 1$, implying $W_{i+1} \subseteq \mathrm{rVal}_\sigma(b_1)$. We thus have $\mathrm{rVal}^*_\sigma(h_{i,j}) \preceq \mathrm{rVal}^*_\sigma(b_1)$. Similarly, if $j = 1$, we have $g_{i+1} \in \mathrm{rVal}^*_\sigma(h_{i,j})$ and $g_{i+1} \notin \mathrm{rVal}^*_\sigma(b_1)$, implying the statement. □

We now begin with our discussion of the application of individual improving switches. This is organized as follows. For each phase, we provide a table that contains a summary of most of the statements related to the corresponding phase. Each row of such a table is then proven by an individual lemma. Both the tables and the proofs are very technical, and it is not obvious why the strategies the strategy improvement algorithm produces have the corresponding properties. We thus defer most of the proofs to Appendix A.2. We do not discuss the execution of the corresponding algorithm in all technical details here, but provide lemmas summarizing several of the more technical lemmas. These lemmas then also relate the application of the individual switches to the occurrence records given in Table 5.6. The formal and exact description of the application of the algorithm is then given in Section 6.3 where the results of this section will be applied.

We refer to Figure 5.10 through 5.15 for visualizations of the strategies at the beginning of the different phases in the graph S_3. To simplify the description of the improving switches, we define $\mathfrak{D}^\sigma := \{(d_{i,j,k}, F_{i,j}) : \sigma(d_{i,j,k}) \neq F_{i,j}\}$ as in Table 5.9.

Improving switches of phase 1

In this phase, cycle edges $(d_{*,*,*}, F_{*,*})$ and edges $(g_*.F_{*,*})$ are applied. As explained previously, we provide an overview describing the application of individual switches during phase 1 in Table 6.4. We interpret each row of this table stating that if a strategy σ fulfills the given conditions, applying the given switch e results in a strategy σe that has the claimed properties. For convenience, conditions specifying the improving switch, resp. the level or cycle center corresponding to the switch, are contained in the second column. Note that we also include one improving switch that technically belongs to phase 2. This is included as Table 6.4 then contains all statements necessary to prove that applying improving switches to $\sigma_{\mathfrak{b}}$ yields the phase-2-strategy that is described in Tables 5.8 and 5.9.

Conditions for σ	Switch e	Properties of σe
$F_{i,j}$ is open and $I_\sigma = \mathfrak{D}^\sigma$	$(d_{i,j,k}, F_{i,j})$	Phase-1-strategy for $\mathfrak{b}$ and $I_{\sigma e} = \mathfrak{D}^{\sigma e}$
$G_n = S_n$ and $I_\sigma = \mathfrak{D}^\sigma$ $\sigma(g_i) = F_{i,1-\mathfrak{b}_{i+1}}$	$(d_{i,1-\mathfrak{b}_{i+1},k}, F_{i,1-\mathfrak{b}_{i+1}})$ $i \neq 1$	Phase-1-strategy for $\mathfrak{b}$ $I_{\sigma e} = \mathfrak{D}^{\sigma e}$
$I_\sigma = \mathfrak{D}^\sigma$ and $\sigma(g_i) = F_{i,\mathfrak{b}_{i+1}}$ $\sigma(d_{i,1-\mathfrak{b}_{i+1},1-k}) = F_{i,1-\mathfrak{b}_{i+1}}$	$(d_{i,1-\mathfrak{b}_{i+1},k}, F_{i,1-\mathfrak{b}_{i+1}})$ $\mathfrak{b}_i = 0$	Phase-1-strategy for $\mathfrak{b}$ $I_{\sigma e} = \mathfrak{D}^{\sigma e} \cup \{(g_i, F_{i,1-\mathfrak{b}_{i+1}})\}$
$I_\sigma = \mathfrak{D}^\sigma \cup \{(g_i, F_{i,1-\mathfrak{b}_{i+1}})\}$ $F_{i,j}$ is closed	$(g_i, F_{i,1-\mathfrak{b}_{i+1}})$ $i \neq 1 \wedge \mathfrak{b}_i = 0$	Phase-1-strategy for $\mathfrak{b}$ $I_{\sigma e} = I_\sigma \setminus \{e\} = \mathfrak{D}^{\sigma e}$
$I_\sigma = \mathfrak{D}^\sigma$ $\sigma(d_{\nu,\mathfrak{b}_{\nu+1},1-k}) = F_{\nu,\mathfrak{b}_{\nu+1}}$	$(d_{\nu,\mathfrak{b}_{\nu+1},k}, F_{\nu,\mathfrak{b}_{\nu+1}})$	$\nu = 1 \Rightarrow$ Phase-3-strategy for $\mathfrak{b}$ $\nu = 1 \wedge \sigma(g_\nu) = F_{\nu,\mathfrak{b}_{\nu+1}}$ imply $I_{\sigma e} = \mathfrak{D}^{\sigma e} \cup \{(b_1, g_1)\} \cup \{(e_{*,*,*}, g_1)\}$ $\nu > 1 \Rightarrow$ Phase-2-strategy for $\mathfrak{b}$ $\nu > 1 \wedge \sigma(g_\nu) = F_{\nu,\mathfrak{b}_{\nu+1}}$ imply $I_{\sigma e} = \mathfrak{D}^{\sigma e} \cup \{(b_\nu, g_\nu), (s_{\nu-1,1}, h_{\nu-1,1})\}$ $\sigma(g_\nu) \neq F_{\nu,\mathfrak{b}_{\nu+1}}$ implies $I_{\sigma e} = \mathfrak{D}^{\sigma e} \cup \{(g_\nu, F_{\nu,\mathfrak{b}_{\nu+1}})\}$ Pseudo phase-2- resp. phase-3-strategy
Pseudo phase-2-strategy and $\nu > 1$ $I_\sigma = \mathfrak{D}^\sigma \cup \{(g_\nu, F_{\nu,\mathfrak{b}_{\nu+1}})\}$	$(g_\nu, F_{\nu,\mathfrak{b}_{\nu+1}})$	Phase-2-strategy for $\mathfrak{b}$ $I_{\sigma e} = \mathfrak{D}^{\sigma e} \cup \{(b_\nu, g_\nu), (s_{\nu-1,1}, h_{\nu-1,1})\}$

Table 6.4.: Improving switches applied during phase 1. For convenience, we always assume $\sigma \in \rho(\sigma_0)$ and that σ is a phase-1-strategy for $\mathfrak{b}$ if not stated otherwise. We thus also always have $\sigma e \in \rho(\sigma_0)$.

The first lemma shows that performing switches at cycle vertices that do not close any cycle centers does not create any new improving switches and does not make existing switches unimproving.

Lemma 6.2.7 (First row of Table 6.4). *Let $\sigma \in \rho(\sigma_0)$ be a well-behaved phase-1-strategy for $\mathfrak{b} \in \mathfrak{B}_n$ with $I_\sigma = \mathfrak{D}^\sigma$. Let $i \in [n], j, k \in \{0,1\}$ such that $e := (d_{i,j,k}, F_{i,j}) \in I_\sigma$ and $\sigma(d_{i,j,1-k}) \neq F_{i,j}$. Then σe is a well-behaved phase-1-strategy for $\mathfrak{b}$ with $\sigma e \in \rho(\sigma_0)$ and $I_{\sigma e} = \mathfrak{D}^{\sigma e}$.*

The next lemma describes what happens when the inactive cycle center $F_{i,1-\beta^\sigma_{i+1}}$ is closed under the assumption that the selector vertex of level i points towards this cycle center. This happens when cycle centers with a low occurrence record have to "catch up". We exclude level 1 here since the edges of the cycle centers in this level switch sufficiently often. Consequently, this behavior does not occur for $i = 1$. Also, we only need to consider this for $G_n = S_n$ since it cannot happen that g_i points towards $F_{i,1-\beta^{\sigma}_{i+1}}$ if $G_n = M_n$.

Lemma 6.2.8 (Second row of Table 6.4). *Let $G_n = S_n$. Let $\sigma \in \rho(\sigma_0)$ be a well-behaved phase-1-strategy for $\mathfrak{b} \in \mathfrak{B}_n$ with $I_\sigma = \mathfrak{D}^\sigma$. Let $i \in [n], j, k \in \{0,1\}$ such that $e := (d_{i,j,k}, F_{i,j}) \in I_\sigma$ and $\sigma(d_{i,j,1-k}) = F_{i,j}, i \neq 1, j \neq \mathfrak{b}_{i+1}$ as well as $\sigma(g_i) = F_{i,j}$. Then σe is a well-behaved phase-1-strategy for $\mathfrak{b}$ with $I_{\sigma e} = \mathfrak{D}^{\sigma e}$ and $\sigma e \in \rho(\sigma_0)$.*

The next lemma describes what happens when the inactive cycle center $F_{i,1-\beta^{\sigma}_{i+1}}$ of some level $i \in [n-1]$ is closed under the assumption that the selector vertex of level i does *not* point towards that cycle center. In this case, the valuation of $F_{i,1-\beta^{\sigma}_{i+1}}$ increases significantly, making the switch $(g_i, F_{i,1-\beta^{\sigma}_{i+1}})$ improving.

Lemma 6.2.9 (Third row of Table 6.4). *Let $\sigma \in \rho(\sigma_0)$ be a well-behaved phase-1-strategy for $\mathfrak{b} \in \mathfrak{B}_n$ with $I_\sigma = \mathfrak{D}^\sigma$. Let $i \in [n-1], j, k \in \{0,1\}$ such that $e := (d_{i,j,k}, F_{i,j}) \in I_\sigma$ and $\sigma(d_{i,j,1-k}) = F_{i,j}, j = 1 - \beta^{\sigma}_{i+1}, \sigma(b_i) = b_{i+1}$ and $\sigma(g_i) = F_{i,1-j}$. Then σe is a well-behaved phase-1-strategy for $\mathfrak{b}$ with $\sigma e \in \rho(\sigma_0)$ and $I_{\sigma e} = \mathfrak{D}^{\sigma e} \cup \{(g_i, F_{i,j})\}$.*

It can thus happen that improving switches $(g_i, F_{i,j})$ are created. We prove that applying this switch again yields a strategy σ with $I_\sigma = \mathfrak{D}^\sigma$.

Lemma 6.2.10 (Fourth row of Table 6.4). *Let $\sigma \in \rho(\sigma_0)$ be a well-behaved phase-1-strategy for $\mathfrak{b} \in \mathfrak{B}_n$ with $I_\sigma = \mathfrak{D}^\sigma \cup \{(g_i, F_{i,1-\mathfrak{b}_{i+1}})\}$ for some index $i \in [n-1]$. Let $e := (g_i, F_{i,1-\mathfrak{b}_{i+1}}) \in I_\sigma$ and $\mathfrak{b}_i = 0, i \neq 1$ and $\bar{\sigma}(d_{i,j})$. Then σe is a well-behaved phase-1-strategy for $\mathfrak{b}$ with $I_{\sigma e} = I_\sigma \setminus \{e\}$.*

This now allows us to formalize the application of the first set of improving switches that are applied during phase 1.

Lemma 6.2.11. *Let $\sigma \in \rho(\sigma_{\mathfrak{b}})$ be a well-behaved phase-1-strategy for $\mathfrak{b}$ with $I_\sigma = \mathfrak{D}^\sigma$. Let $\sigma_{\mathfrak{b}} \in \rho(\sigma_0)$ and let $\sigma_{\mathfrak{b}}$ have the canonical properties. Let $i \in [n], j, k \in \{0,1\}$ such that $e := (d_{i,j,k}, F_{i,j}) \in I_\sigma, I_{\sigma_{\mathfrak{b}}}$ with $\phi^{\sigma}(e) = \phi^{\sigma_{\mathfrak{b}}}(e) = \lfloor(\mathfrak{b}+1)/2\rfloor - 1$. Then σe is a well-behaved phase-1-strategy for $\mathfrak{b}$ with $\sigma e \in \rho(\sigma_0)$. Furthermore, $\sigma(d_{i,j,1-k}) = F_{i,j}, j \neq \mathfrak{b}_{i+1}, \sigma(g_i) = F_{i,1-j}$ and $\sigma(b_i) \neq g_i$ imply $I_{\sigma e} = (I_\sigma \setminus \{e\}) \cup \{(g_i, F_{i,j})\}$. Otherwise, $I_{\sigma e} = I_\sigma \setminus \{e\}$. In addition, the occurrence record of e with respect to σe is described correctly by Table 5.6 when interpreted for $\mathfrak{b}+1$.*

The next lemma now formalizes the last row of Table 6.4. It describes what happens when the cycle center $F_{\nu,\mathfrak{b}_{\nu+1}}$ is closed, concluding phase 1.

Lemma 6.2.12 (Fifth row of Table 6.4). *Let $\sigma \in \rho(\sigma_0)$ be a well-behaved phase-1-strategy for $\mathfrak{b} \in \mathfrak{B}_n$ and $I_\sigma = \mathfrak{D}^\sigma$. Let $\nu := \ell(\mathfrak{b}+1)$ and $j := \mathfrak{b}_{\nu+1}$. Let $e := (d_{\nu,j,k}, F_{\nu,j}) \in I_\sigma$ and $\sigma(d_{\nu,j,1-k}) = F_{\nu,j}$ for some $k \in \{0,1\}$. The following statements hold.*

1. *$\beta^{\sigma e} = \mathfrak{b}+1$.*
2. *σe has Properties (EV1)$_i$ and (EV3)$_i$ for all $i > \nu$. It also has Property (EV2)$_i$ and Property (USV1)$_i$ for all $i \geq \nu$ as well as Property (REL1), and $\mu^{\sigma e} = \mu^\sigma = \nu$.*
3. *σe is well-behaved and $\sigma e \in \rho(\sigma_0)$.*

4. *If $\nu = 1$, then σe is a phase-3-strategy for $\mathfrak{b}$. If $\sigma(g_\nu) = F_{\nu,j}$, then it holds that $I_{\sigma e} = \mathfrak{D}^{\sigma e} \cup \{(b_1, g_1)\} \cup \{(e_{*,*,*}, g_1)\}$. If $\sigma(g_\nu) \neq F_{\nu,j}$, then $I_{\sigma e} = \mathfrak{D}^{\sigma e} \cup \{(g_\nu, F_{\nu,j})\}$ and σe is a pseudo phase-3-strategy.*
5. *If $\nu > 1$, then σe is a phase-2-strategy for $\mathfrak{b}$. If $\sigma(g_\nu) = F_{\nu,j}$, then it holds that $I_{\sigma e} = \mathfrak{D}^{\sigma e} \cup \{(b_\nu, g_\nu)\} \cup \{(s_{\nu-1,1}, h_{\nu-1,1})\}$. If $\sigma(g_\nu) \neq F_{\nu,j}$, then $I_{\sigma e} = \mathfrak{D}^{\sigma e} \cup \{(g_\nu, F_{\nu,j})\}$ and σe is a pseudo phase-2-strategy.*

The final statement contained in Table 6.4 does technically not belong to Phase 1. It considers the case that $\sigma(g_\nu) \neq F_{\nu,\mathfrak{b}_{\nu+1}}$ when the cycle center $F_{\nu,\mathfrak{b}_{\nu+1}}$ is closed. We show that applying $(g_i, F_{\nu,\mathfrak{b}_{\nu+1}})$ then results in the same strategy that would be achieved if $\sigma(g_\nu) = F_{\nu,j}$ already held.

Lemma 6.2.13 (Sixth row of Table 6.4). *Let $\sigma \in \rho(\sigma_0)$ be a well-behaved pseudo phase-2-strategy for $\mathfrak{b} \in \mathfrak{B}_n$ with $\nu > 1$. Let $e := (g_\nu, F_{\nu,\mathfrak{b}_{\nu+1}})$ and $I_\sigma = \mathfrak{D}^\sigma \cup \{(g_\nu, F_{\nu,\mathfrak{b}_{\nu+1}})\}$. Assume that σ has Property (REL1). Then σe is a well-behaved phase-2-strategy for $\mathfrak{b}$ with $\sigma e \in \rho(\sigma_0)$ and $I_{\sigma e} = \mathfrak{D}^{\sigma e} \cup \{(b_\nu, g_\nu), (s_{\nu-1,1}, h_{\nu-1,1})\}$.*

This concludes our discussion of the application of improving switches that potentially yield a phase-2-strategy for $\mathfrak{b}$ as described by Tables 5.8 and 5.9. We next provide the lemmas necessary for proving that the strategy improvement algorithm reaches a phase-3-strategy regardless of whether we have $G_n = S_n$ or $G_n = M_n$ and of the parity of $\mathfrak{b}$. This is done by investigating the improving switches of phase 2 as well as proving how a "real" phase-3-strategy can be obtained by the respective algorithm only yields a pseudo phase-3-strategy at the end of phase 1.

Improving switches of phase 2

During phase 2, the entry vertices b_i of levels $i \in \{2, \dots, \nu\}$ and the upper selection vertices $s_{i,(\mathfrak{b}+1)_{i+1}}$ of levels $i \leq \nu - 1$ are updated. We again provide an overview describing the application of individual improving switches during phase 2 as well as the application of the switch $(g_\nu, F_{\nu,(\mathfrak{b}+1)_{\nu+1}})$ if the algorithm produces a pseudo phase-3-strategy.

We now formalize and prove the statements summarized in Table 6.5. We begin by describing the application of (b_ν, g_ν).

Lemma 6.2.14 (First row of Table 6.5). *Let $\sigma \in \rho(\sigma_0)$ be a well-behaved phase-2-strategy for $\mathfrak{b} \in \mathfrak{B}_n$ with $\nu > 1$. Let $I_\sigma = \mathfrak{D}^\sigma \cup \{(b_\nu, g_\nu), (s_{\nu-1,1}, h_{\nu-1,1})\}$. Let σ have Property (REL1) as well as Property (USV3)$_i$ for all $i < \nu$. Let $e := (b_\nu, g_\nu)$. Then, σe is a well-behaved phase-2-strategy for $\mathfrak{b}$ with $\sigma e \in \rho(\sigma_0)$. In addition ,$\nu \neq 2$ implies*

$$I_{\sigma e} = \mathfrak{D}^{\sigma e} \cup \{(b_{\nu-1}, b_\nu), (s_{\nu-1,1}, h_{\nu-1,1}), (s_{\nu-2,0}, h_{\nu-2,0})\}$$

if $\nu \neq 2$ and $\nu = 2$ implies

$$I_{\sigma e} = \mathfrak{D}^{\sigma e} \cup \{(b_1, b_2), (s_{1,1}, h_{1,1})\} \cup \{(e_{*,*,*}, b_2)\}.$$

Properties of σ	Switch e	Properties of σe
$I_\sigma = \mathfrak{D}^\sigma \cup \{(b_\nu, g_\nu), (s_{\nu-1,1}, h_{\nu-1,1})\}$ Property (REL1) Property (USV3)$_i$ for all $i < \nu$	(b_ν, g_ν)	Phase-2-strategy for $\mathfrak{b}$ $\nu \neq 2$ implies $I_{\sigma e} = (I_\sigma \setminus \{e\}) \cup \{(b_{\nu-1}, b_\nu), (s_{\nu-2,0}, h_{\nu-2,0})\}$ $\nu = 2$ implies $I_{\sigma e} = \mathfrak{D}^{\sigma e} \cup \{(b_1, b_2), (s_{1,1}, h_{1,1})\} \cup \{(e_{*,*,*}, b_2)\}$
$i' < \mu^\sigma \Rightarrow F_{i',\bar{\sigma}(g_{i'})}$ is closed Property (USV3)$_{i'}$ for all $i' \leq i$ Properties (EV1)$_{\mu^\sigma}$, (EV1)$_{i+1}$	$(s_{i,j}, h_{i,j})$ $i < \mu^\sigma$ $j = \beta^\sigma_{i+1}$	$i \neq 1 \Rightarrow$ Phase-2-strategy for $\mathfrak{b}$ $i = 1 \Rightarrow$ Phase-3-strategy for $\mathfrak{b}$ $I_{\sigma e} = I_\sigma \setminus \{e\}$
$i' < \mu^\sigma \Rightarrow F_{i',\bar{\sigma}(g_{i'})}$ is closed Property (USV3)$_{i'}$ for all $i' \leq i$ $i' > i \Rightarrow$ Properties (EV1)$_{i'} \wedge$(EV2)$_{i'}$ $i' > i, i' \neq \mu^\sigma \Rightarrow$Property (EV3)$_{i'}$	(b_i, b_{i+1}) $i > 1$ $i < \mu^\sigma$	Phase-2-strategy for $\mathfrak{b}$ $i \neq 2$ implies $I_{\sigma e} = (I_\sigma \setminus \{e\}) \cup \{(b_{i-1}, b_i), (s_{i-2,0}, h_{i-2,0})\}$ $i = 2$ implies $I_{\sigma e} = (I_\sigma \setminus \{e\}) \cup \{(b_1, b_2)\} \cup \{(e_{*,*,*}, b_2)\}$
Pseudo phase-3-strategy and $\nu = 1$ $I_\sigma = \mathfrak{D}^\sigma \cup \{(g_\nu, F_{\nu,\mathfrak{b}_{\nu+1}})\}$	$(g_\nu, F_{\nu,\mathfrak{b}_{\nu+1}})$	Phase-3-strategy for $\mathfrak{b}$ $I_{\sigma e} = \mathfrak{D}^{\sigma e} \cup \{(b_1, g_1)\} \cup \{(e_{*,*,*}, g_1)\}$

Table 6.5.: Improving switches applied during phase 2. For convenience, we always assume $\sigma \in \rho(\sigma_0)$, that σ is a phase-2-strategy for $\mathfrak{b}$ and that $\nu > 1$ if not stated otherwise. We thus also always have $\sigma e \in \rho(\sigma_0)$. We also include one application here that technically belongs to phase 3.

The following lemma describes the application of switches $(s_{i,j}, h_{i,j})$ for $i \in [\mu^\sigma - 1]$ and $j = \beta^\sigma_{i+1}$. Depending on whether $i \neq 1$ or $i = 1$, applying this switch might conclude phase 2 and thus lead to a phase-3-strategy for $\mathfrak{b}$. As the following lemma describes a strategy that is obtained after the application of several improving switches during phase 2, we include several additional assumptions that encode the application of these previously applied switches.

Lemma 6.2.15 (Second row of Table 6.5). *Let $\sigma \in \rho(\sigma_0)$ be a well-behaved phase-2-strategy for some $\mathfrak{b} \in \mathfrak{B}_n$ with $\nu > 1$. Assume that $\bar{\sigma}(d_{i'}) = 1$ for all $i' < \mu^\sigma$ and that $e = (s_{i,j}, h_{i,j}) \in I_\sigma$ for some $i \in [\mu^\sigma - 1]$ where $j := \beta^\sigma_{i+1}$. Further assume that σ has Property (USV3)$_{i'}$ for all $i' \leq i$. Also, assume that σ has Properties (EV1)$_{\mu^\sigma}$ and (EV1)$_{i+1}$. If $i \neq 1$, then σe is a well-behaved phase-2-strategy for $\mathfrak{b}$. If $i = 1$, then σe is a well-behaved phase-3-strategy for $\mathfrak{b}$. In either case, $I_{\sigma e} = I_\sigma \setminus \{e\}$.*

The following lemma describes the application of an improving switch (b_i, b_{i+1}) for levels $i \in \{2, \ldots, \nu - 1\}$ during phase 2.

Lemma 6.2.16 (Third row of Table 6.5). *Let $\sigma \in \rho(\sigma_0)$ be a well-behaved phase-2-strategy for $\mathfrak{b} \in \mathfrak{B}_n$ with $\nu > 1$. Assume that $\bar{\sigma}(d_{i'}) = 1$ for all $i' < \mu^\sigma$ and $e = (b_i, b_{i+1}) \in I_\sigma$ for some $i \in \{2, \ldots, \mu^\sigma - 1\}$. In addition, assume that σ has Property (USV3)$_{i'}$ for all $i' < i$, Property (EV1)$_{i'}$ and Property (EV2)$_{i'}$ for all $i' > i$ as well as Property (EV3)$_{i'}$ for all $i' > i, i' \neq \mu^\sigma$.*

Then σe is a well-behaved phase-2-strategy for $\mathfrak{b}$. Furthermore, $i \neq 2$ implies

$$I_{\sigma e} = (I_\sigma \setminus \{e\}) \cup \{(b_{i-1}, b_i), (s_{i-2,0}, h_{i-2,0})\}$$

and $i = 2$ implies $I_{\sigma e} = (I_\sigma \setminus \{e\}) \cup \{(b_1, b_2)\} \cup \{(e_{*,*,*}, b_2)\}$.

This concludes our overview related to the improving switches applied during phase 2. The next lemma considers a special case that can occur at the beginning of phase 3. Although we closed the cycle center $F_{\nu,(\mathfrak{b}+1)_{\nu+1}}$ at the end of phase 1, it is not guaranteed that the selection vertex of level ν points towards this cycle center if $\nu = 1$. That is, it is not guaranteed that we immediately obtain a "proper" phase-3-strategy. Such a strategy is then called *pseudo* phase-3-strategy. If the first phase-3-strategy is a pseudo phase-3-strategy, then the improving switch $(g_\nu, F_{\nu,(\mathfrak{b}+1)_{\nu-1}})$ will be applied immediately at the beginning of phase 3. The lemma thus describes the last row of Table 6.5.

Lemma 6.2.17 (Last row of Table 6.5). *Let $\sigma \in \rho(\sigma_0)$ be a well-behaved pseudo phase-3-strategy for some $\mathfrak{b} \in \mathfrak{B}_n$ with $\nu = 1$. Let $I_\sigma = \mathfrak{D}^\sigma \cup \{(g_\nu, F_{\nu,\mathfrak{b}_{\nu+1}})\}$ and $e := (g_\nu, F_{\nu,\mathfrak{b}_{\nu+1}})$. Then σe is a well-behaved phase-3-strategy for $\mathfrak{b}$ with $\sigma e \in \rho(\sigma_0)$ and*

$$I_{\sigma e} = (I_\sigma \setminus \{e\}) \cup \{(b_1, g_1)\} \cup \{(e_{*,*,*}, g_1)\}.$$

These are all lemmas necessary for describing phase 2. We consider the statements related to the application of the improving switches during phase 3 next.

Improving switches of phase 3

We now discuss the application of improving switches during phase 3, which highly depends on whether we have $G_n = S_n$ or $G_n = M_n$ and on the least significant set bit of $\mathfrak{b}+1$. As usual, we provide an overview describing the application of individual improving switches during phase 3. To simplify and unify the arguments, we define $t^\rightarrow := b_2$ if $\nu > 1$ and $t^\rightarrow := g_1$ if $\nu = 1$. Similarly, let $t^\leftarrow := g_1$ if $\nu > 1$ and $t^\leftarrow := b_2$ if $\nu = 1$. We furthermore define $\mathfrak{E}^\sigma := \{(d_{i,j,k}, F_{i,j}), (e_{i,j,k}, t^\rightarrow) \colon \sigma(e_{i,j,k}) = t^\leftarrow\}$.

There are also additional statements describing the application of the improving switches during phase 3. These statements are however more involved and cannot be stated in the same way the statements contained in Table 6.6 can be described. We defer these statements and their discussion for the moment and begin with a lemma characterizing the vertex valuations for several phase-3-strategies. As its proof is rather short and yields some interesting insights regarding phase-3-strategies, it is also given directly here and not deferred to the appendix.

Lemma 6.2.18. *Let $\sigma \in \rho(\sigma_0)$ be a well-behaved phase-3-strategy for $\mathfrak{b} \in \mathfrak{B}_n$.*

1. *If $\nu = 1$, then $\mathrm{rVal}^*_\sigma(b_2) = L^*_2$ and $\mathrm{rVal}^*_\sigma(g_1) = W^*_1 \oplus \mathrm{rVal}^*_\sigma(b_2)$, so in particular $\mathrm{Val}^*_\sigma(g_1) \succ \mathrm{Val}^*_\sigma(b_2)$.*
2. *If $\nu > 1$, then $\mathrm{rVal}^*_\sigma(b_2) = L^*_2$ and $\mathrm{Val}^*_\sigma(b_2) \succ \mathrm{Val}^*_\sigma(g_1) \oplus [\![s_{i,j}]\!]$ for $i \in [n], j \in \{0,1\}$, so in particular $\mathrm{rVal}^*_\sigma(b_2) \succ \mathrm{rVal}^*_\sigma(g_1)$.*

Proof. Let $\nu = 1$. Since σ is a phase-3-strategy, this implies $\mathrm{rVal}^*_\sigma(b_2) = L^*_2$ as $\mu^\sigma = 1$. Let $G_n = S_n$. By Property (EV1)$_2$ and Property (CC2), $\bar{\sigma}(g_1) = \bar{\sigma}(b_2) = \beta^\sigma_2$. Thus, $\beta^\sigma_2 = 0$ implies $\sigma(g_1) = F_{1,0}$ and $\beta_2 = 1$ implies $\sigma(b_2) = g_2$. In either case, $\lambda^{\mathrm{S}}_1 = 1$. We now investigate which of the cases of Corollary 6.1.18 can occur and prove that $\mathrm{rVal}^{\mathrm{S}}_\sigma(g_1) = W^{\mathrm{S}}_1 \cup \mathrm{rVal}^{\mathrm{S}}_\sigma(b_2)$ for the respective cases. The first case cannot occur since

Properties of σ	Switch e	Properties of σe
If $G_n = S_n$: Property (USV2)$_{i',*} \forall i' < \mu^\sigma$ If $G_n = M_n$: $\sigma(s_{i',*}) = b_1$ implies $\bar{\sigma}(eb_{i',*}) \wedge \neg\bar{\sigma}(eg_{i',*}) \mid \forall i' < \mu^\sigma$	$(e_{i,j,k}, t^{\rightarrow})$	Phase-3-strategy for $\mathfrak{b}$ $\sigma(d_{i,j,1-k}) = e_{i,j,1-k}$ $\vee[\sigma(d_{i,j,1-k}) = F_{i,j} \wedge j \neq \beta^\sigma_{i+1}]$ imply $I_{\sigma e} = (I_\sigma \setminus \{e\}) \cup \{(d_{i,j,k}, e_{i,j,k})\}$ $\sigma(d_{i,j,1-k}) = F_{i,j} \wedge j = \beta^\sigma_{i+1}$ imply $I_{\sigma e} = I_\sigma \setminus \{e\}$
$G_n = S_n$ $\bar{\sigma}(d_{i,j}) \implies j \neq \beta^\sigma_{i+1}$ $\sigma(e_{i,j,k}) = t^{\rightarrow}$	$(d_{i,j,k}, e_{i,j,k})$	Phase-3-strategy for $\mathfrak{b}$ $I_{\sigma e} = I_\sigma \setminus \{e\}$
$G_n = M_n$ $\sigma(e_{i,j,k}) = t^{\rightarrow}$	$(d_{i,j,k}, e_{i,j,k})$ $\beta^\sigma_i = 1$ $j = \beta^\sigma_{i+1}$	Phase-3-strategy for $\mathfrak{b}$ $I_{\sigma e} = I_\sigma \setminus \{e\}$
$G_n = M_n$ and $\sigma(g_i) = F_{i,1-j}$ $F_{i,j}$ is $t^{\leftarrow}$-halfopen, $F_{i,1-j}$ is $t^{\rightarrow}$-open $\sigma(e_{i,j,k}) = t^{\rightarrow}$	$(d_{i,j,k}, e_{i,j,k})$ $\beta^\sigma_i = 0$ $j = \beta^\sigma_{i+1}$	Phase-3-strategy for $\mathfrak{b}$ $I_{\sigma e} = I_\sigma \setminus \{e\}$
$G_n = M_n$ $\nu > 1$ $\bar{\sigma}(eb_{i,j}) \wedge \neg\bar{\sigma}(eg_{i,j})$	$(s_{i,j}, b_1)$ $i < \nu$ $j = 1 - \beta^\sigma_{i+1}$	Phase-3-strategy for $\mathfrak{b}$ $I_{\sigma e} = (I_\sigma \setminus \{e\})$
$G_n = M_n$ and $\sigma(e_{i,j,k}) = t^{\rightarrow}$ $F_{i,j}$ is $t^{\rightarrow}$-halfopen $\beta^\sigma_i = 0 \Rightarrow [\sigma(g_i) = F_{i,j} \wedge F_{i,1-j}$ is $t^{\leftarrow}$-halfopen]	$(d_{i,j,k}, e_{i,j,k})$ $j = 1 - \beta^\sigma_{i+1}$	Phase-3-strategy for $\mathfrak{b}$ $I_{\sigma e} = (I_\sigma \setminus \{e\})$
$G_n = S_n$, $\nu > 1$ and $I_\sigma = \mathfrak{E}^\sigma \cup \{(b_1, b_2)\}$ $\sigma(d_{i,j,k}) = F_{i,j} \Leftrightarrow \beta^\sigma_i = 1 \wedge \beta^\sigma_{i+1} = j$ (ESC4)$_{i,j}$ for all $(i,j) \in S_1$ (ESC5)$_{i,j}$ for all $(i,j) \in S_2$ $i < \mu^\sigma \Rightarrow \sigma(s_{i,*}) = h_{i,*}$	(b_1, b_2)	Phase-4-strategy for $\mathfrak{b}$ with $\mu^{\sigma e} = 1$ $I_{\sigma e} = (I_\sigma \setminus \{e\}) \cup \{(s_{\nu-1,0}, b_1)\}$ $\cup\{(s_{i,1}, b_1) : i \leq \nu - 2\} \cup X_0 \cup X_1$
$G_n = M_n$, $\nu > 1$ and $I_\sigma = \mathfrak{E}^\sigma \cup \{(b_1, b_2)\}$ $\sigma(d_{i,j,k}) = F_{i,j} \Leftrightarrow \beta^\sigma_i = 1 \wedge \beta^\sigma_{i+1} = j$ (ESC4)$_{i,j}$ for all $(i,j) \in S_1$ (ESC5)$_{i,j}$ for all $(i,j) \in S_2$ Equation (USV1)$_i$ for all $i \in [n]$	(b_1, b_2)	Phase-5-strategy for $\mathfrak{b}$ with $\mu^{\sigma e} = 1$ $I_{\sigma e} = (I_\sigma \setminus \{e\}) \cup X_0 \cup X_1$ $\cup\{(d_{i,-1,\beta^\sigma_{i+1},*}, F_{i,1-\beta^\sigma_{i+1}}) : i < \nu\}$
$\nu = 1$ and $I_\sigma = \mathfrak{E}^\sigma \cup \{(b_1, g_1)\}$ $\sigma(d_{i,j,k}) = F_{i,j} \Leftrightarrow \beta^\sigma_i = 1 \wedge \beta^\sigma_{i+1} = j$ (ESC3)$_{i,j}$ for all $(i,j) \in S_4$ (ESC5)$_{i,j}$ for all $(i,j) \in S_3$	(b_1, g_1)	Phase-5-strategy with $\mu^{\sigma e} = u$ $I_{\sigma e} = (I_\sigma \setminus \{e\})$ $\cup \bigcup_{\substack{i'=u+1 \\ \beta^\sigma_i = 0}}^{m-1} \{(d_{i,1-\beta^\sigma_{i+1},*}, F_{i,1-\beta^\sigma_{i+1}})\}$

Table 6.6.: Improving switches applied during phase 3. For convenience, we always assume $\sigma \in \rho(\sigma_0)$ and that σ is a phase-3-strategy for $\mathfrak{b}$ if not stated otherwise, implying that $\sigma e \in \rho(\sigma_0)$. The definition of the sets X_k, S_i can be found in Tables 5.5 and 5.9.

$\mu^\sigma = 1$ implies $\sigma(b_1) = b_2$. The second up to the fifth case cannot occur since Property (REL2) and Property (CC2) imply $\bar{\sigma}(d_1)$. In addition, Property (CC2) implies $\bar{\sigma}(g_1) = \beta_2$. Thus, by Property (USV1)$_1$, $\bar{\sigma}(s_{1,\bar{\sigma}(g_1)}) = h_{1,\bar{\sigma}(g_1)}$, so the conditions of the sixth case of Corollary 6.1.18 cannot hold. Consequently, the conditions of one of the last two cases of Corollary 6.1.18 are fulfilled. However, since $\sigma(b_2) = b_3$ if $\sigma(g_1) = F_{1,0}$ by Property (CC2) and Property (EV1)$_2$, the statement follows in either case.

Let $G_n = M_n$ and consider Corollary 6.1.17. If $\beta_2 = 0$, then $\sigma(g_1) = F_{1,0}$ by Property (CC2), implying $\lambda_1^{\mathrm{M}} = 1$. Since $\bar{\sigma}(d_1) \wedge \bar{\sigma}(s_1)$ as shown previously,it follows that $\mathrm{rVal}_\sigma^{\mathrm{M}}(g_1) = W_1^{\mathrm{M}} + \mathrm{rVal}_\sigma^{\mathrm{M}}(b_2)$ since the conditions of the last case are fulfilled. If $\beta_2^{\sigma e} = 1$, then $\lambda_1^{\mathrm{M}} = 2$ by Property (EV2)$_2$. Consequently, $\mathrm{rVal}_{\sigma e}^{\mathrm{M}}(g_1) = W_1^{\mathrm{M}} + \mathrm{rVal}_{\sigma e}^{\mathrm{M}}(b_2)$ since the conditions of the first case are fulfilled.

This concludes the case $\nu = 1$, hence assume $\nu > 1$. We prove $\mathrm{rVal}_\sigma^*(b_2) = L_2^*$ first. If $\sigma(b_2) = b_3$, this follows by definition. Hence assume $\sigma(b_2) = g_2$. Then, by Property (EV1)$_2$ and Property (CC2), $\sigma(g_1) = F_{1,0}$. In addition, $\nu = \mu^\sigma > 1$ implies $\sigma(b_1) = g_1$. Consequently, $1 \in \mathcal{I}^\sigma$. Since σ has Property (EV1)$_i$ and Property (EV2)$_i$ for all $i > 1$, no other index can be contained in $\mathcal{I}^\sigma$. But this implies $\mathcal{I}^\sigma = \{1\}$ and thus, since $\sigma(b_2) = g_2$, $\mu^\sigma = 2$, implying $\mathrm{rVal}_\sigma^*(b_2) = L_2^*$ as claimed.

We now prove that $\nu > 1$ implies $\mathrm{Val}_\sigma^*(b_2) \succ \mathrm{Val}_\sigma^*(g_1) \oplus [\![s_{i,j}]\!]$ for $i \in [n], j \in \{0,1\}$. Since $\nu > 1$ implies $\sigma(b_1) = g_1$, this implies that $\mathrm{rVal}_\sigma^*(g_1) = \mathrm{rVal}_\sigma^*(b_1)$. Furthermore, by $1 < \mu^\sigma$ and $\sigma(b_1) = g_1$, Lemma 6.1.13 and $B_2^* = L_2^*$ imply that either $\mathrm{rVal}_\sigma^*(b_1) = R_1^*$ or $G_n = M_n$ and $\mathrm{rVal}_\sigma^{\mathrm{M}}(b_1) = \langle g_k \rangle + \sum_{j=1}^{k-1} W_j^{\mathrm{M}} + L_2^{\mathrm{M}}$ where $k = \min\{i \geq 1 \colon \neg\bar{\sigma}(d_i)\} < \mu^\sigma$. In the second case the statement follows directly, in the first it follows by Lemma 6.1.10 since $\mathcal{I}^\sigma \neq \emptyset$ implies $\sigma(b_{\mu^\sigma}) = g_{\mu^\sigma}$ by Lemma 6.1.4. □

We now begin with the lemmas describing phase 3. The first lemma describes the application of $(e_{i,j,k}, g_1)$ resp. $(e_{i,j,k}, b_2)$ for the case that $\sigma(d_{i,j,k}) = F_{i,j}$. As all of the following lemmas, this lemma contains some conditions encoding the behavior of the strategy improvement algorithm and the application of previous improving switches. Since phase 3 is not exactly identical for S_n and M_n, there are also some conditions distinguishing between the two.

Lemma 6.2.19 (First row of Table 6.6). *Let $\sigma \in \rho(\sigma_0)$ be a well-behaved phase-3-strategy for $\mathfrak{b} \in \mathfrak{B}_n$. Let $i \in [n], j, k \in \{0,1\}$ such that $(e_{i,j,k}, t^{\rightarrow}) \in I_\sigma$ and $\sigma(d_{i,j,k}) = F_{i,j}$. Further assume the following.*

1. *If $G_n = S_n$, then, σ has Property (USV2)$_{i',j'}$ for all $i' < \mu^\sigma, j' \in \{0,1\}$.*
2. *If $G_n = M_n$, then, $\sigma(s_{i',j'}) = b_1$ implies $\bar{\sigma}(eb_{i',j'}) \wedge \neg\bar{\sigma}(eg_{i',j'})$ for all $i' < \mu^\sigma$ and $j' \in \{0,1\}$.*

Then σe is a well-behaved phase-3-strategy for $\mathfrak{b}$ with $\sigma e \in \rho(\sigma_0)$. If $\sigma(d_{i,j,1-k}) = e_{i,j,1-k}$ or $[\sigma(d_{i,j,1-k}) = F_{i,j}$ and $j \neq \beta_{i+1}^\sigma]$, then $I_{\sigma e} = (I_\sigma \setminus \{e\}) \cup \{(d_{i,j,k}, e_{i,j,k})\}$. If $\sigma(d_{i,j,1-k}) = F_{i,j}$ and $j = \beta_{i+1}^\sigma$, then $I_{\sigma e} = I_\sigma \setminus \{e\}$.

We now want to describe the application of improving switches $(d_{i,j,k}, e_{i,j,k})$ in phase 3. The main challenge regarding these switches is that there are several different cases that need to be considered when a switch of this type is applied. We thus provide several individual lemmas that are combined later to give a lemma summarizing the application of these switches. We first show that we always obtain a well-behaved phase-3-strategy.

Lemma 6.2.20. *Let $\sigma \in \rho(\sigma_0)$ be a well-behaved phase-3-strategy for $\mathfrak{b}$. Let $i \in [n]$ and $j, k \in \{0,1\}$ such that $\sigma(e_{i,j,k}) = t^{\rightarrow}$ and $e := (d_{i,j,k}, e_{i,j,k}) \in I_\sigma$. Let $\sigma(d_{i,j,1-k}) = e_{i,j,1-k}$ or $[\sigma(d_{i,j,1-k}) = F_{i,j}$ and $j \neq \beta^\sigma_{i+1}]$. Then σe is a well-behaved phase-3-strategy for $\mathfrak{b}$ with $\sigma e \in \rho(\sigma_0)$.*

Lemma 6.2.20 justifies to omit the upper index when referring to the induced bit state. For a well-behaved phase-3-strategy σ with $e \in I_\sigma$, we thus define $\beta := \beta^\sigma = \beta^{\sigma e} = \mathfrak{b} + 1$ while we discuss phase 3.

We now describe the application of switches $(d_{*,*,*}, e_{*,*,*})$. While this application is not hard to describe in S_n, it is very complex in M_n. The reason is that applying these switches *always* has an influence on the valuation of the cycle centers in M_n. Thus, we need to carefully investigate the application of these switches and need to pay heavy attention to the exact order of application.

We begin with the application of an improving switch $(d_{*,*,*}, e_{*,*,*})$ during phase 3 in S_n. This lemma is significantly easier than the lemmas for the case $G_n = M_n$ as the valuation of the cycle center $F_{i,j}$ does not change when applying $(d_{i,j,k}, e_{i,j,k}), i \in [n], j, k \in \{0,1\}$ in S_n.

Lemma 6.2.21 (Second row of Table 6.6). *Let $G_n = S_n$. Let $\sigma \in \rho(\sigma_0)$ be a well-behaved phase-3-strategy for $\mathfrak{b} \in \mathfrak{B}_n$. Let $i \in [n], j, k \in \{0,1\}$ such that $e := (d_{i,j,k}, e_{i,j,k}) \in I_\sigma$ and $\sigma(e_{i,j,k}) = t^{\rightarrow}$. Further assume that $\bar{\sigma}(d_{i,j})$ implies $j \neq \beta^\sigma_{i+1}$. Then σe is a well-behaved phase-3-strategy for $\mathfrak{b}$ with $\sigma e \in \rho(\sigma_0)$ and $I_{\sigma e} = I_\sigma \setminus \{e\}$.*

We now focus on the case $G_n = M_n$. The next lemma describes the application of switches $(d_{i,j,k}, e_{i,j,k})$ where $i \in [n], j, k \in \{0,1\}$ within levels i with $\beta^\sigma_i = 1$. We skip the upper index M to denote that we have $G_n = M_n$ since we exclusively consider this case.

Lemma 6.2.22 (Third row of Table 6.6). *Let $G_n = M_n$ and let $\sigma \in \rho(\sigma_0)$ be a well-behaved phase-3-strategy for $\mathfrak{b}$. Let $i \in [n]$ with $\beta_i = 1$ and let $j := 1-\beta_{i+1}$. Let $e := (d_{i,j,k}, e_{i,j,k}) \in I_\sigma$ and $\sigma(e_{i,j,k}) = t^{\rightarrow}$ for some $k \in \{0,1\}$. Then σe is a well-behaved phase-3-strategy for $\mathfrak{b}$ with $\sigma e \in \rho(\sigma_0)$ and $I_{\sigma e} = I_\sigma \setminus \{e\}$.*

The next lemma describes the application of an improving switch $(d_{i,j,k}, e_{i,j,k})$ within a $t^{\rightarrow}$-open cycle center.

Lemma 6.2.23 (Fourth row of Table 6.6). *Let $G_n = M_n$. Let $\sigma \in \rho(\sigma_0)$ be a well-behaved phase-3-strategy for $\mathfrak{b} \in \mathfrak{B}_n$. Let $i \in [n]$ with $\beta^\sigma_i = 0, j := \beta^\sigma_{i+1}$ and let $F_{i,j}$ be $t^{\leftarrow}$-halfopen. Let $F_{i,1-j}$ be $t^{\rightarrow}$-open and $\sigma(g_i) = F_{i,1-j}$. Let $e := (d_{i,j,k}, e_{i,j,k}) \in I_\sigma$ and $\sigma(e_{i,j,k}) = t^{\rightarrow}$ with $k \in \{0,1\}$. Then σe is a well-behaved phase-3-strategy for $\mathfrak{b}$ with $\sigma e \in \rho(\sigma_0)$ and $I_{\sigma e} = I_\sigma \setminus \{e\}$.*

The next lemma describes the application of improving switches within levels $i \in [n]$ in which no cycle center is closed at the beginning of phase 3. The first case describes the first improving switch that is applied in such a level. This switch is applied in the cycle center $F_{i,\bar{\sigma}(g_i)}$ to avoid the creation of an additional improving switch at the selector vertex. The second case describes the second improving switch that is then applied in the cycle center $F_{i,1-\bar{\sigma}(g_i)}$. The statement of this lemma is not included in Table 6.6 as it is to involved and does not fit the framework of the lemmas summarized there.

Lemma 6.2.24. *Let $G_n = M_n$. Let $\sigma \in \rho(\sigma_0)$ be a well-behaved phase-3-strategy for $\mathfrak{b} \in \mathfrak{B}_n$. Let $i \geq \mu^\sigma + 1$ and assume $\bar{\sigma}(g_i) = \beta^\sigma_{i+1}$.*

1. *If both cycle centers of level i are $t^{\leftarrow}$-halfopen, then let $j := \bar{\sigma}(g_i)$.*
2. *If $F_{i,\beta^\sigma_{i+1}}$ is mixed and $F_{i,1-\beta^\sigma_{i+1}}$ is $t^{\leftarrow}$-halfopen, then let $j := 1 - \bar{\sigma}(g_i)$.*

In any case, assume $e := (d_{i,j,k}, e_{i,j,k}) \in I_\sigma$ and $\sigma(e_{i,j,k}) = t^{\rightarrow}$ for $k \in \{0,1\}$. Then, σe is a well-behaved phase-3-strategy for $\mathfrak{b}$ with $\sigma e \in \rho(\sigma_0)$ and $I_{\sigma e} = I_\sigma \setminus \{e\}$.

The next lemma describes the application of a switch $(d_{i,*,*}, e_{i,*,*})$ within a closed but inactive cycle center for the case that $\beta_i = 0$. The lemma requires that the strategy σ fulfills several rather complicated assumptions. As usual, these assumptions somehow "encode" the order of application of the improving switches.

Lemma 6.2.25. *Let $G_n = M_n$. Let σ be a well-behaved phase-3-strategy for $\mathfrak{b} \in \mathfrak{B}_n$ with $\sigma \in \rho(\sigma_0)$. Let $i \in [n]$ and $j := 1 - \beta^\sigma_{i+1}$. Let $e := (d_{i,j,k}, e_{i,j,k}) \in I_\sigma$ and $\sigma(e_{i,j,k}) = t^{\rightarrow}$ for some $k \in \{0,1\}$. Further assume that there is no other triple of indices i', j', k' with $(d_{i',j',k'}, e_{i',j',k'}) \in I_\sigma$, that $F_{i,j}$ is closed and that σ fulfills the following assumptions:*

1. *If $\beta^\sigma_i = 0$, then $\sigma(g_i) = F_{i,j}$ and $F_{i,1-j}$ is $t^{\leftarrow}$-halfopen.*
2. *$i < \mu^\sigma$ implies [$\sigma(s_{i,j}) = h_{i,j}$ and $\sigma(s_{i',j'}) = h_{i',j'} \land \bar{\sigma}(d_{i'})$ for all $i' < i, j' \in \{0,1\}$] and that the cycle center $F_{i',1-\bar{\sigma}(g_{i'})}$ is $t^{\leftarrow}$-halfopen for all $i' < i$. In addition, $i < \mu^\sigma - 1$ implies $\bar{\sigma}(eb_{i+1})$.*
3. *$i' > i$ implies $\sigma(s_{i,1-\beta^\sigma_{i'+1}}) = b_1$.*
4. *$i' > i$ and $\beta^\sigma_{i'} = 0$ imply that either [$\bar{\sigma}(g_{i'}) = \beta^\sigma_{i'+1}$ and $F_{i,0}, F_{i,1}$ are mixed] or [$\bar{\sigma}(g_{i'}) = 1 - \beta^\sigma_{i'+1}$, $F_{i',1-\beta^\sigma_{i'+1}}$ is $t^{\rightarrow}$-open and $F_{i',\beta^\sigma_{i'+1}}$ is mixed] and*
5. *$i' > i$ and $\beta^\sigma_{i'} = 1$ imply that $F_{i',1-\beta^\sigma_{i'+1}}$ is either mixed or $t^{\rightarrow}$-open.*

Then σe is a well-behaved phase-3-strategy for $\mathfrak{b}$ with $\sigma e \in \rho(\sigma_0)$ and $I_{\sigma e} = I_\sigma \setminus \{e\}$ if $i \geq \mu^\sigma$ and $I_{\sigma e} = [I_\sigma \cup \{(s_{i,j}, b_1)\}] \setminus \{e\}$ if $i < \mu^\sigma$.

The next lemma describes the application of an improving switch $(s_{i,j}, b_1)$ that might be unlocked by the application of a switch $(d_{i,j,k}, e_{i,j,k})$. In M_n, these switches are already implied during phase 3 while they are applied during phase 4 in S_n. Thus, the following lemma only considers M_n.

Lemma 6.2.26 (Fifth row of Table 6.6). *Let $G_n = M_n$. Let $\sigma \in \rho(\sigma_0)$ be a well-behaved phase-3-strategy for $\mathfrak{b} \in \mathfrak{B}_n$ with $\nu > 1$. Let $i < \mu^\sigma, j = 1 - \beta^\sigma_{i+1}$ and $e := (s_{i,j}, b_1) \in I_\sigma$. Further assume $\bar{\sigma}(eb_{i,j}) \land \neg\bar{\sigma}(eg_{i,j})$. Then σe is a well-behaved phase-3-strategy for $\mathfrak{b}$ with $I_{\sigma e} = I_\sigma \setminus \{e\}$ and $\sigma e \in \rho(\sigma_0)$.*

The next lemma now describes the application of the second improving switch of the kind $(d_{i,j,k}, e_{i,j,k})$ within a cycle center that was closed in phase 1.

Lemma 6.2.27 (Sixth row of Table 6.6). *Let $G_n = M_n$. Let $\sigma \in \rho(\sigma_0)$ be a well-behaved phase-3-strategy for $\mathfrak{b} \in \mathfrak{B}_n$. Let $i \in [n]$ and $j := 1 - \beta^\sigma_{i+1}$. Let $F_{i,j}$ be $t^{\rightarrow}$-halfopen and assume that $\beta^\sigma_i = 0$ implies that $F_{i,1-j}$ is $t^{\leftarrow}$-halfopen as well as $\sigma(g_i) = F_{i,j}$. Let $e := (d_{i,j,k}, e_{i,j,k}) \in I_\sigma$ and $\sigma(e_{i,j,k}) = t^{\rightarrow}$ for $k \in \{0,1\}$. Then σe is a well-behaved phase-3-strategy for $\mathfrak{b}$ with $\sigma e \in \rho(\sigma_0)$ and $I_{\sigma e} = I_\sigma \setminus \{e\}$.*

This concludes the discussion of the application of switches $(d_{*,*,*}, e_{*,*,*})$. The next lemma describes the end of phase 3 in S_n for $\nu > 1$. In contrast to the Markov decision process M_n, none of the switches $(s_{i,1-\beta_{i+1}}, b_1)$ with $i < \mu^\sigma$ is applied during phase 3. In S_n, these switches only become improving after applying the switch (b_1, b_2). This then starts phase 4 and the beginning of this phase is described by the following lemma. We refer to Table 5.5 resp. Table 5.9 for the definition of the sets S_1 and S_2 resp. X_k that are used in the statement.

Lemma 6.2.28 (Seventh row of Table 6.6). *Let $G_n = S_n$. Let $\sigma \in \rho(\sigma_0)$ be a well-behaved phase-3-strategy for $\mathfrak{b} \in \mathfrak{B}_n$ with $\nu > 1$. Let*

$$I_\sigma = \{(b_1, b_2)\} \cup \{(d_{i,j,k}, F_{i,j}), (e_{i,j,k}, b_2) : \sigma(e_{i,j,k}) = g_1\}$$

and $\sigma(d_{i,j,k}) = F_{i,j} \Leftrightarrow \beta^\sigma_i = 1 \wedge \beta^\sigma_{i+1} = j$ for all $i \in [n], j, k \in \{0,1\}$. Assume that σ has Property (ESC4)$_{i,j}$ for all $(i,j) \in S_1$ and Property (ESC5)$_{i,j}$ for all $(i,j) \in S_2$. Further assume that $\sigma(s_{i,j}) = h_{i,j}$ for all $i < \nu, j \in \{0,1\}$. Let $e := (b_1, b_2)$ and $m := \max\{i : \beta^\sigma_i = 1\}$. Then σe is a well-behaved phase-4-strategy for $\mathfrak{b}$ with $\mu^{\sigma e} = 1$ and

$$I_{\sigma e} = (I_\sigma \setminus \{e\}) \cup \{(s_{\nu-1,0}, b_1)\} \cup \{(s_{i,1}, b_1) : i \leq \nu - 2\} \cup X_0 \cup X_1$$

where X_k is defined as in Table 5.9.

As mentioned earlier, there is no phase 4 if $G_n = M_n$, even for $\nu > 1$. Hence, the application of the improving switch (b_1, b_2) directly yields a phase-5-strategy if all of the switches $(s_{i,j}, b_1)$ have been applied before.

Lemma 6.2.29 (Eighth row of Table 6.6). *Let $G_n = M_n$. Let $\sigma \in \rho(\sigma_0)$ be a well-behaved phase-3-strategy for $\mathfrak{b} \in \mathfrak{B}_n$ with $\nu > 1$. Let*

$$I_\sigma = \{(b_1, b_2)\} \cup \{(d_{i,j,k}, F_{i,j}), (e_{i,j,k}, b_2) : \sigma(e_{i,j,k}) = g_1\}.$$

Let σ have Property (USV1)$_i$ for all $i \in [n]$ and let $\sigma(d_{i,j,k}) = F_{i,j} \Leftrightarrow \beta^\sigma_i = 1 \wedge \beta^\sigma_{i+1} = j$ for all $i \in [n], j, k \in \{0,1\}$. Let σ have Property (ESC4)$_{i,j}$ for all $(i,j) \in S_1$ and Property (ESC5)$_{i,j}$ for all $(i,j) \in S_2$. Further assume that $e := (b_1, b_2) \in I_\sigma$ and let $m := \max\{i : \beta^\sigma_i = 1\}$. Then, σe is a well-behaved phase-5-strategy for $\mathfrak{b}$ with $\mu^{\sigma e} = 1$ and

$$I_{\sigma e} = (I_\sigma \setminus \{e\}) \cup \{(d_{i,1-\beta^\sigma_{i+1},k}, F_{i,1-\beta^\sigma_{i+1}}) : i < \nu\} \cup X_0 \cup X_1$$

where X_k is defined as in Table 5.9.

The next lemma now describes the direct transition from phase 3 to phase 5 for $\mathfrak{b} \in \mathfrak{B}_n$ with $\nu = 1$. In this case, there is no need to distinguish whether $G_n = S_n$ or $G_n = M_n$.

Lemma 6.2.30 (Last row of Table 6.6). *Let $\sigma \in \rho(\sigma_0)$ be a well-behaved phase-3-strategy for $\mathfrak{b} \in \mathfrak{B}_n$ with $\nu = 1$. Let $I_\sigma = \{(b_1, g_1)\} \cup \{(d_{i,j,k}, F_{i,j}), (e_{i,j,k}, g_1) \colon \sigma(e_{i,j,k}) = b_2\}$ and assume that σ has Property (ESC5)$_{i,j}$ for all $(i,j) \in S_3$ and Property (ESC3)$_{i,j}$ for all $(i,j) \in S_4$. Let $\sigma(d_{i,j,k}) = F_{i,j} \Leftrightarrow \beta_i^\sigma = 1 \wedge \beta_{i+1}^\sigma = j$ for all $i \in [n], j,k \in \{0,1\}$. Let $e := (b_1, g_1)$ and define $m := \max\{i \colon \beta_i^\sigma = 1\}$ and $u := \min\{i \colon \beta_i^\sigma = 0\}$. Then σe is a well-behaved phase-5-strategy for $\mathfrak{b}$ with $\mu^{\sigma e} = u, \sigma e \in \rho(\sigma_0)$ and*

$$I_{\sigma e} = (I_\sigma \setminus \{e\}) \cup \bigcup_{\substack{i'=u+1 \\ \beta_i^\sigma = 0}}^{m-1} \{(d_{i,1-\beta_{i+1}^\sigma,0}, F_{i,1-\beta_{i+1}^\sigma}), (d_{i,1-\beta_{i+1}^\sigma,1}, F_{i,1-\beta_{i+1}^\sigma})\}.$$

This concludes our discussion of the application of the improving switches of phase 3. We now discuss the improving switches that are applied during phase 4 if this phase is present.

Improving switches of phase 4

As explained earlier, it is still necessary to apply improving switches $(s_{*,*}, b_1)$ in S_n if $\nu > 1$. These switches are applied during phase 4. Since these switches are the only switches that are applied during phase 4, we do not provide a table summarizing the application of improving switches during this phase. Instead, we provide the following single lemma.

Lemma 6.2.31. *Let $G_n = S_n$. Let $\sigma \in \rho(\sigma_0)$ be a well-behaved phase-4-strategy for $\mathfrak{b} \in \mathfrak{B}_n$ with $\nu > 1$. Assume that there is an index $i < \nu$ such that $e := (s_{i,j}, b_1) \in I_\sigma$ where $j := 1 - \beta_{i+1}^\sigma$. Further assume the following:*

1. *σ has Property (USV1)$_{i'}$ for all $i' > i$.*
2. *For all i', j', k', it holds that $\sigma(d_{i',j',k'}) = F_{i',j'}$ if and only if $\beta_{i'}^\sigma = 1 \wedge \beta_{i'+1}^\sigma = j'$.*
3. *$i' < \nu$ implies $\bar{\sigma}(g_{i'}) = 1 - \beta_{i'+1}^\sigma$.*
4. *$i' < i$ implies $\sigma(s_{i',*}) = h_{i',*}$.*

If there is an index $i' < i$ such that $(s_{i',1-\beta_{i'+1}^\sigma}, b_1) \in I_\sigma$, then σe is a well-behaved phase-4-strategy for $\mathfrak{b}$. Otherwise, it is a well-behaved phase-5-strategy for $\mathfrak{b}$. In either case, it holds that $I_{\sigma e} = (I_\sigma \setminus \{e\}) \cup \{(d_{i,j,0}, F_{i,j}), (d_{i,j,1}, F_{i,j})\}$.

Improving switches of phase 5

We now discuss the improving switches which are applied during phase 5. As usual, we provide a table that contains one row per “type” of improving switch and provide a statement for each row of that table. This overview is given by Table 6.7. There is one more complex statement describing the application of improving switches of the type $(e_{*,*,*}, g_1)$ resp. $(e_{*,*,*}, b_2)$ during phase 5. Due to its complexity, this statement is not contained in Table 6.7.

We begin by providing the lemma describing the application of the improving switches involving the escape vertices. As usual, we define $t^{\rightarrow} := g_1 \wedge t^{\leftarrow} := b_2$ if $\nu = 1$ and $t^{\leftarrow} := b_2 \wedge t^{\rightarrow} := g_1$ if $\nu > 1$.

Lemma 6.2.32. *Let $\sigma \in \rho(\sigma_0)$ be a well-behaved phase-5-strategy for $\mathfrak{b} \in \mathfrak{B}_n$. Let $i \in [n]$ and $j, k \in \{0,1\}$ with $e := (e_{i,j,k}, t^{\rightarrow}) \in I_\sigma$ and $\bar{\sigma}(eb_{i,j}) \wedge \bar{\sigma}(eg_{i,j})$. Furthermore assume that $G_n = S_n$ implies*

$$j = 1 \wedge \nu > 1 \implies \neg\bar{\sigma}(eg_{i,1-j}) \quad \text{and} \quad j = 1 \wedge \nu = 1 \implies \neg\bar{\sigma}(eb_{i,1-j}).$$

Similarly, assume that $G_n = M_n$ implies

$$j = 1 - \beta^\sigma_{i+1} \wedge \nu > 1 \implies \neg\bar{\sigma}(eg_{i,1-j}) \quad \text{and} \quad j = 1 - \beta^\sigma_{i+1} \wedge \nu = 1 \implies \neg\bar{\sigma}(eb_{i,1-j}).$$

Moreover, assume that $\nu = 2$ implies $\sigma(g_1) = F_{1,0}$ if $G_n = S_n$. Then the following hold.

1. *If there are indices $(i', j', k') \neq (i, j, k)$ with $(e_{i',j',k'}, t^{\rightarrow}) \in I_\sigma$ or if there is an index i' such that σ does not have Property (SVG)$_{i'}$/(SVM)$_{i'}$, then σe is a phase-5-strategy for $\mathfrak{b}$.*
2. *The strategy σe is well-behaved.*
3. *If there are no indies $(i', j', k') \neq (i, j, k)$ with $(e_{i',j',k'}, t^{\rightarrow}) \in I_\sigma$ and if σ has Property (SVG)$_{i'}$/(SVM)$_{i'}$ for all $i'[n]$, then σe is a phase-1-strategy for $\mathfrak{b} + 1$.*
4. *If $G_n = S_n$, then*

$$(g_i, F_{i,j}) \in I_{\sigma e} \iff \beta^\sigma_i = 0 \wedge \overline{\sigma e}(g_i) = 1 \wedge j = 0 \wedge \begin{cases} \bar{\sigma}(eb_{i,1-j}), & \nu > 1 \\ \bar{\sigma}(eg_{i,1-j}), & \nu = 1 \end{cases}.$$

If $G_n = M_n$, then

$$(g_i, F_{i,j}) \in I_{\sigma e} \iff \beta^\sigma_i = 0 \wedge \overline{\sigma e}(g_i) = 1 - \beta^\sigma_{i+1} \wedge j = \beta^\sigma_{i+1} \wedge \begin{cases} \bar{\sigma}(eb_{i,1-j}), & \nu > 1 \\ \bar{\sigma}(eg_{i,1-j}), & \nu = 1 \end{cases}.$$

If the corresponding conditions are fulfilled, then

$$I_{\sigma e} = (I_\sigma \setminus \{e\}) \cup \{(d_{i,j,1-k}, F_{i,j}), (g_i, F_{i,j})\}.$$

Otherwise, $I_{\sigma e} = (I_\sigma \setminus \{e\}) \cup \{(d_{i,j,1-k}, F_{i,j})\}$.

Properties of σ	Switch e	Properties of σe
$\sigma(b_i) = b_{i+1}$ $\bar{\sigma}(g_i) = 1 - \beta^\sigma_{i+1}$	$(d_{i,j,k}, F_{i,j})$ $i \neq 1$ $j = 1 - \beta^\sigma_{i+1}$	Phase-5-strategy for $\mathfrak{b}$ $I_{\sigma e} = I_\sigma \setminus \{e\}$
$\beta^\sigma_i = 0$ $\nu = 1 \Rightarrow \bar{\sigma}(eg_{i,j}) \wedge \neg\bar{\sigma}(eg_{i,j})$ $\nu > 1 \Rightarrow \bar{\sigma}(eb_{i,j}) \wedge \neg\bar{\sigma}(eg_{i,j})$ $\mu^\sigma = 1$ implies $[i' \geq i \Rightarrow \bar{\sigma}(d_{i',*}) \vee (\bar{\sigma}(eb_{i',*}) \wedge \neg\bar{\sigma}(eg_{i',*})]$	$(g_i, F_{i,j})$	$\sigma(e_{*,*,*}) = t^{\rightarrow} \wedge$(SVG)$_{i'}$/(SVM)$_{i'}$ $\forall i' \in [n]$ $\Rightarrow$ Phase-1-strategy for $\mathfrak{b} + 1$ Otherwise phase-5-strategy for $\mathfrak{b}$ $I_{\sigma e} = I_\sigma \setminus \{e\}$

Table 6.7.: Improving switches applied during phase 5. For convenience, we always assume $\sigma \in \rho(\sigma_0)$ and that σ is a phase-5-strategy for $\mathfrak{b}$. Note that we thus also always have $\sigma e \in \rho(\sigma_0)$.

The following lemma describes the application of improving switches that involve cycle vertices during phase 5.

Lemma 6.2.33 (First row of Table 6.7). *Let $\sigma \in \rho(\sigma_0)$ be a well-behaved phase-5-strategy for $\mathfrak{b} \in \mathfrak{B}_n$. Let $i \in [n], j = 1 - \beta^\sigma_{i+1}, k \in \{0,1\}$ with $e := (d_{i,j,k}, F_{i,j}) \in I_\sigma$ and assume $\sigma(b_i) = b_{i+1}, \bar{\sigma}(g_i) = 1 - \beta^\sigma_{i+1}$ and $i \neq 1$. Then σe is a well-behaved Phase-5-strategy for $\mathfrak{b}$ with $\sigma e \in \rho(\sigma_0)$ and $I_{\sigma e} = I_\sigma \setminus \{e\}$.*

The next lemma concludes our discussion on the application of the improving switches and the corresponding transition through the phases. It describes the application of switches involving selector vertices during phase 5.

Lemma 6.2.34 (Second row of of Table 6.7). *Let $\sigma \in \rho(\sigma_0)$ be a well-behaved phase-5-strategy for $\mathfrak{b} \in \mathfrak{B}_n$. Let $i \in [n], j \in \{0,1\}$ with $e := (g_i, F_{i,j}) \in I_\sigma$ and $\beta^\sigma_i = 0$. Assume that $\nu = 1$ implies $\bar{\sigma}(eg_{i,j}) \wedge \neg\bar{\sigma}(eb_{i,j})$ and that $\nu > 1$ implies $\bar{\sigma}(eb_{i,j}) \wedge \neg\bar{\sigma}(eg_{i,j})$. Further assume that $\mu^\sigma = 1$ implies that for any $i' \geq i$ and $j' \in \{0,1\}$, either $\bar{\sigma}(d_{i',j'})$ or $\bar{\sigma}(eb_{i',j'}) \wedge \neg\bar{\sigma}(eg_{i',j'})$. If $\sigma(e_{i',j',k'}) = t^{\rightarrow}$ for all $i' \in [n], j', k' \in \{0,1\}$ and if σe has Property (SVG)$_{i'}$/(SVM)$_{i'}$ for all $i' \in [n]$, then σe is a phase-1-strategy for $\mathfrak{b}+1$. Otherwise it is a phase-5-strategy for $\mathfrak{b}$. In either case, σe is well-behaved and $I_{\sigma e} = I_\sigma \setminus \{e\}$.*

This concludes our discussion of the lemmas describing the exact application of individual improving switches in G_n. The next section now applies the results of this section to provide formal proofs of the statements of Section 5.3.

6.3. Proving the Main Statements

In this section, we provide the formal proofs for the statements given in Section 5.3. Before providing these proofs, we briefly explain how this section is organized. We begin by considering a canonical strategy $\sigma_{\mathfrak{b}}$ for some $\mathfrak{b} \in \mathfrak{B}_n$ that has the canonical properties. For each of the k phases, we prove that applying improving switches according to Zadeh's rule yields a phase-k-strategy $\sigma^{(k)}$ as described by Tables 5.8 and 5.9. This is done by considering the phases one after another. At the end, we prove that applying the improving switches of phase 5 to $\sigma^{(5)}$ yields a canonical strategy $\sigma_{\mathfrak{b}+1}$ for $\mathfrak{b}+1$.

When proving these statements, we typically immediately prove that the occurrence record of an edge e is described correctly by Table 5.6 when interpreted for $\mathfrak{b}+1$ after its application. The only kind of edges for which this is not proven immediately are edges of the form $(g_*, F_{*,*})$. The reason is that we need to analyze more than a single transition to properly analyze the occurrence records of these edges. Consequently, we defer the discussion of the occurrence records of these edges to the end of this section.

When proving the statements of this section, we often state smaller statements within the proofs as **Claims.** Using claims allows us to hide several more technical aspects on the macroscopic level, making the important proofs shorter and thus easier to comprehend. The proofs of all claims can however be found in Appendix A.2.

Consider some fixed $\mathfrak{b} \in \mathfrak{B}_n$, let $\nu := \ell(\mathfrak{b}+1)$ denote the least significant set bit of $\mathfrak{b}+1$ and let $\sigma_{\mathfrak{b}} \in \rho(\sigma_0)$ be a canonical strategy for $\mathfrak{b}$ that has the canonical conditions. As a a reminder, for two strategies σ, σ' with $\sigma' \in \rho(\sigma)$, the set $\mathfrak{A}^{\sigma'}_{\sigma}$ describes the set of improving switches applied by the strategy improvement resp. policy iteration algorithm during the transition $\sigma \to \sigma'$. Also, we define the parameter $\mathfrak{m} := \lfloor(\mathfrak{b}+1)/2\rfloor$ as this quantity will often be used when analyzing the occurrence records as it serves as a natural upper bound and is the maximum occurrence record that improving switches have.

Reaching a phase-2-strategy

We begin by proving that Zadeh's pivot rule together with the tie-breaking rule given in Definition 5.3.5 yields a phase-2-strategy $\sigma^{(2)} \in \rho(\sigma_0)$ as described by the corresponding rows of Tables 5.8 and 5.9. That is, we provide the proof of a slightly extended version of Lemma 5.3.14. This extension states that $\sigma^{(2)}$ is well-behaved. This was not included in the original statement as the term "well-behaved" was only introduced in Chapter 6.

Lemma 6.3.1 (Extended version of Lemma 5.3.14). *Let $\sigma_{\mathfrak{b}} \in \rho(\sigma_0)$ be a canonical strategy for $\mathfrak{b} \in \mathfrak{B}_n$ with $\nu = \ell(\mathfrak{b}+1) > 1$ having the canonical properties. After applying finitely many improving switches, the strategy improvement algorithm produces a well-behaved phase-2-strategy $\sigma^{(2)}$ for $\mathfrak{b}$ as described by the corresponding rows of Tables 5.8 and 5.9.*

Proof. By Lemma 6.2.2, $\sigma_{\mathfrak{b}}$ is well-behaved. Let $j := \mathfrak{b}_{\nu+1} = (\mathfrak{b}+1)_{\nu+1}$. Since $\sigma_{\mathfrak{b}}$ is a canonical strategy, we have $\sigma_{\mathfrak{b}}(d_{\nu,j,*}) \neq F_{\nu,j}$. Moreover, $I_{\sigma_{\mathfrak{b}}} = \mathfrak{D}^{\sigma_{\mathfrak{b}}}$ and $\sigma_{\mathfrak{b}}$ is a phase-1-strategy for $\mathfrak{b}$ by Lemma 5.3.12. In particular, $(d_{\nu,j,*}, F_{\nu,j}) \in I_{\sigma_{\mathfrak{b}}}$. By Lemma 6.2.5, $(d_{\nu,j,0}, F_{\nu,j})$ maximizes the occurrence record among all improving switches and $\phi^{\sigma_{\mathfrak{b}}}(d_{\nu,j,1}, F_{\nu,j}) = \mathfrak{m} - 1$. By Property (OR4)$_{*,*,*}$, $I_{\sigma_{\mathfrak{b}}}$ can be partitioned into $I_{\sigma_{\mathfrak{b}}} = I_{\sigma_{\mathfrak{b}}}^{<\mathfrak{m}} \cup I_{\sigma_{\mathfrak{b}}}^{\mathfrak{m}}$ where $e \in I_{\sigma_{\mathfrak{b}}}^{<\mathfrak{m}}$ if $\phi^{\sigma_{\mathfrak{b}}}(e) = \mathfrak{m} - 1$ and $e \in I_{\sigma_{\mathfrak{b}}}^{\mathfrak{m}}$ if $\phi^{\sigma_{\mathfrak{b}}}(e) = \mathfrak{m}$. If $I_{\sigma_{\mathfrak{b}}}^{<\mathfrak{m}} \neq \emptyset$, then a switch contained in this set is applied first as the LeastEntered pivot rule always chooses an improving switch minimizing the occurrence record. By applying Lemma 6.2.11 iteratively, the algorithm applies switches $e \in I_{\sigma_{\mathfrak{b}}}^{<\mathfrak{m}}$ until it either reaches a strategy σ with $I_{\sigma}^{<\mathfrak{m}} = \emptyset$ or until an edge $(g_i, F_{i,j'})$ with $j' \neq j$ becomes improving. By Lemma 6.2.11, Table 5.6 (interpreted for $\mathfrak{b}+1$) correctly describes the occurrence record of all switches applied in the process.

Claim 1. If an edge $(g_i, F_{i,j'})$ with $i \in [n]$ and $j' \neq \mathfrak{b}_{\nu+1}$ becomes improving during the application of improving switches contained in $I^{<\mathfrak{m}}$, then it is applied immediately. Its application is described by row 4 of Table 6.4.

Such an edge $(g_i, F_{i,j'})$ is only applied if $F_{i,j'}$ was closed by the previous applications. This implies that either $(d_{i,j',0}, F_{i,j'}), (d_{i,j',1}, F_{i,j'}) \in I_{\sigma_{\mathfrak{b}}}^{<\mathfrak{m}}$ or $\sigma(d_{i,j',1-k}) = \sigma_{\mathfrak{b}}(d_{i,j',1-k}) = F_{i,j'}$. The first case can only happen for $i = \nu$ and $j' = 1 - \mathfrak{b}_{\nu+1}$. Thus, if a switch $(g_i, F_{i,j'})$ is applied, then either $i = \nu$ or $\sigma_{\mathfrak{b}}(d_{i,j',1-k}) = F_{i,j'}$ and $(d_{i,j',k}, F_{i,j'}) \in I_{\sigma_{\mathfrak{b}}}^{<\mathfrak{m}}$.

The previous arguments can now be applied until a strategy is reached for which no edge has a "low" occurrence record. Thus let σ be a phase-1-strategy σ with $I_{\sigma}^{<\mathfrak{m}} = \emptyset$ and $\mathbb{G} \cap I_{\sigma} = \emptyset$. Further note that $\mathfrak{A}_{\sigma_{\mathfrak{b}}}^{\sigma} \subseteq \mathbb{D}^1 \cup \mathbb{G}$ and that $(g_i, F_{i,j}) \in \mathfrak{A}_{\sigma_{\mathfrak{b}}}^{\sigma}$ implies $\mathfrak{b}_i = 0 \wedge \mathfrak{b}_{i+1} \neq j$ and that the previous arguments hold independent of ν.

We discuss improving switches contained in $I_{\sigma_{\mathfrak{b}}}^{\mathfrak{m}}$ next.

Claim 2. Let $\nu > 1$ and let σ denote the strategy obtained after applying all improving switches contained in $I_{\sigma}^{<\mathfrak{m}}$. For all suitable indices $i \in [n], j' \in \{0,1\}$ it holds that $\sigma(d_{i,j',1}) = F_{i,j'}$, implying that no cycle center is open for σ.

Let $\sigma \in \rho(\sigma_{\mathfrak{b}})$ be a phase-1-strategy with $I_\sigma = \{e = (d_{i,j,k}, F_{i,j})\colon \phi^\sigma(e) = \mathfrak{m}\} = \mathfrak{D}^\sigma$ as described previously. We prove that $e := (d_{\nu,j,0}, F_{\nu,j})$ is applied next. By Lemma 6.2.5, the definition of a canonical strategy and since only edges with an occurrence record less than $\mathfrak{m}$ were applied so far, this implies $e \in I_\sigma$. Since all improving switches have the same occurrence records, it is sufficient to show that no other improving switch is ranked lower by the tie-breaking rule. By Claim 2, there are no open cycle centers. Hence, the ordering of the edges is based on the bits represented by the levels, the index of the levels and whether the cycle center is active. To be precise, the first switch according to the tie-breaking rule is the improving switch contained in the active cycle center of the lowest level with a bit equal to 0. This edge is precisely $e = (d_{\nu,j,0}, F_{\nu,j})$.

We now prove that the occurrence record of e is described by Table 5.6 when interpreted for $\mathfrak{b}+1$ after the application. Since $F_{\nu,j}$ is closed for σ but was open for $\sigma_{\mathfrak{b}}$, we prove

$$\phi^{\sigma e}(e) = \left\lceil \frac{\operatorname{lfn}(\mathfrak{b}+1, \nu, \{(\nu+1, j)\}) + 1}{2} \right\rceil.$$

By the definition of ν, it holds that $\mathfrak{b}+1 = \operatorname{lfn}(\mathfrak{b}+1, \nu, \{(\nu+1, j)\})$. The statement thus follows since $\mathfrak{m}+1 = \lceil (\mathfrak{b}+1+1)/2 \rceil$.

By row 5 of Table 6.4, σe is a well-behaved (potentially pseudo) phase-2-strategy for $\mathfrak{b}$. If $\sigma(g_\nu) \neq F_{\nu,j}$, then $(g_\nu, F_{\nu,j})$ minimizes the occurrence record among all improving switches. Due to the tie-breaking rule, this switch is then applied next, and this application is formalized in row 6 of Table 6.4.

Let σ denote the strategy obtained after applying $(g_\nu, F_{\nu,j})$ if $\sigma(g_\nu) \neq F_{\nu,j}$ resp. after applying $(d_{\nu,j,0}, F_{\nu,j})$ if $\sigma(g_\nu) = F_{\nu,j}$. Then, by row 5 resp. 6 of Table 6.4,

$$I_\sigma = \mathfrak{D}^\sigma \cup \{(b_\nu, g_\nu)\} \cup \{(s_{\nu-1,1}, h_{\nu-1,1})\}.$$

Furthermore, σ has Property (USV3)$_i$ for all $i < \nu$ as $\sigma_{\mathfrak{b}}$ has Property (USV1)$_i$ and $\mathfrak{b}_i = 1 - (\mathfrak{b}+1)_i$ for $i \leq \nu$. In addition, $\sigma(d_{i,j,k}) \neq F_{i,j}$ implies $\phi^{\sigma^{(2)}}(d_{i,j,k}, F_{i,j}) = \mathfrak{m}$ by Corollary 6.3.3. Moreover, since no improving switch $(d_{*,*,*}, e_{*,*,*})$ was applied and $\mathfrak{b}_i = 1 - \beta_{i+1}^\sigma$ for all $i < \nu$, it holds that $\bar{\sigma}(g_i) = 1 - \beta_{i+1}^\sigma$ and $\bar{\sigma}(d_{i,1-\beta_{i+1}^\sigma})$ for all $i < \nu$. □

We henceforth refer to the phase-2-strategy that is described by the corresponding rows of Tables 5.8 and 5.9 and whose existence we just proved by $\sigma^{(2)}$. When proving the existence of this strategy, we furthermore implicitly proved the following three corollaries. We later show that the condition $\nu > 1$ can be dropped in the first corollary.

Corollary 6.3.2. *Let $\sigma_{\mathfrak{b}}$ be a canonical strategy for $\mathfrak{b}$ having the canonical properties and $\nu > 1$. Let $i \in [n]$ and $j \in \{0,1\}$. Then, the edge $(g_i, F_{i,j})$ is applied as improving switch during phase 1 if and only if $F_{i,j}$ is closed during phase 1, $\sigma_{\mathfrak{b}}(g_i) \neq F_{i,j}$ and $i \neq \nu$. A cycle center can only be closed during phase 1 if either $i = \nu$ or if there exists an index $k \in \{0,1\}$ such that $\sigma_{\mathfrak{b}}(d_{i,1-\mathfrak{b}_{i+1},k}) = F_{i,1-\mathfrak{b}_{i+1}}, \phi^{\sigma_{\mathfrak{b}}}(d_{i,1-\mathfrak{b}_{i+1},1-k}, F_{i,\mathfrak{b}_{i+1}}) < \mathfrak{m}$ and $\sigma_{\mathfrak{b}}(b_i) = b_{i+1}$.*

Corollary 6.3.3. *Let $\sigma_{\mathfrak{b}}$ be a canonical strategy for $\mathfrak{b}$ having the canonical properties and let $i \in [n], j, k \in \{0,1\}$ such that $\sigma_{\mathfrak{b}}(d_{i,j,k}) \neq F_{i,j}$. If $\phi^{\sigma_{\mathfrak{b}}}(d_{i,j,k}, F_{i,j}) < \mathfrak{m}$, then $(d_{i,j,k}, F_{i,j})$ is applied during phase 1.*

Corollary 6.3.4. *No cycle center is open with respect to* $\sigma^{(2)}$.

Corollary 6.3.5. *Table 5.6 correctly specifies the occurrence record of every improving switch applied during* $\sigma_{\mathfrak{b}} \to \sigma^{(2)}$ *when interpreted for* $\mathfrak{b}+1$*, excluding switches* $(g_*, F_{*,*})$*. In addition, each switch is applied at most once.*

Reaching a phase-3-strategy

We now prove that the algorithm produces a phase-3-strategy by proving a slightly extended version of Lemma 5.3.15. If $\nu = 1$, then this follows by analyzing phase 1 in a similar fashion as done when proving Lemma 6.3.1. It in fact turns out that nearly the identical arguments can be applied. If $\nu > 1$, then we use that lemma to argue that we obtain a phase-2-strategy. We then investigate phase 2 in detail and prove that we also obtain a phase-3-strategy. The proof uses the statements summarized in Tables 6.4 and 6.5, and we refer to these tables and the corresponding statements in proofs of the related statements.

Lemma 6.3.6 (Extended version of Lemma 5.3.15). *Let* $\sigma_{\mathfrak{b}} \in \rho(\sigma_0)$ *be a canonical strategy for* $\mathfrak{b} \in \mathfrak{B}_n$ *having the canonical properties. After applying a finite number of improving switches, the strategy improvement algorithm produces a well-behaved phase-3-strategy* $\sigma^{(3)} \in \rho(\sigma_0)$ *as described by the corresponding rows of Tables 5.8 and 5.9.*

Proof. Consider the case $\nu = 1$ first. As shown in the proof of Lemma 6.3.1, the set $I_{\sigma_{\mathfrak{b}}}$ can be partitioned into $I_{\sigma_{\mathfrak{b}}}^{<\mathfrak{m}}$ and $I_{\sigma_{\mathfrak{b}}}^{\mathfrak{m}}$. Since Lemma 6.2.11 also applies for $\nu = 1$, the same arguments imply that the algorithms calculate a phase-1-strategy $\sigma \in \rho(\sigma_0)$ with $I_\sigma = \{e = (d_{i,j,k}, F_{i,j}) \colon \phi^{\sigma}(e) = \mathfrak{m}\} = \mathfrak{D}^{\sigma}$. We can again deduce $\mathfrak{A}_{\sigma_{\mathfrak{b}}}^{\sigma} \subseteq \mathbb{D}^1 \cup \mathbb{G}$ and that $(g_i, F_{i,j}) \in \mathfrak{A}_{\sigma_{\mathfrak{b}}}^{\sigma}$ implies $\mathfrak{b}_i = 0 \wedge \mathfrak{b}_{i+1} \neq j$ or $i = \nu$ for all $i \in [n], j \in \{0,1\}$. We can further assume $(g_*, F_{*,*}) \notin I_\sigma$. Also, by Lemma 6.2.11, the occurrence records of edges $(d_{*,*,*}, F_{*,*}) \in \mathfrak{A}_{\sigma_{\mathfrak{b}}}^{\sigma}$ is described by Table 5.6 when interpreted for $\mathfrak{b}+1$.

Since all improving switches now have the same occurrence records, their order of application depends on the tie-breaking rule. Due to the first criterion, improving switches contained in open cycle centers are applied first. Hence, a sequence of strategies is produced until a strategy without open cycle centers is reached. All produced strategies are well-behaved phase-1-strategies for $\mathfrak{b}$, reachable from σ_0 by row 1 of Table 6.4. Also, by the tie-breaking rule, the edge $(d_{*,*,0}, F_{*,*})$ is applied as improving switch in an open cycle center $F_{*,*}$. By the same arguments used when proving Lemma 6.3.1, the second switch of $F_{\nu,\mathfrak{b}_{\nu+1}}$ is applied next and, possibly, $(g_\nu, F_{\nu,\mathfrak{b}_{\nu+1}})$ is applied afterwards.

Let $\sigma^{(3)}$ denote the strategy obtained after closing the cycle center $F_{\nu,\mathfrak{b}_{\nu+1}}$ resp. after applying $(g_\nu, F_{\nu,\mathfrak{b}_{\nu+1}})$ if it becomes improving.

Claim 3. Let $i \in [n], j,k \in \{0,1\}$ such that $(d_{i,j,k}, F_{i,j}) \in \mathfrak{A}_{\sigma_{\mathfrak{b}}}^{\sigma^{(3)}}$. The occurrence records of $(d_{i,j,k}, F_{i,j})$ with respect to $\sigma^{(3)}$ is specified by Table 5.6 when interpreted for $\mathfrak{b}+1$.

Note that the last row of Table 6.5 can be used to describe the application of $(g_\nu, F_{\nu,\mathfrak{b}_{\nu+1}})$. Then, by row 5 of Table 6.4 resp. the last row of Table 6.5 and our previous arguments, $\sigma^{(3)}$ has all properties listed in the respective rows of Tables 5.8 and 5.9. Furthermore, as we used the same arguments, Corollary 6.3.2 is also valid for $\nu = 1$ and we can drop the assumption $\nu > 1$.

Consider the case $\nu > 1$, implying $\mathfrak{b} \geq 1$. By Lemma 6.3.1, applying improving switches to $\sigma_{\mathfrak{b}}$ yields a phase-2-strategy $\sigma = \sigma^{(2)}$ for $\mathfrak{b}$ with $I_\sigma = \mathfrak{D}^\sigma \cup \{(b_\nu, g_\nu), (s_{\nu-1,1}, h_{\nu-1,1})\}$ and $\sigma \in \rho(\sigma_0)$. By Table 5.6 and Lemma 6.2.6,

$$\phi^\sigma(b_\nu, g_\nu) = \mathrm{fl}(\mathfrak{b}, \nu) = \phi^\sigma(s_{\nu-1,1}, h_{\nu-1,1}) \quad \text{and} \quad \mathrm{fl}(\mathfrak{b}, \nu) = \left\lfloor \frac{\mathfrak{b} + 2^{\nu-1}}{2^\nu} \right\rfloor.$$

Since $\nu > 1$ and $\mathfrak{b} \geq 1$, this implies $\mathrm{fl}(\mathfrak{b}, \nu) \leq \lfloor(\mathfrak{b}+2)/4\rfloor \leq \mathfrak{m}$. By Lemma 6.3.1, any improving switch $(d_{*,*,*}, F_{*,*}) \in I_\sigma$ has an occurrence record of $\mathfrak{m}$. Thus, by the tie-breaking rule, (b_ν, g_ν) is applied next. Let σe denote the strategy obtained after applying (b_ν, g_ν). It is easy to verify that σ has the properties of row 1 of Table 6.5. Consequently, σe is a phase-2-strategy for $\mathfrak{b}$ with $\sigma e \in \rho(\sigma_0)$. By Lemma 6.2.6, $\phi^{\sigma e}(b_\nu, g_\nu) = \mathrm{fl}(\mathfrak{b}, \nu) + 1 = \mathrm{fl}(\mathfrak{b}+1, \nu)$, so Table 5.6 describes the occurrence record of (b_ν, g_ν) with respect to $\mathfrak{b}+1$. The set of improving switches for σe now depends on ν, see row 1 of Table 6.5.

Let $\nu = 2$. Then $I_{\sigma e} = \mathfrak{D}^{\sigma e} \cup \{(b_1, b_2), (s_{1,1}, h_{1,1})\} \cup \{(e_{*,*,*}, b_2)\}$. In this case, $(s_{1,1}, h_{1,1})$ is applied next and its application yields the desired phase-3-strategy $\sigma^{(3)}$.

Claim 4. Let $\nu = 2$ and consider the phase-2-strategy σ obtained after the application of (b_ν, g_ν). Then, the edge $(s_{1,1}, h_{1,1})$ is applied next, and the obtained strategy is a well-behaved phase-3-strategy for $\mathfrak{b}$ described by the respective rows of Tables 5.8 and 5.9.

If $\nu > 1$, then we do not obtain the desired strategy yet and we have to consider a longer sequence of improving switches that are applied. Thus, let $\nu > 2$, implying $\mathfrak{b} \neq 1$. Then, the first row of Table 6.5 implies

$$I_{\sigma e} = \mathfrak{D}^{\sigma e} \cup \{(b_{\nu-1}, b_\nu), (s_{\nu-1,1}, h_{\nu-1,1}), (s_{\nu-2,0}, h_{\nu-2,0})\}.$$

By Table 5.6, $\phi^{\sigma e}(b_{\nu-1}, b_\nu) = \mathrm{fl}(\mathfrak{b}, \nu-1) - 1$ and $\phi^{\sigma e}(s_{\nu-1,1}, h_{\nu-1,1}) = \mathrm{fl}(\mathfrak{b}, \nu)$. In addition, $\phi^{\sigma e}(s_{\nu-2,0}, h_{\nu-2,0}) = \mathrm{fl}(\mathfrak{b}, \nu-1) - 1$. Hence, both edges $(b_{\nu-1}, b_\nu)$ and $(s_{\nu-2,0}, h_{\nu-2,0})$ minimize the occurrence record. By the tie-breaking rule, the switch $e' := (b_{\nu-1}, b_\nu)$ is now applied. We show that the application of e' can be described by row 3 of Table 6.5. We thus need to show the following:

- $\overline{\sigma e}(d_{i'})$ **for all** $i' < \mu^{\sigma e}$: This follows from Lemma 6.3.1 as no switch $(d_{*,*,*}, e_{*,*,*})$ was applied during $\sigma_{\mathfrak{b}} \to \sigma^{(2)}$ and no improving switch involving selector vertices was applied in a level $i' < \mu^{\sigma e}$.
- σe **has Property (USV3)**$_{i'}$ **for all** $i' < \nu - 1$: Since no switch $(s_{i',*}, *)$ was applied for $i' < \nu - 1$, this follows since $\sigma_{\mathfrak{b}}$ has Property (USV1)$_{i'}$ for those indices.
- σe **has Property (EV1)**$_{i'}$ **and (EV2)**$_{i'}$ **for all** $i' > \nu - 1$ **and (EV3)**$_{i'}$ **for all** $i' > \nu - 1$ **with** $i' \neq \mu^{\sigma e}$: Since $\mu^{\sigma e} - 1 = \nu - 1$ and σe is a phase-2-strategy for $\mathfrak{b}$, it suffices to prove that σe has Property (EV1)$_\nu$ and Property (EV2)$_\nu$. This however follows since the strategy in which (b_ν, g_ν) was applied had Property (CC2).

For simplicity, we denote the strategy that is obtained by applying e' to σe by σ. By our previous arguments and row 3 of Table 6.5, σ is a well-behaved phase-2-strategy for $\mathfrak{b}$ that has Property (CC2) as well as Properties (EV1)$_i$ and (EV2)$_i$ for all $i \geq \nu - 1$

and Property (EV3)$_i$ for all $i > \nu - 1, i \neq \nu$. In addition, $\bar{\sigma}(d_i)$ for all $i < \nu$ and σ has Property (USV3)$_i$ for all $i < \nu - 1$ Furthermore, Lemma 6.2.6 implies

$$\phi^{\sigma}(e) = \mathrm{fl}(\mathfrak{b}, \nu - 1) - 1 + 1 = \mathrm{fl}(\mathfrak{b}, \nu - 1) = \mathrm{fl}(\mathfrak{b} + 1, \nu - 1) - (\mathfrak{b} + 1)_{\nu-1},$$

so Table 5.6 describes the occurrence record of e with respect to $\mathfrak{b} + 1$. By row 3 of Table 6.5, $\nu - 1 > 2$ implies

$$I_\sigma = \mathfrak{D}^\sigma \cup \{(s_{\nu-1,1}, h_{\nu-1,1}), (s_{\nu-2,0}, h_{\nu-2,0}), (b_{\nu-2}, b_{\nu-1}), (s_{\nu-3,0}, h_{\nu-3,0})\}.$$

Similarly, $\nu - 1 = 2$ implies

$$I_\sigma = \mathfrak{D}^\sigma \cup \{(e_{i,j,k}, b_2)\} \cup \{(b_1, b_2), (s_{2,1}, h_{2,1}), (s_{1,0}, h_{1,0})\}.$$

In both cases, $e := (s_{\nu-1,1}, h_{\nu-1,1}) \in I_\sigma$ is applied next.

Claim 5. After the application of $(b_{\nu-1}, b_\nu)$ in the case $\nu > 2$, the switch $e = (s_{\nu-1,1}, h_{\nu-1,1})$ is applied next. Its application can be described by row 2 of Table 6.5 and Table 5.6 specifies its occurrence record after the application correctly when interpreted for $\mathfrak{b} + 1$.

Let $\nu - 1 > 2$. We argue that applying improving switches according to Zadeh's pivot rule and our tie-breaking rule then results in a sequence of strategies such that we finally obtain a strategy σ' with $I_{\sigma'} = \mathfrak{D}^{\sigma'} \cup \{(e_{*,*,*}, b_2)\} \cup \{(b_1, b_2), (s_{1,0}, h_{1,0})\}$. Note that such a strategy is also obtained after the application of $(s_{\nu-1,1}, h_{\nu-1,1})$ if $\nu - 1 = 2$. For any $x \in \{2, \ldots, \nu - 2\}$, Lemma 6.2.6 implies

$$\begin{aligned} \phi^{\sigma e}(s_{\nu-x,0}, h_{\nu-x,0}) < \phi^{\sigma e}(b_{\nu-x}, b_{\nu-(x+1)}) = \phi^{\sigma e}(s_{\nu-(x-1),0}, h_{\nu-(x-1),0}) \\ < \phi^{\sigma e}(e_{*,*,*}, b_2). \end{aligned} \tag{6.2}$$

Thus, $(s_{\nu-2,0}, h_{\nu-2,0})$ is applied next. It is easy to verify that σe meets the requirements of row 2 of Table 6.5, so it can be used to describe the application of $(s_{\nu-2,0}, h_{\nu-2,0})$.

Let σ' denote the strategy obtained. Then $I_{\sigma'} = \mathfrak{D}^\sigma \cup \{(b_{\nu-2}, b_{\nu-1}), (s_{\nu-3,0}, h_{\nu-3,0})\}$. Also, the occurrence record of $(s_{\nu-2,0}, h_{\nu-2,0})$ is described by Table 5.6 when interpreting the table for $\mathfrak{b} + 1$. By Equation (6.2) and the tie-breaking rule, $(b_{\nu-2}, b_{\nu-1})$ is applied next. Similar to the previous cases, it is easy to check that row 3 of Table 6.5 applies to this switch. We thus obtain a strategy σ such $\nu - 2 \neq 2$ implies

$$I_\sigma = \mathfrak{D}^\sigma \cup \{(s_{\nu-3,0}, h_{\nu-3,0}), (b_{\nu-3}, b_{\nu-2}), (s_{\nu-4,0}, h_{\nu-4,0})\}$$

and $\nu - 2 = 2$ implies

$$I_\sigma = \mathfrak{D}^\sigma \cup \{(e_{*,*,*,}, b_2)\} \cup \{(b_1, b_2), (s_{1,0}, h_{1,0})\}.$$

In either case, a simple calculation implies that the occurrence record of $(b_{\nu-2}, b_{\nu-1})$ is described by Table 5.6 interpreted for $\mathfrak{b} + 1$.

In the first case, we can now apply the same arguments again iteratively as Equation (6.2) remains valid for σ' and $x \in \{2, \ldots, \nu - 3\}$. After applying a finite number of improving switches we thus obtain a phase-2-strategy $\sigma \in \rho(\sigma_0)$ with

$$I_\sigma = \mathfrak{D}^\sigma \cup \{(e_{*,*,*}, b_2)\} \cup \{(b_1, b_2), (s_{1,0}, h_{1,0})\}.$$

Furthermore, σ has Properties (EV1)$_i$, (EV2)$_i$ and (USV2)$_{i,\beta_{i+1}}$ for all $i > 1$ as well as Property (EV3)$_i$ for all $i > 1, i \neq \mu^\sigma$ and Property (CC2). In addition, $\bar{\sigma}(g_i) = 1 - \beta_{i+1}$ and $\bar{\sigma}(d_{i,1-\beta_{i+1}})$ for all $i < \nu$ and the occurrence records of all edges applied so far (with the exception of switches $(g_*, F_{*,*})$) is described by Table 5.6 when being interpreted for $\mathfrak{b}+1$. Note that all of this also holds if $\nu - 1 = 2$.

Consequently, σ meets the requirements of row 2 of Table 6.5. As $\nu > 2$, we have $\beta_2 = 0$. By Table 5.6,

$$\phi^\sigma(s_{1,0}, h_{1,0}) = \mathrm{fl}(\mathfrak{b}, 2) - 1 < \mathrm{fl}(\mathfrak{b}, 1) - 1 = \phi^\sigma(b_1, b_2)$$

as well as

$$\mathrm{fl}(\mathfrak{b}, 2) - 1 = \lfloor (\mathfrak{b}+2)/4 \rfloor - 1 < \lfloor \mathfrak{b}/2 \rfloor = \phi^\sigma(e_{*,*,*}, b_2).$$

Hence, the switch $e = (s_{1,0}, h_{1,0})$ is applied next and by row 2 of Table 6.5, $\sigma^{(3)} := \sigma e$ is a phase-3-strategy for $\mathfrak{b}$ with

$$I_{\sigma^{(3)}} = \mathfrak{D}^{\sigma^{(3)}} \cup \{(e_{i,j,k}, b_2)\} \cup \{(b_1, b_2)\}.$$

We thus obtain a strategy as described by the corresponding rows of Tables 5.8 and 5.9. □

We henceforth use $\sigma^{(3)}$ to refer to the phase-3-strategy described by Lemma 6.3.6. Note that we implicitly proved the following corollaries where the second follows by Corollary 6.3.5.

Corollary 6.3.7. *No cycle center is open with respect to $\sigma^{(3)}$.*

Corollary 6.3.8. *Table 5.6 specifies the occurrence record of every improving switch applied during $\sigma_\mathfrak{b} \to \sigma^{(3)}$ when interpreted for $\mathfrak{b}+1$, excluding switches $(g_*, F_{*,*})$. In addition, each such switch was applied once.*

Reaching a phase-4-strategy or a phase-5-strategy

We now discuss the application of improving switches during phase 3, which highly depends on whether $G_n = M_n$ or $G_n = S_n$ and on the least significant set bit of $\mathfrak{b}+1$. The next lemma now summarizes the application of improving switches during phase 3 and is a generalization of Lemma 5.3.16. Depending on G_n and ν, we then either obtain a phase-4-strategy or a phase-5-strategy for $\mathfrak{b}$. As with the previous lemmas, this lemma is an extension of Lemma 5.3.16. We also use the usual notation and define $t^\rightarrow := b_2$ if $\nu > 1$ and $t^\rightarrow := g_1$ if $\nu = 1$. Similarly, let $t^\leftarrow := g_1$ if $\nu > 1$ and $t^\leftarrow := b_2$ if $\nu = 1$.

Lemma 6.3.9 (Extended version of Lemma 5.3.16). *Let $\sigma_\mathfrak{b} \in \rho(\sigma_0)$ be a canonical strategy for $\mathfrak{b} \in \mathfrak{B}_n$ having the canonical properties. After applying finitely many improving switches, the strategy improvement algorithm produces a well-behaved strategy σ with the following properties: If $\nu > 1$, then σ is a phase-k-strategy for $\mathfrak{b}$, where $k = 4$ if $G_n = S_n$ and $k = 5$ if $G_n = M_n$. If $\nu = 1$, then σ is a phase-5-strategy for $\mathfrak{b}$. In any case, $\sigma \in \rho(\sigma_0)$ and σ is described by the corresponding rows of Tables 5.8 and 5.9.*

Before proving this lemma, we provide an additional lemma that summarizes the application of switches of the type $(d_{*,*,*}, e_{*,*,*})$. Its proof is omitted here and deferred to Appendix A.2.

Lemma 6.3.10. *Let $\sigma \in \rho(\sigma^{(3)})$ be a well-behaved phase-3-strategy for $\mathfrak{b}$ obtained through the application of a sequence $\mathfrak{A}^{\sigma}_{\sigma^{(3)}} \subseteq \mathbb{E}^1 \cup \mathbb{D}^0$ of improving switches. Assume that the conditions of row 1 of Table 6.6 were fulfilled for each intermediate strategy σ' of the transition $\sigma^{(3)} \to \sigma$. Let $t^{\rightarrow} := b_2$ if $\nu > 1$ and $t^{\rightarrow} := g_1$ if $\nu = 1$. Let $i \in [n], j, k \in \{0,1\}$ such that $e := (d_{i,j,k}, e_{i,j,k}) \in I_\sigma$ is applied next and assume $\sigma(e_{i,j,k}) = t^{\rightarrow}, \beta_i^\sigma = 0 \vee \beta_{i+1}^\sigma \neq j$ and $I_\sigma \cap \mathbb{D}^0 = \{e\}$. Further assume that either $i \geq \nu$ or that we consider the case $G_n = S_n$. Then σe is a phase-3-strategy for $\mathfrak{b}$ with $I_{\sigma e} = (I_\sigma \setminus \{e\})$.*

This now enables us to prove Lemma 6.3.9.

Proof of Lemma 6.3.9. By Lemma 6.3.6, applying improving switches according to Zadeh's pivot rule and our tie-breaking rule yields a phase-3-strategy $\sigma^{(3)} \in \rho(\sigma_0)$ described by the corresponding rows of Tables 5.8 and 5.9. As it simplifies the formal proof significantly, we begin by describing phase 3 informally.

For every cycle vertex $d_{*,*,*}$, it either holds that $\sigma^{(3)}(d_{*,*,*}) = F_{*,*}$ or $\sigma^{(3)}(d_{*,*,*}) = e_{*,*,*}$ and $(d_{*,*,*}, F_{*,*}) \in I_{\sigma^{(3)}}$. It will turn out that only switches corresponding to cycle vertices of the first type are applied during phase 3. Consider an arbitrary but fixed such cycle vertex $d_{i,j,k}$ for some suitable indices i, j, k. Then, the switch $(e_{i,j,k}, t^{\rightarrow})$ will be applied. If $(\mathfrak{b}+1)_i = 0$ or $(\mathfrak{b}+1)_{i+1} \neq j$, then $(d_{i,j,k}, e_{i,j,k})$ becomes improving and is applied next. This procedure then continues until all such improving switches have been applied. During this procedure, it might happen that an edge $(s_{i',*}, b_1)$ with $i' < \nu$ becomes improving after applying some switch $(d_{*,*,*}, e_{*,*,*})$ if $\nu > 1$ and $G_n = M_n$. In this case, the corresponding switch is applied immediately. Finally, (b_1, b_2) resp. (b_1, g_1) is applied, resulting in a phase-4-strategy if $\nu > 1$ and $G_n = S_n$ and in a phase-5-strategy otherwise.

We now formalize this behavior. We first show that switches $(e_{*,*,*}, t^{\rightarrow})$ minimize the occurrence record among all improving switches. Consider some indices i, j, k such that $(d_{i,j,k}, F_{i,j}) \in I_{\sigma^{(3)}}$. Then, $\phi^{\sigma^{(3)}}(d_{i,j,k}, F_{i,j}) = \mathfrak{m}$ by Lemma 6.3.6 resp. Table 5.8. If $\nu > 1$, then

$$\phi^{\sigma^{(3)}}(e_{i,j,k}, b_2) = \left\lfloor \frac{\mathfrak{b}}{2} \right\rfloor = \mathfrak{m} - 1 = \mathrm{fl}(\mathfrak{b}, 1) - \mathfrak{b}_1 = \phi^{\sigma^{(3)}}(b_1, b_2)$$

by Table 5.6. Similarly, if $\nu = 1$, then $\phi^{\sigma^{(3)}}(e_{i,j,k}, g_1) = \phi^{\sigma^{(3)}}(b_1, g_1)$. By the tie-breaking rule, a switch of the type $(e_{i',j',k'}, t^{\rightarrow})$ with $\sigma^{(3)}(d_{i',j',k'}) = F_{i',j'}$ for some suitable indices is thus applied next. Since $\sigma^{(3)}(s_{i',*}) = h_{i',*}$ for all $i' < \mu^{\sigma^{(3)}}$ by Lemma 6.3.6, the statement of row 1 of Table 6.6 can be applied.

Let $i \in [n], j, k \in \{0,1\}$ denote the indices such that $e := (e_{i,j,k}, t^{\rightarrow}) \in I_\sigma \cap \mathbb{E}^1$ is the switch that is applied next. We prove that the characterization given in the first row of Table 6.6 implies $I_{\sigma e} = (I_\sigma \setminus \{e\}) \cup \{(d_{i,j,k}, e_{i,j,k})\}$ if $\beta_i = 0 \vee \beta_{i+1} \neq j$ and $I_{\sigma e} = I_\sigma \setminus \{e\}$ else. As explained earlier, the strategy σ fulfills the requirements of the first row of Table 6.6. Consider the strategy σe. By the first row of Table 6.6, $(d_{i,j,k}, e_{i,j,k})$ is improving for σe if and only if either $\sigma(d_{i,j,1-k}) = e_{i,j,1-k}$ or $[\sigma(d_{i,j,1-k}) = F_{i,j}$ and $j \neq \beta_{i+1}]$. It thus suffices to prove that $\beta_i = 0 \vee \beta_{i+1} \neq j$ is equivalent to the disjunction of these two conditions. We do so by showing $\beta_i = 1 \wedge \beta_{i+1} = j \Leftrightarrow \sigma(d_{i,j,1-k}) = F_{i,j} \wedge \beta_{i+1} = j$. The direction "$\Rightarrow$" follows since the cycle center $F_{i,j}$ is then active and closed. The direction "$\Leftarrow$" follows since $\sigma(d_{i,j,1-k}) = F_{i,j}$ implies that $F_{i,j}$ is closed as e being improving for σ implies $\sigma(d_{i,j,k}) = F_{i,j}$. But then, by the definition of β and the choice of j, $\beta_i = 1$.

Consequently, by the tie-breaking rule and row 1 of Table 6.6, improving switches $(e_{*,*,*}, t^{\rightarrow}) \in \mathbb{E}^1$ are applied until a switch of this type with $i \in [n], j \in \{0,1\}$ and $\beta_i = 0 \vee \beta_{i+1} \neq j$ is applied. The occurrence record of each applied switch is described by Table 5.6 when interpreted for $\mathfrak{b}+1$ since $\lfloor \mathfrak{b}/2 \rfloor + 1 = \mathfrak{m}$ if $\mathfrak{b}$ is odd and $\lceil \mathfrak{b}/2 \rceil + 1 = \lceil (\mathfrak{b}+1)/2 \rceil$ if $\mathfrak{b}$ is even. By row 1 of Table 6.6, $(d_{i,j,k}, e_{i,j,k})$ now becomes improving. As $(d_{i,j,k}, e_{i,j,k}) \notin \mathfrak{A}^{\sigma}_{\sigma_{\mathfrak{b}}}$ and since switches of the type $(e_{*,*,*}, t^{\rightarrow})$ minimize the occurrence record, Table 5.6 and the tie-breaking rule imply that $(d_{i,j,k}, e_{i,j,k})$ is applied next. In particular, an edge $(d_{i,j,k}, e_{i,j,k})$ is applied immediately if it becomes improving and this requires that $(e_{i,j,k}, t^{\rightarrow})$ was applied earlier. Therefore, the application of improving switches $(e_{*,*,*}, t^{\rightarrow})$ is described by row 1 of Table 6.6 and whenever an edge $(d_{*,*,*}, e_{*,*,*})$ becomes improving, its application is described by Lemma 6.3.10. In particular, the occurrence record of all these edges is described by Table 5.6 when interpreted for $\mathfrak{b}+1$.

Let $G_n = S_n$. Then, row 1 of Table 6.6 and Lemma 6.3.10 can be applied until reaching a strategy σ such that all improving switches $(e_{i,j,k}, t^{\rightarrow})$ with $i \in [n], j, k \in \{0,1\}$ and $\sigma^{(3)}(d_{i,j,k}) = F_{i,j}$ were applied. Since a fixed improving switch $(d_{i,j,k}, e_{i,j,k})$ was applied if and only if $\beta_i = 0 \vee \beta_{i+1} = j$, this implies that $\sigma(d_{i,j,k}) = F_{i,j}$ is equivalent to $\beta_i = 1 \wedge \beta_{i+1} = j$ for all $i \in [n], j, k \in \{0,1\}$. Consequently, every cycle center is closed or escapes towards $t^{\rightarrow}$. In addition, for suitable indices i, j, k, an edge $(d_{i,j,k}, F_{i,j})$ is an improving switch exactly if the corresponding switch $(e_{i,j,k}, t^{\rightarrow})$ was not applied. Consequently,

$$I_{\sigma} = \{(d_{i,j,k}, F_{i,j}), (e_{i,j,k}, t^{\rightarrow}) \colon \sigma(e_{i,j,k}) = t^{\leftarrow}\} \cup \{(b_1, t^{\rightarrow})\}.$$

Now, as $\phi^{\sigma}(b_1, t^{\rightarrow}) = \phi^{\sigma}(e_{*,*,*}, t^{\rightarrow})$ and $\mathbb{E}^1 = \emptyset$, the switch $e := (b_1, t^{\rightarrow})$ is applied next due to the tie-breaking rule. We prove that we can apply row 7 resp. 9 of Table 6.6, implying the statement for the case $G_n = S_n$ and arbitrary ν and for the case $G_n = M_n$ and $\nu = 1$. The following claim shows that one of the key requirements for the application of the corresponding statements is fulfilled.

Claim 6. Let σ denote the phase-3-strategy in which the improving switch $(b_1, t^{\rightarrow})$ should be applied next. If $\nu > 1$, then $\bar{\sigma}(eb_{i,j}) \wedge \neg\bar{\sigma}(eg_{i,j})$ for all $(i,j) \in S_1$ and, in addition, $\bar{\sigma}(eb_{i,j}) \wedge \bar{\sigma}(eg_{i,j})$ for all $(i,j) \in S_2$. If $\nu = 1$, then $\bar{\sigma}(eg_{i,j}) \wedge \neg\bar{\sigma}(eb_{i,j})$ for all $(i,j) \in S_4$ and $\bar{\sigma}(eb_{i,j}) \wedge \bar{\sigma}(eg_{i,j})$ for all $(i,j) \in S_3$.

In addition to the two statements of the claim, it holds that $\sigma(d_{i,j,*}) = F_{i,j}$ if and only if $\beta_i = 1 \wedge \beta_{i+1} = j$ for all $i \in [n], j \in \{0,1\}$.. Consequently, all requirements of row 7 are met for the case $G_n = S_n$ and $\nu > 1$, implying that the application of $e = (b_1, b_2)$ yields a phase-4-strategy as described by the corresponding rows of Tables 5.8 and 5.9. Analogously, all requirements of row 9 are met for the case that $\nu = 1$, implying that the application of $e = (b_1, g_1)$ yields a phase-5-strategy as described by the corresponding rows of Tables 5.8 and 5.9 in this case.

It remains to consider the case $\nu > 1$ for $G_n = M_n$, implying $t^{\rightarrow} = b_2$ and $t^{\leftarrow} = g_1$. Using the same argumentation as before, row 1 of Table 6.6 and Lemma 6.3.10 imply that improving switches within levels $i \geq \nu$ are applied until we obtain a phase-3-strategy σ for $\mathfrak{b}$ with

$$I_{\sigma} = \{(d_{i,j,k}, F_{i,j}) \colon i < \nu \wedge \sigma(d_{i,j,k}) \neq F_{i,j}\}$$

$$\cup \{(d_{i,j,k}, F_{i,j}), (e_{i,j,k}, b_2): i \geq \nu \wedge \sigma(e_{i,j,k}) = g_1\} \cup \{(b_1, b_2)\}.$$

As no cycle center in any level $i' < \nu$ was opened yet, the switch $e = (e_{i,j,k}, b_2)$ with $i = \nu - 1, j = 1 - \beta_{i+1}$ and $k \in \{0, 1\}$ is applied next. Since $\sigma(d_{i,j,k}) = F_{i,j}$, row 1 of Table 6.6 implies $I_{\sigma e} = (I_\sigma \setminus \{e\}) \cup \{(d_{i,j,k}, e_{i,j,k})\}$. Due to the tie-breaking rule, $(d_{i,j,k}, e_{i,j,k})$ is applied next.

Claim 7. The strategy σe meets the five requirements of Lemma 6.2.25 and the lemma thus describes the application of the improving switch $(d_{i,j,k}, e_{i,j,k})$.

Therefore, applying $(d_{i,j,k}, e_{i,j,k})$ yields a well-behaved phase-3-strategy $\sigma \in \rho(\sigma_0)$ for $\mathfrak{b}$ with $I_\sigma = (I_{\sigma e} \setminus \{(d_{i,j,k}, e_{i,j,k})\}) \cup \{(s_{i,j}, b_1)\}$. We prove $\phi^\sigma(s_{i,j}, b_1) < \phi^\sigma(e_{i,j,k}, b_2) = \lfloor \mathfrak{b}/2 \rfloor$, implying that $(s_{i,j}, b_1)$ is applied next. It is easy to verify that $(s_{i,j}, b_1) \notin \mathfrak{A}^\sigma_{\sigma_\mathfrak{b}}$. Consequently, by Table 5.6 and as $i = \nu - 1$ and $j = 1 - \beta_{i+1} = 0$,

$$\phi^\sigma(s_{i,j}, b_1) = \text{fl}(\mathfrak{b}, i+1) - j \cdot \mathfrak{b}_{i+1} = \text{fl}(\mathfrak{b}, \nu) \leq \left\lfloor \frac{\mathfrak{b}+2}{4} \right\rfloor < \left\lfloor \frac{\mathfrak{b}+1}{2} \right\rfloor$$

if $\mathfrak{b} \geq 3$ since $\nu \geq 2$. If $\mathfrak{b}_1 = 1$, then $(s_{i,j}, b_1)$ is also the next switch applied as the tie-breaking rule then ranks $(s_{i,j}, b_1)$ higher than any switch of the type $(e_{*,*,*}, b_2)$. Since $(e_{i,j,k}, b_2), (d_{i,j,k}, e_{i,j,k}) \in \mathfrak{A}^\sigma_{\sigma_\mathfrak{b}}$ and since the cycle center $F_{i,j}$ was closed when $(e_{i,j,k}, b_2)$ was applied, we have $\bar{\sigma}(eb_{i,j}) \wedge \neg\bar{\sigma}(eg_{i,j})$. Therefore, the fifth row of Table 6.6 describes the application of $e = (s_{i,j}, b_1)$. Consequently, σe is a phase-3-strategy with $I_{\sigma e} = I_\sigma \setminus \{e\}$ and $\phi^{\sigma e}(s_{i,j}, b_1) = \text{fl}(\mathfrak{b}, \nu) + 1 = \text{fl}(\mathfrak{b}+1, \nu)$ by Lemma 6.2.6. Thus, Table 5.6 describes the occurrence record of $(s_{i,j}, b_1)$ when interpreted for $\mathfrak{b}+1$. Since $F_{i,j}$ is b_2-halfopen for σe whereas $F_{i,1-j}$ is g_1-halfopen, $(e_{i,j,1-k}, b_2)$ is applied next. By the first row of Table 6.6, this application unlocks $(d_{i,j,1-k}, e_{i,j,1-k})$. Using our previous arguments and observations, it is easy to verify that $(d_{i,j,1-k}, e_{i,j,1-k})$ is applied next and that its application is described by the second-to-last row of Table 6.6. The tie-breaking rule then chooses to apply $(e_{i,1-j,k}, b_2) \in \mathbb{E}^1$ next. By row 1 of Table 6.6, $(d_{i,1-j,k}, e_{i,1-j,k})$ then becomes improving and is applied next. Its application is described by row 5 of Table 6.6. After applying this switch, we then obtain a strategy σ with

$$\begin{aligned} I_\sigma = &\{(d_{i,j,k}, F_{i,j}): i < \nu - 1 \wedge \sigma(d_{i,j,k}) \neq F_{i,j}\} \\ &\cup \{(d_{i,j,k}, F_{i,j}), (e_{i,j,k}, b_2): i \geq \nu - 1 \wedge \sigma(e_{i,j,k}) = g_1\} \cup \{(b_1, b_2)\}. \end{aligned}$$

It is easy to verify that the same arguments can be applied iteratively as applying a switch $(s_{i',j'}, b_1)$ with $i' < \nu$ always requires to open the corresponding cycle center $F_{i',j'}$ first. Thus, after finitely many iterations, we obtain a strategy σ with

$$I_\sigma = \{(d_{i,j,k}, F_{i,j}), (e_{i,j,k}, b_2): \sigma(e_{i,j,k}) = g_1\} \cup \{(b_1, b_2)\}.$$

By the same arguments as for $G_n = S_n$, the conditions of the row 8 of Table 6.6 are met, so we obtain a strategy as described by the corresponding rows of Tables 5.8 and 5.9. □

Note that we implicitly proved the following which follows from Corollary 6.3.8.

Corollary 6.3.11. *Let $\sigma^{(4)}$ be the phase-4-strategy calculated by the strategy improvement algorithm when starting with a canonical strategy $\sigma_{\mathfrak{b}}$ having the canonical properties as described by Lemma 6.3.9. Then, Table 5.6 specifies the occurrence record of every improving switch applied during $\sigma_{\mathfrak{b}} \to \sigma^{(4)}$ when interpreted for $\mathfrak{b}+1$, excluding switches $(g_*, F_{*,*})$. In addition, each such switch was applied once.*

As indicated by Lemma 6.3.9, we do not always obtain a phase-5-strategy immediately after phase 3 as there might be improving switches involving selection vertices $s_{i,*}$ in levels $i < \nu$ that still need to be applied if $G_n = S_n$. We thus prove that we also reach a phase-5-strategy after applying these switches. Consequently, we always reach a phase-5-strategy. The following lemma generalizes Lemma 5.3.17.

Lemma 6.3.12 (Extended version of Lemma 5.3.17). *Let $\sigma_{\mathfrak{b}} \in \rho(\sigma_0)$ be a canonical strategy for $\mathfrak{b} \in \mathfrak{B}_n$ having the canonical properties. After applying finitely many improving switches, the strategy improvement resp. policy iteration algorithm produces a well-behaved phase-5-strategy $\sigma^{(5)} \in \rho(\sigma_0)$ as described by the corresponding rows of Tables 5.8 and 5.9.*

Proof. By Lemma 6.3.9, it suffices to consider the case $G_n = S_n$ and $\nu > 1$. The same lemma implies that the strategy improvement algorithm calculates a phase-4-strategy σ for $\mathfrak{b}$ with $\sigma \in \rho(\sigma_0)$ and

$$\begin{aligned} I_\sigma = \{&(d_{i,j,k}, F_{i,j}), (e_{i,j,k}, b_2) \colon \sigma(e_{i,j,k}) = g_1\} \cup \{(s_{\nu-1,0}, b_1)\} \\ &\cup \{(s_{i,1}, b_1) \colon i \leq \nu - 2\} \cup X_0 \cup X_1. \end{aligned}$$

Claim 8. Let σ denote the first phase-4-strategy in S_n for $\nu > 1$. Then, the switch $(s_{\nu-1,0}, b_1)$ is applied next and the application of this switch is described by Lemma 6.2.31.

Consider the case $\nu = 2$ first. Then, applying $e = (s_{1,0}, b_1)$ yields a phase-5-strategy and $\phi^{\sigma e}(e) = \mathrm{fl}(\mathfrak{b}, \nu) + 1 = \mathrm{fl}(\mathfrak{b}+1, \nu)$ by Lemma 6.2.6. Hence, Table 5.6 describes the occurrence record of e with respect to $\mathfrak{b}+1$. In addition, we then have

$$\begin{aligned} I_{\sigma e} &= (I_\sigma \setminus \{e\}) \cup \{(d_{1,0,0}, F_{1,0}), (d_{1,0,1}, F_{1,0})\}) \\ &= \{(d_{i,j,k}, F_{i,j}), (e_{i,j,k}, b_2) \colon \sigma e(e_{i,j,k}) = g_1\} \cup \{(d_{i,1-\beta_{i+1},*}, F_{i,1-\beta_{i+1}}) \colon i \leq \nu - 1\} \\ &\quad \cup X_0 \cup X_1. \end{aligned}$$

Since σe is a phase-5-strategy, it has Property (REL1), implying $\mu^{\sigma e} = u = \min\{i \colon \beta_i = 0\}$. Thus, σe has all properties listed in the corresponding rows of Tables 5.8 and 5.9.

Before discussing the case $\nu > 2$, we discuss edges $(d_{i,j,*}, F_{i,j})$ that become improving when a switch $(s_{i,j}, b_1)$ with $i < \nu$ and $j = 1 - \beta_{i+1}$ is applied, see Lemma 6.2.31. Since $i < \nu$ implies $1 - \beta_{i+1} = \mathfrak{b}_{i+1}$, their cycle centers $F_{i,j}$ were closed for $\sigma_{\mathfrak{b}}$. Therefore, their occurrence records might be very low with respect to the current strategy σ. However, their occurrence records are not "too low" in the sense that they interfere with the improving switches applied during phase 4. More precisely, we prove that $i < \nu$ and $j = \mathfrak{b}_{i+1}$ imply $\phi^{\sigma_{\mathfrak{b}}}(d_{i,j,k}, F_{i,j}) > \lfloor (\mathfrak{b}+2)/4 \rfloor - 1$. By Table 5.6,

$$\phi^{\sigma_{\mathfrak{b}}}(d_{i,j,k}, F_{i,j}) = \left\lceil \frac{\mathrm{lfn}(\mathfrak{b}, i, \{(i+1, j)\}) + 1 - k}{2} \right\rceil.$$

Since $i < \nu$, we have $\mathfrak{b}_1 = \cdots = \mathfrak{b}_i = 1$ and, by the choice of j, $\mathfrak{b}_{i+1} = j$ and $\mathfrak{b} \geq 2^{\nu-1} - 1$. This implies $\operatorname{lfn}(\mathfrak{b}, i, \{(i+1, j)\}) = \mathfrak{b} - \sum(\mathfrak{b}, i) = \mathfrak{b} - 2^{i-1} + 1$. Thus

$$\phi^{\sigma_{\mathfrak{b}}}(d_{i,j,k}, F_{i,j}) = \left\lceil \frac{\mathfrak{b} - 2^{i-1} + 2 - k}{2} \right\rceil \geq \left\lceil \frac{\mathfrak{b} - 2^{i-1} + 1}{2} \right\rceil = \left\lfloor \frac{2\mathfrak{b} - 2^i + 4}{4} \right\rfloor.$$

Since $\lfloor(\mathfrak{b}+2)/4\rfloor - 1 = \lfloor(\mathfrak{b}-2)/4\rfloor$, it suffices to prove $2\mathfrak{b} - 2^i + 4 - (\mathfrak{b} - 2) > 4$. This follows as $i \leq \nu - 1$ implies

$$2\mathfrak{b} - 2^i + 4 - \mathfrak{b} + 2 = \mathfrak{b} - 2^i + 6 \geq 2^{\nu-1} - 1 - 2^i + 6 \geq 2^i - 2^i + 5 = 5.$$

Let $\nu > 2$. We obtain $\phi^{\sigma e}(e) = \operatorname{fl}(\mathfrak{b}+1, \nu)$ as before. Furthermore, Lemma 6.2.31 yields

$$\begin{aligned} I_{\sigma e} &= \{(d_{i,j,k}, F_{i,j}), (e_{i,j,k}, b_2) \colon \sigma e(e_{i,j,k}) = g_1\} \\ &\cup \{(s_{i,1}, b_1) \colon i \leq \nu - 2\} \cup \{(d_{\nu-1,0,0}, F_{\nu-1,0}), (d_{\nu-1,0,1}, F_{\nu-1,0})\}. \end{aligned}$$

We show that the switches $(s_{\nu-2,1}, b_1), \dots, (s_{1,1}, b_1)$ are applied next and in this order. To simplify notation, we denote the current strategy by σ. By Table 5.6, it holds that $\phi^{\sigma}(s_{i,1}, b_1) = \operatorname{fl}(\mathfrak{b}, i+1) - 1$ for all $i \leq \nu - 2$. Hence $\phi^{\sigma}(s_{\nu-2,1}, b_1) < \cdots < \phi^{\sigma}(s_{1,1}, b_1)$ by Lemma 6.2.6. It thus suffices to show that the occurrence record of $(s_{1,1}, b_1)$ is smaller than the occurrence record of any switch improving for σ and any improving switch that might be unlocked by applying some switch $(s_{i,1}, b_1)$ for $i \leq \nu - 2$.

The second statement follows since $\phi^{\sigma_{\mathfrak{b}}}(s_{1,1}, b_1) = \operatorname{fl}(\mathfrak{b}, 2) - 1 = \lfloor(\mathfrak{b}+2)/4\rfloor - 1$ and since the occurrence record of any edge that becomes improving is bounded by $\lfloor(\mathfrak{b}+2)/4\rfloor$ as discussed earlier. It thus suffices to show the first statement.

Let $e := (d_{i,j,k}, F_{i,j}) \in I_{\sigma}$ with $i \in [n], j, k \in \{0, 1\}$ and $\sigma(e_{i,j,k}) = g_1$. By Lemma 6.3.6 and Lemma 6.2.6, it then holds that $\phi^{\sigma}(e) = \mathfrak{m} = \operatorname{fl}(\mathfrak{b}, 1)$. In addition, $\nu > 2$ implies $\operatorname{fl}(\mathfrak{b}, 1) > \operatorname{fl}(\mathfrak{b}, \nu - 1)$, hence $\phi^{\sigma}(e) < \phi^{\sigma}(s_{1,1}, b_1)$ follows. Next let $e := (e_{i,j,k}, b_2) \in I_{\sigma}$ with $i \in [n], j, k \in \{0, 1\}$ and $\sigma(e_{i,j,k}) = g_1$. Then, since $\mathfrak{b}$ is odd, Table 5.6 implies $\phi^{\sigma}(e) = \lfloor \mathfrak{b}/2 \rfloor = \lfloor(\mathfrak{b}+1)/2\rfloor - 1 = \operatorname{fl}(\mathfrak{b}, 1) - 1$. Consequently, we have $\phi^{\sigma}(e) > \phi^{\sigma}(s_{1,1}, b_1)$. If $\mathfrak{b} + 1$ is not a power of two, we need to show this estimation for some more improving switches. But this can be shown by easy calculations similar to the calculations necessary when discussing the application of $(s_{\nu-1,0}, b_1)$ which can be found in the proof of Claim 8 in Appendix A.2.

Consequently, the switches $(s_{\nu-1}, b_1), \dots, (s_{1,1}, b_1)$ are applied next, and they are applied in this order. It is easy to verify that the requirements of Lemma 6.2.31 are always met, so this lemma describes the application of these switches. It is also easy to check that the occurrence records of these edges are described by Table 5.6 after applying them. Let σ denote the strategy obtained after applying $(s_{1,1}, b_1)$. Then σ is a well-behaved phase-5-strategy for $\mathfrak{b}$ with $\sigma \in \rho(\sigma_0)$ and $\mu^{\sigma} = \min\{i \colon \beta_i = 0\}$. This further implies

$$\begin{aligned} I_{\sigma} &= \{(d_{i,j,k}, F_{i,j}), (e_{i,j,k}, b_2) \colon \sigma(e_{i,j,k}) = g_1\} \\ &\cup \{(d_{i,1-(\mathfrak{b}-1)_{i+1},*}, F_{i,1-(\mathfrak{b}+1)_{i+1}}) \colon i \leq \nu - 1\} \cup X_0 \cup X_1. \end{aligned}$$

We observe that $\sigma(e_{i,j,k}) = g_1$ still implies $\phi^{\sigma_{\mathfrak{b}}}(d_{i,j,k}, F_{i,j}) = \phi^{\sigma}(d_{i,j,k}, F_{i,j}) = \mathfrak{m}$ for all indices $i \in [n], j, k \in \{0, 1\}$ since the corresponding switches are improving since the end of phase 1. Also, every improving switch was applied at most once and we proved that the occurrence record of every improving switch that was applied is described correctly by Table 5.6 when interpreted for $\mathfrak{b}+1$. Since no improving switches involving cycle vertices were applied, $\sigma(d_{i,j,*}) = F_{i,j}$ if and only if $(\mathfrak{b}+1)_i = 1$ and $(\mathfrak{b}+1)_{i+1} = j$ where $i \in [n], j \in \{0,1\}$. Hence, all conditions listed in the corresponding rows of Tables 5.8 and 5.9 are fulfilled, proving the statement. □

We henceforth use $\sigma^{(5)}$ to refer to the phase-5-strategy described by Lemma 6.3.12. As before, we implicitly proved the following corollary which follows from Corollaries 6.3.8 and 6.3.11.

Corollary 6.3.13. *Let $\sigma^{(5)}$ be the phase-5-strategy calculated by the strategy improvement algorithm when starting with a canonical strategy $\sigma_{\mathfrak{b}}$ having the canonical properties as described by Lemma 6.3.12. Then, Table 5.6 specifies the occurrence record of every improving switch applied during $\sigma_{\mathfrak{b}} \to \sigma^{(5)}$ when interpreted for $\mathfrak{b}+1$, excluding switches $(g_*, F_{*,*})$. In addition, each such switch was applied once.*

Reaching a canonical strategy part I: Everything but the occurrence records

There are two major statements that we still have to prove. First, we have to prove that applying improving switches to $\sigma^{(5)}$ yields a canonical strategy $\sigma_{\mathfrak{b}+1}$ for $\mathfrak{b}+1$ having the canonical properties. Note that this implies Lemma 5.3.18, stating that applying improving switches yields the strategies as described by Tables 5.8 and 5.9. Second, we need to investigate the occurrence records of edges $(g_*, F_{*,*})$ which we ignored until now.

We begin by proving the first statement. We also prove several smaller statements implicitly which will be used when proving that $\sigma_{\mathfrak{b}+1}$ has the canonical properties.

Lemma 6.3.14. *Let $\sigma_{\mathfrak{b}} \in \rho(\sigma_0)$ be a canonical strategy for $\mathfrak{b}$ having the canonical properties. Then, applying improving switches according to Zadeh's pivot rule and the tie-breaking rule produces a canonical strategy $\sigma_{\mathfrak{b}+1} \in \rho(\sigma_0)$ for $\mathfrak{b}+1$ with $I_{\sigma_{\mathfrak{b}+1}} = \mathfrak{D}^{\sigma_{\mathfrak{b}+1}}$.*

Proof. By Lemma 6.3.12, applying improving switches according to Zadeh's pivot rule and our tie-breaking rule yields a phase-5-strategy $\sigma := \sigma^{(5)}$ for $\mathfrak{b}$ with $\sigma^{(5)} \in \rho(\sigma_0)$ and $\mu^\sigma = u = \min\{i : \beta_i = 0\}$. Let $m := \max\{i : \beta_i = 1\}$.

Consider the case $\nu = 1$. We begin by proving that the occurrence records of the improving switches are bounded by $\mathfrak{m}$. We furthermore characterize the improving switches which will be applied next.

Claim 9. For all $e \in I_\sigma$, it holds that $\phi^\sigma(e) \leq \mathfrak{m}$. Let $e \in I_\sigma$ with $\phi^\sigma(e) < \mathfrak{m}$. Then, $e = (d_{i,j,k}, F_{i,j})$ with $i \in \{u+1, \dots, m-1\}, j = 1 - \beta^{i+1}, k \in \{0,1\}$ and $\sigma_{\mathfrak{b}}(d_{i,j,k}) = F_{i,j}$.

Thus, improving switches $(d_{i,j,k}, F_{i,j})$ with $i \in \{u+1, \dots, m-1\}, \beta_i = 0, j = 1 - \beta_{i+1}$ and $k \in \{0,1\}$ are applied first. Let $e = (d_{i,j,k}, F_{i,j})$ denote such a switch with $\phi^\sigma(e) < \mathfrak{m}$ minimizing the occurrence record. Since $\sigma_{\mathfrak{b}}(d_{i,j,k}) = F_{i,j}$, e was not applied during phase 1, it follows that $\phi^\sigma(e) = \phi^{\sigma_{\mathfrak{b}}}(e) = \ell^{\mathfrak{b}}(i,j,k) + 1$.

Claim 10. Let σ denote the phase-5-strategy at the beginning of phase 5 for $\nu = 1$. Let $i, \in [n], j, k \in \{0,1\}$ such that $e = (d_{i,j,k}, F_{i,j}) \in I_\sigma$ and $\phi^\sigma(e) < \mathfrak{m}$. Row 1 of Table 6.7 can be applied to describe the application of e.

Thus, σe is a well-behaved phase-5-strategy for $\mathfrak{b}$ with $\sigma e \in \rho(\sigma_0)$ and $I_{\sigma e} = I_\sigma \setminus \{e\}$. By Lemma 6.2.4 and the choice of i and j, it follows that $\ell^{\mathfrak{b}}(i,j,k) + 1 = \ell^{\mathfrak{b}+1}(i,j,k)$. In particular,

$$\phi^{\sigma e}(e) = \ell^{\mathfrak{b}}(i,j,k) + 1 + 1 = \ell^{\mathfrak{b}+1}(i,j,k) + 1 \leq \left\lfloor \frac{\mathfrak{b}+1}{2} \right\rfloor \leq \left\lfloor \frac{(\mathfrak{b}+1)+1-k}{2} \right\rfloor.$$

Thus, by choosing the parameter $t_{\mathfrak{b}+1} = 1$, which is feasible since $i \neq 1$, the occurrence record of e is described by Table 5.6 when interpreted for $\mathfrak{b}+1$.

Now, the same arguments can be used for all improving switches $e' \in \mathbb{D}^1 \cap I_\sigma$ with $\phi^\sigma(e') < \mathfrak{m}$. All of these switches are thus applied and their occurrence records are specified by Table 5.6 when interpreted for $\mathfrak{b}+1$. After the application of these switches, we obtain a well-behaved phase-5-strategy σ for $\mathfrak{b}$ with $\sigma \in \rho(\sigma_0)$ and

$$\begin{aligned} I_\sigma =& \{(d_{i,j,k}, F_{i,j}), (e_{i,j,k}, g_1) \colon \sigma(e_{i,j,k}) = b_2\} \\ &\cup \bigcup_{\substack{i=u+1 \\ \beta_i = 0}}^{m-1} \{e = (d_{i,1-\beta_{i+1},*}, F_{i,1-\beta_{i+1}}) \colon \phi^\sigma(e) = \mathfrak{m}\}. \end{aligned} \tag{6.3}$$

In particular, all improving switches have an occurrence record of $\mathfrak{m}$. Thus, the tie-breaking rule now applies a switch of the type $(e_{*,*,*}, g_1)$. Let $i \in [n], j, k \in \{0,1\}$ such that $e := (e_{i,j,k}, g_1)$ is the next applied improving switch.

Claim 11. Let $\nu = 1$ and let σ denote the strategy obtained after applying all improving switches with an occurrence record less than $\mathfrak{m}$ during phase 5. Then, Lemma 6.2.32 can be applied to describe the application of $e = (e_{i,j,k}, g_1)$.

In fact, Claim 11 can be applied for any improving switch of the type $(e_{*,*,*}, g_1)$. Furthermore, $\phi^{\sigma e}(e)$ is specified by Table 5.6 when interpreted for $\mathfrak{b}+1$ as $\nu = 1$ implies $\lceil \mathfrak{b}/2 \rceil + 1 = \lceil (\mathfrak{b}+1)/2 \rceil$. Depending on whether the conditions listed in the fourth case of Lemma 6.2.32 are fulfilled, either

$$I_{\sigma e} = (I_{\sigma e} \setminus \{e\}) \cup \{(d_{i,j,1-k}, F_{i,j}), (g_i, F_{i,j})\} \quad \text{or} \quad I_{\sigma e} = (I_{\sigma e} \setminus \{e\}) \cup \{(d_{i,j,1-k}, F_{i,j})\}.$$

In particular, $\tilde{e} := (d_{i,j,1-k}, F_{i,j})$ becomes improving in either case. As formalized by the following corollary, $\tilde{e}$ has an occurrence record of at least $\mathfrak{m}$. This corollary will be used in later arguments, hence it is not a claim as we use the term claim solely for statements that are only relevant within a single proof. Nevertheless, its proof is deferred to Appendix A.2.

Corollary 6.3.15. *Let $\nu = 1$ and $i \in [n], j, k \in \{0,1\}$. If the edge $\tilde{e} = (d_{i,j,1-k}, F_{i,j})$ becomes improving during phase 5 due to the application of $(e_{i,j,k}, g_1)$, then the corresponding strategy has Property (OR4)$_{i,j,1-k}$.*

Now, consider the case that $(g_i, F_{i,j})$ becomes improving when applying $(e_{i,j,k}, g_1)$. We prove that this implies $(g_i, F_{i,j}) \notin \mathfrak{A}_{\sigma_\mathfrak{b}}^{\sigma\!e}$. The conditions stated in Lemma 6.2.32 imply that the switch was not applied previously in phase 5. For the sake of a contradiction, assume that $(g_i, F_{i,j})$ was applied during phase 1 of the current transition. Then, by Corollary 6.3.2, the cycle center $F_{i,j}$ was closed during phase 1. Since $(e_{i,j,k}, g_1)$ was applied immediately before unlocking $(g_i, F_{i,j})$, we have $\phi^{\sigma_\mathfrak{b}}(d_{i,j,k}, F_{i,j}) = \mathfrak{m}$ by Lemma 6.3.12. However, by Corollary 6.3.2, a cycle center can only be closed during phase 1 if either $i = \nu$ or if the occurrence record of both cycle edges is less than $\mathfrak{m}$. We thus need to have $i = \nu = 1$. But then $\beta_i = 1$, implying that $(g_i, F_{i,j})$ cannot become improving. Hence, a switch $(g_i, F_{i,j})$ that is unlocked during phase 5 was not applied earlier in the same transition if $\nu = 1$.

Since $\phi^{\sigma\!e}(g_i, F_{i,j}) = \phi^{\sigma_\mathfrak{b}}(g_i, F_{i,j})$, we have $\phi^{\sigma\!e}(g_i, F_{i,j}) \leq \phi^{\sigma_\mathfrak{b}}(d_{i,j,k}, F_{i,j}) = \mathfrak{m}$ by Table 5.6. By Corollary 6.3.15, $\phi^{\sigma\!e}(d_{i,j,1-k}, F_{i,j}) \geq \mathfrak{m}$. Therefore, the occurrence record of any improving switch except $(g_i, F_{i,j})$ is at least $\mathfrak{m}$. Thus, $(g_i, F_{i,j})$ either uniquely minimizes the occurrence record or has the same occurrence record as all other improving switches. Consequently, by the tie-breaking rule, $(g_i, F_{i,j})$ is applied next in either case.

We prove that row 2 of Table 6.7 applies to this switch. Since $\nu = 1, \mu^{\sigma\!e} = u > 1$ and $\beta_i = 0$, it suffices to prove $\overline{\sigma\!e}(eg_{i,j}) \wedge \neg\overline{\sigma\!e}(eb_{i,j})$. But this follows as we applied $(e_{i,j,k}, g_1)$ earlier and since $F_{i,j}$ was mixed when this switch was applied. Observe that the following corollary holds due to the conditions which specify when a switch $(g_i, F_{i,j})$ is unlocked, independent on ν.

Corollary 6.3.16. *Let $\nu = 1$. If an improving switch $(g_i, F_{i,j})$ is applied during phase 5, then the resulting strategy has Property (SVG)$_i$/(SVM)$_i$.*

Let σ denote the strategy obtained after applying $(e_{i,j,k}, g_1)$ (and potentially $(g_i, F_{i,j})$ if it became improving). Assume that there is an improving switch of the type $(e_{*,*,*}, g_1) \in I_\sigma$. Then, by Lemma 6.2.32 resp. row 2 of Table 6.7, σ is a phase-5-Strategy for $\mathfrak{b}$. By our previous discussion, the occurrence records of all improving switches are at least $\mathfrak{m}$. Among all improving switches with an occurrence record of exactly $\mathfrak{m}$, the tie-breaking rule then decides which switch to apply. There are two types of improving switches. Each switch is either of the form $(d_{*,*,*}, F_{*,*j})$ or of the form $(e_{i',j',k'}, g_1)$ for indices $i' \in [n], j', k' \in \{0,1\}$ with $\sigma(d_{i',j',k'}) = e_{i',j',k'}$. Since every edge $(e_{*,*,*}, g_1)$ minimizes the occurrence record among all improving switches, an edge of this type is chosen. Let $(e_{i',j',k'}, g_1)$ denote this switch. Then, the same arguments used previously can be used again. More precisely, Lemma 6.2.32 applies to this such a switch, making the edge $(d_{i',j',1-k'}, F_{i',j'})$ and eventually also $(g_{i'}, F_{i',j'})$ improving. Also, Corollaries 6.3.15 and 6.3.16 apply to these switches and another switch of the form $(e_{*,*,*}, g_1)$ is applied afterwards. Thus, inductively, all remaining improving switches $(e_{*,*,*}, g_1)$ are applied.

Let σ denote the strategy that is reached before the last improving switch $(e_{*,*,*}, g_1)$ is applied. We argue that this switch is $e := (e_{1,1-\beta_2,k}, g_1)$ for some $k \in \{0,1\}$ and that σ has Property (SVG)$_i$/(SVM)$_i$ for all $i \in [n]$. As the tie-breaking rule applies improving switches in higher levels first, it suffices to prove that there there is a $k \in \{0,1\}$ such that $e \in I_{\sigma^{(5)}}$. This however follows from Lemma 6.3.12 as $\nu = 1$ implies $(1, \beta_2) \in S_3$. It remains to prove that σ has Property (SVG)$_i$/(SVM)$_i$ for all $i \in [n]$.

Claim 12. If $\nu = 1$, then the strategy σ obtained before the application of the switch $e := (e_{1,1-\beta_2,k}, g_1)$ has Property (SVG)$_i$/(SVM)$_i$ for all $i \in [n]$.

Thus, Lemma 6.2.32 applies to $e := (e_{1,\beta_2^\sigma,k}, g_1)$. Let $\sigma_{\mathfrak{b}+1} := \sigma e$ denote the strategy obtained by applying e. Then, as we assume that there are no further indices (i', j', k') such that $(e_{i',j',k'}, g_1) \in I_{\sigma_{\mathfrak{b}+1}}$, Lemma 6.2.32 implies that $\sigma_{\mathfrak{b}+1}$ is a phase-1-strategy for $\mathfrak{b}+1$ with $\sigma_{\mathfrak{b}+1} \in \rho(\sigma_0)$. Since every edge was applied at most once during $\sigma_{\mathfrak{b}} \to \sigma^{(5)}$ by Lemma 6.3.12 and since no edge applied during $\sigma^{(5)} \to \sigma_{\mathfrak{b}+1}$ was applied earlier, every edge was applied at most once as improving switch during $\sigma_{\mathfrak{b}} \to \sigma_{\mathfrak{b}+1}$. We furthermore implicitly proved the following corollary where the second statement follows from Corollary 6.3.15.

Corollary 6.3.17. *Let $\nu = 1$ and let $\sigma_{\mathfrak{b}+1}$ denote the strategy obtained after the application of the final improving switch $(e_{*,*,*}, g_1)$. Let $i \in [n]$ and $j, k \in \{0, 1\}$. Then, $(d_{i,j,k}, F_{i,j}) \in \mathfrak{A}_{\sigma^{(5)}}^{\sigma_{\mathfrak{b}+1}}$ if and only if $\sigma_{\mathfrak{b}}(d_{i,j,k}) = F_{i,j}, \phi^{\sigma_{\mathfrak{b}}}(d_{i,j,k}, F_{i,j}) < \mathfrak{m}, i \in \{u+1, \dots, m-1\}, \beta_i = 0$ and $j = 1 - \beta_{i+1}$. In addition, $\sigma_{\mathfrak{b}+1}$ has Property (OR2)$_{i,j,k}$.*

It remains to prove that $\sigma_{\mathfrak{b}+1}$ is a canonical strategy for $\mathfrak{b}+1$ with $I_{\sigma_{\mathfrak{b}+1}} = \mathfrak{D}^{\sigma_{\mathfrak{b}+1}}$.

We begin with the second statement. This can be proven by using the characterization given in Equation (6.3) and showing $I_\sigma \subseteq \mathfrak{D}^{\sigma_{\mathfrak{b}+1}}$ and $I_\sigma \supseteq \mathfrak{D}^{\sigma_{\mathfrak{b}+1}}$.

Claim 13. It holds that $I_{\sigma_{\mathfrak{b}+1}} = \{(d_{i,j,k}, F_{i,j}) \colon \sigma_{\mathfrak{b}+1}(d_{i,j,k}) \neq F_{i,j}\}$.

To simplify notation, let $\sigma := \sigma_{\mathfrak{b}+1}$. We now prove that σ is a canonical strategy for $\mathfrak{b}$, concluding the case $\nu = 1$. Since σ is a phase-1-strategy for $\mathfrak{b}+1$, it holds that $\mathfrak{b}+1 = \beta$. Consider the conditions listed in Definition 5.1.2 resp. 5.2.1. Condition 1 is fulfilled since $\sigma(e_{*,*,*}) = g_1$ and $\nu = 1$. Condition 2(a) is fulfilled since $\beta_i^\sigma = (\mathfrak{b}+1)_i = 1$ implies $\sigma(b_i) = g_i$ by Property (EV1)$_i$ for every $i \in [n]$. Consider condition 2(b) and let $i \in [n]$. If $(\mathfrak{b}+1)_i = 1$, then $F_{i,(\mathfrak{b}+1)_{i+1}}$ is closed by Property (EV1)$_i$. We prove that $(\mathfrak{b}+1)_i = 1$ implies that $F_{i,j}$ with $j := 1 - (\mathfrak{b}+1)_{i+1}$ cannot be closed.

Consider $\sigma^{(5)}$ and let $k \in \{0, 1\}$. Then, $\sigma^{(5)}(d_{i,j,k}) = F_{i,j}$ if and only if $\beta_i^{\sigma^{(5)}} = 1 \wedge \beta_{i+1}^{\sigma^{(5)}} = j$. Hence, $\sigma^{(5)}(d_{i,j,0}) \neq F_{i,j}$ and it suffices to show that $e := (d_{i,j,0}, F_{i,j})$ was not applied during $\sigma^{(5)} \to \sigma$. By Corollary 6.3.17, it suffices to show $\phi^{\sigma^{(5)}}(e) \geq \mathfrak{m}$. By Lemma 6.2.3, it holds that $\ell^{\mathfrak{b}}(i, j, 0) \geq \mathfrak{m}$. Since $\nu = 1$, Property (OR4)$_{i,j,0}$ implies $\phi^{\sigma_{\mathfrak{b}}}(e) \neq \ell^{\mathfrak{b}}(i, j, 0) - 1$, hence $\phi^{\sigma^{(5)}}(e) \geq \phi^{\sigma_{\mathfrak{b}}}(e) \geq \mathfrak{m}$. Consequently, condition 2(b) is fulfilled. Condition 2(c) is fulfilled by $\beta^\sigma = \mathfrak{b}+1$ and Property (EV2)$_*$.

Conditions 3(a) and 3(b) are fulfilled since σ has Property (EV1)$_*$. Consider condition 3(c) and let $i \in [n]$. We prove that $(\mathfrak{b}+1)_i = 0, j = 1 - (\mathfrak{b}+1)_{i+1}$ and $\bar{\sigma}(d_{i,j})$ imply $\bar{\sigma}(g_i) = F_{i,j}$. Since S_n is a sink game and M_n is weakly unichain, $F_{i,j}$ being closed implies $\mathrm{rVal}_\sigma^*(F_{i,j}) = \mathrm{rVal}_\sigma^*(s_{i,j})$. Thus, $\mathrm{Val}_\sigma^*(F_{i,j}) = [\![s_{i,j}]\!] \oplus \mathrm{Val}_\sigma^*(g_1)$ by the choice of j and since $\nu = 1$. As shown by Lemmas 6.1.15 and 6.1.16, $\mu^\sigma \neq 1, \bar{\sigma}(eg_{i,1-j}), \neg\bar{\sigma}(eb_{i,1-j})$ and $1 - j = \beta_{i+1}$ implies $\mathrm{Val}_\sigma^*(F_{i,1-j}) = \{F_{i,1-j}, d_{i,1-j,k}, e_{i,1-j,k}, b_1\} \cup \mathrm{Val}_\sigma^*(g_1)$ for some $k \in \{0, 1\}$. But this implies $\sigma(g_i) = F_{i,j}$ since $(g_i, F_{i,1-j}) \in I_\sigma$ otherwise, contradicting $I_\sigma = \{(d_{i,j,k}, F_{i,j}) \colon \sigma(d_{i,j,k}) = F_{i,j}\}$. Consider condition 3(d) and let $i \in [n]$ and let $j := 0$ if $G_n = S_n$ and $j := \beta_{i+1}$ if $G_n = M_n$. It suffices to prove $\mathrm{Val}_\sigma^*(F_{i,j}) \succ \mathrm{Val}_\sigma^*(F_{i,1-j})$ if none of the cycle centers are closed. For $G_n = M_n$, this follows from Lemma 6.2.1 or an easy calculation using $i \geq 1 = \nu$. For $G_n = S_n$, this follows from $\Omega(F_{i,0}) > \Omega(F_{i,1})$ and since both priorities are even.

Conditions 4 and 5 follow easily since σ has Property (USV1)$_*$. For condition 6, let $i := \ell(\mathfrak{b}+2), j := (\mathfrak{b}+1)_{i+1}$ and $k \in \{0,1\}$. Since $\ell(\mathfrak{b}+1) = 1$, we have $i \geq 2$ and $\mathfrak{b}_i = (\mathfrak{b}+1)_i = 0$ as well as $\mathfrak{b}_{i+1} = (\mathfrak{b}+1)_{i+1} = j$. We prove $\sigma(d_{i,j,k}) \neq F_{i,j}$. For the sake of a contradiction, let $\sigma(d_{i,j,k}) = F_{i,j}$. Then, by the choice of i and j and Lemma 6.3.12, it holds that$(d_{i,j,k}, F_{i,j}) \in \mathfrak{A}^{\sigma}_{\sigma(5)}$. Thus, by Corollary 6.3.17 and Property (OR2)$_{i,j,k}$, it holds that $\phi^{\sigma^{(5)}}(d_{i,j,k}, F_{i,j}) < \mathfrak{m}$ and $\phi^{\sigma_{\mathfrak{b}}}(d_{i,j,k}, F_{i,j}) = \ell^{\mathfrak{b}}(i,j,k)+1$. But, by Lemma 6.2.3, we have

$$\ell^{\mathfrak{b}}(i,j,k) = \left\lceil \frac{\mathfrak{b} + 2^{i-1} + \sum(\mathfrak{b},i) + 1 - k}{2} \right\rceil \geq \left\lceil \frac{\mathfrak{b}+3-k}{2} \right\rceil = \left\lfloor \frac{\mathfrak{b}+2-k}{2} \right\rfloor,$$

which is a contradiction. Hence, $\sigma(d_{i,j,k}) \neq F_{i,j}$.

This concludes the case $\nu = 1$. We now prove the same statements for the case $\nu > 1$.

Consider the case $\nu > 1$. Then, $\mathfrak{b}$ is odd and $\mathfrak{m} = \lfloor \mathfrak{b}/2 \rfloor + 1$. By Lemma 6.3.12, applying improving switches according to Zadeh's pivot rule and the tie-breaking rule given in Definition 5.3.5 yields a well-behaved phase-5-strategy σ for $\mathfrak{b}$ with $\sigma \in \rho(\sigma_0)$ and $\mu^\sigma = u$. In addition

$$I_\sigma = \{(d_{i,j,k}, F_{i,j}), (e_{i,j,k}, b_2) \colon \sigma(e_{i,j,k}) = g_1\} \cup \bigcup_{i=1}^{\nu-1} \{(d_{i,1-\beta_{i+1},*}, F_{i,1-\beta_{i+1}})\} \cup X_0 \cup X_1, \tag{6.4}$$

where X_k is defined as in Table 5.9.

To deduce which improving switch is applied next, it is necessary to analyze their occurrence records.

Claim 14. Let $\nu > 1$. The occurrence records of the improving switches with respect to the phase-5-strategy σ described by Lemma 6.3.12 is described correctly by Table 6.8.

We partition I_σ into three subsets, based on their occurrence records. An improving switch $e \in I_\sigma$ is called

- *type 1 switch* if $\phi^\sigma(e) = \mathfrak{m}$
- *type 2 switch* if $\phi^\sigma(e) = \mathfrak{m} - 1$ and
- *type 3 switch* if $\phi^\sigma(e) < \mathfrak{m} - 1$.

By Zadeh's pivot rule, type 3 switches are applied first, and we discuss the application of these switches next.

Claim 15. Let $\nu > 1$ and consider the first phase-5-strategy. The application of type 3 switches is described by row 1 of Table 6.7.

Let $i \in [n], j,k \in \{0,1\}$ and let $e = (d_{i,j,k}, F_{i,j})$ denote the type 3 switch that is applied next. We show that Table 5.6 specifies the occurrence record of e after its application when interpreted for $\mathfrak{b}+1$. Consider the case $i \in \{\nu+1, \dots, m-1\}, \beta_i = 0, j = 1 - \beta_{i+1}$ and $k \in \{0,1\}$ first. Since e is a type 3 switch, it holds that $\sigma_{\mathfrak{b}}(d_{i,j,k}) = F_{i,j}$,

Switch e	$(d_{i,j,k},F_{i,j})$	$(e_{i,j,k},b_2)$	$(d_{\nu,1-\mathfrak{b}_{\nu+1},k},F_{\nu,1-\mathfrak{b}_{\nu+1}})$
Condition	$\sigma(e_{i,j,k})=g_1$		–
$\phi^{\sigma}(e)$	$=\mathfrak{m}$	$=\mathfrak{m}-1$	$=\mathfrak{m}$

Switch e	$(d_{i,j,k},F_{i,j})$	
Condition	$i\in\{\nu+1,\dots,m\},\mathfrak{b}_i=0,j=1-\mathfrak{b}_{i+1},k\in\{0,1\}$	
	$\sigma_{\mathfrak{b}}(d_{i,j,k})=F_{i,j}$	$\sigma_{\mathfrak{b}}(d_{i,j,k})\neq F_{i,j}$
$\phi^{\sigma}(e)$	$\leq\mathfrak{m}-1$	$=\mathfrak{m}$

Switch e	$(d_{i,j,k},F_{i,j})$			
Condition	$i\leq\nu-1,j=1-\mathfrak{b}_{i+1}$			
	$i=1$	$i=2$	$i=3$	$i>3$
$\phi^{\sigma}(e)$	$\mathfrak{m}$	$=\mathfrak{m}-k$	$=\mathfrak{m}-1-k$	$<\mathfrak{m}-1$

Table 6.8.: Occurrence records of the improving switches at the beginning of phase 5 for $\nu>1$.

implying $\phi^{\sigma_{\mathfrak{b}}}(e)=\ell^{\mathfrak{b}}(i,j,k)+1$ by Property $(\text{OR2})_{i,j,k}$. Thus, the statement follows since $\ell^{\mathfrak{b}+1}(i,j,k)=\ell^{\mathfrak{b}}(i,j,k)+1$ by Lemma 6.2.6. Now consider the case $i\leq\nu-1$. Then, $F_{i,j}$ was closed with respect to $\sigma_{\mathfrak{b}}$ and $j=\mathfrak{b}_{i+1}=1-\beta_{i+1}$. It is easy to verify that this implies $\phi^{\sigma_{\mathfrak{b}}}(e)=\lceil(\mathfrak{b}-\sum(\mathfrak{b},i)+1-k)/2\rceil$. Since $(\mathfrak{b}+1)_i=0\wedge(\mathfrak{b}+1)_{i+1}\neq j$ and the switch e is applied, it suffices to prove $\ell^{\mathfrak{b}+1}(i,j,k)=\lceil(\mathfrak{b}-\sum(\mathfrak{b},i)+1-k)/2\rceil$ as we can then choose $t^{\mathfrak{b}+1}=1$ as feasible parameter. This however follows directly from

$$\begin{aligned}\ell^{\mathfrak{b}+1}(i,j,k)&=\left\lceil\frac{\mathfrak{b}+1-2^{i-1}+\sum(\mathfrak{b}+1,i)+1-k}{2}\right\rceil=\left\lceil\frac{\mathfrak{b}+1-2^{i-1}+1-k}{2}\right\rceil\\&=\left\lceil\frac{\mathfrak{b}+1+\sum(\mathfrak{b},i)-1+1-k}{2}\right\rceil=\left\lceil\frac{\mathfrak{b}-\sum(\mathfrak{b},i)+1-k}{2}\right\rceil.\end{aligned}$$

Note that we do not prove yet that choosing this parameter is in accordance with Properties $(\text{OR1})_{*,*,*}$ to $(\text{OR4})_{*,*,*}$. Since e is a type 3 switch, this furthermore implies $\phi^{\sigma e}(e)\leq\mathfrak{m}-1=\lfloor(\mathfrak{b}+1+1)/2\rfloor-1$. Hence, σe has Property $(\text{OR1})_{i,j,k}$ and we have implicitly proven the following corollary.

Corollary 6.3.18. *Let $\nu>1$ and $i\in[n],j,k\in\{0,1\}$. Every switch $e=(d_{i,j,k},F_{i,j})$ with $\phi^{\sigma_{\mathfrak{b}}}(e)<\mathfrak{m}-1$ is applied during phase 5, and the resulting strategy has Property $(OR1)_{i,j,k}$.*

Now, the first row of Table 6.7 and the corresponding arguments can be applied for every improving switch of type 3. Thus, we obtain a phase-5-strategy $\sigma\in\rho(\sigma_0)$ such that every improving switch is of type 1 or 2. The next improving switch that is applied has an occurrence record of $\lfloor(\mathfrak{b}+1)/2\rfloor-1$, i.e., it is of type 2, so we discuss the application of these switches next.

Since any improving switch is either of the form $(d_{*,*,*},F_{*,*})$ or $(e_{*,*,*},b_2)$ and since the latter switches are of type 2, some improving switch $(e_{*,*,*},b_2)$ is applied next due to the tie-breaking rule.

Claim 16. Let $\nu > 1$ and let σ denote the strategy obtained after the application of all improving switches of type 3 during phase 5. The application of type 2 switches of the form $(e_{*,*,*}, b_2)$ is described by row 1 of Lemma 6.2.32.

Let $i \in [n], j, k \in \{0,1\}$ and let $e = (e_{i,j,k}, b_2)$ denote the applied improving switch. Then, Table 5.6 describes the occurrence record of e after the application when interpreted for $\mathfrak{b}+1$ since $\phi^{\sigma e}(e) = \phi^{\sigma_{\mathfrak{b}}}(e)+1 = \lfloor \mathfrak{b}/2 \rfloor + 1 = \mathfrak{m}$. Now, by Lemma 6.2.32, $(d_{i,j,1-k}, F_{i,j}) \in I_{\sigma e}$ and the edge $(g_i, F_{i,j})$ might become improving for σe. The strategy σe is now either a phase-5-strategy for $\mathfrak{b}$ or a phase-1-strategy for $\mathfrak{b}+1$. The following corollary which is proven in Appendix A.2 now describes the improving switch $(d_{i,j,1-k}, F_{i,j})$ in more detail.

Corollary 6.3.19. *Let $i \in [n], j, k \in \{0,1\}$ and let σ denote the strategy obtained after the application of an improving switch $(e_{i,j,k}, b_2)$ during phase 5. If $(d_{i,j,1-k}, F_{i,j}) \in I_\sigma$, then$\sigma$ has Property (OR4)$_{i,j,1-k}$ and it holds that $\min_{k' \in \{0,1\}} \phi^{\sigma_{\mathfrak{b}}}(d_{i,j,k'}, F_{i,j}) \leq \mathfrak{m} - 1$.*

We now use Corollary 6.3.19 to prove that $e := (g_i, F_{i,j})$ is applied next if it becomes improving. For simplicity, let σ denote the current strategy that was obtained by applying an improving switch $(e_{i,j,*}, b_2)$ according to Lemma 6.2.32.

By the tie-breaking rule and Corollary 6.3.19, it suffices to prove

$$\phi^{\sigma}(g_i, F_{i,j}) \leq \left\lfloor \frac{\mathfrak{b}+1}{2} \right\rfloor - 1. \tag{6.5}$$

Since Table 5.6 and Corollary 6.3.19 yield

$$\phi^{\sigma_{\mathfrak{b}}}(g_i, F_{i,j}) \leq \min_{k' \in \{0,1\}} \phi^{\sigma_{\mathfrak{b}}}(d_{i,j,k'}, F_{i,j}) \leq \left\lfloor \frac{\mathfrak{b}+1}{2} \right\rfloor - 1,$$

it suffices to prove $(g_i, F_{i,j}) \notin \mathfrak{A}^{\sigma}_{\sigma_{\mathfrak{b}}}$.

Claim 17. Let $i \in [n], j, k \in \{0,1\}$ and let σ denote the strategy obtained after the application of an improving switch $(e_{i,j,k}, b_2)$ during phase 5. If $(g_i, F_{i,j}) \in I_\sigma$, then $(g_i, F_{i,j}) \notin \mathfrak{A}^{\sigma}_{\sigma_{\mathfrak{b}}}$.

Due to the tie-breaking rule, $(g_i, F_{i,j})$ is thus applied next. We prove that row 2 of Table 6.6 applies to the application of e.

First, $\beta_i = 0$ follows from the conditions of Lemma 6.2.32. Second, $\bar{\sigma}(eb_{i,j}) \wedge \neg\bar{\sigma}(eg_{i,j})$ follows as the cycle center $F_{i,j}$ was mixed earlier and since we just applied $(e_{i,j,k}, b_2)$. To prove that $\bar{\sigma}(d_{i',j'}) \vee [\bar{\sigma}(eb_{i',j'}) \wedge \neg\bar{\sigma}(eg_{i',j'})]$ holds for all $i' \geq i$ and $j \in \{0,1\}$, fix some $i' \geq i$ and $j' \in \{0,1\}$. If $\beta_{i'} = 1 \wedge \beta_{i'+1} = j'$, then the statement follows from Property (EV1)$_{i'}$. We may hence assume $\beta_{i'} = 0 \vee j' \neq \beta_{i+1}$ and that $F_{i',j'}$ is not closed. Then, by Lemma 6.3.12, either $\bar{\sigma}(eb_{i',j'}) \wedge \bar{\sigma}(eg_{i',j'})$ or $\bar{\sigma}(eb_{i',j'}) \wedge \neg\bar{\sigma}(eg_{i',j'})$. Assume that the first case was true, implying $i' \neq i$. Then, $\sigma(e_{i',j',k}) = g_1$ and $\sigma(d_{i',j',k}) = e_{i',j',k}$ for some $k \in \{0,1\}$. This in particular implies $(e_{i',j',k}, b_2) \in I_\sigma$. But this is a contradiction to the fact that we apply improving switches according to the tie-breaking rule since $i' > i$ implies that the switch $(e_{i',j',k}, b_2)$ is applied before the switch $(e_{i,j',k}, b_2)$.

Hence, all requirements of the second row of Table 6.7 are met. Further note that the strategy obtained after applying the switch has Property (SVG)$_i$/(SVM)$_i$ due to the conditions described in Lemma 6.2.32. In particular, Corollary 6.3.16 also holds for $\nu > 1$.

After the application of $(e_{i,j,k}, b_2)$ (or $(g_i, F_{i,j})$ if it becomes improving), the tie-breaking rule determines which switch is applied next. Since $(d_{i,j,1-k}, F_{i,j})$ has an occurrence record of at least $\mathfrak{m}-1$, another switch of the type $(e_{*,*,*}, b_2)$ is applied. But then, the same arguments used previously can be applied again. That is, we can apply some switch $(e_{i',j',k'}, b_2)$, making $(d_{i',j',1-k'}, F_{i',j'})$ improving, and eventually making $(g_{i'}, F_{i',j'})$ improving as well. The switch $(g_{i'}, F_{i',j'})$ is applied immediately (if it becomes improving) whereas the other switch is not applied. Then, inductively, all remaining switches of the form $(e_{*,*,*}, b_2)$ are applied.

Let σ denote the strategy that is reached after applying the final improving switch of the type $(e_{*,*,*}, b_2)$. We prove that σ has Property (SV*)$_1$ if $(g_1, F_{1,j})$ does not become improving and Property (SV*)$_i$ for all $i \geq 2$. We first determine which is the last switch of the form $(e_{*,*,*}, b_2)$ that will be applied. It holds that $(1, \beta_2) \in S_2$, implying $(e_{1,\beta_2,k}, b_2) \in I_{\sigma^{(5)}}$ for some $k \in \{0, 1\}$ by Lemma 6.3.12. Due to the tie-breaking rule, this is thus the last switch of the form $(e_{*,*,*}, b_2)$ that will be applied. This might also unlock the corresponding improving switch (g_1, F_{1,β_2}). Let σ denote the strategy obtained after the application of the switch $(e_{1,\beta_2,k}, b_2)$ resp. after the application of the switch (g_1, F_{1,β_2}) if it becomes improving.

Claim 18. Let $\nu > 1$. The strategy σ obtained after the application of the final improving switch of phase 5 has Property (SV*)$_i$ for all $i \in [n]$.

Thus, by Lemma 6.2.32 resp. the row 2 of Table 6.7, σ is a well-behaved phase-1-strategy for $\mathfrak{b}+1$ with $\sigma \in \rho(\sigma_0)$. It remains to show that σ is a canonical strategy for $\mathfrak{b}+1$ with $I_\sigma = \{(d_{i,j,k}, F_{i,j}) \colon \sigma(d_{i,j,k}) \neq F_{i,j}\}$. This is formalized by the two following claims whose proofs can be found in Appendix A.2. The first statement is again shown by proving that the two sets are contained in each other. The proof of the second statement is analogous to the corresponding statement for $\nu = 1$ and is thus deferred to the appendix.

Claim 19. Let σ denote the strategy obtained after applying the final improving switch of phase 5 for $\nu > 1$. Then $I_\sigma = \{(d_{i,j,k}, F_{i,j}) \colon \sigma(d_{i,j,k}) \neq F_{i,j}\}$.

Claim 20. Let σ denote the strategy obtained after applying the final improving switch of phase 5 for $\nu > 1$. Then σ is a canonical strategy for $\mathfrak{b}+1$.

This concludes the case $\nu > 1$ and hence proves the statement. □

Using the previous similar corollaries of this type, it follows that we also implicitly proved the following corollary.

Corollary 6.3.20. *Let $\sigma_{\mathfrak{b}+1}$ be the canonical strategy for $\mathfrak{b}+1$ calculated by the strategy improvement algorithm as described by Lemma 6.3.14. Then, Table 5.6 specifies the occurrence record of every improving switch applied until reaching $\sigma_{\mathfrak{b}+1}$, excluding switches $(g_*, F_{*,*})$, when interpreted for $\mathfrak{b}+1$. In addition, each such switch was applied once.*

It remains to prove that the canonical strategy $\sigma_{\mathfrak{b}+1}$ fulfills the canonical conditions and to investigate the occurrence records of edges of the type $(g_*, F_{*,*})$. By Corollary 6.3.20, it suffices to prove that $\sigma_{\mathfrak{b}+1}$ has Properties (OR1)$_{*,*,*}$ to (OR4)$_{*,*,*}$ and that Table 5.6 specifies the occurrence records of all edges that were not applied during $\sigma_{\mathfrak{b}} \to \sigma_{\mathfrak{b}+1}$.

We begin by investigating the canonical properties. The following statement is required when discussing Properties (OR1)$_{*,*,*}$ to (OR4)$_{*,*,*}$. It states that the occurrence record of the cycle edges of $F_{\ell(\mathfrak{b}+2),1-(\mathfrak{b}+2)}$ are large if $\mathfrak{b}$ is even and will be used repeatedly. This is useful as the canonical properties with respect to $\sigma_{\mathfrak{b}+1}$ depend on $\mathfrak{b}+2$. Its proof can be found in Appendix A.2.

Lemma 6.3.21. *Let $\mathfrak{b} \in \mathfrak{B}_n$ be even, $i := \ell(\mathfrak{b}+2)$ and $j := 1-(\mathfrak{b}+2)_{i+1}$. If $\mathfrak{b}+2$ is a power of 2, then $\phi^{\sigma_{\mathfrak{b}}}(d_{i,j,*}, F_{i,j}) = \mathfrak{m}$. Otherwise, $\phi^{\sigma_{\mathfrak{b}}}(d_{i,j,0}, F_{i,j}) = \lfloor(\mathfrak{b}+1)/2\rfloor$ and $\phi^{\sigma_{\mathfrak{b}}}(d_{i,j,1}, F_{i,j}) = \mathfrak{m}-1$. In any case, $\sigma_{\mathfrak{b}}(d_{i,j,k}) \neq F_{i,j}$ for both $k \in \{0,1\}$.*

We now prove that $\sigma_{\mathfrak{b}+1}$ has Properties (OR1)$_{*,*,*}$ to (OR4)$_{*,*,*}$.

Lemma 6.3.22. *Let $\sigma_{\mathfrak{b}+1}$ denote the canonical strategy calculated by the strategy improvement algorithm as described by Lemma 6.3.14. Then $\sigma_{\mathfrak{b}+1}$ has Properties (OR1)$_{*,*,*}$ to (OR4)$_{*,*,*}$.*

Proof. To simplify notation, let $\sigma := \sigma_{\mathfrak{b}+1}$. We first prove that σ has Properties (OR1)$_{*,*,*}$, (OR2)$_{*,*,*}$ and (OR4)$_{*,*,*}$ and discuss Property (OR3)$_{*,*,*}$ at the end.

Consider the case $\nu > 1$ first. Let $i \in [n], j, k \in \{0,1\}$ and consider Property (OR4)$_{i,j,k}$. We prove that any improving switch has an occurrence record of either $\mathfrak{m}$ or $\mathfrak{m}-1$ as $\mathfrak{m} = \lfloor(\mathfrak{b}+1+1)/2\rfloor$ due to $\nu > 1$. Any $e \in I_\sigma$ was either improving for $\sigma^{(5)}$ or became improving when transitioning from $\sigma^{(5)}$ to σ. As shown in the proof of Lemma 6.3.14, all improving switches not applied during phase 5 had an occurrence record of at least $\mathfrak{m}-1$. More precisely, this was shown implicitly when giving the characterization of the improving switches. Also, the occurrence records of these edges are at most $\mathfrak{m}$, proving the statement for these edges. For improving switches that were unlocked during phase 5, the statement follows by Corollary 6.3.19. Hence, σ has Property (OR4)$_{i,j,k}$.

We prove that σ has Property (OR2)$_{*,*,*}$ and Property (OR1)$_{*,*,*}$. Consider some indices $i \in [n], j \in \{0,1\}$ with $\beta_i = 0 \vee \beta_{i+1} \neq j$ and let $k \in \{0,1\}$. We prove

$$\sigma(d_{i,j,k}) = F_{i,j} \iff \phi^{\sigma}(d_{i,j,k}, F_{i,j}) = \ell^{\mathfrak{b}+1}(i,j,k) + 1. \tag{6.6}$$

Let $\sigma(d_{i,j,k}) = F_{i,j}$. Then, since $\sigma^{(5)}(d_{i,j,k}) \neq F_{i,j}$ by the choice of i and j and Table 5.8, the switch was applied during $\sigma^{(5)} \to \sigma$. Consequently, the edge $(d_{i,j,k}, F_{i,j})$ was not applied as improving switch before phase 5 as switches are applied at most once by Corollary 6.3.20. Thus, $\phi^{\sigma_{\mathfrak{b}}}(d_{i,j,k}, F_{i,j}) = \phi^{\sigma^{(5)}}(d_{i,j,k}, F_{i,j}) < \mathfrak{m}-1$. But this implies $\sigma_{\mathfrak{b}}(d_{i,j,k}) = F_{i,j}$ since the switch would have been applied in phase 1 otherwise. Consequently, by Lemma 6.2.6,

$$\phi^{\sigma}(d_{i,j,k}, F_{i,j}) = \phi^{\sigma_{\mathfrak{b}}}(d_{i,j,k}, F_{i,j}) + 1 = \ell^{\mathfrak{b}}(i,j,k) + 1 + 1 = \ell^{\mathfrak{b}+1}(i,j,k) + 1 \leq \mathfrak{m}-1.$$

This implies both "$\Rightarrow$" of the equivalence (6.6) as well as Property (OR1)$_{i,j,k}$.

Now, let $\phi^{\sigma}(d_{i,j,k}, F_{i,j}) = \ell^{\mathfrak{b}+1}(i,j,k)+1$. We prove that this implies $\sigma(d_{i,j,k}) = F_{i,j}$. First, $\phi^{\sigma}(d_{i,j,k}, F_{i,j}) = \ell^{\mathfrak{b}+1}(i,j,k) + 1 \leq \lfloor(\mathfrak{b}+1+1-k)/2\rfloor$ implies $\ell^{\mathfrak{b}+1}(i,j,k) \leq \lfloor(\mathfrak{b}-k)/2\rfloor$. By Lemma 6.2.3, this implies that $\beta_{i+1} = 1-j$. Consider the case $\mathfrak{b}_i = 0 \wedge \mathfrak{b}_{i+1} \neq j$. Then, $\phi^{\sigma_{\mathfrak{b}}}(d_{i,j,k}, F_{i,j}) = \min(\lfloor(\mathfrak{b}+1-k)/2\rfloor, \ell^{\mathfrak{b}}(i,j,k) + t_{\mathfrak{b}})$ for some $t_{\mathfrak{b}}$ feasible for $\mathfrak{b}$.

Assume $\phi^{\sigma_{\mathfrak{b}}}(d_{i,j,k}, F_{i,j}) \neq \ell^{\mathfrak{b}}(i,j,k) + t_{\mathfrak{b}}$ for all feasible parameters and note that this implies $\phi^{\sigma_{\mathfrak{b}}}(d_{i,j,k}, F_{i,j}) = \lfloor(\mathfrak{b}+1-k)/2\rfloor$. Then $\phi^{\sigma}(d_{i,j,k}, F_{i,j}) < \ell^{\mathfrak{b}}(i,j,k)+1$, implying

$$\ell^{\mathfrak{b}+1}(i,j,k) = \ell^{\mathfrak{b}}(i,j,k) + 1 > \left\lfloor \frac{\mathfrak{b}+1-k}{2} \right\rfloor + 1 = \left\lfloor \frac{\mathfrak{b}+3-k}{2} \right\rfloor \geq \left\lfloor \frac{\mathfrak{b}+1+1-k}{2} \right\rfloor$$

which is a contradiction. Consequently, $\phi^{\sigma_{\mathfrak{b}}}(d_{i,j,k}, F_{i,j}) = \ell^{\mathfrak{b}}(i,j,k)+t_{\mathfrak{b}}$ for some feasible $t_{\mathfrak{b}}$. Assume $\phi^{\sigma}(d_{i,j,k}, F_{i,j}) = \ell^{\mathfrak{b}}(i,j,k)$. Then

$$\phi^{\sigma}(d_{i,j,k}, F_{i,j}) = \ell^{\mathfrak{b}+1}(i,j,k) + 1 = \ell^{\mathfrak{b}}(i,j,k) + 2 = \phi^{\sigma_{\mathfrak{b}}}(d_{i,j,k}, F_{i,j}) + 2,$$

implying that the switch would have been applied twice during $\sigma_{\mathfrak{b}} \to \sigma$. This is a contradiction. The same contradiction follows if we assume $\phi^{\sigma_{\mathfrak{b}}}(d_{i,j,k}, F_{i,j}) = \ell^{\mathfrak{b}}(i,j,k)-1$. Hence, it holds that $\phi^{\sigma_{\mathfrak{b}}}(d_{i,j,k}, F_{i,j}) = \ell^{\mathfrak{b}}(i,j,k)+1$, implying $\sigma_{\mathfrak{b}}(d_{i,j,k}) = F_{i,j}$. Since $\ell^{\mathfrak{b}}(i,j,k) = \ell^{\mathfrak{b}+1}(i,j,k)-1$, this also implies that the switch was indeed applied during the transition. However, $\sigma_{\mathfrak{b}}(d_{i,j,k}) = F_{i,j}$ implies that the switch was not applied during phase 1 of that transition. But then it must have been applied in phase 5, implying $\sigma(d_{i,j,k}) = F_{i,j}$.

We now show that the same holds if $\mathfrak{b}_i = 1$ and $\mathfrak{b}_{i+1} = j$, implying $i < \nu$. This then yields

$$\begin{aligned}\phi^{\sigma_{\mathfrak{b}}}(d_{i,j,k}, F_{i,j}) &= \left\lceil \frac{\operatorname{lfn}(\mathfrak{b}, i, \{(i+1,j)\}) + 1 - k}{2} \right\rceil = \left\lceil \frac{\mathfrak{b} - 2^{i-1} + 1 + 1 - k}{2} \right\rceil \\ &= \left\lceil \frac{\mathfrak{b} + 1 - 2^{i-1} + 1 - k}{2} \right\rceil = \ell^{\mathfrak{b}+1}(i,j,k).\end{aligned}$$

Since $\phi^{\sigma}(d_{i,j,k}, F_{i,j}) = \ell^{\mathfrak{b}+1}(i,j,k)+1$, this implies that the switch was applied during phase 5 of $\sigma_{\mathfrak{b}} \to \sigma$. Consequently, $\sigma(d_{i,j,k}) = F_{i,j}$. This proves "$\Leftarrow$" and hence the equivalence (6.6). Most importantly, σ thus has Property (OR2)$_{i,j,k}$.

Now assume $\nu = 1$. Let $i \in [n], j,k \in \{0,1\}$ and consider Property (OR4)$_{i,j,k}$. We prove that $e := (d_{i,j,k}, F_{i,j}) \in I_{\sigma}$ implies that e has an occurrence record of $\lfloor(\mathfrak{b}+2)/2\rfloor - 1 = \mathfrak{m}$ or $\lfloor(\mathfrak{b}+2)/2\rfloor = \mathfrak{m}+1$. It is easy to verify that for such an edge e, one of the following cases holds.

- $e \in I_{\sigma'}$ for all $\sigma' \in \rho(\sigma_{\mathfrak{b}})$, i.e., the switch was improving during the complete transition. Then, $\phi^{\sigma}(e) = \phi^{\sigma_{\mathfrak{b}}}(e) = \mathfrak{m}$ by Corollary 6.3.3.
- There is a strategy $\sigma' \in \rho(\sigma^{(5)})$ with $(d_{i,j,k}, F_{i,j}) \in I_{\sigma'}$ but $(d_{i,j,k}, F_{i,j}) \notin I_{\sigma^{(5)}}$. That is, the switch became improving during phase 5. Then, σ has Property (OR4)$_{i,j,k}$ by Corollary 6.3.15.
- The edge e became an improving switch when applying (b_1, g_1) at the end of phase 3. Then $i \in \{u+1, \dots, m-1\}, j = 1 - \beta_{i+1}$ and $\beta_i = 0$. Thus, by the characterization of I_{σ} given in the beginning of the proof of Lemma 6.3.14, $\phi^{\sigma}(d_{i,j,k}, F_{i,j}) = \mathfrak{m}$ since the switch would have been applied during phase 5 otherwise.

Thus, σ has Property (OR4)$_{i,j,k}$.

Now let $i \in [n]$ and $j \in \{0,1\}$ with $\beta_i = 0 \vee \beta_{i+1} \neq j$, let $k \in \{0,1\}$ and consider Property (OR2)$_{i,j,k}$. Then $\sigma^{(5)}(d_{i,j,k}) \neq F_{i,j}$ by Lemma 6.3.12. We again prove that σ fulfills the equivalence (6.6) and that σ has Property (OR1)$_{i,j,k}$ simultaneously.

Let $\sigma(d_{i,j,k}) = F_{i,j}$. By Lemma 6.3.12, $(d_{i,j,k}, F_{i,j}) \in \mathfrak{A}^{\sigma}_{\sigma^{(5)}}$. Since improving switches are applied at most once per transition, this implies

$$\phi^{\sigma^{(5)}}(d_{i,j,k}, F_{i,j}) = \phi^{\sigma_{\mathfrak{b}}}(d_{i,j,k}, F_{i,j}) < \mathfrak{m}$$

and $\sigma_{\mathfrak{b}}(d_{i,j,k}) = F_{i,j}$ by Corollary 6.3.17. Thus, by Property $(\text{OR2})_{i,j,k}$ and Lemma 6.2.6, $\phi^{\sigma_{\mathfrak{b}}}(d_{i,j,k}, F_{i,j}) = \ell^{\mathfrak{b}}(i,j,k) + 1 = \ell^{\mathfrak{b}+1}(i,j,k)$. Hence

$$\phi^{\sigma}(d_{i,j,k}, F_{i,j}) = \phi^{\sigma^{(5)}}(d_{i,j,k}, F_{i,j}) + 1 = \ell^{\mathfrak{b}+1}(i,j,k) + 1 < \mathfrak{m} + 1 = \left\lfloor \frac{\mathfrak{b}+2}{2} \right\rfloor$$

by integrality. Thus, "$\Rightarrow$" as well as Property $(\text{OR1})_{i,j,k}$ follow.

Let $\phi^{\sigma}(d_{i,j,k}, F_{i,j}) = \ell^{\mathfrak{b}+1}(i,j,k) + 1$. By Lemma 6.3.12. $\sigma^{(5)}(d_{i,j,k}) = F_{i,j}$ if and only if $\beta_i = 1 \wedge \beta_{i+1} = j$. It thus suffices to prove $(d_{i,j,k}, F_{i,j}) \in \mathfrak{A}^{\sigma}_{\sigma^{(5)}}$. By Corollary 6.3.17, we thus need to show that

1. $\phi^{\sigma_{\mathfrak{b}}}(d_{i,j,k}, F_{i,j}) < \mathfrak{m} \wedge \sigma_{\mathfrak{b}}(d_{i,j,k}) = F_{i,j}$,
2. $\beta_i = 0 \wedge \beta_{i+1} \neq j$ and
3. $i \in \{u+1, \dots, m-1\}$.

Since

$$\phi^{\sigma}(d_{i,j,k}, F_{i,j}) = \ell^{\mathfrak{b}+1}(i,j,k) + 1 \leq \left\lfloor \frac{\mathfrak{b}+1+1-k}{2} \right\rfloor, \tag{6.7}$$

Lemma 6.2.3 implies that $\beta_i = 0 \wedge \beta_{i+1} = 1 - j,$. Consequently since $\nu = 1$ implies that no bit switches from 1 to 0, it follows that $\mathfrak{b}_i = 0 \wedge \mathfrak{b}_{i+1} = 1 - j$. This implies that there is a feasible $t_{\mathfrak{b}}$ with $\phi^{\sigma_{\mathfrak{b}}}(d_{i,j,k}, F_{i,j}) = \min(\lfloor(\mathfrak{b}+1-k)/2\rfloor, \ell^{\mathfrak{b}}(i,j,k) + t_{\mathfrak{b}})$. Note that $t_{\mathfrak{b}} \neq -1$ due to the parity of $\mathfrak{b}$ and Property $(\text{OR3})_{i,j,k}$. We prove that $\phi^{\sigma_{\mathfrak{b}}}(d_{i,j,k}, F_{i,j}) = \ell^{\mathfrak{b}}(i,j,k)+1$ by ruling out the other possible cases.

- Assume $\phi^{\sigma_{\mathfrak{b}}}(d_{i,j,k}, F_{i,j}) = \lfloor(\mathfrak{b}+1-k)/2\rfloor$ and that neither 0 nor 1 are feasible parameters. As $i \neq \nu$, this implies $\ell^{\mathfrak{b}+1}(i,j,k) = \ell^{\mathfrak{b}}(i,j,k) + 1 > \lfloor(\mathfrak{b}+1-k)/2\rfloor$. But then $\ell^{\mathfrak{b}+1}(i,j,k) + 1 > \lfloor(\mathfrak{b}+1+1-k)/2\rfloor$, contradicting Equation (6.7).
- Next assume $\phi^{\sigma_{\mathfrak{b}}}(d_{i,j,k}, F_{i,j}) = \ell^{\mathfrak{b}}(i,j,k)$. Then, since $\ell^{\mathfrak{b}}(i,j,k) = \ell^{\mathfrak{b}+1}(i,j,k) - 1$, the switch $(d_{i,j,k}, F_{i,j})$ would have been switched twice during $\sigma_{\mathfrak{b}} \to \sigma$. This is a contradiction.

Hence $\phi^{\sigma_{\mathfrak{b}}}(d_{i,j,k}, F_{i,j}) = \ell^{\mathfrak{b}}(i,j,k) + 1$. It remains to prove $i \in \{u+1, \dots, m-1\}$. Since $i \geq m$ implies $\ell^{\mathfrak{b}}(i,j,k) \geq \mathfrak{b}$, this implies that we need to have $i < m$ as we have $\phi^{\sigma_{\mathfrak{b}}}(d_{i,j,k}, F_{i,j}) = \ell^{\mathfrak{b}}(i,j,k) + 1 < \mathfrak{m}$. Also, assuming $i = u$ yields $\phi^{\sigma_{\mathfrak{b}}}(d_{i,j,k}, F_{i,j}) \geq \mathfrak{m}$ as discussed earlier. Consequently, all of the three necessary conditions hold, so Corollary 6.3.17 implies the direction "$\Leftarrow$" of the equivalence (6.6). Thus, σ has Properties $(\text{OR1})_{*,*,*}$, $(\text{OR2})_{*,*,*}$ and $(\text{OR4})_{*,*,*}$ if $\nu = 1$.

It remains to prove that σ has Property $(\text{OR3})_{*,*,*}$. Property $(\text{OR3})_{i,j,k}$ states that $\phi^{\sigma}(d_{i,j,k}, F_{i,j}) = \ell^{\mathfrak{b}+1}(i,j,k) - 1 \wedge \phi^{\sigma}(d_{i,j,k}, F_{i,j}) \neq \lfloor(\mathfrak{b}+1+1-k)/2\rfloor$ if and only if $\mathfrak{b}+1$ is odd, $\mathfrak{b}+2$ is not a power of 2, $i = \ell(\mathfrak{b}+2)$, $j \neq (\mathfrak{b}+2)_{i+1}$ and $k = 0$. We first prove the "if" part. Since $\mathfrak{b}+1$ is odd, $\mathfrak{b}$ is even. As $\mathfrak{b}+2$ is not a power of 2 by assumption, $\phi^{\sigma_{\mathfrak{b}}}(d_{i,j,0}, F_{i,j}) = \mathfrak{m}$ and $\phi^{\sigma_{\mathfrak{b}}}(d_{i,j,1}, F_{i,j}) = \mathfrak{m} - 1$ as well $\sigma_{\mathfrak{b}}(d_{i,j,k}) \neq F_{i,j}$ for both $k \in \{0,1\}$ by Lemma 6.3.21. Consider phase 1 of $\sigma_{\mathfrak{b}} \to \sigma$. Then, $(d_{i,j,1}, F_{i,j})$ is applied in this phase by Corollary 6.3.3. Thus, by the tie breaking rule, $(d_{i,j,0}, F_{i,j})$ is not applied during phase 1. Since no switch with an occurrence record of $\mathfrak{m}$ is applied during phase 5, the switch is also not applied during phase 5. Consequently,

$$\phi^{\sigma}(d_{i,j,0}, F_{i,j}) = \phi^{\sigma_{\mathfrak{b}}}(d_{i,j,0}, F_{i,j}) = \left\lfloor \frac{\mathfrak{b}+1}{2} \right\rfloor = \left\lfloor \frac{\mathfrak{b}+1+1}{2} \right\rfloor - 1$$

since $\mathfrak{b}+1$ is odd. It remains to show $\ell^{\mathfrak{b}+1}(i,j,0) = \lfloor (\mathfrak{b}+1+1)/2 \rfloor$. Since $\mathfrak{b}+1$ is odd, $\ell(\mathfrak{b}+2) \neq \nu$ and $\mathfrak{b}_i = 0$. Hence, $\ell^{\mathfrak{b}+1}(i,j,0) = \ell^{\mathfrak{b}}(i,j,0) + 1 = \lfloor \mathfrak{b}/2 \rfloor + 1 = \lfloor (\mathfrak{b}+1+1)/2 \rfloor$ by Lemma 6.2.6. Thus, the "if" part is fulfilled.

The "only if" part can be show using contraposition by dividing the proof into several small statements, each proving that one of the conditions is necessary. We state all of the statements here and defer their proofs to Appendix A.2. More precisely, the following statements imply the "only if" part:

Claim 21. Let $i \in [n], j, k \in \{0,1\}$ and consider the two equations

$$\phi^{\sigma}(e) \neq \ell^{\mathfrak{b}+1}(i,j,k) - 1, \tag{6.8}$$

$$\phi^{\sigma}(e) = \left\lfloor \frac{\mathfrak{b}+1+1-2}{2} \right\rfloor. \tag{6.9}$$

1. If $j = (\mathfrak{b}+2)_{i+1}$, then either Equation (6.8) or Equation (6.9) holds.
2. If $i \neq \ell(\mathfrak{b}+2)$ and $j \neq (\mathfrak{b}+2)_{i+1}$, then either Equation (6.8) or Equation (6.9) holds.
3. If $\mathfrak{b}+1$ is even, $i = \ell(\mathfrak{b}+2)$ and $j \neq (\mathfrak{b}+2)_{i+1}$, then Equation (6.9) holds.
4. If $\mathfrak{b}+1$ is odd, $i = \ell(\mathfrak{b}+2), j = 1 - (\mathfrak{b}+2)_{i+1}, k \in \{0,1\}$ and $\mathfrak{b}+2$ is a power of two, then Equation (6.9) holds.
5. If $\mathfrak{b}$ is even, $i = \ell(\mathfrak{b}+2), j \neq (\mathfrak{b}+2)_{i+1}, k = 1$ and $\mathfrak{b}+2$ is not a power of two, then Equation (6.9) holds.

This show that σ has Property (OR3)$_{*,*,*}$ and thus yields the statement. □

Reaching a canonical strategy part II: The occurrence records

It now remains to prove that Table 5.6 specifies the occurrence records with respect to the canonical strategy $\sigma_{\mathfrak{b}+1}$ for $\mathfrak{b}+1$ when it is interpreted for $\mathfrak{b}+1$. This then implies that $\sigma_{\mathfrak{b}+1}$ has the canonical properties which can then be used to give inductive proofs of the main statements of Section 5.3.

As in particular the investigation of edges of the type $(g_*, F_{*,*})$ is rather involved, we show two separate statements and consider all other edges first.

Lemma 6.3.23. *Let $\sigma_{\mathfrak{b}+1}$ be the canonical strategy for $\mathfrak{b}+1$ calculated by the strategy improvement resp. policy iteration algorithm when starting with a canonical strategy $\sigma_{\mathfrak{b}}$ having the canonical properties as described by Lemma 6.3.14. Then, Table 5.6 specifies the occurrence records of all edges $e \in E_0$ but edges of the type $(g_*, F_{*,*})$ that were applied during $\sigma_{\mathfrak{b}} \to \sigma_{\mathfrak{b}+1}$.*

Proof. There are two types of edges. Each edge was either applied as improving switch when transitioning from $\sigma_{\mathfrak{b}}$ to $\sigma_{\mathfrak{b}+1}$ or was not applied as an improving switch. We already proved that Table 5.6 specifies the occurrence records of all improving switch that were applied, with the exception of switches $(g_*, F_{*,*})$. It thus suffices to consider switches that were not applied when transitioning from $\sigma_{\mathfrak{b}}$ to $\sigma_{\mathfrak{b}+1}$. We thus identify edges that were not applied as improving switches and prove that their occurrence records are described by Table 5.6. To simplify notation, let $\sigma := \sigma_{\mathfrak{b}+1}$.

Let $\nu > 1$ and let $i \in [n], j, k \in \{0,1\}$ be suitable indices. We first prove the statement for all edges that are not of the type $(d_{*,*,*}, F_{*,*})$.

1. Consider edges of the type $(b_i, *)$. Since $\nu > 1$, the edges (b_i, b_{i+1}) for $i \in [\nu-1]$ as well as the edge (b_ν, g_ν) were applied as improving switches. Let $e = (b_i, b_{i+1})$ and $i \geq \nu$. Then $\phi^\sigma(e) = \mathrm{fl}(\mathfrak{b}, i) - \mathfrak{b}_i = \mathrm{fl}(\mathfrak{b}+1, i) - (\mathfrak{b}+1)_i$ since either $\mathrm{fl}(\mathfrak{b}, i) = \mathrm{fl}(\mathfrak{b}+1, i)$ and $\mathfrak{b}_i = (\mathfrak{b}+1)_{i+1}$ (if $i > \nu$) or $\mathrm{fl}(\mathfrak{b}+1, i) = \mathrm{fl}(\mathfrak{b}, i) + 1$ and $\mathfrak{b}_i = 0, (\mathfrak{b}+1)_i = 1$ (if $i = \nu$). Let $e = (b_i, g_i)$ for $i \neq \nu$. Then, by Lemma 6.2.6, $\phi^\sigma(e) = \mathrm{fl}(\mathfrak{b}, i) = \mathrm{fl}(\mathfrak{b}+1, i)$.
2. Consider some edge $(g_i, F_{i,j})$ that was not applied during $\sigma_{\mathfrak{b}} \to \sigma$. Then, the upper bound remains valid as it can only increase.
3. Consider some vertex $s_{i,j}$. Since $\nu > 1$, the edges $(s_{\nu-1,1}, h_{\nu-1,1}), (s_{\nu-1,0}, b_1)$ as well as the edges $(s_{i,0}, h_{i,0}), (s_{i,1}, b_1)$ for $i \in [\nu-2]$ were switched. It thus suffices to consider indices $i \geq \nu$. For these edges, the choice of i implies

$$\phi^\sigma(s_{i,j}, b_1) = \mathrm{fl}(\mathfrak{b}, i+1) - j \cdot \mathfrak{b}_{i+1} = \mathrm{fl}(\mathfrak{b}+1, i+1) - j \cdot (\mathfrak{b}+1)_{i+1},$$
$$\phi^\sigma(s_{i,j}, h_{i,j}) = \mathrm{fl}(\mathfrak{b}, i+1) - (1-j)\mathfrak{b}_{i+1} = \mathrm{fl}(\mathfrak{b}+1, i+1) - (1-j)(\mathfrak{b}+1)_{i+1}.$$

4. For $e = (e_{i,j,k}, g_1)$, Table 5.6 implies $\phi^\sigma(e_{i,j,k}, g_1) = \lceil \mathfrak{b}/2 \rceil = \lceil (\mathfrak{b}+1)/2 \rceil$ since $\nu > 1$.
5. For $e = (d_{i,j,k}, e_{i,j,k})$, we need to prove $\phi^\sigma(e) \leq \phi^\sigma(e_{i,j,k}, g_1) = \lceil (\mathfrak{b}+1)/2 \rceil$ since $\mathfrak{b}$ is odd. This follows from $\phi^\sigma(e) \leq \phi^{\sigma_{\mathfrak{b}}}(e) + 1 \leq \lfloor \mathfrak{b}/2 \rfloor + 1 = \lfloor (\mathfrak{b}+2)/2 \rfloor = \lceil (\mathfrak{b}+1)/2 \rceil$.

Let $i \in [n], j, k \in \{0,1\}$ and consider some edge $e = (d_{i,j,k}, F_{i,j})$ that was not switched during $\sigma_{\mathfrak{b}} \to \sigma$. We distinguish the following cases.

1. Let $(\mathfrak{b}_i = 1 \wedge \mathfrak{b}_{i+1} = j)$ and $((\mathfrak{b}+1)_i = 1 \wedge (\mathfrak{b}+1)_{i+1} = j)$. Then, since any intermediate strategy had Property (EV1)$_i$, the cycle $F_{i,j}$ was always closed during $\sigma_{\mathfrak{b}} \to \sigma$. Thus $i \neq \nu$, implying $\mathrm{lfn}(\mathfrak{b}, i, \{(i+1, j)\}) = \mathrm{lfn}(\mathfrak{b}+1, i, \{(i+1, j)\})$. Therefore, $\phi^\sigma(e)$ is described by Table 5.6.
2. Let $(\mathfrak{b}_i = 1 \wedge \mathfrak{b}_{i+1} = j)$ and $(\mathfrak{b}+1)_i = 0$, implying $i < \nu$. Then bit $i+1$ also switched, so $(\mathfrak{b}+1)_i = 0 \wedge (\mathfrak{b}+1)_{i+1} \neq j$. Consequently, e was not switched during phase 1 since $F_{i,j}$ was closed with respect to any intermediate strategy due to Property (EV1)$_i$. It is however possible that such a switch is applied during phase 5. Since $i \leq \nu - 1$, this switch is applied if and only if $\phi^{\sigma_{\mathfrak{b}}}(e) < \mathfrak{m} - 1$. We may thus assume $\phi^{\sigma_{\mathfrak{b}}}(e) \geq \mathfrak{m} - 1 = \lfloor (\mathfrak{b}-1)/2 \rfloor$ and only need to consider the edge e if $\lfloor (\mathrm{lfn}(\mathfrak{b}, i, \{(i+1, j)\}) + 2 - k)/2 \rfloor \geq \lfloor (\mathfrak{b}-1)/2 \rfloor$. This inequality holds if and only if one of the following three cases applies:
   - $\mathrm{lfn}(\mathfrak{b}, i, \{(i+1, j)\}) + 2 - k \geq \mathfrak{b} - 1$.
   - $\mathrm{lfn}(\mathfrak{b}, i, \{(i+1, j)\}) + 2 - k$ is even and $\mathrm{lfn}(\mathfrak{b}, i, \{(i+1, j)\}) + 2 - k = \mathfrak{b} - 2$.

- $\operatorname{lfn}(\mathfrak{b}, i, \{(i+1, j)\}) + 2 - k$ is odd and $\operatorname{lfn}(\mathfrak{b}, i, \{(i+1, j)\}) + 2 - k = \mathfrak{b}$.

These assumptions can only hold if $i \in \{1, 2\} \vee (i = 3 \wedge k = 0)$. It thus suffices to consider three more cases.

For $i = 1$, it holds that

$$\ell^{\mathfrak{b}+1}(i, j, k) = \left\lceil \frac{\mathfrak{b} + 1 - 1 + 1 - k}{2} \right\rceil = \left\lceil \frac{\mathfrak{b} + 1 - k}{2} \right\rceil = \phi^{\sigma}(e).$$

Similarly, for $i = 2$, it holds that

$$\ell^{\mathfrak{b}+1}(i, j, k) = \left\lceil \frac{\mathfrak{b} + 1 - 2 + 1 - k}{2} \right\rceil = \left\lceil \frac{\mathfrak{b} - k}{2} \right\rceil = \left\lfloor \frac{\mathfrak{b} + 1 - k}{2} \right\rfloor = \phi^{\sigma}(e).$$

Finally, for $i = 3$ and $k = 0$,it holds that

$$\ell^{\mathfrak{b}+1}(i, j, k) = \left\lceil \frac{\mathfrak{b} + 1 - 4 + 1 - k}{2} \right\rceil = \left\lceil \frac{\mathfrak{b} - 2 - k}{2} \right\rceil = \phi^{\sigma}(e).$$

Hence, the parameter $t_{\mathfrak{b}-1} = 0$ can be chosen in all three cases.

3. Let $(\mathfrak{b}_i = 0 \wedge \mathfrak{b}_{i+1} \neq j)$ and $((\mathfrak{b}+1)_i = 0 \wedge (\mathfrak{b}+1)_{i+1} \neq j)$, implying $i > \nu$. First let $\mathbf{1}_{j=0}\operatorname{lfn}(\mathfrak{b}, i+1) + \mathbf{1}_{j=1}\operatorname{lufn}(\mathfrak{b}, i+1) = 0$. Then $\ell^{\mathfrak{b}}(i, j, k) \geq \mathfrak{b}$ by Lemma 6.2.3, implying $\phi^{\sigma_{\mathfrak{b}}}(e) = \lfloor (\mathfrak{b} + 1 - k)/2 \rfloor$. Since $\mathfrak{b}$ is odd, $\lfloor (\mathfrak{b} + 1 - 1)/2 \rfloor < \mathfrak{m}$. Hence, $(d_{i,j,1}, F_{i,j})$ was applied during phase 1 of $\sigma_{\mathfrak{b}} \to \sigma$ and $e = (d_{i,j,0}, F_{i,j}) \notin \mathfrak{A}^{\sigma}_{\sigma_{\mathfrak{b}}}$. Thus, since $\ell^{\mathfrak{b}+1}(i, j, k) \geq \mathfrak{b} + 1$ by the choice of i, choosing $t_{\mathfrak{b}+1} = 0$ yields the desired characterization.

   Let $\mathbf{1}_{j=0}\operatorname{lfn}(\mathfrak{b}, i+1) + \mathbf{1}_{j=1}\operatorname{lufn}(\mathfrak{b}, i+1) \neq 0$, implying $i < m = \max\{i : \beta_i = 1\}$. Using $i > \nu \geq 2$, this yields

$$\begin{aligned}
\ell^{\mathfrak{b}}(i, j, k) &= \left\lceil \frac{\mathfrak{b} - 2^{i-1} + \sum(\mathfrak{b}, i) + 1 - k}{2} \right\rceil \\
&= \left\lceil \frac{\mathfrak{b} - 2^{i-1} + \sum_{l=1}^{\nu-1} 2^{l-1} + \sum_{l=\nu+1}^{i-1} \mathfrak{b}_l 2^{l-1} + 1 - k}{2} \right\rceil \\
&\leq \left\lceil \frac{\mathfrak{b} - 2^{i-1} + 2^{\nu-1} - 1 + 2^{i-1} - 2^{\nu} + 1 - k}{2} \right\rceil = \left\lceil \frac{\mathfrak{b} - 2^{\nu-1} - k}{2} \right\rceil \\
&\leq \left\lceil \frac{\mathfrak{b} - 2 - k}{2} \right\rceil = \left\lfloor \frac{\mathfrak{b} - 1 - k}{2} \right\rfloor \leq \left\lfloor \frac{\mathfrak{b} + 1 - k}{2} \right\rfloor - 1.
\end{aligned}$$

   If $\sigma_{\mathfrak{b}}(d_{i,j,k}) \neq F_{i,j}$, this implies $\phi^{\sigma_{\mathfrak{b}}}(d_{i,j,k}, F_{i,j}) \leq \ell^{\mathfrak{b}}(i, j, k) \leq \lfloor (\mathfrak{b} + 1 - k)/2 \rfloor - 1$. Then, by Corollary 6.3.3 the switch was applied during phase 1. We may hence assume $\sigma_{\mathfrak{b}}(d_{i,j,k}) = F_{i,j}$, implying $\phi^{\sigma_{\mathfrak{b}}}(e) = \ell^{\mathfrak{b}}(i, j, k) + 1 \leq \lfloor (\mathfrak{b} + 1 - k)/2 \rfloor$ as well as $\phi^{\sigma_{\mathfrak{b}}}(e) \leq \mathfrak{m} - 1$ by Property $(\text{OR1})_{i,j,k}$. As we assume $e \notin \mathfrak{A}^{\sigma}_{\sigma_{\mathfrak{b}}}$, it suffices to consider the case $\phi^{\sigma}(e) = \phi^{\sigma_{\mathfrak{b}}}(e) = \lfloor (\mathfrak{b} + 1)/2 \rfloor - 1$ since e is applied during phase 5 otherwise (see Corollary 6.3.19). Since $\ell^{\mathfrak{b}+1}(i, j, k) = \ell^{\mathfrak{b}}(i, j, k) + 1$ by Lemma 6.2.6, choosing $t_{\mathfrak{b}+1} = 0$ yields the desired characterization.

4. Let $\mathfrak{b}_i = 0 \wedge \mathfrak{b}_{i+1} \neq j$ and $(\mathfrak{b}+1)_i = 1 \wedge (\mathfrak{b}+1)_{i+1} \neq j$, so $i = \nu$. The statement follows by the same argument used earlier if $\mathbf{1}_{j=0}\mathrm{lfn}(\mathfrak{b}, i+1) + \mathbf{1}_{j=1}\mathrm{lufn}(\mathfrak{b}, i+1) = 0$. Hence consider the case $\mathbf{1}_{j=0}\mathrm{lfn}(\mathfrak{b}, i+1) + \mathbf{1}_{j=1}\mathrm{lufn}(\mathfrak{b}, i+1) \neq 0$. This implies that we have $\ell^{\mathfrak{b}}(i,j,k) = \lfloor(\mathfrak{b}+1-k)/2\rfloor$. Since $\sigma_{\mathfrak{b}}$ is a canonical strategy for $\mathfrak{b}$, we have $\sigma_{\mathfrak{b}}(d_{i,j,k}) \neq F_{i,j}$. If $\phi^{\sigma_{\mathfrak{b}}}(e) = \ell^{\mathfrak{b}}(i,j,k)$, then $\phi^{\sigma_{\mathfrak{b}}}(e) = \lfloor(\mathfrak{b}+1-k)/2\rfloor$ and the same arguments used in the third case can be used to show the statement. If $\phi^{\sigma_{\mathfrak{b}}}(e) = \ell^{\mathfrak{b}}(i,j,k) - 1$, then $\phi^{\sigma_{\mathfrak{b}}}(e) = \mathfrak{m} - 1$ since we need to have $k = 0$ by Property (OR3)$_{i,j,k}$. But this implies that e was switched during phase 1 and that we do not need to consider it here.
5. Finally, let $\mathfrak{b}_i = 0 \wedge \mathfrak{b}_{i+1} = j$. It suffices to consider the case $(\mathfrak{b}+1)_i = 0 \wedge (\mathfrak{b}+1)_{i+1} = j$, implying $i > \nu$. If $\mathbf{1}_{j=0}\mathrm{lfn}(\mathfrak{b}, i+1) + \mathbf{1}_{j=1}\mathrm{lufn}(\mathfrak{b}, i+1) = 0$, then the statement follows by the same arguments made earlier. Otherwise, we can also use the previous same arguments since $\ell^{\mathfrak{b}}(i,j,k) > \lfloor(\mathfrak{b}+1-k)/2\rfloor$ implies $\phi^{\sigma}(e) = \lfloor(\mathfrak{b}+1-k)/2\rfloor$.

Now let $\nu = 1$ and $i \in [n], j, k \in \{0,1\}$. We again begin by proving the statement for all edges that are not of the type $(d_{*,*,*}, F_{*,*})$.

1. Consider edges of the type $(b_i, *)$. Since $\nu = 1$, the only such edge that was applied was (b_1, g_1). Let $e = (b_i, g_i)$ and $i \neq 1$. Then, $\phi^{\sigma}(e) = \phi^{\sigma_{\mathfrak{b}}}(e) = \mathrm{fl}(\mathfrak{b}, i) = \mathrm{fl}(\mathfrak{b}+1, i)$ by Table 5.6 and Lemma 6.2.6 as required.

   For $e = (b_i, b_{i+1})$ and $i \in [n]$, we have $\phi^{\sigma}(e) = \phi^{\sigma_{\mathfrak{b}}}(e) = \mathrm{fl}(\mathfrak{b}, i) - \mathfrak{b}_i$. If $i \neq 1$, then $\mathrm{fl}(\mathfrak{b}+1, i) = \mathrm{fl}(\mathfrak{b}, i)$ and $\mathfrak{b}_i = (\mathfrak{b}+1)_i$. If $i = 1$, then $\mathrm{fl}(\mathfrak{b}+1, i) = \mathrm{fl}(\mathfrak{b}, i) + 1$ and $\mathfrak{b}_i = 0$ as well as $(\mathfrak{b}+1)_i = 1$. In both cases, the occurrence record is described by Table 5.6.
2. Consider some edge $(g_i, F_{i,j})$ that was not applied. Then, the upper bound can only increase and thus remains valid.
3. Consider some vertex $s_{i,j}$. Then, since $\nu = 1$, no edge $(s_{*,*}, *)$ was switched. The statement then follows since $\mathrm{fl}(\mathfrak{b}, i+1) - (1-j)\mathfrak{b}_{i+1} = \mathrm{fl}(\mathfrak{b}+1, i+1) - (1-j)(\mathfrak{b}+1)_{i+1}$ and $\mathrm{fl}(\mathfrak{b}, i+1) - j \cdot \mathfrak{b}_{i+1} = \mathrm{fl}(\mathfrak{b}+1, i+1) - j(\mathfrak{b}+1)_{i+1}$.
4. Consider some edge $e = (e_{i,j,k}, b_2)$. Then, the statement follows directly as $\nu = 1$ implies $\phi^{\sigma}(e) = \phi^{\sigma_{\mathfrak{b}}}(e) = \lfloor \mathfrak{b}/2 \rfloor = \mathfrak{m}$.
5. Consider some edge of the type $e = (d_{i,j,k}, e_{i,j,k})$ that was not applied. Then, $\nu = 1$ implies that $\phi^{\sigma}(e) = \phi^{\sigma_{\mathfrak{b}}}(e) \leq \lceil \mathfrak{b}/2 \rceil = \lfloor(\mathfrak{b}+1)/2\rfloor = \lfloor \mathfrak{b}/2 \rfloor$. The upper bound is thus valid for σ since $\phi^{\sigma_{\mathfrak{b}}}(e_{i,j,k}, b_2) = \phi^{\sigma}(e_{i,j,k}, b_2)$,.

It remains to consider edges of the type $e = (d_{i,j,k}, F_{i,j})$ that were not applied. As the arguments used for proving this are similar to the ones used for the case $\nu > 1$, we defer the discussion of these edges to Appendix A.2.

Claim 22. Let $\nu = 1$. The occurrence records of edges of the type $(d_{*,*,*}, F_{*,*})$ not applied during $\sigma_{\mathfrak{b}} \to \sigma$ is described correctly by Table 5.6.

□

We now investigate the occurrence records of edges of the type $(g_*, F_{*,*})$ that were applied during $\sigma_{\mathfrak{b}} \to \sigma$. Determining the exact occurrence records of these edges is challenging as it is challenging to describe the exact conditions under which edges of these type become improving. In particular, these conditions depend on whether we consider the sink game S_n or the Markov decision process M_n, making it even harder to describe these in terms of the unified framework G_n. For these reasons, we prove that the upper bound on the occurrence records of these edges given in Table 5.6 remains valid for σ by an inductive argument as follows. We begin by determining the exact set of conditions under which the application of an improving switch $(g_*, F_{*,*})$ might yield an occurrence record that could violate Table 5.6. That is, we explicitly determine properties that σ needs to have for the upper bound to hold with equality. The idea of the proof is then to show that these conditions imply that there is an earlier canonical strategy σ' in which there was a slack between the upper bound of the occurrence record of $(g_*, F_{*,*})$ and the actual occurrence record of this edge. We then prove that this slack is still present when the strategy σ is reached, implying that the upper bound cannot hold with equality and remains valid.

For this reason, the proof itself is an inductive proof that requires us to consider up to 4 previous transitions and uses the statements of this section. As we always excluded the occurrence records of edges of the type $(g_*, F_{*,*})$ in these statements, we can in fact use them within the following induction. For example, we heavily use that each improving switch is applied at most once per transition. To keep the proof more readable and not refer to one of the previous lemmas in every second sentence, we do not always explicitly mention the lemma proving such a statement.

Lemma 6.3.24. *Let $\sigma_{\mathfrak{b}} \in \rho(\sigma_0)$ be a canonical strategy for $\mathfrak{b} \in \mathfrak{B}_n$ calculated by the strategy improvement resp. policy iteration algorithm having the canonical properties. Then $\phi^{\sigma_{\mathfrak{b}}}(g_i, F_{i,j}) \leq \min_{k\in\{0,1\}} \phi^{\sigma_{\mathfrak{b}}}(d_{i,j,k}, F_{i,j})$. In particular, Table 5.6 specifies the occurrence records of all edges of the type $(g_*, F_{*,*})$.*

Proof. Let $i \in \{n\}, j \in \{0,1\}$ be fixed and let $e := (g_i, F_{i,j})$ be an arbitrary but fixed edge of the type $(g_*, F_{*,*})$. We prove the statement via induction on $\mathfrak{b}$. We first consider the case $i \neq 1$ and discuss the case $i = 1$ later.

We begin by investigating the first transition in which e could have been applied. Thus, let $\mathfrak{b} \leq 2^i =: \mathfrak{t}$. Then $\mathfrak{t}_{i+1} = 1$ and for all $\mathfrak{d} \leq \mathfrak{t}$, it holds that $\mathfrak{d}_{i+1} = 0$. We prove that e was applied at most once when transitioning from σ_0 to $\sigma_{\mathfrak{t}}$ and that this application can only happen during $\sigma_{\mathfrak{t}-1} \to \sigma_{\mathfrak{t}}$. The statement then follows since the occurrence records of the cycle edges of $F_{i,j}$ are both at least one.

Since $\sigma_0(g_i) = F_{i,0}$, the switch e cannot have been applied during phase 1 of any transition encountered during the sequence $\sigma_0 \to \sigma_{\mathfrak{t}}$ as the choice of $\mathfrak{t}$ implies that there is no $\mathfrak{d} \leq \mathfrak{t}$ with $\mathfrak{d}_i = 1 \wedge \mathfrak{d}_{i+1} = 1$. It is also easy to show that this implies that it cannot happen that the cycle center $F_{i,j}$ was closed during phase 1 if $j = 1-\mathfrak{d}_{i+1}$ as the occurrence record of the cycle edges is too low. The switch $(g_i, F_{i,j})$ can thus only have been applied during some phase 5. However, since $\sigma_0(g_i) = F_{i,0}$ and due to the choice of $\mathfrak{t}$, this can only happen when transitioning from $\sigma_{\mathfrak{t}-1}$ to $\sigma_{\mathfrak{t}}$.

Thus, the statement holds for all canonical strategies $\sigma_{\mathfrak{b}}$ representing numbers $\mathfrak{b} \leq 2^i$. Now, assume that it holds for all $\mathfrak{b}' < \mathfrak{b}$ for an arbitrary but fixed $\mathfrak{b} > 2^i$. We prove that the statement also holds for $\sigma_{\mathfrak{b}}$. Consider the strategy $\sigma_{\mathfrak{b}-1}$. We begin by arguing that several cases do not need to be considered.

First of all, every improving switch is applied at most once in a single transition. If $\min_{k\in\{0,1\}} \phi^{\sigma_{\mathfrak{b}}}(d_{i,j,k}, F_{i,j}) > \min_{k\in\{0,1\}} \phi^{\sigma_{\mathfrak{b}-1}}(d_{i,j,k}, F_{i,j})$, then the statement thus follows by the induction hypothesis. We thus assume

$$\min_{k\in\{0,1\}} \phi^{\sigma_{\mathfrak{b}}}(d_{i,j,k}, F_{i,j}) = \min_{k\in\{0,1\}} \phi^{\sigma_{\mathfrak{b}-1}}(d_{i,j,k}, F_{i,j}). \tag{6.10}$$

Similarly, if e is not applied during $\sigma_{\mathfrak{b}-1} \to \sigma_{\mathfrak{b}}$, then the statement also follows by the induction hypothesis. We thus assume $e \in \mathfrak{A}_{\sigma_{\mathfrak{b}-1}}^{\sigma_{\mathfrak{b}}}$.

These observations give first structural insights on $\mathfrak{b}-1$ and $\mathfrak{b}$. First, if $\mathfrak{b}_i = 1 \wedge (\mathfrak{b}-1)_i = 1$, then it is not possible to apply e during $\sigma_{\mathfrak{b}-1} \to \sigma_{\mathfrak{b}}$. Second, if $\mathfrak{b}_i = 1 \wedge (\mathfrak{b}-1)_i = 0$, then $i = \ell(\mathfrak{b})$. By Definition 5.1.2 resp. 5.2.1, both cycle centers of level $\ell(\mathfrak{b})$ are open for $\sigma_{\mathfrak{b}-1}$. Hence, Corollary 6.3.2 implies that $F_{i,j}$ is closed during $\sigma_{\mathfrak{b}-1} \to \sigma_{\mathfrak{b}}$ by applying both switches $(d_{i,j,0}, F_{i,j})$ and $(d_{i,j,1}, F_{i,j})$. But then, Equation (6.10) is not fulfilled and the statement follows. This implies that it suffices to consider the case $\mathfrak{b}_i = 0$.

We now show that these three conditions imply that the occurrence record of the edges $(d_{i,j,*}, F_{i,j})$ is "large". To be precise, we prove that Equation (6.10), $e \in \mathfrak{A}_{\sigma_{\mathfrak{b}-1}}^{\sigma_{\mathfrak{b}}}$ and $\mathfrak{b}_i = 0$ imply

$$\min_{k\in\{0,1\}} \phi^{\sigma_{\mathfrak{b}-1}}(d_{i,j,k}, F_{i,j}) \geq \left\lfloor \frac{\mathfrak{b}}{2} \right\rfloor - 1. \tag{6.11}$$

It then suffices to prove $\phi^{\sigma_{\mathfrak{b}-1}}(g_i, F_{i,j}) < \lfloor \frac{\mathfrak{b}}{2} \rfloor - 1$ to complete the proof.

It is easy but tedious to verify that these conditions either already imply the desired inequality or give additional structural insights.

Claim 23. Equation (6.10), $e = (g_i, F_{i,j}) \in \mathfrak{A}_{\sigma_{\mathfrak{b}-1}}^{\sigma_{\mathfrak{b}}}$ and $\mathfrak{b}_i = 0$ either imply Inequality (6.11) directly or that exactly one of the cycle edges of $F_{i,j}$ is switched during $\sigma_{\mathfrak{b}-1} \to \sigma_{\mathfrak{b}}$.

Consequently, it suffices to consider the case that exactly one of the two cycle edges $(d_{i,j,0}, F_{i,j}), (d_{i,j,1}, F_{i,j})$ is applied during $\sigma_{\mathfrak{b}-1} \to \sigma_{\mathfrak{b}}$. However, by Equation (6.10), this implies that the the occurrence record of both edges $(d_{i,j,0}, F_{i,j}), (d_{i,j,1}, F_{i,j})$ is the same with respect to $\sigma_{\mathfrak{b}-1}$.

Now assume that $F_{i,j}$ is open or halfopen for $\sigma_{\mathfrak{b}-1}$. Then, there is at least one $k \in \{0,1\}$ with $\sigma_{\mathfrak{b}-1}(d_{i,j,k}) \neq F_{i,j}$. The statement thus follows since Property $(\text{OR4})_{i,j,k}$ implies $\phi^{\sigma_{\mathfrak{b}-1}}(d_{i,j,k}, F_{i,j}) \geq \lfloor \mathfrak{b}/2 \rfloor - 1$. Thus assume that $F_{i,j}$ is closed for $\sigma_{\mathfrak{b}-1}$. This implies that either $(\mathfrak{b}-1)_i = 1 \wedge (\mathfrak{b}-1)_{i+1} = j$ or $(\mathfrak{b}-1)_i = 0 \wedge (\mathfrak{b}-1)_{i+1} \neq j$ holds. In the first case, $\phi^{\sigma_{\mathfrak{b}-1}}(d_{i,j,k}, F_{i,j}) = \lfloor (\mathfrak{b} - 2^{i-1} + 2 - k)/2 \rfloor$ and $\ell(\mathfrak{b}) > 1$ need to hold. This implies that $\mathfrak{b} - 2^{i-1} + 2$ is even as we have $i \neq 1$ by assumption. But this implies that we have $\phi^{\sigma_{\mathfrak{b}-1}}(d_{i,j,1}, F_{i,j}) < \phi^{\sigma_{\mathfrak{b}-1}}(d_{i,j,0}, F_{i,j})$ which is a contradiction.

We thus need to have $(\mathfrak{b}-1)_i = 0 \wedge (\mathfrak{b}-1)_{i+1} \neq j$. Then, since $F_{i,j}$ is closed, Property $(\text{OR2})_{i,j,*}$ implies $\ell^{\mathfrak{b}-1}(i,j,0) = \ell^{\mathfrak{b}-1}(i,j,1)$ since

$$\phi^{\sigma_{\mathfrak{b}-1}}(d_{i,j,0}, F_{i,j}) = \ell^{\mathfrak{b}-1}(i,j,0) + 1 = \ell^{\mathfrak{b}-1}(i,j,1) + 1 = \phi^{\sigma_{\mathfrak{b}-1}}(d_{i,j,1}, F_{i,j}).$$

But this implies $\lceil (\mathfrak{b} - 1 - 2^{i-1} + \sum(\mathfrak{b}-1, i))/2 \rceil = \lceil (\mathfrak{b} - 2^{i-1} + \sum(\mathfrak{b}-1, i))/2 \rceil$. Since $\mathfrak{b} - 2^{i-1} + \sum(\mathfrak{b}-1, i)$ is always odd due to $i \neq 1$, this is however not possible.

This concludes this part of the proof. Since $\min_{k\in\{0,1\}} \phi^{\sigma_{\mathfrak{b}-1}}(d_{i,j,k}, F_{i,j}) \geq \lfloor \mathfrak{b}/2 \rfloor - 1$, it suffices to prove $\phi^{\sigma_{\mathfrak{b}-1}}(g_i, F_{i,j}) < \lfloor \mathfrak{b}/2 \rfloor - 1$ under the given conditions.

We begin by stating one more structural insight.

Claim 24. Assume that Equation (6.10), $e = (g_i, F_{i,j}) \in \mathfrak{A}^{\sigma_{\mathfrak{b}}}_{\sigma_{\mathfrak{b}-1}}, \mathfrak{b}_i = 0$ hold and that exactly one of the two cycle edges $(d_{i,j,0}, F_{i,j}), (d_{i,j,1}, F_{i,j})$ is applied during $\sigma_{\mathfrak{b}-1} \to \sigma_{\mathfrak{b}}$. Then $(\mathfrak{b} - 1)_i = 0$.

To simplify notation, we denote the binary number obtained by subtracting 1 from a binary number $(\mathfrak{b}'_n, \dots, \mathfrak{b}'_1)$ by $[\mathfrak{b}'_n, \dots, \mathfrak{b}'_1] - 1$. Then, $\mathfrak{b}$ and $\mathfrak{b} - 1$ can be represented as

$$\begin{aligned} \mathfrak{b} &= (\mathfrak{b}_n, \dots, \mathfrak{b}_{i+1}, \mathbf{0}, \mathfrak{b}_{i-1}, \dots, \mathfrak{b}_1), \\ \mathfrak{b} - 1 &= (\mathfrak{b}_n, \dots, \mathfrak{b}_{i+1}, \mathbf{0}, [\mathfrak{b}_{i-1}, \dots, \mathfrak{b}_1] - 1) \end{aligned}$$

where bit i is marked in **bold**. The idea of the proof is now the following. We define two smaller numbers that are relevant for the application of $(g_i, F_{i,j})$. We use these numbers and the induction hypothesis to prove that even if $(g_i, F_{i,j})$ was applied during the maximum number of transitions, the claimed bound still holds.

We define

$$\begin{aligned} \bar{\mathfrak{b}} &:= ([\mathfrak{b}_n, \dots, \mathfrak{b}_{i+1}] - 1, \mathbf{1}, 1 \dots, 1) \\ \widehat{\mathfrak{b}} &:= ([\mathfrak{b}_n, \dots, \mathfrak{b}_{i+1}] - 1, \mathbf{1}, 0, \dots, 0) \end{aligned}$$

where bit i is again marked in **bold**. These numbers are well-defined since $\mathfrak{b} \geq 2^i$.

Consider $\widehat{\mathfrak{b}}$. Let $\mathfrak{N}(\widehat{\mathfrak{b}}, \mathfrak{b} - 1)$ denote the number of applications of $(g_i, F_{i,j})$ when transitioning from $\sigma_{\widehat{\mathfrak{b}}}$ to $\sigma_{\mathfrak{b}-1}$. Then, since $\mathfrak{b}'_i = 1$ for all $\mathfrak{b}' \in \{\widehat{\mathfrak{b}}, \dots, \bar{\mathfrak{b}}\}$, we have $\mathfrak{N}(\widehat{\mathfrak{b}}, \mathfrak{b} - 1) = \mathfrak{N}(\bar{\mathfrak{b}}, \mathfrak{b} - 1)$. We thus can describe the occurrence record of $(g_i, F_{i,j})$ as

$$\phi^{\sigma_{\mathfrak{b}-1}}(g_i, F_{i,j}) = \mathfrak{N}(0, \mathfrak{b} - 1) = \mathfrak{N}(0, \widehat{\mathfrak{b}}) + \mathfrak{N}(\widehat{\mathfrak{b}}, \mathfrak{b} - 1) = \phi^{\sigma_{\widehat{\mathfrak{b}}}}(g_i, F_{i,j}) + \mathfrak{N}(\bar{\mathfrak{b}}, \mathfrak{b} - 1).$$

Our goal is to bound the two terms on the right-hand side. Using the induction hypothesis and that $\widehat{\mathfrak{b}}$ is even, it is easy to verify that the first term can be bounded by $\lfloor \widehat{\mathfrak{b}}/2 \rfloor$. Since every improving switch is applied at most once per transition by Corollary 6.3.20, we have $\mathfrak{N}(\bar{\mathfrak{b}}, \mathfrak{b} - 1) \leq (\mathfrak{b} - 1) - \bar{\mathfrak{b}}$. However, this upper bound is not strong enough. To improve this bound, we now distinguish between when exactly $(g_i, F_{i,j})$ is applied during $\sigma_{\mathfrak{b}-1} \to \sigma_{\mathfrak{b}}$. Note that we refer to even earlier transitions in the last statement of the following claim.

Claim 25. Assume that Equation (6.10), $e = (g_i, F_{i,j}) \in \mathfrak{A}^{\sigma_{\mathfrak{b}}}_{\sigma_{\mathfrak{b}-1}}, \mathfrak{b}_i = 0$ hold and that exactly one of the two cycle edges $(d_{i,j,0}, F_{i,j}), (d_{i,j,1}, F_{i,j})$ is applied during $\sigma_{\mathfrak{b}-1} \to \sigma_{\mathfrak{b}}$. If $(g_i, F_{i,j})$ is applied during phase 1 of $\sigma_{\mathfrak{b}-1} \to \sigma_{\mathfrak{b}}$, then

1. $\mathfrak{b}$ is even and $i \neq 2$,
2. $\sum(\mathfrak{b}, i) = 2^{i-1} - 2$ and
3. if $(g_i, F_{i,j}) \in \mathfrak{A}^{\sigma_{\mathfrak{b}-1}}_{\sigma_{\mathfrak{b}-2}}$, then $(g_i, F_{i,j}) \notin \mathfrak{A}^{\sigma_{\mathfrak{b}-2}}_{\sigma_{\mathfrak{b}-3}}$.

If $(g_i, F_{i,j})$ was applied during phase 1, then the last statement of Claim 25 implies $\mathfrak{N}(\bar{\mathfrak{b}}, \mathfrak{b}-1) \leq (\mathfrak{b}-1) - \bar{\mathfrak{b}} - 1$. Combining these results and using $\bar{\mathfrak{b}} = \mathfrak{b} - \sum(\mathfrak{b}, i) - 1$ and $\widehat{\mathfrak{b}} = \mathfrak{b} - \sum(\mathfrak{b}, i) - 2^{i-1}$ yields the statement as

$$\begin{aligned}
\phi^{\sigma_{\mathfrak{b}-1}}(g_i, F_{i,j}) &= \phi^{\sigma_{\widehat{\mathfrak{b}}}}(g_i, F_{i,j}) + \mathfrak{N}(\bar{\mathfrak{b}}, \mathfrak{b}-1) \\
&\leq \left\lfloor \frac{\widehat{\mathfrak{b}}}{2} \right\rfloor + (\mathfrak{b}-1) - \bar{\mathfrak{b}} - 1 = \left\lfloor \frac{\widehat{\mathfrak{b}}}{2} \right\rfloor + \mathfrak{b} - \bar{\mathfrak{b}} - 2 \\
&= \left\lfloor \frac{\mathfrak{b} - \sum(\mathfrak{b}, i) - 2^{i-1}}{2} \right\rfloor + \mathfrak{b} - \mathfrak{b} + \sum(\mathfrak{b}, i) + 1 - 2 \\
&= \left\lfloor \frac{\mathfrak{b} - 2^{i-1} + \sum(\mathfrak{b}, i)}{2} \right\rfloor - 1 = \left\lfloor \frac{\mathfrak{b} - 2^{i-1} + 2^{i-1} - 2}{2} \right\rfloor - 1 \\
&= \left\lfloor \frac{\mathfrak{b} - 2}{2} \right\rfloor - 1 = \left\lfloor \frac{\mathfrak{b}}{2} \right\rfloor - 2 < \left\lfloor \frac{\mathfrak{b}}{2} \right\rfloor - 1.
\end{aligned}$$

This proves the statement if $(g_i, F_{i,j})$ was applied during phase 1 of $\sigma_{\mathfrak{b}-1} \to \sigma_{\mathfrak{b}}$.

Hence assume that $e = (g_i, F_{i,j})$ was applied during phase 5 of $\sigma_{\mathfrak{b}-1} \to \sigma_{\mathfrak{b}}$. Let σ denote the phase-5-strategy in which e is applied. Then, $\bar{\sigma}(g_i) = 1 - j$ needs to hold. Consequently, either $\bar{\sigma}_{\mathfrak{b}-1}(g_i) = 1 - j$ or $\bar{\sigma}_{\mathfrak{b}-1}(g_i) = j$ and $(g_i, F_{i,1-j}) \in \mathfrak{A}_{\sigma_{\mathfrak{b}-1}}^{\sigma}$. We thus distinguish between these two cases. In the first case, the following statement similar to Claim 25 can be shown.

Claim 26. Assume that Equation (6.10), $e = (g_i, F_{i,j}) \in \mathfrak{A}_{\sigma_{\mathfrak{b}-1}}^{\sigma_{\mathfrak{b}}}, \mathfrak{b}_i = 0$ hold and that exactly one of the two cycle edges $(d_{i,j,0}, F_{i,j}), (d_{i,j,1}, F_{i,j})$ is applied during $\sigma_{\mathfrak{b}-1} \to \sigma_{\mathfrak{b}}$. If $(g_i, F_{i,j})$ is applied during phase 5 of $\sigma_{\mathfrak{b}-1} \to \sigma_{\mathfrak{b}}$ and $\bar{\sigma}_{\mathfrak{b}-1}(g_i) = 1 - j$, then

1. $i \neq 2$,
2. $\sum(\mathfrak{b}, i) = 2^{i-1} - 2$ and
3. if $(g_i, F_{i,j}) \in \mathfrak{A}_{\sigma_{\mathfrak{b}-2}}^{\sigma_{\mathfrak{b}-1}}$, then$(g_i, F_{i,j}) \notin \mathfrak{A}_{\sigma_{\mathfrak{b}-3}}^{\sigma_{\mathfrak{b}-2}}$.

The statement thus follows analogously.

Thus, assume $\bar{\sigma}_{\mathfrak{b}-1}(g_i) = j$ and $(g_i, F_{i,1-j}) \in \mathfrak{A}_{\sigma_{\mathfrak{b}-1}}^{\sigma_{\mathfrak{b}}}$. Since at most one improving switch involving a selection vertex is applied during phase 5, this implies that $(g_i, F_{i,1-j})$ was applied during phase 1 of $\sigma_{\mathfrak{b}-1} \to \sigma_{\mathfrak{b}}$. It could technically also be applied at the beginning of phase 2 resp. 3 when closing the final cycle center only creates a pseudo phase-2 resp. pseudo phase-3-strategy. For clarity of presentation, we include this case in the second case and interpret the application of this improving switch as being part of phase 1. In particular, we thus have $1 - j = 1 - \mathfrak{b}_{i+1}$ resp. $j = \mathfrak{b}_{i+1}$ as $(\mathfrak{b}-1)_i = \mathfrak{b}_i = 0$ implies $i \neq \ell(\mathfrak{b})$. We prove that we need to have $i \neq 2$.

For the sake of a contradiction, assume $i = 2$. Then, since $(\mathfrak{b}-1)_i = 0$, we have $\mathfrak{b}_1 = 1$ and $\mathfrak{b} - 2 = \bar{\mathfrak{b}}$. Consequently, $1 - j = 1 - \mathfrak{b}_3 = (\mathfrak{b}-2)_3$. Thus, the cycle center $F_{i,1-j}$ is active and closed with respect to $\sigma_{\mathfrak{b}-2}$. As $\mathfrak{b}$ is odd, this implies

$$\phi^{\sigma_{\mathfrak{b}-2}}(d_{i,1-j,k}, F_{i,1-j}) = \left\lfloor \frac{\operatorname{lfn}(\mathfrak{b}-2, i, \{(i+1, (\mathfrak{b}-2)_3)\}) - k}{2} \right\rfloor + 1$$

$$= \left\lfloor \frac{\mathfrak{b}-2-1-k}{3} \right\rfloor + 1 = \left\lfloor \frac{\mathfrak{b}-1-k}{2} \right\rfloor = \left\lfloor \frac{\mathfrak{b}-1}{2} \right\rfloor - k.$$

Since the cycle center is closed with respect to $\sigma_{\mathfrak{b}-2}$, none of these two edges is applied as improving switch during phase 1 of $\sigma_{\mathfrak{b}-2} \to \sigma_{\mathfrak{b}-1}$. However, since $\ell(\mathfrak{b}-1) > 1$, the switches are also not applied during phase 5 of that transition. But this implies $\sigma_{\mathfrak{b}-1}(d_{i,1-j,k}) \neq F_{i,1-j}$ for both $k \in \{0,1\}$, contradicting that $(g_i, F_{i,1-j})$ is applied during phase 1 of $\sigma_{\mathfrak{b}-1} \to \sigma_{\mathfrak{b}}$. Note that this argument further proves that we cannot have $\mathfrak{b}-2 = \bar{\mathfrak{b}}$.

We can thus assume $i > 2$ and $\mathfrak{b}-3 \geq \bar{\mathfrak{b}}$. Since we apply $(g_i, F_{i,1-j})$ during phase 1 of $\sigma_{\mathfrak{b}-1} \to \sigma_{\mathfrak{b}}$, we can use the same arguments used when proving Claim 25 resp. 26 to prove $\sum(\mathfrak{b}, i) \leq 2^{i-1} - 2$. Similar to the last cases, we prove that there is at least one transition between $\bar{\mathfrak{b}}$ and $\mathfrak{b}-1$ in which the switch $(g_i, F_{i,j})$ is not applied. As this follows if $(g_i, F_{i,j}) \notin \mathfrak{A}_{\sigma_{\mathfrak{b}-2}}^{\sigma_{\mathfrak{b}-1}}$, assume $(g_i, F_{i,j}) \in \mathfrak{A}_{\sigma_{\mathfrak{b}-2}}^{\sigma_{\mathfrak{b}-1}}$.

First, since $\mathfrak{b}-3 \geq \bar{\mathfrak{b}}$ and since we apply $(g_i, F_{i,j})$ in phase 5 of $\sigma_{\mathfrak{b}-1} \to \sigma_{\mathfrak{b}}$, it holds that $j = \mathfrak{b}_{i+1} = (\mathfrak{b}-1)_{i+1} = (\mathfrak{b}-2)_{i+1}$. This implies that $(g_i, F_{i,j})$ was not applied during phase 1 of $\sigma_{\mathfrak{b}-3} \to \sigma_{\mathfrak{b}-2}$. The reason is that this could only happen if $i = \ell(\mathfrak{b}-2)$, contradicting $(\mathfrak{b}-2)_i = 0$, or if $(\mathfrak{b}-3)_i = 0 \wedge j \neq (\mathfrak{b}-3)_{i+1}$. However, this then contradicts the previous identities regarding j. Thus, $(g_i, F_{i,j})$ is applied during phase 5 of $\sigma_{\mathfrak{b}-3} \to \sigma_{\mathfrak{b}-2}$.

For the sake of a contradiction, assume $(\mathfrak{b}-3)_i = 1$. Then, $\mathfrak{b}-3 = \bar{\mathfrak{b}}$, implying $(\mathfrak{b}-3)_{i+1} \neq j$. This further implies $(\mathfrak{b}-3)_{i'} = 1$ for all $i' \leq i$. Then, since $F_{i,1-j}$ is closed with respect to $\sigma_{\mathfrak{b}-3}$ and $i \geq 3$, it is easy to calculate that we then need to have $\phi^{\sigma_{\mathfrak{b}-3}}(d_{i,1-j,k}, F_{i,1-j}) < \lfloor(\mathfrak{b}-3+1)/2\rfloor - 1$. But, since $\ell(\mathfrak{b}-2) > 1$, this implies that both of these edges are applied at the beginning of phase 5 of $\sigma_{\mathfrak{b}-3} \to \sigma_{\mathfrak{b}-2}$. Thus, $F_{i,1-j}$ is closed at the beginning of phase 5 of $\sigma_{\mathfrak{b}-3} \to \sigma_{\mathfrak{b}-2}$, contradicting the assumption that $(g_i, F_{i,j})$ is applied during phase 5 of $\sigma_{\mathfrak{b}-3} \to \sigma_{\mathfrak{b}-2}$, see Lemma 6.2.32.

Thus assume $(\mathfrak{b}-3)_i = 0 \wedge (\mathfrak{b}-3)_{i+1} = j$. This implies $\mathfrak{b}-4 \geq \bar{\mathfrak{b}}$ and that the transition from $\sigma_{\mathfrak{b}-4}$ to $\sigma_{\mathfrak{b}-3}$ is thus part of the currently considered sequence of transitions. Then, since $(g_i, F_{i,j})$ is applied during both $\sigma_{\mathfrak{b}-3} \to \sigma_{\mathfrak{b}-2}$ and $\sigma_{\mathfrak{b}-2} \to \sigma_{\mathfrak{b}-1}$, the improving switch $(g_i, F_{i,1-j})$ has to be applied in between. This switch can only be applied during phase 1 of $\sigma_{\mathfrak{b}-2} \to \sigma_{\mathfrak{b}-1}$. It is easy to see that this implies that there is a $k \in \{0,1\}$ such that $\phi^{\sigma_{\mathfrak{b}-2}}(d_{i,1-j,k}, F_{i,1-j}) = \ell^{\mathfrak{b}-2}(i, 1-j, k) + 1 \leq \lfloor(\mathfrak{b}-1)/2\rfloor - 1$ and $\phi^{\sigma_{\mathfrak{b}-2}}(d_{i,1-j,1-k}, F_{i,1-j}) = \lfloor(\mathfrak{b}-1)/2\rfloor - 1$. Using $\sum(\mathfrak{b}, i) \leq 2^{i-1} - 2$, it is then easy to verify that $\ell^{\mathfrak{b}-3}(i, 1-j, k') \leq \lfloor(\mathfrak{b}-k')/2\rfloor - 3$ for $k' \in \{0,1\}$.

This implies that we need to have $\phi^{\sigma_{\mathfrak{b}-3}}(d_{i,1-j,1}, F_{i,1-j}) = \ell^{\mathfrak{b}-3}(i, 1-j, 1) + 1$ and that the edge $(d_{i,1-j,1}, F_{i,1-j})$ is applied as improving switch during phase 5 of $\sigma_{\mathfrak{b}-3} \to \sigma_{\mathfrak{b}-2}$. We thus need to have $k = 1$. Consider $(d_{i,1-j,0}, F_{i,1-j})$. Then,

$$\ell^{\mathfrak{b}-3}(i, 1-j, 0) = \left\lfloor \frac{\mathfrak{b}}{2} \right\rfloor - 3 = \left\lfloor \frac{\mathfrak{b}-2}{2} \right\rfloor - 2 < \left\lfloor \frac{\mathfrak{b}-2}{2} \right\rfloor - 1 = \left\lfloor \frac{\mathfrak{b}-3+1}{2} \right\rfloor - 1.$$

By Property $(\text{OR4})_{i,1-j,0}$, we thus need to have $\sigma_{\mathfrak{b}-3}(d_{i,1-j,0}) = F_{i,1-j}$. In particular, it implies that $F_{i,1-j}$ is closed for $\sigma_{\mathfrak{b}-3}$ and thus $\sigma_{\mathfrak{b}-3}(g_i) = F_{i,1-j}$.

This now enables us to show that the edge $(g_i, F_{i,j})$ was not applied as improving switch during the transition $\sigma_{\mathfrak{b}-4} \to \sigma_{\mathfrak{b}-3}$. Independent on whether $\mathfrak{b}-4 = \bar{\mathfrak{b}}$ or $\mathfrak{b}-4 \neq \bar{\mathfrak{b}}$, the switch was not applied during phase 5 of $\sigma_{\mathfrak{b}-4} \to \sigma_{\mathfrak{b}-3}$ as we have $\sigma_{\mathfrak{b}-3}(g_i) = F_{i,1-j}$. If $(\mathfrak{b}-4)_i = 0 \wedge (\mathfrak{b}-4)_{i+1} = j$, then it also follows directly that the switch was not applied during phase 1 of that transition. Thus consider the case $(\mathfrak{b}-4)_i = 1 \wedge (\mathfrak{b}-4)_{i+1} \neq j$, implying $\mathfrak{b}-4 = \bar{\mathfrak{b}}$. But this immediately implies that the switch is not applied during phase 1.

This concludes the proof for the case that $(g_i, F_{i,j})$ was applied during phase 5 of $\sigma_{\mathfrak{b}-1} \to \sigma_{\mathfrak{b}}$ and thus concludes the proof for $i \geq 2$.

It remains to consider the case $i = 1$. We prove the statement again via induction on $\mathfrak{b}$. It is easy to verify that the statement hold for both $G_n = S_n$ and $G_n = M_n$ for the canonical strategies $\sigma_0, \sigma_1, \sigma_2$. Hence let $\mathfrak{b} > 2$ and assume that the statement holds for all $\mathfrak{b}' < \mathfrak{b}$. We show that the statement then also holds for $\mathfrak{b}$.

An improving switch $(g_1, F_{1,j})$ can only be applied during phase 1 if $1 = \ell(\mathfrak{b})$ and $j = \mathfrak{b}_2$. Since we then have $(\mathfrak{b}-1)_i = 0$, both edges $(d_{1,\mathfrak{b}_2,0}, F_{1,\mathfrak{b}_2})$ and $(d_{1,\mathfrak{b}_2,1}, F_{1,\mathfrak{b}_2})$ are switched during phase 1 of $\sigma_{\mathfrak{b}-1} \to \sigma_{\mathfrak{b}}$. Thus, the statement follows by the induction hypothesis.

Thus consider the case $(\mathfrak{b}-1)_i = 1$. Then, a switch $(g_1, F_{1,j})$ can only be applied in phase 5. Consider the case $G_n = S_n$ first. Then, by the conditions of the application of such a switch in phase 5, we need to have $j = 0$ and $\sigma_{\mathfrak{b}-1}(g_1) = 1$. Since $(\mathfrak{b}-1)_1 = 1$ this implies $\mathfrak{b} = (\dots, 0, 0), \mathfrak{b}-1 = (\dots, 1, 1)$ and $\mathfrak{b}-2 = (\dots, 1, 0)$. It follows directly that $(g_1, F_{1,0})$ is not applied during the transition $\sigma_{\mathfrak{b}-2} \to \sigma_{\mathfrak{b}-1}$. However, during phase 1 of both transitions $\sigma_{\mathfrak{b}-2} \to \sigma_{\mathfrak{b}-1}$, exactly one of the cycle edges of $F_{1,0}$ is switched. Using the induction hypothesis, this implies the statement. If $G_n = M_n$, then we then need to have $j = \mathfrak{b}_2$ and $\sigma_{\mathfrak{b}-1}(g_1) = 1 - \mathfrak{b}_2$. If $j = 0$, then the statement follows by the exact same arguments used for the case $G_n = S_n$. If $j = 1$, it follows by similar arguments.

□

We can now prove the statements of Section 5.3. For convenience, we restate these statements before proving them.

We begin by showing that applying the improving switches according to Zadeh's pivot rule and the tie-breaking rule of Definition 5.3.5 yields the strategies described by Tables 5.8 and 5.9.

Lemma 5.3.18. *Let $\sigma_{\mathfrak{b}} \in \rho(\sigma_0)$ be a canonical strategy for $\mathfrak{b} \in \mathfrak{B}_n$ having the canonical properties. Let σ be a strategy obtained by applying a sequence of improving switches to $\sigma_{\mathfrak{b}}$. Let σ and I_σ have the properties of row k of Table 5.8 and 5.9 for some $k \in [5]$. Then, applying improving switches according to Zadeh's pivot rule and the tie-breaking rule of Definition 5.3.5 produces a strategy σ' that is described by the next feasible rows of Tables 5.8 and 5.9.*

Proof. By Lemma 6.3.14, applying improving switches to σ produces a canonical strategy $\sigma_{\mathfrak{b}+1}$ for $\mathfrak{b}+1$. When proving this lemma, we proved that the algorithm produces the intermediate strategies as described by the corresponding rows of Tables 5.8 and 5.9. More precisely, this follows from Lemmas 6.3.1, 6.3.6, 6.3.9 and 6.3.12.

By Lemma 6.3.14, $I_{\sigma_{\mathfrak{b}+1}} = \{(d_{i,j,k}, F_{i,j}) : \sigma_{\mathfrak{b}+1} \neq F_{i,j}\}$. In particular, this set is described as specified by Table 5.9. It thus remains to prove that $\sigma_{\mathfrak{b}+1}$ has the canonical properties. More precisely, we prove the following three statements:

1. The occurrence records $\phi^{\sigma_{\mathfrak{b}+1}}$ are described correctly by Table 5.6: This follows from Lemmas 6.3.23 and 6.3.24.
2. $\sigma_{\mathfrak{b}+1}$ has Properties (OR1)$_{*,*,*}$ to (OR4)$_{*,*,*}$: This follows from Lemma 6.3.22.
3. Any improving switch was applied at most once per previous transition between canonical strategies: This follows from Corollary 6.3.20.

Thus, $\sigma_{\mathfrak{b}+1}$ is a canonical strategy for $\mathfrak{b}+1$ and has the canonical properties. □

This now immediately implies the following theorem of Section 5.3, stating that applying improving switches to a canonical strategy for $\mathfrak{b}$ having the canonical properties produces such a strategy for $\mathfrak{b}+1$.

Theorem 5.3.19. *Let $\sigma_{\mathfrak{b}} \in \rho(\sigma_0)$ be a canonical strategy for $\mathfrak{b} \in \mathfrak{B}_n$ having the canonical properties. After applying finitely many improving switches according to Zadeh's pivot rule and the tie-breaking rule of Definition 5.3.5, the strategy improvement algorithm calculates a strategy $\sigma_{\mathfrak{b}+1}$ with the following properties.*

1. *$I_{\sigma_{\mathfrak{b}+1}} = \{(d_{i,j,k}, F_{i,j}) : \sigma_{\mathfrak{b}+1}(d_{i,j,k}) \neq F_{i,j}\}$.*
2. *The occurrence records are described by Table 5.6 when interpreted for $\mathfrak{b}+1$.*
3. *$\sigma_{\mathfrak{b}+1}$ is a canonical strategy for $\mathfrak{b}+1$ and has Properties (OR1)$_{*,*,*}$ to (OR4)$_{*,*,*}$.*
4. *When transitioning from $\sigma_{\mathfrak{b}}$ to $\sigma_{\mathfrak{b}+1}$, every improving switch is applied at most once.*

In particular, $\sigma_{\mathfrak{b}+1}$ has the canonical properties.

This now enables us to prove the remaining statements of Section 5.3 simultaneously.

Theorem 5.3.10. *Let $\sigma_{\mathfrak{b}}$ be a canonical strategy for $\mathfrak{b} \in \mathfrak{B}_n$ and assume that the improving switches are applied as described in Section 5.3.3. Then, Table 5.6 describes the occurrence records of all edges $e \in E_0$ with respect to $\sigma_{\mathfrak{b}}$.*

Theorem 5.3.20. *Using Zadeh's pivot rule and the tie-breaking rule of Definition 5.3.5 when applying*

1. *the strategy improvement algorithm of [VJ00] to S_n,*
2. *the policy iteration algorithm of [How60] to M_n,*
3. *the simplex algorithm of [Dan51] to the linear program induced by M_n*

to the game G_n or the induced linear program requires at least 2^n iterations for finding the optimal strategy resp. solution when using σ_0 as initial strategy.

Proof. By Observation 5.3.2 and Lemma 5.3.13, the initial strategy σ_0 is a canonical strategy representing 0 having the canonical properties. In addition, by Lemma 5.3.3 it is a sink strategy for S_n and a weak unichain policy for M_n.

By Theorem 5.3.19, applying improving switches to σ_0 yields a canonical strategy σ_1 representing the number 1. Also, the occurrence record of the edges is described correctly by Table 5.6 for σ_1. In particular, Theorem 5.3.19 can be applied to σ_1 again, yielding a canonical strategy σ_2 representing the number 2.

This argument can now be applied iteratively. Thus, applying improving switches according to Zadeh's pivot rule and the tie-breaking rule defined in Definition 5.3.5

produces the strategies $\sigma_0, \sigma_1, \ldots, \sigma_{2^n-1}$. By Theorem 5.3.19, Table 5.6 describes the occurrence records for each of these strategies, implying Theorem 5.3.10. Since these are 2^n different strategies, since G_n has size $\mathcal{O}(n)$ and all rewards and probabilities (for $G_n = M_n$) and priorities (for $G_n = S_n$) can be encoded using a polynomial number of bits, this implies the exponential lower bound for the strategy improvement and policy iteration algorithm. By Corollary 3.3.5, there is a linear program such that the simplex algorithm using the same pivot and tie-breaking rule requires the same number of iterations as it requires for M_n. Consequently, the lower bound also applies to the simplex algorithm. □

7. Conclusion

In this thesis, we considered the general framework of strategy improvement and its application to parity games and Markov decision processes. We discussed the connection between two subclasses of these frameworks, sink games and weakly unichain Markov decision processes, and investigated the connection of the latter to linear programming. Our main focus was Zadeh's LeastEntered pivot rule, and we considered the strategy improvement algorithm, the policy iteration algorithm and the simplex algorithm when parameterized with this pivot rule.

We began by introducing parity games, Markov decision processes and linear programs. We introduced the abstract concepts of valuations for parity games and Markov decision processes and considered special subclasses of parity games and Markov decision processes afterwards. More precisely, we analyzed the classes of sink games and weakly unichain Markov decision processes in detail, redeveloped previously used definitions and provided a clean framework for using these two classes. In addition, we revisited the connection between Markov decision processes and linear programs by providing a linear program and discussing the connection between Howard's policy iteration algorithm and Dantzig's simplex algorithm.

The first major contribution of this thesis was a detailed exposition of Friedmann's subexponential lower bound construction for Zadeh's pivot rule. This famous construction was discussed in detail, and we showed that it belongs to a new class of lower bound constructions based on the connection between Markov decision processes and linear programs. We then pointed out that there are several smaller and one major issue regarding the original analysis. Although the major issue requires a significant change regarding the application of the improving switches, we resolved all of the issues without affecting the macroscopic structure of Friedmann's example, and were able to retain his original result.

The second main and major contribution was the development of a new lower bound construction, based on the same connection between Markov decision processes and linear programs. More precisely, we presented the first exponential lower bound for Zadeh's LeastEntered pivot rule for all of the discussed algorithms. This in particular answered the question whether one of the classic deterministic pivot rule admits a subexponential worst-case running time, a question that remained open even after Friedmann's result. This construction implements the same key ideas as Friedmann's example but only requires linear space. The example was designed such that a single construction and description could be used for both sink games and weakly unichain Markov decision process, allowing us to unify most of the proofs and statements. We believe that our framework is applicable to previous and future constructions as well, and that it might even be possible to define a class of sink games and weakly unichain Markov decision processes on which the strategy improvement and policy iteration algorithm behave exactly the same. In addition, our findings were verified for sink games and small instances using an implementation of Friedmann.

Of course, there are still many open questions and mysteries. It is still unknown whether there is a pivot rule guaranteeing a polynomial number of iterations for any of the algorithms. Also, the construction presented in this thesis highly depends on the chosen tie-breaking rule. Although it would be extremely challenging to find an example that provides a superpolynomial lower bound independent of the chosen tie-breaking rule, it might be possible to find at least examples that rule out full classes of tie-breaking rules.

On a more abstract level, it is still an open question whether there is an even closer connection between sink games and weakly unichain Markov decision processes. Although our own, as well as previously developed constructions, use well-known similarities and connections between these frameworks, there is still no canonical transformation between subclasses of parity games and Markov decision processes. Such a common subclass would allow to phrase and analyze several algorithmic problems and questions from both game theory and linear programming in a single context.

In addition, it remains unclear whether the observation that all of the lower bound constructions using Markov decision processes can be formalized. As all of them implement the same key idea of implementing a binary counter and share several similarities, it might be possible to prove that all of these constructions are just special cases of some general binary counting Markov decision process. Such a general lower bound construction would not only simplify and unify all of the new lower bound constructions, but would potentially also allow for easily constructing new lower bound examples.

Bibliography

[AC78] D. Avis and V. Chvátal. Notes on Bland's pivoting rule. In M. L. Balinski and A. J. Hoffman, editors, *Polyhedral Combinatorics: Dedicated to the memory of D.R. Fulkerson*, volume 8 of *Mathematical Programming Studies*, pages 24–34. Springer-Verlag Berlin Heidelberg, 1978.

[AF17] D. Avis and O. Friedmann. An exponential lower bound for Cunningham's rule. *Mathematical Programming*, 161(1–2):271–305, January 2017. Springer-Verlag Berlin Heidelberg.

[APR14] I. Adler, C. H. Papadimitriou, and A. Rubinstein. On Simplex Pivoting Rules and Complexity Theory. In J. Lee and J. Vygen, editors, *Proceedings of the 17th International Conference on Integer Programming and Combinatoral Optimization (IPCO)*, volume 8494 of *Lecture Notes in Computer Science*, pages 13–24. Springer-Verlag Berlin Heidelberg, 2014.

[Avi09] D. Avis. Postscript to "What is the Worst Case Behavior of the Simplex Algorithm?". In D. Avis, D. Bremner, and A. Deza, editors, *Polyhedral Computation*, volume 48 of *CRM Proceedings & Lecture notes*, pages 145–148. American Mathematical Society, 2009.

[AVW03] A. Arnold, A. Vincent, and I. Walukiewicz. Games for synthesis of controllers with partial observation. *Theoretical Computer Science*, 303(1):7–34, June 2003.

[AZ98] N. Amenta and G. M. Ziegler. Deformed products and maximal shadows of polytopes. In B. Chazelle, J. E. Goodman, and R. Pollack, editors, *Advances in Discrete and Computational Geometry*, volume 223 of *Contemporary Mathematics*, pages 57–90. American Mathematical Society, 1998.

[Bel57] R. Bellman. *Dynamic Programming*. Princeton University Press, Princeton, NJ, USA, 1st edition, 1957.

[Bla77] R. G. Bland. New finite pivoting rules for the simplex method. *Mathematics of Operations Research*, 2:103–107, 1977.

[Bor87] K. H. Borgwardt. The shadow-vertex algorithm. In *The Simplex Method: A Probabilistic Analysis*, pages 62–111. Springer-Verlag Berlin Heidelberg, 1987.

[BT97] D. Bertsimas and J. N. Tsitsiklis. *Introduction to Linear Optimization*, volume 6 of *Athena Scientific Series in Optimization and Neural Computation*. Athena Scientific, Belmont, Massachusetts, 1997.

[BvD17] R. Boucherie and M. van Dijk, editors. *Markov Decision Processes in Practice*, volume 248 of *International Series in Operations Research & Management Science*. Springer International Publishing, 2017.

[CDF+19] W. Czerwiński, L. Daviaud, N. Fijalkow, M. Jurdziński, R. Lazić, and P. Parys. Universal trees grow inside separating automata: Quasi-polynomial lower bounds for parity games. In T. M. Chan, editor, *Proceedings of the 2019 Annual ACM-SIAM Symposium on Discrete Algorithms (SODA)*, pages 2333–2349, USA, January 2019. Society for Industrial and Applied Mathematics.

[CJK+17] C. S. Calude, S. Jain, B. Khoussainov, W. Li, and F. Stephan. Deciding parity games in quasipolynomial time. In *Proceedings of the 49th Annual ACM SIGACT Symposium on Theory of Computing (STOC)*, pages 252–263, New York, NY, USA, 2017. Association for Computing Machinery.

[Con92] A. Condon. The complexity of stochastic games. *Information and Computation*, 96(2):203–224, 1992.

[Cun79] W. H. Cunningham. Theoretical Properties of the Network Simplex Method. *Mathematics of Operations Research*, 4(2):196–208, 1979.

[Dan51] G. B. Dantzig. Maximization of a Linear Function of Variables Subject to Linear Inequalities. In T. C. Koopmans, editor, *Activity Analysis of Production and Allocation*, pages 339–347, New York, 1951. John Wiley & Sons.

[Dan63] G. B. Dantzig. *Linear Programming and Extensions*. United States Air Force Project RAND. The RAND Corporation, August 1963. Available at `http://www.rand.org/pubs/reports/R366.html`. Accessed on 5 July 2020.

[d'E63] F. d'Epenoux. A probabilistic production and inventory problem. *Management Science*, 10(1):98–108, October 1963.

[DFH19] Y. Disser, O. Friedmann, and A. V. Hopp. An Exponential Lower Bound for Zadeh's pivot rule. `https://arxiv.org/abs/1911.01074`, November 2019.

[DGP09] C. Daskalakis, P. W. Goldberg, and C. H. Papadimitriou. The complexity of computing a nash equilibrium. *SIAM Journal on Computing*, 39(1):195–259, 2009.

[DH18] Y. Disser and A. V. Hopp. On Friedmann's subexponential lower bound for Zadeh's pivot rule. `http://tuprints.ulb.tu-darmstadt.de/7557/`, July 2018. Full version. Accessed 5 July 2020.

[DH19] Y. Disser and A. V. Hopp. On Friedmann's Subexponential Lower Bound for Zadeh's Pivot Rule. In *Proceedings of the 20th International Conference on Integer Programming and Combinatorial Optimization (IPCO)*, volume 11480 of *Lecture Notes in Computer Science*, pages 168–180, Ann Arbor, MI, USA, May 2019. Springer-Verlag Berlin Heidelberg.

[DOW55] G. B. Dantzig, A. Orden, and P. Wolfe. The generalized simplex method for minimizing a linear form under linear inequality restraints. *Pacific Journal of Mathematics*, 5(2):183–195, 1955.

[DP11] C. Daskalakis and C. Papadimitriou. Continuous Local Search. In D. Randall, editor, *Proceedings of the 2011 Annual ACM-SIAM Symposium on Discrete Algorithms (SODA)*, pages 790–804, USA, January 2011. Society for Industrial and Applied Mathematics.

[DS15] Y. Disser and M. Skutella. The simplex algorithm is NP-mighty. In *Proceedings of the 26th Annual ACM-SIAM Symposium on Discrete Algorithms (SODA)*, pages 858–872, 2015.

[EJ91] E. A. Emerson and C. S. Jutla. Tree automata, μ-calculus and determinacy. In *Proceedings of the 32nd Annual Symposium of Foundations of Computer Science (FOCS)*, pages 368–377, San Juan, Puerto Rico, October 1991. IEEE Computer Society Press.

[EJS93] E. A. Emerson, C. S. Jutla, and A. P. Sistla. On model-checking for fragments of μ-calculus. In C. Courcoubetis, editor, *Proceedings of the 5th International Conference on Computer Aided Verification (CAV 1993)*, volume 697 of *Lecture Notes in Computer Science*, pages 385–396, Elounda, Greece, June and July 1993. Springer-Verlag Berlin Heidelberg.

[Fea10a] J. Fearnley. Exponential Lower Bounds for Policy Iteration. In S. Abramsky, C. Gavoille, C. Kirchner, F. Meyer auf der Heide, and P. G. Spirakis, editors, *Proceedings of the 37th International Colloquium on Automata, Languages and Programming (ICALP)*, volume 6199 of *Lecture Notes in Computer Science*, pages 551–562, Bordeaux, France, July 2010. Springer-Verlag Berlin Heidelberg.

[Fea10b] J. Fearnley. *Strategy Iteration Algorithms for Games and Markov Decision Processes*. PhD thesis, The University of Warwick, Department of Computer Science, August 2010.

[FG92] J. J. Forrest and D. Goldfarb. Steepest-edge simplex algorithms for linear programming. *Mathematical Programming*, 57(1–3):341–374, 1992.

[FHZ11a] O. Friedmann, T. D. Hansen, and U. Zwick. A subexponential lower bound for the random facet algorithm for parity games. In D. Randall, editor, *Proceedings of the 2011 Annual ACM-SIAM Symposium on Discrete Algorithms (SODA)*, pages 202–216, San Francisco, CA, USA, January 2011. Association for Computing Machinery, New York, NY, USA and the Society for Industiral and Applied Mathematics, Philadelphia, PA, USA.

[FHZ11b] O. Friedmann, T. D. Hansen, and U. Zwick. Subexponential lower bounds for randomized pivoting rules for the simplex algorithm. In *Proceedings of the*

43rd Annual ACM Symposium on Theory of Computing (STOC), pages 283–292, San Jose, CA, USA, June 2011. Association for Computing Machinery, New York, NY, USA.

[FJdK+19] J. Fearnley, S. Jain, B. de Keijzer, S. Schewe, F. Stephan, and D. Wojtczak. An ordered approach to solving parity games in quasi-polynomial time and quasi-linear space. *International Journal on Software Tools for Technology Transfer*, 21:325–349, 2019.

[FL17] O. Friedmann and M. Lange. PGSolver, version 4.1, June 2017. Available at `https://github.com/tcsprojects/pgsolver`.

[FLL10] O. Friedmann, M. Latte, and M. Lange. A Decision Procedure for CTL* Based on Tableaux and Automata. In J. Giesl and R. Hähnle, editors, *Automated Reasoning*, volume 6173 of *Lecture Notes in Artificial Intelligence*, pages 331–345. Springer-Verlag Berlin Heidelberg, July 2010.

[Fri09] O. Friedmann. An exponential lower bound for the parity game strategy improvement algorithm as we know it. In *Proceedings of the 24th Annual IEEE Symposium on Logic in Computer Science*, pages 145–156. IEEE Computer Society, August 2009.

[Fri11a] O. Friedmann. An Exponential Lower Bound for the Latest Deterministic Strategy Iteration Algorithms. *Logical Methods in Computer Science*, 7, 2011.

[Fri11b] O. Friedmann. *Exponential Lower Bounds for Solving Infinitary Payoff Games and Linear Programs*. PhD thesis, Ludwig-Maximilians-Universität München, Department of Computer Science, July 2011.

[Fri11c] O. Friedmann. A subexponential lower bound for Zadeh's pivoting rule for solving linear programs and games. In O. Günlük and G. J. Woeginger, editors, *Proceedings of the 15th International Conference on Integer Programming and Combinatoral Optimization (IPCO)*, volume 6655 of *Lecture Notes in Computer Science*, pages 192–206, New York, NY, USA, June 2011. Springer-Verlag Berlin Heidelberg.

[Fri19] O. Friedmann. Animation of the exponential sink game construction for 3 and 4 levels, 2019.
`https://oliverfriedmann.com/downloads/add/zadeh-exponential-animation-3.pdf`
`https://oliverfriedmann.com/downloads/add/zadeh-exponential-animation-4.pdf`. Accessed on 5 July 2020.

[FS02] E. A. Feinberg and A. Shwartz, editors. *Handbook of Markov Decision Processes: Methods and Applications*, volume 40 of *International Series in Operations Research & Management Science*. Springer US, 2002.

[FS15] J. Fearnley and R. Savani. The Complexity of the Simplex Method. In *Proceedings of the 47th Annual ACM on Symposium on Theory of Computing (STOC)*, pages 201–208, 2015.

[GS55] S. Gass and T. Saaty. The computational algorithm for the parametric objective function. *Naval Research Logistics Quarterly*, 2(1–2):39–45, 1955.

[GS79] D. Goldfarb and W. Y. Sit. Worst case behavior of the steepest edge simplex method. *Discrete Applied Mathematics*, 1(4):277–285, 1979.

[GTW02] E. Grädel, W. Thomas, and T. Wilke, editors. *Automata Logics, and Infinite Games: A Guide to Current Research*, volume 2500 of *Lecture Notes in Computer Science*. Springer-Verlag Berlin Heidelberg, 1st edition, 2002.

[Han12] T. D. Hansen. *Worst-case Analysis of Strategy Iteration and the Simplex Method*. PhD thesis, Aarhus Universitet, Department of Computer Science, July 2012.

[HK66] A. J. Hoffman and R. M. Karp. On nonterminating stochastic games. *Management Science*, 12(5):359–370, January 1966.

[HM18] S. Haddad and B. Monmege. Interval iteration algorithm for MDPs and IMDPs. *Theoretical Computer Science*, 735:111–131, 2018. Reachability Problems 2014: Special Issue.

[How60] R. A. Howard. *Dynamic Programming and Markov Processes*. The M.I.T. Press, Cambridge, MA, USA, 1960.

[Jer73] R. G. Jeroslow. The simplex algorithm with the pivot rule of maximizing criterion improvement. *Discrete Mathematics*, 4(4):367–377, 1973.

[JPZ08] M. Jurdziński, M. Paterson, and U. Zwick. A deterministic subexponential algorithm for solving parity games. *SIAM Journal on Computing*, 38(4):1519–1532, 2008.

[Jur98] M. Jurdziński. Deciding the winner in parity games is in $\mathsf{UP} \cap \mathsf{co\text{-}UP}$. *Information Processing Letters*, 68(3):119 – 124, 1998.

[Jur00] M. Jurdziński. Small progress measures for solving parity games. In *Proceedings of the 17th Annual Symposium on Theoretical Aspects of Computer Science (STACS)*, volume 1770 of *Lecture Notes in Computer Science*, pages 290–301. Springer-Verlag Berlin Heidelber, 2000.

[Kal91] G. Kalai. The diameter of graphs of convex polytopes and f-vector theory. In P. Gritzmann and B. Sturmfels, editors, *Applied Geometry And Discrete Mathematics: The Victor Klee Festschrift*, Series in Discrete Mathematics and Theoretical Computer Science (DIMACS), pages 387–412. American Mathematical Society, 1991.

[Kal92] G. Kalai. A subexponential randomized simplex algorithm (extended abstract). In *Proceedings of the 24th Annual ACM Symposium on Theory of Computing (STOC)*, pages 475–482, New York, NY, USA, 1992. Association for Computing Machinery.

[Kal97] G. Kalai. Linear programming, the simplex algorithm and simple polytopes. *Mathematical Programming*, 79(1–3):217–233, October 1997.

[Kar84] N. Karmarkar. A new polynomial-time algorithm for linear programming. *Combinatorica*, 4:373–395, December 1984.

[Kha80] L. G. Khachiyan. Polynomial algorithms in linear programming. *USSR Computational Mathematics and Mathematical Physics*, 20(1):53–72, 1980.

[KM72] V. Klee and G. J. Minty. How good is the simplex algorithm? In *Inequalities, III (Proc. Third Sympos., Univ. California, Los Angeles, Calif., 1969; dedicated to the memory of Theodore S. Motzkin)*, pages 159–175. Academic Press, New York, 1972.

[Man60] A. S. Manne. Linear programming and sequential decisions. *Management Science*, 6(3):259–267, April 1960.

[Mat94] J. Matoušek. Lower bounds for a subexponential optimization algorithm. *Random Structures & Algorithms*, 5(4):591–607, 1994.

[McM70] P. McMullen. The maximum numbers of faces of a convex polytope. *Mathematika*, 17(2):179–184, 1970.

[MSW96] J. Matoušek, M. Sharir, and E. Welzl. A subexponential bound for linear programming. *Algorithmica*, 16(4–5):498–516, 1996.

[Mur80] K. G. Murty. Computational complexity of parametric linear programming. *Mathematical Programming*, 19(1):213–219, 1980.

[Pap94] C. H. Papadimitriou. On the complexity of the parity argument and other inefficient proofs of existence. *Journal of Computer and System Sciences*, 48(3):498–532, 1994.

[Par19] P. Parys. Parity Games: Zielonka's Algorithm in Quasi-Polynomial Time. In P. Rossmanith, P. Heggernes, and J.-P. Katoen, editors, *44th International Symposium on Mathematical Foundations of Computer Science (MFCS 2019)*, volume 138 of *Leibniz International Proceedings in Informatics (LIPIcs)*, pages 10:1–10:13, Dagstuhl, Germany, 2019. Schloss Dagstuhl–Leibniz-Zentrum fuer Informatik.

[Pow19] W. B. Powell. A unified framework for stochastic optimization. *European Journal of Operational Research*, 275(3):795–821, 2019.

[Pur95] A. Puri. *Theory of Hybrid Systems and Discrete Event Systems*. PhD thesis, EECS Department, University of California, Berkeley, December 1995.

[Put05] M. L. Puterman. *Markov Decision Processes: Discrete Stochastic Dynamic Programming*. Wiley Series in Probability and Statistics. John Wiley & Sons, Hoboken, NJ, USA, 2005.

[PY15] I. Post and Y. Ye. The simplex method is strongly polynomial for deterministic markov decision processes. *Mathematics of Operations Research*, 40(4):859–868, November 2015.

[San12] F. Santos. A counterexample to the Hirsch conjecture. *Annals of mathematics*, 176:383–412, 2012.

[SB18] R. S. Sutton and A. G. Barto. *Reinforcement Learning: An Introduction*. Adaptivecomputation and machine learning series. The MIT Press, 2nd edition, 2018.

[Sch17] S. Schewe. Solving parity games in big steps. *Journal of Computer and System Sciences*, 84:243–262, March 2017.

[Sha53] L. S. Shapley. Stochastic games. *Proceedings of the National Academy of Sciences*, 39(10):1095–1100, 1953.

[Sti95] C. Stirling. Local model checking games (extended abstract). In I. Lee and S. A. Smolka, editors, *CONUR' 95: Concurrency Theory*, volume 962 of *Lecture Notes in Computer Science*, pages 1–11, Philadephia, PA, USA, August 1995. Springer-Verlag Berlin Heidelberg.

[SW92] M. Sharir and E. Welzl. A combinatorial bound for linear programming and related problems. In A. Finkel and M. Jantzen, editors, *STACS 92*, pages 567–579, Berlin, Heidelberg, 1992. Springer-Verlag Berlin Heidelberg.

[Szp30] E. Szpilrajn. Sur l'extension de l'ordre partiel. *Fundamenta Mathematicae*, 16:386–389, 1930.

[Ter01a] T. Terlaky. Criss-Cross Pivoting Rules. In C.A. Floudas and P.M. Pardalos, editors, *Encyclopedia of Optimization*, pages 361–366. Springer US, Boston, MA, 2001.

[Ter01b] T. Terlaky. Lexicographic pivoting rules. In C. A. Floudas and P. M. Pardalos, editors, *Encyclopedia of Optimization*, pages 1263–1267. Springer US, Boston, MA, 2001.

[Tho17] A. Thomas. Exponential lower bounds for history-based simplex pivot rules on abstract cubes. In *Proceedings of the 25th Annual European Symposium on Algorithms (ESA)*, pages 69:1–69:14, 2017.

[TZ93] T. Terlaky and S. Zhang. Pivot rules for linear programming: A survey on recent theoretical developments. *Annals of Operations Research*, 46(1):203–233, 1993.

[VJ00] J. Vöge and M. Jurdziński. A Discrete Strategy Improvement Algorithm for Solving Parity Games. In E. A. Emerson and A. P. Sistla, editors, *Proceedings of the 12th International Conference on Computer Aided Verification (CAV 2000)*, volume 1855 of *Lecture Notes in Computer Science*, pages 202–215. Springer-Verlag Berlin Heidelberg, July 2000.

[Ye11] Y. Ye. The simplex and policy-iteration methods are strongly polynomial for the markov decision problem with a fixed discount rate. *Mathematics of Operations Research*, 36(4):593–603, November 2011.

[Zad80] N. Zadeh. What is the Worst Case Behavior of the Simplex Algorithm? Technical Report 27, Departments of Operations Research, Stanford, 1980.

[Zie98] W. Zielonka. Infinite games on finitely coloured graphs with applications to automata on infinite trees. *Theoretical Computer Science*, 200(1):135 – 183, 1998.

[Zie04] Günter M. Ziegler. Typical and extremal linear programs. In M. Grötschel, editor, *The Sharpest Cut: The Impact of Manfred Padberg and His Work*, volume 4 of *MPS-SIAM Series on Optimization*, chapter 14, pages 217–230. SIAM, Philadelphia, PA, 2004.

A. Proofs

This appendix contains all proofs that were omitted in the main part of the thesis. For convenience, we restate the proven statements.

A.1. Proofs of Chapter 4

Lemma 4.1.4. *Let $\mathfrak{b} \in \mathfrak{B}_n$ and $i, j \in [n]$. Then the following hold:*

1. *Let S, S' be schemes and $S \subseteq S'$. Then $M(\mathfrak{b}, S') \subseteq M(\mathfrak{b}, S)$.*
2. *Let S, S' be schemes and $S \subseteq S'$. Then $f(\mathfrak{b}, i, S') \leq f(\mathfrak{b}, i, S)$.*
3. *It holds that $f(\mathfrak{b}, j) = f(\mathfrak{b}, j, \{(i, 0)\}) + f(\mathfrak{b}, j, \{(i, 1)\})$ and $f(\mathfrak{b}, j) = \lfloor (\mathfrak{b} + 2^{j-1})/2^j \rfloor$.*
4. *Let $i \leq j$ and S be a scheme. Then $f(\mathfrak{b}, j, S) \leq f(\mathfrak{b}, i, S)$ and thus $f(\mathfrak{b}, j) \leq f(\mathfrak{b}, i)$.*
5. *Let $i < j$. Then $F(\mathfrak{b}, j) = F(\mathfrak{b}, j, \{(i, 0)\})$ and thus $f(\mathfrak{b}, j, \{(i, 0)\}) = f(\mathfrak{b}, j)$.*

Proof. We prove the statements one after another.

1. Let S, S' be schemes and $S \subseteq S'$. Since every number matching S' also matches S, it follows that $M(\mathfrak{b}, S') \subseteq M(\mathfrak{b}, S)$ for all $\mathfrak{b} \in \mathfrak{B}_n$.
2. This follows directly from (1) and the definition of $f(\mathfrak{b}, i, S')$.
3. The first statement follows since either $\mathfrak{b}_i = 0$ or $\mathfrak{b}_i = 1$ for every $\mathfrak{b} \in \mathfrak{B}_n$ and $i \in [n]$.

   It remains to prove $f(\mathfrak{b}, j) = \lfloor (\mathfrak{b} + 2^{j-1})/2^j \rfloor$ for $\mathfrak{b} \in \mathfrak{B}_n$ and $j \in [n]$. The smallest number matching the scheme $S_j := \{(j, 1), (j-1, 0), \ldots, (1, 0)\}$ is 2^{j-1}. This implies the statement for $\mathfrak{b} < 2^{j-1}$. Let m_i denote the i-th number matching the scheme S_j. Then, by the previous argument, $m_1 = 2^{j-1}$. As only numbers ending on the subsequence $(1, 0, \ldots, 0)$ of length j match S_j, we have $m_i = (i-1) \cdot 2^j + 2^{j-1}$. This implies $f(m_i, j) = \lfloor (m_i + 2^{j-1})/2^j \rfloor$ since $f(m_i, j) = i$ by definition and

   $$\left\lfloor \frac{m_i + 2^{j-1}}{2^j} \right\rfloor = \left\lfloor \frac{(i-1) \cdot 2^j + 2^{j-1} + 2^{j-1}}{2^j} \right\rfloor = \left\lfloor \frac{i \cdot 2^j}{2^j} \right\rfloor = i.$$

   Now let $\mathfrak{b} \in \mathfrak{B}_n$ and choose $i \in \mathbb{N}$ such that $\mathfrak{b} \in [m_i, m_{i+1})$. Then, $f(\mathfrak{b}, j) = i$ by the definition of $f(\mathfrak{b}, j)$. In addition, by the choice of i,

   $$\left\lfloor \frac{\mathfrak{b} + 2^{j-1}}{2^j} \right\rfloor \geq \left\lfloor \frac{m_i + 2^{j-1}}{2^j} \right\rfloor = f(m_i, j) = i \tag{A.1}$$

   and

   $$\left\lfloor \frac{\mathfrak{b} + 2^{j-1}}{2^j} \right\rfloor < \left\lfloor \frac{m_{i+1} + 2^{j-1}}{2^j} \right\rfloor = f(m_{i+1}, j) = i + 1. \tag{A.2}$$

By integrality, Equations (A.1) and (A.2) imply $\lfloor (\mathfrak{b}+2^{j-1})/2^j \rfloor = i$ and thus the statement.

4. Let $i \leq j$ and $\mathfrak{b} \in \mathfrak{B}_n$. Let $S_j := \{(j,1),(j-1,0),\dots,(1,0)\}$ and define S_i analogously. Consider any $\mathfrak{b}' \leq \mathfrak{b}$ matching S_j and S. Then, since $i \leq j$ there needs to be at least one $\hat{\mathfrak{b}} \leq \mathfrak{b}'$ matching S_i and S. This immediately implies $f(\mathfrak{b},j,S) \leq f(\mathfrak{b},i,S)$.
   The second inequality follows immediately when setting $S := \emptyset$.
5. Let $i < j$ and define $S_j := \{(j,1),(j-1,0),\dots,(1,0)\}$. Since $i < j$, we have $(i,0) \in S_j$, immediately implying $F(\mathfrak{b},j) = F(\mathfrak{b},j,\{(i,0)\})$. □

Lemma 4.3.2. *None of the edges $(b_{i,k}^1, A_i^1)$ for $i \in [n]$ and $k \in \{0,1\}$ is an improving switch with respect to σ^*.*

Proof. To simplify the calculations, let $\mathrm{Val} := \mathrm{Val}_{\sigma^*}$. Let $i \in [n]$ and $k \in \{0,1\}$. Then, the definition of σ^* implies $\sigma^*(b_{i,k}^1) = t$. Therefore, $\mathrm{Val}(b_{i,k}^1) = \mathrm{Val}(\sigma^*(b_{i,k}^1)) = \mathrm{Val}(t) = 0$ since $r(b_{i,k}^1, t) = 0$. Analogously, $\mathrm{Val}(b_{i,1-k}^1) = 0$. This implies that $(b_{i,k}^1, A_i^1)$ is an improving switch if and only if $\mathrm{Val}(A_i^1) > 0$. But, due to $\sigma^*(k_{i+1}) = t$, it holds that

$$\begin{aligned}
\mathrm{Val}(A_i^1) &= \varepsilon\,\mathrm{Val}(d_i^1) + \frac{1-\varepsilon}{2}\mathrm{Val}(b_{i,k}^1) + \frac{1-\varepsilon}{2}\mathrm{Val}(b_{i,1-k}^1) \\
&= \varepsilon\,\mathrm{Val}(d_i^1) = \varepsilon\,(-N)^6 + \mathrm{Val}(\sigma^\star(d_i^1))] = \varepsilon\left[N^6 + \mathrm{Val}(h_i^1)\right] \\
&= \varepsilon\left[N^6 + (-N)^{2i+8} + \mathrm{Val}(k_{i+1})\right] \\
&= \varepsilon\left[N^6 + N^{2i+8} + (-N)^{2(i+1)+7} + \mathrm{Val}(t)\right] = \varepsilon\left[N^6 + N^{2i+8} - N^{2i+9}\right] < 0,
\end{aligned}$$

as $N = 7n+1 \geq 8$ and $i \geq 1$. Consequently, $(b_{i,r}^1, A_i^1)$ is not an improving switch. □

Lemma 4.3.9. *Let σ be a phase 3 strategy and let $e \in L_\sigma^3$. Then $L_{\sigma e}^3 = L_\sigma \setminus \{e\}$.*

Proof. We only discuss the case $e \in L_\sigma^{3,1}$ as the cases $e \in L_\sigma^{3,2}$ and $e \in L_\sigma^{3,3}$ follow from similar arguments. Let $e \in L_\sigma^{3,1}$. Then, $e = (k_i, k_\nu)$ for some $i \in [n]$ with $\sigma(k_i) \neq k_\nu$ and $(\mathfrak{b}+1)_i = 0$. As any edge in L_σ^3 is improving for σ by [Fri11c, Lemma 4], (k_i, k_ν) is improving for σ. Thus, $\sigma e(k_i) = k_\nu$, implying $e \notin L_{\sigma e}^{3,1}$ and $e \notin L_{\sigma e}^3$.

Let $\tilde{e} \in L_\sigma^3$ and $\tilde{e} \neq e$. We prove $\tilde{e} \in L_{\sigma e}^3$. Since $\tilde{e} \in L_\sigma^3$, we have $\tilde{e} = (v, k_\nu)$ where either $v = k_{i'}$ or $v = b_{i',k}^j$ for some $i' \in [n]$ and $k,j \in \{0,1\}$. In addition, since $\tilde{e} \in L_\sigma^3$, we have $\sigma(v) \neq k_\nu$. The switch (k_i, k_ν) is the only switch applied in σ. Therefore, $\sigma(v) \neq k_\nu$ implies $\sigma e(v) \neq k_\nu$ as the target of no vertex other than k_i changes. As furthermore the conditions $(\mathfrak{b}+1)_i = 0$ and $(\mathfrak{b}+1)_{i+1} \neq j$ remain valid, it follows that $\tilde{e} \in L_{\sigma e}^3$. This implies $L_\sigma^3 \subseteq L_{\sigma e}^3 \cup \{e\}$.

For the sake of a contradiction, assume that there is some edge $\tilde{e} \in L_{\sigma e}^3 \cup \{e\}$ but $\tilde{e} \notin L_\sigma^3$. Then, $e \in L_\sigma^3$ implies $e \neq \tilde{e}$. Thus, $\tilde{e} = (v, k_\nu)$ for some v as previously and $\sigma e(v) \neq k_\nu$. Since (k_i, k_ν) is the only switch applied in σ, this implies $\sigma(v) \neq k_\nu$. But then, $e \in L_\sigma^3$ which is a contradiction. Consequently, $L_{\sigma e}^3 \cup \{e\} \subseteq L_\sigma^3$ and thus $L_{\sigma e}^3 \cup \{e\} = L_\sigma^3$. □

Lemma 4.3.11. *Let $i \in \{2,\dots,n-2\}$ and $l < i$. Then, there is a number $\mathfrak{b} \in \mathfrak{B}_n$ with $\ell(\mathfrak{b}+1) = \nu = l$ such that for all $j \in \{i+2,\dots,n\}$, it holds that $\phi^{\sigma_\mathfrak{b}}(k_i, k_\nu) < \phi^{\sigma_\mathfrak{b}}(k_j, k_\nu)$ and $(k_i, k_\nu), (k_j, k_\nu) \in L_{\sigma_\mathfrak{b}}^3$.*

Proof. Let $\mathfrak{b} := 2^i + 2^{l-1} - 1$ and $j \in \{i+2, \dots, n\}$. Then, $\ell(\mathfrak{b}+1) = \ell(2^i + 2^{l-1}) = l$ since $l < i$. Furthermore, $j \geq i+2, i > l$ and $i \geq 2$ imply

$$\mathfrak{b} + 1 = 2^i + 2^{l-1} \leq 2^i + 2^{i-2} \leq 2^{j-2} + 2^{j-4} < 2^{j-1} - 1.$$

Now consider the flip set $F(\mathfrak{b}, l)$ containing all $\tilde{\mathfrak{b}} \leq \mathfrak{b}$ with $\ell(\tilde{\mathfrak{b}}) = l$. Since $\mathfrak{b} < 2^{j-1}$, it holds that $\tilde{\mathfrak{b}}_j = 0$ for all $\tilde{\mathfrak{b}} \leq \mathfrak{b}$, hence $F(\mathfrak{b}, l) = F(\mathfrak{b}, l, \{(j, 0)\})$. Thus, by Table 4.5,

$$\phi^{\sigma_\mathfrak{b}}(k_j, k_\nu) = \phi^{\sigma_\mathfrak{b}}(k_j, k_l) = f(\mathfrak{b}, l, \{(j, 0)\}) = f(\mathfrak{b}, l).$$

In addition, since $\mathfrak{b} + 1 < 2^{j-1} - 1$ and thus $(\mathfrak{b}+1)_j = 0$ and $\sigma_\mathfrak{b}(k_j) = k_\ell \neq k_\nu$ due to the invariants discussed in Section 4.1, we have $(k_j, k_\nu) \in L^3_{\sigma_\mathfrak{b}}$. However, since $\mathfrak{b} > 2^i, i \geq 2$ and $i > l$, it holds that $\tilde{\mathfrak{b}} := 2^{i-1} + 2^{l-1} \in F(\mathfrak{b}, l)$ since $\tilde{\mathfrak{b}} \leq \mathfrak{b}$. Furthermore $\tilde{\mathfrak{b}}_i = 1$. As a consequence, $\tilde{\mathfrak{b}} \notin F(\mathfrak{b}, l, \{(i, 0)\}))$. But this implies $F(\mathfrak{b}, l, \{(i, 0)\}) \subsetneq F(\mathfrak{b}, l)$. Since $\phi^{\sigma_\mathfrak{b}}(k_i, k_l) = f(\mathfrak{b}, l, \{(i, 0)\})$ and $|F(\mathfrak{b}, l\{(i, 0)\})| = f(\mathfrak{b}, l, \{(i, 0)\})$ by Table 4.5, this implies

$$\phi^{\sigma_\mathfrak{b}}(k_i, k_\nu) = \phi^{\sigma_\mathfrak{b}}(k_i, k_l) = f(\mathfrak{b}, l, \{(i, 0)\}) < f(\mathfrak{b}, l) = \phi^{\sigma_\mathfrak{b}}(k_j, k_l) = \phi^{\sigma_\mathfrak{b}}(k_j, k_\nu).$$

As $i > l = \ell(\mathfrak{b}+1)$ and $\sigma_\mathfrak{b}(k_i) = k_\ell \neq k_\nu$ imply $(\mathfrak{b}+1)_i = \mathfrak{b}_i = 0$, we also have $(k_i, k_\nu) \in L^3_{\sigma_\mathfrak{b}}$ as claimed. □

Lemma 4.3.13. *Assume that all edges of $L^3_{\sigma_\mathfrak{b}}$ are applied during phase 3 of the transition from $\sigma_\mathfrak{b}$ to $\sigma_{\mathfrak{b}+1}$ for all $\mathfrak{b} \in \mathfrak{B}_n$. Let $i \in \{2, \dots, n-2\}$ and $l < i$ be fixed. Then, there is a $\mathfrak{b} \in \mathfrak{B}_n$ with $\ell(\mathfrak{b}+1) = l$ such that $\phi^{\sigma_\mathfrak{b}}(k_{i+1}, k_\nu) < \phi^{\sigma_\mathfrak{b}}(b^1_{i,k}, k_\nu)$ for some $k \in \{0, 1\}$ and $(k_{i+1}, k_\nu), (b^1_{i,k}, k_\nu) \in L^3_{\sigma_\mathfrak{b}}$.*

Proof. We begin by observing that Table 4.5 can be used as we assume that all edges of $L^3_{\sigma_\mathfrak{b}}$ are applied during phase 3, and since this set is exactly the set of edges that should be applied during phase 3 by Lemma 4.3.8.

Consider some $\mathfrak{b} \in \mathfrak{B}_n$ with $\ell(\mathfrak{b}+1) = \nu = l$, its exact value will be fixed later. By Table 4.5, since $\nu = l$, and by Lemma 4.1.4 it holds that

$$\begin{aligned} \phi^{\sigma_\mathfrak{b}}(k_{i+1}, k_\nu) &= f(\mathfrak{b}, \nu, \{(i+1, 0)\}), \\ \phi^{\sigma_\mathfrak{b}}(b^1_{i,k}, k_\nu) &= f(\mathfrak{b}, \nu, \{(i, 0)\}) + f(\mathfrak{b}, \nu, \{(i, 1), (i+1, 0)\}), \\ f(\mathfrak{b}, \nu, \{(i, 0)\}) &= f(\mathfrak{b}, \nu, \{(i, 0), (i+1, 0)\}) + f(\mathfrak{b}, \nu, \{(i, 0), (i+1, 1)\}). \end{aligned}$$

These equations imply that $\phi^{\sigma_\mathfrak{b}}(b^1_{i,k}, k_\nu)$ can be expressed as

$$f(\mathfrak{b}, \nu, \{(i, 0), (i+1, 0)\}) + f(\mathfrak{b}, \nu, \{(i, 0), (i+1, 1)\}) + f(\mathfrak{b}, \nu, \{(i, 1), (i+1, 0)\}).$$

Since $f(\mathfrak{b}, \nu, \{(i, 1), (i+1, 0)\}) + f(\mathfrak{b}, \nu, \{(i, 0), (i+1, 0)\}) = f(\mathfrak{b}, \nu, \{(i+1, 0)\})$, the inequality $\phi^{\sigma_\mathfrak{b}}(k_{i+1}, k_\nu) < \phi^{\sigma_\mathfrak{b}}(b^1_{i,k}, k_\nu)$ can be formulated equivalently as

$$f(\mathfrak{b}, \nu, \{(i+1, 0)\}) < f(\mathfrak{b}, \nu, \{(i+1, 0)\}) + f(\mathfrak{b}, \nu, \{(i, 0), (i+1, 1)\}).$$

It thus suffices to find a $\mathfrak{b} \in \mathfrak{B}_n$ such that $f(\mathfrak{b}, \nu, \{(i, 0), (i+1, 1)\}) > 0, \ell(\mathfrak{b}+1) = l$ and $(k_{i+1}, k_\nu), (b^1_{i,k}, k_\nu) \in L^3_{\sigma_\mathfrak{b}}$.

We prove that $\mathfrak{b} := 2^{i+1} + 2^{l-1} - 1$ has all of these properties. Since $l < i$, we have $\ell(\mathfrak{b}+1) = \ell(2^{i+1} + 2^{l-1}) = l$. In addition, $l < i$ implies $\mathfrak{b}_{i+1} = 0$, so $\sigma_{\mathfrak{b}}(k_{i+1}) = k_\ell \neq k_\nu$. Furthermore, $(\mathfrak{b}+1)_{i+1} = 0$, implying $(k_{i+1}, k_\nu) \in L^3_{\sigma_{\mathfrak{b}}}$. Also, since $(\mathfrak{b}+1)_{i+1} = 0 \neq 1$ and $\sigma_{\mathfrak{b}}(b^1_{i,k}) = k_\ell \neq k_\nu$, we also have $(b^1_{i,k}, k_\nu) \in L^3_{\sigma_{\mathfrak{b}}}$.

Now, consider $\mathfrak{b}' := 2^i + 2^{l-1} \in \mathfrak{B}_n$. This number fulfills $\mathfrak{b}' < \mathfrak{b}, \mathfrak{b}'_i = 0$ and $\mathfrak{b}'_{i+1} = 1$. But this implies $f(\mathfrak{b}, \nu, \{(i,0),(i+1,1)\}) \geq 1$ and thus concludes the proof. □

Lemma 4.4.6. *Let σ be a phase 3 strategy. Then $\max_{e \in L^3_\sigma} \phi^\sigma(e) \leq f(\mathfrak{b}, \nu)$.*

Proof. As discussed in Section 4.3, the set L^3_σ can be partitioned into three subsets $L^{3,1}_\sigma$, $L^{3,2}_\sigma$ and $L^{3,3}_\sigma$, so we distinguish three cases. The last two cases can be discussed together as the occurrence records of edges contained in $L^{3,2}_\sigma$ and $L^{3,3}_\sigma$ are the same (cf. Table 4.5).

1. $\boldsymbol{e \in L^{3,1}_\sigma}$. Then, $e = (k_i, k_\nu)$, where $\sigma(k_i) \neq k_\nu$ and $(\mathfrak{b}+1)_i = 0$. The first condition implies that the switch e was not applied yet during $\sigma_{\mathfrak{b}} \to \sigma_{\mathfrak{b}+1}$. Consequently, $\phi^{\sigma_{\mathfrak{b}}}(k_i, k_\nu) = \phi^{\sigma_{\mathfrak{b}}}(k_i, k_\nu)$. Since $\phi^{\sigma_{\mathfrak{b}}}(e) = f(\mathfrak{b}, \nu, \{(i,0)\})$ by Table 4.5, this implies $\phi^\sigma(e) = f(\mathfrak{b}, \nu, \{(i,0)\})$. By Lemma 4.1.4 (3), this yields
$$\phi^\sigma(e) = f(\mathfrak{b}, \nu, \{(i,0)\}) = f(\mathfrak{b}, \nu) - f(\mathfrak{b}, \nu, \{i, 1\}) \leq f(\mathfrak{b}, \nu).$$
2. $\boldsymbol{e \in L^{3,2}_\sigma}$ **or** $\boldsymbol{e \in L^{3,3}_\sigma}$. Then, $e = (b^j_{i,r}, k_\nu)$ for some $r \in \{0,1\}$ where $\sigma(b^j_{i,r}) \neq k_\nu$ and either $(\mathfrak{b}+1)_i = 0$ or $(\mathfrak{b}+1)_{i+1} \neq j$. The first condition implies that e was not applied yet during $\sigma_{\mathfrak{b}} \to \sigma_{\mathfrak{b}+1}$. Hence $\phi^\sigma(b^j_{i,r}, k_\nu) = \phi^{\sigma_{\mathfrak{b}}}(b^j_{i,r}, k_\nu)$. Thus, Table 4.5 implies
$$\phi^\sigma(e) = f(\mathfrak{b}, \nu, \{(i,0)\}) + f(\mathfrak{b}, \nu, \{(i,1),(i+1,1-j)\}).$$
By Lemma 4.1.4 (2), it also holds that $f(\mathfrak{b}, \nu, \{(i,1),(i+1,1-j)\}) \leq f(\mathfrak{b}, \nu, \{(i,1)\})$. Thus, $\phi^\sigma(e) \leq f(\mathfrak{b}, \nu, \{(i,0)\}) + f(\mathfrak{b}, \nu, \{(i,1)\}) = f(\mathfrak{b}, \nu)$. □

Lemma 4.4.9. *Let σ be a phase 3 strategy. Assume that the strategy iteration algorithm is started with the initial strategy $\sigma^\star$. Then $\min_{e \in L^4_\sigma \cup L^5_\sigma \cup L^5_\sigma} \phi^{\sigma_{\mathfrak{b}}}(e) \geq f(\mathfrak{b}, \nu)$.*

Proof. Since σ is calculated after $\sigma_{\mathfrak{b}}$, we have $\phi^\sigma(e) \geq \phi^{\sigma_{\mathfrak{b}}}(e)$ for all edges e. It therefore suffices to show $\phi^{\sigma_{\mathfrak{b}}}(e) \geq f(\mathfrak{b}, \nu)$ for all $e \in L^4_\sigma \cup L^5_\sigma \cup L^6_\sigma$. We distinguish three cases.

1. $\boldsymbol{e \in L^4_\sigma}$: Then, by Table 4.4, $e = (h^0_i, k_{\ell_{i+2}(\mathfrak{b}+1)})$ for some $i \in [n]$ and, in addition, $\sigma(h^0_i) \notin \{k_{\ell_{i+2}(\mathfrak{b}+1)}, t\}$. Since $\sigma(h^0_i) \neq t$, there needs to be a next bit equal to 1 with an index of at least $i+2$ as a switch (h^0_i, k_l) is only applied when a number $\mathfrak{b}'$ with $\ell(\mathfrak{b}') = l$ is calculated.

   By the definition of $\nu := \ell(\mathfrak{b}+1)$, it holds that $\mathfrak{b}_j = (\mathfrak{b}+1)_j$ for all $j \in \{\nu+1, \dots, n\}$. Therefore, the first bit equal to 1 with an index of at least $i+2$ does not change if $i \geq \nu - 1$, so $\ell_{i+2}(\mathfrak{b}) = \ell_{i+2}(\mathfrak{b}+1)$. This implies $i \leq \nu - 2$ since $\sigma(h^0_i) = k_{\ell_{i+2}(\mathfrak{b}+1)}$ otherwise. But this contradicts that $(h^0_i, k_{\ell_{i+2}(\mathfrak{b}+1)})$ is an improving switch. As also $(\mathfrak{b}+1)_j = 0$ for $j < \nu$, it follows that $\ell_{j+2}(\mathfrak{b}+1) = \nu$ for all $j \in [\nu - 2]$. Thus, $e = (h^0_i, k_\nu)$ for some $i \in [\nu-2]$, and $\phi^{\sigma_{\mathfrak{b}}}(e) = f(\mathfrak{b}, \nu)$ by Table 4.5.
2. $\boldsymbol{e \in L^5_\sigma}$: Then, by Table 4.4, $e = (s, k_\nu)$. Therefore, since $\phi^{\sigma_{\mathfrak{b}}}(s, k_\nu) = f(\mathfrak{b}, \nu)$ by Table 4.5, it holds that $\phi^{\sigma_{\mathfrak{b}}}(e) = f(\mathfrak{b}, \nu)$.

3. $\boldsymbol{e \in L_\sigma^6}$: We have $L_{\sigma_\mathfrak{b}}^6 = \{(d_{\nu-1}^1, h_{\nu-1}^1), (d_{\nu-1}^0, s)\} \cup \{(d_i^0, h_i^0), (d_i^1, s) : i \in [\nu-2]\}$ by Theorem 4.4.8. Since $L_\sigma^6 \subseteq L_{\sigma_\mathfrak{b}}^6$ can be proven analogously to Lemma 4.3.8, it suffices to prove the inequality for all $e \in L_{\sigma_\mathfrak{b}}^6$.

   Let $e = (d_{\nu-1}^0, s)$. Then, $\phi^{\sigma_\mathfrak{b}}(d_{\nu-1}^0, s) = f(\mathfrak{b}, \nu-1+1) + 0 \cdot \mathfrak{b}_{i+1} = f(\mathfrak{b}, \nu)$ by Table 4.5. Analogously, $\phi^{\sigma_\mathfrak{b}}(d_{\nu-1}^1, h_{\nu-1}^1) = f(\mathfrak{b}, \nu)$ for $e = (d_{\nu-1}^1, h_{\nu-1}^1)$. Therefore, it holds that $\phi^{\sigma_\mathfrak{b}}(e) \geq f(\mathfrak{b}, \nu)$ for $e \in \{(d_{\nu-1}^0, s), (d_{\nu-1}^1, h_{\nu-1}^1)\}$.

   Let $e = (d_i^1, s)$ for some $i \in [\nu-2]$. Then, e is improving if and only bit $i+1$ switches from 1 to 0. The first transition in which (d_i^1, s) becomes improving is therefore the transition from $\sigma_{2^{i+1}-1}$ to $\sigma_{2^{i+1}}$. As the strategy iteration algorithm is initialized with the strategy representing the number 0, the number $\mathfrak{b} \in \mathfrak{B}_n$ is represented after $\mathfrak{b}$ many transitions. Therefore, e is an improving switch every 2^{i+1}-th transition.

   We now interpret $\phi^{\sigma_\mathfrak{b}}(e)$ and $f(\mathfrak{b}, \nu)$ as "counters", which increase during the execution of the algorithm. As argued previously, $\phi^{\sigma_\mathfrak{b}}(e)$ increases every 2^{i+1} transitions. In contrast to this, $f(\mathfrak{b}, \nu)$ (for *fixed* ν with increasing $\mathfrak{b}$) increases the first time when $2^{\nu-1}$ is reached. But then, after another $2^{\nu-1}$ transitions, the number 2^ν is reached and $\ell(2^\nu) = \nu+1$. Therefore, it takes another $2^{\nu-1}$ transitions until the "counter" $f(\mathfrak{b}, \nu)$ increases again. To summarize, the counter $f(\mathfrak{b}, \nu)$ increases every 2^ν iterations, excluding the first increase which happens after $2^{\nu-1}$ iterations. Since $i+1 \leq \nu-1$ as $i \leq \nu-2$, this proves that whenever the counter $f(\mathfrak{b}, \nu)$ is increased, the counter $\phi^{\sigma_\mathfrak{b}}(e)$ must have been increased at least once before or in the same iteration. Therefore, $\phi^{\sigma_\mathfrak{b}}(e) \geq f(\mathfrak{b}, \nu)$.

   For $e = (d_i^0, h_i^0)$, the statement follows by the same arguments as follows. The switch (d_i^1, s) is applied whenever bit $i+1$ is no longer equal to 1. The switch (d_i^0, h_i^0) is applied whenever bit $i+1$ becomes 0. Both of these happen whenever bit $i+1$ switches from 1 to 0 and thus, the same arguments used before can be applied. □

Lemma 4.4.10. *Let σ be a phase 3 strategy. Let $e_1 \in L_\sigma^3$ and $e_2 \in I_\sigma \cap (U_\sigma^{3,4} \cup \cdots \cup U_\sigma^{3,9})$. Then $\phi^\sigma(e_1) \leq \phi^\sigma(e_2)$.*

Proof. Let σ be a phase 3 strategy and let $e_1 \in L_\sigma^3$. Then, $\phi^\sigma(e_1) \leq f(\mathfrak{b}, \nu)$ by Lemma 4.4.6. It thus suffices to show $\phi^\sigma(e) \geq f(\mathfrak{b}, \nu)$ for all $e \in I_\sigma \cap (U_\sigma^{3,4} \cup \cdots \cup U_\sigma^{3,9})$. We distinguish the following 5 cases.

1. $\boldsymbol{e \in U_\sigma^{3,4}}$: Then, $e = (h_i^0, k_l)$ for some $l \leq \ell_{i+2}(\mathfrak{b}+1)$. By Table 4.5, $\phi^{\sigma_\mathfrak{b}}(e) = f(\mathfrak{b}, l)$ and since σ is calculated after $\sigma_\mathfrak{b}$, also $\phi^\sigma(\tilde{e}) \geq \phi^{\sigma_\mathfrak{b}}(e) = f(\mathfrak{b}, l)$. Let $l \leq \nu$. Then, by Lemma 4.1.4 (4), $f(\mathfrak{b}, l) \geq f(\mathfrak{b}, \nu)$, implying $\phi^\sigma(e) \geq f(\mathfrak{b}, \nu)$.

   For the sake of a contradiction, let $l > \nu$. We prove that (h_i^0, k_l) is not an improving switch in this case by showing $\operatorname{Val}_\sigma(\sigma(h_i^0)) \geq \operatorname{Val}_\sigma(k_l)$. Let $\ell_{i+2} := \ell_{i+2}(\mathfrak{b}+1)$. By construction, $\sigma(h_i^0) \in \{t, k_{i+2}, \ldots, k_n\}$. Assume $\ell_{i+2} \neq n+1$ first. Then, by the definition of ℓ_{i+2} and the invariants discussed in Section 4.1, $\sigma(h_i^0) = k_{\ell_{i+2}}$. We thus prove $\operatorname{Val}_\sigma(k_{\ell_{i+2}}) \geq \operatorname{Val}_\sigma(k_l)$.

   Since $l > \nu$ and $\ell_{i+2} \geq l$ by the choice of e, also $\ell_{i+2} > \nu$. Therefore, $\mathfrak{b}_j = (\mathfrak{b}+1)_j$ for all $j \geq \ell_{i+2}$. This implies that no bicycle of one of these levels was opened during phase 1. It furthermore implies that the target of none of the vertices

$k_{\ell_{i+2}}, \dots, k_n$ was changed during phase 2 as only the target of k_ν is switched during phase 2. Therefore, by Lemma 4.3.6, $\operatorname{Val}_\sigma(k_{\ell_{i+2}}) = S_{\ell_{i+2}}$. By the same lemma, also $\operatorname{Val}_\sigma(k_l) \leq T_l$. Thus, using that $\mathfrak{b}_j = (\mathfrak{b}+1)_j$ for all $j > \nu$ and $l > \nu$, we obtain,

$$\begin{aligned}\operatorname{Val}_\sigma(k_l) \leq T_l &= \sum_{j \geq l : (\mathfrak{b}+1)_j = 1} [N^{2j+8} - N^{2j+7} - N^7 + N^6] \\ &= \sum_{j \geq l : \mathfrak{b}_j = 1} [N^{2j+8} - N^{2j+7} - N^7 + N^6].\end{aligned}$$

By definition, ℓ_{i+2} is the smallest index larger than or equal to $i+2$ such that the corresponding bit of $\mathfrak{b}+1$ is equal to 1. By constriction, $\sigma(h_i^0) \in \{t, k_{i+2}, \dots, k_n\}$, implying $l \geq i+2$. Therefore, $\mathfrak{b}_l = \mathfrak{b}_{l+1} = \dots = \mathfrak{b}_{\ell_{i+2}-1} = 0$ since $l \leq \ell_{i+2}$. Using the previous inequality, we thus obtain

$$\begin{aligned}\operatorname{Val}_\sigma(k_l) &\leq \sum_{j \geq l : \mathfrak{b}_j = 1} [N^{2j+8} - N^{2j+7} - N^7 + N^6] \\ &= \sum_{j \geq \ell_{i+2} : \mathfrak{b}_j = 1} [N^{2j+8} - N^{2j+7} - N^7 + N^6] = S_{\ell_{i+2}} = \operatorname{Val}_\sigma(k_{\ell_{i+2}}).\end{aligned}$$

Consequently, e is not an improving switch if $\ell_{i+2} \neq n+1$.

Now assume $\ell_{i+2} = n+1$. Then, $(\mathfrak{b}+1)_{i'} = \mathfrak{b}_{i'} = 0$ for all $i' \geq i+2$. In particular, we then have $\sigma(h_i^0) = t$. As $\operatorname{Val}_\sigma(t) = 0$, it suffices to prove $\operatorname{Val}_\sigma(k_l) \leq 0$. This however follows immediately since $l > i+2$ implies $\sigma(k_l) = k_\ell$ and since Lemma 4.3.6 yields $\operatorname{Val}_\sigma(k_k) = r(k_l, k_\ell) + \operatorname{Val}_\sigma(k_\ell) < 0$.

Thus, $l > \nu$ implies $e \notin I_\sigma$ in every case so, $\phi^\sigma(e) \geq f(\mathfrak{b}, \nu)$ for all $e \in I_\sigma \cap U^{3,4}$.

2. **$e \in U_\sigma^{3,5}$:** Then, $e = (s, k_i)$ for some $i < \nu$ and $\sigma(s) \neq k_i$. Therefore, by Table 4.5, $\phi^{\sigma_\mathfrak{b}}(s, k_i) = f(\mathfrak{b}, i)$. Since σ is calculated after $\sigma_\mathfrak{b}$, also $\phi^\sigma(s, k_i) \geq \phi^{\sigma_\mathfrak{b}}(s, k_i)$. Since $i < \nu$ and by Lemma 4.1.4 (4), this implies $\phi^\sigma(s, k_i) \geq \phi^{\sigma_\mathfrak{b}}(s, k_i) = f(\mathfrak{b}, i) \geq f(\mathfrak{b}, \nu)$.
3. **$e \in U_\sigma^{3,6}$:** Then, $e = (d_i^j, v)$ for $v \in \{s, h_i^j\}$ where $i \in [\nu], j \in \{0,1\}$ and $\sigma(d_i^j) \neq v$. First, assume that $v = s$. Then $\sigma(d_i^j) = h_i^j$. Since $i < \nu$, it holds that $\mathfrak{b}_{i+1} = 1$ for $i \neq \nu - 1$ and $\mathfrak{b}_{i+1} = 0$ for $i = \nu - 1$. Furthermore, the target vertex of d_i^j can only be changed during phase 6 and was thus not changed yet. This implies that either $d_i^j = d_i^1$ and $\sigma(d_i^1) = h_i^1$ and $i < \nu - 1$ or $d_i^j = d_{\nu-1}^0$ and $\sigma(d_i^j) = h_{i-1}^0$ if $i = \nu - 1$. For these switches we however already showed in the proof of Lemma 4.4.9 that $\phi^{\sigma_\mathfrak{b}}(e) \geq f(\mathfrak{b}, \nu)$, implying $\phi^\sigma(e) \geq f(\mathfrak{b}, \nu)$.

   Now assume $v = h_i^j$, so $e = (d_i^j, h_i^j)$. Analogously to the case $v = s$ it is easy to verify that either $h_i^j = h_i^0$ and $\sigma(d_i^0) = h_i^0$ for $i < \nu - 1$ or $h_i^j = h_i^1$ and $\sigma(d_i^1) = h_i^1$ for $i = \nu - 1$. Again, these edges have already been investigated in the proof of Lemma 4.4.9 and the inequality $\phi^\sigma(e) \geq f(\mathfrak{b}, \nu)$ was shown there.
4. **$e \in U_\sigma^{3,7}$ or $e \in U_\sigma^{3,8}$:** By Lemma 4.4.9, $\phi^\sigma(e) \geq f(\mathfrak{b}, \nu)$ for all $e \in L_\sigma^6$. Since $U_\sigma^{3,7}, U_\sigma^{3,8} \subseteq L_\sigma^6$, this implies the statement.

5. $e \in U_\sigma^{3,9}$: The set $U_\sigma^{3,9}$ contains edges that are improving switches since phase 1. We thus refer to the rules listed in Section 4.4 describing the application of these edges. We need to investigate the occurrence record of switches that could have been applied during phase 1 but were not applied. By the rules 1 to 5 and Theorem 4.4.3, only one instead of two improving switches are switched in a bicycle A_i^j when $\phi^{\sigma_\mathfrak{b}}(A_i^j) = \mathfrak{b}$. Since we always chose to switch the edge with the lower occurrence record in a bicycle and their occurrence records differ at most by one by Equation (4.3), this implies $\phi^{\sigma_\mathfrak{b}}(b_{i,l}^j, A_i^j) = \lceil \mathfrak{b}/2 \rceil = \lfloor (\mathfrak{b}+1)/2 \rfloor$ for any $e = (b_{i,l}^j, A_i^j) \in U_\sigma^{3,9}$ with $\sigma(b_{i,l}^j) \neq A_i^j$. By Lemma 4.1.4 (3) resp. (4), we obtain $\lfloor (\mathfrak{b}+1)/2 \rfloor = f(\mathfrak{b}, 1)$ and consequently $\phi^\sigma(e) \geq \phi^{\sigma_\mathfrak{b}}(e) = f(\mathfrak{b}, 1) \geq f(\mathfrak{b}, \nu)$. □

Lemma 4.4.12. *Let σ be a phase 3 strategy and let e denote the switch that is applied in σ. Let σ' denote an arbitrary phase 3 strategy of $\sigma_\mathfrak{b} \to \sigma_{\mathfrak{b}+1}$ calculated after the strategy σ.*

1. *If $e = (k_i, k_\nu)$, then $I_{\sigma'} \cap S_{i,\sigma'}^{3,1} = \emptyset$.*
2. *If $e = (b_{i,l}^j, k_\nu)$ with $\sigma(b_{i,l}^j) \neq k_\nu$ and $(\mathfrak{b}+1)_i = 0$, then $I_{\sigma'} \cap S_{i,j,l,\sigma'}^{3,2} = \emptyset$.*
3. *If $e = (b_{i,l}^j, k_\nu)$ with $\sigma(b_{i,l}^j) \neq k_\nu$ and $(\mathfrak{b}+1)_{i+1} \neq j$, then $I_{\sigma'} \cap S_{i,j,l,\sigma'}^{3,3} = \emptyset$.*

Proof. We prove the first statement in detail and only sketch the proof of the other two statements since all of them use the same arguments.

1. Let $\tilde{e} \in S_{i,\sigma'}^{3,1}$. We show that $\tilde{e}$ is not an improving switch with respect to σ'. Due to the application of e in σ and since σ' is reached after σ, it holds that $\sigma'(k_i) = k_\nu$. Since $\tilde{e} \in S_{i,\sigma'}^{3,1}$, we have $\tilde{e} = (k_i, k_z)$ for some $z \leq \nu$ and $\sigma'(k_i) \neq k_z$. It thus suffices to show that $\mathrm{Val}_{\sigma'}(k_\nu) \geq \mathrm{Val}_{\sigma'}(k_z)$. Since σ' is a phase 3 policy, Lemma 4.3.6 implies

$$\mathrm{Val}_{\sigma'}(k_z) \leq \sum_{j \geq z : (\mathfrak{b}+1)_j = 1} \left[(-N)^{2j+8} + (-N)^{2j+7} + (-N)^7 + (-N)^6\right]. \tag{A.3}$$

Since σe is also a phase 3 policy, the active bicycle of level ν was already closed during phase 1 and $\sigma'(k_\nu) = c_\nu^{j'}$ where $j' = (\mathfrak{b}+1)_{i+1}$. In addition, no active bicycle contained in a level $j > \nu$ was opened as $\mathfrak{b}_j = (\mathfrak{b}+1)_j$ for these indices. This implies

$$\mathrm{Val}_{\sigma e}(k_\nu) = \sum_{j \geq \nu : (\mathfrak{b}+1)_j = 1} \left[(-N)^{2j+8} + (-N)^{2j+7} + (-N)^7 + (-N)^6\right]. \tag{A.4}$$

As the valuations are non-decreasing, $\mathrm{Val}_{\sigma'}(k_\nu) \geq \mathrm{Val}_{\sigma e}(k_\nu)$. Since $(\mathfrak{b}+1)_j = 0$ for all $j < \nu$, combining Equations (A.3) and (A.4) yields

$$\begin{aligned}\mathrm{Val}_{\sigma'}(k_\nu) \geq \mathrm{Val}_{\sigma e}(k_\nu) &= \sum_{j \in \geq \nu : (\mathfrak{b}+1)_j = 1} \left[(-N)^{2j+8} + (-N)^{2j+7} + (-N)^7 + (-N)^6\right] \\ &= \sum_{j \in [n] : (\mathfrak{b}+1)_j = 1} \left[(-N)^{2j+8} + (-N)^{2j+7} + (-N)^7 + (-N)^6\right]\end{aligned}$$

$$\geq \sum_{j\geq z:(\mathfrak{b}+1)_j=1} [(-N)^{2j+8}+(-N)^{2j+7}+(-N)^7+(-N)^6] \geq \mathrm{Val}_{\sigma'}(k_z)$$

Thus, $\mathrm{Val}_{\sigma'}(k_\nu) \geq \mathrm{Val}_{\sigma'}(k_z)$ and $\tilde{e}=(k_i,k_z)$ is not improving for σ'.

2. We need to show that for every phase 3 policy σ' reached after applying $e=(b_{i,r}^j,k_\nu)$ in σ, no switch contained in $S_{i,j,r,\sigma'}^{3,2}$ is an improving switch.

   Let σ' be a phase 3 policy of $\sigma_\mathfrak{b} \to \sigma_{\mathfrak{b}+1}$ reached after σ. Then, $\sigma e(b_{i,r}^j)=k_\nu$ and thus $\mathrm{Val}_{\sigma e}(b_{i,r}^j)=\mathrm{Val}_{\sigma e}(k_\nu)$. Since any edge $\tilde{e}\in S_{i,j,r,\sigma'}^{3,2}$ is of the form $\tilde{e}=(b_{i,r}^j,k_z)$ for some $z\leq\nu$, it therefore suffices to show $\mathrm{Val}_{\sigma'}(k_\nu)\geq\mathrm{Val}_{\sigma'}(k_z)$. This however follows by the same estimations used in the first case.

3. This is proven analogously to 2. □

Lemma 4.4.13. *Let σ be a phase 3 strategy. Then $L_\sigma^3 \cap \arg\min_{e'\in I_\sigma} \phi^\sigma(e') \neq \emptyset$.*

Proof. Since a policy is optimal if and only if the set of improving switches is empty, we have $I_\sigma\neq\emptyset$ as σ is a phase 3 strategy. Let $e\in\arg\min_{\tilde{e}\in I_\sigma}\phi^\sigma(e)$.

Since σ is a phase 3 strategy, $L_\sigma^3\neq\emptyset$. By [Fri11c, Lemma 4], $e\in L_\sigma^3$ or $e\in U_\sigma^3\setminus L_\sigma^3$. Let $e\in U_\sigma^3\setminus L_\sigma^3$, since the statement follows directly in the first case. Since $U_\sigma^{3,1},\dots,U_\sigma^{3,9}$ define a partition of U_σ^3, there is exactly one $k\in\{1,\dots,9\}$ with $e\in U_\sigma^{3,k}$.

Let $k\in\{4,\dots,9\}$. Then, by Lemma 4.4.10, $\phi^\sigma(e)\geq\phi^\sigma(\tilde{e})$ for all $\tilde{e}\in L_\sigma^3$ since $e\in I_\sigma$. Since e minimizes the occurrence record, this implies $\phi^\sigma(e)=\phi^\sigma(\tilde{e})$ for all $\tilde{e}\in L_\sigma^3$. This implies there is at least one $\tilde{e}\in L_\sigma^3$ minimizing the occurrence record, so $\tilde{e}\in\arg\min_{\tilde{e}\in I_\sigma}\phi^\sigma(\tilde{e})\cap L_\sigma^3$.

Now let $k\in\{1,2,3\}$ and distinguish the following cases.

1. $\boldsymbol{e\in U_\sigma^{3,1}}$: Then, $e=(k_i,k_z)$ for some $i\in[n]$ and $z\in[\ell]$ with $\sigma(k_i)\notin\{k_z,k_\nu\}$ and $(\mathfrak{b}+1)_i=0$. Thus, $e\in S_{i,\sigma}^{3,1}$. First assume that (k_i,k_ν) was not applied yet in the current transition. Then, $\phi^\sigma(k_i,k_\nu)=\phi^{\sigma_\mathfrak{b}}(k_i,k_\nu)$ and, by Table 4.5, $\phi^{\sigma_\mathfrak{b}}(k_i,k_\nu)=f(b,\nu,\{(i,0)\})$. Using $z\leq\nu$ and Lemma 4.1.4 (4), this implies

$$\phi^\sigma(k_i,k_\nu)=\phi^{\sigma_\mathfrak{b}}(k_i,k_\nu)=f(\mathfrak{b},\nu,\{(i,0)\})\leq f(\mathfrak{b},z,\{(i,0)\})=\phi^{\sigma_\mathfrak{b}}(e)\leq\phi^\sigma(e).$$

   Since e minimizes the occurrence records among all improving switches, it holds that $\phi^\sigma(k_i,k_\nu)=\phi^\sigma(e)$. This implies $(k_i,k_\nu)\in\arg\min_{\tilde{e}\in I_\sigma}\phi^\sigma(\tilde{e})$, so the statement follows from $(k_i,k_\nu)\in L_\sigma^3$.

   It remains to prove that (k_i,k_ν) was not applied yet. For the sake of a contradiction, assume that it was applied previously during this transition. Then there was a phase 3 strategy σ' reached before σ such that (k_i,k_ν) was applied in σ'. But then, by Lemma 4.4.12, $I_\sigma\cap S_{i,\sigma}^{3,1}=\emptyset$. This is a contradiction since $e\in I_\sigma$ and $e\in S_{i,\sigma}^{3,1}$.

2. $\boldsymbol{e\in U_\sigma^{3,2}}$: Then, $e=(b_{i,r}^j,k_z)$ for some $i\in[n]$ and $z[\nu]$ with $\sigma(b_{i,l}^j)\notin\{k_z,k_\nu\}$ and $(\mathfrak{b}+1)_i=0$. Hence, $e\in S_{i,j,r,\sigma}^{3,2}$. Assume that $(b_{i,r}^j,k_\nu)$ was not applied yet in the current transition. Then, since $z\leq\nu$, by Table 4.5 and by Lemma 4.1.4 (4),

$$\phi^\sigma(b_{i,r}^j,k_\nu)=\phi^{\sigma_\mathfrak{b}}(b_{i,l}^j,k_\nu)=f(\mathfrak{b},\nu,\{(i,0)\})+f(\mathfrak{b},\nu,\{(i,1),(i+1,1-j)\})$$

$$\leq f(\mathfrak{b}, z, \{(i,0)\}) + f(\mathfrak{b}, z, \{(i,1),(i+1,1-j)\}) = \phi^{\sigma_\mathfrak{b}}(e) \leq \phi^\sigma(e).$$

Since e minimizes the occurrence records among all improving switches, it holds that $\phi^\sigma(b_{i,r}^j, k_\nu) = \phi^\sigma(e)$. This implies that $(b_{i,r}^j, k_\nu) \in \arg\min_{\tilde{e} \in I_\sigma} \phi^\sigma(\tilde{e})$, so the statement follows from $(b_{i,r}^j, k_\nu) \in L_\sigma^3$.

It remains to show that $(b_{i,r}^j, k_\nu)$ was not applied yet. However, assuming that this switch was applied before results in the same contradiction as in the last case.

3. $\boldsymbol{e \in U_\sigma^{3,3}}$: This follows analogously to the previous case. □

Lemma 4.4.16. *Let $p \in \{1,2,4,5,6\}$ and let σ be a phase p strategy. Then, there is an improving switch $e \in L_\sigma^p$ such that $\phi^\sigma(e) \leq \min_{e' \in U_\sigma^p \cap I_\sigma} \phi^\sigma(e')$.*

Proof. We distinguish between the five possible choices for p. Let σ denote a phase p policy for the corresponding p.

- Let $p=1$. Then, since Table 4.4 implies $L_\sigma = I_\sigma = U_\sigma$ for any phase 1 policy σ, the statement follows directly.
- Let $p=2$ and $e \in L_\sigma^2$. Then, by Table 4.4, $e = (k_\nu, c_\nu^j)$ where $j = (\mathfrak{b}+1)_{\nu+1}$. Since $U_\sigma^2 = L_\sigma^1 \cup L_\sigma^2$ and $I_\sigma \subset U_\sigma^2$, it suffices to prove $\phi^\sigma(e) \leq \min_{e' \in L_\sigma^1} \phi^\sigma(e')$.

  Let $e' \in L_\sigma^1$. Then, $e' = (b_{i,k}^j, A_i^j)$ for some $i \in [n]$ and $j,k \in \{0,1\}$ and $\sigma(b_{i,k}^j) \neq A_i^j$. This implies that e' was not applied during phase 1. As we have already discussed in the proof of Lemma 4.4.10, we thus have $\phi^\sigma(e') = f(\mathfrak{b}, 1)$. But then, since $\phi^\sigma(e) = f(\mathfrak{b}, \nu, \{(\nu+1, j)\})$ by Table 4.5 and Lemma 4.1.4 (2,4), this implies that $\phi^\sigma(e) = f(\mathfrak{b}, \nu, \{(\nu+1, j)\}) \leq f(\mathfrak{b}, \nu) \leq f(\mathfrak{b}, 1) = \phi^\sigma(e')$.
- Let $p=4$ and $e \in L_{\sigma^4}$. As proven in Lemma 4.4.9, Case 1, $\phi^\sigma(e) = f(\mathfrak{b}, \nu)$. It therefore suffices to prove $\phi^\sigma(e) \geq f(\mathfrak{b}, \nu)$ for all $e \in U_\sigma^4 \cap I_\sigma$.

  By definition, $U_\sigma^4 = U_\sigma^{3,4} \cup \dots \cup U_\sigma^{3,9}$. We already proved $\phi^{\sigma'}(e') \geq f(\mathfrak{b}, \nu)$ for all $e' \in I_{\sigma'} \cap U_{\sigma'}^4$ in the proof of Lemma 4.4.10 when σ' is a phase 3 policy. The statement follows for phase 4 policies by applying the same arguments.
- Let $p=5$ and $e \in L_\sigma^5$, implying $e = (s, k_\nu)$. Thus, by Table 4.5, $\phi^\sigma(e) = f(\mathfrak{b}, \nu)$, so it suffices to prove $\phi^\sigma(e') \geq f(\mathfrak{b}, \nu)$ for all $e' \in U_\sigma^5$. This can be shown by the same arguments used in the proof of Lemma 4.4.10 since $U_\sigma^5 = U_\sigma^{3,5} \cup \dots \cup U_\sigma^{3,9}$.
- Let $p=6$ and $e \in L_\sigma^6$. Since $U_\sigma^6 = L_\sigma^1 \cup L_\sigma^6$, it suffices to prove $\phi^\sigma(e) \leq \phi^\sigma(e')$ for all $e' \in L_\sigma^1$. Let $e' \in L_{\sigma^1}$. As shown in the proof of Lemma 4.4.10, $\phi^\sigma(e') = f(\mathfrak{b}, 1)$ since $e' \in L_\sigma^6$ and e' was not applied during phase 1. Since $e \in L_\sigma^6$, either $e = (d_i^j, s)$ for some $i \in [n]$ and $j \in \{0,1\}$, implying $\phi^\sigma(e) = f(\mathfrak{b}, i+1) - j \cdot \mathfrak{b}_{i+1}$ or $e = (d_i^j, h_i^j)$ for some $i \in [n], j \in \{0,1\}$, implying $\phi^\sigma(e) = f(\mathfrak{b}, i+1) - (1-j) \cdot \mathfrak{b}_{i+1}$. Using Lemma 4.1.4 (4), this yields $\phi^\sigma(e) \leq f(\mathfrak{b}, i+1) \leq f(\mathfrak{b}, 2) \leq f(\mathfrak{b}, 1) = \phi^\sigma(e')$.

□

A.2. Proofs of Chapter 6

This part of the appendix contains all proofs related to the exponential lower bound construction. We begin by providing proofs for the statements of Section 6.1.

Omitted proofs of Section 6.1

Lemma 6.1.8. *Let $P = \{g_*, s_{*,*}, h_{i*,*}\}$ be the set of vertices with priorities in M_n. Let $S, S', P \subseteq P$ be non-empty subsets, let $\sum(S) := \sum_{v \in S} \langle v \rangle$ and define $\sum(S')$ analogously.*

1. *$|\sum(S)| < N^{2n+11}$ and $\varepsilon \cdot |\sum(S)| < 1$ for every subset $S \subseteq P$, and*
2. *$|\max_{v\in S} \langle v \rangle| < |\max_{v \in S'} \langle v \rangle|$ if and only if $|\sum(S)| < |\sum(S')|$.*

Proof. The highest priority of any vertex is $2n + 10$ and there are no more then $5n$ vertices with priorities. Let v, w be two vertices with $\Omega(v) > \Omega(w)$. Then, by construction, $\langle v \rangle \geq N \cdot \langle w \rangle$. In other words, if two vertices v, w do not have the same priority, then the rewards associated with the vertices are apart by at least a factor of N. Thus, for $S \subseteq P$,

$$\left|\sum(S)\right| \leq \left|\sum(P)\right| \leq |P| \cdot \left|\max_{v\in P} \langle v \rangle\right| \leq |P| \cdot N^{2n+10} = 5n \cdot N^{2n+10} < N^{2n+11},$$

implying the first statement since $\varepsilon = N^{-(2n+11)}$ by definition.

Let $S, S' \subseteq P$ be non-empty. Let $|\max_{v\in S} \langle v \rangle| < |\max_{v\in S'} \langle v \rangle|$. Then

$$\left|\sum(S)\right| \leq |S| \cdot \left|\max_{v\in S} \langle v \rangle\right| < 5n \left|\max_{v\in S} \langle v \rangle\right| \leq 5n \frac{|\max_{v\in S'} \langle v \rangle|}{N} < \left|\max_{v\in S'} \langle v \rangle\right| \leq \left|\sum(S')\right|.$$

Now let $|\sum(S)| < |\sum(S')|$. Then

$$\left|\max_{v\in S} \langle v \rangle\right| \leq \left|\sum(S)\right| < \left|\sum(S')\right| < 5n \left|\max_{v\in S'} \langle v \rangle\right|,$$

so $|\max_{v\in S} \langle v \rangle| < 5n\, |\max_{v\in S'} \langle v \rangle|$. Since $N \geq 7n$, this implies the statement. □

Lemma 6.1.10. *Let $\sigma \in \rho(\sigma_0)$ be well-behaved.*

1. *Let $\sigma(b_{\mu^\sigma}) = b_{\mu^\sigma+1}$. Then $L_i^* \preceq R_i^*$ for all $i \in [n]$ and $L_i^* \prec R_j^*$ for $j < i \leq \mu^\sigma$.*
2. *Let $\sigma(b_{\mu^\sigma}) = g_{\mu^\sigma}$. Then $L_i^* \succeq R_i^*$ for all $i \in [n]$ and $L_i \succ R_j^*$ for $i \leq \mu^\sigma$ and $j \in [n]$ and $L_i \oplus [\![g_j]\!] \succ R_j$ for $i \leq \mu^\sigma$ and $j < \mu^\sigma$.*
3. *Let $i \geq \mu^\sigma > j$. Then $R_j^* \prec [\![s_{i,j}, h_{i,j}]\!] \oplus L_{i+1}^*$.*
4. *For all $i \in [n]$, it holds that $[\![g_i, s_{i,*}, h_{i,*}]\!] \succ \bigoplus_{i'<i} W_{i'}^*$ and $L_1^* \prec [\![s_{i,j}, h_{i,j}]\!] \oplus L_{i+1}^*$.*

Proof. We prove the statements one after another.

1. The first statement follows directly if $i \geq \mu^\sigma$ since this implies $R_i^* = L_i^*$. Thus assume $i < \mu^\sigma$. Then, $\sigma(b_{\mu^\sigma}) = b_{\mu^\sigma+1}$ implies $L_i^* = \bigoplus_{\ell=i}^{\mu^\sigma-1} \{W_\ell^* \colon \sigma(b_\ell) = g_\ell\} \oplus L_{\mu^\sigma+1}^*$. The first statement follows since

$$\bigoplus_{\ell=i}^{\mu^\sigma-1} \{W_\ell^* \colon \sigma(b_\ell) = g_\ell\} \preceq \bigoplus_{\ell=i}^{\mu^\sigma-1} W_\ell^* \quad \text{and} \quad R_i^* = \bigoplus_{\ell=i}^{\mu^\sigma-1} W_\ell^* \oplus L_{\mu^\sigma+1}.$$

The second statement follows since $j < i \leq \mu^\sigma$ implies $\bigoplus_{\ell=i}^{\mu^\sigma-1} W_\ell^* \prec \bigoplus_{\ell=j}^{\mu^\sigma-1} W_\ell^*$.

2. The first statement follows directly if $i > \mu^\sigma$ since this implies $R_i^* = L_i^*$. Thus assume $i \leq \mu^\sigma$. Since $\sigma(b_{\mu^\sigma}) = g_{\mu^\sigma}$, the statement then follows since

$$\begin{aligned} L_i^* &= \bigoplus_{\ell \geq i} \{W_\ell^* : \sigma(b_\ell) = g_\ell\} = \bigoplus_{\ell=i}^{\mu^\sigma - 1} \{W_\ell^* : \sigma(b_\ell) = g_\ell\} \oplus W_{\mu^\sigma} \oplus L_{\mu^\sigma+1}^* \\ &\succ \bigoplus_{\ell=1}^{\mu^\sigma-1} W_\ell^* \oplus L_{\mu^\sigma+1}^* = R_1^* \succeq R_j^*. \end{aligned}$$

The same calculation implies the third statement as the estimations remain correct if $j < \mu^\sigma$.

3. By $i \geq \mu^\sigma > j$, we have

$$\begin{aligned} R_j^* &= \bigoplus_{\ell=j}^{\mu^\sigma-1} W_\ell \oplus \bigoplus_{\ell \geq \mu^\sigma+1} \{W_\ell \colon \sigma(b_\ell) = g_\ell\} \prec \bigoplus_{\ell=1}^{\mu^\sigma} W_\ell \oplus \bigoplus_{\ell \geq \mu^\sigma+1} \{W_\ell \colon \sigma(b_\ell) = g_\ell\} \\ &\prec \bigoplus_{\ell=1}^{i} W_\ell \oplus \bigoplus_{\ell \geq i+1} \{W_\ell \colon \sigma(b_\ell) = g_\ell\} \prec [\![s_{i,j}, h_{i,j}]\!] \oplus L_{i+1}^*. \end{aligned}$$

4. For $G_n = S_n$, the statement follows since the most significant difference is the vertex h_{i*} and since the priority of this vertex is even. For $G_n = M_n$, this follows intuitively since the $\langle h_{i,*}\rangle$ has the largest exponent of all terms in the expression and since it is the only vertex that has this exponent. Thus, $\langle h_{i,*}\rangle$ is by a factor of N larger than all other quantities in the given expression, and as N is sufficiently large, this term dominates. Formally, this can be shown by an easy but tedious calculation.

   The second statement follows from the first. □

Lemma 6.1.12. *Let $\sigma \in \rho(\sigma_0)$ be well-behaved and $i < \mu^\sigma$. Then $\mathrm{rVal}_\sigma^{\mathrm{S}}(g_i) = R_i^{\mathrm{S}}$ and*

$$\mathrm{rVal}_\sigma^{\mathrm{M}}(g_i) = \begin{cases} B_2^{\mathrm{M}} + \sum_{j=i}^{k-1} W_j^{\mathrm{M}} + \langle g_k \rangle, & \text{if } k := \min\{k \geq i \colon \neg\bar{\sigma}(d_k)\} < \mu^\sigma \\ \mathrm{rVal}_\sigma^{\mathrm{M}}(g_i) = R_i^{\mathrm{M}}, & \text{otherwise.} \end{cases}$$

Proof. This statement is shown by backwards induction on i. Let $i = \mu^\sigma - 1$, implying $\sigma(g_i) = F_{i,0}$ by Property (BR1).

- Let $\bar{\sigma}(d_i)$. Since $\sigma(s_{i,0}) = h_{i,0}$ by Property (S2), Lemma 6.1.11 yields

$$\mathrm{rVal}_\sigma^*(g_i) = W_i^* \oplus \mathrm{rVal}_\sigma^*(b_{i+2}) = W_i^* \oplus B_{i+2}^* = W_i^* \oplus L_{i+2}^* = R_i^*.$$

- Let $\neg\bar{\sigma}(d_i)$. Consider $G_n = S_n$ first. By Property (BR2), either $\tau^\sigma(F_{i,0}) = d_{i,0,k}$ where $\sigma(d_{i,0,k}) = e_{i,0,k}$ and $\sigma(e_{i,0,k}) = b_2$ for some $k \in \{0,1\}$ or $\tau^\sigma(F_{i,0}) = s_{i,0}$. Since player 1 chooses $\tau^\sigma(F_{i,0})$ such that the valuation of g_i is minimized we need to compare $B_2^{\mathrm{S}} \cup \{g_i\}$ (if player 1 chooses $d_{i,0,k}$) and R_i^{S} (if player 1 chooses $s_{i,0}$). Note that $\sigma(b_2) = b_3$ if $\mu^\sigma > 2$ by Property (EB6) and that $\mu^\sigma = 2$ implies $B_2^{\mathrm{S}} = L_2^{\mathrm{S}}$. We prove that this implies $R_i^{\mathrm{S}} \lhd L_2^{\mathrm{S}} \cup \{g_i\}$ and thus $\mathrm{rVal}_\sigma^{\mathrm{S}}(g_i) = R_i^{\mathrm{S}}$. As mentioned before, $\sigma(g_i) = F_{i,0}$. In addition, we have $\bar{\sigma}(eb_{i,0}) \wedge \neg\bar{\sigma}(eg_{i,0})$. Thus, by Property (EB1), we have $\bar{\sigma}(b_{i+1}) = \bar{\sigma}(b_{\mu^\sigma}) \neq 0$. Hence $\sigma(b_{\mu^\sigma}) = g_{\mu^\sigma}$ and $\sigma(b_{i+1}) = g_{i+1}$. The statement thus follows from Lemma 6.1.10 (2). Consider $G_n = M_n$ next. By the choice of i and assumption, it holds that $i = \min\{k \geq i\colon \neg\bar{\sigma}(d_k)\} < \mu^\sigma$. By Property (BR2), this implies $\neg\bar{\sigma}(eg_{i,0})$. Thus $\mathrm{rVal}_\sigma^{\mathrm{M}}(F_{i,j}) = \mathrm{rVal}_\sigma^{\mathrm{M}}(b_2)$. Therefore, $\mathrm{rVal}_\sigma^{\mathrm{M}}(g_i) = \langle g_i\rangle + \mathrm{rVal}_\sigma^{\mathrm{M}}(b_2) = \langle g_i\rangle + B_2^{\mathrm{M}}$ by Lemma 6.1.11.

Now let $i < \mu^\sigma - 1$, implying $\sigma(g_i) = F_{i,1}$ by Property (BR1) and $\mu^\sigma \geq 3$.

1. Let $\bar{\sigma}(d_i)$. Consider $G_n = S_n$ first. Then, $\tau^\sigma(F_{i,1}) = s_{i,1}$ and $\sigma(s_{i,1}) = h_{i,1}$ by Property (S2). Using the induction hypotheses, this yields

$$\mathrm{rVal}_\sigma^{\mathrm{S}}(g_i) = W_i^{\mathrm{S}} \cup \mathrm{rVal}_\sigma^{\mathrm{S}}(g_{i+1}) = W_i^{\mathrm{S}} \cup R_{i+1}^{\mathrm{S}} = R_i^{\mathrm{S}}.$$

   If $G_n = M_n$, then the same property implies $\mathrm{rVal}_\sigma^{\mathrm{M}}(F_{i,1}) = \mathrm{rVal}_\sigma^{\mathrm{M}}(s_{i,1})$ as well as $\sigma(s_{i,1}) = h_{i,1}$. Thus $\mathrm{rVal}_\sigma^{\mathrm{M}}(g_i) = W_i^{\mathrm{M}} + \mathrm{rVal}_\sigma^{\mathrm{M}}(g_{i+1})$. Applying the induction hypotheses to $\mathrm{rVal}_\sigma^{\mathrm{M}}(g_{i+1})$ yields the result.

2. Let $\neg\bar{\sigma}(d_i)$. Consider $G_n = S_n$ first. By the same arguments used for $i = \mu^\sigma - 1$, we need to show $R_i^{\mathrm{S}} \lhd B_2^{\mathrm{S}} \cup \{g_i\}$. By Properties (BR2) and (EB2) we thus have $\sigma(b_2) = b_3$ and $B_2 = L_2$ as $\mu^\sigma \geq 3$. It thus suffices to prove $R_i^{\mathrm{S}} \lhd L_2^{\mathrm{S}} \cup \{g_i\}$. Let, for the sake of contradiction, $\mu^\sigma = \min\{i'\colon \sigma(b_{i'}) = b_{i'+1}\}$. By Property (BR2) and $\neg\bar{\sigma}(d_i)$, we have $\bar{\sigma}(eb_{i,1}) \wedge \neg\bar{\sigma}(eg_{i,1})$. Thus, by Property (EB1), $\sigma(b_{i+1}) = b_{i+2}$. But this implies $\mu^\sigma \leq i+1$, contradicting $i < \mu^\sigma$. Hence $\mathcal{I}^\sigma \neq \emptyset$, implying $\sigma(b_{\mu^\sigma}) = g_{\mu^\sigma}$ by Lemma 6.1.4. But then, the statement follows from Lemma 6.1.10 (2).

   Consider $G_n = M_n$ next. Then $i = \min\{k \geq i\colon \neg\bar{\sigma}(d_k)\} < \mu^\sigma$. As in the case $i = \mu^\sigma - 1$, we have $\mathrm{rVal}_\sigma^{\mathrm{M}}(F_{i,j}) = \mathrm{rVal}_\sigma^{\mathrm{M}}(b_2)$. Therefore,

$$\mathrm{rVal}_\sigma^{\mathrm{M}}(g_i) = \langle g_i\rangle + \mathrm{rVal}_\sigma^{\mathrm{M}}(b_2) = \langle g_i\rangle + B_2^{\mathrm{M}}$$

   by Lemma 6.1.11.

□

Lemma 6.1.14. *Let $\mu^\sigma = 1$ and $m := \min\{\overline{m}_g^\sigma, \overline{m}_s^\sigma\}$. Then*

$$\mathrm{rVal}_\sigma^*(g_1) = \begin{cases} \langle g_1\rangle + \mathrm{rVal}_\sigma^{\mathrm{M}}(b_2), & \text{if } m_b^\sigma \leq \overline{m}_s^\sigma, \overline{m}_g^\sigma \wedge G_n = M_n \wedge \neg\bar{\sigma}(d_1), \\ W_1^* \oplus \mathrm{rVal}_\sigma^*(b_2), & \text{if } m_b^\sigma \leq \overline{m}_s^\sigma, \overline{m}_g^\sigma, \\ & \quad \wedge (G_n = S_n \vee [G_n = M_n \wedge \bar{\sigma}(d_1)]), \\ \bigoplus_{i'=1}^{m} W_{i'}^* \oplus \mathrm{rVal}_\sigma^*(b_{\overline{m}_g^\sigma+2}) & \text{if } \overline{m}_g^\sigma < \overline{m}_s^\sigma, m_b^\sigma, \\ & \quad \wedge [(\bar{\sigma}(b_{\overline{m}_g^\sigma+1}) \wedge G_n = S_n) \vee \neg\bar{\sigma}(eb_{\overline{m}_g^\sigma})], \\ \bigoplus_{i'=1}^{m-1} W_{i'}^* \oplus [\![g_m]\!] \oplus \mathrm{rVal}_\sigma^*(b_2) & \text{otherwise.} \end{cases}$$

Proof. This statement is proven by distinguishing between several cases. Most of the cases are proven by backwards induction, some are proven directly.

1. $\boldsymbol{m_b^\sigma \leq \overline{m}_s^\sigma, \overline{m}_g^\sigma, G_n = M_n}$ **and** $\boldsymbol{\neg\bar{\sigma}(d_1)}$**:** We prove that this implies that we have $\mathrm{rVal}_\sigma^{\mathrm{M}}(g_1) = \langle g_1 \rangle + \mathrm{rVal}_\sigma^{\mathrm{M}}(b_2)$. By Lemma 6.1.6, $m_b^\sigma \leq \overline{m}_s^\sigma, \overline{m}_g^\sigma$ implies $m_b^\sigma = 2$. Thus, $\sigma(b_2) = g_2, \sigma(g_1) = F_{1,1}$ and it suffices to prove $\bar{\sigma}(eb_{1,1}) \wedge \neg\bar{\sigma}(eg_{1,1})$. By Property (EG1), it cannot hold that $\bar{\sigma}(eg_{1,1}) \wedge \neg\bar{\sigma}(eb_{1,1})$ as this would imply $\sigma(s_{1,1}) = b_1$. This however contradicts $\sigma(s_{1,1}) = h_{1,1}$ which follows from $1 < \overline{m}_s^\sigma$ and $\sigma(g_1) = F_{1,1}$. By Property (EBG3), we cannot have $\bar{\sigma}(eb_{1,1}) \wedge \bar{\sigma}(eg_{1,1})$ as this would imply $\bar{\sigma}(d_1)$, contradicting the current assumptions. Thus, $\bar{\sigma}(eb_{1,1}) \wedge \neg\bar{\sigma}(eg_{1,1})$.
2. $\boldsymbol{m_b^\sigma \leq \overline{m}_s^\sigma, \overline{m}_g^\sigma, G_n = M_n}$ **and** $\boldsymbol{\bar{\sigma}(d_1)}$**:** By Lemma 6.1.6, it holds that $m_b^\sigma = 2$. This implies $\sigma(b_2) = g_2, \sigma(g_1) = F_{1,1}$ and $\sigma(s_{1,1}) = h_{1,1}$. Thus, the chose cycle center of level 1 is closed, implying $\mathrm{rVal}_\sigma^{\mathrm{M}}(g_1) = W_1^{\mathrm{M}} + \mathrm{rVal}_\sigma^{\mathrm{M}}(b_2)$.
3. $\boldsymbol{m_b^\sigma \leq \overline{m}_s^\sigma, \overline{m}_g^\sigma}$ **and** $\boldsymbol{G_n = S_n}$**:** By the same argument used in the last case, it suffices to prove $\bar{\sigma}(d_1)$. This however follows since $\neg\bar{\sigma}(eb_1)$ by Property (MNS1) and $\neg\bar{\sigma}(eg_1)$ by Property (EG1).
4. $\boldsymbol{\overline{m}_g^\sigma < \overline{m}_s^\sigma, m_b^\sigma \wedge \neg\bar{\sigma}(eb_{\overline{m}_g^\sigma}) \wedge [G_n = S_n \implies \neg\bar{\sigma}(b_{\overline{m}_g^\sigma+1})]}$**:** We prove that

$$\mathrm{rVal}_\sigma^*(g_i) = \bigoplus_{i'=i}^{\overline{m}_g^\sigma} W_{i'}^* \oplus \mathrm{rVal}_\sigma^*(b_{\overline{m}_g^\sigma+2}) \tag{A.5}$$

for all $i \leq \overline{m}_g^\sigma$ by backwards induction. Let $i = \overline{m}_g^\sigma$. Then, by the choice of i, $\sigma(g_i) = F_{i,0}$ and $\sigma(s_{i,0}) = h_{i,0}$. Since we assume $\neg\bar{\sigma}(eb_{\overline{m}_g^\sigma}) = \neg\bar{\sigma}(eb_{i,0})$, $F_{i,0}$ does not escape towards b_2. In addition, by Property (EG1), it cannot be the case that $\bar{\sigma}(eg_{i,0}) \wedge \neg\bar{\sigma}(eb_{i,0})$ as this would imply $\sigma(s_{i,0}) = b_1$. Hence $F_{i,0}$ is closed, so $\mathrm{rVal}_\sigma^*(F_{i,0}) = \mathrm{rVal}_\sigma^*(s_{i,0})$. Consequently, $\mathrm{rVal}_\sigma^*(g_i) = W_i^* \oplus \mathrm{rVal}_\sigma^*(b_{i+2})$ which is exactly Equation (A.5) by the choice of i.

Let $i < \overline{m}_g^\sigma$. By $i < \overline{m}_g^\sigma$, it holds that $\sigma(g_i) = F_{i,1}$. By Property (MNS2), it also holds that $\neg\bar{\sigma}(eb_i)$. Using Property (EG1) as before, we conclude that $F_{i,1}$ is closed, so $\mathrm{rVal}_\sigma^*(F_{i,1}) = \mathrm{rVal}_\sigma^*(s_{i,1})$. Also, $\sigma(s_{i,1}) = h_{i,1}$ since $i < \overline{m}_s^\sigma$. This implies $\mathrm{rVal}_\sigma^*(g_i) = W_i^* \oplus \mathrm{rVal}_\sigma^*(g_{i+1})$, so Equation (A.5) follows from the induction hypothesis.
5. $\boldsymbol{\overline{m}_g^\sigma < \overline{m}_s^\sigma, m_b^\sigma \wedge \bar{\sigma}(b_{\overline{m}_g^\sigma+1})}$ **and** $\boldsymbol{G_n = S_n}$**:** We prove that

$$\mathrm{rVal}_\sigma^{\mathrm{S}}(g_i) = \bigcup_{i'=i}^{\overline{m}_g^\sigma} W_{i'}^{\mathrm{S}} \cup \mathrm{rVal}_\sigma^{\mathrm{S}}(b_{\overline{m}_g^\sigma+2}) \tag{A.6}$$

for all $i \leq \overline{m}_g^\sigma$ by backwards induction. Let $i = \overline{m}_g^\sigma$. Then, by the choice of i, it holds that $\sigma(g_i) = F_{i,0}$ and $\sigma(s_{i,0}) = h_{i,0}$. By Property (EG1), it cannot be the case that $\bar{\sigma}(eg_{i,0}) \wedge \neg\bar{\sigma}(eb_{i,0})$ as this would imply $\sigma(s_{i,0}) = b_1$. By Property (EBG1), it cannot be the case that $\bar{\sigma}(eg_{i,0}) \wedge \bar{\sigma}(eb_{i,0})$ as this would imply $\bar{\sigma}(b_{i+1}) = 0$, contradicting the assumption. In particular, this implies $\neg\bar{\sigma}(eg_{i,0})$. Thus, depending on the choice of player 1, either $\mathrm{rVal}_\sigma^{\mathrm{S}}(F_{i,0}) = \mathrm{rVal}_\sigma^{\mathrm{S}}(b_2)$ or $\mathrm{rVal}_\sigma^{\mathrm{S}}(F_{i,0}) = \mathrm{rVal}_\sigma^{\mathrm{S}}(s_{i,0})$. It is now easy to see that $\bar{\sigma}(b_{\overline{m}_g^\sigma+1})$ implies $\mathrm{rVal}_\sigma^{\mathrm{S}}(s_{i,0}) \lhd \mathrm{rVal}_\sigma^{\mathrm{S}}(b_2)$ and thus $\tau^\sigma(F_{i,0}) = s_{i,0}$. Hence $\mathrm{rVal}_\sigma^{\mathrm{S}}(g_i) = W_i^{\mathrm{S}} \cup \mathrm{rVal}_\sigma^{\mathrm{S}}(b_{i+2})$ as required.

Let $i < \overline{m}_g^\sigma$. By $i < \overline{m}_g^\sigma$, it follows that $\sigma(g_i) = F_{i,1}$ and consequently $\sigma(s_{i,1}) = h_{i,1}$. Using Properties (EG1), (EBG1) and $i < m_b^\sigma$, we can again conclude that $\neg\bar{\sigma}(eg_{i,1})$. Consequently, either $\mathrm{rVal}_\sigma^{\mathrm{S}}(F_{i,1}) = \mathrm{rVal}_\sigma^{\mathrm{S}}(b_2)$ or $\mathrm{rVal}_\sigma^{\mathrm{S}}(F_{i,1}) = \mathrm{rVal}_\sigma^{\mathrm{S}}(s_{i,1})$. Using $\bar{\sigma}(b_{m_g^\sigma+1})$ and the induction hypotheses, it is again follows that$\mathrm{rVal}_\sigma^{\mathrm{S}}(s_{i,1}) \lhd \mathrm{rVal}_\sigma^{\mathrm{S}}(b_2)$. Thus, the statement again follows by applying the induction hypotheses.

6. $\boldsymbol{\overline{m}_g^\sigma < \overline{m}_s^\sigma, m_b^\sigma \wedge \neg\bar{\sigma}(b_{\overline{m}_g^\sigma+1}) \wedge \bar{\sigma}(eb_{\overline{m}_g^\sigma})}$ **and** $\boldsymbol{G_n = S_n}$**:** Let, for the sake of contradiction, $\overline{m}_g^\sigma > 1$. Then, by Lemma 6.1.6, $m_b^\sigma = \overline{m}_g^\sigma + 1$ and in particular $\bar{\sigma}(b_{\overline{m}_g^\sigma+1})$, contradicting the assumption. Thus, $\overline{m}_g^\sigma = 1$. This implies $\sigma(g_1) = F_{1,0}$ and $\sigma(s_{1,0}) = h_{1,0}$. Let, for the sake of contradiction, $\bar{\sigma}(eb_{1,0}) \wedge \bar{\sigma}(eg_{1,0})$. Then, $\bar{\sigma}(g_1) = \bar{\sigma}(b_2)$ as $\neg\bar{\sigma}(b_2)$ by assumption and $\overline{m}_g^\sigma = 1$. But then, Property (EBG3) implies $\bar{\sigma}(d_1)$ which is a contradiction. Consequently, $\bar{\sigma}(eb_{i,0}) \wedge \neg\bar{\sigma}(eg_{i,0})$. Since

$$\mathrm{rVal}_\sigma(b_2) \lhd \{s_{1,0}, h_{1,0}\} \cup \mathrm{rVal}_\sigma(b_2) = \langle s_{1,0}, h_{1,0}\rangle \cup \mathrm{rVal}_\sigma(b_3),$$

this yields $\mathrm{rVal}_\sigma^{\mathrm{S}}(g_1) = \{g_1\} \cup \mathrm{rVal}_\sigma(b_3)$.

7. $\boldsymbol{\overline{m}_g^\sigma < \overline{m}_s^\sigma, m_b^\sigma \wedge \bar{\sigma}(eb_{\overline{m}_g^\sigma})}$ **and** $\boldsymbol{G_n = M_n}$**:** We prove that

$$\mathrm{rVal}_\sigma^{\mathrm{M}}(g_i) = \sum_{i'=i}^{\overline{m}_g^\sigma - 1} W_i^{\mathrm{M}} + \left\langle g_{\overline{m}_g^\sigma} \right\rangle + \mathrm{rVal}_\sigma^{\mathrm{M}}(b_2) \tag{A.7}$$

for all $i \leq \overline{m}_g^\sigma$ by backwards induction. Let $i = \overline{m}_g^\sigma$. Then, by construction, $\overline{m}_g^\sigma \neq n$ as $m_b^\sigma \leq n$. We prove $\neg\bar{\sigma}(eg_{i,0})$. Assume otherwise, implying $\bar{\sigma}(eb_{i,0}) \wedge \bar{\sigma}(eg_{i,0})$. Assume $\overline{m}_g^\sigma > 1$. Then, by Lemma 6.1.6, $\bar{\sigma}(b_{\overline{m}_g^\sigma+1}) = 1$, contradicting Property (EBG1). Hence assume $\overline{m}_g^\sigma = 1$. If $\bar{\sigma}(b_2) = \bar{\sigma}(g_1)$, then Property (EBG3) implies $\bar{\sigma}(d_1)$, contradicting the assumption. If $\bar{\sigma}(b_2) \neq \bar{\sigma}(g_1)$, then $\bar{\sigma}(b_2) = 1$, again contradicting Property (EBG1). Thus, $\neg\bar{\sigma}(eg_{i,0})$ needs to hold. Consequently, as $\bar{\sigma}(eb_{i,0}) \wedge \neg\bar{\sigma}(eg_{i,0})$, this yields $\mathrm{rVal}_\sigma^{\mathrm{M}}(g_{\overline{m}_g^\sigma}) = \langle g_{\overline{m}_g^\sigma}\rangle + \mathrm{rVal}_{\sigma e}^{\mathrm{M}}(b_2)$.

Let $i < \overline{m}_g^\sigma$. Then $\sigma(g_i) = F_{i,1}$ and $\sigma(s_{i,1}) = h_{i,1}$. In addition, by Property (MNS2), $\neg\bar{\sigma}(eb_{i,1})$. Since $\bar{\sigma}(eg_{i,1})$ would imply $\sigma(s_{i,1}) = b_1$ by Property (EB1), we have $\neg\bar{\sigma}(eb_i) \wedge \neg\bar{\sigma}(eg_i)$, implying $\bar{\sigma}(d_i)$. This implies $\mathrm{rVal}_\sigma^{\mathrm{M}}(g_i) = W_i^{\mathrm{M}} + \mathrm{rVal}_\sigma^{\mathrm{M}}(g_{i+1})$ and the statement then follows by using the induction hypotheses.

8. $\boldsymbol{\overline{m}_s^\sigma \leq \overline{m}_g^\sigma < m_b^\sigma}$ **or** $\boldsymbol{\overline{m}_s^\sigma < m_b^\sigma \leq \overline{m}_g^\sigma}$**:** Let $m := \min\{\overline{m}_g^\sigma, \overline{m}_s^\sigma\}$. We prove that

$$\mathrm{rVal}_\sigma^*(g_i) = \bigoplus_{i'=i}^{m-1} W_{i'}^* \oplus [\![g_m]\!] \oplus \mathrm{rVal}_\sigma^*(b_2) \tag{A.8}$$

for all $i \leq m$. Let $i = m$ and $j := \bar{\sigma}(g_i)$. We can assume $i = \overline{m}_s^\sigma$ in both cases, implying $i = \overline{m}_s^\sigma \leq n$ in both cases. By either Property (MNS4) or Property (MNS6), we have $\bar{\sigma}(eb_i) \wedge \neg\bar{\sigma}(eg_i)$. This implies $\mathrm{rVal}_\sigma^{\mathrm{M}}(F_{i,j}) = \mathrm{rVal}_\sigma^{\mathrm{M}}(b_2)$, hence the statement follows for $G_n = M_n$. Consider the case $G_n = S_n$. Since $i = \overline{m}_s^\sigma$, we have $\sigma(s_{i,j}) = b_1$. Therefore, using $\sigma(b_1) = b_2$, we obtain $\mathrm{rVal}_\sigma^{\mathrm{S}}(s_{i,j}) = \{s_{i,j}\} \cup \mathrm{rVal}_\sigma^{\mathrm{S}}(b_2) \rhd \mathrm{rVal}_\sigma^{\mathrm{S}}(b_2)$. Thus $\tau^\sigma(F_{i,j}) = d_{i,j,k}$ and therefore $\mathrm{rVal}_\sigma^{\mathrm{S}}(g_m) = \{g_m\} \cup \mathrm{rVal}_\sigma^{\mathrm{S}}(b_2)$.

Let $i < m$. Since $i < \overline{m}_s^\sigma \leq \overline{m}_g^\sigma$ in both cases, $\sigma(g_i) = F_{i,1}$ and $\sigma(s_{i,1}) = h_{i,1}$. Let $G_n = S_n$. By Property (EG1), it cannot be the case that $\bar{\sigma}(eg_{i,1}) \wedge \neg\bar{\sigma}(eb_{i,1})$

as this would imply $\sigma(s_{i,1}) = b_1$. By Property (EBG1), it cannot be the case that $\bar{\sigma}(eb_{i,1}) \wedge \bar{\sigma}(eg_{i,1})$ as this would imply $\sigma(b_{i+1}) = g_{i+1}$, contradicting the choice of i. Hence, either $\bar{\sigma}(d_{i,j})$ or $\bar{\sigma}(eb_{i,1}) \wedge \neg\bar{\sigma}(eg_{i,1})$. We prove that $\tau^\sigma(F_{i,j}) = s_{i,j}$ holds in any case. It suffices to consider the second case as this follows directly in the first case. Since $\bar{\sigma}(eb_{i,1}) \wedge \neg\bar{\sigma}(eg_{i,1})$, it suffices to prove $\mathrm{rVal}^{\mathrm{S}}_\sigma(s_{i,1}) \lhd \mathrm{rVal}^{\mathrm{S}}_\sigma(b_2)$. This however follows by the induction hypotheses and $\mathrm{rVal}^{\mathrm{S}}_\sigma(s_{i,1}) = \{s_{i,1}, h_{i,1}\} \cup \mathrm{rVal}^{\mathrm{S}}_\sigma(g_{i+1})$. Thus, $\tau^\sigma(F_{i,j}) = s_{i,j}$ for $G_n = S_n$ in any case.

If $G_n = M_n$, then either Property (MNS3) or Property (MNS5) implies the cycle center $F_{i,1}$ is closed. Hence, using the induction hypotheses, Equation (A.8) follows from $\mathrm{rVal}^*_\sigma(g_i) = W^*_i \oplus \mathrm{rVal}^*_\sigma(g_{i+1})$.

Note that the cases listed here suffice, i.e., every possible relation between the three parameters $\overline{m}^\sigma_s, \overline{m}^\sigma_g$ and m^σ_b is covered by exactly one of the cases. □

Lemma 6.1.16. *Let $G_n = S_n$. Let $\sigma \in \rho(\sigma_0)$ be well-behaved and $i \in [n], j \in \{0,1\}$. Then*

$$\mathrm{rVal}^{\mathrm{S}}_\sigma(F_{i,j}) = \begin{cases} \mathrm{rVal}^{\mathrm{S}}_\sigma(s_{i,j}), & \text{if } \bar{\sigma}(d_{i,j}), \\ \{s_{i,j}\} \cup \mathrm{rVal}^{\mathrm{S}}_\sigma(b_2), & \text{if } \bar{\sigma}(eg_{i,j}) \wedge \neg\bar{\sigma}(eb_{i,j}) \wedge \mu^\sigma = 1, \\ \mathrm{rVal}^{\mathrm{S}}_\sigma(g_1), & \text{if } \bar{\sigma}(eg_{i,j}) \wedge \neg\bar{\sigma}(eb_{i,j}) \wedge \mu^\sigma \neq 1, \\ \mathrm{rVal}^{\mathrm{S}}_\sigma(b_2), & \text{if } \bar{\sigma}(eb_{i,j}) \wedge \neg\bar{\sigma}(eg_{i,j}) \wedge \mu^\sigma = 1 \\ & \quad \wedge (\neg\bar{\sigma}(s_{i,j}) \vee \bar{\sigma}(b_{i+1}) = j), \\ \mathrm{rVal}^{\mathrm{S}}_\sigma(s_{i,j}), & \text{if } \bar{\sigma}(eb_{i,j}) \wedge \neg\bar{\sigma}(eg_{i,j}) \\ & \quad \wedge (\mu^\sigma \neq 1 \vee (\bar{\sigma}(s_{i,j}) \wedge \bar{\sigma}(b_{i+1}) \neq j)), \\ \mathrm{rVal}^{\mathrm{S}}_\sigma(g_1), & \text{if } \bar{\sigma}(eb_{i,j}) \wedge \bar{\sigma}(eg_{i,j}) \wedge \bar{\sigma}(g_1) \neq \bar{\sigma}(b_2), \\ \mathrm{rVal}^{\mathrm{S}}_\sigma(b_2), & \text{if } \bar{\sigma}(eb_{i,j}) \wedge \bar{\sigma}(eg_{i,j}) \wedge \bar{\sigma}(g_1) = \bar{\sigma}(b_2). \end{cases}$$

Proving this statement requires the following additional lemma.

Lemma A.2.1. *Let $G_n = S_n$ and let $\sigma \in \rho(\sigma_0)$. Let $i \in [n]$ such that $\bar{\sigma}(eg_i)$. Then there is some $i' < i$ such that either $\sigma(g_{i'}) = F_{i',0}$ or $\sigma(s_{i',\bar{\sigma}(g_{i'})}) = b_1$.*

Proof. Let, for the sake of contradiction, $\sigma(g_{i'}) = F_{i',1}$ and $\sigma(s_{i',\bar{\sigma}(g_{i'})}) = h_{i',\bar{\sigma}(g_{i'})}$ for all $i' < i$. Then, player 1 can create a cycle by setting $\tau^\sigma(F_{k,1}) = s_{k,1}$ for all $k < i$ and $\tau(F_{i,\bar{\sigma}(g_i)}) = d_{i,\bar{\sigma}(g_i),k}$ where k is chosen such that the cycle center escapes towards g_1. But this contradicts the fact that S_n is a sink game. □

Proof of Lemma 6.1.16. Consider a cycle center $F_{i,j}$. We distinguish the following cases:

1. **$\bar{\sigma}(d_{i,j})$:** Then $F_{i,j}$ is closed and $\mathrm{rVal}^{\mathrm{S}}_\sigma(F_{i,j}) = \mathrm{rVal}^{\mathrm{S}}_\sigma(s_{i,j})$ since S_n is a sink game.
2. **$\bar{\sigma}(eg_{i,j}), \neg\bar{\sigma}(eb_{i,j})$ and $\mu^\sigma = 1$:** We prove $\mathrm{rVal}^{\mathrm{S}}_\sigma(F_{i,j}) = \{s_{i,j}\} \cup \mathrm{rVal}^{\mathrm{S}}_\sigma(b_2)$. Since $\neg\bar{\sigma}(eb_{i,j})$, player 1 choose a cycle vertex escaping towards g_1 or $s_{i,j}$. As player 1 minimizes the vertex valuations, it suffices to prove $\mathrm{rVal}^{\mathrm{S}}_\sigma(s_{i,j}) \lhd \mathrm{rVal}^{\mathrm{S}}_\sigma(g_1)$. Property (EG2) implies $\bar{\sigma}(d_1)$ and hence $\mathrm{rVal}^{\mathrm{S}}_\sigma(g_1) = \{g_1\} \cup \mathrm{rVal}^{\mathrm{S}}_\sigma(s_{1,\bar{\sigma}(g_1)})$. By Property (EG3), also $\sigma(s_{1,\bar{\sigma}(g_1)}) = h_{1,\bar{\sigma}(g_1)}$. Since $\bar{\sigma}(g_1) = \bar{\sigma}(b_2)$ by Property (EG4), it then follows that $\mathrm{rVal}^{\mathrm{S}}_\sigma(h_{1,\bar{\sigma}(g_1)}) = \mathrm{rVal}^{\mathrm{S}}_\sigma(b_2)$. Hence $\mathrm{rVal}^{\mathrm{S}}_\sigma(g_1) = W^{\mathrm{S}}_1 \cup \mathrm{rVal}^{\mathrm{S}}_\sigma(b_2)$. Furthermore, $\mathrm{rVal}^{\mathrm{S}}_\sigma(s_{i,j}) = \{s_{i,j}\} \cup \mathrm{rVal}^{\mathrm{S}}_\sigma(b_1) = \{s_{i,j}\} \cup \mathrm{rVal}^{\mathrm{S}}_\sigma(b_2)$ by Property (EG1) and $\mu^\sigma = 1$, and consequently $\mathrm{rVal}^{\mathrm{S}}_\sigma(s_{i,j}) \lhd \mathrm{rVal}^{\mathrm{S}}_\sigma(g_1)$.

3. **$\bar{\sigma}(eg_{i,j}), \neg\bar{\sigma}(eb_{i,j})$ and $\mu^\sigma \neq 1$:** We prove $\mathrm{rVal}_\sigma^\mathrm{S}(F_{i,j}) = \mathrm{rVal}_\sigma^\mathrm{S}(g_1)$. Assume $\sigma(s_{i,j}) = b_1$. Since $\sigma(b_1) = g_1$ by Lemma 6.1.3,
$$\mathrm{rVal}_\sigma^\mathrm{S}(g_1) = \mathrm{rVal}_\sigma^\mathrm{S}(b_1) \lhd \{s_{i,j}\} \cup \mathrm{rVal}_\sigma^\mathrm{S}(b_1) = \{s_{i,j}\} \cup \mathrm{rVal}_\sigma^\mathrm{S}(g_1) = \mathrm{rVal}_\sigma^\mathrm{S}(s_{i,j}),$$
implying $\tau^\sigma(F_{i,j}) = s_{i,j}$ since $\neg\bar{\sigma}(eb_{i,j})$. Assume $\sigma(s_{i,j}) = h_{i,j}$. By Property (EG5), it then holds that $j = \bar{\sigma}(b_{i+1})$. Since also $\sigma(b_1) = g_1$ and $\mu^\sigma \neq 1$, it suffices to prove $R_1^\mathrm{S} \lhd \{s_{i,j}, h_{i,j}\} \cup \mathrm{rVal}_\sigma(b_{i+2-j})$. This can be shown by the following case distinction based on j and the relation between $i+1$ and μ^σ.
   a) Let $j = 1$ and $i+1 < \mu^\sigma$. Then $\mathrm{rVal}_\sigma^\mathrm{S}(b_{i+2-j}) = \mathrm{rVal}_\sigma^\mathrm{S}(b_{i+1}) = R_{i+1}^\mathrm{S}$ since $\sigma(b_{i+1}) = g_{i+1}$ by Property (EG5). It thus suffices to prove $R_1^\mathrm{S} \lhd \{s_{i,j}, h_{i,j}\} \cup R_{i+1}^\mathrm{S}$ which follows from $\Delta(R_{i+1}^\mathrm{S} \cup \{s_{i,j}, h_{i,j}\}, R_1^\mathrm{S}) = g_i$.
   b) Let$j = 1$ and $i+1 \geq \mu^\sigma$. Then $\mathrm{rVal}_\sigma^\mathrm{S}(b_{i+1}) = L_{i+1}^\mathrm{S}$ and it suffices to prove $R_1^\mathrm{S} \lhd \{s_{i,j}, h_{i,j}\} \cup L_{i+1}^\mathrm{S}$. This follows from Lemma 6.1.10 if $i \geq \mu^\sigma$ and is easy to verify for $i+1 = \mu^\sigma$.
   c) Let $j = 0$ and $i+2 \leq \mu^\sigma$. We first show $\mathrm{rVal}_\sigma^\mathrm{S}(b_{i+2}) = L_{i+2}^\mathrm{S}$. If $i+2 = \mu^\sigma$, then this follows by definition. Thus let $i+1 < \mu^\sigma - 1$. Since $\sigma(s_{i,j}) = h_{i,j}$, Property (EG5) implies $\sigma(b_{i+1}) = b_{i+2}$. This implies $\sigma(b_{i+2}) = b_{i+3}$ by Property (B1). Consequently, also $\mathrm{rVal}_\sigma^\mathrm{S}(b_{i+2}) = L_{i+2}^\mathrm{S}$ in this case. As usual, we have $\mathrm{rVal}_\sigma(b_1) = R_1^\mathrm{S}$ and thus prove $R_1^\mathrm{S} \lhd \{s_{i,j}, h_{i,j}\} \cup L_{i+2}^\mathrm{S}$. But this follows from Lemma 6.1.10 since $\sigma(b_{i+1}) = b_{i+2}$ implies $L_{i+1}^\mathrm{S} = L_{i+2}^\mathrm{S}$ as follows: By $\mu^\sigma \geq i+2$ and $\sigma(b_{i+1}) = b_{i+2}$, we have $\mu^\sigma \neq \min\{i' \in [n] \colon \sigma(b_{i'}) = b_{i'+1}\}$. Hence $\mathcal{I}^\sigma \neq \emptyset$ and $\sigma(b_{\mu^\sigma}) = g_{\mu^\sigma}$ by Lemma 6.1.4. Thus $i+2 \leq \mu^\sigma$ implies $R_1^\mathrm{S} \lhd L_{i+2}^\mathrm{S} \lhd L_{i+2}^\mathrm{S} \cup \{s_{i,j}, h_{i,j}\}$ by Lemma 6.1.10.
   d) Let $j = 0$ and $i+2 > \mu^\sigma$. Then $\mathrm{rVal}_\sigma^\mathrm{S}(b_{i+2}) = L_{i+2}^\mathrm{S} = L_{i+1}^\mathrm{S}$ since $\sigma(b_{i+1}) = b_{i+2}$ by Property (EG5). Thus, $R_1^\mathrm{S} \lhd \{s_{i,j}, h_{i,j}\} \cup L_{i+1}^\mathrm{S}$ by Lemma 6.1.10.
4. **$\bar{\sigma}(eb_{i,j}), \neg\bar{\sigma}(eg_{i,j}), \mu^\sigma = 1$ and $(\sigma(s_{i,j}) = b_1 \vee \bar{\sigma}(b_{i+1}) = j)$:** We prove $\mathrm{rVal}_\sigma^\mathrm{S}(F_{i,j}) = \mathrm{rVal}_\sigma^\mathrm{S}(b_2)$. Due to $\neg\bar{\sigma}(eg_{i,j})$, it suffices to show $\mathrm{rVal}_\sigma^\mathrm{S}(b_2) \lhd \mathrm{rVal}_\sigma^\mathrm{S}(s_{i,j})$. If $\sigma(s_{i,j}) = b_1$, then the statement follows directly since $\mu^\sigma = 1$ implies $\sigma(b_1) = b_2$). Hence let $\sigma(s_{i,j}) = h_{i,j} \wedge \bar{\sigma}(b_{i+1}) = j$. As $\mu^\sigma = 1$, it holds that $\mathrm{rVal}_\sigma^\mathrm{S}(b_i) = L_i^\mathrm{S}$ for all $i \in [n]$. Hence, $\mathrm{rVal}_\sigma^\mathrm{S}(b_2) = L_2^\mathrm{S}$ and $\mathrm{rVal}_\sigma^\mathrm{S}(s_{i,j}) = L_{i+2-j}^\mathrm{S} \cup \{s_{i,j}, h_{i,j}\}$ since $\bar{\sigma}(b_{i+1}) = j$. We thus prove $L_2^\mathrm{S} \lhd L_{i+2-j}^\mathrm{S} \cup \{s_{i,j}, h_{i,j}\}$. It is sufficient to show $\sigma(b_i) = b_{i+1}$ since this implies $W_i \not\subseteq L_2$. For the sake of contradiction let $\sigma(b_i) = g_i$. Since $\mu^\sigma = 1$, Property (D1) implies that the chosen cycle center of level i is closed. However, $\bar{\sigma}(b_{i+1}) = j$ and $\mathcal{I}^\sigma = \emptyset$ then imply $\bar{\sigma}(g_i) = j$. Hence this cycle center is $F_{i,j}$, contradicting $\bar{\sigma}(eb_{i,j})$.
5. **$\bar{\sigma}(eb_{i,j}), \neg\bar{\sigma}(eg_{i,j})$ and $\mu^\sigma \neq 1 \vee (\sigma(s_{i,j}) = h_{i,j} \wedge \bar{\sigma}(b_{i+1}) \neq j)$:** By Property (EB1), we can assume $\bar{\sigma}(b_{i+1}) \neq j$ in either case. We prove that it holds that $\mathrm{rVal}_\sigma^\mathrm{S}(F_{i,j}) = \mathrm{rVal}_\sigma^\mathrm{S}(s_{i,j})$ by proving $\mathrm{rVal}_\sigma^\mathrm{S}(s_{i,j}) \lhd \mathrm{rVal}_\sigma^\mathrm{S}(b_2)$.
   a) Let $\sigma(s_{i,j}) = h_{i,j}$ and $\sigma(b_1) = b_2$, implying $\mu^\sigma = 1$. Let $j = 0$. Since none of the cycle vertices of $F_{i,j}$ escapes to g_1 and $\sigma(s_{i,j}) = \sigma(s_{i,0}) = h_{i,0}$, we have to prove $\{s_{i,0}, h_{i,0}\} \cup \mathrm{rVal}_\sigma^\mathrm{S}(L_{i+2}) \lhd \mathrm{rVal}_\sigma^\mathrm{S}(L_2)$ since $\mu^\sigma = 1$. This however follows as $\bar{\sigma}(b_{i+1}) \neq j = 0$ implies $\sigma(b_{i+1}) = g_{i+1}$. Now let $j = 1$. We then need to

prove $\mathrm{rVal}_\sigma^{\mathrm{S}}(s_{i,1}) \lhd \mathrm{rVal}_\sigma^{\mathrm{S}}(b_2)$. However, the exact valuation of $s_{i,1}$ is not clear in this case and depends on several vertices of level 1 and $i+1$. To be precise we can have the following paths:

$$\begin{array}{ccccccccccccc}
 & & & & & & \boxed{b_2} & \longleftarrow & b_1 & & \boxed{b_{i+3}} & & \\
 & & & & & & \uparrow & & \uparrow & & \uparrow & & \\
s_{i,1} & \longrightarrow & h_{i,1} & \longrightarrow & g_{i+1} & \rightarrow & F_{i+1,*} & \rightarrow & s_{i+1,*} & \rightarrow & h_{i+1,*} & \rightarrow & \boxed{g_{i+2}} \\
 & & & & & & \downarrow & & & & & & \\
 & & & & & & g_1 & \longrightarrow & F_{1,*} & \longrightarrow & s_{1,*} & \longrightarrow & h_{1,*} \longrightarrow \boxed{b_3} \\
 & & & & & & & & \downarrow & & \downarrow & & \downarrow \\
 & & & & & & & & \boxed{b_2} & \longleftarrow & b_1 & & \boxed{g_2}
\end{array}$$

We show that $\mathrm{rVal}_\sigma^{\mathrm{S}}(s_{i,1}) \lhd \mathrm{rVal}_\sigma^{\mathrm{S}}(b_2)$ holds for all marked "endpoints" that could be reached by $s_{i,1}$. In all cases, $j=1$ and $\bar{\sigma}(b_{i+1}) \neq j$ imply $\sigma(b_{i+1}) = b_{i+2}$.

- $\boldsymbol{b_2}$: Then, $\mathrm{rVal}_\sigma^{\mathrm{S}}(s_{i,1}) \lhd \mathrm{rVal}_\sigma^{\mathrm{S}}(b_2)$ follows as it is easy to verify that we then have $\Delta(\mathrm{rVal}_\sigma^{\mathrm{S}}(b_2), \mathrm{rVal}_\sigma^{\mathrm{S}}(s_{i,1})) = g_{i+1}$ in each possible case.
- $\boldsymbol{b_{i+3}}$: Then $\mathrm{rVal}_\sigma^{\mathrm{S}}(s_{i,1}) = \{s_{i,1}, h_{i,1}, g_{i+1}, s_{i+1,0}, h_{i+1,0}\} \cup \mathrm{rVal}_\sigma^{\mathrm{S}}(b_{i+3})$ and $\sigma(g_{i+1}) = F_{i+1,0}$. Since $\sigma(s_{i,1}) = h_{i,1}$, Property (B3) implies $\bar{\sigma}(g_{i+1}) \neq \bar{\sigma}(b_{i+2})$, hence $\sigma(b_{i+2}) = g_{i+2}$. Since $\mu^\sigma = 1$ implies $\mathrm{rVal}_\sigma^{\mathrm{S}}(b_2) = L_2^{\mathrm{S}}$ we therefore have $W_{i+2}^{\mathrm{S}} \subseteq \mathrm{rVal}_\sigma^{\mathrm{S}}(b_2)$. This yields the statement.
- $\boldsymbol{b_3}$: Then $\mathrm{rVal}_\sigma^{\mathrm{S}}(s_{i,1}) = \{s_{i,1}, h_{i,1}, g_{i+1}\} \cup W_1^{\mathrm{S}} \cup \mathrm{rVal}_\sigma^{\mathrm{S}}(b_3)$ and thus,

$$\begin{aligned}
\mathrm{rVal}_\sigma^{\mathrm{S}}(s_{i,1}) &= \{s_{i,1}, h_{i,1}, g_{i+1}\} \cup W_1^{\mathrm{S}} \cup L_3^{\mathrm{S}} \\
&\unlhd \{s_{i,1}, h_{i,1}, g_{i+1}\} \cup W_1^{\mathrm{S}} \cup L_2^{\mathrm{S}} \lhd \mathrm{rVal}_\sigma^{\mathrm{S}}(b_2).
\end{aligned}$$

- $\boldsymbol{g_2}$: Then, it holds that $\mathrm{rVal}_\sigma^{\mathrm{S}}(s_{i,1}) = \{s_{i,1}, h_{i,1}, g_{i+1}\} \cup W_1^{\mathrm{S}} \cup \mathrm{rVal}_\sigma^{\mathrm{S}}(g_2)$ and $\mathrm{rVal}_\sigma^{\mathrm{S}}(b_2) = L_2^{\mathrm{S}}$. As before we need to show $\mathrm{rVal}_\sigma^{\mathrm{S}}(s_{i,1}) \lhd \mathrm{rVal}_\sigma^{\mathrm{S}}(b_2)$. Note that we can assume $i \geq 2$ since S_n is a sink game and the valuation of $s_{i,1}$ contains a cycle for $i=1$.

  First let $\sigma(b_2) = g_2$. Then $\mathrm{rVal}_\sigma^{\mathrm{S}}(g_2) = \mathrm{rVal}_\sigma^{\mathrm{S}}(b_2)$. This implies that we have $\Delta(\mathrm{rVal}_\sigma^{\mathrm{S}}(s_{i,1}), \mathrm{rVal}_\sigma^{\mathrm{S}}(b_2)) = g_{i+1}$ since $i \geq 2$ and $W_{i+1}^{\mathrm{S}} \nsubseteq \mathrm{rVal}_\sigma^{\mathrm{S}}(b_2)$ due to $\sigma(b_{i+1}) = b_{i+2}$ and $\mu^\sigma = 1$. Since $g_{i+1} \subseteq \mathrm{rVal}_\sigma^{\mathrm{S}}(s_{i,1})$, this implies $\mathrm{rVal}_\sigma^{\mathrm{S}}(s_{i,1}) \lhd \mathrm{rVal}_\sigma^{\mathrm{S}}(b_2)$.

  Thus let $\sigma(b_2) = b_3$, implying $\mathrm{rVal}_\sigma^{\mathrm{S}}(b_2) = \mathrm{rVal}_\sigma^{\mathrm{S}}(b_3)$ and let $k := \bar{\sigma}(g_2)$. Similar to the picture showing the "directions" to which the vertex $s_{i,1}$ can lead, there are several possibilities towards which vertex the path starting in $F_{2,k}$ leads. For all of the following cases, the main argument will be the following. No matter what choices are made in the lower levels and no matter how many levels the path starting in g_2 might traverse, the vertex g_{i+1} contained in $\mathrm{rVal}_\sigma^{\mathrm{S}}(s_{i,1})$ will always ensure $\mathrm{rVal}_\sigma^{\mathrm{S}}(s_{i,1}) \lhd \mathrm{rVal}_\sigma^{\mathrm{S}}(b_2)$. We distinguish the following cases.

  i. $F_{2,k}$ escapes towards b_2. Then $\mathrm{rVal}_\sigma^{\mathrm{S}}(g_2) = \{g_2\} \cup \mathrm{rVal}_\sigma^{\mathrm{S}}(b_2)$ and thus

$$\mathrm{rVal}_\sigma^{\mathrm{S}}(s_{i,1}) = \{s_{i,1}, h_{i,1}, g_{i+1}, g_2\} \cup W_1^{\mathrm{S}} \cup \mathrm{rVal}_\sigma^{\mathrm{S}}(b_2) \lhd \mathrm{rVal}_\sigma^{\mathrm{S}}(b_2).$$

ii. $F_{2,k}$ escapes towards g_1. Depending on the configuration of level 1, the path can end in different vertices. As S_n is a sink game, it cannot end in g_1 or g_2 since this would constitute a cycle. It can thus either end in b_1, b_2 or b_3. Since $\sigma(b_1) = b_2$ and $\sigma(b_2) = b_3$, we then have $\mathrm{rVal}^{\mathrm{S}}_{\sigma}(g_2) = \{g_2\} \cup W^{\mathrm{S}}_1 \cup \mathrm{rVal}^{\mathrm{S}}_{\sigma}(b_3)$ in either case and thus

$$\mathrm{rVal}^{\mathrm{S}}_{\sigma}(s_{i,1}) = \{s_{i,1}, h_{i,1}, g_{i+1}, g_2\} \cup W^{\mathrm{S}}_1 \cup \mathrm{rVal}^{\mathrm{S}}_{\sigma}(b_2) \lhd \mathrm{rVal}^{\mathrm{S}}_{\sigma}(b_2).$$

iii. $F_{2,k}$ does not escape level 2 and $k = 0$. In this case, $\tau(F_{2,k}) = s_{2,0}$. If $\sigma(s_{2,0}) = b_1$, then the statement follows by the same arguments used in the last case. Thus consider the case $\sigma(s_{2,0}) = h_{2,0}$, implying $\mathrm{rVal}^{\mathrm{S}}_{\sigma}(g_2) = W^{\mathrm{S}}_2 \cup \mathrm{rVal}^{\mathrm{S}}_{\sigma}(b_4)$. Then, since $i \geq 2$,

$$\begin{aligned}\mathrm{rVal}^{\mathrm{S}}_{\sigma}(s_{i,1}) &= \{s_{i,1}, h_{i,1}, g_{i+1}\} \cup W^{\mathrm{S}}_1 \cup W^{\mathrm{S}}_2 \cup \mathrm{rVal}^{\mathrm{S}}_{\sigma}(b_4) \\ &\lhd \mathrm{rVal}^{\mathrm{S}}_{\sigma}(b_4) \unlhd \mathrm{rVal}^{\mathrm{S}}_{\sigma}(b_2).\end{aligned}$$

iv. $F_{2,k}$ does not escape level 2 and $k = 1$. In this case we can use the exact same arguments to show that either $\mathrm{rVal}_{\sigma}(s_{i,1}) \lhd \mathrm{rVal}_{\sigma}(b_2)$ or that the path reaches g_4. In fact, the same arguments can be used until vertex g_{i-1} is reached. We now show that once this vertex is reached the inequality $\mathrm{rVal}^{\mathrm{S}}_{\sigma}(s_{i,1}) \lhd \mathrm{rVal}^{\mathrm{S}}_{\sigma}(b_2)$ is fulfilled. Let $k' := \bar{\sigma}(g_{i-1})$. If $F_{i-1,k'}$ escapes towards b_2, then $\mathrm{rVal}^{\mathrm{S}}_{\sigma}(s_{i,1}) \lhd \mathrm{rVal}^{\mathrm{S}}_{\sigma}(b_2)$ follows from

$$\mathrm{rVal}^{\mathrm{S}}_{\sigma}(g_2) = \bigcup_{i'=2}^{i-2} W^{\mathrm{S}}_{i'} \cup \{g_{i-1}\} \cup \mathrm{rVal}^{\mathrm{S}}_{\sigma}(b_2).$$

If $F_{i-1,k'}$ escapes towards b_1 via $s_{i-1,k'}$, then the statement follows analogously since $\sigma(b_1) = b_2$. Thus assume that the cycle center escapes towards g_1. By the same arguments used before, it can be shown that level 1 needs to escape towards either b_1, b_2 or b_3. However, the same calculation used before can be applied in each of these cases.

Next assume that the cycle center $F_{i-1,k'}$ does not escape level $i-1$ but traverses the level and reaches vertex b_{i+1}. Then,

$$\mathrm{rVal}^{\mathrm{S}}_{\sigma}(g_2) = \bigcup_{i'=2}^{i-1} W^{\mathrm{S}}_{i'} \cup \mathrm{rVal}^{\mathrm{S}}_{\sigma}(b_{i+1})$$

and $\mathrm{rVal}^{\mathrm{S}}_{\sigma}(s_{i,1}) \lhd \mathrm{rVal}^{\mathrm{S}}_{\sigma}(b_{i+1}) \unlhd \mathrm{rVal}^{\mathrm{S}}_{\sigma}(b_2)$. The last case we need to consider is if level $i-1$ is traversed and g_i is reached. In this case we need to have $\sigma(g_i) = F_{i,0}$ since player 1 could create a cycle otherwise, contradicting that S_n is a sink game. If the cycle center $F_{i,0}$ escapes towards g_1 or b_2 the statement follows by the same arguments used before. If it reaches b_{i+2}, then the statement follows from

$$\mathrm{rVal}^{\mathrm{S}}_{\sigma}(s_{i,1}) = \{s_{i,1}, h_{i,1}, g_{i+1}\} \cup \bigcup_{i'=1}^{i} W^{\mathrm{S}}_{i'} \cup \mathrm{rVal}^{\mathrm{S}}_{\sigma}(b_{i+2}) \unlhd \mathrm{rVal}^{\mathrm{S}}_{\sigma}(b_2).$$

- $\boldsymbol{g_{i+2}}$: This implies $\mathrm{rVal}_\sigma^{\mathrm{S}}(s_{i,1}) = \{s_{i,1}, h_{i,1}\} \cup W_{i+1}^{\mathrm{S}} \cup \mathrm{rVal}_\sigma^{\mathrm{S}}(g_{i+2})$ and we prove $\mathrm{rVal}_\sigma^{\mathrm{S}}(s_{i,1}) \lhd \mathrm{rVal}_\sigma^{\mathrm{S}}(b_2)$. This case is organized similarly to the last case. We prove that the statement follows for all but one possible configurations of the levels $i+2$ to $n-1$. It then turns out that this missing configuration contradicts Property (DN1). We distinguish the following cases.

  i. Level $i+2$ escapes towards b_2 via some cycle center $F_{i+2,*}$. Then $\mathrm{rVal}_\sigma^{\mathrm{S}}(g_{i+2}) = \{g_{i+2}\} \cup \mathrm{rVal}_\sigma^{\mathrm{S}}(b_2)$ implies the statement.

  ii. Level $i+2$ escapes towards b_1 via some upper selection vertex $s_{i+2,*}$. Then, $\mathrm{rVal}_\sigma^{\mathrm{S}}(g_{i+2}) = \{g_{i+2}, s_{i+2,*}\} \cup \mathrm{rVal}_\sigma^{\mathrm{S}}(b_2)$ as $\sigma(b_1) = b_2$, implying the statement.

  iii. Level $i+2$ is traversed completely and reaches b_{i+4} directly. In this case, $\sigma(g_{i+2}) = F_{i+2,0}$. Since $\sigma(s_{i+1,1}) = h_{i+1,1}$ and $\sigma(b_{i+2}) = b_{i+3}$ by Property (B3), this implies $\sigma(b_{i+3}) = g_{i+3}$. Thus, it holds that $\mathrm{rVal}_\sigma^{\mathrm{S}}(g_{i+2}) = W_{i+2}^{\mathrm{S}} \cup \mathrm{rVal}_\sigma^{\mathrm{S}}(b_{i+4})$ which implies the statement due to $\sigma(b_{i+3}) = g_{i+3}$.

  iv. Level $i+2$ escapes towards g_1 via some cycle center $F_{i+2,*}$. Then, $\mathrm{rVal}_\sigma^{\mathrm{S}}(g_{i+2}) = \{g_{i+2}\} \cup \mathrm{rVal}_\sigma^{\mathrm{S}}(g_1)$. Consider level 1. If $F_{1,\bar{\sigma}(g_1)}$ escapes towards b_1, b_2 or b_3, the statement follows by the same arguments used in the last two cases since $\mathrm{rVal}_\sigma^{\mathrm{S}}(b_3) \unlhd \mathrm{rVal}_\sigma^{\mathrm{S}}(b_2)$. Since S_n is a sink game, the cycle center cannot escape towards g_1. Thus assume that it escapes towards g_2. By the same arguments used previously, the statement either holds or level 3 is traversed and the path reaches g_4. We now iterate this argument until we reach a level $k < i+2$ such that either $\sigma(g_k) = F_{k,0}$ or $\sigma(s_{k,\bar{\sigma}(g_k)}) = b_1$. Such a level exists by Lemma A.2.1. We only consider the second case here since the statement follows by calculations similar to the previous ones if $\sigma(g_k) = F_{k,0}$. Then

  $$\mathrm{rVal}_\sigma^{\mathrm{S}}(g_{i+2}) = \{g_{i+2}\} \cup \bigcup_{i'=1}^{k-1} W_{i'}^{\mathrm{S}} \cup \{g_k, s_{k,*}\} \cup \mathrm{rVal}_\sigma^{\mathrm{S}}(b_1),$$

  implying the statement since $k \leq i+1$.

  v. Level $i+2$ is traversed completely and reaches g_{i+3}. This implies that $\sigma(s_{i+1,1}) = h_{i+1,1}, \sigma(b_{i+2}) = b_{i+3}$ and $\sigma(g_{i+2}) = F_{i+2,1}$. Thus, by Property (B3), also $\sigma(b_{i+3}) = b_{i+4}$. We can therefore use the same arguments used before since $\mathrm{rVal}_\sigma^{\mathrm{S}}(g_{i+2}) = W_{i+2}^{\mathrm{S}} \cup \mathrm{rVal}_\sigma^{\mathrm{S}}(g_{i+3})$. That is, the statement either holds or we reach the vertex g_{n-1}. If level $n-1$ escapes towards b_1, b_2 or g_1, the statement follows by the same arguments used for level $i+2$. We thus assume that level $n-1$ is traversed completely. Note that $\sigma(s_{n-2,1}) = h_{n-2,1}$ and $\sigma(b_{n-1}) = b_n$ (apply Property (B3) iteratively). Consider the case $\sigma(g_{n-1}) = F_{n-1,0}$

first. Then, $\sigma(b_n) = g_n$ by Property (B3) and thus

$$\mathrm{rVal}_\sigma^{\mathrm{S}}(s_{i,1}) = \{s_{i,1}, h_{i,1}\} \cup \bigcup_{i'=i+1}^{n-1} W_{i'}^{\mathrm{S}} \lhd W_n^{\mathrm{S}} = \mathrm{rVal}_\sigma^{\mathrm{S}}(b_n) \unlhd \mathrm{rVal}_\sigma^{\mathrm{S}}(b_2).$$

Consider the case $\sigma(g_{n-1}) = F_{n-1,1}$ next. Since we assume that level $n-1$ does not escape towards one of the vertices b_1, b_2 or g_1, we traverse this level and reach g_n. If level n escapes towards b_1, b_2 or g_1 the statement follows as usual. We thus assume that the level n is traversed completely. We observe that the vertex $h_{n,1}$ has the highest even priority among all vertices in the parity game. Thus, player 1 would avoid this vertex if this was possible. We thus need to have $\bar{\sigma}(d_n)$. But this is a contradiction to Property (DN1) since $\sigma(b_n) = t$ by Property (B3) and $\sigma(b_1) = b_2$ by assumption.

b) Let $\sigma(s_{i,j}) = h_{i,j}$ and $\sigma(b_1) = g_1$, implying $\mu^\sigma \neq 1$ by Lemma 6.1.3.
   i. Let $j = 0$ and $i = 1$. Then, $\mathrm{rVal}_\sigma^{\mathrm{S}}(s_{i,j}) = \{s_{1,0}, h_{1,0}\} \cup \mathrm{rVal}_\sigma^{\mathrm{S}}(b_3)$. By Property (EB2), it holds that $\mu^\sigma = 2$, implying $\mathrm{rVal}_\sigma^{\mathrm{S}}(b_2) = L_2^{\mathrm{S}}$ and $\mathrm{rVal}_\sigma^{\mathrm{S}}(b_3) = L_3^{\mathrm{S}}$. It thus suffices to prove $L_2^{\mathrm{S}} \rhd \{s_{1,0}, h_{1,0}\} \cup L_3^{\mathrm{S}}$ which follows from Property (EB1) as this implies $\sigma(b_{i+1}) \neq j$, so $\sigma(b_2) = g_2$.
   ii. Let $j = 0$ and $i > 1$. Then $\mathrm{rVal}_\sigma^{\mathrm{S}}(s_{i,j}) = \{s_{i,0}, h_{i,0}\} \cup \mathrm{rVal}_\sigma^{\mathrm{S}}(b_{i+2})$. By Property (EB2), it follows that $\mu^\sigma = i+1$, implying $\mathrm{rVal}_\sigma^{\mathrm{S}}(b_{i+2}) = L_{i+2}^{\mathrm{S}}$. In addition, $\sigma(b_2) = b_3$ by Property (EB3), implying $\mathrm{rVal}_\sigma^{\mathrm{S}}(b_2) = L_2^{\mathrm{S}}$. We thus prove $L_2^{\mathrm{S}} \rhd \{s_{i,0}, h_{i,0}\} \cup L_{i+2}^{\mathrm{S}}$. This follows since Property (EB1) implies $\sigma(b_{i+1}) = \sigma(b_{\mu^\sigma}) \neq j = 0$ and thus $\sigma(b_{\mu^\sigma}) = g_{\mu^\sigma}$
   iii. Let $j = 1$. Then $\mathrm{rVal}_\sigma^{\mathrm{S}}(s_{i,j}) = \{s_{i,1}, h_{i,1}\} \cup \mathrm{rVal}_\sigma^{\mathrm{S}}(g_{i+1})$. By Property (EB4), we have $i+1 < \mu^\sigma$ and, by Property (EB1), also $\sigma(b_{i+1}) = b_{i+2}$. Therefore, $\mathrm{rVal}_\sigma^{\mathrm{S}}(g_{i+1}) = R_{i+1}^{\mathrm{S}}$. This also implies $\mu^\sigma \neq \min\{i' \in [n] \colon \sigma(b_{i'}) = b_{i'+1}\}$ since this would contradict $\mu^\sigma > i+1$. In particular it holds that $\mathcal{I}^\sigma \neq \emptyset$, implying $\sigma(b_{\mu^\sigma}) = g_{\mu^\sigma}$ by Lemma 6.1.4. Since $\mu^\sigma > i+1 \geq 2$, we can apply Property (EB6), implying $\sigma(b_2) = b_3$. Hence $\mathrm{rVal}_\sigma^{\mathrm{S}}(b_2) = L_2^{\mathrm{S}}$. It thus suffices to prove $L_2^{\mathrm{S}} \rhd \{s_{i,1}, h_{i,1}\} \cup R_{i+1}^{\mathrm{S}}$ which follows from $\sigma(b_{\mu^\sigma}) = g_{\mu^\sigma}$ and $i+1 < \mu^\sigma$.

c) Next let $\sigma(s_{i,j}) = b_1$ and $\sigma(b_1) = g_1$, implying $\mu^\sigma \neq 1$. Consider the case $\mu^\sigma > 2$ first. Then, by Property (EB6), $\sigma(b_2) = b_3$, so $\mathrm{rVal}_\sigma^{\mathrm{S}}(b_2) = L_2^{\mathrm{S}}$ and $\mu^\sigma \neq \min\{i'[n] \colon \sigma(b_{i'}) = b_{i'+1}\}$. Hence $\mathcal{I}^\sigma \neq \emptyset$ and thus $\sigma(b_{\mu^\sigma}) = g_{\mu^\sigma}$. Since $\mathrm{rVal}_\sigma^{\mathrm{S}}(b_1) = R_1^{\mathrm{S}}$, we prove $\mathrm{rVal}_\sigma^{\mathrm{S}}(s_{i,j}) = \{s_{i,j}\} \cup R_1^{\mathrm{S}} \lhd L_2^{\mathrm{S}}$ which follows from $\sigma(b_{\mu^\sigma}) = g_{\mu^\sigma}$. Now consider the case $\mu^\sigma = 2$, implying $\mathrm{rVal}_\sigma^{\mathrm{S}}(b_2) = L_2^{\mathrm{S}}$. Since $\sigma(b_2) = g_2$ by Property (EB5), the statement follows by the same arguments.

6. **$\bar{\sigma}(eb_{i,j}), \bar{\sigma}(eg_{i,j})$ and $\bar{\sigma}(g_1) \neq \bar{\sigma}(b_2)$:** We prove $\mathrm{rVal}_\sigma^{\mathrm{S}}(F_{i,j}) = \mathrm{rVal}_\sigma^{\mathrm{S}}(g_1)$. To simplify the proof, we show $\mathrm{rVal}_\sigma^{\mathrm{S}}(g_1) \lhd \mathrm{rVal}_\sigma^{\mathrm{S}}(b_2)$ first. If $\mathrm{rVal}_\sigma^{\mathrm{S}}(F_{1,\bar{\sigma}(g_1)}) = \mathrm{rVal}_\sigma^{\mathrm{S}}(b_2)$, then the statement follows from $\mathrm{rVal}_\sigma^{\mathrm{S}}(g_1) = \{g_1\} \cup \mathrm{rVal}_\sigma^{\mathrm{S}}(F_{1,\bar{\sigma}(g_1)}) = \{g_1\} \cup \mathrm{rVal}_\sigma^{\mathrm{S}}(b_2)$. In addition, since S_n is a sink game, the chosen cycle center of level 1 cannot escape

towards g_1 since this would close a cycle. If $\mathrm{rVal}^{\mathrm{S}}_\sigma(F_{i,\bar{\sigma}(g_1)} = \mathrm{rVal}^{\mathrm{S}}_\sigma(g_1)$, then the claim also follows as player 1 always minimizes the valuations and could choose vertex b_2 but prefers g_1. Thus let $\tau^\sigma(F_{1,\bar{\sigma}(g_1)}) = s_{1,\bar{\sigma}(g_1)}$ and assume $\sigma(s_{1,\bar{\sigma}(g_1)}) = b_1$. Then, since S_n is a sink game, we need to have $\sigma(b_1) = b_2$ since there would be a cycle otherwise. But then $\mathrm{rVal}^{\mathrm{S}}_\sigma(g_1) = \{g_1, s_{1,\bar{\sigma}(g_1)}\} \cup \mathrm{rVal}^{\mathrm{S}}_\sigma(b_2) \lhd \mathrm{rVal}^{\mathrm{S}}_\sigma(b_2)$. We can therefore assume $\sigma(s_{1,\bar{\sigma}(g_1)}) = h_{1,\bar{\sigma}(g_1)}$ and distinguish two cases.

- Let $\sigma(g_1) = F_{1,0}$. Then $\sigma(b_2) = g_2$. Therefore, by Property (EBG4), $\mu^\sigma \leq 2$. Thus $\mathrm{rVal}^{\mathrm{S}}_\sigma(b_2) = L^{\mathrm{S}}_2$ and $\mathrm{rVal}^{\mathrm{S}}_\sigma(b_3) = L^{\mathrm{S}}_3$, hence

$$\mathrm{rVal}^{\mathrm{S}}_\sigma(g_1) = W^{\mathrm{S}}_1 \cup \mathrm{rVal}^{\mathrm{S}}_\sigma(b_3) = W^{\mathrm{S}}_1 \cup L^{\mathrm{S}}_3 \lhd W^{\mathrm{S}}_2 \cup L^{\mathrm{S}}_3 = L^{\mathrm{S}}_2 = \mathrm{rVal}^{\mathrm{S}}_\sigma(b_2).$$

- Let $\sigma(g_1) = F_{1,1}$. Then, $\mathrm{rVal}^{\mathrm{S}}_\sigma(g_1) = W^{\mathrm{S}}_1 \cup \mathrm{rVal}^{\mathrm{S}}_\sigma(g_2)$ and $\sigma(b_2) = b_3$. This implies $\mu^\sigma \neq 2$ by Property (EBG5). Consider the case $\mu^\sigma > 2$ first. Lemma 6.1.12 then implies $\mathrm{rVal}^{\mathrm{S}}_\sigma(b_2) = L^{\mathrm{S}}_2$ as well as $\mathrm{rVal}^{\mathrm{S}}_\sigma(g_2) = R^{\mathrm{S}}_2$. Also, $\sigma(b_2) = b_3$ and $\mu^\sigma > 2$ together imply $\mu^\sigma \neq \min\{i' \in [n] \colon \sigma(b_{i'}) = b_{i'+1}\}$. Thus, $\mathcal{I}^\sigma \neq \emptyset$ and $\sigma(b_{\mu^\sigma}) = g_{\mu^\sigma}$ by Lemma 6.1.4. Combining all of this then yields

$$\mathrm{rVal}^{\mathrm{S}}_\sigma(g_1) = W^{\mathrm{S}}_1 \cup R^{\mathrm{S}}_2 \lhd W^{\mathrm{S}}_{\mu^\sigma} \cup L_{\mu^\sigma+1} = L^{\mathrm{S}}_{\mu^\sigma} \trianglelefteq L^{\mathrm{S}}_2 = \mathrm{rVal}^{\mathrm{S}}_\sigma(b_2).$$

  Now consider the case $\mu^\sigma = 1$. Then again $\mathrm{rVal}^{\mathrm{S}}_\sigma(b_2) = L^{\mathrm{S}}_2$. We apply Lemma 6.1.14 to give the exact valuation of g_1. The case $m^\sigma_b \leq \overline{m}^\sigma_s, \overline{m}^\sigma_g$ cannot occur as this would imply $\sigma(b_2) = g_2$ by Lemma 6.1.6.

  Consider the case $\overline{m}^\sigma_g < \overline{m}^\sigma_s, \overline{m}^\sigma_b$. As $\sigma(g_1) = F_{1,1}$, it holds that $\overline{m}^\sigma_g \neq 1$. Let $i := \overline{m}^\sigma_g$. Thus, by assumption, $\sigma(g_{i-1}) = F_{i-1,1}$ and consequently $\sigma(s_{i-1,1}) = h_{i-1,1}$ as well as $\sigma(b_i) = b_{i+1}$. Thus, Property (B3) implies $0 = \bar{\sigma}(g_i) \neq \bar{\sigma}(b_{i+1})$, hence $\bar{\sigma}(b_{i+1}) = 1$. But then, $\bar{\sigma}(b_{\overline{m}^\sigma_g+1})$, so Lemma 6.1.14 yields

$$\begin{aligned}\mathrm{rVal}_\sigma(g_1) &= \bigcup_{i'=1}^{\overline{m}^\sigma_g} W^{\mathrm{S}}_{i'} \cup \mathrm{rVal}^{\mathrm{S}}_\sigma(b_{\overline{m}^\sigma_g+2}) \\ &\lhd W_{\overline{m}^\sigma_g+1} \cup \mathrm{rVal}^{\mathrm{S}}_\sigma(b_{\overline{m}^\sigma_g+2}) = \mathrm{rVal}^{\mathrm{S}}_\sigma(b_{\overline{m}^\sigma_g+1}) \trianglelefteq \mathrm{rVal}^{\mathrm{S}}_\sigma(b_2).\end{aligned}$$

Thus $\mathrm{rVal}^{\mathrm{S}}_\sigma(g_1) \lhd \mathrm{rVal}^{\mathrm{S}}_\sigma(b_2)$. We next prove that we have $\mathrm{rVal}^{\mathrm{S}}_\sigma(g_1) \lhd \mathrm{rVal}^{\mathrm{S}}_\sigma(s_{i,j})$, implying that player 1 chooses to escape to g_1.

a) Let $\sigma(s_{i,j}) = b_1$. If $\sigma(b_1) = g_1$, then $\mathrm{rVal}^{\mathrm{S}}_\sigma(s_{i,j}) = \{s_{i,j}\} \cup \mathrm{rVal}^{\mathrm{S}}_\sigma(g_1)$ implies the statement. If $\sigma(b_1) = b_2$ we have $\mathrm{rVal}^{\mathrm{S}}_\sigma(b_1) = \mathrm{rVal}^{\mathrm{S}}_\sigma(b_2)$. The statement then follows since $\mathrm{rVal}^{\mathrm{S}}_\sigma(g_1) \lhd \mathrm{rVal}^{\mathrm{S}}_\sigma(b_2)$ and $\mathrm{rVal}^{\mathrm{S}}_\sigma(s_{i,j}) = \{s_{i,j}\} \cup \mathrm{rVal}^{\mathrm{S}}_\sigma(b_1)$.

b) Let $\sigma(s_{i,j}) = h_{i,j}$. Then, Property (EBG1) implies $\bar{\sigma}(b_{i+1}) = j$ and thus $\mathrm{rVal}^{\mathrm{S}}_\sigma(s_{i,j}) = \{s_{i,j}, h_{i,j}\} \cup \mathrm{rVal}^{\mathrm{S}}_\sigma(b_{i+1})$. Let $\mu^\sigma = 1$. Then, $\sigma(b_1) = b_2$, implying $\mathrm{rVal}^{\mathrm{S}}_\sigma(b_1) = \mathrm{rVal}^{\mathrm{S}}_\sigma(b_2)$ and $\mathrm{rVal}^{\mathrm{S}}_\sigma(b_{i+1}) = L^{\mathrm{S}}_{i+1}$. Combining this with $\mathrm{rVal}^{\mathrm{S}}_\sigma(g_1) \lhd \mathrm{rVal}^{\mathrm{S}}_\sigma(b_2)$ yields the statement as

$$\mathrm{rVal}^{\mathrm{S}}_\sigma(g_1) \lhd \mathrm{rVal}^{\mathrm{S}}_\sigma(b_2) = L^{\mathrm{S}}_2 = \lhd \{s_{i,j}, h_{i,j}\} \cup L^{\mathrm{S}}_{i+1} = \mathrm{rVal}^{\mathrm{S}}_\sigma(s_{i,j}).$$

Now let $\mu^\sigma \neq 1$, implying $\mathrm{rVal}^{\mathrm{S}}_\sigma(g_1) = R^{\mathrm{S}}_1$ by Lemma 6.1.12. Consider the case $\mu^\sigma \geq i+1$ and $B^{\mathrm{S}}_{i+1} = L^{\mathrm{S}}_{i+1}$ first. Then, since $\sigma(b_{i+1}) = b_{i+2}$ by $B^{\mathrm{S}}_{i+1} = L^{\mathrm{S}}_{i+1}$, we have $\mu^\sigma \neq i+1$. This implies $\mu^\sigma \neq \min(\{i' : \sigma(b_{i'}) = b_{i'+1}\})$ and consequently $\sigma(b_{\mu^\sigma}) = g_{\mu^\sigma}$. The statement then follows from Lemma 6.1.10 (2) and $\mathrm{rVal}^{\mathrm{S}}_\sigma(s_{i,j}) \rhd L^{\mathrm{S}}_{i+1}$. If $\mu^\sigma \geq i+1$ and $B^{\mathrm{S}}_{i+1} = R^{\mathrm{S}}_{i+1}$ then the statement follows since $R^{\mathrm{S}}_1 \lhd \{s_{i,j}, h_{i,j}\} \cup R^{\mathrm{S}}_{i+1}$ in this case. If $\mu^\sigma < i+1$, then the statement follows from Lemma 6.1.10 (3).

7. $\boldsymbol{\bar{\sigma}(eb_{i,j}), \bar{\sigma}(eg_{i,j})}$ **and** $\boldsymbol{\bar{\sigma}(g_1) = \bar{\sigma}(b_2)}$**:** We prove $\mathrm{rVal}^{\mathrm{S}}_\sigma(F_{i,j}) = \mathrm{rVal}^{\mathrm{S}}_\sigma(b_2)$. Similar to the last case we prove $\mathrm{rVal}^{\mathrm{S}}_\sigma(b_2) \lhd \mathrm{rVal}^{\mathrm{S}}_\sigma(g_1)$ and $\mathrm{rVal}^{\mathrm{S}}_\sigma(b_2) \lhd \mathrm{rVal}^{\mathrm{S}}_\sigma(s_{i,j})$.

   The assumption $\bar{\sigma}(g_1) = \bar{\sigma}(b_2)$ implies $\sigma(h_{1,\bar{\sigma}(g_1)}) = \sigma(b_2)$. By Property (EBG3), the chosen cycle center of level 1 is closed. In addition, $\bar{\sigma}(s_1)$ by Property (EBG2). Hence, $\mathrm{rVal}^{\mathrm{S}}_\sigma(g_1) = W^{\mathrm{S}}_1 \cup \mathrm{rVal}^{\mathrm{S}}_\sigma(b_2)$, implying $\mathrm{rVal}^{\mathrm{S}}_\sigma(b_2) \lhd \mathrm{rVal}^{\mathrm{S}}_\sigma(g_1)$.

   It remains to show $\mathrm{rVal}^{\mathrm{S}}_\sigma(b_2) \lhd \mathrm{rVal}^{\mathrm{S}}_\sigma(s_{i,j})$. Let $\sigma(s_{i,j}) = b_1$ first. If $\sigma(b_1) = b_2$, then the statement follows from $\mathrm{rVal}^{\mathrm{S}}_\sigma(s_{i,j}) = \{s_{i,j}\} \cup \mathrm{rVal}^{\mathrm{S}}_\sigma(b_2)$. Thus let $\sigma(b_1) = g_1$, implying $\mathrm{rVal}^{\mathrm{S}}_\sigma(s_{i,j}) = \{s_{i,j}\} \cup \mathrm{rVal}^{\mathrm{S}}_\sigma(g_1)$. But this implies $\mathrm{rVal}^{\mathrm{S}}_\sigma(s_{i,j}) \rhd \mathrm{rVal}^{\mathrm{S}}_\sigma(g_1)$ and consequently also $\mathrm{rVal}^{\mathrm{S}}_\sigma(s_{i,j}) \rhd \mathrm{rVal}^{\mathrm{S}}_\sigma(b_2)$. Thus, let $\sigma(s_{i,j}) = h_{i,j}$, implying $\bar{\sigma}(b_{i+1}) = j$ by Property (EBG1). We distinguish two cases.

   a) Let $j = 0$. Then $\mathrm{rVal}^{\mathrm{S}}_\sigma(s_{i,j}) = \{s_{i,0}, h_{i,0}\} \cup B^{\mathrm{S}}_{i+2}$. We first consider the case that $B^{\mathrm{S}}_{i+2} = L^{\mathrm{S}}_{i+2}$ and show $\{s_{i,0}, h_{i,0}\} \cup L^{\mathrm{S}}_{i+2} \rhd L^{\mathrm{S}}_2, R^{\mathrm{S}}_2$ since this suffices to show $\mathrm{rVal}_\sigma(s_{i,j}) \rhd \mathrm{rVal}_\sigma(b_2)$. Since $\sigma(b_{i+1}) = b_{i+2}$ by Property (EBG1) and $j = 0$, $i \geq 2$ implies

   $$\{s_{i,0}, h_{i,0}\} \cup L^{\mathrm{S}}_{i+2} = \rhd \bigcup_{i'=2}^{i} \{W^{\mathrm{S}}_{i'} : \sigma(b_{i'}) = g_{i'}\} \cup \bigcup_{i' \geq i+1} \{W^{\mathrm{S}}_{i'} : \sigma(b_{i'}) = g_{i'}\} = L^{\mathrm{S}}_2$$

   and $i = 1$ implies $\{s_{i,0}, h_{i,0}\} \cup L^{\mathrm{S}}_{i+2} = \{s_{1,0}, h_{1,0}\} \cup L^{\mathrm{S}}_3 = \{s_{1,0}, h_{1,0}\} \cup L^{\mathrm{S}}_2 \rhd L^{\mathrm{S}}_2$.

   Thus let $\mathrm{rVal}^{\mathrm{S}}_\sigma(b_2) = R^{\mathrm{S}}_2$ and consider the case $\mu^\sigma \leq i+1$ first. Then, by Lemma 6.1.10 and since $\sigma(b_{i+1}) = b_{i+2}$, it holds that $L^{\mathrm{S}}_{i+1} = L^{\mathrm{S}}_{i+2}$ and thus $R^{\mathrm{S}}_2 \lhd R^{\mathrm{S}}_1 \lhd \{s_{i,0}, h_{i,0}\} \cup L^{\mathrm{S}}_{i+1}$. Now assume $\mu^\sigma > i+1$, implying $\mu^\sigma \neq 1$ and $i+2 \leq \mu^\sigma$. This implies $\mu^\sigma \neq \min\{i' : \sigma(b_{i'}) = b_{i'+1}\}$ since Property (EBG1) implies $\sigma(b_{i+1}) = b_{i+2}$. Thus $\sigma(b_{\mu^\sigma}) = g_{\mu^\sigma}$ by Lemma 6.1.4. Then, the statement follows from $\{s_{i,0}, h_{i,0}\} \cup L^{\mathrm{S}}_{i+2} \rhd L^{\mathrm{S}}_{i+2}$ and $L^{\mathrm{S}}_{i+2} \rhd R^{\mathrm{S}}_2$ which follows from Lemma 6.1.10 (2).

   Next, let $B^{\mathrm{S}}_{i+2} = R^{\mathrm{S}}_{i+2}$. We show that this results in a contradiction. First, $B^{\mathrm{S}}_{i+2} = R^{\mathrm{S}}_{i+2}$ implies $i+2 < \mu^\sigma$ and $\sigma(b_{i+2}) = g_{i+2}$. In particular we have $\mu^\sigma \geq 4$, implying $\mu^\sigma - 1 \geq 3$. But then Property (BR1) implies $\sigma(g_1) = F_{1,1}$ which implies $\sigma(b_2) = g_2$ by assumption. Now consider level i. Again, by Property (BR1), $\sigma(g_i) = F_{i,1}$. Now, combining all of this and using Property (S2) yields $\sigma(s_{i,1}) = h_{i,1}$. But then, since we have $\sigma(b_{i+1}) = b_{i+2}$ by assumption, Property (B3) now implies $\bar{\sigma}(g_{i+1}) \neq \bar{\sigma}(b_{i+2})$. Since $i+1 < \mu^\sigma - 1$, Property (BR1) now implies $\sigma(g_{i+1}) = F_{i+1,1}$, i.e., we have $\bar{\sigma}(g_{i+1}) = 1$. But this now implies $\bar{\sigma}(b_{i+2}) = 0$, i.e., $\sigma(b_{i+2}) = b_{i+3}$ which is a contradiction since $\sigma(b_{i+2}) = g_{i+2}$ by $B^{\mathrm{S}}_{i+2} = R^{\mathrm{S}}_{i+2}$.

b) Let $j = 1$. Then, $\sigma(b_{i+1}) = g_{i+1}$, implying $\mathrm{rVal}_\sigma^{\mathrm{S}}(s_{i,j}) = \{s_{i,1}, h_{i,1}\} \cup B_{i+1}^{\mathrm{S}}$. We now show $\{s_{i,1}, h_{i,1}\} \cup B_{i+1}^{\mathrm{S}} \rhd B_2^{\mathrm{S}}$ for all possible "choices" of B_{i+1}^{S} and B_2^{S}. Let $B_{i+1}^{\mathrm{S}} = L_{i+1}^{\mathrm{S}}$ and $B_2^{\mathrm{S}} = L_2^{\mathrm{S}}$. Then $\{s_{i,1}, h_{i,1}\} \cup L_{i+1}^{\mathrm{S}} \rhd L_2^{\mathrm{S}}$, so the statement holds. Now consider the case $B_2^{\mathrm{S}} = R_2^{\mathrm{S}}$, implying that $2 < \mu^\sigma$. First assume that $\mu^\sigma \leq i$. Then, $\{s_{i,1}, h_{i,1}\} \cup L_{i+1}^{\mathrm{S}} \rhd R_2^{\mathrm{S}}$ follows from Lemma 6.1.10 (3). It cannot happen that$\mu^\sigma > i$, as this would yield $\mu^\sigma \geq i+1$. But this is a contradiction as this would imply $B_{i+1}^{\mathrm{S}} = R_{i+1}^{\mathrm{S}}$ as we currently assume $B_{i+2}^{\mathrm{S}} = L_{i+2}^{\mathrm{S}}$.

Now consider the case $B_{i+1}^{\mathrm{S}} = R_{i+1}^{\mathrm{S}}$ and $B_2^{\mathrm{S}} = R_2^{\mathrm{S}}$. Then $i+1 < \mu^\sigma$, hence the statement follows from $\{s_{i,1}, h_{i,1}\} \rhd \bigcup_{i'<i} W_{i'}$. Finally assume $B_{i+1}^{\mathrm{S}} = R_{i+1}^{\mathrm{S}}$ and $B_2^{\mathrm{S}} = L_2^{\mathrm{S}}$. Since $\mu^\sigma > i+1 \geq 2$ it holds that $\sigma(b_2) = b_3$, implying $B_2^{\mathrm{S}} = B_3^{\mathrm{S}}$. Applying Property (B1) repeatedly thus yields $B_2^{\mathrm{S}} = B_k^{\mathrm{S}} = R_k^{\mathrm{S}}$ where $k = \min\{i' \in \{2, \dots, i+1\} : \sigma(b_{i'}) = g_{i'}\} \leq i+1$. Thus, the statement follows from $\{s_{i,1}, h_{i,1}\} \rhd \bigcup_{i'<i} W_{i'}$ resp. $\{s_{i,j}, h_{i,1}\} \rhd \emptyset$. □

Omitted proofs of Section 6.2

Here, we provide the formal proofs of all statements of Section 6.2 that have not been proven there.

Lemma 6.2.1. *Let $G_n = M_n$. Let $\sigma \in \rho(\sigma_0)$ be a well-behaved phase-k-strategy for some $\mathfrak{b} \in \mathfrak{B}_n$ having Property (USV1)$_i$ and Property (EV1)$_{i+1}$ for some $i \in [n]$ where $k \in [5]$. If $F_{i,0}$ and $F_{i,1}$ are in the same state and if either $i \geq \nu$ or σ has Property (REL1), then* $\mathrm{Val}_\sigma^{\mathrm{M}}(F_{i,\beta_{i+1}}) > \mathrm{Val}_\sigma^{\mathrm{M}}(F_{i,1-\beta_{i+1}})$.

Proof. To simplify notation let $j := \beta_{i+1}$. Since both cycle centers are in the same state, it suffices to prove $\mathrm{rVal}_\sigma^{\mathrm{M}}(s_{i,j}) > \mathrm{rVal}_\sigma^{\mathrm{M}}(s_{i,1-j})$. By Property (USV1)$_i$ and Property (EV1)$_{i+1}$, $\mathrm{rVal}_\sigma^{\mathrm{M}}(s_{i,1-j}) = \langle s_{i,1-j}\rangle + \mathrm{rVal}_\sigma^{\mathrm{M}}(b_1)$ and $\mathrm{rVal}_\sigma^{\mathrm{M}}(s_{i,j}) = \langle s_{i,j}, h_{i,j}\rangle + \mathrm{rVal}_\sigma^{\mathrm{M}}(b_{i+1})$. Let $\mu^\sigma = 1$. Then, $\sigma(b_1) = b_2$ and thus $\mathrm{rVal}_\sigma^{\mathrm{M}}(b_1) = \mathrm{rVal}_\sigma^{\mathrm{M}}(b_2) = L_2^{\mathrm{M}}$ and $\mathrm{rVal}_\sigma^{\mathrm{M}}(b_{i+1}) = L_{i+1}^{\mathrm{M}}$. The statement then follows from $\langle s_{i,j}, h_{i,j}\rangle > \sum_{\ell \leq i} W_\ell^{\mathrm{M}} + \langle s_{i,1-j}\rangle$.

Hence let $\mu^\sigma > 1$, implying $\sigma(b_1) = g_1$. We distinguish the following cases.

1. Let $\mathrm{rVal}_\sigma^{\mathrm{M}}(b_1) = R_1^{\mathrm{M}}$ and $\mathrm{rVal}_\sigma^{\mathrm{M}}(b_{i+1}) = R_{i+1}^{\mathrm{M}}$. This implies $i+1 < \mu^\sigma$ and the statement thus again follows from $\langle s_{i,j}, h_{i,j}\rangle > \sum_{\ell \leq i} W_\ell^{\mathrm{M}} + \langle s_{i,1-j}\rangle$.
2. Let $\mathrm{rVal}_\sigma^{\mathrm{M}}(b_1) = R_1^{\mathrm{M}}$ and $\mathrm{rVal}_\sigma^{\mathrm{M}}(b_{i+1}) = L_{i+1}^{\mathrm{M}}$. Property (EV1)$_{i+1}$ implies

$$\mathrm{Val}_\sigma^{\mathrm{M}}(s_{i,j}) = \langle s_{i,j}, h_{i,j}\rangle + L_{i+1}^{\mathrm{M}} > \sum_{\ell \leq i} W_\ell^{\mathrm{M}} + \langle s_{i,1-j}\rangle + L_{i+1}^{\mathrm{M}}.$$

If $\mu^\sigma \leq i$, then

$$\sum_{\ell \leq i} W_\ell^{\mathrm{M}} = \sum_{\ell \leq \mu^\sigma} W_\ell^{\mathrm{M}} + \sum_{\ell = \mu^\sigma + 1}^{i} W_\ell^{\mathrm{M}} > \sum_{\ell < \mu^\sigma} W_\ell^{\mathrm{M}} + \sum_{\ell = \mu^\sigma + 1}^{i} \{W_\ell^{\mathrm{M}} \colon \sigma(b_i) = g_i\},$$

implying $\mathrm{Val}_\sigma^{\mathrm{M}}(s_{i,j}) > \mathrm{Val}_\sigma^{\mathrm{M}}(s_{i,1-j})$. If $\mu^\sigma = i+1$, then $\sum_{\ell \leq i} W_\ell^{\mathrm{M}} = \sum_{\ell < \mu^\sigma} W_\ell^{\mathrm{M}}$, again implying $\mathrm{Val}_\sigma^{\mathrm{M}}(s_{i,j}) > \mathrm{Val}_\sigma^{\mathrm{M}}(s_{i,1-j})$. Let $\mu^\sigma > i+1$. Then, by assumption, it needs to hold that $\sigma(b_{i+1}) = b_{i+2}$. Thus $\mu^\sigma \neq \min\{i' \colon \sigma(b_{i'}) = b_{i'+1}\}$, implying $\sigma(b_{\mu^\sigma}) = g_{\mu^\sigma}$ by Lemma 6.1.4. The statement the follows from

$$\begin{aligned}\sum_{\ell\leq i} W_\ell^{\mathrm{M}} + L_{i+1}^{\mathrm{M}} &= \sum_{\ell\leq i} W_\ell^{\mathrm{M}} + \sum_{\ell=i+1}^{\mu^\sigma-1}\{W_\ell^{\mathrm{M}} \colon \sigma(b_\ell) = g_\ell\} + W_{\mu^\sigma}^{\mathrm{M}} + L_{\mu^\sigma+1}^{\mathrm{M}}\\ &> \sum_{\ell<\mu^\sigma} W_\ell^{\mathrm{M}} + L_{\mu^\sigma+1}^{\mathrm{M}} = R_1^{\mathrm{M}}.\end{aligned}$$

3. Let $\mathrm{rVal}_\sigma^{\mathrm{M}}(b_1) = \langle g_k\rangle + \sum_{\ell<k} W_\ell^{\mathrm{M}} + \mathrm{rVal}_\sigma^{\mathrm{M}}(b_2), k = \min\{i' \colon \neg\bar{\sigma}(d_{i'})\} < \mu^\sigma$, and $\mathrm{rVal}_\sigma^{\mathrm{M}}(b_{i+1}) = R_{i+1}^{\mathrm{M}}$. We show that these assumptions yield a contradiction. The second equality implies that $i+1 < \mu^\sigma$. We now prove that σ has Property (REL1) in any case, so assume that $i \geq \nu$. This implies $\nu < i+1 < \mu^\sigma$ and thus $\nu \neq \mu^\sigma$. Consequently, σ cannot be a phase-2-strategy or phase-3-strategy for $\mathfrak{b}$ as it then had Property (REL2), implying $\mu^\sigma = \nu$. Therefore, by the definition of the phases, σ has Property (REL1) in any case, so $\mu^\sigma = \min\{i' \colon \sigma(b_{i'}) = b_{i'+1}\}$. Consequently, it holds that $\mathcal{I}^\sigma = \emptyset$. But then $i' < \mu^\sigma$ implies $\bar{\sigma}(d_{i'})$ by Corollary 6.1.5. This contradicts the characterization of $\mathrm{rVal}_\sigma^{\mathrm{M}}(b_1)$.
4. Let $\mathrm{rVal}_\sigma^{\mathrm{M}}(b_1) = \langle g_k\rangle + \sum_{\ell<k} W_\ell^{\mathrm{M}} + \mathrm{rVal}_\sigma^{\mathrm{M}}(b_2), k = \min\{i' \colon \neg\bar{\sigma}(d_{i'})\} < \mu^\sigma$, and $\mathrm{rVal}_\sigma^{\mathrm{M}}(b_{i+1}) = L_{i+1}^{\mathrm{M}}$. Then

$$\mathrm{rVal}_\sigma^{\mathrm{M}}(s_{i,j}) = \langle s_{i,j}, h_{i,j}\rangle + L_{i+1}^{\mathrm{M}} > \sum_{\ell\leq i} W_\ell^{\mathrm{M}} + L_{i+1}^{\mathrm{M}} + \langle s_{i,1-j}\rangle \geq \langle s_{i,1-j}\rangle + L_2^{\mathrm{M}}.$$

If $\mathrm{rVal}_\sigma^{\mathrm{M}}(b_2) = L_2^{\mathrm{M}}$, the statement thus follows since $\mathrm{rVal}_\sigma^{\mathrm{M}}(b_1) < \mathrm{rVal}_\sigma^{\mathrm{M}}(b_2)$ in this case. Thus assume $\mathrm{rVal}_\sigma^{\mathrm{M}}(b_2) = R_2^{\mathrm{M}}$, implying $\sigma(b_2) = g_2$ and $\mu^\sigma > 2$. If $\sigma(b_{\mu^\sigma}) = g_{\mu^\sigma}$, then $\langle s_{i,1-j}\rangle + L_2^{\mathrm{M}} > R_2^{\mathrm{M}}$, implying the statement. Hence assume $\sigma(b_{\mu^\sigma}) = b_{\mu^\sigma+1}$, implying $\mu^\sigma = \min\{i' \colon \sigma(b_{i'}) = b_{i'+1}\}$. In particular, $\sigma(b_{i'}) = g_{i'}$ for all $i' \in \{1, \dots, \mu^\sigma - 1\}$. Thus, since $\mathrm{rVal}_\sigma^{\mathrm{M}}(b_{i+1}) = L_{i+1}^{\mathrm{M}}$, we need to have $i+1 \geq \mu^\sigma$ and in particular $i \geq \mu^\sigma - 1$. Then, the statement follows from

$$\begin{aligned}\mathrm{rVal}_\sigma^{\mathrm{M}}(s_{i,j}) &> \sum_{\ell\leq i} W_\ell^{\mathrm{M}} + L_{i+1}^{\mathrm{M}} + \langle s_{i,1-j}\rangle\\ &\geq \sum_{\ell<\mu^\sigma} W_\ell^{\mathrm{M}} + L_{\mu^\sigma+1}^{\mathrm{M}} + \langle s_{i,1-j}\rangle = R_2^{\mathrm{M}} + \langle s_{i,1-j}\rangle.\end{aligned}$$

□

Lemma 6.2.3. *Let $\mathfrak{b} \in \mathfrak{B}_n$. If $\mathbf{1}_{j=0}\mathrm{lfn}(\mathfrak{b}, i+1) + \mathbf{1}_{j=1}\mathrm{lufn}(\mathfrak{b}, i+1) = 0$ for $i \in [n], j \in \{0,1\}$, then $\ell^{\mathfrak{b}}(i,j,k) \geq \mathfrak{b}$ for , $k \in \{0,1\}$. Otherwise, the following hold:*

Setting of bits	$\mathfrak{b}_i = 1 \wedge \mathfrak{b}_{i+1} = 1-j$	$\mathfrak{b}_i = 0 \wedge \mathfrak{b}_{i+1} = j$	$\mathfrak{b}_i = 0 \wedge \mathfrak{b}_{i+1} = 1-j$
$\ell^{\mathfrak{b}}(i,j,k) =$	$\left\lceil\frac{\mathfrak{b}+\sum(\mathfrak{b},i)+1-k}{2}\right\rceil$	$\left\lceil\frac{\mathfrak{b}+2^{i-1}+\sum(\mathfrak{b},i)+1-k}{2}\right\rceil$	$\left\lceil\frac{\mathfrak{b}-2^{i-1}+\sum(\mathfrak{b},i)+1-k}{2}\right\rceil$

Proof. Let $m := \mathbf{1}_{j=0}\mathrm{lfn}(\mathfrak{b}, i+1) + \mathbf{1}_{j=1}\mathrm{lufn}(\mathfrak{b}, i+1) \neq 0$. Then, $\mathrm{lfn}(\mathfrak{b}, i, \{(i+1, j)\}) \neq 0$ and we distinguish three cases.

1. Let $\mathfrak{b}_i = 1$ and $\mathfrak{b}_{i+1} = 1-j$. We prove $\mathrm{lfn}(\mathfrak{b}, i, \{(i+1,j)\}) = \mathfrak{b} - 2^i - \sum(\mathfrak{b}, i)$ and $m = \mathfrak{b} - 2^{i-1} - \sum(\mathfrak{b}, i)$. By definition, $\mathfrak{b}' := \mathrm{lfn}(\mathfrak{b}, i, \{(i+1, j)\})$ is the largest number smaller than $\mathfrak{b}$ such that $\nu(\mathfrak{b}') = i$ and $\mathfrak{b}'_{i+1} = j$. Since $\mathfrak{b}_{i+1} = 1-j$, subtracting 2^i switches bit $i+1$ and only bit $i+1$. By subtracting $\sum(\mathfrak{b}, i)$, all bits below bit i that are equal to 1 are set to 0. Therefore, $\mathfrak{b}' = \mathfrak{b} - 2^i - \sum(\mathfrak{b}, i)$. Note that $\mathfrak{b}' > 0$.

   Assume $j = 1$, implying $m = \mathrm{lfn}(\mathfrak{b}, i+1) \neq 0$. Since m is the largest number smaller than $\mathfrak{b}$ with least significant set bit equal to 1 being bit $i+1$ and since $\mathfrak{b}_{i+1} = 1$, we have $m = \mathfrak{b} - \sum(\mathfrak{b}, i+1) = \mathfrak{b} - 2^{i-1} - \sum(\mathfrak{b}, i)$. Consequently,

$$\begin{aligned} \ell^{\mathfrak{b}}(i,j,k) &= \left\lceil \frac{\mathfrak{b} - 2^i - \sum(\mathfrak{b}, i) + 1 - k}{2} \right\rceil + \mathfrak{b} - \mathfrak{b} + 2^{i-1} + \sum(\mathfrak{b}, i) \\ &= \left\lceil \frac{\mathfrak{b} + \sum(\mathfrak{b}, i) + 1 - k}{2} \right\rceil. \end{aligned}$$

2. By similar arguments, it can be shown that $\mathrm{lfn}(\mathfrak{b}, i, \{(i+1, j)\}) = \mathfrak{b} - 2^i - 2^{i-1} - \sum(\mathfrak{b}, i)$ and $m = \mathfrak{b} - 2^i - \sum(\mathfrak{b}, i)$ in this case, implying the statement analogously.
3. By similar arguments, it can be shown that $\mathrm{lfn}(\mathfrak{b}, i, \{(i+1, j)\}) = \mathfrak{b} - 2^{i-1} - \sum(\mathfrak{b}, i)$ and $m = \mathfrak{b} - \sum(\mathfrak{b}, i)$ in this case, implying the statement analogously.

If $\mathbf{1}_{j=0}\mathrm{lfn}(\mathfrak{b}, i+1) + \mathbf{1}_{j=1}\mathrm{lufn}(\mathfrak{b}, i+1) = 0$, the statement follows immediately. □

Lemma 6.2.4. *Let $\mathfrak{b} \in \mathfrak{B}_n$ and $i \in [n]$ and $j \in \{0,1\}$ such that $\mathfrak{b}_i = 0$ or $\mathfrak{b}_{i+1} \neq j$. Then,*

$$\mathbf{1}_{j=0}\mathrm{lfn}(\mathfrak{b}, i+1) - \mathbf{1}_{j=1}\mathrm{lufn}(\mathfrak{b}, i+1) = \mathbf{1}_{j=0}\mathrm{lfn}(\mathfrak{b}+1, i+1) - \mathbf{1}_{j=1}\mathrm{lufn}(\mathfrak{b}+1, i+1).$$

Moreover, if $i \neq \nu$, then $\ell^{\mathfrak{b}}(i,j,k) + 1 = \ell^{\mathfrak{b}+1}(i,j,k)$.

Proof. Consider the first statement. Assume $\mathrm{lufn}(\mathfrak{b}, i+1) \neq \mathrm{lufn}(\mathfrak{b}+1, i+1)$. This can only occur if $\mathfrak{b}+1 = \mathrm{lufn}(\mathfrak{b}+1, i+1)$, implying $(\mathfrak{b}+1)_{i+1} = \dots = (\mathfrak{b}+1)_1 = 0$. But this implies $\nu(\mathfrak{b}+1) \geq i+2$, hence $\mathfrak{b}_1 = \dots = \mathfrak{b}_i = \mathfrak{b}_{i+1} = 1$. Since $\mathfrak{b}_i = 0 \vee \mathfrak{b}_{i+1} \neq j$ by assumption, it thus needs to hold that $j = 0$. This proves that $\mathrm{lufn}(\mathfrak{b}, i+1) \neq \mathrm{lufn}(\mathfrak{b}+1, i+1)$ implies $j = 0$. In a similar way it can be proven that $\mathrm{lfn}(\mathfrak{b}, i+1) \neq \mathrm{lfn}(\mathfrak{b}+1, i+1)$ implies $j = 1$. Consequently, it is impossible that both $\mathrm{lufn}(\mathfrak{b}, i+1) \neq \mathrm{lufn}(\mathfrak{b}+1, i+1)$ and $\mathrm{lfn}(\mathfrak{b}, i+1) \neq \mathrm{lufn}(\mathfrak{b}+1, i+1)$ hold. If $\mathrm{lufn}(\mathfrak{b}, i+1) \neq \mathrm{lufn}(\mathfrak{b}+1, i+1)$, then it holds that $j = 0$ and $\mathrm{lfn}(\mathfrak{b}, i+1) = \mathrm{lfn}(\mathfrak{b}+1, i+1)$. If $\mathrm{lfn}(\mathfrak{b}, i+1) \neq \mathrm{lfn}(\mathfrak{b}+1, i+1)$, we have $j = 1$ and $\mathrm{lufn}(\mathfrak{b}, i+1) = \mathrm{lufn}(\mathfrak{b}+1, i+1)$. But this implies

$$\mathbf{1}_{j=0}\mathrm{lfn}(\mathfrak{b}, i+1) - \mathbf{1}_{j=1}\mathrm{lufn}(\mathfrak{b}, i+1) = \mathbf{1}_{j=0}\mathrm{lfn}(\mathfrak{b}+1, i+1) - \mathbf{1}_{j=1}\mathrm{lufn}(\mathfrak{b}+1, i+1).$$

Now let also $i \neq \nu$. It suffices to prove $\mathrm{lfn}(\mathfrak{b}, i, \{(i+1, j)\}) = \mathrm{lfn}(\mathfrak{b}+1, i, \{(i+1, j)\})$. But this follows directly since the choice of i implies $\mathrm{lfn}(\mathfrak{b}+1, i, \{(i+1, j)\}) \neq \mathfrak{b}+1$. □

Lemma 6.2.5. *Let $\sigma_{\mathfrak{b}}$ be a canonical strategy for $\mathfrak{b}$ such that its occurrence records are described by Table 5.6. Assume that $\sigma_{\mathfrak{b}}$ has Properties (OR1)$_{*,*,*}$ to (OR4)$_{*,*,*}$. Then, the following hold.*

1. *Let $i \in [n]$ and $j \in \{0,1\}$ and assume that either $\mathfrak{b}_i = 0$ or $\mathfrak{b}_{i+1} \neq j$. Then, it holds that $\phi^{\sigma_\mathfrak{b}}(d_{i,j,*}, F_{i,j}) \leq \lfloor (\mathfrak{b}+1)/2 \rfloor$.*
2. *Let $j := \mathfrak{b}_{\nu+1}$. Then, $\phi^{\sigma_\mathfrak{b}}(d_{\nu,j,0}, F_{\nu,j}) = \lfloor (\mathfrak{b}+1)/2 \rfloor$. In addition, $\nu = 1$ implies $\phi^{\sigma_\mathfrak{b}}(d_{\nu,j,1}, F_{\nu,j}) = \lfloor (\mathfrak{b}+1)/2 \rfloor$ and $\nu > 1$ implies $\phi^{\sigma}(d_{\nu,j,1}, F_{\nu,j}) = \lfloor (\mathfrak{b}+1)/2 \rfloor - 1$.*
3. *If $i = 1$, then $\sigma_\mathfrak{b}(d_{1,1-\mathfrak{b}_2,*}) \neq F_{1,1-\mathfrak{b}_2}$ and $\phi^{\sigma_\mathfrak{b}}(d_{1,1-\mathfrak{b}_2,0}, F_{1,1-\mathfrak{b}_2}) = \lfloor (\mathfrak{b}+1)/2 \rfloor$.*

Proof. The first statement follows immediately since $\mathfrak{b}_i = 0 \vee \mathfrak{b}_{i+1} \neq j$ imply

$$\phi^{\sigma_\mathfrak{b}}(d_{i,j,k}, F_{i,j}) = \min\left(\left\lfloor \frac{\mathfrak{b}+1-k}{2} \right\rfloor, \ell^\mathfrak{b}(i,j,k) + t_\mathfrak{b}\right) \leq \left\lfloor \frac{\mathfrak{b}+1-k}{2} \right\rfloor \leq \left\lfloor \frac{\mathfrak{b}+1}{2} \right\rfloor.$$

Consider the second statement and observe that it suffices to prove

$$\phi^{\sigma_\mathfrak{b}}(d_{\nu,j,k}, F_{\nu,j}) = \left\lfloor \frac{\mathfrak{b}+1-k}{2} \right\rfloor \tag{A.9}$$

for $k \in \{0,1\}$. Let $\mathbf{1}_{j=0}\mathrm{lfn}(\mathfrak{b},\nu+1) - \mathbf{1}_{j=1}\mathrm{lufn}(\mathfrak{b},\nu+1) = 0$. Then, by Lemma 6.2.3, $\ell^\mathfrak{b}(\nu,j,k) \geq \mathfrak{b}$. In order to show Equation (A.9), it thus suffices to prove that either $\mathfrak{b} - 1 \geq \lfloor (\mathfrak{b}+1-k)/2 \rfloor$ or that the parameter $t_\mathfrak{b} = -1$ is not feasible. Since it holds that $\mathfrak{b} - 1 \geq \lfloor (\mathfrak{b}+1)/2 \rfloor$ for $\mathfrak{b} \geq 2$, it suffices to show that $t_\mathfrak{b} = -1$ is not feasible for $\mathfrak{b} = 0, 1$. By Table 5.6, the parameter -1 can only be feasible if $\mathfrak{b}_1 = 1 \wedge \nu \neq 1$. It is therefore not feasible for $\mathfrak{b} = 0$. Let $\mathfrak{b} = 1$ and $\phi^{\sigma_\mathfrak{b}}(d_{\nu,j,k}, F_{\nu,j}) = \ell^\mathfrak{b}(\nu,j,k) - 1$. Since $\mathfrak{b}+1 = 2$ is a power of two and since $\sigma_\mathfrak{b}$ has Properties (OR1)$_{*,*,*}$ to (OR4)$_{*,*,*}$, Property (OR3)$_{\nu,j,k}$ implies $\phi^{\sigma_\mathfrak{b}}(d_{\nu,j,k}, F_{\nu,j}) = \lfloor (\mathfrak{b}+1-k)/2 \rfloor$. Consequently, $\phi^{\sigma_\mathfrak{b}}(d_{\nu,j,k}, F_{\nu,j}) = \lfloor (\mathfrak{b}+1-k)/2 \rfloor$. Now let $\mathbf{1}_{j=0}\mathrm{lfn}(\mathfrak{b},\nu+1) - \mathbf{1}_{j=1}\mathrm{lufn}(\mathfrak{b},\nu+1) \neq 0$. Then, by the definition of ν and j and Lemma 6.2.3,

$$\ell^\mathfrak{b}(\nu,j,k) = \left\lceil \frac{\mathfrak{b} + 2^{\nu-1} + \sum(\mathfrak{b},\nu) + 1 - k}{2} \right\rceil \geq \left\lceil \frac{\mathfrak{b} + 2^{\nu-1} + 1 - k}{2} \right\rceil \geq \left\lfloor \frac{\mathfrak{b}+1-k}{2} \right\rfloor + 1.$$

Since $-1, 0$ and 1 are the only feasible parameters, this implies Equation (A.9).

Consider the third statement and let $i = 1, j = 1 - \mathfrak{b}_2$. Then, independent of whether $\mathfrak{b}_1 = 0$ or $\mathfrak{b}_1 = 1$, $\ell^\mathfrak{b}(i,j,k) \geq \lceil (\mathfrak{b}-k)/2 \rceil = \lfloor (\mathfrak{b}+1-k)/2 \rfloor$ by Lemma 6.2.3. By the first statement and by Property (OR1)$_{i,j,k}$ and Property (OR2)$_{i,j,k}$, this implies $\sigma(d_{i,j,k}) \neq F_{i,j}$ for both $k \in \{0,1\}$ as $\phi^{\sigma_\mathfrak{b}}(d_{i,j,k}, F_{i,j}) = \ell^\mathfrak{b}(i,j,k) + 1$ otherwise. Furthermore, this implies $\ell^\mathfrak{b}(i,j,0) = \lfloor (\mathfrak{b}+1)/2 \rfloor$. Assume that $\phi^{\sigma}(d_{i,j,0}, F_{i,j}) = \ell^\mathfrak{b}(i,j,0) - 1$. Then, $\phi^{\sigma_\mathfrak{b}}(d_{i,j,0}, F_{i,j}) \neq \lfloor (\mathfrak{b}+1)/2 \rfloor$. Hence, by Property (OR4)$_{i,j,k}$, it holds that $\mathfrak{b}$ is odd and $i = \nu(\mathfrak{b}+1)$. But this contradicts $i = 1$.

Consequently, $\phi^{\sigma_\mathfrak{b}}(d_{i,j,0}, F_{i,j}) = \ell^\mathfrak{b}(i,j,0) = \lfloor (\mathfrak{b}+1)/2 \rfloor$. □

Lemma 6.2.6. *Let $\mathfrak{b} \in \mathfrak{B}_n$ and $i \in [n]$. It holds that $\mathrm{fl}(\mathfrak{b},i) = \lfloor (\mathfrak{b} + 2^{i-1})/2^i \rfloor$ and $\mathrm{fl}(\mathfrak{b}+1,i) = \mathrm{fl}(\mathfrak{b},i) + \mathbf{1}_{i=\nu}$. In addition, for indices $i_1, i_2 \in [n]$ with $i_1 < i_2$ and $\mathfrak{b} \geq 2^{i_1-1}$ imply $\mathrm{fl}(\mathfrak{b},i_1) > \mathrm{fl}(\mathfrak{b},i_2)$. Furthermore, if $k := \frac{\mathfrak{b}+1}{2^{\nu-1}}$ and $x \in [\nu-1]$, then $\mathrm{fl}(\mathfrak{b},\nu-x) = k \cdot 2^{x-1}$.*

Proof. As a reminder, a binary number $\mathfrak{b}$ *matches* the pair (i,q) if $\mathfrak{b}_i = q$. It matches a set S if $\mathfrak{b}$ matches every $(i,q) \in S$. Consider the first two statements. By definition, it holds that $\mathrm{fl}(\mathfrak{b}+1,i) = \mathrm{fl}(\mathfrak{b},i) + \mathbf{1}_{i=\nu}$. Let $S_i := \{(i,1),(i-1,0),\ldots,(1,0)\}$. By definition,

$\text{fl}(\mathfrak{b}, i)$ is the number of numbers smaller than or equal to $\mathfrak{b}$ matching S_i. Since 2^{i-1} is the smallest number matching S_i, the statement follows if $\mathfrak{b} < 2^{i-1}$. Let m_k denote the k-th number matching the scheme S_i. Then $m_1 = 2^{i-1}$. As only numbers ending on the subsequence $(1, 0, \ldots, 0)$ of length i match S_i, we have $m_k = (k-1) \cdot 2^i + 2^{i-1}$. Since $\text{fl}(m_k, i) = k$ by definition and

$$\left\lfloor \frac{m_k + 2^{i-1}}{2^i} \right\rfloor = \left\lfloor \frac{(k-1) \cdot 2^i + 2^{i-1} + 2^{i-1}}{2^i} \right\rfloor = \left\lfloor \frac{k \cdot 2^k}{2^k} \right\rfloor = k,$$

this implies $\text{fl}(m_k, i, =) \lfloor (m_k + 2^{i-1})/2^i \rfloor$. Let $\mathfrak{b} \in \mathfrak{B}_n$ and $k \in \mathbb{N}$ such that $\mathfrak{b} \in [m_k, m_{k+1})$. Then, by the definition of $\text{fl}(\mathfrak{b}, i)$, we have $\text{fl}(\mathfrak{b}, i) = k$. In addition,

$$\left\lfloor \frac{\mathfrak{b} + 2^{i-1}}{2^i} \right\rfloor \geq \left\lfloor \frac{m_k + 2^{i-1}}{2^i} \right\rfloor = \text{fl}(m_k, i) = k$$

by the choice of k and

$$\left\lfloor \frac{\mathfrak{b} + 2^{i-1}}{2^i} \right\rfloor < \left\lfloor \frac{m_{k+1} + 2^{i-1}}{2^i} \right\rfloor = \text{fl}(m_{k+1}, i) = k+1.$$

Integrality thus implies $\lfloor (\mathfrak{b} + 2^{i-1})/2^i \rfloor r = k$, hence $\text{fl}(\mathfrak{b}, i) = k = \lfloor (\mathfrak{b} + 2^{i-1})/2^i \rfloor$.

Now let $i_1, i_2 \in [n]$ with $i_1 < i_2$ and $\mathfrak{b} \geq 2^{i_1 - 1}$. Then, $\text{fl}(\mathfrak{b}, i_1) = \lfloor (\mathfrak{b} + 2^{i_1-1})/2^{i_1} \rfloor$ and similarly $\text{fl}(\mathfrak{b}, i_2) = \lfloor (\mathfrak{b} + 2^{i_2-1})/2^{i_2} \rfloor$. If $\mathfrak{b} = 2^{i_1-1}$, then $\mathfrak{b} < 2^{i_2-1}$, implying

$$\text{fl}(\mathfrak{b}, i_1) = \left\lfloor \frac{\mathfrak{b}}{2^{i_1}} + \frac{1}{2} \right\rfloor = \left\lfloor \frac{2^{i_1-1}}{2^{i_1}} + \frac{1}{2} \right\rfloor = 1 > 0 = \left\lfloor \frac{2^{i_1-1}}{2^{i_2}} + \frac{1}{2} \right\rfloor = \text{fl}(\mathfrak{b}, i_2).$$

Thus let $\mathfrak{b} > 2^{i_1-1}$. Choose $k \in \mathbb{N}$ such that $k \cdot 2^{i_1-1} < \mathfrak{b} \leq (k+1)2^{i_1-1}$. Then

$$\text{fl}(\mathfrak{b}, i_1) = \left\lfloor \frac{\mathfrak{b}}{2^{i_1}} + \frac{1}{2} \right\rfloor > \left\lfloor \frac{k \cdot 2^{i_1-1}}{2^{i_1}} + \frac{1}{2} \right\rfloor = \left\lfloor \frac{k+1}{2} \right\rfloor \geq \left\lfloor \frac{k}{2} \right\rfloor$$

and thus $\text{fl}(\mathfrak{b}, i_1) \geq \lfloor k/2 \rfloor + 1$ by integrality. In addition, $\mathfrak{b} \leq (k+1)2^{i_1-1} < (k+1)2^{i_2-1}$ implies

$$\text{fl}(\mathfrak{b}, i_2) = \left\lfloor \frac{\mathfrak{b}}{2^{i_2}} + \frac{1}{2} \right\rfloor < \left\lfloor \frac{(k+1)2^{i_2-1}}{2^{i_2}} + \frac{1}{2} \right\rfloor = \left\lfloor \frac{k+2}{2} \right\rfloor = \left\lfloor \frac{k}{2} \right\rfloor + 1,$$

hence $\text{fl}(\mathfrak{b}, i_2) \leq \lfloor k/2 \rfloor$, implying the statement.

Now consider the third statement. By definition, ν is the least significant set bit of $\mathfrak{b} + 1$. Consequently, $\mathfrak{b} + 1$ is dividable by $2^{\nu-1}$, hence $k \in \mathbb{N}$ and in particular $\mathfrak{b} = k \cdot 2^{\nu-1} - 1$. Using Lemma 6.2.3 and $\frac{1}{2} - \frac{1}{2^{\nu-x}} \in (0, \frac{1}{2})$, this implies

$$\begin{aligned} \text{fl}(\mathfrak{b}, \nu - x) &= \left\lfloor \frac{\mathfrak{b} + 2^{\nu-x-1}}{2^{\nu-x}} \right\rfloor = \left\lfloor \frac{k \cdot 2^{\nu-1} - 1 + 2^{\nu-x-1}}{2^{\nu-x}} \right\rfloor = \left\lfloor \frac{k \cdot 2^{\nu-1}}{2^{\nu-x}} - \frac{1}{2^{\nu-x}} + \frac{1}{2} \right\rfloor \\ &= \left\lfloor \frac{k \cdot 2^{\nu-1}}{2^{\nu-x}} \right\rfloor = \lfloor k \cdot 2^{x-1} \rfloor = k \cdot 2^{x-1}. \end{aligned}$$

□

Lemma 6.2.7 (First row of Table 6.4). *Let $\sigma \in \rho(\sigma_0)$ be a well-behaved phase-1-strategy for $\mathfrak{b} \in \mathfrak{B}_n$ with $I_\sigma = \mathfrak{D}^\sigma$. Let $i \in [n], j, k \in \{0,1\}$ such that $e := (d_{i,j,k}, F_{i,j}) \in I_\sigma$ and $\sigma(d_{i,j,1-k}) \neq F_{i,j}$. Then σe is a well-behaved phase-1-strategy for $\mathfrak{b}$ with $\sigma e \in \rho(\sigma_0)$ and $I_{\sigma e} = \mathfrak{D}^{\sigma e}$.*

Proof. Since $F_{i,j}$ is open for σ, Property (ESC1) implies $\bar{\sigma}(eb_{i,j}) = \overline{\sigma e}(eb_{i,j}), \bar{\sigma}(eg_{i,j}) = \overline{\sigma e}(eg_{i,j})$ and $\bar{\sigma}(d_{i,j}) = \overline{\sigma e}(d_{i,j}) = 0$. Hence, σ being well-behaved implies that σe is well-behaved. By the same arguments, σe is a phase-1-strategy for $\mathfrak{b}$ and it suffices to prove $I_{\sigma[e]} = \mathfrak{D}^{\sigma e}$.

Consider the case $G_n = S_n$. By Property (ESC1), it holds that $\bar{\sigma}(eb_{i,j}) = \overline{\sigma e}(eb_{i,j})$ and $\bar{\sigma}(eg_{i,j}) = \overline{\sigma e}(eg_{i,j})$. Since σe is a phase-1-strategy for $\mathfrak{b}$, also $\mu^\sigma = \mu^{\sigma e}$ by the choice of e. Thus, $\mathrm{rVal}_\sigma^{\mathrm{S}}(F_{i,j}) = \mathrm{rVal}_{\sigma e}^{\mathrm{S}}(F_{i,j})$ by Lemma 6.1.16. In particular, the valuation of $F_{i,j}$ does not change. Since $F_{i,j}$ is the only vertex that has an edge towards $d_{i,j,k}$, this implies that the valuation of no other vertex but $d_{i,j,k}$ changes, hence $I_{\sigma e} = I_\sigma \setminus \{e\}$ if $G_n = S_n$.

Consider the case $G_n = M_n$ and let $j = \bar{\sigma}(g_i)$. Then, $F_{i,\bar{\sigma}(g_i)}$ is not closed with respect to either σ or σe. Therefore, the valuations of $F_{i,j}$ and g_i increase, but only by terms of size $o(1)$. Now, Property (EV1)$_i$ and Property (EV2)$_i$ imply $\sigma(b_i) = b_{i+1}$, hence $\sigma(s_{i-1,1}) = b_1$ by Property (USV1)$_{i-1}$. In particular, the valuation of no other vertex than $d_{i,j,k}, F_{i,j}, g_i$ and $h_{i-1,1}$ increases. It is now easy to calculate that $(b_i, g_i), (s_{i-1,1}, h_{i-1,1}) \notin I_{\sigma e}$ as the change of the valuation of g_i is only of size $o(1)$, implying the statement.

Let $j \neq \bar{\sigma}(g_i)$ and let $t^{\rightarrow} := g_1$ if $\mathfrak{b}$ is odd and $t^{\rightarrow} := b_2$ if $\mathfrak{b}$ is even. Then, $d_{i,j,k}$ and $F_{i,j}$ are the only vertices whose valuation increases by applying e. Since $(g_i, F_{i,j}) \notin I_\sigma$ by assumption, it thus suffices to prove $(g_i, F_{i,j}) \notin I_{\sigma e}$. By the choice of e, it holds that $\mathrm{Val}_{\sigma e}^{\mathrm{M}}(F_{i,j}) = \frac{1-\varepsilon}{1+\varepsilon} \mathrm{Val}_{\sigma e}^{\mathrm{M}}(t^{\rightarrow}) + \frac{2\varepsilon}{1+\varepsilon} \mathrm{Val}_\sigma^{\mathrm{M}}(s_{i,j})$. First assume that $F_{i,1-j}$ is $t^{\rightarrow}$-open. Then, by Lemma 6.2.1 and since $(g_i, F_{i,j}) \notin I_\sigma$, we have $j = 1 - \mathfrak{b}_{i+1}$. We prove that $\sigma(s_{i,j}) = b_1$ (Property (USV1)$_i$), $\sigma(b_1) = t^{\rightarrow}$ (Property (EV1)$_1$ and Property (ESC1)) as well as $\sigma(s_{i,1-j}) = h_{i,1-j}$ and $\bar{\sigma}(b_{i+1}) = 1 - j$ (Property (USV1)$_i$) imply the statement. We have

$$
\begin{aligned}
\mathrm{Val}_{\sigma e}^{\mathrm{M}}&(F_{i,1-j}) - \mathrm{Val}_{\sigma e}^{\mathrm{M}}(F_{i,j}) \\
&= \frac{\varepsilon(1-\varepsilon)}{1+\varepsilon} \mathrm{Val}_{\sigma e}^{\mathrm{M}}(t^{\rightarrow}) + \varepsilon \mathrm{Val}_{\sigma e}^{\mathrm{M}}(s_{i,1-j}) - \frac{2\varepsilon}{1+\varepsilon} \left(\langle s_{i,j}\rangle + \mathrm{Val}_{\sigma e}^{\mathrm{M}}(t^{\rightarrow})\right) \\
&= \varepsilon \left(\mathrm{Val}_{\sigma e}^{\mathrm{M}}(s_{i,1-j}) - \mathrm{Val}_{\sigma e}^{\mathrm{M}}(t^{\rightarrow}) - \frac{2}{1+\varepsilon} \langle s_{i,j}\rangle \right) \\
&> \varepsilon \left(\mathrm{Val}_{\sigma e}^{\mathrm{M}}(s_{i,1-j}) - \mathrm{Val}_{\sigma e}^{\mathrm{M}}(t^{\rightarrow}) - 2N^{10} \right) \\
&= \varepsilon \left(\langle s_{i,1-j}, h_{i,1-j}\rangle + \mathrm{Val}_{\sigma e}^{\mathrm{M}}(b_{i+1}) - \mathrm{Val}_{\sigma e}^{\mathrm{M}}(t^{\rightarrow}) - 2N^{10} \right).
\end{aligned}
$$

It thus suffices to prove $\langle s_{i,1-j}, h_{i,1-j}\rangle + \mathrm{Val}_{\sigma e}^{\mathrm{M}}(b_{i+1}) - \mathrm{Val}_{\sigma e}^{\mathrm{M}}(t^{\rightarrow}) - 2N^{10} \geq 0$. We distinguish three cases.

1. Let $t^{\rightarrow} = b_2$. Then, $\sigma(b_1) = b_2$ and $\mu^\sigma = 1$. In particular, $\mathrm{Val}_{\sigma e}^{\mathrm{M}}(b_{i+1}) = L_{i+1}^{\mathrm{M}}$ and $\mathrm{Val}_{\sigma e}^{\mathrm{M}}(t^{\rightarrow}) = L_2^{\mathrm{M}}$. Consequently,

$$
\begin{aligned}
\langle s_{i,1-j}, &h_{i,1-j}\rangle + \mathrm{Val}_{\sigma e}^{\mathrm{M}}(b_{i+1}) - \mathrm{Val}_{\sigma e}^{\mathrm{M}}(t^{\rightarrow}) - 2N^{10} \\
&= \langle s_{i,1-j}, h_{i,1-j}\rangle + L_{i+1}^{\mathrm{M}} - L_2^{\mathrm{M}} - 2N^{10} = \langle s_{i,1-j}, h_{i,1-j}\rangle + L_{2,i}^{\mathrm{M}} - 2N^{10}
\end{aligned}
$$

$$
\begin{aligned}
&> \langle s_{i,1-j}, h_{i,1-j}\rangle - \sum_{\ell=2}^{i} W_\ell^{\mathrm{M}} - 2N^{10} \\
&\geq N^{2i+10} + N^8 - \sum_{\ell=1}^{i}(N^{2\ell+10} - N^{2\ell+9} + N^{10}) - 2N^{10} \\
&= N^{2i+10} + N^8 - \frac{N^{2i+11} - N^{11}}{N+1} - (i+2)N^{10} \\
&> N^{2i+10} + N^8 - \frac{N^{2i+11} - N^{11}}{N+1} - N^{11}.
\end{aligned}
$$

This term is larger than 0 if $(2N+1)N^{2i+2} + N + 1 > N^3(N+2)$ which holds since $i \geq 1$ and N is sufficiently large.

2. Let $t^{\rightarrow} = g_1$ and $\mathrm{Val}_{\sigma}^{\mathrm{M}}(b_{i+1}) = R_{i+1}^{\mathrm{M}}$. Then $\sigma(b_1) = g_1$ and $\mathrm{Val}_{\sigma}^{\mathrm{M}}(t^{\rightarrow}) = R_1^{\mathrm{M}}$. In particular, since σ is a phase-1-strategy and $i+1 < \mu^{\sigma} \wedge \sigma(b_{i+1}) = g_{i+1}$ by assumption, it holds that $\mathfrak{b}_1 = \cdots = \mathfrak{b}_{i+1} = 1$. This then implies

$$
\begin{aligned}
&\langle s_{i,1-j}, h_{i,1-j}\rangle + \mathrm{Val}_{\sigma}^{\mathrm{M}}(b_{i+1}) - \mathrm{Val}_{\sigma}^{\mathrm{M}}(t^{\rightarrow}) - 2N^{10} \\
&\quad= \langle s_{i,1-j}, h_{i,1-j}\rangle + R_{i+1}^{\mathrm{M}} - R_1^{\mathrm{M}} - 2N^{10} = \langle s_{i,1}, h_{i,1}\rangle - \sum_{\ell=1}^{i} W_\ell^{\mathrm{M}} - 2N^{10} \\
&\quad= \langle s_{i,1}, h_{i,1}\rangle - \sum_{\ell=1}^{i}(N^{2\ell+10} - N^{2\ell+9} + N^8) - 2N^{10} \\
&\quad> N^{2i+10} + N^8 - \sum_{\ell=1}^{i}(N^{2\ell+10} - N^{2\ell+9}) - N^{11}
\end{aligned}
$$

which is larger than 0 as shown above.

3. Let $t^{\rightarrow} = g_1$ and $\mathrm{Val}_{\sigma}^{\mathrm{M}}(b_{i+1}) = L_{i+1}^{\mathrm{M}}$. It cannot hold that $i+1 < \mu^{\sigma}$ since this implies $\sigma(b_{i+1}) = g_{i+1}$ and thus $\mathrm{Val}_{\sigma}^{\mathrm{M}}(b_{i+1}) = R_{i+1}^{\mathrm{M}}$. Consequently, $i+1 \geq \mu^{\sigma}$. In addition, $\mathrm{Val}_{\sigma}^{\mathrm{M}}(b_1) = R_1^{\mathrm{M}}$ as before. Consequently,

$$
\begin{aligned}
&\langle s_{i,1-j}, h_{i,1-j}\rangle + \mathrm{Val}_{\sigma}^{\mathrm{M}}(b_{i+1}) - \mathrm{Val}_{\sigma}^{\mathrm{M}}(t^{\rightarrow}) - 2N^{10} \\
&\qquad= \langle s_{i,1-j}, h_{i,1-j}\rangle + L_{i+1}^{\mathrm{M}} - R_1^{\mathrm{M}} - 2N^{10} \\
&\qquad= \langle s_{i,1-j}, h_{i,1-j}\rangle - \sum_{\ell=1}^{\mu^{\sigma}-1} W_\ell^{\mathrm{M}} - \sum_{\ell=\mu^{\sigma}+1}^{i} \{W_\ell^{\mathrm{M}} \colon \sigma(b_\ell) = g_\ell\} - 2N^{10} \\
&\qquad> \langle s_{i,1-j}, h_{i,1-j}\rangle - \sum_{\ell=1}^{i} W_\ell^{\mathrm{M}} - 2N^{10}
\end{aligned}
$$

which is larger than 0 as proven before.

This concludes the case that $F_{i,1-j}$ is $t^{\rightarrow}$-open. If it is not $t^{\rightarrow}$-open, then it has to be closed or $t^{\rightarrow}$-halfopen by Property (ESC1). Assume that it is closed. If $1-j = \mathfrak{b}_{i+1}$, then $\mathrm{rVal}_{\sigma}^{\mathrm{M}}(F_{i,1-j}) = \langle s_{i,1-j}, h_{i,1-j}\rangle + \mathrm{rVal}_{\sigma}^{\mathrm{M}}(b_{i+1})$ by Properties (USV1)$_i$ and (EV1)$_{i+1}$. Since

$\langle s_{i,1-j}, h_{i,1-j}\rangle > \sum_{\ell\in[i]} W_\ell^{\mathrm{M}}$, this implies $\mathrm{rVal}_{\sigma e}^{\mathrm{M}}(F_{i,1-j}) > \mathrm{rVal}_{\sigma e}(F_{i,j}) = \mathrm{rVal}_{\sigma e}^{\mathrm{M}}(t^{\rightarrow})$. If $j = \mathfrak{b}_{i+1}$, then the same properties imply $\mathrm{rVal}_{\sigma e}(F_{i,1-j}) = \langle s_{i,1-j}\rangle + \mathrm{rVal}_{\sigma e}^{\mathrm{M}}(b_1)$. Since $\mathrm{rVal}_{\sigma e}^{\mathrm{M}}(b_1) = \mathrm{rVal}_{\sigma e}^{\mathrm{M}}(t^{\rightarrow}) = \mathrm{rVal}_{\sigma e}^{\mathrm{M}}(F_{i,j})$, this implies the statement.

Hence let $F_{i,1-j}$ be $t^{\rightarrow}$-halfopen. Then $\mathrm{Val}_{\sigma e}^{\mathrm{M}}(F_{i,1-j}) = \frac{1-\varepsilon}{1+\varepsilon}\mathrm{Val}_{\sigma}^{\mathrm{M}}(t^{\rightarrow}) + \frac{2\varepsilon}{1+\varepsilon}\mathrm{Val}_{\sigma e}^{\mathrm{M}}(s_{i,1-j})$. We prove that $\mathrm{Val}_{\sigma e}^{\mathrm{M}}(s_{i,1-j}) > \mathrm{Val}_{\sigma e}^{\mathrm{M}}(t^{\rightarrow})$ in this case. If $1-j \neq \mathfrak{b}_{i+1}$, then this follows from Property (USV1)$_i$ as $\mathrm{Val}_{\sigma e}^{\mathrm{M}}(s_{i,1-j}) = \langle s_{i,1-j}\rangle + \mathrm{Val}_{\sigma e}^{\mathrm{M}}(t^{\rightarrow})$ in that case. If $1-j = \mathfrak{b}_{i+1}$, then $\mathrm{Val}_{\sigma e}^{\mathrm{M}}(s_{i,1-j}) = \langle s_{i,1-h}, h_{i,1-j}\rangle + \mathrm{Val}_{\sigma e}^{\mathrm{M}}(b_{i+1})$ by Properties (USV1)$_i$ and (EV1)$_{i+1}$. The statement then follows since $\langle h_{i,1-j}\rangle > \sum_{\ell\in[i]} W_\ell^{\mathrm{M}}$.

Consequently, $\mathrm{Val}_{\sigma e}^{\mathrm{M}}(s_{i,1-j}) > \mathrm{Val}_{\sigma e}^{\mathrm{M}}(t^{\rightarrow})$. This implies $\mathrm{Val}_{\sigma e}^{\mathrm{M}}(F_{i,1-j}) > (1-\varepsilon)\mathrm{Val}_{\sigma e}^{\mathrm{M}}(t^{\rightarrow}) + \varepsilon\,\mathrm{Val}_{\sigma e}^{\mathrm{M}}(s_{i,1-j})$ which yields $\mathrm{Val}_{\sigma e}^{\mathrm{M}}(F_{i,1-j}) > \mathrm{Val}_{\sigma e}^{\mathrm{M}}(F_{i,j})$ as proven earlier. □

Lemma 6.2.8 (Second row of Table 6.4). *Let $G_n = S_n$. Let $\sigma \in \rho(\sigma_0)$ be a well-behaved phase-1-strategy for $\mathfrak{b} \in \mathfrak{B}_n$ with $I_\sigma = \mathfrak{D}^\sigma$. Let $i \in [n], j, k \in \{0,1\}$ such that $e := (d_{i,j,k}, F_{i,j}) \in I_\sigma$ and $\sigma(d_{i,j,1-k}) = F_{i,j}, i \neq 1, j \neq \mathfrak{b}_{i+1}$ as well as $\sigma(g_i) = F_{i,j}$. Then σe is a well-behaved phase-1-strategy for $\mathfrak{b}$ with $I_{\sigma e} = \mathfrak{D}^{\sigma e}$ and $\sigma e \in \rho(\sigma_0)$.*

Proof. Since $\sigma(g_i) = F_{i,j}$ and $j \neq \mathfrak{b}_{i+1} = \beta_{i+1}^\sigma$, we have $\sigma e(b_i) = \sigma(b_i) = b_{i+1}$ by Property (EV2)$_i$. This implies $i \geq \mu^\sigma$ by Property (REL1). As $\mu^\sigma = \mu^{\sigma e}$, this implies that σe has Property (CC1)$_{i'}$ for all indices i'. This further implies that σe has Property (ESC1),(EV1)$_{i'}$ and (USV1)$_{i'}$ for all $i' \in [n]$. Furthermore, since σ has all other properties defining a phase-1-strategy, σe has them as well. As we do not perform changes within the cycle center F_{i,β_{i+1}^σ}, also $\beta^\sigma = \beta^{\sigma e} =: \beta$. Since σ has Property (SVG)$_i$ and since the cycle center $F_{i,j}$ is not closed for σ by the choice of e, we have $j = 0$. This implies that σe has Property (SVG)$_i$ as well. Hence σe is a phase-1-strategy for $\mathfrak{b}$.

Proving that σe is well-behaved follows by re-evaluating Properties (D1), (MNS4), (MNS6), (EG2), (DN1) and (DN2). This set of properties is sufficient as we do not need to verify properties where the conclusion might become true or the premise might become false since the implication is then already true.

- (D1) By the premise of this property, $\sigma e(b_i) = g_i$, contradicting $\sigma e(b_i) = b_{i+1}$. Property (D2) holds by the same argument since $i < \mu^{\sigma e}$ implies $\sigma e(b_i) = g_i$.
- (MNS4) Since σ is well-behaved, this only needs to be reevaluated if $i = \overline{m}_s^{\sigma e}$. Since $i \neq 1$ cannot occur by assumption, let $i > 1$. Then, $1 < \overline{m}_s^{\sigma e} \leq \overline{m}_g^{\sigma e} < m_b^{\sigma e}$. Thus, in particular $\sigma e(b_1) = b_2, \sigma e(g_1) = F_{1,1}$ and $\sigma e(s_{1,1}) = h_{1,1}$. By Property (USV1)$_1$ and Property (EV1)$_2$, $\sigma e(b_2) = g_2$, implying $m_b^{\sigma e} = 2$. But this contradicts the premise since $1 < \overline{m}_s^{\sigma e} < m_b^{\sigma e}$ implies $m_b^{\sigma e} \geq 3$.
- (MNS6) Since σ is well-behaved, this only needs to be reevaluated if $i = \overline{m}_s^{\sigma e}$. Since $i = 1$ cannot occur by assumption, let $i > 1$. Then, $1 < \overline{m}_s^{\sigma e} \leq \overline{m}_g^{\sigma e} < m_b^{\sigma e}$, implying the same contradiction as in the last case.
- (EG2) The cycle center $F_{i,j}$ is closed with respect to σe, so the premise is incorrect.
- (DN*) Since the only cycle center in level n is $F_{n,0}$ and since we always have $\mathfrak{b}_{n+1} = 0$ by definition, the choice of j implies that we cannot have $i = n$.

We next prove $I_{\sigma e} = \{(d_{i,j,k}, F_{i,j}) \colon \sigma e(d_{i,j,k}) \neq F_{i,j}\}$. The only vertices that have an edge towards $F_{i,j}$ are $d_{i,j,*}$ and g_i. Since closing $F_{i,j}$ increases its valuation, the valuation of these vertices might increase as well. Since no player 0 vertex has an edge to either $d_{i,j,0}$ or $d_{i,j,1}$, no new improving switch involving these vertices can emerge. However, the valuation of g_i might increase due to $\sigma(g_i) = \sigma e(g_i) = F_{i,j}$. We now prove that this increase does not create new improving switches and that all switches but e that are improving for σ stay improving for σe.

It suffices to prove that $\sigma e(b_i) = b_{i+1}$ and $\mathrm{Val}^{\mathrm{S}}_{\sigma e}(g_i) \trianglelefteq \mathrm{Val}^{\mathrm{S}}_{\sigma e}(b_{i+1})$ as well as$\sigma e(s_{i-1,1}) = b_1$ and $\mathrm{Val}^{\mathrm{S}}_{\sigma e}(h_{i-1,1}) \trianglelefteq \mathrm{Val}^{\mathrm{S}}_{\sigma e}(b_1)$. Since $\sigma(b_i) = \sigma e(b_i) = b_{i+1}$ and $i \geq \mu^{\sigma e}$, we have $\mathrm{rVal}^{\mathrm{S}}_{\sigma e}(b_{i+1}) = L^{\mathrm{S}}_{i+1}$. Since $\overline{\sigma e}(d_{i,j})$ implies $\mathrm{rVal}^{\mathrm{S}}_{\sigma e}(F_{i,j}) = \mathrm{rVal}^{\mathrm{S}}_{\sigma e}(s_{i,j})$ by Lemma 6.1.16, we have $\mathrm{rVal}^{\mathrm{S}}_{\sigma e}(g_i) = \{g_i, s_{i,j}\} \cup \mathrm{rVal}^{\mathrm{S}}_{\sigma e}(b_1) = \{g_i, s_{i,j}\} \cup B^{\mathrm{S}}_1$ by the choice of j and Property (USV1)$_i$. Thus, $\{g_i, s_{i,j}\} \lhd \bigcup_{\ell \in [i]} W^{\mathrm{S}}_\ell$ and $\{g_i, s_{i,j}\} \lhd \bigcup_{\ell \in [i]} \{W^{\mathrm{S}}_\ell \colon \sigma e(b_\ell) = g_\ell\}$ yield $\mathrm{rVal}^{\mathrm{S}}_{\sigma e}(g_i) \lhd \mathrm{rVal}^{\mathrm{S}}_{\sigma e}(b_{i+1})$. For the second statement, we observe that $\mathfrak{b}_i = 0$ implies $\sigma e(s_{i-1,1}) = b_1$ by Property (USV1)$_{i-1}$. The second part then follows using similar calculations as before since $\mathrm{rVal}^{\mathrm{S}}_{\sigma e}(h_{i-1,1}) = \{h_{i-1,1}, g_i, s_{i,j}\} \cup \mathrm{rVal}^{\mathrm{S}}_{\sigma e}(b_1)$. □

Lemma 6.2.9 (Third row of Table 6.4). *Let $\sigma \in \rho(\sigma_0)$ be a well-behaved phase-1-strategy for $\mathfrak{b} \in \mathfrak{B}_n$ with $I_\sigma = \mathfrak{D}^\sigma$. Let $i \in [n-1], j, k \in \{0,1\}$ such that $e := (d_{i,j,k}, F_{i,j}) \in I_\sigma$ and $\sigma(d_{i,j,1-k}) = F_{i,j}, j = 1 - \beta^\sigma_{i+1}, \sigma(b_i) = b_{i+1}$ and $\sigma(g_i) = F_{i,1-j}$. Then σe is a well-behaved phase-1-strategy for $\mathfrak{b}$ with $\sigma e \in \rho(\sigma_0)$ and $I_{\sigma e} = \mathfrak{D}^{\sigma e} \cup \{(g_i, F_{i,j})\}$.*

Proof. By similar arguments used in the proof of Lemma 6.2.8, Properties (ESC1), (REL1) and (USV1)$_{i'}$, (CC1)$_{i'}$, (EV1)$_{i'}$, (EV2)$_{i'}$ and (EV3)$_{i'}$ for $i' \in [n]$ are valid for σe. Consider Property (SVM)$_i$ and let $G_n = M_n$. Since $1 - j = \beta^\sigma_{i+1}$, the premise of this property is incorrect, hence σe has Property (SVM)$_i$. Consider Property (SVG)$_i$ and let $G_n = S_n$. If $1 - j = \beta^\sigma_{i+1} = 0$, then σe has Property (SVG)$_i$ as well. Hence assume $1 - j = \beta^\sigma_{i+1} = 1$. Then, since σ has Property (SVG)$_i$, it follows that $\bar{\sigma}(d_{i,1})$, implying $\overline{\sigma e}(d_{i,1})$. Thus, σe has Property (SVG)$_i$ resp. Property (SVM)$_i$, implying that σe is a phase-1-strategy for $\mathfrak{b}$.

Since $\sigma(g_i) = F_{i,1-j}$, applying e does not close the chosen cycle center. It is thus not necessary to reevaluate the assumptions of Table 6.1 and thus, σ being well-behaved implies that σe is well-behaved. It hence remains to prove $I_{\sigma e} = \mathfrak{D}^{\sigma e} \cup \{(g_i, F_{i,j})\}$.

By Property (EV1), $\sigma(b_i) = \sigma e(b_i) = b_{i+1}$ implies that $F_{i,1-j}$ is not closed with respect to both σ and σe. Hence, $\mathrm{rVal}^*_{\sigma e}(F_{i,1-j}) = \mathrm{rVal}^*_{\sigma e}(b_1)$ by Lemma 6.1.19. Since Property (USV1)$_i$ and the choice of j imply $\sigma e(s_{i,j}) = \sigma(s_{i,j}) = b_1$, it holds that

$$\mathrm{rVal}^*_{\sigma e}(F_{i,1-j}) = \mathrm{rVal}_{\sigma e}(b_1)^* \lhd [\![s_{i,j}]\!] \oplus \mathrm{rVal}^*_{\sigma e}(b_1) = \mathrm{rVal}^*_{\sigma e}(F_{i,j}),$$

implying $(g_i, F_{i,j}) \in I_{\sigma e}$.

Since $\sigma(g_i) = F_{i,1-j}$, the only vertices whose valuations change by applying e are $d_{i,j,0}, d_{i,j,1}$ and $F_{i,j}$. This implies that no new switches besides the switch $(g_i, F_{i,j})$ are created and that all improving switches for σ but e stay improving for σe. □

Lemma 6.2.10 (Fourth row of Table 6.4). *Let $\sigma \in \rho(\sigma_0)$ be a well-behaved phase-1-strategy for $\mathfrak{b} \in \mathfrak{B}_n$ with $I_\sigma = \mathfrak{D}^\sigma \cup \{(g_i, F_{i,1-\mathfrak{b}_{i+1}})\}$ for some index $i \in [n-1]$. Let $e := (g_i, F_{i,1-\mathfrak{b}_{i+1}}) \in I_\sigma$ and $\mathfrak{b}_i = 0, i \neq 1$ and $\bar{\sigma}(d_{i,j})$. Then σe is a well-behaved phase-1-strategy for $\mathfrak{b}$ with $I_{\sigma e} = I_\sigma \setminus \{e\}$.*

Proof. Let $j := 1 - \mathfrak{b}_{i+1} = 1 - \beta^\sigma_{i+1}$. Since σ is a phase-1-strategy for $\mathfrak{b}$, $\mathfrak{b}_i = 0$ implies $\sigma(b_i) = b_{i+1}$. Since no cycle center is closed when applying e, σe has Properties (ESC1), (EV1)$_{i'}$, (EV3)$_{i'}$, (CC1)$_{i'}$ and (USV1)$_{i'}$ for all $i' \in [n]$. Also, since $\sigma e(b_i) = b_{i+1}$, the premise of Property (EV2)$_i$ is incorrect with respect to σe, hence it has the property for all indices. Since σ has Property (REL1) and $\sigma e(b_i) = b_{i+1}$, it also has Property (REL1). This implies $\mathcal{I}^{\sigma e} = \emptyset$, and thus $i \geq \mu^{\sigma e} = \mu^\sigma$. Also, since $\bar{\sigma}(d_{i,j})$ by assumption, σe has Property (SVG)$_i$ resp. Property (SVM)$_i$.

By the choice of e, by $i \geq \mu^{\sigma e}$ and since σ is well-behaved, it suffices to investigate Properties (B3), (MNS4), (MNS6) and (EG4) in order to prove that σe is well-behaved.

(B3) Since σ has Property (USV1)$_i$, the premise of this property is incorrect.

(MNS4) Let the premise be correct, i.e., let $\mu^{\sigma e} = 1 \wedge \overline{m}^{\sigma e}_s \leq \overline{m}^{\sigma e}_g < m^{\sigma e}_b$. Let $i' := \overline{m}^{\sigma e}_s$. For the sake of a contradiction, let $i' = 1$. Then $\sigma e(b_1) = b_2$ since $\mu^{\sigma e} = 1$. If $\sigma e(g_1) = F_{1,1}$, then $\sigma e(s_{1,1}) = b_1$. Thus, $\beta_2 = \mathfrak{b}_2 = 0$ by Property (USV1)$_1$. If also $\sigma(g_1) = F_{1,1}$, Property (SVG)$_1$ resp. Property (SVM)$_1$ would imply $\bar{\sigma}(d_{1,1})$, contradicting Property (MNS4) for σ. If $\sigma(g_1) = F_{1,0}$, then $\overline{m}^\sigma_g = 1$. If also $\overline{m}^{\sigma e}_g = 1$, then the statement follows by applying Property (MNS4) to σ. Otherwise, we need to have $e = (g_1, F_{1,1})$, contradicting the assumption $i \neq 1$. Hence consider the case $i' = \overline{m}^{\sigma e}_s > 1$. Then $1 < \overline{m}^{\sigma e}_s \leq \overline{m}^{\sigma e}_g < m^{\sigma e}_b$, implying $\sigma e(g_1) = F_{1,1}, \sigma e(s_{1,1}) = h_{1,1}$ and $m^{\sigma e}_b \geq 3$. By Property (USV1)$_1$, this implies $\beta_2 = 1$, hence $\sigma e(b_2) = g_2$ by Property (EV1)$_2$. But then $m^{\sigma e}_b = 2$ which is a contradiction. Therefore the premise cannot be correct, hence the implication is correct.

(MNS6) Assume the premise is correct, i.e., assume $\mu^{\sigma e} = 1 \wedge \overline{m}^{\sigma e}_s < m^{\sigma e}_b \leq \overline{m}^{\sigma e}_g$. Let $i' := \overline{m}^{\sigma e}_s$ and assume $i' = 1$. Then $\sigma e(g_1) = F_{1,1}, \sigma e(s_{1,1}) = b_1$ and $\sigma e(b_1) = b_2$. If also $\sigma(g_1) = F_{1,1}$, then Property (SVG)$_1$ resp. Property (SVM)$_1$ would imply $\bar{\sigma}(d_{1,1})$ as in the last case, contradicting Property (MNS6) for σ. However, since $\sigma(g_1) = F_{1,0}$ implies $e = (g_1, F_{1,1})$, this again contradicts the assumption $i \neq 1$. Hence consider the case $i' = \overline{m}^{\sigma e}_s > 1$. Then $1 < \overline{m}^{\sigma e}_s < m^{\sigma e}_b \leq \overline{m}^{\sigma e}_g$ which implies the same contradiction that occurred when discussing Property (MNS4).

(EG4) By Property (ESC1), the premise of this property is always incorrect, hence the implication is correct.

Note that the other Properties (MNS*) do not need to be considered since their conclusion is correct for level i by assumption. In addition, none of the properties (EBG*) needs to be checked due to Property (ESC1).

It remains to prove $I_{\sigma e} = I_\sigma \setminus \{e\}$. This follows by proving $\sigma e(b_i) = b_{i+1}$ and $\operatorname{Val}^*_{\sigma e}(g_i) \prec \operatorname{Val}^*_{\sigma e}(b_{i+1})$ as well as $\sigma e(s_{i-1,1}) = b_1$ and $\operatorname{Val}^*_{\sigma e}(h_{i-1,1}) \prec \operatorname{Val}^*_{\sigma e}(b_1)$. This can be proven in the same way as it was proven in the proof of Lemma 6.2.8. □

Lemma 6.2.11. *Let $\sigma \in \rho(\sigma_\mathfrak{b})$ be a well-behaved phase-1-strategy for $\mathfrak{b}$ with $I_\sigma = \mathfrak{D}^\sigma$. Let $\sigma_\mathfrak{b} \in \rho(\sigma_0)$ and let $\sigma_\mathfrak{b}$ have the canonical properties. Let $i \in [n], j, k \in \{0,1\}$ such that $e := (d_{i,j,k}, F_{i,j}) \in I_\sigma, I_{\sigma_\mathfrak{b}}$ with $\phi^\sigma(e) = \phi^{\sigma_\mathfrak{b}}(e) = \lfloor(\mathfrak{b}+1)/2\rfloor - 1$. Then σe is a well-behaved phase-1-strategy for $\mathfrak{b}$ with $\sigma e \in \rho(\sigma_0)$. Furthermore, $\sigma(d_{i,j,1-k}) = F_{i,j}, j \neq \mathfrak{b}_{i+1}, \sigma(g_i) = F_{i,1-j}$ and $\sigma(b_i) \neq g_i$ imply $I_{\sigma e} = (I_\sigma \setminus \{e\}) \cup \{(g_i, F_{i,j})\}$. Otherwise, $I_{\sigma e} = I_\sigma \setminus \{e\}$. In* *addition, the occurrence record of* e *with respect to* σe *is described correctly by Table 5.6 when interpreted for* $\mathfrak{b}+1$.

Proof. By transitivity, $\sigma \in \rho(\sigma_0)$. Since $e = (d_{i,j,k}, F_{i,j}) \in I_\sigma$, the cycle center $F_{i,j}$ cannot be closed. Hence, since σ is a phase-1-strategy for $\mathfrak{b}$, either $\mathfrak{b}_i = 0$ or $\mathfrak{b}_{i+1} \neq j$. Consequently, exactly one of the following cases is true:

1. $\sigma(d_{i,j,1-k}) \neq F_{i,j}$
2. $\sigma(d_{i,j,1-k}) = F_{i,j} \wedge j \neq \mathfrak{b}_{i+1} \wedge \sigma(g_i) = F_{i,j}$
3. $\sigma(d_{i,j,1-k}) = F_{i,j} \wedge j \neq \mathfrak{b}_{i+1} \wedge \sigma(g_i) = F_{i,1-j} \wedge \sigma(b_i) \neq g_i$
4. $\sigma(d_{i,j,1-k}) = F_{i,j} \wedge j \neq \mathfrak{b}_{i+1} \wedge \sigma(g_i) = F_{i,1-j} \wedge \sigma(b_i) = g_i$
5. $\sigma(d_{i,j,1-k}) = F_{i,j} \wedge j = \mathfrak{b}_{i+1}$

We prove that case four and five cannot occur. Assume that the conditions of the fourth case were true. Then, since σ is a phase-1-strategy for $\mathfrak{b}$, Property (EV1)$_i$ and Property (EV2)$_i$, imply $\mathfrak{b}_i = 1$ and $\mathfrak{b}_{i+1} = 1-j$. This also implies that $t_\mathfrak{b} = 0$ is the only feasible parameter for $(d_{i,j,k}, F_{i,j})$ and $(d_{i,j,1-k}, F_{i,j})$. Now, by assumption, $\sigma(d_{i,j,1-k}) = F_{i,j}$. If $\sigma_\mathfrak{b}(d_{i,j,1-k}) = F_{i,j}$, then Property (OR1)$_{i,j,k}$ and Property (OR2)$_{i,j,k}$ imply $\phi^{\sigma_\mathfrak{b}}(d_{i,j,1-k}, F_{i,j}) = \ell^\mathfrak{b}(i,j,1-k)+1$, contradicting that $t_\mathfrak{b} = 0$ is the only feasible parameter. Thus, assume $\sigma_\mathfrak{b}(d_{i,j,1-k}) \neq F_{i,j}$. Then $(d_{i,j,1-k}, F_{i,j})$ was applied during $\sigma_\mathfrak{b} \to \sigma$ and in particular before $(d_{i,j,k}, F_{i,j})$. Thus, $\sigma_\mathfrak{b}(d_{i,j,1-k}) \neq F_{i,j}$ and Property (OR4)$_{i,j,1-k}$ implies $\phi^{\sigma_\mathfrak{b}}(d_{i,j,1-k}, F_{i,j}) = \lfloor\frac{\mathfrak{b}+1}{2}\rfloor - 1$. Since $\mathfrak{b}_i = 1 \wedge \mathfrak{b}_{i+1} = 1-j$, Lemma 6.2.3 implies

$$\ell^\mathfrak{b}(i,j,1-k) = \left\lceil \frac{\mathfrak{b} + \sum(\mathfrak{b},i) + 1 - k}{2} \right\rceil \geq \left\lfloor \frac{\mathfrak{b}+1}{2} \right\rfloor.$$

As $t_\mathfrak{b} = 0$ is the only feasible parameter, it thus needs to hold that

$$\phi^{\sigma_\mathfrak{b}}(d_{i,j,1-k}, F_{i,j}) = \left\lfloor \frac{\mathfrak{b}+1}{2} \right\rfloor - 1 = \left\lfloor \frac{\mathfrak{b}+1-(1-k)}{2} \right\rfloor.$$

This implies $k = 0$ and that $\mathfrak{b}$ is odd. But then $\phi^\sigma(d_{i,j,k}, F_{i,j}) = \lfloor(\mathfrak{b}+1)/2\rfloor$, contradicting the assumptions.

Consider the fifth case. Then, $\mathfrak{b}_i = 0$ as $j = \mathfrak{b}_{i+1}$. If $\sigma_\mathfrak{b}(d_{i,j,1-k}) = \sigma(d_{i,j,1-k}) = F_{i,j}$, then $i \neq \nu$ by the definition of a canonical strategy. If $\sigma_\mathfrak{b}(d_{i,j,1-k}) \neq F_{i,j}$, then the switch $(d_{i,j,1-k}, F_{i,j})$ was applied during $\sigma_\mathfrak{b} \to \sigma$. This implies $\phi^\sigma(d_{i,j,1-k}, F_{i,j}) < \lfloor(\mathfrak{b}+1)/2\rfloor$. By Lemma 6.2.5, there can be at most one improving switch in level ν with an occurrence record strictly smaller than $\lfloor(\mathfrak{b}+1)/2\rfloor$. This implies $i \neq \nu$. Since $i = 1$ would imply $i = \nu$ due to $\mathfrak{b}_i = 0$, we thus have $i \geq 2$.

Assume $\sigma_\mathfrak{b}(d_{i,j,1-k}) = F_{i,j}$. Then $\phi^\sigma(d_{i,j,1-k}, F_{i,j}) = \ell^\mathfrak{b}(i,j,1-k) + 1 < \lfloor(\mathfrak{b}+1)/2\rfloor$ by Property (OR2)$_{i,j,1-k}$ and Property (OR1)$_{i,j,-1-k}$, hence $\ell^\mathfrak{b}(i,j,1-k) < \lfloor(\mathfrak{b}+1)/2\rfloor - 1$. However, since $\mathfrak{b}_i = 0$ and $\mathfrak{b}_{i+1} = j$, Lemma 6.2.3 implies that either $\ell^\mathfrak{b}(i,j,1-k) \geq \mathfrak{b}$ or

$$\ell^\mathfrak{b}(i,j,1-k) = \left\lceil \frac{\mathfrak{b} + 2^{i-1} + \sum(\mathfrak{b},i) + 1 - (1-k)}{2} \right\rceil \geq \left\lfloor \frac{\mathfrak{b}+3}{2} \right\rfloor = \left\lfloor \frac{\mathfrak{b}+1}{2} \right\rfloor + 1$$

which is a contradiction in either case. The case $\sigma_\mathfrak{b}(d_{i,j,1-k}) \neq F_{i,j}$ yields the same contradiction developed for case four.

Thus one of the first three listed cases needs to be true. In the first resp. third case, we can apply Lemma 6.2.7 resp. 6.2.9 to prove the part of the statement regarding the improving switches. In order to apply Lemma 6.2.8, we need to prove that the conditions of the second case can only occur if $G_n = S_n$ and $i \neq 1$.

Thus assume that the conditions of the second case were true. Assume $i = 1$. Then, since $\sigma(g_1) = F_{1,1-\mathfrak{b}_2}$, wit holds that $\mathfrak{b}_1 = 0$ by Property (EV1)$_1$ and Property (EV2)$_1$. By the choice of j and Lemma 6.2.3, this implies $\ell^{\mathfrak{b}}(i,j,k) = \lfloor(\mathfrak{b}+1-k)/2\rfloor$. By Property (OR3)$_{i,j,k}$ and Property (OR4)$_{i,j,k}$, it thus needs to hold that $\phi^{\sigma_{\mathfrak{b}}}(d_{i,j,k}, F_{i,j}) = \ell^{\mathfrak{b}}(i,j,k) = \lfloor(\mathfrak{b}+1)/2\rfloor - 1$. This can only happen if $k = 1$ and if $\mathfrak{b}$ is odd, contradicting $\mathfrak{b}_1 = 0$. Consequently, $i \neq 1$. Proving that the conditions can only occur if $G_n = S_n$ can be done by proving that we have $(g_i, F_{i,1-j}) \in I_\sigma$ if $G_n = M_n$, contradicting $I_\sigma = \mathfrak{D}^\sigma$. As proving this is rather tedious, we omit this part here.

It remains to show that there is a feasible parameter $t_{\mathfrak{b}+1}$ for $\mathfrak{b}+1$ such that

$$\phi^{\sigma e}(d_{i,j,k}, F_{i,j}) = \min\left(\left\lfloor\frac{(\mathfrak{b}+1)+1-k}{2}\right\rfloor, \ell^{\mathfrak{b}+1}(i,j,k) + t_{\mathfrak{b}+1}\right).$$

Since $\phi^{\sigma}(e) = \phi^{\sigma_{\mathfrak{b}}}(e)$, we have $\phi^{\sigma e}(e) = \phi^{\sigma_{\mathfrak{b}}}(e) + 1$. Also, there is a parameter $t_{\mathfrak{b}}$ feasible for $\mathfrak{b}$ such that $\phi^{\sigma}(e) = \min(\lfloor(\mathfrak{b}+1-k)/2\rfloor, \ell^{\mathfrak{b}}(i,j,k) + t_{\mathfrak{b}}) = \lfloor(\mathfrak{b}+1)/2\rfloor - 1$ by the choice of e. Consequently, $\phi^{\sigma e}(e) = \lfloor(\mathfrak{b}+1)/2\rfloor$. We distinguish two cases.

1. $i = \nu \wedge j = \mathfrak{b}_{\nu+1}$. Since we have one of the first three cases discussed earlier, this implies $\sigma(d_{i,j,1-k}) \neq F_{i,j}$. Moreover, $\mathfrak{b}$ needs to be odd since both cycle edges of $F_{\nu,\mathfrak{b}_{\nu+1}}$ have an occurrence record of $\lfloor(\mathfrak{b}+1)/2\rfloor$ if $\mathfrak{b}$ is even. Consequently, $\nu > 1$. Thus, by the choice of e and Lemma 6.2.5, it holds that $k = 1$. It therefore suffices to show $\lfloor(\mathfrak{b}+1)/2\rfloor = \lceil \mathrm{lfn}(\mathfrak{b}+1, i, \{(i+1,j)\})/2\rceil$. This however follows immediately from the choice of i and j and the fact that $\mathfrak{b}$ is odd.
2. $i \neq \nu \vee j \neq \mathfrak{b}_{i+1}$. This implies $\mathfrak{b}_i = 0 \vee j \neq \mathfrak{b}_{i+1}$, hence $(\mathfrak{b}+1)_i = 0 \vee (\mathfrak{b}+1)_{i+1} \neq j$. We thus need to show that there is a parameter $t_{\mathfrak{b}+1}$ feasible for $\mathfrak{b}+1$ such that

$$\left\lfloor\frac{\mathfrak{b}+1}{2}\right\rfloor = \min\left(\left\lfloor\frac{(\mathfrak{b}+1)+1-k}{2}\right\rfloor, \ell^{\mathfrak{b}+1}(i,j,k) + t_{\mathfrak{b}+1}\right).$$

   By Lemma 6.2.4, $\ell^{\mathfrak{b}}(i,j,k) + 1 = \ell^{\mathfrak{b}+1}(i,j,k)$. We distinguish the following cases.
   a) Let $\lfloor(\mathfrak{b}+1)/2\rfloor - 1 = \lfloor(\mathfrak{b}+1-k)/2\rfloor$. This implies $k = 1$ and $\mathfrak{b} \bmod 2 = 1$. Consequently, $\phi^{\sigma e}(e) = \lfloor(\mathfrak{b}+1)/2\rfloor = \lfloor(\mathfrak{b}+1+1-k)/2\rfloor$. It remains to define a feasible parameter $t_{\mathfrak{b}+1}$. Since $\phi^{\sigma_{\mathfrak{b}}}(e) = \lfloor(\mathfrak{b}+1-k)/2\rfloor$, there is a feasible $t_{\mathfrak{b}}$ for $\mathfrak{b}$ such that $\lfloor(\mathfrak{b}+1-k)/2\rfloor \leq \ell^{\mathfrak{b}}(i,j,k) + t_{\mathfrak{b}}$. Since $\sigma_{\mathfrak{b}}(d_{i,j,k}) \neq F_{i,j}$ due to $e \in I_{\sigma_{\mathfrak{b}}}$, Property (OR2)$_{i,j,k}$ implies $t_{\mathfrak{b}} \neq 1$. Hence we can choose $t_{\mathfrak{b}+1} = 0$ as $\phi^{\sigma e}(e) = \lfloor(\mathfrak{b}+1-k)/2\rfloor + 1 \leq \ell^{\mathfrak{b}}(i,j,k) + 1 = \ell^{\mathfrak{b}+1}(i,j,k)$.
   b) Let $\lfloor(\mathfrak{b}+1)/2\rfloor - 1 = \ell^{\mathfrak{b}}(i,j,k) + t_{\mathfrak{b}}$ for some parameter $t_{\mathfrak{b}}$ feasible for $\mathfrak{b}$ but $\lfloor(\mathfrak{b}+1)/2\rfloor - 1 \neq \lfloor(\mathfrak{b}+1-k)/2\rfloor$. Then, Property (OR2)$_{i,j,k}$ implies $t_{\mathfrak{b}} \neq 1$. Consider the case $t_{\mathfrak{b}} = 0$ first. Then $\phi^{\sigma e}(e) = \ell^{\mathfrak{b}}(i,j,k) + 1 = \ell^{\mathfrak{b}+1}(i,j,k)$ and $\phi^{\sigma}(e) = \lfloor(\mathfrak{b}+1)/2\rfloor \leq \lfloor(\mathfrak{b}+1+1-k)/2\rfloor$. Thus, choosing $t_{\mathfrak{b}+1} = 0$ is a feasible choice giving the correct characterization. Thus consider the case $t_{\mathfrak{b}} = -1$. Then, by Property (OR3)$_{i,j,k}$, $\mathfrak{b}$ is odd and $k = 0$. This then

implies that $\phi^{\sigma e}(e) = \ell^{\mathfrak{b}}(i,j,k) = \lfloor(\mathfrak{b}+1)/2\rfloor = \lfloor(\mathfrak{b}+1+1-k)/2\rfloor$ as well as $\phi^{\sigma e}(e) = \ell^{\mathfrak{b}+1}(i,j,k) - 1$. We thus choose $t_{\mathfrak{b}+1} = 0$ which is a feasible choice, does not contradict Property (OR3) for $\mathfrak{b}+1$ and yields the desired characterization. □

Lemma 6.2.12 (Fifth row of Table 6.4). *Let $\sigma \in \rho(\sigma_0)$ be a well-behaved phase-1-strategy for $\mathfrak{b} \in \mathfrak{B}_n$ and $I_\sigma = \mathfrak{D}^\sigma$. Let $\nu := \ell(\mathfrak{b}+1)$ and $j := \mathfrak{b}_{\nu+1}$. Let $e := (d_{\nu,j,k}, F_{\nu,j}) \in I_\sigma$ and $\sigma(d_{\nu,j,1-k}) = F_{\nu,j}$ for some $k \in \{0,1\}$. The following statements hold.*

1. *$\beta^{\sigma e} = \mathfrak{b}+1$.*
2. *σe has Properties (EV1)$_i$ and (EV3)$_i$ for all $i > \nu$. It also has Property (EV2)$_i$ and Property (USV1)$_i$ for all $i \geq \nu$ as well as Property (REL1), and $\mu^{\sigma e} = \mu^\sigma = \nu$.*
3. *σe is well-behaved and $\sigma e \in \rho(\sigma_0)$.*
4. *If $\nu = 1$, then σe is a phase-3-strategy for $\mathfrak{b}$. If $\sigma(g_\nu) = F_{\nu,j}$, then it holds that $I_{\sigma e} = \mathfrak{D}^{\sigma e} \cup \{(b_1, g_1)\} \cup \{(e_{*,*,*}, g_1)\}$. If $\sigma(g_\nu) \neq F_{\nu,j}$, then $I_{\sigma e} = \mathfrak{D}^{\sigma e} \cup \{(g_\nu, F_{\nu,j})\}$ and σe is a pseudo phase-3-strategy.*
5. *If $\nu > 1$, then σe is a phase-2-strategy for $\mathfrak{b}$. If $\sigma(g_\nu) = F_{\nu,j}$, then it holds that $I_{\sigma e} = \mathfrak{D}^{\sigma e} \cup \{(b_\nu, g_\nu)\} \cup \{(s_{\nu-1,1}, h_{\nu-1,1})\}$. If $\sigma(g_\nu) \neq F_{\nu,j}$, then $I_{\sigma e} = \mathfrak{D}^{\sigma e} \cup \{(g_\nu, F_{\nu,j})\}$ and σe is a pseudo phase-2-strategy.*

Proof. We have $\nu = \mu^\sigma$ as σ has Property (REL1) and Property (EV1)$_{i'}$ for all $i' < \nu$. Also, $\mu^\sigma = \mu^{\sigma e}$ by the choice of e. Since we do not close any cycle centers in any level below $\mu^{\sigma e}$, σe has Property (CC1)$_{i'}$ for all $i' \in [n]$.

1. Since the cycle centers of levels $i > \nu$ are not changed, $\beta_i^{\sigma e} = \beta_i^\sigma = \mathfrak{b}_i = (\mathfrak{b}+1)_i$ for all $i > \nu$. Moreover, $\beta_i^{\sigma e} = \overline{\sigma e}(d_{\nu,j}) = 1 = (\mathfrak{b}+1)_\nu$ by the definition of ν and the choice of e. It remains to show $\beta_i^{\sigma e} = 0$ for all $i < \nu$. This is proven by backwards induction. Hence let $i = \nu - 1$ and consider $\beta_i^{\sigma e} = \overline{\sigma e}(d_{i,\beta_{i+1}^{\sigma e}})$.

   Since $\beta_{i+1}^{\sigma e} = 1$, we prove $\overline{\sigma e}(d_{i-1,1}) = 0$. We have $\beta_{\nu-1}^\sigma = 1$ and $\beta_\nu^\sigma = 0$. Thus $\sigma(b_{\nu-1}) = g_{\nu-1}$ by Property (EV1)$_{\nu-1}$, so $0 = \bar\sigma(d_{\nu-1,1-\beta_\nu^\sigma}) = \bar\sigma(d_{\nu-1,1}) = \overline{\sigma e}(d_{\nu-1,1})$ by Property (EV3)$_{\nu-1}$.

   Now consider some $i < \nu - 1$. By the induction hypotheses, $\beta_{i+1}^{\sigma e} = 0$. We hence prove $\overline{\sigma e}(d_{i,\beta_{i+1}^{\sigma e}}) = \overline{\sigma e}(d_{i,0}) = 0$. By the definition of ν, $\beta_i^\sigma = \beta_{i+1}^\sigma = 1$. Hence, $\sigma(b_i) = g_i$ by Property (EV1)$_i$, implying $\bar\sigma(d_{i,0}) = 0$ by Property (EV3)$_i$.
2. We prove that σe has the listed properties. Since $\beta_i^{\sigma e} = \beta_i^\sigma$ for all $i > \nu$, σe has Property (EV1)$_i$ for all $i \geq \nu$. This also implies that it has Property (EV2)$_i$ and (USV1)$_i$ for all $i \geq \nu$. In addition, it has Property (EV3)$_i$ for all $i \geq \nu$ and thus in particular for all $i > \nu$. As Property (REL1) does not consider cycle centers, it remains valid for σe. Since $\beta_1^\sigma = 1$ if and only if $\beta_1^{\sigma e} = 0$ and since σ has Property (ESC1), σe has Property (ESC2) if $\nu = 0$. Thus σe has all properties for the bound $\mu^{\sigma e}$ if $\nu > 1$ resp. for the bound 1 if $\nu = 1$ as specified in Table 5.5.
3. Since σ is well-behaved, it suffices to reevaluate Properties (MNS4), (MNS6), (DN1) and (DN2).

(MNS4) By the choice of e, the premise of this property is true for σe if and only if it is true for σ. In particular, $\overline{m}_s^{\sigma e} = \overline{m}_s^{\sigma}, \overline{m}_g^{\sigma e} = \overline{m}_g^{\sigma}$ and $m_b^{\sigma} = m_b^{\sigma e}$. In addition, $\mu^{\sigma} = \mu^{\sigma e} = 1$ implies that we close the cycle center $F_{1,\mathfrak{b}_2}$. If $\overline{m}_s^{\sigma} \neq 1$, then the conclusion is correct for σe if and only if it is correct for σe, hence σe has Property (MNS4). It thus suffices to consider the case $\overline{m}_s^{\sigma e} = 1$. Assume the conditions of the premise were fulfilled and let $j' := \bar{\sigma}(g_1)$. Then, by assumption, $\overline{\sigma e}(s_{\overline{m}_s^{\sigma e}}) = \overline{\sigma e}(s_{1,j'}) = b_1$. Thus, by the choice of j, it follows that we do not close the cycle center $F_{1,j'}$. Hence, since $\bar{\sigma}(eb_{\overline{m}_s^{\sigma}}) \wedge \neg\bar{\sigma}(eg_{\overline{m}_s^{\sigma}})$ by Property (MNS4), also $\overline{\sigma e}(eb_{\overline{m}_g^{\sigma e}}) \wedge \neg\bar{\sigma}(eg_{\overline{m}_s^{\sigma e}})$.

(MNS6) This follows by the same arguments used for Property (MNS4).

(DN1) Since $i = n$ in this case, $\sigma e(b_1) = \sigma(b_1) = g_1$ by the definition of ν.

(DN2) This statement only needs to be considered if $\neg\bar{\sigma}(d_n) \wedge \overline{\sigma e}(d_n)$, hence, only if $\nu = n$. Then, $\beta_1^{\sigma} = \cdots = \beta_{n-1}^{\sigma} = 1$. But then Property (EV1)$_i$ implies $\sigma e(b_i) = g_i$ for all $i \leq n-1$.

Since σ is well-behaved, σe is thus well-behaved.

4. We prove that $\sigma(g_\nu) \neq F_{\nu,j}$ and $\nu = 1$ imply $I_{\sigma e} = \mathfrak{D}^{\sigma e} \cup \{(g_\nu, F_{\nu,j})\}$.

We first prove $(g_\nu, F_{\nu,j}) \in I_{\sigma e}$. Since $\nu = \mu^{\sigma e} = 1$ and by Property (ESC1) and Property (USV1)$_i$, either $\mathrm{rVal}^*_{\sigma e}(F_{\nu,1-j}) = \mathrm{rVal}^*_{\sigma e}(s_{\nu,1-j}) = [\![s_{\nu,1-j}]\!] \oplus \mathrm{rVal}^*_{\sigma e}(b_1)$ or $\mathrm{rVal}^*_{\sigma e}(F_{\nu,1-j}) = \mathrm{rVal}^*_{\sigma e}(b_2)$. By Property (USV1)$_i$ and Property (EV1)$_{\nu+1}$, it also holds that $\mathrm{rVal}^*_{\sigma e}(F_{\nu,j}) = [\![s_{\nu,j}, h_{\nu,j}]\!] \oplus \mathrm{rVal}^*_{\sigma e}(b_{\nu+1})$. The statement thus follows in either case since $[\![h_{\nu,j}]\!] \succ [\![s_{\nu,1-j}]\!] \oplus L^*_{1,\nu} \succ L^*_{1,\nu}$ and $\mathrm{rVal}^*_{\sigma}(b_1) = \mathrm{rVal}^*_{\sigma}(b_2) = L^*_{1,\nu} \oplus L^*_{\nu+1}$ as well as $\mathrm{rVal}^*_{\sigma}(b_{\nu+1}) = L^*_{\nu+1}$, implying that the valuation of $F_{\nu,j}$ is higher than the valuation of $F_{\nu,1-j}$.

Since $\sigma(g_\nu) = F_{\nu,1-j}$, the valuation of g_ν does not change. Hence, only the valuations of the cycle vertices $d_{\nu,j,0}, d_{\nu,j,1}$ can change. Since $F_{\nu,j}$ is the only vertex with an edge to these vertices, the valuations of all other vertices remain the same. Thus, all switches improving with respect to σ but e stay improving with respect to σe and no further improving switches are created.

Next, let $\sigma(g_\nu) = F_{\nu,j}$ and $\nu = 1$. We prove $I_{\sigma e} = \mathfrak{D}^{\sigma e} \cup \{(b_1, g_1)\} \cup \{(e_{*,*,*}, g_1)\}$. We first prove $(b_1, g_1) \in I_{\sigma e}$.

By Property (EV1)$_1$, $\sigma e(b_1) = b_2$, and it suffices to show $\mathrm{rVal}^*_{\sigma e}(g_1) \succ \mathrm{rVal}^*_{\sigma e}(b_2)$. Since $\mu^{\sigma e} = 1$, we have $\mathrm{rVal}^*_{\sigma e}(b_2) = L^*_2$. Let $G_n = S_n$. We use Corollary 6.1.18 to determine the valuation of g_1. We hence need to analyze λ_1^{S}. If $\sigma e(b_2) = g_2$, then $\lambda_1^{\mathrm{S}} = 1$. If $\sigma e(b_2) = b_3$, then $j = \mathfrak{b}_2 = 0$ by Property (EV1)$_2$ and thus $\sigma e(g_j) = F_{j,0}$ by assumption. Thus, $\lambda_1^{\mathrm{S}} = 1$ in either case. Consider the different cases listed in Corollary 6.1.18. Since $\sigma e(b_1) = b_2$, the first case cannot occur. In addition, since $\sigma e(g_1) = F_{1,j}$ and the cycle center $F_{1,j}$ is closed, the cases 2 to 5 cannot occur. Hence consider the sixth case. As before, $\overline{\sigma e}(g_1) = j = \mathfrak{b}_2$ by assumption, implying $\sigma e(s_{1,\overline{\sigma e}(g_1)}) = \sigma e(s_{1,j}) = h_{1,j}$ by Property (USV1)$_1$. Thus, the sixth case cannot occur. As a consequence, by applying either the seventh or eighth case of Corollary 6.1.18, Property (USV1)$_1$ implies $\mathrm{rVal}^{\mathrm{S}}_{\sigma e}(g_1) = W_1^{\mathrm{S}} \cup \mathrm{rVal}^{\mathrm{S}}_{\sigma e}(b_2) \rhd \mathrm{rVal}^{\mathrm{S}}_{\sigma e}(b_2)$ since $j = \overline{\sigma e}(b_2)$. This also implies that any edge $(e_{*,*,*}, g_1)$ is an improving switch as claimed.

Now consider the case $G_n = M_n$. We use Corollary 6.1.17 to evaluate $\mathrm{rVal}^{\mathrm{M}}_{\sigma}(g_1)$ and thus determine λ^{M}_1. If $\sigma e(b_2) = b_3$, then $\lambda^{\mathrm{M}}_1 = 1$ by the same arguments used when analyzing λ^{S}_1. Since $\overline{\sigma e}(d_1) \wedge \overline{\sigma e}(s_1)$ in this case, the conditions of the last case of Corollary 6.1.17 are fulfilled. Consequently, $\sigma e(b_2) = b_3$ implies that $\mathrm{rVal}^{\mathrm{M}}_{\sigma}(g_1) = W_1 + \mathrm{rVal}^{\mathrm{M}}_{\sigma}(b_2) > \mathrm{rVal}^{\mathrm{M}}_{\sigma}(b_2)$. If $\sigma e(b_2) = g_2$, we have $\lambda^{\mathrm{M}}_1 = 2$. However, by Corollary 6.1.17, case 1, $\mathrm{rVal}^{\mathrm{M}}_{\sigma}(g_1) = W_1 + \mathrm{rVal}^{\mathrm{M}}_{\sigma}(b_2) > \mathrm{rVal}^{\mathrm{M}}_{\sigma}(b_2)$ holds also in this case. This again implies that any edge $(e_{*,*,*}, g_1)$ is improving for σe.

We now show that no further improving switches are created and that existing improving switches remain improving. The only vertices having edges towards g_1 are the vertices b_1 and $e_{*,*,*}$. It thus suffices to show that the valuations of these vertices does not change. This however follows from $\sigma e(b_1) = b_2$ and $\sigma e(e_{i,j,k}) \neq g_1$.

It remains to show that σe is a phase-3-strategy for $\mathfrak{b}$ in either case. By the first two statements, it suffices to show that σe has Property $(\mathrm{USV2})_{i,\mathfrak{b}_{i+1}}$ for all $i < \nu$. But, since $\nu = 1$, there is no such i. Also, by the definition of a pseudo phase-3-strategy, it directly follows that σe is a such a strategy if $\sigma(g_\nu) \neq F_{\nu,\mathfrak{b}_{\nu+1}}$.

5. Since σ is a phase-1-strategy for $\mathfrak{b}$, $\sigma(s_{i,\beta^\sigma_{i+1}}) = h_{i,\beta^\sigma_{i+1}}$ and $\sigma(s_{i,1-\beta^\sigma_{i+1}}) = b_1$ by Property $(\mathrm{USV1})_i$ for all $i < \nu$. As $\mathfrak{b}_i = \beta^\sigma_i = 1 - \beta^{\sigma e}_i = 1 - (\mathfrak{b}+1)_{i+1}$ for all $i \leq \nu$, this implies that σe has Property $(\mathrm{USV3})_i$ for all $i < \nu$.

   We prove that $\sigma(g_\nu) \neq F_{\nu,j}$ and $\nu > 1$ imply $I_{\sigma e} = \mathfrak{D}^{\sigma e} \cup \{(g_\nu, F_{\nu,j})\}$. We observe that either $\mathrm{rVal}^*_{\sigma e}(F_{\nu,1-j}) = \mathrm{rVal}^*_{\sigma e}(s_{\nu,1-j})$ or $\mathrm{rVal}^*_{\sigma e}(F_{\nu,1-j}) = \mathrm{rVal}^*_{\sigma e}(g_1)$. In addition, $\bigoplus_{\ell\in[\nu-1]} W_{\ell^*} \prec [\![s_{\nu,1-j}]\!] \oplus \bigoplus_{\ell\in[\nu-1]} W^*_\ell \prec [\![h_{\nu,j}]\!]$. The statement can thus be shown by the same arguments used in the case $\nu > 1$.

   Let $\sigma(g_\nu) = F_{\nu,j}$ and $\nu > 1$. We prove $I_{\sigma e} = \mathfrak{D}^{\sigma e} \cup \{(b_\nu, g_\nu)\} \cup \{(s_{\nu-1,1}, h_{\nu-1,1})\}$. We first show that $(s_{\nu-1,1}, h_{\nu-1,1})$ is improving for σe. Since $\sigma e(s_{\nu-1,1}) = b_1$ by Property $(\mathrm{USV1})_{\nu-1}$, we prove $\mathrm{rVal}^*_{\sigma e}(h_{\nu-1,1}) \succ \mathrm{rVal}^*_{\sigma e}(b_1)$.

   It holds that $\mathrm{rVal}^*_{\sigma e}(h_{\nu-1,1}) = [\![h_{\nu-1,1}]\!] \oplus \mathrm{rVal}^*_{\sigma e}(g_\nu)$. Since $F_{\nu,j}$ is closed for σe, Properties $(\mathrm{USV1})_\nu$ and $(\mathrm{EV1})_{\nu+1}$ imply $\mathrm{rVal}^*_{\sigma e}(g_\nu) = W^*_\nu \oplus \mathrm{rVal}^*_{\sigma e}(b_{\nu+1})$. As it also holds that $\mathrm{rVal}^*_{\sigma}(b_{\nu+1}) = L^*_{\nu+1}$, it hence follows that

$$\mathrm{rVal}^*_{\sigma e}(h_{\nu-1,1}) = [\![h_{\nu-1,1}]\!] \oplus W^*_\nu \cup L^*_{\nu+1} \succ \bigoplus_{i=1}^{\nu-1} W^*_i \oplus L^*_{\nu+1} = R^*_1 = \mathrm{rVal}^*_{\sigma e}(b_1).$$

   Thus $(s_{\nu-1,1}, h_{\nu-1,1}) \in I_{\sigma e}$. Also,

$$\mathrm{rVal}^*_{\sigma e}(g_\nu) = W^*_\nu \oplus \mathrm{rVal}^*_{\sigma e}(b_{\nu+1}) \rhd \mathrm{rVal}^*_{\sigma e}(b_{\nu+1}) = \mathrm{rVal}^*_{\sigma e}(b_\nu)$$

   since $\sigma e(b_\nu) = b_{\nu+1}$, implying $(b_\nu, g_\nu) \in I_{\sigma e}$.

   We argue why no further improving switches are created and that existing improving switches remain improving. The only vertices with edges to g_ν are $s_{\nu-1,1}$ and b_ν. It thus suffices to show that their valuations does not change. But this follows from $\sigma e(b_\nu) = b_{\nu+1}$ and $\sigma e(s_{\nu-1,1}) = b_1$.

   It remains to prove that σe is a phase-2-strategy. By the first two statements, it suffices to show that there is some $i < \nu$ such that Property $(\mathrm{USV3})_i$ and the negations of both Property (EV2)$_i$ and Property (EV3)$_i$ are fulfilled as $\nu = \mu^{\sigma e}$. Choose any $i < \nu$. Then, by our previous arguments, σe has Property (USV3)$_i$. We next show that σe does not have Property (EV2)$_i$. This follows from $\beta^{\sigma}_{i+1} = 1$, Property (EV1)$_i$, Property (EV2)$_i$ (both applied to σ) and $1 - \beta^{\sigma e}_{i+1} = \beta^{\sigma}_{i+1}$. We finally show that σe does not have Property (EV3)$_i$. But this also immediately follows from $1 - \beta^{\sigma e}_{i+1} = \beta^{\sigma}_{i+1}$ and by applying Property (EV1)$_i$ and Property (EV3)$_i$ to σ. By definition, this also implies that σe is a pseudo phase-2-strategy if $\sigma(g_\nu) \neq F_{\nu,j}$. □

Lemma 6.2.13 (Sixth row of Table 6.4). *Let $\sigma \in \rho(\sigma_0)$ be a well-behaved pseudo phase-2-strategy for $\mathfrak{b} \in \mathfrak{B}_n$ with $\nu > 1$. Let $e := (g_\nu, F_{\nu,\mathfrak{b}_{\nu+1}})$ and $I_\sigma = \mathfrak{D}^\sigma \cup \{(g_\nu, F_{\nu,\mathfrak{b}_{\nu+1}})\}$. Assume that σ has Property (REL1). Then σe is a well-behaved phase-2-strategy for $\mathfrak{b}$ with $\sigma e \in \rho(\sigma_0)$ and $I_{\sigma e} = \mathfrak{D}^{\sigma e} \cup \{(b_\nu, g_\nu), (s_{\nu-1,1}, h_{\nu-1,1})\}$.*

Proof. Let $j := \mathfrak{b}_{\nu+1}$. We prove that σe is a phase-2-strategy for $\mathfrak{b}$. By the choice of e, $\beta^{\sigma e} = \beta^{\sigma} = \mathfrak{b} + 1 =: \beta$. As σ has Property (REL2), $\nu = \mu^\sigma$. Since $e \in I_\sigma$ implies $\sigma(b_\nu) = b_{\nu+1}$ by Property (EV2)$_\nu$, we have $\mathcal{I}^\sigma = \emptyset$ as σ has Property (REL1). By the choice of e and $\sigma e(b_\nu) = \sigma(b_\nu) = b_{\nu+1}$, this implies $\mathcal{I}^{\sigma e} = \mathcal{I}^\sigma = \emptyset$ and $\mu^{\sigma e} = \mu^\sigma = \nu$. Hence σe has Properties (REL1) and (REL2). By the choice of e, $\sigma e(g_\nu) = F_{\nu,j}$. Hence Property (EV2)$_i$ remains valid for all $i \geq \nu$. It remains to show that there is an $i < \nu$ such that σe has Property (USV3)$_i$ but not Property (EV2)$_i$ and Property (EV3)$_i$. Since σ is a pseudo phase-2-strategy for $\mathfrak{b}$, there is such an index fulfilling these conditions with respect to σ. This index also fulfills these conditions with respect to σe. As σ being a pseudo phase-2-strategy implies that σe has the remaining properties, σe is a phase-2-strategy for $\mathfrak{b}$.

Since σ is well-behaved, $\mu^\sigma = \mu^{\sigma e} = \nu \neq 1$ and a switch involving a selector vertex is applied we need to reevaluate the following properties.

(B3) Assume that the premise was fulfilled by σe. Then, by Property (USV1)$_\nu$ and Property (EV1)$_{\nu+1}$, $\sigma e(s_{\nu,1}) = h_{\nu,1}$ implies $j = \beta_{\nu+1} = 1$. Consequently, it holds that $\sigma e(b_{\nu+1}) = g_{\nu+1}$, contradicting $\sigma(b_{\nu+1}) = b_{\nu+2}$.

(EG4) Since $\nu > 1$, the target of g_1 is not changed.

(EBG*) Any premise requires a cycle center to escape towards both g_1 and b_2, contradicting Property (ESC2).

(DN2) Since σe is a pseudo phase-2−strategy for $\mathfrak{b}$ there is some index i such that Property (EV2)$_i$ is not fulfilled. This implies $\sigma e(b_i) = g_i$.

We prove $I_{\sigma e} = \mathfrak{D}^{\sigma e} \cup \{(b_\nu, g_\nu), (s_{\nu-1,1}, h_{\nu-1,1})\}$ and prove $(s_{\nu-1,1}, h_{\nu-1,1}) \in I_{\sigma e}$ first. By Property (USV3)$_{\nu-1}$, $\sigma e(s_{\nu-1,1}) = b_1$. It thus suffices to prove $\mathrm{rVal}^*_{\sigma e}(h_{\nu-1,1}) \succ \mathrm{rVal}^*_{\sigma e}(b_1)$. It holds that

$$\mathrm{rVal}^*_{\sigma e}(h_{\nu-1,1}) = [\![h_{\nu-1,1}]\!] \oplus \mathrm{rVal}^*_{\sigma e}(g_\nu) = [\![h_{\nu-1,1}]\!] \oplus W^*_\nu \oplus \mathrm{rVal}^*_{\sigma e}(b_{\nu+1})$$

since $\overline{\sigma e}(g_\nu) = \beta_{\nu+1}$ and σe has Property (USV1)$_\nu$. Since $\mu^{\sigma e} = \nu$, we also have that $\mathrm{rVal}^*_{\sigma e}(b_{\nu+1}) = L^*_{\nu+1}$ and $\overline{\sigma e}(b_{\mu^{\sigma e}}) = b_{\mu^{\sigma e}+1}$. The statement then follows since Corollary 6.1.5 implies $\mathrm{rVal}^*_\sigma(b_1) = R^*_1$.

We next show $(b_\nu, g_\nu) \in I_{\sigma e}$. Since $\sigma e(b_\nu) = b_{\nu+1}$, we prove$\mathrm{rVal}^*_{\sigma e}(g_\nu) \rhd \mathrm{rVal}^*_{\sigma e}(b_{\nu+1})$. This however follows since $\mathrm{rVal}^*_{\sigma e}(g_\nu) = W^*_\nu \oplus \mathrm{rVal}^*_{\sigma e}(b_{\nu+1})$ as discussed previously.

It remains to show that improving switches remain improving and that no new improving switches are created. By the choice of e, the valuation of g_ν increases. However, as discussed before, $\sigma e(b_\nu) = b_{\nu+1}$ and $\sigma e(s_{\nu-1,1}) = b_1$. Since b_ν and $s_{\nu-1,1}$ are the only vertices that have an edge towards g_ν, the vertex g_ν is the only vertex whose valuation changes when transitioning from σ to σe, implying the statement. □

Lemma 6.2.14 (First row of Table 6.5). *Let $\sigma \in \rho(\sigma_0)$ be a well-behaved phase-2-strategy for $\mathfrak{b} \in \mathfrak{B}_n$ with $\nu > 1$. Let $I_\sigma = \mathfrak{D}^\sigma \cup \{(b_\nu, g_\nu), (s_{\nu-1,1}, h_{\nu-1,1})\}$. Let σ have Property (REL1) as well as Property (USV3)$_i$ for all $i < \nu$. Let $e := (b_\nu, g_\nu)$. Then, σe is a well-behaved phase-2-strategy for $\mathfrak{b}$ with $\sigma e \in \rho(\sigma_0)$. In addition ,$\nu \neq 2$ implies*

$$I_{\sigma e} = \mathfrak{D}^{\sigma e} \cup \{(b_{\nu-1}, b_\nu), (s_{\nu-1,1}, h_{\nu-1,1}), (s_{\nu-2,0}, h_{\nu-2,0})\}$$

if $\nu \neq 2$ and $\nu = 2$ implies

$$I_{\sigma e} = \mathfrak{D}^{\sigma e} \cup \{(b_1, b_2), (s_{1,1}, h_{1,1})\} \cup \{(e_{*,*,*}, b_2)\}.$$

Proof. We first show that σe is a phase-2-strategy for $\mathfrak{b}$. Since the same set of cycle centers is closed for σ and σe, $\beta^{\sigma e} = \beta^\sigma = \mathfrak{b} + 1 =: \beta$. Thus Property (USV1)$_{i'}$ remains valid for all $i \geq \mu^\sigma$ and Property (CC1)$_i$ remains valid for all $i \in [n]$. We next show $\mu^\sigma = \mu^{\sigma e}$. By the choice of e, $\sigma e(b_\nu) = g_\nu$. In addition, since σ has Property (REL2), $\mu^\sigma = \nu$. Thus $\sigma e(b_{\nu-1}) = \sigma(b_{\nu-1}) = g_{\nu-1}$ by Lemma 6.1.4 as Property (REL1) applied to σ implies $\mathcal{I}^\sigma = \emptyset$. Note that Lemma 6.1.4 is applied to σ which is well-behaved. Since σ is well behaved and $\nu - 1 < \mu^\sigma$ we have $\sigma e(g_{i-1}) = \sigma(g_{i-1}) = F_{i,0}$ by Property (BR1). But then, since $\overline{\sigma e}(b_i) = \bar{\sigma}(b_i)$ for all $i \in [n], i \neq \nu$ and $\overline{\sigma e}(g_i) = \bar{\sigma}(g_i)$ for all $i \in [n]$, we have $\mathcal{I}^\sigma = \{\nu - 1\}$. Since $\overline{\sigma e}(g_\nu) = \overline{\sigma e}(b_{\nu+1})$ by Property (CC2), it therefore follows that $\mu^{\sigma e} = \nu$. Thus, since σ is phase-2-strategy, any statement regarding a level larger than $\nu = \mu^\sigma = \mu^{\sigma e}$ remains valid. Property (EV1)$_\nu$ and Property (EV2)$_\nu$ follow directly from Property (CC2) and the choice of e. It remains to show that there is some $i < \mu^{\sigma e}$ such that Property (USV3)$_i$ as well as the negations of both Property (EV2)$_i$ and Property (EV3)$_i$ hold. However, since σ is a phase-2-strategy, there exists such an index for σ, so the same index can be used for σe.

Since we switched the target of b_ν and $\nu = \mu^{\sigma e} \neq 1$ we need to reevaluate the following assumptions to prove that σe is well-behaved.

(S1) Since $\sigma(g_\nu) = F_{\nu,\beta_{\nu+1}}$ by Property (EV2)$_\nu$, the premise and the conclusion are true.

(S2) This property only needs to be checked if $\mu^{\sigma e} = 2$. Then, the only index for which the premise might become true is $i = 1$. But then, it cannot hold that $\sigma e(b_2) = g_2 \wedge i > 1$. Thus, the premise is either incorrect for $i = 1$, implying that the implication is correct for σe, or one of the other two conditions of the premise is true for σe. But then, these conditions were also already true for σ, and hence $\bar{\sigma}(s_i) = \overline{\sigma e}(s_i) = 1$ follows.

(B3) As discussed earlier, $\sigma e(b_{\nu-1}) = g_{\nu-1}$, hence the premise is incorrect.

(D1) By Property (EV1)$_\nu$ and Property (EV2)$_\nu$, the conclusion is true, hence the implication.

(D2) Again, this property only needs to be checked if $\mu^{\sigma e} = 2$. But then, there is no $i \geq 2$ with $i < \mu^{\sigma e}$, hence the premise is incorrect.

(EG5) We only need to show that the premise is not true for $j = 0$. It thus suffices to show that the cycle center $F_{\nu-1,0}$ is closed. If $\mu^{\sigma e} > 2$, then $\nu - 1 > 1$. By Lemma 6.1.4, it then holds that $\sigma e(b_{\nu-1}) = \sigma(b_{\nu-1}) = g_{\nu-1}$. Hence, by Properties (D1) and (BR1), $\bar{\sigma e}(d_{\nu-1}) = \bar{\sigma e}(d_{\nu-1,0})$. This in particular implies $\neg\bar{\sigma e}(eb_{\nu-1,0})$, so the premise is incorrect of $\mu^{\sigma e} > 2$. Now consider the case $\mu^{\sigma e} = 2$. Then, by the definition of a phase-2-strategy, the negation of Property (EV3)$_1$ holds. Thus, since $\beta_2 = 1$ in this case, we have $\bar{\sigma e}(d_{\nu,1-\beta_2}) = \bar{\sigma e}(d_{1,0})$.

We next prove that $I_{\sigma e} = \mathfrak{D}^{\sigma e} \cup \{(b_{\nu-1}, b_\nu), (s_{\nu-2,0}, h_{\nu-2,0}), (s_{\nu-1,1}, h_{\nu-1,1})\}$ if $\nu \neq 2$. We first show that $(b_{\nu-1}, b_\nu) \in I_{\sigma e}$. Since $\sigma e(b_\nu) = g_\nu$ and $\sigma e(b_{\nu-1}) = g_{\nu-1}$ it suffices to show $\operatorname{Val}^*_{\sigma e}(b_\nu) \succ \operatorname{Val}^*_{\sigma e}(b_{\nu-1})$. This follows since $\operatorname{rVal}^*_{\sigma e}(b_\nu) = L^*_\nu \succ R^*_{\nu-1} = \operatorname{rVal}^*_{\sigma e}(b_{\nu-1})$ by Lemma 6.1.10.

We next show $(s_{\nu-2,0}, h_{\nu-2,0}) \in I_{\sigma e}$. By Property (USV3)$_{\nu-2}$, $\sigma e(s_{\nu-2,0}) \neq h_{\nu-2,0}$. Using $\sigma e(b_\nu) = g_\nu, \nu = \mu^{\sigma e}, \beta_\nu = 1$, Property (USV1)$_\nu$, $(s_{\nu-2,0}, h_{\nu-2,0}) \in I_{\sigma e}$ follows from

$$\operatorname{rVal}^*_{\sigma e}(h_{\nu-2,0}) = [\![h_{\nu-2,0}]\!] \oplus W^*_\nu \oplus L^*_{\nu+1} \succ \bigoplus_{i<\nu} W^*_i \cup L^*_{\nu+1} = R^*_1 = \operatorname{Val}^*_{\sigma e}(b_1).$$

Using the same arguments yields $(s_{\nu-1,1}, h_{\nu-1,1}) \in I_{\sigma e}$. Since the valuation of all other vertices is unchanged, no other switch becomes improving and improving switches stay improving.

We prove that $I_{\sigma e} = \mathfrak{D}^{\sigma e} \cup \{(b_1, b_2), (s_{1,1}, h_{1,1})\} \cup \{(e_{*,*,*}, g_1)\}$ if $\nu = 2$. All of the equations developed for the case $\nu \neq 2$ are also valid for $\nu = 2$. In particular we have $\operatorname{rVal}^*_{\sigma e}(b_2) \succ \operatorname{rVal}^*_{\sigma e}(g_1)$ and $\sigma e(b_1) = g_1$, implying $(b_1, b_2) \in I_{\sigma e}$. In addition, we have $\sigma e(e_{i,j,k}) = g_1$ for all $i \in [n]$ and $j, k \in \{0, 1\}$, hence $(e_{i,j,k}, b_2) \in I_{\sigma e}$ for these indices. By the usual arguments, no other new improving switches are created and improving switches stay improving (with the exception of e). □

Lemma 6.2.15 (Second row of Table 6.5). *Let $\sigma \in \rho(\sigma_0)$ be a well-behaved phase-2-strategy for some $\mathfrak{b} \in \mathfrak{B}_n$ with $\nu > 1$. Assume that $\bar{\sigma}(d_{i'}) = 1$ for all $i' < \mu^\sigma$ and that $e = (s_{i,j}, h_{i,j}) \in I_\sigma$ for some $i \in [\mu^\sigma - 1]$ where $j := \beta^\sigma_{i+1}$. Further assume that σ has Property (USV3)$_{i'}$ for all $i' \leq i$. Also, assume that σ has Properties (EV1)$_{\mu^\sigma}$ and (EV1)$_{i+1}$. If $i \neq 1$, then σe is a well-behaved phase-2-strategy for $\mathfrak{b}$. If $i = 1$, then σe is a well-behaved phase-3-strategy for $\mathfrak{b}$. In either case, $I_{\sigma e} = I_\sigma \setminus \{e\}$.*

Proof. We first observe that $\sigma(s_{1,\beta^\sigma_2}) = b_1$ by Property (USV3)$_1$. Since σ has Properties (REL2) and (EV1)$_{\mu^\sigma}$, it follows that $\sigma(b_{\mu^\sigma}) = g_{\mu^\sigma}$. Thus, by Lemma 6.1.4, $\mathcal{I}^\sigma \neq \emptyset$. By the choice of e, $\beta^\sigma = \beta^{\sigma e} =: \beta, \mu^\sigma = \mu^{\sigma e}, \mathcal{I}^{\sigma e} = \mathcal{I}^\sigma \neq \emptyset$ and $\sigma e(b_{\mu^{\sigma e}}) = g_{\mu^{\sigma e}}$. In particular, σe has Properties (EV1)$_{\mu^{\sigma e}}$ and (EV1)$_{i+1}$. Let $i \neq 1$. We prove that σe is a phase-2-strategy. Since $i < \mu^{\sigma e}$, it suffices to check the special conditions of phase 2 since all other properties of Table 5.5 remain valid for σe. We show that the index 1 fulfills these special conditions. Since $\mu^{\sigma e} \neq 1$, we have $\sigma e(b_1) = g_1$. As the choice of $i \neq 1$ implies $\mu^\sigma = \mu^{\sigma e} > 2$, applying Property (BR1) to σ yields $\bar{\sigma e}(g_1) = \bar{\sigma}(g_1) = 1$. For the sake of a contradiction, assume that σe had Property (EV2)$_1$. Then, $1 = \bar{\sigma}(g_1) = \beta_2$, implying $\nu = \mu^\sigma = 2$, contradicting the choice of i. Consequently, Property (EV2)$_1$ does not hold for σe. Now, for the sake of contradiction, assume that σe had Property (EV3)$_1$. Then, since $\sigma(b_1) = g_1$ and $\nu = \mu^\sigma > 2$, the cycle center $F_{1,1-\beta_2} = F_{1,1}$ is not closed. By Property (ESC2), this implies $\bar{\sigma}(eg_{1,1}) \wedge \neg\bar{\sigma}(eb_{1,1})$. Since $\bar{\sigma}(g_1) = 1$, Property (EG3) then implies $\bar{\sigma}(s_1) = \bar{\sigma}(s_{1,1}) = 1$. Consequently, by Property (EG5), this implies $\bar{\sigma}(b_2) = 1$, so $\sigma(b_2) = g_2$. But then, $\sigma(b_2) = g_2 \Leftrightarrow \mu^\sigma > 2$ as both statements are true. Thus, since $\sigma(b_1) = g_1$, Property (D1) implies that $F_{1,\bar{\sigma}(g_1)} = F_{1,1}$ is closed which is a contradiction. Hence σe does not have Property (EV3)$_1$. Finally, we have $\sigma e(s_{1,0}) = \sigma(s_{1,0}) = \sigma(s_{1,\beta_2}) = b_1$ by assumption and $\overline{\sigma e}(s_{1,1}) = \bar{\sigma}(s_{1,1}) = \bar{\sigma}(s_{1,1-\beta_2}) = 1$ by Property (S2). Hence the index 1 fulfills all of the special conditions of the definition of a phase-2-strategy, so σe is a phase-2-strategy for $\mathfrak{b}$.

If $i = 1$, then the assumptions imposed on σ and the choice of e directly imply that σe is a phase-3-strategy for $\mathfrak{b}$.

We prove that σe is well-behaved. Note that $\overline{\sigma e}(g_i) = \bar{\sigma}(g_i)$ and thus, by Property (BR1), $\bar{\sigma}(g_i) = 1$ if and only if $i \neq \mu^{\sigma e} - 1$, implying $j = 1 - \bar{\sigma}(g_i)$. By the usual arguments, it suffices to investigate the following properties.

- (B3) We only need to consider this property if $j = 1$, i.e., if $\beta_{i+1} = 1$. Since $i < \mu^{\sigma e}$ this implies that $i = \mu^{\sigma e} - 1$. But then $\sigma e(b_{i+1}) = \sigma e(b_{\mu^{\sigma e}}) = g_{\mu^{\sigma e}}$, so the premise is incorrect.
- (EG5) Since σe fulfills Property (EV1)$_{i+1}$, $\bar{\sigma}(b_{i+1}) = \beta_{i+1} = j$. Thus, the conclusion of Property (EG5) is correct, implying that the implication is correct.

It remains to show $I_{\sigma e} = I_\sigma \setminus \{e\}$. The vertex $F_{i,j}$ is the only vertex that has an edge to $s_{i,j}$. Let $G_n = S_n$ first. Since $\mathrm{Val}^{\mathrm{S}}_\sigma(s_{i,j}) \trianglelefteq \mathrm{Val}^{\mathrm{S}}_{\sigma e}(s_{i,j})$, proving $\tau^\sigma(F_{i,j}) \neq s_{i,j}$ implies $\tau^{\sigma e}(F_{i,j}) \neq s_{i,j}$. This then implies that the valuation of no other vertex than $s_{i,j}$ changes, implying $I_{\sigma e} = I_\sigma \setminus \{e\}$.

For the sake of a contradiction, assume $\tau^\sigma(F_{i,j}) = s_{i,j}$. Then, by Lemma 6.1.16, one of three cases holds. Since $\mu^{\sigma e} \neq 1$, it cannot hold that $[\bar{\sigma}(eb_{i,j}) \wedge \neg\bar{\sigma}(eb_{i,j}) \wedge \mu^\sigma = 1]$. As $\bar{\sigma}(d_{i,1-j})$ by Property (BR1) and assumption, Property (CC1)$_i$ implies $\neg\bar{\sigma}(d_{i,j})$. Since σe has Property (ESC2), $\bar{\sigma}(eb_{i,j}) \wedge \neg\bar{\sigma}(eg_{i,j}) \wedge [\mu^\sigma \neq 1 \vee (\bar{\sigma}(s_{i,j}) \wedge \bar{\sigma}(b_{i+1}) \neq j)]$ also cannot hold. Consequently, by Lemma 6.1.16, $\tau^\sigma(F_{i,j}) \neq s_{i,j}$.

Now let $G_n = M_n$. Again, as $\overline{\sigma e}(d_{i,1-j})$ by assumption, Property (CC2) implies that $F_{i,j}$ is not closed. By Lemma 6.1.15 and Property (ESC2), this implies $\mathrm{rVal}^{\mathrm{M}}_{\sigma e}(F_{i,j}) = \mathrm{rVal}^{\mathrm{M}}_{\sigma e}(g_1)$ and in particular $\mathrm{rVal}^{\mathrm{M}}_{\sigma e}(F_{i,j}) \neq \mathrm{rVal}^{\mathrm{M}}_{\sigma e}(s_{i,j})$. The only vertices that have an edge to $F_{i,j}$ are $d_{i,j,0}, d_{i,j,1}$ and g_i. We prove that $\sigma(d_{i,j,k}) \neq F_{i,j}$ implies $\mathrm{Val}^{\mathrm{M}}_\sigma(F_{i,j}) > \mathrm{Val}^{\mathrm{M}}_\sigma(e_{i,j,k})$, so $(d_{i,j,k}, F_{i,j}) \in I_\sigma$. Since $\sigma(d_{i,j,k}) = F_{i,j}$ implies $(d_{i,j,k}, F_{i,j}) \notin I_\sigma$, this then proves that $\sigma(d_{i,j,k}) \neq F_{i,j} \Leftrightarrow (d_{i,j,k}, F_{i,j}) \in I_\sigma$. We then argue why the same arguments can be applied to σe which proves $(d_{i,j,k}, F_{i,j}) \in I_\sigma \Leftrightarrow (d_{i,j,k}, F_{i,j}) \in I_{\sigma e}$.

Hence assume $\sigma(d_{i,j,k}) \neq F_{i,j}$, implying $\sigma(d_{i,j,k}) = e_{i,j,k}$. By Property (ESC2), all escape vertices escape to g_1, hence $F_{i,j}$ is either g_1-open or g_1-halfopen. Also, $\sigma(s_{i,j}) = b_1$ and $\sigma(b_1) = g_1$ imply $\mathrm{Val}^{\mathrm{M}}_\sigma(s_{i,j}) = \langle s_{i,j}\rangle + \mathrm{Val}^{\mathrm{M}}_\sigma(g_1)$. Thus,

$$\mathrm{Val}^{\mathrm{M}}_\sigma(F_{i,j}) - \mathrm{Val}^{\mathrm{M}}_\sigma(e_{i,j,k}) = q[\mathrm{Val}^{\mathrm{M}}_\sigma(s_{i,j}) - \mathrm{Val}^{\mathrm{M}}_\sigma(g_1)],$$

where the exact value of $q > 0$ depends on whether $F_{i,j}$ is open or halfopen. But then, $\mathrm{Val}^{\mathrm{M}}_\sigma(s_{i,j}) = \langle s_{i,j}\rangle + \mathrm{Val}^{\mathrm{M}}_\sigma(g_1)$ implies $\mathrm{Val}^{\mathrm{M}}_\sigma(F_{i,j}) > \mathrm{Val}^{\mathrm{M}}_\sigma(e_{i,j,k})$. Since

$$\mathrm{Val}^{\mathrm{M}}_{\sigma e}(s_{i,j}) = \langle s_{i,j}\rangle + \mathrm{Val}^{\mathrm{M}}_{\sigma e}(h_{i,j}) > \langle s_{i,j}\rangle + \mathrm{Val}^{\mathrm{M}}_{\sigma e}(g_1)$$

as the edge $(s_{i,j}, b_1)$ would otherwise be improving for σe which cannot happen, the same argument implies $\mathrm{Val}^{\mathrm{M}}_{\sigma e}(F_{i,j}) > \mathrm{Val}^{\mathrm{M}}_{\sigma e}(e_{i,j,k})$.

No vertex but $F_{i,j}$ has an edge towards $d_{i,j,k}$. Thus, although the valuation of $d_{i,j,k}$ increases due to the application of e, it is impossible to have an improving switch $(*, d_{i,j,k})$ for either σ or σe. Consequently, we do not need to consider this vertex when investigating whether new improving switches are created.

It thus remains to prove $\sigma(g_i) = \sigma e(g_i) = F_{i,1-j}$ and $(g_i, F_{i,j}) \notin I_\sigma, I_{\sigma e}$. Once this statement is proven, combining all of the previous statements yields $I_{\sigma e} = I_\sigma \setminus \{e\}$. Since Property (BR1)$_i$ and the choice of j imply that $\sigma e(g_i) = F_{i,1-j}$, it suffices to prove $\mathrm{Val}^{\mathrm{M}}_{\sigma e}(F_{i,1-j}) > \mathrm{Val}^{\mathrm{M}}_{\sigma e}(F_{i,j})$. Since $\sigma e(b_1) = g_1$, the assumption $\overline{\sigma e}(d_{i'})$ for all $i' < \mu^{\sigma e}$ implies $\mathrm{rVal}^{\mathrm{M}}_{\sigma e}(F_{i,j}) = \mathrm{rVal}^{\mathrm{M}}_{\sigma e}(g_1) = R^{\mathrm{M}}_1$. This furthermore yields $\mathrm{rVal}^{\mathrm{M}}_{\sigma e}(F_{i,1-j}) = \mathrm{rVal}^{\mathrm{M}}_{\sigma e}(s_{i,1-j})$. Property (USV2)$_{i,1-j}$ implies that $\sigma e(s_{i,1-j}) = h_{i,1-j}$. If $j = \beta_{i+1} = 1$, then

$$\mathrm{rVal}^{\mathrm{M}}_{\sigma e}(F_{i,1-j}) = [\![s_{i,1-j}, h_{i,1-j}]\!] + \mathrm{rVal}^{\mathrm{M}}_{\sigma e}(b_{i+2}).$$

But this implies $i = \nu - 1$, so

$$\mathrm{rVal}^{\mathrm{M}}_{\sigma e}(F_{i,1-j}) = [\![s_{\nu-1,1-j}, h_{\nu-1,1-j}]\!] + \mathrm{rVal}^{\mathrm{M}}_{\sigma e}(b_{\nu+1}) > \sum_{\ell=1}^{\nu-1} W^{\mathrm{M}}_\ell + L^{\mathrm{M}}_{\nu+1} = R^{\mathrm{M}}_1.$$

If $j = \beta_{i+1} = 0$, then $i < \nu - 1$ and $\mathrm{rVal}^{\mathrm{M}}_{\sigma e}(F_{i,1-j}) = [\![s_{i,1-j}, h_{i,1-j}]\!] + \mathrm{rVal}^{\mathrm{M}}_{\sigma e}(g_{i+1})$. Using Lemma 6.1.12, this implies

$$\begin{aligned}\mathrm{rVal}^{\mathrm{M}}_{\sigma e}(F_{i,1-j}) &= [\![s_{i,1-j}, h_{i,1-j}]\!] + R^{\mathrm{M}}_{i+1}\\ &= [\![s_{i,1-j}, h_{i,1-j}]\!] + \sum_{\ell=i+1}^{\nu-1} W^{\mathrm{M}}_\ell + L^{\mathrm{M}}_{\nu+1} > \sum_{\ell=1}^{\nu-1} W^{\mathrm{M}}_\ell + L^{\mathrm{M}}_{\nu+1} = R^{\mathrm{M}}_1.\end{aligned}$$

□

Lemma 6.2.16 (Third row of Table 6.5). *Let $\sigma \in \rho(\sigma_0)$ be a well-behaved phase-2-strategy for $\mathfrak{b} \in \mathfrak{B}_n$ with $\nu > 1$. Assume that $\bar{\sigma}(d_{i'}) = 1$ for all $i' < \mu^\sigma$ and $e = (b_i, b_{i+1}) \in I_\sigma$ for some $i \in \{2, \ldots, \mu^\sigma - 1\}$. In addition, assume that σ has Property (USV3)$_{i'}$ for all $i' < i$, Property (EV1)$_{i'}$ and Property (EV2)$_{i'}$ for all $i' > i$ as well as Property (EV3)$_{i'}$ for all $i' > i, i' \neq \mu^\sigma$.*

Then σe is a well-behaved phase-2-strategy for $\mathfrak{b}$. Furthermore, $i \neq 2$ implies

$$I_{\sigma e} = (I_\sigma \setminus \{e\}) \cup \{(b_{i-1}, b_i), (s_{i-2,0}, h_{i-2,0})\}$$

and $i = 2$ implies $I_{\sigma e} = (I_\sigma \setminus \{e\}) \cup \{(b_1, b_2)\} \cup \{(e_{*,*,*}, b_2)\}$.

Proof. By the choice of e and by assumption, σe has Property (USV3)$_{i'}$ for all $i' < i$. In particular, $\sigma(s_{i-1,0}) = \sigma(s_{i-2,0}) = b_1$ since $i < \mu^\sigma = \nu$.

We prove that σe is a phase-2-strategy for $\mathfrak{b}$. Since $e = (b_i, b_{i+1})$ and $i < \mu^\sigma = \nu$, it suffices to prove $\mu^\sigma = \mu^{\sigma e}$ and that there is an index $i' < i$ fulfilling the special conditions of a phase-2-strategy. Since $\mu^\sigma > i, \mu^\sigma = \nu$ and since σ has Property (EV1)$_{\mu^\sigma}$, $\sigma(b_{\mu^\sigma}) = g_{\mu^\sigma}$. Hence, by Lemma 6.1.4, $\mathcal{I}^\sigma \neq \emptyset$. We show $i = \max\{i' \in \mathcal{I}^\sigma\}$. By the choice of e,

$\bar{\sigma}(b_i) = 1$. If $i+1 = \mu^\sigma$, then $\bar{\sigma}(b_{i+1}) = \bar{\sigma}(b_{\mu^\sigma}) = 1$ and $\bar{\sigma}(g_i) = 0$ by Property (BR1). If $i+1 < \mu^\sigma$, then $i+1 \leq \mu^\sigma - 1$, so $\bar{\sigma}(b_{i+1}) = \beta^\sigma_{i+1} = (\mathfrak{b}+1)_{i+1} = 0$ by Property (EV1)$_{i+1}$ and $\bar{\sigma}(g_i) = 1$ by Property (BR1). In either case $\bar{\sigma}(g_i) \neq \bar{\sigma}(b_{i+1})$, hence $i \in \mathcal{I}^\sigma$. For any $i' \in \{i+1, \dots \nu-1\}$, Property (EV1)$_{i'}$ and $\mu^\sigma = \nu$ imply $\bar{\sigma}(b_{i'}) = 0$. Thus $i = \max\{i' \in \mathcal{I}^\sigma\}$. We now prove $i-1 \in \mathcal{I}^{\sigma e}$ since this suffices to prove $\mu^{\sigma e} = \mu^\sigma$ as $\sigma e(b_{i'}) = b_{i'+1}$ for all $i' \in \{i, \dots, \mu^\sigma - 1\}$.

By Property (EV1)$_{i-1}$, it holds that $\overline{\sigma e}(b_{i-1}) = \bar{\sigma}(b_{i-1}) = 1$ since Property (B2) would imply $\sigma(b_i) = b_{i+1}$ otherwise. By Property (BR1) and $i-1 < \mu^\sigma - 1$, it follows that $\overline{\sigma e}(g_{i-1}) = \bar{\sigma}(g_{i-1}) = 1$. Also, $\overline{\sigma e}(b_i) = 0$ by the choice of e. Hence $i-1 \in \mathcal{I}^{\sigma e}$, implying $i-1 = \max\{i' \in \mathcal{I}^{\sigma e}\}$. Consequently, $\mu^{\sigma e} = \mu^\sigma = \nu$, so σe has Property (REL2).

We show that $i-1$ fulfills the special conditions of Table 5.5 for phase-2-strategies. As shown previously, $\overline{\sigma e}(b_{i-1}) = 1$ and $\overline{\sigma e}(g_{i-1}) = 1 = 1 - \beta_i$. Thus Property (EV2)$_{i-1}$ does not hold for σe. If $i > 2$, Lemma 6.1.4 implies $\sigma(b_2) = g_2$ as $i = \max\{i' \in \mathcal{I}^\sigma\}$. If $i = 2$ then $\sigma(b_2) = g_2$ since $(b_2, b_3) \in I_\sigma$. Thus, by applying Property (D2) to σ, it follows that $\bar{\sigma}(d_{i-1}) = \overline{\sigma e}(d_{i-1}) = \overline{\sigma e}(d_{i-1,1-\beta_i}) = 1$. Thus σe fulfills the negation of Property (EV3)$_{i-1}$. Finally, σe also has Property (USV3)$_{i-1}$ by assumption. Thus the index $i-1$ fulfills the special conditions of Table 5.5, so σe is phase-2-strategy for $\mathfrak{b}$.

Since σ is a phase-2-strategy, it suffices to check the following properties:

(B1) If $i < \mu^{\sigma e} - 1$, then $i+1 < \mu^{\sigma e}$. Since σ has Property (EV1)$_{i+1}$ by assumption, Property (REL2) implies $\sigma e(b_{i+1}) = \sigma(b_{i+1}) = b_{i+2}$.

(B3) For this property, it might happen that either the premise becomes true with respect to σe or that it is true while the conclusion becomes false for σe. Consider the first case first. Then, $\sigma e(s_{i-1,1}) = h_{i-1,1}$ and $\sigma e(b_i) = b_{i+1}$. However, since $i = \max\{i' \in \mathcal{I}^{\sigma e}\}$, we have $\overline{\sigma e}(g_i) = \bar{\sigma}(g_i) \neq \bar{\sigma}(b_{i+1}) = \overline{\sigma e}(b_{i+1})$, hence the conclusion is true as well. Now assume that the premise is correct for σe while the conclusion became false by the application of e. Then $\sigma e(s_{i-2,1}) = h_{i-2,1}$ and $\sigma e(b_{i-1}) = b_i$. But this cannot happen since $\sigma e(b_{i-1}) = g_{i-1}$ as proven earlier.

(EG5) Since $\overline{\sigma e}(b_i) = 0$ we need to show $\overline{\sigma e}(d_{i-1,1}) = 1$ which was already shown earlier.

We now show that $i \neq 2$ implies $I_{\sigma e} = (I_\sigma \setminus \{e\}) \cup \{(b_{i-1}, b_i), (s_{i-2,0}, h_{i-2,0})\}$ and that $i = 2$ implies $I_{\sigma e} = (I_\sigma \setminus \{e\}) \cup \{(b_1, b_2)\} \cup \{(e_{*,*,*}, g_1)\}$. By Lemma 6.1.4, $\sigma e(b_{i-1}) = g_{i-1}$ and also, by assumption, $\sigma(s_{i-2,0}) = \sigma e(s_{i-2,0}) = b_1$ if $i \neq 2$. We hence need to show $\mathrm{rVal}^*_{\sigma e}(b_i) \succ \mathrm{rVal}^*_{\sigma e}(g_{i-1})$ and $[\![h_{i-2,0}]\!] \oplus \mathrm{rVal}^*_{\sigma e}(b_i) \succ \mathrm{rVal}^*_{\sigma e}(b_1)$. This in particular implies that any edge $(e_{*,*,*}, b_2)$ is improving for σe if $i = 2$ since $\sigma(e_{*,*,*}) = g_1$ by Property (ESC2). Since either $i+1 < \mu^{\sigma e}$ and $\sigma e(b_{i+1}) = b_{i+2}$ by Property (B1) or $i+1 = \mu^{\sigma e}$, we have $\mathrm{rVal}^*_{\sigma e}(b_i) = L^*_{i+1}$ since $\sigma e(b_i) = b_{i+1}$. By assumption, $\bar{\sigma}(d_{i'}) = \overline{\sigma e}(d_{i'}) = 1$ for all $i' < \mu^{\sigma e}$. Thus, $\mathrm{rVal}^*_{\sigma e}(g_{i-1}) = R^*_{i-1}$ by Lemma 6.1.12. Therefore, $\sigma e(b_{\mu^{\sigma e}}) = g_{\mu^{\sigma e}}$ implies $\mathrm{rVal}^*_{\sigma e}(b_i) \succ \mathrm{rVal}^*_{\sigma e}(g_{i-1})$ by Lemma 6.1.10.Since $\mathrm{rVal}^*_{\sigma e}(b_1) = R^*_1$, Lemma 6.1.10 further implies $[\![h_{i-2,0}]\!] \oplus \mathrm{rVal}^*_{\sigma e}(b_i) \succ \mathrm{rVal}^*_{\sigma e}(b_1)$ if $i \neq 2$. Thus $(b_{i-1}, b_i), (s_{i-2,0}, h_{i-2,0}) \in I_{\sigma e}$ if $i \neq 2$ and $(b_1, b_2), (e_{*,*,*}, b_2) \in I_\sigma$ if $i = 2$. □

Lemma 6.2.17 (Last row of Table 6.5). *Let $\sigma \in \rho(\sigma_0)$ be a well-behaved pseudo phase-3-strategy for some $\mathfrak{b} \in \mathfrak{B}_n$ with $\nu = 1$. Let $I_\sigma = \mathfrak{D}^\sigma \cup \{(g_\nu, F_{\nu,\mathfrak{b}_{\nu+1}})\}$ and $e := (g_\nu, F_{\nu,\mathfrak{b}_{\nu+1}})$.*

Then σe *is a well-behaved phase-3-strategy for* $\mathfrak{b}$ *with* $\sigma e \in \rho(\sigma_0)$ *and*

$$I_{\sigma e} = (I_\sigma \setminus \{e\}) \cup \{(b_1, g_1)\} \cup \{(e_{*,*,*}, g_1)\}.$$

Proof. Let $j := \mathfrak{b}_{\nu+1}$. We first show $\mu^{\sigma e} = \mu^\sigma$. Since σ is a pseudo phase-3-strategy for $\mathfrak{b}$, $\mu^\sigma = \nu = 1$. Hence, $\sigma(b_1) = b_2$, implying $\sigma e(b_1) = b_2$, so $\mu^{\sigma e} = 1$. Note that this implies that σe has Property (CC2) as the cycle center $F_{\nu,j}$ is closed due to $\beta^\sigma = \beta^{\sigma e} = \mathfrak{b} + 1 =: \beta$.

We next show that σe is a phase-3-strategy for $\mathfrak{b}$. The only properties other than Property (CC2) involving $e = (g_\nu, F_{\nu,j})$ are Properties (REL1) and $(\text{EV2})_1$. These do not need to be fulfilled for a phase-3-strategy so, σ being a pseudo phase-3-strategy implies that σe is a phase-3-strategy for $\mathfrak{b}$.

We now show that σe is well-behaved. Since σ is a well-behaved pseudo phase-3-strategy for $\mathfrak{b}$, $\mu^{\sigma e} = 1$ and by the choice of e, it suffices to investigate the following properties:

(MNS1) Since the cycle center $F_{1,j}$ is closed with respect to σe due to $\beta^{\sigma e} = \mathfrak{b} + 1$ and since $j = \overline{\sigma e}(g_i)$, the conclusion of this property is true for σe.

(MNS2) If $\sigma e(b_2) = g_2$, then $m_b^{\sigma e} = 2$ and there cannot be an index fulfilling the conditions of the premise. If $\sigma e(b_2) = b_3$, then $\beta_2 = j = 0$ by Property $(\text{EV1})_2$, implying $\sigma e(g_1) = F_{1,0}$. But then $\overline{m}_g^{\sigma e} = 1$, hence there cannot be an index such that the conditions of the premise are fulfilled.

(MNS3) This follows by the same arguments used for Property (MNS2).

(MNS4) By the choice of e, the definition of j and Property $(\text{EV1})_2$, we either have $\overline{m}_g^{\sigma e} = 1 \land \overline{m}_b^{\sigma e} > 2$ or $\overline{m}_g^{\sigma e} > 1 \land m_b^{\sigma e} = 2$. Since the second case contradicts the conditions of the premise, assume $\overline{m}_g^{\sigma e} = 1 \land m_b^{\sigma e} > 2$. Then, $\sigma e(b_2) = b_3$, hence $j = \beta_2^{\sigma e} = 0$, implying $\sigma e(g_1) = F_{1,0}$. In addition, $\overline{m}_s^{\sigma e} = 1$ as $\overline{m}_s^{\sigma e} \leq \overline{m}_g^{\sigma e}$. Thus, by the definition of $\overline{m}_s^{\sigma e}$, we have $\sigma e(s_{1,0}) = b_1$. But this contradicts Property $(\text{USV1})_1$ as this implies $\sigma e(s_{1,\beta_2}) = \sigma e(s_{1,0}) = h_{1,0}$.

(MNS5) This follows by the same arguments used for Property (MNS2).

(MNS6) By the choice of e, the definition of j and Property $(\text{EV1})_2$, we either have $\overline{m}_g^{\sigma e} = 1 \land \overline{m}_b^{\sigma e} > 2$ or $\overline{m}_g^{\sigma e} > 1 \land m_b^{\sigma e} = 2$. Since the first case contradicts the conditions of the premise, assume $\overline{m}_g^{\sigma e} > 1 \land m_b^{\sigma e} = 2$. Then, $\sigma e(b_2) = g_2$, hence $j = \beta_2 = 1$, implying $\sigma e(g_1) = F_{1,1}$. In addition, $\overline{m}_s^{\sigma e} = 1$ as $\overline{m}_s^{\sigma e} < m_b^{\sigma e} = 2$. Thus, by the definition of $\overline{m}_s^{\sigma e}$, we have $\sigma e(s_{1,1}) = b_1$. But this contradicts Property $(\text{USV1})_1$ as this implies $\sigma e(s_{1,\beta_2^{\sigma e}}) = \sigma e(s_{1,1}) = h_{1,1}$.

(EG4) Since the conclusion is true for σe by the choice of e, the implication is true.

(EBG2) Since $\overline{\sigma e}(g_1) = \overline{\sigma e}(b_2) = \beta_2 = j$, $\mu^{\sigma e} = 1$ and Property $(\text{USV1})_1$ together imply $\overline{\sigma e}(s_1) = 1$.

(EBG3) Since $\overline{\sigma e}(g_1) = \overline{\sigma e}(b_2) = \beta_2 = j$, $\nu = 1$ and $\beta = \mathfrak{b} + 1$ imply $\overline{\sigma e}(d_1) = 1$.

It thus remains to show that $I_{\sigma e} = (I_\sigma \setminus \{e\}) \cup \{(b_1, g_1)\} \cup \{(e_{*,*,*}, g_1)\}$. Since $\sigma e(b_1) = b_2$ and $\mu^{\sigma e} = 1$, this can be shown by using the same arguments used in the proof of Lemma 6.2.12 (4). □

Lemma 6.2.19 (First row of Table 6.6). *Let* $\sigma \in \rho(\sigma_0)$ *be a well-behaved phase-3-strategy for* $\mathfrak{b} \in \mathfrak{B}_n$. *Let* $i \in [n], j, k \in \{0, 1\}$ *such that* $(e_{i,j,k}, t^\rightarrow) \in I_\sigma$ *and* $\sigma(d_{i,j,k}) = F_{i,j}$. *Further assume the following.*

1. *If $G_n = S_n$, then, σ has Property (USV2)$_{i',j'}$ for all $i' < \mu^\sigma, j' \in \{0,1\}$.*
2. *If $G_n = M_n$, then, $\sigma(s_{i',j'}) = b_1$ implies $\bar{\sigma}(eb_{i',j'}) \wedge \neg\bar{\sigma}(eg_{i',j'})$ for all $i' < \mu^\sigma$ and $j' \in \{0,1\}$.*

Then σe is a well-behaved phase-3-strategy for $\mathfrak{b}$ with $\sigma e \in \rho(\sigma_0)$. If $\sigma(d_{i,j,1-k}) = e_{i,j,1-k}$ or $[\sigma(d_{i,j,1-k}) = F_{i,j}$ and $j \neq \beta^\sigma_{i+1}]$, then $I_{\sigma e} = (I_\sigma \setminus \{e\}) \cup \{(d_{i,j,k}, e_{i,j,k})\}$. If $\sigma(d_{i,j,1-k}) = F_{i,j}$ and $j = \beta^\sigma_{i+1}$, then $I_{\sigma e} = I_\sigma \setminus \{e\}$.

Proof. Since a phase-3-strategy does not need to fulfill Property (ESC1) or (ESC2), σ being a phase-3-strategy for $\mathfrak{b}$ implies that σe is a phase-3-strategy for $\mathfrak{b}$. We prove that σe is well-behaved. By assumption, $\sigma e(d_{i,j,k}) = \sigma(d_{i,j,k}) = F_{i,j}$. Hence $F_{i,j}$ escapes towards g_1 resp. b_2 with respect to σ if and only if it escapes towards the same vertex with respect to σe. Since there are no other conditions on escape vertices except the escape of cycle centers in Table 6.1, σe is well-behaved since σ is well-behaved.

It remains to show the statements related to the improving switches. We first prove that $\sigma(d_{i,j,1-k}) = e_{i,j,1-k}$ implies $I_{\sigma e} = (I_\sigma \setminus \{e\}) \cup \{(d_{i,j,k}, e_{i,j,k})\}$. As $d_{i,j,k}$ is the only vertex having an edge to $e_{i,j,k}$, it suffices to prove $(d_{i,j,k}, e_{i,j,k}) \in I_{\sigma e}$. We distinguish two cases.

1. Let $\mu^{\sigma e} = 1$. Then $\mathrm{rVal}^*_{\sigma e}(d_{i,j,k}) = \mathrm{rVal}^*_{\sigma e}(F_{i,j})$ and $\mathrm{rVal}^*_{\sigma e}(e_{i,j,k}) = \mathrm{rVal}^*_{\sigma e}(g_1)$. Moreover, $\mathrm{rVal}^*_{\sigma e}(g_1) = W_1^* \oplus \mathrm{rVal}^*_{\sigma e}(b_2)$ and $\mathrm{rVal}^*_{\sigma e}(b_2) = L_2^*$ by Lemma 6.2.18. We thus prove $\mathrm{rVal}^*_{\sigma e}(g_1) \succ \mathrm{rVal}^*_{\sigma e}(F_{i,j})$ and distinguish two further cases.
   a) Let $\sigma e(e_{i,j,1-k}) = g_1$. Since $\sigma e(d_{i,j,1-k}) = e_{i,j,1-k}$ and $\sigma e(d_{i,j,k}) = F_{i,j}$ we then have $\overline{\sigma e}(eg_{i,j}) \wedge \neg\overline{\sigma e}(eb_{i,j})$ and, by assumption, $\mu^{\sigma e} = 1$. Let $G_n = S_n$. Then, Lemma 6.1.16 implies $\mathrm{rVal}^{\mathrm{S}}_{\sigma e}(F_{i,j}) = \mathrm{rVal}^{\mathrm{S}}_{\sigma e}(s_{i,j})$. Since Property (EG1) implies $\sigma e(s_{i,j}) = b_1$ and $\sigma e(b_1) = b_2$ follows from $\mu^{\sigma e} = 1$, it holds that $\mathrm{rVal}^{\mathrm{S}}_{\sigma e}(F_{i,j}) = \{s_{i,j}\} \cup \mathrm{rVal}^{\mathrm{S}}_{\sigma e}(b_2)$. Hence, by Lemma 6.2.18, $\mathrm{rVal}^{\mathrm{S}}_{\sigma e}(g_1) \rhd \mathrm{rVal}^{\mathrm{S}}_{\sigma e}(F_{i,j})$, so $(d_{i,j,k}, e_{i,j,k}) \in I_{\sigma e}$. Now let $G_n = M_n$. Then, since $F_{i,j}$ is g_1-halfopen,
   $$\mathrm{Val}^{\mathrm{M}}_{\sigma e}(g_1) - \mathrm{Val}^{\mathrm{M}}_{\sigma e}(F_{i,j}) = \frac{2\varepsilon}{1+\varepsilon}[\mathrm{Val}^{\mathrm{M}}_{\sigma}(g_1) - \mathrm{Val}^{\mathrm{M}}_{\sigma}(s_{i,j})],$$
   so it suffices to prove $\mathrm{Val}^{\mathrm{M}}_{\sigma e}(s_{i,j}) < \mathrm{Val}^{\mathrm{M}}_{\sigma e}(g_1)$. This follows by the previous arguments since $\mathrm{Val}^{\mathrm{M}}_{\sigma e}(s_{i,j}) = \langle s_{i,j}\rangle + \mathrm{Val}^{\mathrm{M}}_{\sigma e}(b_2) < W_1^{\mathrm{M}} + \mathrm{Val}^{\mathrm{M}}_{\sigma e}(b_2) = \mathrm{Val}^{\mathrm{M}}_{\sigma e}(g_1)$.
   b) Now assume $\sigma e(e_{i,j,1-k}) = b_2$. By the same arguments used in case 1(a) this implies $\overline{\sigma e}(eb_{i,j}) \wedge \neg\overline{\sigma e}(eg_{i,j})$. Let $G_n = S_n$ and consider Lemma 6.1.16. Either the conditions of case four or of case five are then fulfilled. If the conditions of case four are true, then $\mathrm{rVal}^{\mathrm{S}}_{\sigma e}(F_{i,j}) = \mathrm{rVal}^{\mathrm{S}}_{\sigma e}(b_2)$. Since $\mathrm{rVal}^{\mathrm{S}}_{\sigma e}(g_1) = W_1^{\mathrm{S}} \cup \mathrm{rVal}^{\mathrm{S}}_{\sigma e}(b_2)$, this implies $\mathrm{rVal}^{\mathrm{S}}_{\sigma e}(g_1) \rhd \mathrm{rVal}^{\mathrm{S}}_{\sigma e}(F_{i,j})$. For the sake of a contradiction, assume that the conditions of case five were true. Then $\overline{\sigma e}(s_{i,j})$ and $\overline{\sigma e}(b_{i+1}) \neq j$. But then Property (USV1)$_i$ implies $j = \beta^{\sigma e}_{i+1}$ and Property (EV1)$_{i+1}$ implies $j \neq \beta_{i+1}$ which is a contradiction. If $G_n = M_n$, then the statement follows directly since $\mathrm{rVal}^{\mathrm{M}}_{\sigma e}(F_{i,j}) = \mathrm{rVal}^{\mathrm{M}}_{\sigma e}(b_2)$.
2. Let $\mu^{\sigma e} \neq 1$. Then, $\mathrm{rVal}^*_{\sigma e}(e_{i,j,k}) = \mathrm{rVal}^*_{\sigma e}(b_2)$ and we prove $\mathrm{rVal}^*_{\sigma e}(F_{i,j}) \prec \mathrm{rVal}^*_{\sigma e}(b_2)$.
   a) Assume $\sigma e(e_{i,j,1-k}) = b_2$. Then $\overline{\sigma e}(eb_{i,j}) \wedge \neg\overline{\sigma e}(eg_{i,j})$ and $\mu^{\sigma e} \neq 1$ by assumption. From Property (EB1) and Property (EV1)$_{i+1}$, it follows that $j \neq \beta_{i+1}$. Let $G_n = S_n$. By Lemma 6.1.16, $\mathrm{rVal}^{\mathrm{S}}_{\sigma e}(F_{i,j}) = \mathrm{rVal}^{\mathrm{S}}_{\sigma e}(s_{i,j})$. Consider the case

$\sigma e(s_{i,j}) = b_1$ first. Then $\mathrm{rVal}^{\mathrm{S}}_{\sigma e}(F_{i,j}) = \{s_{i,j}\} \cup \mathrm{rVal}^{\mathrm{S}}_{\sigma e}(b_1) = \{s_{i,j}\} \cup R^{\mathrm{S}}_1$. Since $\sigma e(b_{\mu^{\sigma e}}) = g_{\mu^{\sigma e}}$ and $W^{\mathrm{S}}_{\mu^{\sigma e}} \rhd \bigcup_{\ell<\mu^{\sigma e}} W^{\mathrm{S}}_\ell \cup \{s_{i,j}\}$, this implies

$$\begin{aligned}\mathrm{rVal}^{\mathrm{S}}_{\sigma e}(F_{i,j}) &= \{s_{i,j}\} \cup R^{\mathrm{S}}_1 \lhd W^{\mathrm{S}}_{\mu^{\sigma e}} \cup \bigcup_{\ell \geq \mu^{\sigma e}+1} \{W^{\mathrm{S}}_\ell \colon \sigma e(b_\ell) = g_\ell\} \\ &= L^{\mathrm{S}}_{\mu^{\sigma e}} = L^{\mathrm{S}}_2 = \mathrm{rVal}^{\mathrm{S}}_{\sigma e}(b_2).\end{aligned}$$

Next let $\sigma e(s_{i,j}) = h_{i,j}$. Then, since $j = 1 - \beta_{i+1}$ and since σe has Property (USV1)$_\ell$ for all $\ell \geq \mu^{\sigma e}$, we must have $i < \mu^{\sigma e}$. Consequently,

$$\mathrm{rVal}^{\mathrm{S}}_{\sigma e}(F_{i,j}) = \{s_{i,j}, h_{i,j}\} \cup \begin{cases}\mathrm{rVal}^{\mathrm{S}}_{\sigma e}(g_{i+1}), & j = 1 \\ \mathrm{rVal}^{\mathrm{S}}_{\sigma e}(b_{i+2}), & j = 0\end{cases}.$$

We now focus on the case $j = 1$ and continue considering the case $G_n = S_n$. Then, $\mathrm{rVal}^{\mathrm{S}}_{\sigma e}(F_{i,j}) = \{s_{i,j}, h_{i,j}\} \cup \mathrm{rVal}^{\mathrm{S}}_{\sigma e}(g_{i+1})$ as Property (BR1) implies $i < \mu^{\sigma e} - 1$. We now determine $\mathrm{rVal}^{\mathrm{S}}_{\sigma e}(g_{i+1})$ using Corollary 6.1.18. By Property (BR1), it holds that $\sigma(g_{i'}) = F_{i',1}$ for all $i' < \mu^{\sigma e} - 1$ as well as $\sigma e(g_{\mu^{\sigma e}-1}) = F_{\mu^{\sigma e}-1,0}$. By assumption, we also have $\sigma e(s_{i',j'}) = h_{i',j'}$ for all $i' < \mu^{\sigma e}$ and $j' \in \{0,1\}$. Since $\sigma(b_{i'}) = b_{i'+1}$ for all $i' < \mu^{\sigma e}$, we obtain $\lambda^{\mathrm{S}}_{i+1} = \mu^{\sigma e} - 1$. By Property (BR2), $\neg\overline{\sigma e}(eg_{\lambda^{\mathrm{S}}_{i+1}})$. But this implies that the conditions of the sixth case of Corollary 6.1.18 are fulfilled, hence

$$\begin{aligned}\mathrm{rVal}^{\mathrm{S}}_{\sigma e}(F_{i,j}) &= \{s_{i,j}, h_{i,j}\} \cup \bigcup_{\ell=i+1}^{\mu^{\sigma e}-1} W^{\mathrm{S}}_\ell \cup \mathrm{rVal}^{\mathrm{S}}_{\sigma e}(b_{\mu^{\sigma e}+1}) \\ &\lhd W^{\mathrm{S}}_{\mu^{\sigma e}} \cup \mathrm{rVal}^{\mathrm{S}}_{\sigma e}(b_{\mu^{\sigma e}+1}) = \mathrm{rVal}^{\mathrm{S}}_{\sigma e}(b_2).\end{aligned}$$

Let $G_n = M_n$ and $j = 1$. It then suffices to prove $\mathrm{Val}^{\mathrm{M}}_{\sigma e}(s_{i,j}) < \mathrm{Val}^{\mathrm{M}}_{\sigma e}(b_2)$. Consider the case $\sigma e(s_{i,j}) = b_1$, implying $\mathrm{rVal}^{\mathrm{M}}_{\sigma e}(s_{i,j}) = \langle s_{i,j}\rangle + \mathrm{rVal}^{\mathrm{M}}_{\sigma e}(b_1)$. If $\mathrm{rVal}^{\mathrm{M}}_{\sigma e}(b_1) = B^{\mathrm{M}}_1$, then $\mathrm{rVal}^{\mathrm{M}}_{\sigma e}(b_1) = R^{\mathrm{M}}_1$ and the arguments follows by the same arguments used for the case $G_n = S_n$. If $\mathrm{rVal}^{\mathrm{M}}_{\sigma e}(b_1) = B^{\mathrm{M}}_2 + \sum_{j'<k} W^{\mathrm{M}}_{j'} + \langle g_k\rangle$ where $k = \min\{i' \geq 1 \colon \neg\overline{\sigma e}(d_{i'})\} < \mu^{\sigma}$, then

$$\mathrm{rVal}^{\mathrm{M}}_{\sigma e}(s_{i,j}) = \langle s_{i,j}, g_k\rangle + \sum_{j'<k} W^k_{j'} + B^{\mathrm{M}}_2 < B^{\mathrm{M}}_2 = \mathrm{rVal}^{\mathrm{M}}_{\sigma e}(b_2).$$

Let $\sigma e(s_{i,j}) = h_{i,j}$. As $j = 1$, it holds that $\mathrm{rVal}^{\mathrm{M}}_{\sigma e}(F_{i,j}) = \langle s_{i,j}, h_{i,j}\rangle + \mathrm{rVal}^{\mathrm{M}}_{\sigma e}(g_{i+1})$. We use Corollary 6.1.17 to evaluate $\mathrm{rVal}^{\mathrm{M}}_{\sigma e}(g_{i+1})$ and thus determine $\lambda^{\mathrm{M}}_{i+1}$. Assume there was some index $i' \in \{i+1, \dots, \mu^{\sigma e}-1\}$ such that $\sigma e(s_{i',\bar{\sigma}(g_{i'})}) = b_1$. Then, by assumption, $\overline{\sigma e}(eb_{i'}) \wedge \neg\overline{\sigma e}(eg_{i'})$. As $\sigma e(b_{i'}) = b_{i'+1}$ by Property (EV1)$_{i'}$, Corollary 6.1.17 then implies the statement.

Hence assume there was no such index. By Property (BR1), $\lambda^{\mathrm{M}}_{i+1} \leq \mu^{\sigma e} - 1$. The case $\overline{\sigma e}(d_{\lambda^{\mathrm{M}}_{i+1}}) \wedge \neg\overline{\sigma e}(s_{\lambda^{\mathrm{M}}_{i+1}})$ cannot happen by assumption. Also, by Property (BR2), $\neg\overline{\sigma e}(eg_{\lambda^{\mathrm{M}}_{i+1}})$. Consequently, either $\overline{\sigma e}(eb_{\lambda^{\mathrm{M}}_{i+1}}) \wedge \neg\overline{\sigma e}(eg_{\lambda^{\mathrm{M}}_{i+1}})$ and the statement follows by the same arguments used before or $\overline{\sigma e}(d_{\lambda^{M}_{i+1}})$. This however implies $\lambda^{M}_{i+1} = \mu^{\sigma e} - 1$ and thus

$$\begin{aligned}\mathrm{rVal}^{M}_{\sigma e}(s_{i,j}) &= \langle s_{i,j,h_{i,j}} \rangle + \sum_{\ell=i+1}^{\mu^{\sigma e}-1} W^{M}_{\ell} + \mathrm{rVal}^{M}_{\sigma e}(b_{\mu^{\sigma e}+1}) \\ &< W^{M}_{\mu^{\sigma e}} + \mathrm{rVal}^{M}_{\sigma e}(b_{\mu^{\sigma e}+1}) = \mathrm{rVal}^{M}_{\sigma e}(b_2).\end{aligned}$$

This concludes the case $j = 1$, hence let $j = 0$. Then, $\beta_{i+1} = 1$, so $i = \mu^{\sigma e} - 1$ and the statement follows from $\mathrm{rVal}^{*}_{\sigma e}(F_{i,j}) = [\![s_{\mu^{\sigma e}-1,0}, h_{\mu^{\sigma e}-1,0}]\!] \oplus L^{*}_{\mu^{\sigma e}+1}$ and $[\![s_{\mu^{\sigma e}-1,0}, h_{\mu^{\sigma e}-1,0}]\!] \prec W^{*}_{\mu^{\sigma e}}$.

b) Now assume $\sigma e(e_{i,j,1-k}) = g_1$. Then $\overline{\sigma e}(eg_{i,j}) \wedge \neg\overline{\sigma e}(eb_{i,j}) \wedge \mu^{\sigma e} \neq 1$. Thus, by Lemmas 6.1.15 and 6.1.16, $\mathrm{rVal}^{*}_{\sigma e}(F_{i,j}) = \mathrm{rVal}^{*}_{\sigma e}(g_1)$. But then Lemma 6.2.18 implies $\mathrm{rVal}^{*}_{\sigma}(F_{i,j}) \prec \mathrm{rVal}^{*}_{\sigma}(b_2)$.

We now show that $\sigma e(d_{i,j,1-k}) = F_{i,j}$ and $j \neq \beta_{i+1}$ also imply $(d_{i,j,k}, e_{i,j,k}) \in I_{\sigma e}$ by proving

$$\mathrm{rVal}^{*}_{\sigma e}(F_{i,j}) \prec \mathrm{rVal}^{*}_{\sigma e}(e_{i,j,k}). \tag{A.10}$$

In this case, $F_{i,j}$ is closed. Thus, by Lemmas 6.1.15 and 6.1.16, $\mathrm{rVal}^{*}_{\sigma e}(F_{i,j}) = \mathrm{rVal}^{*}_{\sigma e}(s_{i,j})$. If $\sigma e(s_{i,j}) = b_1$, then this implies $\mathrm{rVal}^{*}_{\sigma e}(F_{i,j}) = [\![s_{i,j}]\!] \oplus \mathrm{rVal}^{*}_{\sigma e}(b_1)$. If $\mu^{\sigma e} = 1$, then $\mathrm{rVal}^{*}_{\sigma e}(e_{i,j,k}) = \mathrm{rVal}^{*}_{\sigma e}(g_1)$. We then obtain (A.10) as in case 1(a).

If $\mu^{\sigma e} \neq 1$, then $\mathrm{rVal}^{*}_{\sigma e}(e_{i,j,k}) = \mathrm{rVal}^{*}_{\sigma e}(b_2)$ and $\mathrm{rVal}^{*}_{\sigma e}(F_{i,j}) = [\![s_{i,j}]\!] \oplus \mathrm{rVal}^{*}_{\sigma e}(g_1)$. By the same arguments used in case 2 (b), this implies (A.10). Hence let $\sigma e(s_{i,j}) = h_{i,j}$. By the choice of j, this implies that we need to have $\mu^{\sigma e} \neq 1$ due to Property $(\mathrm{USV1})_i$. But then, (A.10) can be shown by the same arguments used for case 2 (a).

Finally, we show that $\sigma e(d_{i,j,1-k}) = F_{i,j}$ and $j = \beta_{i+1}$ imply $I_{\sigma e} = I_{\sigma} \setminus \{e\}$. As only the valuation of $e_{i,j,k}$ can increase, it suffices to show that $(d_{i,j,k}, e_{i,j,k}) \notin I_{\sigma}, I_{\sigma e}$. As $F_{i,j} = F_{i,\beta_{i+1}}$ is closed for σ, this implies $\beta_i = 1$ by the definition of the induced bit state. Thus, by Property (REL2), it follows that $i \geq \nu = \mu^{\sigma} = \mu^{\sigma e}$. Consequently, by Property $(\mathrm{USV1})_i$, Property $(\mathrm{EV1})_{i+1}$ and $\sigma e(d_{i,j,1-k}) = F_{i,j}$, it follows that

$$\mathrm{rVal}^{*}_{\sigma e}(F_{i,j}) = \mathrm{rVal}^{*}_{\sigma e}(s_{i,j}) = [\![s_{i,j}]\!] \oplus \mathrm{rVal}^{*}_{\sigma e}(h_{i,j}) = [\![s_{i,j}, h_{i,j}]\!] \oplus \mathrm{rVal}^{*}_{\sigma e}(b_{i+1}).$$

We distinguish the following cases.

1. Let $\mu^{\sigma e} = 1$. Then $\sigma e(e_{i,j,k}) = g_1$, hence $\mathrm{rVal}^{*}_{\sigma e}(e_{i,j,k}) = W^{*}_1 \oplus \mathrm{rVal}^{*}_{\sigma e}(b_2) = W^{*}_1 \oplus L^{*}_2$ by Lemma 6.2.18. In this case, $\mu^{\sigma e} = 1$ implies

$$\mathrm{rVal}^{*}_{\sigma e}(F_{i,j}) = [\![s_{i,j}, h_{i,j}]\!] \oplus L^{*}_{i+1} \succ \bigoplus_{i'=1}^{i} W^{*}_{i'} \oplus L^{*}_{i+1} \succeq W^{*}_1 \oplus L^{*}_2 = \mathrm{rVal}^{*}_{\sigma e}(e_{i,j,k})$$

if $i \neq 1$ whereas $i = 1$ implies

$$\mathrm{rVal}^{*}_{\sigma e}(F_{i,j}) = [\![s_{1,j}, h_{1,j}]\!] \oplus \mathrm{rVal}^{*}_{\sigma e}(b_2) \succ W^{*}_1 \oplus \mathrm{rVal}^{*}_{\sigma e}(b_2) = \mathrm{rVal}^{*}_{\sigma e}(e_{i,j,k}).$$

Hence $\mathrm{rVal}^{*}_{\sigma e}(F_{i,j}) \succ \mathrm{rVal}_{\sigma e^{*}}(e_{i,j,k})$ in either case, so $(d_{i,j,k}, e_{i,j,k}) \notin I_{\sigma e}$.

2. Let $\mu^{\sigma e} \neq 1$. Then, $\sigma e(e_{i,j,k}) = b_2$, hence $\operatorname{rVal}^*_{\sigma e}(e_{i,j,k}) = \operatorname{rVal}^*_{\sigma e}(b_2) = L^*_2$ by Lemma 6.2.18. If $\operatorname{rVal}^*_{\sigma e}(b_{i+1}) = L^*_{i+1}$, then the statement follows from

$$\operatorname{rVal}^*_{\sigma e}(F_{i,j}) = [\![s_{i,j}, h_{i,j}]\!] \oplus L^*_{i+1} \succ \bigoplus_{\ell=2}^{i} W_\ell \oplus L^*_{i+1} \succ L^*_2.$$

If $\operatorname{rVal}^*_{\sigma e}(b_{i+1}) = R^*_{i+1}$, then $\sigma e(b_{i+1}) = g_{i+1}$ and $i+1 < \mu^{\sigma e}$. But then, by Property (EV1)$_{i+1}$, $\beta_{i+1} = 1$ contradicting $\mu^{\sigma e} = \nu$.

Since these arguments can be applied to σ analogously, $(d_{i,j,k}, e_{i,j,k}) \notin I_\sigma, I_{\sigma e}$ and the statement follows. □

Lemma 6.2.20. *Let $\sigma \in \rho(\sigma_0)$ be a well-behaved phase-3-strategy for $\mathfrak{b}$. Let $i \in [n]$ and $j,k \in \{0,1\}$ such that $\sigma(e_{i,j,k}) = t^{\rightarrow}$ and $e := (d_{i,j,k}, e_{i,j,k}) \in I_\sigma$. Let $\sigma(d_{i,j,1-k}) = e_{i,j,1-k}$ or $[\sigma(d_{i,j,1-k}) = F_{i,j}$ and $j \neq \beta^\sigma_{i+1}]$. Then σe is a well-behaved phase-3-strategy for $\mathfrak{b}$ with $\sigma e \in \rho(\sigma_0)$.*

Proof. We first show that σe is a phase-3-strategy for $\mathfrak{b}$. If $F_{i,j}$ is halfopen, then the application of e can only influence Properties (ESC1) and (ESC2). In that case, there is nothing to show as σe does not need to fulfill these properties. If $F_{i,j}$ is closed for σ, then $j \neq \beta^\sigma_{i+1}$ by assumption. As $j \neq \beta^{\sigma e}_{i+1}$, Property (EV1)$_i$ remains valid. Consequently, σe is a phase-3-strategy for $\mathfrak{b}$ and in particular $\mu^{\sigma e} = \nu = \mu^\sigma$. We thus skip the upper index σ resp. σe when referring to the induced bit status as $\mathfrak{b}+1 = \beta^\sigma = \beta^{\sigma e}$.

We prove that σe is well-behaved. We need to consider all properties related to escape vertices where the premise might become true or where the conclusion might become false.

(BR2) This property only needs to be checked if the conclusion becomes false. Thus, assume $i < \mu^{\sigma e}$, implying $\mu^{\sigma e} > 1$. But then, $e = (e_{i,j,k}, b_2)$, hence it cannot happen that the conclusion becomes false by applying the switch e.

(D2) By Property (REL2),$\sigma e(b_2) = g_2$ implies $\mu^{\sigma e} \leq 2$, hence the premise cannot be fulfilled.

(MNS1) Assume $\mu^{\sigma e} = 1 \wedge m^{\sigma e}_b \leq \overline{m}^{\sigma e}_s, \overline{m}^{\sigma e}_g$ and $G_n = S_n$. By Property (REL2), it holds that $\mu^{\sigma e} = \nu = 1$. Since σe has Property (CC2), this implies that the cycle center $d_{1,\overline{\sigma e}(g_1)}$ is closed and thus $\neg\overline{\sigma e}(eb_1)$.

(MNS2) Assume $\mu^{\sigma e} = 1$ and let $i < \overline{m}^{\sigma e}_g < \overline{m}^{\sigma e}_s, m^{\sigma e}_b$. If $\sigma e(b_2) = g_2$, then $m^{\sigma e}_b = 2$ by definition and there cannot be an index fulfilling the conditions of the premise. If $\sigma e(b_2) = b_3$, then $\beta_2 = 0$ by Property (EV1)$_2$, implying $\sigma e(g_1) = F_{1,0}$ by Property (CC2). But then $\overline{m}^{\sigma e}_g = 1$, hence there cannot be an index such that the conditions of the premise are fulfilled.

(MNS3) This follows by the same arguments used for Property (MNS2).

(MNS4) Assume $\mu^{\sigma e} = 1 \wedge \overline{m}^{\sigma e}_s \leq \overline{m}^{\sigma e}_g < m^{\sigma e}_b$. If $m^{\sigma e}_b = 2$, then $\overline{m}^{\sigma e}_s = \overline{m}^{\sigma e}_g = 1$, implying $\sigma e(g_1) = F_{1,0}$ and $\sigma e(s_{1,0}) = b_1$. But this contradicts Property (CC2) since Property (EV1)$_2$ implies $\overline{\sigma e}(g_1) = \beta_2 = \overline{\sigma e}(b_2) = 1$. Thus assume $m^{\sigma e}_b > 3$, implying $\sigma e(b_2) = b_3$. Then, by Property (CC2), $\sigma e(g_1) = F_{1,0}$ and $\sigma e(s_{1,0}) = h_{1,0}$ by Property (USV1)$_1$. But this implies $1 = \overline{m}^{\sigma e}_g < \overline{m}^{\sigma e}_s$, contradicting the premise.

(MNS5) Assume $\mu^{\sigma e} = 1 \wedge i < \overline{m}_s^{\sigma e} < m_b^{\sigma e} \leq \overline{m}_g^{\sigma e}$. Since no such i can exist if $m_b^{\sigma e} = 2$, assume $m_b^{\sigma e} > 2$. This in particular implies $\sigma e(b_2) = b_3$, hence $\sigma e(g_1) = F_{1,0}$ by Property (CC2). This implies $\overline{m}_g^{\sigma e} = 1$, contradicting the assumption.

(MNS6) If $m_b^{\sigma e} > 2$, then the same arguments used for Property (MNS5) can be applied again. Hence assume $m_b^{\sigma e} = 2$, implying $\sigma e(b_2) = g_2$. Then, by Property (CC2), $\sigma e(g_1) = F_{1,1}$ and $\sigma e(s_{1,1}) = h_{1,1}$. But this implies $\overline{m}_s^{\sigma e} > 2$, contradicting $\overline{m}_s^{\sigma e} < m_b^{\sigma e}$.

(EG1) This property only needs to be considered if its premise is incorrect for σ but correct for σe. Therefore, since $\mu^{\sigma} = \mu^{\sigma e} = 1$ implies that the switch $(e_{i,j,k}, g_1)$ was applied, we need to have $\neg\overline{\sigma}(eg_{i,j}) \wedge \neg\overline{\sigma}(eb_{i,j})$. This implies $\overline{\sigma}(d_{i,j})$, hence $j \neq \beta_{i+1}$ by assumption, implying $\overline{\sigma e}(s_{i,j}) = b_1$ by Property (USV1)$_i$.

(EG2) If $\mu^{\sigma e} = 1$, then $\overline{\sigma e}(d_1) = \overline{\sigma e}(d_{\mu^{\sigma e}})$ by Property (CC2) and the choice of e, so the implication is correct.

(EG3) The premise of this property can only become true if $\mu^{\sigma e} = 1$ as we need to apply the switch $(e_{i,j,k}, g_1)$. Thus, Property (CC2) implies that $\overline{\sigma e}(d_1), \overline{\sigma e}(g_1) = \beta_{i+1}$ and hence, by Property (USV1)$_1$, $\overline{\sigma e}(s_1)$.

(EG4) By Property (EV1)$_2$, it holds that $\overline{\sigma e}(b_2) = \beta_2$ and by Property (CC2) and the premise we have $\overline{\sigma e}(g_{\mu^{\sigma e}}) = \overline{\sigma e}(g_1) = \beta_2$.

(EG5) The premise of this implication cannot become correct since $\overline{\sigma e}(eg_{i,j}) \wedge \neg\overline{\sigma e}(eb_{i,j})$ imply $\mu^{\sigma e} = 1$.

(EB1) Assume that the conditions of the premise were fulfilled. Then $\overline{\sigma e}(eb_{i,j}) \wedge \neg\overline{\sigma e}(eg_{i,j})$. If also $\overline{\sigma}(eb_{i,j}) \wedge \neg\overline{\sigma}(eg_{i,j})$, the statement follows since σ is well-behaved. Hence suppose that this is not the case. Then, $\overline{\sigma}(d_{i,j})$, implying that $j \neq \beta_{i+1} = \overline{\sigma e}(b_{i+1})$ by assumption and Property (EV1)$_{i+1}$.

(EB2) If the premise is true for σ, then there is nothing to prove. Hence assume that it is incorrect for σ. Then, either $\overline{\sigma}(eb_{i,0}) \wedge \overline{\sigma}(eg_{i,0})$ or $\overline{\sigma}(d_{i,0})$. In the first case, Property (EBG1) (applied to σ) yields $\overline{\sigma}(b_{i+1}) = 0$. This is a contradiction to Property (EB1) (applied to σe) since it implies $\overline{\sigma}(b_{i+1}) = \overline{\sigma e}(b_{i+1}) \neq 0$. Consequently, $\overline{\sigma}(d_{i,0})$, implying $0 \neq \beta_{i+1}$ and thus $1 = \beta_{i+1}$. We now show that $i < \mu^{\sigma e}$, implying $\mu^{\sigma e} = i+1$ since $\mu^{\sigma e} = \nu$ by Property (REL2). For the sake of a contradiction assume $i \geq \mu^{\sigma e}$. Then, by Property (USV1)$_i$ and Property (EV1)$_{i+1}$, $\overline{\sigma e}(b_{i+1}) = \beta_{i+1} = 0$. However, by Property (EB1), also $\overline{\sigma e}(b_{i+1}) \neq 0$ which is a contradiction.

(EB3) As before, there is nothing to prove if the premise is also correct for σ. By the same arguments used for Property (EB2), we can deduce that assuming $\overline{\sigma}(eb_{i,j}) \wedge \overline{\sigma}(eg_{i,j})$ yields a contradiction. Hence, $\overline{\sigma}(d_{i,j})$, implying $j \neq \beta_{i+1}^{\sigma e}$ by assumption. If $\mu^{\sigma e} > 2$, then $\sigma e(b_2) = b_3$ follows by Property (EV1)$_2$. Hence assume $\mu^{\sigma e} = 2$, implying $i \geq \mu^{\sigma e}$. Consequently, σe has Property (USV1)$_i$, implying $j = \beta_{i+1}$ since $\sigma e(s_{i,j}) = h_{i,j}$ by assumption. This is however a contradiction to the choice of j.

(EB4) If the premise is true for σ, then there is nothing to prove. Hence assume that it is incorrect for σ. By the same arguments used earlier, we deduce $\overline{\sigma}(d_{i,1})$, implying $1 \neq \beta_{i+1}$ and thus $0 = \beta_{i+1}$. We now show that $i < \mu^{\sigma e}$, implying

$\mu^{\sigma e} > i+1$ since $\mu^{\sigma e} = \nu$ by Property (REL2). Towards a contradiction assume $i \geq \mu^{\sigma e}$. Then, by Property (USV1)$_i$ and Property (EV1)$_{i+1}$, $\overline{\sigma e}(b_{i+1}) = \beta_{i+1} = 0$. However, by Property (EB1), also $\overline{\sigma e}(b_{i+1}) \neq 0$ which is a contradiction.

(EB6) By Property (REL2), $\mu^{\sigma e} > 2$ implies $\nu > 2$. Hence $\beta_2 = 0$, implying $\sigma e(b_2) = b_3$ by Property (EV1)$_2$.

(EB5) By Lemma 6.1.4 it suffices to show $\mathcal{I}^{\sigma e} \neq \emptyset$. Consider the case $\mu^{\sigma e} > 2$ first. Then $\sigma e(b_2) = b_3$ by Property (EB6), implying $\mu^{\sigma e} \neq \min\{i' \colon \sigma e(b_{i'}) = b_{i'+1}\}$. Hence $\mathcal{I}^{\sigma e} \neq \emptyset$ in this case. Now consider the case $\mu^{\sigma e} = 2$. Then, by assumption, $\sigma e(b_1) = g_1$ and $\sigma e(b_2) = g_2$ by Property (REL2) and Property (EV1)$_2$. Furthermore, by Property (BR1), $\overline{\sigma e}(g_1) = F_{1,0}$, hence $\mathcal{I}^{\sigma e} \neq \emptyset$.

(EBG1) If $i \geq \mu^{\sigma e}$, then Property (USV1)$_i$ and $\sigma e(s_{i,j}) = h_{i,j}$ imply that $j = \beta_{i+1}$. Hence $j = \overline{\sigma e}(b_{i+1})$ by Property (EV1)$_{i+1}$. Thus assume $i < \mu^{\sigma e}$, implying $\mu^{\sigma e} > 1$. Therefore, the switch $(e_{i,j,k}, b_2)$ was applied, implying $\bar{\sigma}(eg_{i,j}) \wedge \neg\bar{\sigma}(eb_{i,j})$. But then, $\overline{\sigma e}(b_{i+1}) = \bar{\sigma}(b_{i+1}) = j$ by Property (EG5).

(EBG2) By assumption and Property (EV2)$_2$, it follows that $\overline{\sigma e}(g_1) = \overline{\sigma e}(b_2) = \beta_2$. Thus, $\overline{\sigma e}(s_1) = \overline{\sigma e}(s_{1,\overline{\sigma e}(g_1)}) = \overline{\sigma e}(s_{1,\beta_2})$ and hence, by either Property (USV2)$_{1,\beta_2}$ or Property (USV1), $\overline{\sigma e}(s_1) = 1$.

(EBG3) If $\mu^{\sigma e} = 1$, then $\overline{\sigma e}(d_1)$ follows from Property (CC2). Thus assume $\mu^{\sigma e} > 1$, implying $\sigma e(b_1) = g_1$. Towards a contradiction, assume that the cycle center $F_{1,\beta_2} = F_{1,\overline{\sigma e}(g_1)}$ was not closed. Since the game is a sink game resp. weakly unichain, the cycle center cannot escape towards g_1 since player 1 could then create a cycle in S_n resp. since M_n would not have the weak unichain condition. Thus, by assumption, $\overline{\sigma e}(eb_{1,\overline{\sigma e}(g_1)}) \wedge \neg\overline{\sigma e}(eg_{1,\overline{\sigma e}(g_1)}) \wedge \sigma e(b_1) = g_1$. But then, Property (EB1) implies $\overline{\sigma e}(g_1) \neq \overline{\sigma e}(b_2)$, contradicting the assumption.

(EBG4) The assumptions $\overline{\sigma e}(b_2) = g_2$ and $\overline{\sigma e}(g_1) = F_{1,0}$ imply $\mu^{\sigma e} = 2$ if $\sigma e(b_1) = g_1$. If this is not the case, we have $\overline{\sigma e}(b_1) = b_2$, implying $\mu^{\sigma e} = 1$.

(EBG5) By assumption $\overline{\sigma e}(g_1) = F_{1,1}$ and $\overline{\sigma e}(b_2) = b_3$. Towards a contradiction assume $\mu^{\sigma e} = 2$. Then we need to have $\sigma e(b_1) = g_1$ and $\mu^{\sigma e} = \min\{i' \colon \sigma e(b_{i'}) = b_{i'+1}\}$. But then $\overline{\sigma e}(b_1) \wedge \overline{\sigma e}(g_1) \neq \overline{\sigma e}(b_2)$, implying $\mathcal{I}^{\sigma e} \neq \emptyset$, contradicting Lemma 6.1.4.

(D1) Assume $i \neq 1$. Then, by Property (EV1)$_i$, $\sigma e(b_i) = g_i$ implies $\beta_i = 1$ and $\overline{\sigma e}(g_i) = \beta_{i+1}$ by Property (EV2)$_i$. Since we only open inactive cycle centers, there is nothing to show in this case. Hence assume $i = 1$. If $\mu^{\sigma e} = 1$, then $\sigma e(b_1) = b_2$, hence the premise is incorrect. Thus assume $\mu^{\sigma e} > 1$. This implies $t^{\rightarrow} = b_2$. In particular, $\overline{\sigma e}(eb_{i,j}) \wedge \neg\overline{\sigma e}(eg_{i,j})$, hence $\mu^{\sigma e} > 2$ implies $\sigma e(b_2) = b_3$ by Property (EB6). Thus, $\mu^{\sigma e} > 2 \not\Leftrightarrow \sigma e(b_2) = g_2$, so the premise is incorrect.

Hence σe is well-behaved. □

Lemma 6.2.21 (Second row of Table 6.6). *Let $G_n = S_n$. Let $\sigma \in \rho(\sigma_0)$ be a well-behaved phase-3-strategy for $\mathfrak{b} \in \mathfrak{B}_n$. Let $i \in [n], j,k \in \{0,1\}$ such that $e := (d_{i,j,k}, e_{i,j,k}) \in I_\sigma$ and $\sigma(e_{i,j,k}) = t^{\rightarrow}$. Further assume that $\bar{\sigma}(d_{i,j})$ implies $j \neq \beta^\sigma_{i+1}$. Then σe is a well-behaved phase-3-strategy for $\mathfrak{b}$ with $\sigma e \in \rho(\sigma_0)$ and $I_{\sigma e} = I_\sigma \setminus \{e\}$.*

Proof. By Lemma 6.2.20, it suffices to prove $I_{\sigma e} = I_\sigma \setminus \{e\}$. By construction, $d_{i,j,k}$ is the only vertex having an edge to $e_{i,j,k}$ and $F_{i,j}$ is the only vertex having an edge towards

$d_{i,j,k}$. It thus suffices to show that the valuation of $F_{i,j}$ does not change when applying e, so we prove $\mathrm{Val}^{\mathrm{S}}_{\sigma}(F_{i,j}) = \mathrm{Val}^{\mathrm{S}}_{\sigma e}(F_{i,j})$. Since $\sigma(d_{i,j,k}) \neq e_{i,j,k}$ implies that we cannot have $\bar{\sigma}(eg_{i,j}) \wedge \bar{\sigma}(eb_{i,j})$, it suffices to distinguishing the following cases.

- Let $\bar{\sigma}(d_{i,j})$, implying $\mathrm{rVal}^{\mathrm{S}}_{\sigma e}(F_{i,j}) = \mathrm{rVal}^{\mathrm{S}}_{\sigma e}(s_{i,j})$ by Lemma 6.1.16. If $t^{\rightarrow} = b_2$, then $\overline{\sigma e}(eb_{i,j}) \wedge \neg\overline{\sigma e}(eg_{i,j})$ and $\mu^{\sigma e} \neq 1$. If $t^{\rightarrow} = g_1$, then $\overline{\sigma e}(eg_{i,j}) \wedge \neg\overline{\sigma e}(eb_{i,j})$ and $\mu^{\sigma e} = 1$. This however yields $\tau^{\sigma e}(F_{i,j}) = s_{i,j}$ in either case by Lemma 6.1.16, implying the statement.
- Let $\bar{\sigma}(eg_{i,j}) \wedge \neg\bar{\sigma}(eb_{i,j})$. Since the target of $F_{i,j}$ does not change when applying e if $\overline{\sigma e}(eg_{i,j}) \wedge \neg\overline{\sigma e}(eb_{i,j})$ we can assume $\overline{\sigma e}(eg_{i,j}) \wedge \overline{\sigma e}(eb_{i,j})$. This implies $t^{\rightarrow} = b_2$, so $\mu^{\sigma e} \neq 1$, implying $\tau^{\sigma e}(F_{i,j}) = g_1$ by Lemma 6.1.16. It now suffices to show $\overline{\sigma e}(g_1) \neq \overline{\sigma e}(b_2)$. For the sake of contradiction, assume $\overline{\sigma e}(g_1) = \overline{\sigma e}(b_2)$. Then, by Property (EBG3), $F_{1,\overline{\sigma e}(g_1)}$ is closed. Thus, by the definition of β, it holds that$1 = \overline{\sigma e}(d_{1,\beta_2}) = \beta_1$, implying $\nu = 1$. But this contradicts $\mu^{\sigma e} \neq 1$ by Property (REL2).
- Let $\bar{\sigma}(eb_{i,j}) \wedge \neg\overline{\sigma e}(eg_{i,j})$. By the same arguments used in the last case we can assume $\overline{\sigma e}(eb_{i,j}) \wedge \overline{\sigma e}(eg_{i,j})$. Hence, $t^{\rightarrow} = g_1$ and $\mu^{\sigma} = \mu^{\sigma e} = 1$. By Property $(\mathrm{USV1})_i$ and Property $(\mathrm{EV1})_{i+1}$, we have $\bar{\sigma}(s_{i,j}) = b_1$ if $j \neq \beta_{i+1}$ or $\bar{\sigma}(b_{i+1}) = j$ if $j = \beta_{i+1}$. In either case, $\tau^{\sigma}(F_{i,j}) = b_2$ by Lemma 6.1.16. It hence suffices to show $\overline{\sigma e}(g_1) = \overline{\sigma e}(b_2)$. This however follows from Property (CC2) since $\mu^{\sigma e} = 1$. □

Lemma 6.2.22 (Third row of Table 6.6). *Let $G_n = M_n$ and let $\sigma \in \rho(\sigma_0)$ be a well-behaved phase-3-strategy for $\mathfrak{b}$. Let $i \in [n]$ with $\beta_i = 1$ and let $j := 1 - \beta_{i+1}$. Let $e := (d_{i,j,k}, e_{i,j,k}) \in I_\sigma$ and $\sigma(e_{i,j,k}) = t^{\rightarrow}$ for some $k \in \{0,1\}$. Then σe is a well-behaved phase-3-strategy for $\mathfrak{b}$ with $\sigma e \in \rho(\sigma_0)$ and $I_{\sigma e} = I_\sigma \setminus \{e\}$.*

Proof. By Lemma 6.2.20, it suffices to prove $I_{\sigma e} = I_\sigma \setminus \{e\}$.

By Property (REL2), $\nu = \mu^{\sigma} = \mu^{\sigma e}$, so in particular $i \geq \mu^{\sigma e}$. By Property $(\mathrm{EV2})_i$, we have $\bar{\sigma}(g_i) = 1 - j = \beta_{i+1}$. We thus begin by showing $(g_i, F_{i,j}) \notin I_\sigma, I_{\sigma e}$. This is done by showing

$$\begin{aligned} \mathrm{Val}_{\sigma}(F_{i,1-j}) &= \mathrm{Val}_{\sigma e}(F_{i,1-j}), \\ \mathrm{Val}_{\sigma e}(F_{i,1-j}) &> \mathrm{Val}_{\sigma e}(F_{i,j}) \end{aligned} \tag{A.11}$$

which suffices as $\mathrm{Val}_{\sigma e}(F_{i,j}) \geq \mathrm{Val}_{\sigma}(F_{i,j})$. Since $\beta_i = 1$ implies $i \geq \nu$, we have $\bar{\sigma}(d_{i,1-j})$ by Property $(\mathrm{EV1})_i$ and $\sigma(s_{i,1-j}) = h_{i,1-j}$ by Property $(\mathrm{USV1})_i$. By Property $(\mathrm{EV1})_{i+1}$, this yields

$$\mathrm{Val}_{\sigma}(F_{i,1-j}) = \langle s_{i,1-j}, h_{i,1-j} \rangle + \mathrm{Val}_{\sigma}(b_{i+1}) = \langle s_{i,1-j}, h_{i,1-j} \rangle + L_{i+1}.$$

This implies $\mathrm{Val}_{\sigma}(F_{i,1-j}) = \mathrm{Val}_{\sigma e}(F_{i,1-j})$. Since $F_{i,j}$ is not closed for σe, it either escapes to $t^{\rightarrow}$ or is mixed. Consequently, it either holds that $\mathrm{rVal}_{\sigma e}(F_{i,j}) = \mathrm{rVal}_{\sigma e}(t^{\rightarrow})$ or $\mathrm{rVal}_{\sigma e}(F_{i,j}) = \frac{1}{2}\mathrm{rVal}_{\sigma e}(t^{\rightarrow}) + \frac{1}{2}\mathrm{rVal}_{\sigma e}(t^{\leftarrow})$. As $\mathrm{rVal}_{\sigma e}(t^{\rightarrow}) > \mathrm{rVal}_{\sigma e}(t^{\leftarrow})$ by Lemma 6.2.18, we have $\mathrm{rVal}_{\sigma e}(F_{i,j}) \leq \mathrm{rVal}_{\sigma e}(t^{\rightarrow})$. Let $t^{\rightarrow} = b_2$. Since $\sigma(b_2) = g_2$ implies $\nu = 2$ by Property $(\mathrm{EV1})_2$, we have $\mathrm{rVal}_{\sigma e}(b_2) = L_2$ in any case. Consequently,

$$\mathrm{rVal}_{\sigma e}(F_{i,j}) \leq \mathrm{rVal}_{\sigma e}(b_2) = L_2 = L_{2,i} + L_{i+1} < \langle s_{i,1-j}, h_{i,1-j} \rangle + L_{i+1} = \mathrm{rVal}_{\sigma e}(F_{i,1-j}).$$

Let $t^{\rightarrow} = g_1$. Then, $\mathrm{rVal}_{\sigma e}(g_1) = W_1 + \mathrm{rVal}_{\sigma e}(b_2)$ by Lemma 6.2.18. Consequently,

$$\begin{aligned}\mathrm{rVal}_{\sigma e}(F_{i,j}) &\leq W_1 + \mathrm{rVal}_{\sigma e}(b_2) = W_1 + L_2 = W_1 + L_{2,i} + L_{i+1}\\ &< \langle s_{i,1-j}, h_{i,1-j}\rangle + L_{i+1} = \mathrm{rVal}_{\sigma e}(F_{i,1-j}).\end{aligned}$$

Thus, $(g_i, F_{i,j}) \notin I_\sigma, I_{\sigma e}$ as claimed.

We now consider the cycle edges of $F_{i,j}$. First, $(d_{i,j,k}, F_{i,j}) \notin I_{\sigma e}$ as $(d_{i,j,k}, e_{i,j,k})$ was just applied. If $\sigma(d_{i,j,1-k}) = F_{i,j}$, then also $(d_{i,j,1-k}, F_{i,j}) \notin I_\sigma, I_{\sigma e}$, implying $I_{\sigma e} = I_\sigma \setminus \{e\}$ since the valuation of no further vertex changes due to $\sigma(g_i) = F_{i,1-j}$. Hence assume $\sigma(d_{i,j,1-k}) = e_{i,j,1-k}$. We prove $(d_{i,j,1-k}, F_{i,j}) \in I_\sigma \Leftrightarrow (d_{i,j,1-k}, F_{i,j}) \in I_{\sigma e}$ which suffices to prove the statement since no other vertex but $F_{i,j}$ has an edge to $d_{i,j,1-k}$.

Let $v := \sigma(e_{i,j,1-k}) = \sigma e(e_{i,j,1-k})$. We prove that $i \geq \mu^{\sigma e}$ and $\sigma(g_i) = \sigma e(g_i) = F_{i,1-j}$ imply $\mathrm{Val}_\sigma(v) = \mathrm{Val}_{\sigma e}(v)$. If $v = b_2$, then $\mathrm{Val}_\sigma(v) = \mathrm{Val}_{\sigma e}(v) = L_2$ by Lemma 6.2.18. If $v = g_1$, then either $\mathrm{rVal}_\sigma(v) = \mathrm{rVal}_{\sigma e}(v) = \langle g_\ell\rangle + \sum_{i' \in [\ell-1]} W_{i'} + \mathrm{rVal}_\sigma(b_2)$ where $\ell = \min\{i' \geq 1 \colon \neg\bar{\sigma}(d_{i'})\} < \mu^\sigma$ or $\mathrm{rVal}_\sigma(v) = \mathrm{rVal}_{\sigma e}(v) = R_1$. We furthermore observe that $\bar{\sigma}(g_i) = \overline{\sigma e}(g_i) \neq j$ implies $\min\{i' \geq 1 \colon \neg\bar{\sigma}(d_{i'})\} = \min\{i' \geq 1 \colon \neg\overline{\sigma e}(d_{i'})\}$. Thus, if $\mathrm{Val}_\sigma(F_{i,j}) > \mathrm{Val}_\sigma(v)$, then also $\mathrm{Val}_{\sigma e}(F_{i,j}) > \mathrm{Val}_{\sigma e}(v)$, so

$$(d_{i,j,1-k}, F_{i,j}) \in I_\sigma \Rightarrow (d_{i,j,1-k}, F_{i,j}) \in I_{\sigma e}.$$

Hence assume $(d_{i,j,1-k}, F_{i,j}) \notin I_\sigma$, implying $\mathrm{Val}_\sigma(F_{i,j}) \leq \mathrm{Val}_\sigma(v)$. Since $\sigma(d_{i,j,k}) = F_{i,j}$ and $\sigma(d_{i,j,1-k}) = e_{i,j,1-k}$, we have

$$\mathrm{Val}_\sigma(F_{i,j}) - \mathrm{Val}_\sigma(v) = \frac{2\varepsilon}{1+\varepsilon}[\langle s_{i,j}\rangle + \mathrm{Val}_\sigma(b_1) - \mathrm{Val}_\sigma(v)] \leq 0.$$

Thus, $\mathrm{Val}_\sigma(b_1) + \langle s_{i,j}\rangle \leq \mathrm{Val}_\sigma(v)$, hence $\mathrm{Val}_\sigma(b_1) < \mathrm{Val}_\sigma(v)$. Since $\mathrm{Val}_\sigma(t^{\leftarrow}) < \mathrm{Val}_\sigma(t^{\rightarrow})$ by Lemma 6.2.18, this implies that $v = t^{\rightarrow}$ and $\sigma(b_1) = t^{\leftarrow}$ have to hold. As it holds that $\mathrm{Val}_\sigma(t^{\leftarrow}) = \mathrm{Val}_{\sigma e}(t^{\leftarrow})$, this then implies

$$\begin{aligned}\mathrm{Val}_{\sigma e}(F_{i,j}) - \mathrm{Val}_{\sigma e}(v) &= (1-\varepsilon)\,\mathrm{Val}_{\sigma e}(t^{\rightarrow}) + \varepsilon\,\mathrm{Val}_{\sigma e}(s_{i,j}) - \mathrm{Val}_{\sigma e}(t^{\rightarrow})\\ &= \varepsilon[\langle s_{i,j}\rangle + \mathrm{Val}_{\sigma e}(b_1) - \mathrm{Val}_{\sigma e}(t^{\rightarrow})] \leq 0,\end{aligned}$$

hence $(d_{i,j,1-k}, F_{i,j}) \notin I_{\sigma e}$. □

Lemma 6.2.23 (Fourth row of Table 6.6). *Let $G_n = M_n$. Let $\sigma \in \rho(\sigma_0)$ be a well-behaved phase-3-strategy for $\mathfrak{b} \in \mathfrak{B}_n$. Let $i \in [n]$ with $\beta_i^\sigma = 0, j := \beta_{i+1}^\sigma$ and let $F_{i,j}$ be $t^{\leftarrow}$-halfopen. Let $F_{i,1-j}$ be $t^{\rightarrow}$-open and $\sigma(g_i) = F_{i,1-j}$. Let $e := (d_{i,j,k}, e_{i,j,k}) \in I_\sigma$ and $\sigma(e_{i,j,k}) = t^{\rightarrow}$ with $k \in \{0,1\}$. Then σe is a well-behaved phase-3-strategy for $\mathfrak{b}$ with $\sigma e \in \rho(\sigma_0)$ and $I_{\sigma e} = I_\sigma \setminus \{e\}$.*

Proof. Since $F_{i,j}$ is $t^{\leftarrow}$-halfopen, the choice of e implies $\sigma(d_{i,j,1-k}) = e_{i,j,1-k}$. Consequently, by Lemma 6.2.20, it suffices to prove $I_{\sigma e} = I_\sigma \setminus \{e\}$.

Since $\sigma(g_i) = F_{i,1-j}$, the application of e can only increase the valuation of $F_{i,j}, d_{i,j,0}$ and $d_{i,j,1}$. In addition, and since there are no player 0 vertices v with $(v, d_{i,j,*}) \in E_0$. It thus suffices to prove $(g_i, F_{i,j}) \notin I_\sigma, I_{\sigma e}$. This however follows from Lemma 6.2.18 since

$$\mathrm{rVal}_\sigma(F_{i,1-j}) = \mathrm{rVal}_\sigma(t^{\rightarrow}) > \mathrm{rVal}_\sigma(t^{\leftarrow}) = \mathrm{rVal}_\sigma(F_{i,j})$$

and

$$\mathrm{rVal}_{\sigma e}(F_{i,1-j}) = \mathrm{rVal}_{\sigma e}(t^{\rightarrow}) > \frac{1}{2}\mathrm{rVal}_{\sigma e}(t^{\rightarrow}) + \frac{1}{2}\mathrm{rVal}_{\sigma e}(t^{\leftarrow}) = \mathrm{rVal}_{\sigma e}(F_{i,j}).$$

□

Lemma 6.2.24. *Let $G_n = M_n$. Let $\sigma \in \rho(\sigma_0)$ be a well-behaved phase-3-strategy for $\mathfrak{b} \in \mathfrak{B}_n$. Let $i \geq \mu^\sigma + 1$ and assume $\bar{\sigma}(g_i) = \beta^\sigma_{i+1}$.*

1. If both cycle centers of level i are $t^{\leftarrow}$-halfopen, then let $j := \bar{\sigma}(g_i)$.

2. If $F_{i,\beta^\sigma_{i+1}}$ is mixed and $F_{i,1-\beta^\sigma_{i+1}}$ is $t^{\leftarrow}$-halfopen, then let $j := 1 - \bar{\sigma}(g_i)$.

In any case, assume $e := (d_{i,j,k}, e_{i,j,k}) \in I_\sigma$ and $\sigma(e_{i,j,k}) = t^{\rightarrow}$ for $k \in \{0,1\}$. Then, σe is a well-behaved phase-3-strategy for $\mathfrak{b}$ with $\sigma e \in \rho(\sigma_0)$ and $I_{\sigma e} = I_\sigma \setminus \{e\}$.

Proof. In both cases, $e = (d_{i,j,k}, e_{i,j,k})$ is applied within a $t^{\leftarrow}$-halfopen cycle center. This implies $\sigma(d_{i,j,1-k}) = e_{i,j,1-k}$, hence, by Lemma 6.2.20, it suffices to prove $I_{\sigma e} = I_\sigma \setminus \{e\}$.

Let both cycle centers be $t^{\leftarrow}$-halfopen for σ and let $j := \bar{\sigma}(g_i) = \beta_{i+1}$. By Lemma 6.2.1, $\mathrm{Val}_\sigma(F_{i,j}) > \mathrm{Val}_\sigma(F_{i,1-j})$ and by Lemma 6.2.18, also $\mathrm{Val}_{\sigma e}(F_{i,j}) > \mathrm{rVal}_{\sigma e}(F_{i,1-j})$. Thus, $(g_i, F_{i,1-j}) \notin I_\sigma, I_{\sigma e}$. Note that Lemma 6.2.1 can be applied since $i \geq \mu^\sigma + 1 = \nu + 1$ by Property (REL2) and since it has Property (USV1)$_i$ and Property (EV1)$_{i+1}$. Due to the application of the switch e, the valuation of g_i increases. We prove that this does not create new improving switches. We thus first prove $\sigma(b_i) \neq g_i$ and $(b_i, g_i) \notin I_\sigma, I_{\sigma e}$. Since $i \geq \mu^\sigma + 1$, no cycle centers being closed implies $\beta_i = 0$, hence $\sigma(b_i) = \sigma e(b_i) = b_{i+1}$ by Property (EV1)$_i$. Furthermore, if $\mu^\sigma > 1$ and thus $t^{\rightarrow} = b_2$, then

$$\mathrm{rVal}_{\sigma e}(b_{i+1}) = L_{i+1} > \langle g_i \rangle + \sum_{\ell=1}^{i-1} W_\ell + L_{i+1} \geq \langle g_i \rangle + L_2 = \langle g_i \rangle + \mathrm{rVal}_{\sigma e}(b_2) = \mathrm{rVal}_{\sigma e}(g_i)$$

and, by Lemma 6.2.18, $\mathrm{rVal}_\sigma(b_{i+1}) > \langle g_i \rangle + \mathrm{rVal}_\sigma(g_1) = \mathrm{rVal}_\sigma(g_i)$ follows by the same estimation. Consequently, $(b_i, g_i) \notin I_\sigma, I_{\sigma e}$ if $\mu^\sigma > 1$. The statement follows by a similar argument if $\mu^\sigma = 1$ by $\mathrm{rVal}_\sigma(g_1) = W_1 + \mathrm{rVal}_\sigma(b_2)$.

We now prove that $\sigma(s_{i-1,1}) = b_1$ and $(s_{i-1,1}, h_{i-1,1}) \notin I_\sigma, I_{\sigma e}$. It cannot happen that $i = 1$ due to $i \geq \mu^\sigma + 1$, hence we do not need to consider a possible increase of the valuation of g_1. Since $i \geq \mu^\sigma + 1$ implies $i - 1 \geq \mu^\sigma$ and since $\beta_{i+1} = 0$, Property (USV1)$_{i-1}$ implies $\sigma(s_{i-1,1}) = \sigma e(s_{i-1,1}) = b_1$. It remains to prove $\mathrm{rVal}_\sigma(b_1) > \mathrm{rVal}_\sigma(h_{i-1,1})$ and $\mathrm{rVal}_{\sigma e}(b_1) > \mathrm{rVal}_{\sigma e}(h_{i-1,1})$. We only prove the second statement since it implies the first statement due to $\mathrm{rVal}_{\sigma e}(b_1) = \mathrm{rVal}_\sigma(b_1)$ and $\mathrm{rVal}_{\sigma e}(h_{i-1,1}) > \mathrm{rVal}_\sigma(h_{i-1,1})$. If $\mu^{\sigma e} = 1$, then $\sigma e(b_1) = b_2, t^{\rightarrow} = g_1$ and $\mathrm{rVal}_{\sigma e}(g_1) = W_1 + \mathrm{rVal}_{\sigma e}(b_2)$. Consequently,

$$\begin{aligned}\mathrm{rVal}_{\sigma e}(b_1) = \mathrm{rVal}_{\sigma e}(b_2) &> \langle h_{i-1,1}, g_i \rangle + W_1 + \mathrm{rVal}_{\sigma e}(b_2)\\ &= \langle h_{i-1,1}, g_i \rangle + \mathrm{rVal}_{\sigma e}(g_1) = \mathrm{rVal}_{\sigma e}(h_{i-1,1}).\end{aligned}$$

If $\mu^{\sigma e} > 1$, then $\sigma e(b_1) = g_1$ and $t^{\rightarrow} = b_2$. The statement can then be shown by similar arguments as $i > \mu^\sigma$ implies $\langle g_i \rangle < \sum_{\ell \in [i-1]} W_\ell$.

This concludes the case that both cycle centers of level i are $t^{\leftarrow}$-open. Consider the case that $F_{i,\beta_{i+1}}$ is mixed and that $F_{i,1-\beta_{i+1}}$ is $t^{\leftarrow}$-halfopen for σ. Then, $j = \beta_{i+1} = 1 - j$ implies that no other edge but $(g_i, F_{i,j})$ can become improving. However, after the application of e, both cycle centers are mixed. Hence, $\mathrm{Val}_{\sigma e}(F_{i,\beta_{i+1}}) > \mathrm{Val}_{\sigma e}(F_{i,1-\beta_{i+1}})$ by Lemma 6.2.1. □

Lemma 6.2.25. *Let $G_n = M_n$. Let σ be a well-behaved phase-3-strategy for $\mathfrak{b} \in \mathfrak{B}_n$ with $\sigma \in \rho(\sigma_0)$. Let $i \in [n]$ and $j := 1 - \beta^\sigma_{i+1}$. Let $e := (d_{i,j,k}, e_{i,j,k}) \in I_\sigma$ and $\sigma(e_{i,j,k}) = t^\rightarrow$ for some $k \in \{0,1\}$. Further assume that there is no other triple of indices i', j', k' with $(d_{i',j',k'}, e_{i',j',k'}) \in I_\sigma$, that $F_{i,j}$ is closed and that σ fulfills the following assumptions:*

1. *If $\beta^\sigma_i = 0$, then $\sigma(g_i) = F_{i,j}$ and $F_{i,1-j}$ is $t^\leftarrow$-halfopen.*
2. *$i < \mu^\sigma$ implies $[\sigma(s_{i,j}) = h_{i,j}$ and $\sigma(s_{i',j'}) = h_{i',j'} \wedge \bar{\sigma}(d_{i'})$ for all $i' < i, j' \in \{0,1\}]$ and that the cycle center $F_{i',1-\bar{\sigma}(g_{i'})}$ is $t^\leftarrow$-halfopen for all $i' < i$. In addition, $i < \mu^\sigma - 1$ implies $\bar{\sigma}(eb_{i+1})$.*
3. *$i' > i$ implies $\sigma(s_{i,1-\beta^\sigma_{i'+1}}) = b_1$.*
4. *$i' > i$ and $\beta^\sigma_{i'} = 0$ imply that either $[\bar{\sigma}(g_{i'}) = \beta^\sigma_{i'+1}$ and $F_{i,0}, F_{i,1}$ are mixed] or $[\bar{\sigma}(g_{i'}) = 1 - \beta^\sigma_{i'+1}$, $F_{i',1-\beta^\sigma_{i'+1}}$ is $t^\rightarrow$-open and $F_{i',\beta^\sigma_{i'+1}}$ is mixed] and*
5. *$i' > i$ and $\beta^\sigma_{i'} = 1$ imply that $F_{i',1-\beta^\sigma_{i'+1}}$ is either mixed or $t^\rightarrow$-open.*

Then σe is a well-behaved phase-3-strategy for $\mathfrak{b}$ with $\sigma e \in \rho(\sigma_0)$ and $I_{\sigma e} = I_\sigma \setminus \{e\}$ if $i \geq \mu^\sigma$ and $I_{\sigma e} = [I_\sigma \cup \{(s_{i,j}, b_1)\}] \setminus \{e\}$ if $i < \mu^\sigma$.

Proof. By Lemma 6.2.20, it suffices to prove the statements related to the set of improving switches. We first prove $(g_i, F_{i,*}) \notin I_\sigma, I_{\sigma e}$. Let $\beta_i = 0$ first. Then, $\sigma(g_i) = F_{i,j}$ by assumption 1., implying $(g_i, F_{i,j}) \notin I_\sigma, I_{\sigma e}$. We thus prove $(g_i, F_{i,1-j}) \notin I_\sigma, I_{\sigma e}$. By assumption 1., $\bar{\sigma}(d_{i,j})$, hence $\mathrm{rVal}_\sigma(F_{i,j}) = \mathrm{rVal}_\sigma(s_{i,j})$ and $\mathrm{rVal}_{\sigma e}(F_{i,j}) = \mathrm{rVal}_{\sigma e}(t^\rightarrow)$ by Lemma 6.1.15. Since $F_{i,1-j}$ is $t^\leftarrow$-halfopen by assumption, $\mathrm{rVal}_{\sigma e}(F_{i,1-j}) < \mathrm{rVal}_{\sigma e}(F_{i,j})$ since $\mathrm{Val}_{\sigma e}(t^\leftarrow) < \mathrm{Val}_{\sigma e}(t^\rightarrow)$ by Lemma 6.2.18. Consequently, $(g_i, F_{i,1-j}) \notin I_{\sigma e}$ and it remains to prove $\mathrm{rVal}_\sigma(F_{i,1-j}) < \mathrm{rVal}_\sigma(F_{i,j})$.

If $\sigma(s_{i,j}) = b_1$, then the equivalences $\sigma(b_1) = b_2 \Leftrightarrow \mu^\sigma = 1 \Leftrightarrow t^\leftarrow = b_2$ implies

$$\mathrm{rVal}_\sigma(F_{i,j}) = \langle s_{i,j} \rangle + \mathrm{rVal}_\sigma(b_1) = \langle s_{i,j} \rangle + \mathrm{rVal}_\sigma(t^\leftarrow) > \mathrm{Val}_\sigma(t^\leftarrow) = \mathrm{Val}_\sigma(F_{i,1-j}).$$

Hence assume $\sigma(s_{i,j}) = h_{i,j}$, implying $i < \mu^\sigma$ by Property (USV1)$_i$ as $j = 1 - \beta_{i+1}$. This implies $\mu^\sigma > 1$, so $t^\leftarrow = g_1$ and $t^\rightarrow = b_2$. Let $i = \mu^\sigma - 1$. Then $j = 1 - \beta_\nu = 0$, hence $\mathrm{rVal}_\sigma(F_{i,j}) = \langle s_{i,j}, h_{i,j} \rangle + \mathrm{rVal}_\sigma(b_{\nu+1})$. By assumption 2, $\bar{\sigma}(d_{i'})$ for all $i' < i$. Consequently,

$$\mathrm{rVal}_\sigma(F_{i,1-j}) = \mathrm{rVal}_\sigma(g_1) = R_1 < \langle s_{\nu-1,j}, h_{\nu-1,j} \rangle + L_{\nu+1} = \mathrm{rVal}_\sigma(F_{i,j}).$$

Let $i < \mu^\sigma - 1$, implying $j = 1$ and thus $\mathrm{rVal}_\sigma(F_{i,j}) = \langle s_{i,j}, h_{i,j} \rangle + \mathrm{rVal}_\sigma(g_{i+1})$. Since $\bar{\sigma}(eb_{i+1})$ by assumption 2. and $\neg\bar{\sigma}(eg_{i+1})$ by Property (BR2) and thus in particular $\neg\bar{\sigma}(d_{i+1})$, we have $\mathrm{rVal}_\sigma(g_{i+1}) = \langle g_{i+1} \rangle + \mathrm{rVal}_\sigma(b_2)$ by Corollary 6.1.17. Furthermore, as $\bar{\sigma}(d_{i'})$ for all $i' < i$, it holds that $\mathrm{rVal}_\sigma(g_1) = \sum_{\ell<i} W_\ell + \langle g_{i+1} \rangle + \mathrm{rVal}_\sigma(b_2)$. Consequently, as $\mathrm{rVal}_\sigma(g_1) = \mathrm{rVal}_\sigma(F_{i,1-j})$, it follows that

$$\mathrm{rVal}_\sigma(F_{i,j}) = \langle s_{i,j}, h_{i,j}, g_{i+1} \rangle + \mathrm{rVal}_\sigma(b_2) > \sum_{\ell \in [i]} W^{\mathrm{M}}_\ell + \langle g_{i+1} \rangle + \mathrm{rVal}_\sigma(b_2) = \mathrm{rVal}_\sigma(F_{i,1-j}).$$

Hence, $\mathrm{rVal}_\sigma(F_{i,1-j}) < \mathrm{rVal}_\sigma(F_{i,j})$ holds in any case, so $(g_i, F_{i,1-j}) \notin I_\sigma, I_{\sigma e}$. As also $(g_i, F_{i,j}) \notin I_\sigma, I_{\sigma e}$, this proves that $\beta_i = 0$ implies $(g_i, F_{i,*}) \notin I_\sigma, I_{\sigma e}$.

Now let $\beta_i = 1$. Then, by Property (REL2), $i \geq \mu^\sigma = \mu^{\sigma e} = \nu$. By Property (CC2), $\sigma(g_i) = F_{i,\beta_{i+1}} = F_{i,1-j}$. We hence prove $(g_i, F_{i,j}) \notin I_\sigma, I_{\sigma e}$. Since $\beta_i = 1$, the cycle center $F_{i,1-j}$ is closed. By Property (USV1)$_i$, Property (EV1)$_{i+1}$ and since $i \geq \mu^\sigma, \mu^{\sigma e}$ we thus obtain $\text{rVal}_\sigma(F_{i,1-j}) = \langle s_{i,1-j}, h_{i,1-j}\rangle + \text{rVal}_\sigma(b_{i+1}) = \langle s_{i,1-j}, h_{i,1-j}\rangle + L_{i+1}$ and the corresponding equality for $\text{rVal}_{\sigma e}(F_{i,1-j})$.

As $F_{i,j}$ is closed with respect to σ and escapes towards $t^\rightarrow$ with respect to σe, Property (USV1)$_i$ yields $\text{rVal}_\sigma(F_{i,j}) = \langle s_{i,j}\rangle + \text{rVal}_\sigma(t^\leftarrow)$ and $\text{rVal}_{\sigma e}(F_{i,j}) = \text{rVal}_{\sigma e}(t^\rightarrow)$. It is now easy to see that $\langle s_{i,1-j}, h_{i,1-j}\rangle > \sum_{\ell = \in [i]} W_\ell$ implies $\text{rVal}_\sigma(F_{i,1-j}) > \text{rVal}_\sigma(F_{i,j})$ and $\text{rVal}_{\sigma e}(F_{i,1-j}) > \text{rVal}_{\sigma e}(F_{i,j})$. Therefore, $(g_i, F_{i,j}) \notin I_\sigma, I_{\sigma e}$ if $\beta_i = 1$. As also $(g_i, F_{i,1-j}) \notin I_\sigma, I_{\sigma e}$ in this case due to $\sigma(g_i) = F_{i,1-j}$, this proves $(g_i, *) \notin I_\sigma, I_{\sigma e}$ in any case. By the choice of e, we also have $(d_{i,j,k}, F_{i,j}) \notin I_\sigma, I_{\sigma e}$ and $\sigma(d_{i,j,1-k}) = F_{i,j}$ as we assume $F_{i,j}$ to be closed with respect to σ. Thus also $(d_{i,j,1-k}, F_{i,j}) \notin I_\sigma, I_{\sigma e}$.

If $\beta_i = 1$, the increase of the valuation of $F_{i,j}$ can only have an immediate effect on the vertices $g_i, d_{i,j,0}$ and $d_{i,j,1}$. However, as $\beta_i = 1$ implies $\sigma(g_i) = F_{i,1-j}$ and since there are no player 0 vertices edges towards $d_{i,j,*}$, we immediately obtain $I_{\sigma e} = I_\sigma \setminus \{e\}$. We thus only consider the case $\beta_i = 0$ for the remainder of this proof, implying $\sigma(g_i) = F_{i,j}$.

Since $\sigma(g_i) = F_{i,j}$, the valuation of g_i increases due to the increase of the valuation of $F_{i,j}$. We investigate how this increase influences the set of improving switches. We first prove that $i \neq 1$ implies $\sigma(b_i) = b_{i+1}$ and $(b_i, g_i) \notin I_\sigma, I_{\sigma e}$. If $i = 1$, then $\mu^\sigma > 1$ as $\beta_i = 0$ by assumption, implying $\sigma(b_1) = g_1$ and thus $(b_i, g_i) \notin I_\sigma, I_{\sigma e}$. Hence let $i \neq 1$. Then, $\sigma(b_i) = b_{i+1}$ by Property (EV1)$_i$. If $\sigma(b_{i+1}) = g_{i+1}$, then $\beta_{i+1} = 1$ and thus $i + 1 \geq \mu^\sigma = \nu$, implying $\text{rVal}_\sigma(b_{i+1}) = L_{i+1}$ in any case. The same holds for σe, so in particular $\text{rVal}_\sigma(b_{i+1}) = \text{rVal}_{\sigma e}(b_{i+1})$. It hence suffices to prove $\text{rVal}_{\sigma e}(b_{i+1}) > \text{rVal}_{\sigma e}(g_i)$ as $\text{rVal}_{\sigma e}(g_i) \geq \text{rVal}_\sigma(g_i)$.

By the choice of e and our assumptions, $\text{rVal}_{\sigma e}(g_i) = \langle g_i\rangle + \text{rVal}_{\sigma e}(t^\rightarrow)$. If $t^\rightarrow = b_2$, then $\text{rVal}_{\sigma e}(t^\rightarrow) = \text{rVal}_{\sigma e}(b_2)$. If $t^\rightarrow = g_1$, then $\text{rVal}_{\sigma e}(t^\rightarrow) = \text{rVal}_{\sigma e}(g_1) = W_1 + \text{rVal}_{\sigma e}(b_2)$ by Lemma 6.2.18 as $\mu^{\sigma e} = 1$. This in particular yields

$$\begin{aligned}\text{rVal}_{\sigma e}(g_i) &= \langle g_i\rangle + \text{rVal}_{\sigma e}(t^\rightarrow) \leq \langle g_i\rangle + W_1 + \text{rVal}_{\sigma e}(b_2) = \langle g_i\rangle + W_1 + L_2\\ &= \langle g_i\rangle + W_1 + L_{2,i-1} + L_{i+1} < L_{i+1} = \text{rVal}_{\sigma e}(b_{i+1})\end{aligned}$$

as $\sigma(b_i) = b_{i+1}$. Thus $(b_i, g_i) \notin I_\sigma, I_{\sigma e}$ for all i.

Now let $i > \mu^{\sigma e}$. We prove that this implies $\sigma(s_{i-1,1}) = b_1, \text{Val}_\sigma(b_1) > \text{Val}_\sigma(h_{i-1,1})$ and $\text{Val}_{\sigma e}(b_1) > \text{Val}_{\sigma e}(h_{i-1,1})$. When proving these statements, we will also prove that $\text{Val}_\sigma(b_1) = \text{Val}_{\sigma e}(b_1), \text{Val}_\sigma(g_1) = \text{Val}_{\sigma e}(g_1)$ and $\text{Val}_\sigma(b_2) = \text{Val}_{\sigma e}(b_2)$. We then argue why this suffices to prove the statement for $\mu^{\sigma e} = 1$ and then consider the case $\mu^{\sigma e} > 1$ and $i < \mu^{\sigma e}$. It is not necessary to consider the case $i = \mu^{\sigma e}$ as $\beta_i = 0$.

Since $\beta_i = 0$ and $i > \mu^{\sigma e}$ implies $i - 1 \geq \mu^{\sigma e}$, Property (USV1)$_{i-1}$ implies $\sigma(s_{i-1,1}) = \sigma e(s_{i-1,1}) = b_1$. This implies that the valuation of no further vertex than $h_{i-1,1}$ and the vertices discussed previously can change when transitioning from σ to σe. None of these vertices are part of the valuation of b_1, g_1 and b_2 since $i > \mu^{\sigma e}, \sigma(b_i) = \sigma e(b_i) = b_{i+1}$ and $\sigma(s_{i-1,1}) = \sigma(s_{i-1,1}) = b_1$, implying that their valuations do not change. In particular, $\text{Val}_\sigma(b_1) = \text{Val}_{\sigma e}(b_1)$. If we can show $(s_{i-1,1}, b_1) \notin I_\sigma, I_{\sigma e}$, this thus proves $I_{\sigma e} = I_\sigma \setminus \{e\}$ for the case $\mu^{\sigma e} = 1$. Since $\text{Val}_\sigma(b_1) = \text{Val}_{\sigma e}(b_1)$, it suffices to prove $\text{Val}_{\sigma e}(b_1) > \text{Val}_{\sigma e}(h_{i-1,1})$ as $\text{Val}_{\sigma e}(h_{i-1,1}) \geq \text{Val}_\sigma(h_{i-1,1})$. Consider the case $\mu^{\sigma e} = 1$ first, implying $t^\rightarrow = g_1$ and $\text{rVal}_{\sigma e}(g_1) = W_1 + \text{rVal}_{\sigma e}(b_2)$ by Lemma 6.2.18. Thus, since $\text{rVal}_{\sigma e}(g_i) = \langle g_i\rangle + \text{rVal}_{\sigma e}(t^\rightarrow)$

and $\mathrm{rVal}_{\sigma e}(b_2) = \mathrm{Val}_{\sigma e}(b_1)$ since $\sigma e(b_1) = b_2$ due to $\mu^{\sigma e} = 1$, it follows that

$$\mathrm{rVal}_{\sigma e}(h_{i-1,1}) = \langle h_{i-1,1}, g_i\rangle + \mathrm{rVal}_{\sigma e}(g_i) = \langle h_{i-1,1}, g_i\rangle + W_1 + \mathrm{rVal}_{\sigma e}(b_2) < \mathrm{rVal}_{\sigma e}(b_1).$$

Consider the case $\mu^{\sigma e} > 1$ next. Then $t^{\rightarrow} = b_2$ and $\sigma e(b_1) = g_1$. Now, either $\mathrm{rVal}_{\sigma e}(b_1) = R_1$ or $\mathrm{rVal}_{\sigma e}(b_1) = g_{i'} + \sum_{\ell<i'} W_\ell + \mathrm{rVal}_{\sigma e}(b_2)$ where $i' = \min\{\ell \geq 1 \colon \neg\overline{\sigma e}(d_{i'})\} < \mu^{\sigma e}$. In the first case, $i > \mu^{\sigma e}$ implies

$$\mathrm{rVal}_{\sigma e}(h_{i-1,1}) = \langle h_{i-1,1}, g_i\rangle + \mathrm{rVal}_{\sigma e}(b_2) < L_{\mu^{\sigma e}+1} < R_1 = \mathrm{rVal}_{\sigma e}(b_1).$$

In the second case, $i > \mu^{\sigma e} > i'$ implies

$$\mathrm{rVal}_{\sigma e}(h_{i-1,1}) = \langle h_{i-1,1}, g_i\rangle + \mathrm{rVal}_{\sigma e}(b_2) < \langle g_{i'}\rangle + \sum_{\ell<i'} W_\ell + \mathrm{rVal}_{\sigma e}(b_2) = \mathrm{rVal}_{\sigma e}(b_1).$$

Hence $i > \mu^{\sigma e}$ implies $(s_{i-1,1}, h_{i-1,1}) \notin I_\sigma, I_{\sigma e}$, proving the statement for $\mu^{\sigma e} = 1$.

It remains to investigate the case $i < \mu^{\sigma e}$, implying $\mu^{\sigma e} > 1$ and $t^{\rightarrow} = b_2, t^{\leftarrow} = g_1$. In this case, opening the cycle center $F_{i,j}$ changes the valuation of g_1. Since $\mu^{\sigma e} > 1$ implies $\sigma(b_1) = g_1$, this also changes the valuation of b_1 and of possibly every vertex that has an edge to either one of these vertices. These are in particular upper selection vertices, escape vertices and cycle centers. We begin by observing that

$$\begin{aligned}
\mathrm{Val}_\sigma(F_{i,j}) = \mathrm{Val}_\sigma(s_{i,j}) &= \langle s_{i,j}, h_{i,j}\rangle + \begin{cases} \mathrm{Val}_\sigma(b_{i+2}), & j = 0 \\ \mathrm{Val}_\sigma(g_{i+1}), & j = 1 \end{cases} \\
&= \langle s_{i,j}, h_{i,j}\rangle + \begin{cases} \mathrm{Val}_\sigma(b_{i+2}), & i = \mu^{\sigma e} - 1 \\ \mathrm{Val}_\sigma(g_{i+1}), & i < \mu^{\sigma e} - 1 \end{cases} \\
\mathrm{Val}_{\sigma e}(F_{i,j}) &= \frac{1-\varepsilon}{1+\varepsilon}\mathrm{Val}_{\sigma e}(b_2) + \frac{2\varepsilon}{1+\varepsilon}\mathrm{Val}_{\sigma e}(s_{i,j}) \\
\mathrm{rVal}_\sigma(g_1) &= \begin{cases} R_1, & i = \mu^{\sigma e} - 1 \\ \langle g_{i+1}\rangle + \sum_{\ell<i+1} W_\ell + \mathrm{Val}_\sigma(b_2), & i < \mu^{\sigma e} - 1 \end{cases}, \\
\mathrm{rVal}_{\sigma e}(g_1) &= \langle g_i\rangle + \sum_{\ell<i} W_\ell + \mathrm{Val}_{\sigma e}(b_2).
\end{aligned}$$

Furthermore, $\mathrm{Val}_\sigma(b_2) = \mathrm{Val}_{\sigma e}(b_2) = L_2$ and $i \neq 1$ implies $\sigma(b_i) = \sigma e(b_i) = b_{i+1}$. Note that we have $\mathrm{rVal}_\sigma(g_1) < \mathrm{rVal}_\sigma(b_2)$ and $\mathrm{rVal}_{\sigma e}(g_1) < \mathrm{rVal}_{\sigma e}(b_2)$ by Lemma 6.2.18. We begin by investigating upper selection vertices and prove that $(s_{i,j}, b_1)$ is improving for σe. Since $\sigma(s_{i,j}) = \sigma e(s_{i,j}) = h_{i,j}$ by assumption, it suffices to prove $\mathrm{Val}_{\sigma e}(h_{i,j}) < \mathrm{Val}_{\sigma e}(b_1)$. Consider the case $i = \mu^{\sigma e} - 1$ first, implying $j = 0$. Then, since Property (EV1)$_{i'}$ implies $\sigma(b_{i'}) = b_{i'+1}$ for all $i' \in \{2, \ldots, \mu^{\sigma e} - 1\}$,

$$\begin{aligned}
\mathrm{rVal}_{\sigma e}(h_{i,j}) &= \langle h_{i,j}\rangle + \mathrm{Val}_{\sigma e}(b_{i+2}) < \langle g_i\rangle + \sum_{\ell<i} W_\ell + W_{i+1} + L_{i+2} \\
&= \langle g_i\rangle + \sum_{\ell<i} W_\ell + L_{i+1} = \langle g_i\rangle + \sum_{\ell<i} W_\ell + L_2 = \mathrm{rVal}_{\sigma e}(g_1) = \mathrm{rVal}_{\sigma e}(b_1).
\end{aligned}$$

Therefore, $(s_{i,j}, b_1) \in I_{\sigma e}$ if $i = \mu^{\sigma e} - 1$. Consider the case $i < \mu^{\sigma e} - 1$, implying $j = 1$. Then, since $\bar{\sigma}(eb_{i+1}) \wedge \neg\bar{\sigma}(eg_{i+1})$ by assumption 2 and Property (BR2),

$$\begin{aligned}\mathrm{rVal}_{\sigma e}(h_{i,j}) &= \langle h_{i,j}\rangle + \mathrm{rVal}_{\sigma e}(g_{i+1}) = \langle h_{i,j}, g_{i+1}\rangle + \mathrm{rVal}_{\sigma e}(b_2)\\ &< \langle g_i\rangle + \sum_{\ell<i} W_\ell + \mathrm{rVal}_{\sigma e}(b_2) = \mathrm{rVal}_{\sigma e}(g_1),\end{aligned}$$

hence $(s_{i,j}, b_1) \in I_{\sigma e}$ if $i < \mu^{\sigma e}$. It remains to prove that no further improving switch is created.

First, we prove that for all $i' \in [n]$ and $j' \in \{0,1\}$ with $(i',j') \neq (i,j)$, $\sigma(s_{i',j'}) = h_{i',j'}$ implies $(s_{i',j'}, b_1) \notin I_\sigma, I_{\sigma e}$. Hence let i', j' be such a pair of indices and $i' \geq \mu^{\sigma e}$ first. Then, $\mathrm{rVal}_\sigma(h_{i',j'}) = \langle h_{i',j'}\rangle + \mathrm{rVal}_\sigma(b_{i'+1}) = \langle h_{i',j'}\rangle + L_{i'+1}$ follows from Property (USV1)$_{i'}$ and Property (EV1)$_{i'+1}$ and $\mathrm{rVal}_{\sigma e}(h_{i',j'}) = \langle h_{i',j'}\rangle + L_{i'+1}$ follows analogously. This implies $\mathrm{rVal}_\sigma(h_{i',j'}), \mathrm{rVal}_{\sigma e}(h_{i',j'}) > \mathrm{rVal}_\sigma(b_2) = \mathrm{rVal}_{\sigma e}(b_2)$. Since $\mathrm{rVal}_\sigma(g_1) < \mathrm{rVal}_\sigma(b_2)$ as well as $\mathrm{rVal}_{\sigma e}(g_1) < \mathrm{rVal}_{\sigma e}(b_2)$ by Lemma 6.2.18, $\sigma(s_{i',j'}) = h_{i',j'}$ thus implies $(s_{i',j'}, b_1) \notin I_\sigma, I_{\sigma e}$ for $i' \geq \mu^{\sigma e}$. Next let $i' < \mu^{\sigma e}$ and $i < i' < \mu^{\sigma e}$. Then, by assumption 3, $j' = \beta_{i'+1}$, so $\mathrm{Val}_{\sigma e}(h_{i',j'}) = \langle h_{i',j'}\rangle + \mathrm{Val}_{\sigma e}(b_{i'+1}) = \langle h_{i',j'}\rangle + L_{i'+1}$ by Property (EV1)$_{i'+1}$. Furthermore,

$$\mathrm{rVal}_{\sigma e}(b_1) = \mathrm{rVal}_{\sigma e}(g_1) = \langle g_i\rangle + \sum_{\ell<i} W_\ell + \mathrm{rVal}_{\sigma e}(b_2) = \langle g_i\rangle + \sum_{\ell<i} W_\ell + L_2.$$

Since $\beta_2 = \cdots = \beta_{i'} = 0$ due to $i' < \mu^{\sigma e} = \nu$ and $\langle g_i\rangle + \sum_{\ell<i} W_\ell < 0$, this implies

$$\mathrm{rVal}_{\sigma e}(b_1) < L_2 = L_{i'+1} < \langle h_{i',j'}\rangle + L_{i'+1} = \mathrm{rVal}_{\sigma e}(h_{i',j'})$$

and thus $(s_{i',j'}, b_1) \notin I_{\sigma e}$. The same calculations also yield $(s_{i',j'}, b_1) \notin I_\sigma$.

Now let $i' < i < \mu^{\sigma e}$. If $j' = \beta_{i'+1}$, then the same arguments used previously can be applied again. Hence let $j' = 1 - \beta_{i'+1}$. Since $i' < i < \mu^{\sigma e}$ implies $i' < \mu^{\sigma e} - 1$, we have $\beta_{i'+1} = 0$ and thus $j' = 1$. By assumption 2, the cycle center $F_{\ell,\overline{\sigma e}(g_\ell)}$ is closed and $\overline{\sigma e}(s_\ell)$ for all $\ell < i$. Since $\ell < i < \mu^{\sigma e}$ furthermore implies $\sigma e(g_\ell) = F_{\ell,1}$ by Property (BR1), we have $\lambda_{i'+1} = i$. Consequently, as the last case of Corollary 6.1.17 is fulfilled,

$$\begin{aligned}\mathrm{rVal}_{\sigma e}(h_{i',j'}) &= \langle h_{i',j'}\rangle + \mathrm{rVal}_{\sigma e}(g_{i'+1}) = \langle h_{i',j'}\rangle + \langle g_i\rangle + \sum_{\ell=i'+1}^{i-1} W_\ell + \mathrm{rVal}_{\sigma e}(b_2)\\ &> \sum_{\ell=1}^{i'} W_\ell + \langle g_i\rangle + \sum_{\ell=i'+1}^{i-1} W_\ell + \mathrm{rVal}_{\sigma e}(b_2)\\ &= \langle g_i\rangle + \sum_{\ell<i} W_\ell + \mathrm{rVal}_{\sigma e}(b_2) = \mathrm{rVal}_{\sigma e}(b_1),\end{aligned}$$

implying $(s_{i',j'}, b_1) \notin I_{\sigma e}$. The same arguments can also be used to show $(s_{i',j'}, b_1) \notin I_\sigma$. If $i' = i < \mu^{\sigma e}$, then $j' = \beta_{i'+1}$ as we consider indices $(i',j') \neq (i,j)$, implying the statement by the same arguments as before.

We next investigate escape vertices $e_{i',j',k'}$. If $\sigma(e_{i',j',k'}) = \sigma e(e_{i',j',k'}) = b_2$, then Lemma 6.2.18 implies $\mathrm{rVal}_\sigma(b_2) > \mathrm{rVal}_\sigma(g_1)$, hence $(e_{i',j',k'}, g_1) \notin I_\sigma$. Analogously,

$(e_{i',j',k'}, g_1) \notin, I_{\sigma e}$ and, by definition, $(e_{i',j',k'}, b_2) \notin I_\sigma, I_{\sigma e}$. Using the same arguments, it follows that $\sigma(e_{i',j',k'}) = g_1$ implies $(e_{i',j',k'}, b_2) \in I_\sigma, I_{\sigma e}$ as well as $(e_{i',j',k'}, g_1) \notin I_\sigma, I_{\sigma e}$. In particular, we have $(e_{i',j',k'}, *) \in I_\sigma \Leftrightarrow (e_{i',j',k'}, *) \in I_{\sigma e}$.

We next investigate the selector vertices $g_{i'}$. We do not need to consider the case $i' = i$ as we already proved $(g_i, *) \notin I_\sigma, I_{\sigma e}$. Consider the case $\beta^\sigma_{i'} = 1$ first, implying $i' \geq \mu^\sigma > i$ by Property (REL2). Since $i' \geq \mu^\sigma > 1$, we have $\bar{\sigma}(g_{i'}) = \beta_{i'+1}, \bar{\sigma}(d_{i'})$ and $\bar{\sigma}(s_{i'})$. We thus prove $\mathrm{Val}_\sigma(F_{i',\beta_{i'+1}}) \geq \mathrm{Val}_\sigma(F_{i',1-\beta_{i'+1}})$. By the previously shown properties,

$$\mathrm{Val}_\sigma(F_{i',\beta_{i'+1}}) = \left\langle s_{i',\beta_{i'+1}}, h_{i',\beta_{i'+1}} \right\rangle + \mathrm{Val}_\sigma(b_{i'+1}).$$

Now, either $\mathrm{rVal}_\sigma(F_{i',1-\beta_{i'+1}}) = \frac{1}{2}\mathrm{rVal}_\sigma(g_1) + \frac{1}{2}\mathrm{rVal}_\sigma(b_2)$ or $\mathrm{rVal}_\sigma(F_{i',1-\beta^\sigma_{i'+1}}) = \mathrm{rVal}_\sigma(b_2)$ by assumption 4. As $\mathrm{Val}_\sigma(b_2) > \mathrm{Val}_\sigma(g_1)$ by Lemma 6.2.18, it suffices to consider the second case. The statement then follows since

$$\begin{aligned}\mathrm{rVal}_\sigma(F_{i',\beta_{i'+1}}) &= \left\langle s_{i',\beta_{i'+1}}, h_{i',\beta_{i'+1}} \right\rangle + \mathrm{rVal}_\sigma(b_{i'+1}) = \left\langle s_{i',\beta_{i'+1}}, h_{i',\beta_{i'+1}} \right\rangle + L_{i'+1} \\ &\geq L_{2,i'} + L_{i'+1} = L_2 = \mathrm{rVal}_{\sigma e}(b_2) \geq \mathrm{rVal}_\sigma(F_{i',1-\beta_{i'+1}}).\end{aligned}$$

Consequently, $(g_{i'}, *) \notin I_\sigma$. As $i' \neq i$, these estimations can also be applied to σe, implying $(g_{i'}, *) \notin I_{\sigma e}$. Hence $\beta^\sigma_{i'} = 1$ implies $(g_{i'}, *) \notin I_\sigma, I_{\sigma e}$.

Next assume $\beta_{i'} = 0$ and $i' > i$. Then, by assumption 4, either [$\bar{\sigma}(g_{i'}) = \beta_{i'+1}$ and $F_{i',0}, F_{i',1}$ are mixed] or [$\bar{\sigma}(g_{i'}) = 1 - \beta_{i'+1}$, $F_{i',1-\beta_{i'+1}}$ is b_2-open and $F_{i',\beta^\sigma_{i'+1}}$ is mixed]. In the first case, both cycle centers are in the same state with respect to both σ and σe. Consequently, it suffices to prove $\mathrm{Val}_\sigma(s_{i',\beta_{i'+1}}) > \mathrm{Val}_\sigma(s_{i',1-\beta_{i'+1}})$. But this follows as $\sigma(s_{i',1-\beta_{i'+1}}) = b_1, \sigma(s_{i',\beta_{i'+1}}) = h_{i',\beta_{i'+1}}$ and $\sigma(b_{\mu^\sigma}) = g_{\mu^\sigma}$. As these arguments also apply to σe, it follows that $(g_{i'}, *) \notin I_\sigma, I_{\sigma e}$. In the second case, the argument follows since $\mathrm{rVal}_\sigma(b_2) > \mathrm{rVal}_\sigma(g_1)$ by Lemma 6.2.18. By the same argument, $(g_{i'}, *) \notin I_{\sigma e}$. This concludes the case $\beta_{i'} = 0$ and $i' > i$.

Consider the case $\beta_{i'} = 0 \wedge i' < i$ next. Then, since $i' < i \leq \mu^\sigma - 1$, Property (BR1) implies $\bar{\sigma}(g_{i'}) = 1$. We thus prove $\mathrm{rVal}_\sigma(F_{i',1}) > \mathrm{rVal}_\sigma(F_{i',0})$. By assumption 2, it holds that $\sigma(s_{i',0}) = h_{i',0}$ and $\sigma(s_{i',1}) = h_{i',1}$. Since $F_{i'\bar{\sigma}(g_{i'})}$ is closed by assumption 2, this implies $\mathrm{rVal}_\sigma(F_{i',1}) = \langle s_{i',1}, h_{i',1} \rangle + \mathrm{rVal}_\sigma(g_{i'+1})$. Using Corollary 6.1.17, it follows that $\mathrm{rVal}_\sigma(g_{i'+1}) = \sum_{\ell=i'+1}^{\mu^\sigma-1} W_\ell + \mathrm{rVal}_\sigma(b_{\mu^\sigma+1})$. By assumption 2., the other cycle center $F_{i',0}$ of level i' is g_1-halfopen. Let $i = \mu^\sigma - 1$, implying $\mathrm{rVal}_\sigma(g_1) = R_1$. Then

$$\begin{aligned}\mathrm{rVal}_\sigma(F_{i',1}) &= \langle s_{i',1}, h_{i',1} \rangle + \mathrm{rVal}_\sigma(g_{i'+1}) = \langle s_{i',1}, h_{i',1} \rangle + \sum_{\ell=i'+1}^{\mu^\sigma-1} W_\ell + \mathrm{rVal}_\sigma(b_{\mu^\sigma+1}) \\ &> \sum_{\ell=1}^{i'} W_\ell + \sum_{\ell=i'+1}^{\mu^\sigma-1} W_\ell + \mathrm{rVal}_\sigma(b_{\mu^\sigma+1}) = R_1 = \mathrm{rVal}_\sigma(g_1) = \mathrm{rVal}_\sigma(F_{i',0}).\end{aligned}$$

Let $i < \mu^\sigma - 1$, implying $\mathrm{rVal}_\sigma(g_1) = \langle g_{i+1} \rangle + \sum_{\ell<i+1} W_\ell + \mathrm{rVal}_\sigma(b_2)$. Then, due to $\bar{\sigma}(eb_{i+1}) \wedge \neg\bar{\sigma}(eg_{i+1})$, it follows that $\mathrm{rVal}_\sigma(g_{i'+1}) = \langle g_{i+1} \rangle + \sum_{\ell=i'+1}^{i} W_\ell + \mathrm{rVal}_\sigma(b_2)$. Consequently,

$$\mathrm{rVal}_\sigma(F_{i',1}) = \langle s_{i',1}, h_{i',1} \rangle + \mathrm{rVal}_\sigma(g_{i'+1}) = \langle s_{i',1}, h_{i',1}, g_{i+1} \rangle + \sum_{\ell=i'+1}^{i} W_\ell + \mathrm{rVal}_\sigma(b_2)$$

$$> \langle g_{i+1}\rangle + \sum_{\ell=1}^{i'} W_\ell + \sum_{\ell=i'+1}^{i} W_\ell + \mathrm{rVal}_\sigma(b_2) = \mathrm{rVal}_\sigma(g_1) = \mathrm{rVal}_\sigma(F_{i',0}).$$

Hence, $(g_{i'},*) \notin I_\sigma$. Since it holds that $\mathrm{rVal}_{\sigma\!e}(g_{i'+1}) = \langle g_i\rangle + \sum_{\ell=i'+1}^{i-1} W_\ell + \mathrm{rVal}_\sigma(b_2)$ and $\mathrm{rVal}_{\sigma\!e}(g_1) = \langle g_i\rangle + \sum_{\ell<i} W_\ell + \mathrm{rVal}_{\sigma\!e}(b_2)$, the same calculation can be used to obtain $\mathrm{rVal}_{\sigma\!e}(F_{i',1}) > \mathrm{rVal}_{\sigma\!e}(F_{i',0})$. Hence, also $(g_{i'},*) \notin I_{\sigma\!e}$.

This covers all cases, hence $(g_{i'}, F_{i',*}) \notin I_\sigma, I_{\sigma\!e}$ for any index $i' \in [n]$.

We next consider entry vertices $b_{i'}$ for $i' \in [n]$. First of all, since $\sigma(b_1) = \sigma\!e(b_1) = g_1$ and $\mathrm{Val}_\sigma(b_2) > \mathrm{Val}_\sigma(g_1)$ as well as $\mathrm{Val}_{\sigma\!e}(b_2) > \mathrm{Val}_{\sigma\!e}(g_1)$ by Lemma 6.2.18, we have $(b_1,b_2) \in I_\sigma, I_{\sigma\!e}$. Thus consider some edge $(b_{i'}, b_{i'+1})$ for $i' \neq 1$. Since $(b_{i'}, b_{i'+1}) \notin I_\sigma, I_{\sigma\!e}$ if $\sigma(b_{i'}) = b_{i'+1}$, assume $\sigma(b_{i'}) = g_{i'}$. Then $\beta_{i'} = 1$, implying $i' \geq \mu^\sigma$. This directly implies $\mathrm{Val}_\sigma(b_{i'}) = L_{i'} = W_{i'} + L_{i'+1} > L_{i'+1} = \mathrm{Val}_\sigma(b_{i'+1})$, hence $(b_{i'}, b_{i'+1}) \notin I_\sigma$. The same argument can be used to prove $(b_{i'}, b_{i'+1}) \notin I_{\sigma\!e}$.

Next, consider some edge $(b_{i'}, g_{i'})$. If $\sigma(b_{i'}) = g_{i'}$, then $(b_{i'}, g_{i'}) \notin I_\sigma, I_{\sigma\!e}$, hence assume $\sigma(b_{i'}) = b_{i'+1}$, implying $i' > 1$. By Property $(\mathrm{EV1})_{i'}$, it then holds that $\beta_{i'} = 0$. Let $i' > i$. By assumption 4, the cycle center $F_{i',\bar\sigma(g_{i'})}$ is then either mixed or b_2-open. Since $\mathrm{Val}_\sigma(b_2) > \mathrm{Val}_\sigma(g_1)$, it suffices to consider the case that it is b_2-open. Thus,

$$\mathrm{rVal}_\sigma(g_{i'}) \leq \langle g_{i'}\rangle + \mathrm{Val}_\sigma(b_2) = \langle g_{i'}\rangle + L_2 = \langle g_{i'}\rangle + L_{2,i'-1} + L_{i'+1} < L_{i'+1} = \mathrm{rVal}_\sigma(b_{i'+1}),$$

hence $(b_{i'}, g_{i'}) \notin I_\sigma$ and, by the same arguments, also $(b_{i'}, g_{i'}) \notin I_{\sigma\!e}$. Now consider the case $i' < i$. Then, by assumption 2, the cycle center $F_{i',\bar\sigma(g_{i'})}$ is closed. Depending on whether $i = \mu^\sigma - 1$ or $i < \mu^\sigma - 1$, we then have

$$\mathrm{rVal}_\sigma(g_{i'}) = \sum_{\ell=i'}^{\mu^\sigma-1} W_\ell + \mathrm{rVal}_\sigma(b_{\mu^\sigma+1}) \quad \text{or} \quad \mathrm{rVal}_\sigma(g_{i'}) = \langle g_{i+1}\rangle + \sum_{\ell=i'}^{i} W_\ell + \mathrm{rVal}_\sigma(b_2).$$

The statement follows in either case since $i' < i < \mu^\sigma$ implies $\mathrm{rVal}_\sigma(b_{i'+1}) = \mathrm{rVal}_\sigma(b_2)$ and since $\sigma(b_{\mu^\sigma}) = g_{\mu^\sigma}$. As the same arguments can be applied to $\sigma\!e$, this implies $(b_{i'}, g_{i'}) \notin I_\sigma, I_{\sigma\!e}$. The case $i' = i$ can be shown by similar arguments since $\mathrm{rVal}_\sigma(b_{i+1}) = L_2$ and in particular $W_{\mu^\sigma} \subset L_2$ due to $\sigma(b_{\mu^\sigma}) = g_{\mu^\sigma}$. Hence, $(b_{i'},*) \in I_\sigma \Leftrightarrow (b_{i'},*) \in I_{\sigma\!e}$ for all $i' \in [n]$.

We next consider upper selection vertices $s_{i',j'}$ for arbitrary i',j'. We already proved that $(i',j') \neq (i,j)$ implies $(s_{i',j'}, b_1) \notin I_\sigma, I_{\sigma\!e}$. We thus only prove $(s_{i',j}, h_{i',j'}) \notin I_\sigma, I_{\sigma\!e}$ for arbitrary i',j'. This is immediate if $\sigma(s_{i',j'}) = h_{i',j'}$, so let $\sigma(s_{i',j'}) = b_1$. This implies $i' > i$ due to assumption 2 as well as $\mathrm{Val}_\sigma(s_{i',j'}) = \langle s_{i',j'}\rangle + \mathrm{Val}_\sigma(g_1)$ and $j' = 1 - \beta_{i'+1}$. We now distinguish several cases. First assume $i = \mu^\sigma - 1$, implying $\mathrm{rVal}_\sigma(g_1) = R_1$. Also, since $i' > i = \mu^\sigma - 1$, we have $i' \geq \mu^\sigma$. Assume $\beta_{i'+1} = 0$, implying $j' = 1 - \beta_{i'+1} = 1$. Thus, $\mathrm{rVal}_\sigma(h_{i',j'}) = \langle h_{i',j'}\rangle + \mathrm{rVal}_\sigma(g_{i'+1})$. By assumption 2, $F_{i'+1,\bar\sigma(g_{i'+1})}$ is either mixed or b_2-open. Since the valuation of the cycle center is larger if it is b_2-open, it suffices to consider this case. Using $\beta_{i'+1} = 0$, we thus have

$$\begin{aligned}\mathrm{rVal}_\sigma(h_{i',j'}) &\leq \langle h_{i',j'}, g_{i'+1}\rangle + \mathrm{rVal}_\sigma(b_2) = \langle h_{i',j'}, g_{i'+1}\rangle + L_2 \\ &< L_{i'+2} = L_{i'+1} \leq L_{\mu^\sigma+1} < \sum_{\ell<\mu^\sigma} W_\ell + L_{\mu^\sigma+1} = \mathrm{rVal}_\sigma(g_1),\end{aligned}$$

implying $(s_{i',j'}, h_{i',j'}) \notin I_\sigma$ if $\beta_{i'+1} = 0$. Consider the case $\beta_{i'+1} = 1$, implying $j' = 0$. Then

$$\mathrm{rVal}_\sigma(h_{i',j'}) < L_{i'+1} \leq L_{\mu^\sigma+1} < \sum_{\ell<\mu^\sigma} W_\ell + L_{\mu^\sigma+1} = \mathrm{rVal}_\sigma(g_1)$$

implying $(s_{i',j'}, h_{i',j'}) \notin I_\sigma$ if $\beta_{i'+1} = 1$. This concludes the case $i = \mu^\sigma - 1$. Hence assume $i < \mu^\sigma - 1$, implying $\mathrm{rVal}_\sigma(g_1) = \langle g_{i+1}\rangle + \sum_{\ell<i+1} W_\ell + \mathrm{rVal}_\sigma(b_2)$. Note that it thus might happen that $i' \leq \mu^\sigma$. Consider the case $\beta_{i'+1} = 0$. By the same arguments used for the case $i = \mu^\sigma - 1$, we then have $\mathrm{rVal}_\sigma(h_{i',j'}) < L_{i'+1}$. If $i' > \mu^\sigma$, then $i+1 < \mu^\sigma$ thus implies

$$\begin{aligned}\mathrm{rVal}_\sigma(h_{i',j'}) < L_{i'+1} \leq L_{\mu^\sigma+1} &< \langle g_{i+1}\rangle + \sum_{\ell<i+1} W_\ell + W_{\mu^\sigma} + L_{\mu^\sigma+1}\\ &= \langle g_{i+1}\rangle + \sum_{\ell<i+1} W_\ell + \mathrm{rVal}_\sigma(b_2) = \mathrm{rVal}_\sigma(g_1),\end{aligned}$$

hence $(s_{i',j'}, h_{i',j'}) \notin I_\sigma$. If $i' < \mu^\sigma$, then $i < i'$ implies

$$\begin{aligned}\mathrm{rVal}_\sigma(h_{i',j'}) &\leq \langle h_{i',j'}, g_{i'+1}\rangle + L_{2,i'} + L_{i'+1} = \langle h_{i',j'}, g_{i'+1}\rangle + L_{\mu^\sigma}\\ &< \langle g_{i+1}\rangle + \sum_{\ell<i+1} W_\ell + \mathrm{rVal}_\sigma(b_2) = \mathrm{rVal}_\sigma(g_1),\end{aligned}$$

hence $(s_{i',j'}, h_{i',j'}) \notin I_\sigma$. Thus consider the case $\beta_{i'+1} = 1$, implying $i' \geq \mu^\sigma - 1$. Since we consider the case $i < \mu^\sigma - 1$, it holds that $j = 1 - \beta_{i'+1} = 0$, implying

$$\begin{aligned}\mathrm{rVal}_\sigma(h_{i',j'}) &= \langle h_{i',j'}\rangle + \mathrm{rVal}_\sigma(b_{i'+2}) = \langle h_{i',j'}\rangle + L_{i'+2}\\ &< \langle g_{i+1}\rangle + \sum_{\ell<i+1} W_\ell + W_{i'+1} + L_{i'+2}\\ &= \langle g_{i+1}\rangle + \sum_{\ell<i+1} W_\ell + L_{i'+1} \leq \langle g_{i+1}\rangle + \sum_{\ell<i+1} W_\ell + L_{\mu^\sigma} = \mathrm{rVal}_\sigma(g_1),\end{aligned}$$

hence $(s_{i',j'}, h_{i',j'}) \notin I_\sigma$. Thus, under all circumstances, $(s_{i',j'}, h_{i',j'}) \notin I_\sigma$.

We now prove $(s_{i',j'}, h_{i',j'}) \notin I_{\sigma e}$. We have $\mathrm{rVal}_{\sigma e}(g_1) = \langle g_i\rangle + \sum_{\ell<i} W_\ell + \mathrm{rVal}_{\sigma e}(b_2)$. Let $\beta_{i'+1} = 0$ first. Then, using the same arguments used for σ as well as $i' > i$, we obtain

$$\mathrm{rVal}_{\sigma e}(h_{i',j'}) = \langle h_{i',j'}\rangle + \mathrm{rVal}_{\sigma e}(g_{i'+1}) < \langle g_i\rangle + \sum_{\ell<i} W_\ell + \mathrm{rVal}_{\sigma e}(b_2) = \mathrm{rVal}_{\sigma e}(g_1),$$

so $(s_{i',j'}, h_{i',j'}) \notin I_{\sigma e}$. Thus let $\beta_{i'+1} = 1$. Then, by the same arguments used before and $i' > i$, it follows that

$$\begin{aligned}\mathrm{rVal}_{\sigma e}(h_{i',j'}) &= \langle h_{i',j'}\rangle + \mathrm{rVal}_{\sigma e}(b_{i'+2}) = \langle h_{i',j'}\rangle + L_{i'+2} < \langle g_i\rangle + \sum_{\ell<i} W_\ell + W_{i'+1} + L_{i'+2}\\ &= \langle g_i\rangle + \sum_{\ell<i} W_\ell + L_{i'+1} \leq \langle g_i\rangle + \sum_{\ell<i} W_\ell + L_2 = \mathrm{rVal}_{\sigma e}(b_2),\end{aligned}$$

implying $(s_{i',j'}, h_{i',j'}) \notin I_{\sigma e}$. We thus have $(s_{i',j'}, *) \notin I_\sigma, I_{\sigma e}$ for all indices i', j' with the exception of the edge $(s_{i,j}, b_1)$.

Since there are no indices i', j', k' besides i, j, k such that $(d_{i',j',k'}, e_{i',j',k'}) \in I_\sigma$ by assumption, it suffices to prove $(d_{i',j',k'}, e_{i',j',k'}) \notin I_{\sigma e}$ for all such indices. The statement follows if $\sigma(d_{i',j',k'}) = e_{i',j',k'}$, hence assume $\sigma(d_{i',j',k'}) = F_{i',j'}$. Let $\sigma(e_{i',j',k'}) = b_2$. Then, since $\mathrm{Val}_\sigma(b_2) = \mathrm{Val}_{\sigma e}(b_2) = L_2$, the valuation of $e_{i',j',k'}$ does not increase by the application of e. As $(d_{i',j',k'}, e_{i',j',k'}) \notin I_\sigma$ implies $\mathrm{Val}_\sigma(F_{i',j'}) \geq \mathrm{Val}_\sigma(e_{i',j',k'})$ and the valuation of $F_{i',j'}$ can only increase, this implies $(d_{i',j',k'}, e_{i',j',k'}) \notin I_{\sigma e}$. Thus let $\sigma(e_{i',j',k'}) = g_1$, implying $\mathrm{rVal}_{\sigma e}(e_{i',j',k'}) = \langle g_i \rangle + \sum_{\ell<i} W_\ell + \mathrm{rVal}_{\sigma e}(b_2)$. Assume that $F_{i',j'}$ is not closed with respect to σ. Then the assumption $\sigma e(d_{i',j',k'}) = F_{i',j'}$ implies that the cycle center is halfopen with respect to σ. Due to the assumptions of this lemma, it is easy to see that this implies that $F_{i',j'}$ is g_1-halfopen with respect to both σ and σe and that we either have $i' = i$ and $j' = 1 - j$ or $i' < i < \mu^{\sigma e}$ and $j' = 1 - \overline{\sigma e}(g_{i'})$. In both cases, we have $j' = 1 - \overline{\sigma e}(g_{i'}) = \beta_{i'+1}$ by Property (BR1) and $\sigma e(s_{i',j'}) = h_{i',j'}$ by Property $(\mathrm{USV2})_{i',\beta_{i'+1}}$. Consequently, by Property $(\mathrm{EV1})_{i'+1}$ and $i' + 1 \leq i + 1 \leq \mu^{\sigma e}$, we obtain

$$\mathrm{rVal}_{\sigma e}(s_{i',j'}) = \langle s_{i',j'}, h_{i',j'} \rangle + \mathrm{rVal}_{\sigma e}(b_{i'+1}) = \langle s_{i',j'}, h_{i',j} \rangle + \mathrm{rVal}_{\sigma e}(b_{\mu^\sigma}) > \mathrm{rVal}_{\sigma e}(b_2),$$

implying $\mathrm{Val}_{\sigma e}(F_{i',j'}) > \mathrm{Val}_{\sigma e}(g_1)$ as $\mathrm{rVal}_{\sigma e}(b_2) > \mathrm{rVal}_{\sigma e}(g_1)$. Consequently, it holds that $(d_{i',j',k'}, e_{i',j',k'}) \notin I_{\sigma e}$. Now let $F_{i',j'}$ be closed with respect to σ. Consider the case $\beta_{i'} = 1 \wedge \beta_{i'+1} = j'$ first, implying $i' > i$ as $i < \mu^{\sigma e} = \nu$ by assumption. Then, Property $(\mathrm{USV1})_{i'}$ implies

$$\begin{aligned}\mathrm{rVal}_{\sigma e}(F_{i',j'}) &= \langle s_{i',j'}, h_{i',j'} \rangle + \mathrm{Val}_{\sigma e}(b_{i'+1}) = \langle s_{i',j'}, h_{i',j'} \rangle + L_{i'+1} > L_{2,i'} + L_{i'+1} \\ &> \langle g_i \rangle + \sum_{\ell<i} W_\ell + L_{2,i'} + L_{i'+1} = \langle g_i \rangle + \sum_{\ell<i} W_\ell + L_2 = \mathrm{rVal}_{\sigma e}(g_1),\end{aligned}$$

hence $(d_{i',j',k'}, e_{i',j',k'}) \notin I_{\sigma e}$. Next assume $\beta_{i'} = 1 \wedge \beta_{i'+1} \neq j'$. Then, Property $(\mathrm{USV1})_{i'}$ implies $\sigma e(s_{i',j'}) = b_1$ as $i' \geq \mu^{\sigma e} = \nu$. Since $F_{i',j'}$ is closed, we then have

$$\mathrm{rVal}_{\sigma e}(F_{i',j'}) = \mathrm{rVal}_{\sigma e}(s_{i',j'}) = \langle s_{i',j'} \rangle + \mathrm{rVal}_{\sigma e}(g_1) > \mathrm{rVal}_{\sigma e}(g_1),$$

hence $(d_{i',j',k'}, e_{i',j',k'}) \notin I_{\sigma e}$. Since $\beta_{i'} = 0 \wedge \beta_{i'+1} = j'$ is impossible if $F_{i',j'}$ is closed, it remains to consider the case $\beta_{i'} = 0 \wedge \beta_{i'+1} \neq j'$. By assumption 4, we then need to have $i' \leq i < \mu^{\sigma e}$ as well as $\sigma e(s_{i',j'}) = h_{i',j'}$. Consider the case $i' = i$ first, implying $j' = j$. As we just applied the switch $(d_{i,j,k}, e_{i,j,k})$, it is clear that this switch is not improving for σe. Hence consider $(d_{i,j,1-k}, e_{i,j,1-k})$. We have

$$\mathrm{Val}_{\sigma e}(F_{i,j}) = \frac{1-\varepsilon}{1+\varepsilon}\mathrm{Val}_{\sigma e}(b_2) + \frac{2\varepsilon}{1+\varepsilon}\mathrm{Val}_{\sigma e}(s_{i,j}).$$

If $\sigma(e_{i,j,1-k}) = g_1$, then $(d_{i,j,1-k}, e_{i,j,1-k}) \notin I_{\sigma e}$ due to $\mathrm{rVal}_{\sigma e}(b_2) > \mathrm{rVal}_{\sigma e}(g_1)$. It cannot happen that $\sigma(e_{i,j,1-k}) = b_2$ since this would imply $(d_{i,j,1-k}, e_{i,j,1-k}) \in I_\sigma$, contradicting the assumption. The reason is that W_{μ^σ} is not part of the valuation of $F_{i,j}$ which results in $\mathrm{rVal}_\sigma(F_{i,j}) = \mathrm{rVal}_\sigma(s_{i,j}) < \mathrm{rVal}_\sigma(b_2) = \mathrm{rVal}_\sigma(e_{i,j,1-k})$. Hence $(d_{i,j,1-k}, e_{i,j,1-k}) \notin I_{\sigma e}$ and we consider the case $i' < i$ next. Then, since $\overline{\sigma e}(d_{i'})$ by assumption 2 and since $i' < i < \mu^{\sigma e}$, we need to have $j' = \overline{\sigma e}(g_{i'}) = 1 - \beta_{i'+1} = 1$. Consequently,

$$\mathrm{rVal}_{\sigma e}(F_{i',j'}) = \mathrm{rVal}_{\sigma e}(s_{i',1}) = \langle s_{i',1}, h_{i',1} \rangle + \mathrm{rVal}_{\sigma e}(g_{i'+1})$$

$$= \langle s_{i',1}, h_{i',1} \rangle + \langle g_i \rangle + \sum_{\ell=i'+1}^{i-1} W_\ell + \mathrm{rVal}_{\sigma e}(b_2)$$

$$> \sum_{\ell=1}^{i'} W_\ell + \sum_{\ell=i'+1}^{i-1} W_\ell + \langle g_i \rangle + \mathrm{rVal}_{\sigma e}(b_2) = \mathrm{rVal}_{\sigma e}(g_1).$$

Thus, if $\sigma(e_{i',j',k'}) = g_1$, then $(d_{i',j',k'}, e_{i',j',k'}) \notin I_{\sigma e}$. Since

$$\mathrm{rVal}_\sigma(F_{i',j'}) \le \mathrm{rVal}_{\sigma e}(F_{i',j'}) < \mathrm{rVal}_{\sigma e}(b_2) = \mathrm{rVal}_\sigma(b_2),$$

also $\sigma(e_{i',j',k'}) = g_1$ has to hold since $\sigma(e_{i',j',k'}) = b_2$ implies $(d_{i',j',k'}, e_{i',j',k'}) \in I_\sigma$, contradicting our assumption. Consequently, $(d_{i',j',k'}, e_{i',j',k'}) \notin I_{\sigma e}$ for all indices i', j', k'.

It remains to consider edges $(d_{*,*,*}, F_{*,*})$. We prove that $(d_{i',j',k'}, F_{i',j'}) \in I_\sigma \Leftrightarrow (d_{i',j',k'}, F_{i',j'}) \in I_{\sigma e}$. If $\sigma(d_{i',j',k'}) = F_{i',j'}$, then $(d_{i',j',k'}, F_{i',j'}) \notin I_\sigma, I_{\sigma e}$ and the statement follows. Note that also $(d_{i,j,k}, F_{i,j}) \notin I_\sigma, I_{\sigma e}$. Hence fix some indices i', j', k' with $\sigma(d_{i',j',k'}) = e_{i',j',k'}$. Then, the cycle center $F_{i',j'}$ is not closed with respect to σ or σe, implying $\beta_{i'} = 0 \vee \beta_{i'+1} \neq j'$. Consider the case $\beta_{i'} = 0$ first and assume $i' > i$. By assumption 4., $F_{i',j'}$ is ether mixed or b_2-open. Consider the case that $F_{i',j'}$ is mixed. Then

$$\mathrm{rVal}_\sigma(F_{i',j'}) = \frac{1}{2}\mathrm{rVal}_\sigma(b_2) + \frac{1}{2}\mathrm{rVal}_\sigma(g_1),$$
$$\mathrm{rVal}_{\sigma e}(F_{i',j'}) = \frac{1}{2}\mathrm{rVal}_{\sigma e}(b_2) + \frac{1}{2}\mathrm{rVal}_{\sigma e}(g_1).$$

If $\sigma(e_{i',j',k'}) = g_1$, then $\mathrm{rVal}_\sigma(e_{i',j',k'}) < \mathrm{rVal}_\sigma(F_{i',j'})$ and $\mathrm{rVal}_{\sigma e}(e_{i',j',k'}) < \mathrm{rVal}_{\sigma e}(F_{i',j'})$ by Lemma 6.2.18, implying $(d_{i',j',k'}, F_{i',j'}) \in I_\sigma, I_{\sigma e}$. If $\sigma(e_{i',j',k'}) = b_2$, then Lemma 6.2.18 implies $\mathrm{rVal}_\sigma(e_{i',j',k'}) > \mathrm{rVal}_\sigma(F_{i',j'})$ and $\mathrm{rVal}_{\sigma e}(e_{i',j',k'}) > \mathrm{rVal}_{\sigma e}(F_{i',j'})$ and it follows that $(d_{i',j',k'}, F_{i',j'}) \notin I_\sigma, I_{\sigma e}$. Next assume that $F_{i',j'}$ is b_2-open. If $\beta_{i'+1} \neq j'$, then assumption 3 implies $\sigma(s_{i',j'}) = b_1$ and thus $\mathrm{Val}_\sigma(s_{i',j'}) = \langle s_{i',j'} \rangle + \mathrm{Val}_\sigma(g_1)$. By Lemma 6.2.18, it holds that $\langle s_{i',j'} \rangle + \mathrm{Val}_\sigma(g_1) < \mathrm{Val}_\sigma(b_2)$, implying $\mathrm{Val}_\sigma(F_{i',j'}) < \mathrm{Val}_\sigma(e_{i',j',k'})$. Since the same holds for σe, this implies $(d_{i',j',k'}, F_{i',j'}) \notin I_\sigma, I_{\sigma e}$. If $\beta_{i'+1} = j'$, then Property (EV1)$_{i'+1}$ implies that $\mathrm{Val}_\sigma(s_{i',j'}) = \langle s_{i',j'}, h_{i',j'} \rangle + \mathrm{Val}_\sigma(b_{i'+1})$ and thus $\mathrm{Val}_\sigma(s_{i',j'}) > \mathrm{Val}_\sigma(b_2)$ as $\langle s_{i',j'}, h_{i',j'} \rangle > L_{2,i'}$. Since the same holds for σe, we thus have $(d_{i',j',k'}, F_{i',j'}) \in I_\sigma, I_{\sigma e}$.

Next, assume $\beta_{i'+1} = 0$ and $i' \le i < \mu^{\sigma e}$. Then, $F_{i',j'}$ is closed for σ if $j' = \bar{\sigma}(g_{i'})$ and g_1-halfopen if $j' = 1 - \bar{\sigma}(g_{i'})$ by either assumption 1 or assumption 2. Since we assume $\sigma(d_{i',j',j'}) = e_{i',j',k'}$, it suffices to consider the second case. Then $\sigma(e_{i',j',k'}) = g_1$ and

$$\mathrm{Val}_\sigma(F_{i',j'}) = \frac{1-\varepsilon}{1+\varepsilon}\mathrm{Val}_\sigma(g_1) + \frac{2\varepsilon}{1+\varepsilon}\mathrm{Val}_\sigma(s_{i',j'}).$$

Now, by Property (BR1), $\bar{\sigma}(g_{i'}) = 1 - \beta_{i'+1}$, implying $1 - \bar{\sigma}(g_{i'}) = j' = \beta_{i'+1}$. Consequently, by Property (USV2)$_{i'}$ and Property (EV1)$_{i'+1}$, we have

$$\mathrm{Val}_\sigma(s_{i',j'}) = \langle s_{i',j'}, h_{i',j'} \rangle + \mathrm{Val}_\sigma(b_{i'+1}) > \mathrm{Val}_\sigma(b_2) > \mathrm{Val}_\sigma(g_1).$$

This implies $\mathrm{Val}_\sigma(F_{i',j'}) > \mathrm{Val}_\sigma(g_1) = \mathrm{Val}_\sigma(e_{i',j',k'})$, hence $(d_{i',j',k'}, F_{i',j'}) \in I_\sigma$. Since the same arguments can be used for σe, we also obtain $(d_{i',j',k'}, F_{i',j'}) \in I_{\sigma e}$.

Finally, consider the case $\beta_{i'} = 1 \wedge \beta_{i'+1} \neq j'$. Then $i' \geq \mu^\sigma > i$, hence $F_{i',j'}$ is either mixed or b_2-open by assumption 5. We however already showed that this implies $(d_{i',j',k'}, F_{i',j'}) \in I_\sigma \Leftrightarrow (d_{i',j',k'}, F_{i',j'}) \in I_{\sigma e}$. □

Lemma 6.2.26 (Fifth row of Table 6.6). *Let $G_n = M_n$. Let $\sigma \in \rho(\sigma_0)$ be a well-behaved phase-3-strategy for $\mathfrak{b} \in \mathfrak{B}_n$ with $\nu > 1$. Let $i < \mu^\sigma, j = 1 - \beta_{i+1}^\sigma$ and $e := (s_{i,j}, b_1) \in I_\sigma$. Further assume $\bar{\sigma}(eb_{i,j}) \wedge \neg\bar{\sigma}(eg_{i,j})$. Then σe is a well-behaved phase-3-strategy for $\mathfrak{b}$ with $I_{\sigma e} = I_\sigma \setminus \{e\}$ and $\sigma e \in \rho(\sigma_0)$.*

Proof. By the choice of e, σe is a phase-3-strategy for $\mathfrak{b}$ with $\sigma e \in \rho(\sigma_0)$. Consequently, by the choice of i and Property (REL2), we have $\nu = \mu^\sigma = \mu^{\sigma e} > 1$. We prove that σe is well-behaved. Since $\mu^{\sigma e} > 1$ and since $\overline{\sigma e}(eb_{i,j}) \wedge \neg\overline{\sigma e}(eg_{i,j})$, we only need to reevaluate Property (S2). We show that the premise of this property cannot be fulfilled. By Property (BR1), we have $\bar{\sigma}(g_i) = j$, hence $\neg\overline{\sigma e}(d_i)$. As $\mu^{\sigma e} > 1$ implies $\sigma e(b_1) = g_1$, assume $\sigma e(b_2) = g_2$. Then, by Property (BR1), $\overline{\sigma e}(g_1) = 0 \neq 1 = \overline{\sigma e}(b_1)$ and $\overline{\sigma e}(b_2)$. Consequently, $1 \in \mathcal{I}^{\sigma e}$. However, since $\overline{\sigma e}(b_{i'})$ implies $\overline{\sigma e}(g_{i'}) = \overline{\sigma e}(b_{i'+1})$ by Property (EV2)$_{i'}$ for $i' > 1$, this implies $\mathcal{I}^{\sigma e} = \{1\}$ and thus $\mu^{\sigma e} = 2$. But then, $i = 1$, hence the premise of the property cannot be fulfilled.

It remains to show that $I_{\sigma e} = I_\sigma \setminus \{e\}$. Applying $e = (s_{i,j}, b_1)$ increases the valuation of $F_{i,j}$. However, since $\overline{\sigma e}(eb_{i,j}) \wedge \neg\overline{\sigma e}(eg_{i,j})$, the valuation is only changed by terms of order $o(1)$. It is now easy but tedious to prove that the increase of the valuation of $F_{i,j}$ by terms of order $o(1)$ neither creates further improving switches nor makes improving switches unimproving. This implies the statement. □

Lemma 6.2.27 (Sixth row of Table 6.6). *Let $G_n = M_n$. Let $\sigma \in \rho(\sigma_0)$ be a well-behaved phase-3-strategy for $\mathfrak{b} \in \mathfrak{B}_n$. Let $i \in [n]$ and $j := 1 - \beta_{i+1}^\sigma$. Let $F_{i,j}$ be $t^\rightarrow$-halfopen and assume that $\beta_i^\sigma = 0$ implies that $F_{i,1-j}$ is $t^\leftarrow$-halfopen as well as $\sigma(g_i) = F_{i,j}$. Let $e := (d_{i,j,k}, e_{i,j,k}) \in I_\sigma$ and $\sigma(e_{i,j,k}) = t^\rightarrow$ for $k \in \{0,1\}$. Then σe is a well-behaved phase-3-strategy for $\mathfrak{b}$ with $\sigma e \in \rho(\sigma_0)$ and $I_{\sigma e} = I_\sigma \setminus \{e\}$.*

Proof. Since $F_{i,j}$ is $t^\rightarrow$-halfopen and by the choice of e, it holds that $\sigma(d_{i,j,1-k}) = e_{i,j,1-k}$. We can thus apply Lemma 6.2.20 and only need to prove $I_{\sigma e} = I_\sigma \setminus \{e\}$.

Since $F_{i,j}$ is $t^\rightarrow$-halfopen with respect to σ and $t^\rightarrow$-open with respect to σe, the valuation of $F_{i,j}$ only changes by terms of order $o(1)$ when applying the switch e. It is easy but tedious to verify that this implies $I_{\sigma e} = I_\sigma \setminus \{e\}$. □

Lemma 6.2.28 (Seventh row of Table 6.6). *Let $G_n = S_n$. Let $\sigma \in \rho(\sigma_0)$ be a well-behaved phase-3-strategy for $\mathfrak{b} \in \mathfrak{B}_n$ with $\nu > 1$. Let*

$$I_\sigma = \{(b_1, b_2)\} \cup \{(d_{i,j,k}, F_{i,j}), (e_{i,j,k}, b_2) : \sigma(e_{i,j,k}) = g_1\}$$

and $\sigma(d_{i,j,k}) = F_{i,j} \Leftrightarrow \beta_i^\sigma = 1 \wedge \beta_{i+1}^\sigma = j$ for all $i \in [n], j, k \in \{0,1\}$. Assume that σ has Property (ESC4)$_{i,j}$ for all $(i,j) \in S_1$ and Property (ESC5)$_{i,j}$ for all $(i,j) \in S_2$. Further assume that $\sigma(s_{i,j}) = h_{i,j}$ for all $i < \nu, j \in \{0,1\}$. Let $e := (b_1, b_2)$ and $m := \max\{i : \beta_i^\sigma = 1\}$. Then σe is a well-behaved phase-4-strategy for $\mathfrak{b}$ with $\mu^{\sigma e} = 1$ and

$$I_{\sigma e} = (I_\sigma \setminus \{e\}) \cup \{(s_{\nu-1,0}, b_1)\} \cup \{(s_{i,1}, b_1) : i \leq \nu - 2\} \cup X_0 \cup X_1$$

where X_k is defined as in Table 5.9.

Proof. We first show $\mu^{\sigma e} = 1$. Since σ is a phase-3-strategy for $\mathfrak{b}$, it has Property (EV1)$_i$ and Property (EV2)$_i$ for all $i > 1$. This implies $i \notin \mathcal{I}^{\sigma}$ for all $i > 1$ and thus, by the choice of e, $i \notin \mathcal{I}^{\sigma e}$ for all $i > 1$. Since $\sigma e(b_1) = b_2$, also $1 \notin \mathcal{I}^{\sigma e}$, hence $\mathcal{I}^{\sigma e} = \emptyset$, implying $\mu^{\sigma e} = \min\{i : \sigma e(b_i) = b_{i+1}\} = 1$.

We now show that σe is a phase-4-strategy for $\mathfrak{b}$. By the choice of e and the induced bit state, $\beta^{\sigma} = \beta^{\sigma e} = \mathfrak{b} + 1 =: \beta$. Since σ is a phase-3-strategy and by the choice of e it suffices to show that σ has Properties (EV1)$_1$, (EV2)$_2$, (EV3)$_1$, (EV3)$_\nu$, (CC2) and (REL1). Furthermore, we need to show that there is an index $i < \nu$ with $\sigma(s_{i,1-\beta_{i+1}}) = h_{i,1-\beta_{i+1}}$.

First, σe has Property (EV3)$_\nu$ since $\bar{\sigma}(d_{i,j}) \Leftrightarrow \beta_i = 1 \wedge \beta_{i+1} = j$. Second, the special condition as well as Property (ESC4)$_{i,j}$ and Property (ESC5)$_{i,j}$ are fulfilled for the relevant indices by assumption as well. In addition, $0 = \beta_1 = \overline{\sigma e}(d_{1,\beta_2})$ by definition. Thus, since $\sigma e(b_1) = b_2$, σe has Property (EV1)$_1$ and consequently also Property (EV2)$_1$ and Property (EV3)$_1$. In addition, σe has Property (CC2) since σ is a phase-3-strategy. Since $\mathcal{I}^{\sigma e} = \emptyset$, it also has Property (REL1). Hence, σe is a phase-4-strategy.

We show that σe is well-behaved. Since $\mu^{\sigma e} = 1$ and since the target of the vertex b_1 changed when transitioning to σe, the following assumptions need to be reevaluated.

(S1) Let $i \geq \mu^{\sigma e} = 1$ and $\sigma e(b_i) = g_i$. By Property (EV1)$_i$ this implies $i \geq \nu$ and thus, by Property (USV1)$_i$, $\overline{\sigma e}(s_i)$.

(D1) $\sigma e(b_i) = g_i$ implies $\overline{\sigma e}(d_i)$ by Property (EV1)$_i$ and Property (EV2)$_i$.

(MNS1) Assume that the premise was correct. Then, by Lemma 6.1.6, $m_b^{\sigma e} = 2$. This implies $\sigma e(b_2) = \sigma(b_2) = g_2$ and thus in particular $\mu^{\sigma} = \nu = 2$. But then, by Property (BR1) (applied to σ), this implies $\sigma(g_1) = \sigma e(g_1) = F_{1,0}$ and thus $\overline{m}_g^{\sigma e} = 1$. This however contradicts the premise.

(MNS2) Assume there was some $i < \overline{m}_g^{\sigma e} < \overline{m}_s^{\sigma e}, m_b^{\sigma e}$ and that $\neg\overline{\sigma e}(b_{\overline{m}_g^{\sigma e}+1})$. Then $m_b^{\sigma e} = \nu = \mu^{\sigma}$, implying $\overline{m}_g^{\sigma e} = m_b^{\sigma e} - 1 = \mu^{\sigma} - 1$ by Property (BR1). But then $\sigma e(b_{\overline{m}_g^{\sigma e}+1}) = \sigma e(b_{m_b^{\sigma e}}) = \sigma e(b_\nu) = g_\nu$ by Property (EV1)$_\nu$, contradicting the assumption.

(MNS3) Assume there was some index $i < \overline{m}_s^{\sigma e} \leq \overline{m}_g^{\sigma e} < m_b^{\sigma e}$ and let $\ell := \overline{m}_s^{\sigma e}$. Then, $\sigma e(s_{\ell,\overline{\sigma e}(g_\ell)}) = b_1$ and $\ell < m_b^{\sigma e} = \nu$. But this contradicts the assumption that $\sigma e(s_{i',j'}) = h_{i',j'}$ for all $i' < \nu, j' \in \{0,1\}$. This argument also applies to Properties (MNS4), (MNS5) and (MNS6), hence σe has all of these properties.

(EG*) It can easily be checked that for all indices $i \in [n], j \in \{0,1\}$ not listed in either of the sets S_1 or S_2, $\bar{\sigma}(d_{i,j})$ and thus $\overline{\sigma e}(d_{i,j})$ holds. Hence $\overline{\sigma e}(eg_{i,j}) \implies \overline{\sigma e}(eb_{i,j})$, so the premise of any of any of the assumptions (EG*) is incorrect.

(DN1) By Property (EV1)$_n$, $\overline{\sigma e}(d_n)$ implies $\overline{\sigma e}(b_n)$.

(DN2) We only need to consider this assumption if $\overline{m}_g^{\sigma e} = n$. Since $\mu^{\sigma} \neq 1$, this implies $\overline{\sigma e}(g_i) = \bar{\sigma}(g_i) = 1$ for all $i \in [n-1]$ by Property (BR1) (applied to σ). Thus, by assumption, $i \neq \mu^{\sigma} - 1$ for all of those i, hence $n = \mu^{\sigma} - 1$. But this implies $\mu^{\sigma} = n + 1$, contradicting the definition of μ^{σ}.

We now show the statements regarding the improving switches. First, $(s_{\nu-1,0}, b_1) \in I_{\sigma e}$ follows since $\sigma(s_{\nu-1,0}) = h_{\nu-1,0}$ by assumption and since $\mu^{\sigma e} = 1$ and $\sigma e(b_1) = b_2$ imply

$$\mathrm{rVal}^{\mathrm{S}}_{\sigma e}(b_1) = L_1^{\mathrm{S}} = L_\nu^{\mathrm{S}} = W_\nu^{\mathrm{S}} \cup L_{\nu+1}^{\mathrm{S}} \rhd \{h_{\nu-1,0}\} \cup L_{\nu+1}^{\mathrm{S}} = \mathrm{rVal}^{\mathrm{S}}_{\sigma e}(h_{\nu-1,0}).$$

We now show $(s_{i,1},b_1)\in I_{\sigma e}$ for all $i\leq\nu-2$. Fix some $i\leq\nu-2$. Then, since $i+1<\nu=\mu^\sigma$,

$$\begin{aligned}\mathrm{rVal}_\sigma^{\mathrm{S}}(h_{i,1})&=\{h_{i,1}\}\cup\mathrm{rVal}_\sigma^{\mathrm{S}}(g_{i+1})=\{h_{i,1}\}\cup R_{i+1}^{\mathrm{S}}\\&\rhd\bigcup_{i'=1}^{i}W_{i'}^{\mathrm{S}}\cup\bigcup_{i'=i+1}^{\mu^\sigma-1}W_{i'}^{\mathrm{S}}\cup L_{\mu^\sigma+1}^{\mathrm{S}}=R_1^{\mathrm{S}}=\mathrm{rVal}_\sigma^{\mathrm{S}}(b_1).\end{aligned}$$

Since $\sigma(s_{i,1})=\sigma e(s_{i,1})=h_{i,1}$, the statement then follows from

$$\mathrm{rVal}_{\sigma e}^{\mathrm{S}}(b_1)=L_1^{\mathrm{S}}=L_{\mu^\sigma}^{\mathrm{S}}=W_{\mu^\sigma}^{\mathrm{S}}\cup L_{\mu^\sigma+1}^{\mathrm{S}}\rhd\bigcup_{i'=i+1}^{\mu^\sigma-1}W_{i'}^{\mathrm{S}}\cup\{h_{i,1}\}\cup L_{\mu^\sigma+1}^{\mathrm{S}}=\mathrm{rVal}_{\sigma e}^{\mathrm{S}}(h_{i,1}).$$

We now show that the edges contained in the sets X_0 and X_1 are improving switches if β^σ is not a power of 2 and that no other edge is an improving switch otherwise. We distinguish the following cases.

1. Let $\beta=2^k$ for some $k\in\mathbb{N}$, implying $\mathrm{rVal}_{\sigma e}^{\mathrm{S}}(b_1)=L_1^{\mathrm{S}}=W_{\mu^\sigma}^{\mathrm{S}}$. By applying the improving switch $e=(b_1,b_2)$ the valuation of b_1 increased. The only vertices with edges towards b_1 are upper selection vertices. We hence show that for any vertex $s_{i,j}$, one of the following statements is true:
   a) $\sigma e(s_{i,j})=h_{i,j}$ and $\mathrm{Val}_{\sigma e}^{\mathrm{S}}(h_{i,j})\unrhd\mathrm{Val}_{\sigma e}^{\mathrm{S}}(b_1)$.
   b) $\sigma e(s_{i,j})=h_{i,j}$ and $(s_{i,j},b_1)\in I_{\sigma e}$.
   c) $\sigma e(s_{i,j})=b_1$ and $\tau^\sigma(F_{i,j}),\tau^{\sigma e}(F_{i,j})\neq s_{i,j}$.

   We distinguish the following cases:
   - **$i\leq\nu-2$ and $j=0$:** Then, $\sigma e(s_{i,0})=h_{i,0}$. Also, $\sigma e(h_{i,0})=b_{i+2}$, so $i+2\leq\mu^\sigma$ implies $\mathrm{rVal}_{\sigma e}^{\mathrm{S}}(h_{i,0})=\{h_{i,0}\}\cup\mathrm{rVal}_{\sigma e}^{\mathrm{S}}(b_{i+2})=\{h_{i,0}\}\cup W_{\mu^\sigma}^{\mathrm{S}}\rhd\mathrm{rVal}_{\sigma e}^{\mathrm{S}}(b_1)$.
   - **$i\leq\nu-2$ and $j=1$:** As proven before, all of these edges are improving.
   - **$i=\nu-1$ and $j=0$:** As proven before, $(s_{\mu^\sigma-1,0},b_1)$ is improving for σe.
   - **$i=\nu-1$ and $j=1$:** By assumption, it holds that $\sigma e(s_{i,1})=h_{i,1}$. Thus, by the choice of i, $\mathrm{rVal}_{\sigma e}^{\mathrm{S}}(h_{i,1})=\{h_{i,1}\}\cup\mathrm{rVal}_{\sigma e}^{\mathrm{S}}(g_{\mu^{\sigma e}})=\{h_{i,1}\}\cup W_{\mu^{\sigma e}}^{\mathrm{S}}\rhd\mathrm{rVal}_{\sigma e}^{\mathrm{S}}(b_1)$.
   - **$i=\nu$ and $j=0$:** Since $\beta^\sigma=2^k$, this then implies $\sigma e(s_{\nu,0})=h_{\nu,0}$ by Property (USV1)$_\nu$. But then, $\mathrm{rVal}_{\sigma e}^{\mathrm{S}}(h_{\nu,0})=\{h_{\nu,0}\}\rhd W_\nu^{\mathrm{S}}=\mathrm{rVal}_{\sigma e}^{\mathrm{S}}(b_1)$.
   - **$i=\nu$ and $j=1$:** Then, by Property (USV1)$_\nu$, $\sigma e(s_{\nu,1})=b_1$. We need to show $\tau^\sigma(F_{i,j})\neq s_{i,j}$ and $\tau^{\sigma e}(F_{i,j})\neq s_{i,j}$. This is done by showing that the first, second and fifth case of Lemma 6.1.16 cannot occur. The first case cannot occur since $j=1=1-\beta_{i+1}=1-\beta_{i+1}$ and both σ and σe have Property (EV3)$_\nu$. The second case cannot occur with respect to both σ and σe since there is no cycle center $F_{i,j}$ with $\bar\sigma(eg_{i,j})\wedge\neg\bar\sigma(eb_{i,j})$ by Property (ESC4)$_{i,j}$ and Property (ESC5)$_{i,j}$. The fifth case cannot occur for σe since $\mu^{\sigma e}=1$ and $\sigma e(s_{i,j})=b_1$. It can also not occur for σ since $\beta=2^k$ implies $\bar\sigma(eb_{\mu^\sigma,1})\wedge\bar\sigma(eg_{\mu^\sigma,1})$ by Property (ESC5)$_{\mu^\sigma,1}$.
   - **$i>\nu$ and $j=0$:** Since $\beta_{i'}=0$ for all $i'\neq\nu$, $i>\nu$ implies $\beta_i=\beta_{i+1}=0$. Hence, by Property (USV1)$_i$, $\sigma e(s_{i,0})=h_{i,0}$ and consequently $\mathrm{rVal}_{\sigma e}^{\mathrm{S}}(h_{i,0})=\{h_{i,0}\}\rhd W_{\mu^\sigma}^{\mathrm{S}}=\mathrm{rVal}_{\sigma e}^{\mathrm{S}}(b_1)$.

- **$i > \nu$ and $j = 1$:** Then $\sigma e(s_{i,j}) = \sigma e(s_{i,1}) = b_1$ by Property (USV1)$_i$, hence it suffices to show $\tau^{\sigma}(F_{i,1}), \tau^{\sigma e}(F_{i,1}) \neq s_{i,1}$. This is again proven by showing that the first, second and fifth case of Lemma 6.1.16 cannot be fulfilled. Since $\beta = 2^k$ for some $k \in \mathbb{N}$ by assumption, $m = \max\{i : \sigma e(b_i) = g_i\} = \nu$. Hence, by Property (ESC5)$_{i,1}$ (resp. by assumption), we have $\overline{\sigma e}(eb_{i,1}) \wedge \overline{\sigma e}(eg_{i,j})$ and $\bar{\sigma}(eb_{i,j}) \wedge \bar{\sigma}(eg_{i,j})$. Consequently, either the sixth or the seventh case of Lemma 6.1.16 is true, both implying $\tau^{\sigma e}(F_{i,1}), \tau^{\sigma}(F_{i,1}) \neq s_{i,1}$.

2. Now assume that there is no $k \in \mathbb{N}$ such that $\beta = 2^k$. We prove $X_0, X_1 \subseteq I_{\sigma e}$ and that the edges contained in $I_{\sigma e}$ according to the lemma are indeed improving. Fix some $k \in \{0, 1\}$. We prove $X_k \subseteq I_{\sigma e}$.
   - We first show $(d_{i,j,k}, F_{i,j}) \in I_{\sigma e}$ where $i = \mu^{\sigma} = \nu$ and $j = 1 - \beta_{i+1}$. By assumption, $\sigma(d_{i,j,k}) = \sigma e(d_{i,j,k}) \neq F_{i,j}$. Hence $\sigma e(d_{i,j,k}) = \sigma(d_{i,j,k}) = e_{i,j,k}$ and it suffices to show $\mathrm{Val}^{\mathrm{S}}_{\sigma e}(F_{i,j}) \rhd \mathrm{Val}^{\mathrm{S}}_{\sigma e}(e_{i,j,k})$. Since $\sigma e(d_{i,j,k}) = e_{i,j,k}$, Property (ESC4)$_{i,j}$ implies $\sigma e(e_{i,j,k}) = b_2$, so $\mathrm{Val}^{\mathrm{S}}_{\sigma e}(e_{i,j,k}) = \{e_{i,j,k}\} \cup \mathrm{Val}^{\mathrm{S}}_{\sigma e}(b_2)$. Since $\neg\overline{\sigma e}(s_{i,j})$ by the choice of j and Property (USV1)$_i$, $\mu^{\sigma e} = 1$, $\overline{\sigma e}(eb_{i,j})$ and $\neg\overline{\sigma e}(eg_{i,j})$, Lemma 6.1.16 thus implies

   $$\mathrm{Val}^{\mathrm{S}}_{\sigma e}(F_{i,j}) = \{F_{i,j}, d_{i,j,k'}, e_{i,j,k'}\} \cup \mathrm{Val}^{\mathrm{S}}_{\sigma e}(b_2) \rhd \{e_{i,j,k}\} \cup \mathrm{Val}^{\mathrm{S}}_{\sigma e}(b_2) = \mathrm{Val}^{\mathrm{S}}_{\sigma e}(e_{i,j,k})$$

   for some $k' \in \{0, 1\}$. Hence $(d_{i,j,k}, F_{i,j}) \in I_{\sigma e}$.
   - Let $i \in \{\nu+1, \ldots, m-1\}$ with $\beta_i = 0$ and $j = 1 - \beta_{i+1}$. We prove $\sigma e(d_{i,j,k}) \neq F_{i,j}$ and $\mathrm{Val}^{\mathrm{S}}_{\sigma e}(F_{i,j}) \rhd \mathrm{Val}^{\mathrm{S}}_{\sigma e}(e_{i,j,k})$. However, since $\overline{\sigma e}(eb_{i,j}) \wedge \neg\overline{\sigma e}(eg_{i,j})$ this can be shown by the same arguments used before.

   We prove that no other edge becomes an improving switch. Let (i, j) be a pair of indices for which the edge $(s_{i,j}, b_1)$ does not become improving for σe. By our assumptions on I_{σ}, it then suffices to prove that one of the following three cases is true.

   a) $\sigma e(s_{i,j}) = h_{i,j}$ and $\mathrm{Val}^{\mathrm{S}}_{\sigma e}(h_{i,j}) \unrhd \mathrm{Val}^{\mathrm{S}}_{\sigma e}(b_1)$ or
   b) $\sigma e(s_{i,j}) = b_1$ and $\tau^{\sigma}(F_{i,j}), \tau^{\sigma e}(F_{i,j}) \neq s_{i,j}$ or
   c) $\sigma e(s_{i,j}) = b_1, j = 1 - \bar{\sigma}(g_i)$ and $\mathrm{Val}^{\mathrm{S}}_{\sigma e}(F_{i,1-j}) > \mathrm{Val}^{\mathrm{S}}_{\sigma e}(F_{i,j})$.

   We distinguish the following cases:
   - **$i \leq \nu - 1$ and $j \in \{0, 1\}$:** Then, the statement follows by the same arguments used for the corresponding cases for $\beta = 2^k, k \in \mathbb{N}$.
   - **$i = \nu$ and $j = \beta_{\nu+1}$:** Then, $\sigma e(s_{i,j}) = h_{i,j}$ by Property (USV1)$_{\nu}$. Hence, by Property (EV1)$_{i+1}$ and since $\{h_{\nu,j}\} \rhd L_{1,\nu}$,

   $$\mathrm{rVal}^{\mathrm{S}}_{\sigma e}(h_{i,j}) = \{h_{\nu,j}\} \cup \mathrm{rVal}^{\mathrm{S}}_{\sigma e}(b_{\nu+1}) = \{h_{\nu,j}\} \cup L_{\nu+1} \rhd L^{\mathrm{S}}_1 = \mathrm{rVal}^{\mathrm{S}}_{\sigma e}(b_1).$$

   - **$i = \nu$ and $j = 1 - \beta_{\nu+1}$:** Then, $\sigma e(s_{i,j}) = b_1$ by Property (USV1)$_{\nu}$ and $\sigma e(g_i) = F_{i,1-j}$ by Property (EV2)$_{\nu}$. We prove $\mathrm{rVal}^{\mathrm{S}}_{\sigma e}(F_{i,1-j}) \rhd \mathrm{rVal}^{\mathrm{S}}_{\sigma e}(F_{i,j})$. Note that we do not need to consider the cycle vertices here as we proved that the corresponding edges become improving for σe. Since $(i, j) \in S_1$, σe has Property (ESC4)$_{i,j}$. Thus, $\overline{\sigma e}(eb_{i,j}) \wedge \neg\overline{\sigma e}(eg_{i,j}) \wedge \mu^{\sigma e} = 1$, implying $\mathrm{rVal}_{\sigma e}(F_{i,j}) =$

$\mathrm{rVal}_{\sigma e}(b_2)$. Since $F_{i,1-j}$ is closed by Property (EV1)$_\nu$, $\sigma e(s_{i,1-j}) = h_{i,1-j}$ by Property (USV1)$_i$, Property (EV1)$_{i+1}$ and the choice of i imply

$$\begin{aligned}\mathrm{rVal}^{\mathrm{S}}_{\sigma e}(F_{i,1-j}) &= \{s_{i,1-j}, h_{i,1-j}\} \cup \mathrm{rVal}^{\mathrm{S}}_{\sigma e}(b_{i+1}) \rhd W^{\mathrm{S}}_i \cup \mathrm{rVal}^{\mathrm{S}}_{\sigma e}(b_{i+1})\\ &= \mathrm{rVal}^{\mathrm{S}}_{\sigma e}(b_i) = \mathrm{rVal}^{\mathrm{S}}_{\sigma e}(b_\nu) = \mathrm{rVal}^{\mathrm{S}}_{\sigma e}(b_2) = \mathrm{rVal}^{\mathrm{S}}_{\sigma e}(F_{i,j}).\end{aligned}$$

- $\boldsymbol{i \in \{\nu+1,\dots,m-1\}, \beta_i = 0}$ **and** $\boldsymbol{j = \beta_{i+1}}$: Again, $\sigma e(s_{i,j}) = h_{i,j}$ by Property (USV1)$_i$ in this case. By Property (EV1)$_{i+1}$ we then obtain

$$\mathrm{rVal}^{\mathrm{S}}_{\sigma e}(h_{i,j}) = \{h_{i,j}\} \cup \mathrm{rVal}^{\mathrm{S}}_{\sigma e}(b_{i+1}) = \{h_{i,j}\} \cup L^{\mathrm{S}}_{i+1} \rhd L^{\mathrm{S}}_1 = \mathrm{rVal}^{\mathrm{S}}_{\sigma e}(b_1).$$

- $\boldsymbol{i \in \{\nu+1,\dots,m-1\}, \beta_i = 0}$ **and** $\boldsymbol{j = 1-\beta_{i+1}}$: Then, by Property (USV1)$_i$, we have $\sigma e(s_{i,j}) = b_1$. In addition, $(i,j) \in S_1$ and $(i,1-j) \in S_2$, implying $\bar\sigma(eb_{i,j}) \wedge \neg\bar\sigma(eg_{i,j})$ as well as $\bar\sigma(eb_{i,1-j}) \wedge \neg\bar\sigma(eg_{i,1-j})$. By Property (EBG3) and since only cycle centers $F_{i',\beta_{i'+1}}$ are closed by assumption, $\nu > 1$ implies $\bar\sigma(g_1) \neq \bar\sigma(b_2)$. Consequently, by Lemma 6.1.16 and since player 1 always chooses the vertex minimizing the valuation,

$$\mathrm{rVal}^{\mathrm{S}}_{\sigma e}(F_{i,j}) = \mathrm{rVal}^{\mathrm{S}}_{\sigma e}(b_2) > \mathrm{rVal}^{\mathrm{S}}_{\sigma e}(g_1) = \mathrm{rVal}^{\mathrm{S}}_{\sigma e}(F_{i,1-j}).$$

By our assumptions on I_σ, this implies that it holds that $\sigma(g_i) = \sigma e(g_i) = F_{i,j}$. We thus prove $\sigma e(b_i) = b_{i+1}$ and $\mathrm{rVal}^{\mathrm{S}}_{\sigma e}(b_{i+1}) > \mathrm{rVal}^{\mathrm{S}}_{\sigma e}(g_i)$ to prove $(b_i, g_i) \notin I_\sigma$ and $\sigma e(s_{i-1,1}) = b_1$ and $\mathrm{rVal}^{\mathrm{S}}_{\sigma e}(b_1) > \mathrm{rVal}^{\mathrm{S}}_{\sigma e}(h_{i-1,1})$ to prove $(s_{i-1,1}, h_{i-1,1}) \notin I_{\sigma e}$.

First, $\sigma e(b_i) = b_{i+1}$ follows from Property (EV1)$_i$ whereas $\sigma e(s_{i-1,1}) = b_1$ follows from $i-1 \geq \nu$ and Property (USV1)$_{i-1}$. Since we need to analyze $\mathrm{rVal}^{\mathrm{S}}_{\sigma e}(g_i)$ using Corollary 6.1.18, we determine λ^{S}_i. However, since $\sigma e(s_{i,j}) = b_1$ and $\overline{\sigma e}(g_i) = j$, this lemma implies $\mathrm{rVal}^{\mathrm{S}}_{\sigma e}(g_i) = \{g_i\} \cup \mathrm{rVal}^{\mathrm{S}}_{\sigma e}(g_1)$. Since the conditions of the third case of Lemma 6.1.14 are fulfilled (by Property (BR1) applied to σ, Property (EV1)$_{i'}$ for $i' \leq \nu$ and our assumption),

$$\begin{aligned}\mathrm{rVal}^{\mathrm{S}}_{\sigma e}(g_i) &= \{g_i\} \cup \mathrm{rVal}^{\mathrm{S}}_{\sigma e}(g_1) = \{g_i\} \cup \bigcup_{i'=1}^{\nu-1} W^{\mathrm{S}}_{i'} \cup \mathrm{rVal}^{\mathrm{S}}_{\sigma}(b_{\nu+1})\\ &= \{g_i\} \cup \bigcup_{i'=1}^{\nu-1} W^{\mathrm{S}}_{i'} \cup L^{\mathrm{S}}_{\nu+1,i-1} + L^{\mathrm{S}}_{i+1} \lhd L^{\mathrm{S}}_{i+1} = \mathrm{rVal}^{\mathrm{S}}_{\sigma e}(b_{i+1}),\end{aligned}$$

$$\mathrm{rVal}^{\mathrm{S}}_{\sigma e}(h_{i-1,1}) = \{h_{i-1,1}, g_i\} \cup \mathrm{rVal}^{\mathrm{S}}_{\sigma e}(g_1) \lhd \mathrm{rVal}^{\mathrm{S}}_{\sigma e}(g_1) \lhd \mathrm{rVal}^{\mathrm{S}}_{\sigma e}(b_2) = \mathrm{rVal}^{\mathrm{S}}_{\sigma e}(b_1).$$

- $\boldsymbol{i \in \{\nu+1,\dots,m-1\}, \beta_i = 1}$ **and** $\boldsymbol{j = \beta_{i+1}}$: As before, $\sigma e(s_{i,j}) = h_{i,j}$ by Property (USV1)$_i$. The statement thus follows by the same arguments used before.
- $\boldsymbol{i \in \{\nu+1,\dots,m-1\}, \beta_i = 1}$ **and** $\boldsymbol{j = 1-\beta_{i+1}}$: Then, $\sigma e(s_{i,j}) = b_1$ by Property (USV1)$_i$. We prove $\tau^\sigma(F_{i,j}), \tau^{\sigma e}(F_{i,j}) \neq s_{i,j}$. By Property (ESC5)$_{i,j}$, both $\overline{\sigma e}(eg_{i,j}) \wedge \overline{\sigma e}(eb_{i,j})$ and $\bar\sigma(eg_{i,j}) \wedge \bar\sigma(eb_{i,j})$ hold. Hence, by Lemma 6.1.16, $\tau^\sigma(F_{i,j}), \tau^{\sigma e}(F_{i,j}) \neq s_{i,j}$.

- **$i \geq m$ and $j = \beta_{i+1}$:** By the choice of i, we then have $\beta_i = 0$. For $i \neq n$, the statements follows similar to the last cases. For $i = n$, we have
$$\mathrm{rVal}^{\mathrm{S}}_{\sigma e}(h_{i,0}) = \{h_{n,0}\} \rhd \bigcup_{i' \geq 1} \{W^{\mathrm{S}}_{i'} \colon \sigma e(b_{i'}) = g_{i'}\} = L^{\mathrm{S}}_1 = \mathrm{rVal}^{\mathrm{S}}_{\sigma e}(b_1).$$
- **$i \geq m$ and $j = 1 - \beta_{i+1}$:** Then, it holds that $\sigma e(s_{i,j}) = b_1$ by Property (USV1)$_i$. Hence we need to show $\tau^{\sigma}(F_{i,j}), \tau^{\sigma e}(F_{i,j}) \neq s_{i,j}$. However, this follows immediately from Lemma 6.1.16 since Property (ESC5) implies $\bar{\sigma}(eb_{i,j}) \wedge \bar{\sigma}(eg_{i,j})$ as well as $\overline{\sigma e}(eb_{i,j}) \wedge \overline{\sigma e}(eg_{i,j})$. □

Lemma 6.2.29 (Eighth row of Table 6.6). *Let $G_n = M_n$. Let $\sigma \in \rho(\sigma_0)$ be a well-behaved phase-3-strategy for $\mathfrak{b} \in \mathfrak{B}_n$ with $\nu > 1$. Let*
$$I_{\sigma} = \{(b_1, b_2)\} \cup \{(d_{i,j,k}, F_{i,j}), (e_{i,j,k}, b_2) \colon \sigma(e_{i,j,k}) = g_1\}.$$
Let σ have Property (USV1)$_i$ for all $i \in [n]$ and let $\sigma(d_{i,j,k}) = F_{i,j} \Leftrightarrow \beta^{\sigma}_i = 1 \wedge \beta^{\sigma}_{i+1} = j$ for all $i \in [n], j, k \in \{0,1\}$. Let σ have Property (ESC4)$_{i,j}$ for all $(i,j) \in S_1$ and Property (ESC5)$_{i,j}$ for all $(i,j) \in S_2$. Further assume that $e := (b_1, b_2) \in I_{\sigma}$ and let $m := \max\{i \colon \beta^{\sigma}_i = 1\}$. Then, σe is a well-behaved phase-5-strategy for $\mathfrak{b}$ with $\mu^{\sigma e} = 1$ and
$$I_{\sigma e} = (I_{\sigma} \setminus \{e\}) \cup \{(d_{i,1-\beta^{\sigma}_{i+1},k}, F_{i,1-\beta^{\sigma}_{i+1}}) \colon i < \nu\} \cup X_0 \cup X_1$$
where X_k is defined as in Table 5.9.

Proof. We begin by proving that σe is a phase-5-strategy. Since $\beta^{\sigma e} = \beta^{\sigma} = \mathfrak{b} + 1 =: \beta$ and $\nu > 1$, σe has Properties (EV1)$_i$, (EV2)$_i$ and (EV3)$_i$ for all $i \in [n]$. Also, σe does not have Property (ESC1) as it has Property (ESC5)$_{i,j}$ for all $(i,j) \in S_2$ and $S_2 \neq \emptyset$. Therefore, as σe has Property (USV1)$_i$ for all $i \in [n]$ by assumption, it is a phase-5-strategy for $\mathfrak{b}$. We next prove that σe is well-behaved. Since $\mu^{\sigma} \neq 1$ but $\mu^{\sigma e} = 1$ as $\mathcal{I}^{\sigma e} = \emptyset$ due to the choice of e, we need to reevaluate the following properties.

(S1) By Properties (USV1)$_i$ and (EV2)$_i$, $\sigma e(b_i) = g_i$ implies $\overline{\sigma e}(s_i)$ for all $i \geq 1$.

(D1) By Properties (EV1)$_i$ and (EV2)$_i$, $\sigma e(b_i) = g_i$ implies $\overline{\sigma e}(d_i)$.

(MNS2) Assume there was some $i < \overline{m}^{\sigma e}_g < \overline{m}^{\sigma e}_s, m^{\sigma e}_b$. Then $1 < \overline{m}^{\sigma e}_g$, implying $\sigma e(g_1) = F_{1,1}$. By the choice of i, it holds that $m^{\sigma e}_b \geq 3$, hence $\sigma e(b_2) = b_3$. But then, Property (USV1)$_1$ implies $\sigma e(s_{1,\overline{\sigma e}(g_1)}) = \sigma e(s_{1,1}) = b_1$, contradicting $\overline{m}^{\sigma e}_g < \overline{m}^{\sigma e}_s$.

(MNS3) Assume there was some $i < \overline{m}^{\sigma e}_s \leq \overline{m}^{\sigma e}_g < m^{\sigma e}_b$. Then $1 < \overline{m}^{\sigma e}_s \leq \overline{m}^{\sigma e}_g$, implying $\sigma e(g_1) = F_{1,1}$ and $\sigma e(s_{1,1}) = h_{1,1}$. Hence, by Property (USV1)$_1$, it holds that $\sigma e(b_2) = g_2$ and thus $m^{\sigma e}_b = 2$. But this is a contradiction as the premise implies $m^{\sigma e}_b > 3$.

(MNS4) If $\overline{m}^{\sigma e}_s > 1$, then the same arguments used for Property (MNS3) can be used again. Hence consider the case $\overline{m}^{\sigma e}_s = 1$. Then, $\sigma e(s_{1,\bar{\sigma}(g_1)}) = b_1$. In particular, Property (USV1)$_1$ implies $\bar{\sigma}(g_1) \neq \beta_2$, hence $\overline{\sigma e}(g_1) = 1 - \beta^{\sigma e}_2$. But then, by Property (ESC4)$_{1,1-\beta_2}$, we have $\overline{\sigma e}(eb_{\overline{m}^{\sigma e}_s}) \wedge \neg\overline{\sigma e}(eg_{\overline{m}^{\sigma e}_s})$.

(MNS5) Assuming that there was some $i < \overline{m}_s^{\sigma e} < m_b^{\sigma e} \leq \overline{m}_b^{\sigma e}$ yields the same contradiction devised for Property (MNS3).

(MNS6) If $\overline{m}_s^{\sigma e} > 1$, then the same arguments used for Equation (MNS3) can be used to show that the premise cannot hold. Hence assume $\overline{m}_s^{\sigma e} = 1$. But then, the same arguments used for Property (MNS4) can be used to prove the statement.

(EG*) It is easy to verify that each cycle center is either closed, escapes only to b_2 or to both b_2 and g_1. In particular, there is no cycle center $F_{i,j}$ with $\bar{\sigma}(eg_{i,j}) \wedge \neg\bar{\sigma}(eb_{i,j})$.

(DN1) By Property $(\text{EV1})_n$, $\overline{\sigma e}(d_n)$ implies $\sigma e(b_n) = g_n$.

(DN2) We only need to consider this assumption if $\overline{m}_g^{\sigma e} = n$. Since $\mu^\sigma \neq 1$, this implies $\overline{\sigma e}(g_i) = \bar{\sigma}(g_i) = 1$ for all $i \in \{1, \ldots, n-1\}$ by Property (BR1) (applied to σ). Thus, by assumption, $i \neq \mu^\sigma - 1$ for all of those i, hence $n = \mu^\sigma - 1$. But this implies $\mu^\sigma = n + 1$, contradicting Property (REL2) for σ.

It remains to prove the statement regarding the improving switches. We observe that $\text{Val}_{\sigma e}^{\text{M}}(g_1) < \text{rVal}_{\sigma e}^{\text{M}}(b_2)$ since $\sigma(e_{i,j,k}) = g_1$ implies $(e_{i,j,k}, b_2) \in I_\sigma$.

Let $i < \nu, j := 1 - \beta_{i+1}$ and $k \in \{0, 1\}$. We prove $(d_{i,j,k}, F_{i,j}) \in I_{\sigma e}$. By assumption, the cycle center $F_{i,j}$ is open, so in particular $\sigma e(d_{i,j,k}) \neq F_{i,j}$. By the choice of i and j, Property $(\text{ESC4})_{i,j}$ implies $\overline{\sigma e}(eb_{i,j}) \wedge \neg\overline{\sigma e}(eg_{i,j})$. Consequently, by Lemma 6.1.15, $\text{Val}_{\sigma e}^{\text{M}}(F_{i,j}) = (1 - \varepsilon)\,\text{Val}_{\sigma e}^{\text{M}}(b_2) + \varepsilon \cdot \text{Val}_{\sigma e}^{\text{M}}(s_{i,j})$. Since $\sigma e(b_1) = b_2$, the choice of j and Property $(\text{USV1})_i$ imply $\text{Val}_{\sigma e}^{\text{M}}(s_{i,j}) = \langle s_{i,j} \rangle + \text{Val}_{\sigma e}^{\text{M}}(b_2)$. Thus, $(d_{i,j,k}, F_{i,j}) \in I_{\sigma e}$ follows from $\text{Val}_{\sigma e}^{\text{M}}(F_{i,j}) = (1 - \varepsilon)\,\text{Val}_{\sigma e}^{\text{M}}(b_2) + \varepsilon\,\text{Val}_{\sigma e}^{\text{M}}(s_{i,j}) > \text{Val}_{\sigma e}^{\text{M}}(b_2) = \text{Val}_{\sigma e}^{\text{M}}(e_{i,j,k})$.

We prove that X_0, X_1 are improving for σe if β is not a power of two. Fix $k \in \{0, 1\}$ and let $i := \nu, j := 1 - \beta_{\nu+1}$. We begin by proving $(d_{i,j,k}, F_{i,j}) \in I_{\sigma e}$. By the choice of j and our assumptions, $\sigma e(d_{i,j,k}) \neq F_{i,j}$. In addition Property $(\text{ESC4})_{i,j}$ implies that $\overline{\sigma e}(eb_{i,j}) \wedge \neg\overline{\sigma e}(eg_{i,j})$. Since this implies $\text{Val}_{\sigma e}(F_{i,j}) = (1-\varepsilon)\,\text{Val}_{\sigma e}^{\text{M}}(b_2) + \varepsilon\,\text{Val}_{\sigma e}^{\text{M}}(s_{i,j})$ as well as $\text{Val}_{\sigma e}^{\text{M}}(e_{i,j,k}) = \text{Val}_{\sigma e}^{\text{M}}(b_2)$, it suffices to prove $\text{Val}_{\sigma e}^{\text{M}}(s_{i,j}) > \text{Val}_{\sigma e}^{\text{M}}(b_2)$. This however follows directly since Property $(\text{USV1})_i$ and the choice of j imply $\text{Val}_{\sigma e}^{\text{M}}(s_{i,j}) = \langle s_{i,j} \rangle \cup \text{Val}_{\sigma e}^{\text{M}}(b_2)$. By applying the same arguments, we also obtain $(d_{i,j,k}, F_{i,j}) \in I_{\sigma e}$ for $i \in \{\nu + 1, \ldots, m - 1\}$ with $\beta_i = 0$ and $j = 1 - \beta_{i+1}$ as $(i, j) \in S_1$ for these indices.

We now prove that no further improving switch is created. Note that no additional improving switches $(d_{i,j,k}, F_{i,j})$ but the ones discussed earlier are created in any case. The reason is that the only indices (i, j) with $(i, j) \in S_1$ are $i < \nu$ and $j = 1 - \beta_{i+1}$ if β is a power of 2. All other indices (i, j) are contained in S_2 since $\nu = m$. Consequently, $\text{rVal}_{\sigma e}^{\text{M}}(F_{i,j}) = \frac{1}{2}\,\text{rVal}_{\sigma e}^{\text{M}}(b_2) + \text{rVal}_{\sigma e}^{\text{M}}(g_1) < \text{rVal}_{\sigma e}^{\text{M}}(b_2)$. By the same argument, no further improving switch $(d_{i,j,k}, F_{i,j})$ besides the ones discussed earlier is created for the case that β is not a power of 2.

The application of e increases the valuation of the vertex b_1. The only vertices that have an edge towards b_1 are upper selection vertices $s_{i,j}$. As we fully covered the cycle vertices, it now suffices to prove that the following statements hold:

1. If $\sigma e(s_{i,j}) = h_{i,j}$, then $(s_{i,j}, b_1) \notin I_\sigma, I_{\sigma e}$.
2. If $\sigma e(s_{i,j}) = b_1$ and $\overline{\sigma e}(g_i) \neq j$, then $(g_i, F_{i,j}) \notin I_\sigma, I_{\sigma e}$.
3. If $\sigma e(s_{i,j}) = b_1$ and $\overline{\sigma e}(g_i) = j$, then $\text{Val}_{\sigma e}^{\text{M}}(g_i) - \text{Val}_\sigma^{\text{M}}(g_i) \in o(1)$.

Consider the case $\sigma e(s_{i,j}) = h_{i,j}$ first. Then, by Property (USV1)$_i$, $j = \beta_{i+1}$. Consequently, since $\sigma e(b_1) = b_2$ and since $\langle h_{i,j}\rangle > \sum_{\ell\in[i]} W_\ell^{\mathrm{M}}$, Property (EV1)$_{i+1}$ yields

$$\mathrm{rVal}_{\sigma e}^{\mathrm{M}}(h_{i,j}) = \langle h_{i,j}\rangle + \mathrm{rVal}_{\sigma e}^{\mathrm{M}}(b_{i+1}) = \langle h_{i,j}\rangle + L_{i+1}^{\mathrm{M}} > L_{1,i}^{\mathrm{M}} + L_{i+1}^{\mathrm{M}} = L_1^{\mathrm{M}} = \mathrm{rVal}_{\sigma e}^{\mathrm{M}}(b_1).$$

Since $\mathrm{rVal}_\sigma(h_{i,j}) = \mathrm{rVal}_{\sigma e}(h_{i,j})$ and $\mathrm{rVal}_{\sigma e}^{\mathrm{M}}(b_1) > \mathrm{rVal}_\sigma^{\mathrm{M}}(b_1)$, this implies $(s_{i,j}, b_1) \notin I_\sigma, I_{\sigma e}$. Now consider the case $\sigma e(s_{i,j}) = b_1$ and $\overline{\sigma e}(g_i) \neq j$. Then, by Property (USV1)$_i$, it holds that $j = 1 - \beta_{i+1}$ and thus, by Property (EV1)$_{i+1}$

$$\begin{aligned}\mathrm{rVal}_{\sigma e}^{\mathrm{M}}(s_{i,j}) &= \langle s_{i,j}\rangle + \mathrm{rVal}_{\sigma e}^{\mathrm{M}}(b_1) = \langle s_{i,j}\rangle + L_\nu^{\mathrm{M}},\\ \mathrm{rVal}_{\sigma e}^{\mathrm{M}}(h_{i,1-j}) &= \langle h_{i,1-j}\rangle + \mathrm{rVal}_{\sigma e}^{\mathrm{M}}(b_{i+1}) = \langle h_{i,1-j}\rangle + L_{i+1}^{\mathrm{M}} > \langle s_{i,j}\rangle + L_\nu^{\mathrm{M}}.\end{aligned}$$

We prove that this implies $\mathrm{Val}_{\sigma e}^{\mathrm{M}}(F_{i,1-j}) > \mathrm{Val}_{\sigma e}^{\mathrm{M}}(F_{i,j})$ in any case. Let $F_{i,1-j}$ be closed. Then, $\beta_i = 1 \wedge \beta_{i+1} = 1-j$. Consequently,

$$\mathrm{rVal}_{\sigma e}^{\mathrm{M}}(F_{i,1-j}) = \langle s_{i,1-j}, h_{i,1-j}\rangle + \mathrm{rVal}_{\sigma e}^{\mathrm{M}}(b_{i+1}) = \langle s_{i,1-j}, h_{i,1-j}\rangle + L_{i+1}^{\mathrm{M}} > L_2^{\mathrm{M}} = \mathrm{rVal}_{\sigma e}^{\mathrm{M}}(b_2).$$

Since either $\mathrm{rVal}_{\sigma e}^{\mathrm{M}}(F_{i,j}) = \mathrm{rVal}_{\sigma e}^{\mathrm{M}}(b_2)$ or $\mathrm{rVal}_{\sigma e}^{\mathrm{M}}(F_{i,j}) = \frac{1}{2}\mathrm{rVal}_{\sigma e}(b_2) + \frac{1}{2}\mathrm{rVal}_{\sigma e}^{\mathrm{M}}(g_1)$ and $\mathrm{rVal}_{\sigma e}^{\mathrm{M}}(g_1) < \mathrm{rVal}_{\sigma e}^{\mathrm{M}}(b_2)$, this implies the statement. Thus assume that $F_{i,1-j}$ is not closed, implying $\beta_i = 0$. For the sake of a contradiction, assume $i < \nu$. Then, $\overline{\sigma e}(g_i) = 1-j = \beta_{i+1}$. However, since $\nu = \mu^\sigma$, applying Property (BR1) to σ implies $\bar\sigma(g_i) = \overline{\sigma e}(g_i) = 1 - \beta_{i+1}$ which is a contradiction. Since $F_{i,1-j}$ is not closed, it suffices to consider the case $i > \nu$. If $i < m$, then $(i, 1-j) \in S_1$ and $(i,j) \in S_2$. Then, by Properties (ESC4)$_{i,1-j}$, (ESC5)$_{i,j}$ and (EV1)$_{i+1}$, we have

$$\mathrm{rVal}_{\sigma e}^{\mathrm{M}}(F_{i,1-j}) = \mathrm{rVal}_{\sigma e}^{\mathrm{M}}(b_2) \quad \text{and} \quad \mathrm{rVal}_{\sigma e}^{\mathrm{M}}(F_{i,j}) = \frac{1}{2}\mathrm{rVal}_{\sigma e}^{\mathrm{M}}(b_2) + \frac{1}{2}\mathrm{rVal}_{\sigma e}^{\mathrm{M}}(g_1),$$

implying the statement. If $i > m$, then $(i,j), (i, 1-j) \in S_2$ and the statement follows from $\mathrm{rVal}_{\sigma e}^{\mathrm{M}}(h_{i,1-j}) > \mathrm{rVal}_{\sigma e}^{\mathrm{M}}(s_{i,j})$. As it holds that $\mathrm{rVal}_\sigma^{\mathrm{M}}(s_{i,j}) < \mathrm{rVal}_{\sigma e}^{\mathrm{M}}(s_{i,j})$ and $\mathrm{rVal}_\sigma^{\mathrm{M}}(h_{i,1-j}) = \mathrm{rVal}_{\sigma e}^{\mathrm{M}}(h_{i,1-j})$ we thus have $(g_i, F_{i,j}) \notin I_\sigma, I_{\sigma e}$.

Finally, assume $\sigma e(s_{i,j}) = b_1$ and $\overline{\sigma e}(g_i) = j$. Since $\mathrm{Val}_{\sigma e}^{\mathrm{M}}(g_i) - \mathrm{Val}_\sigma^{\mathrm{M}}(g_i) \geq 0$, we prove that this difference is smaller than 1. By Property (USV1)$_i$, $j = 1 - \beta_{i+1}$. By the assumptions of the lemma, this implies that $F_{i,j}$ is neither closed with respect to σ nor to σe. Consequently, either $\mathrm{Val}_{\sigma e}^{\mathrm{M}}(g_i) - \mathrm{Val}_\sigma^{\mathrm{M}}(g_i) = \varepsilon[\mathrm{Val}_{\sigma e}^{\mathrm{M}}(s_{i,j}) - \mathrm{Val}_\sigma^{\mathrm{M}}(s_{i,j})]$ or $\mathrm{Val}_{\sigma e}^{\mathrm{M}}(g_i) - \mathrm{Val}_\sigma^{\mathrm{M}}(g_i) = \frac{2\varepsilon}{1+\varepsilon}[\mathrm{Val}_{\sigma e}^{\mathrm{M}}(s_{i,j}) - \mathrm{Val}_\sigma^{\mathrm{M}}(s_{i,j})]$. In either case, the difference is smaller than 1 by the choice of ε. □

Lemma 6.2.30 (Last row of Table 6.6). *Let $\sigma \in \rho(\sigma_0)$ be a well-behaved phase-3-strategy for $\mathfrak{b} \in \mathfrak{B}_n$ with $\nu = 1$. Let $I_\sigma = \{(b_1, g_1)\} \cup \{(d_{i,j,k}, F_{i,j}), (e_{i,j,k}, g_1) \colon \sigma(e_{i,j,k}) = b_2\}$ and assume that σ has Property (ESC5)$_{i,j}$ for all $(i,j) \in S_3$ and Property (ESC3)$_{i,j}$ for all $(i,j) \in S_4$. Let $\sigma(d_{i,j,k}) = F_{i,j} \Leftrightarrow \beta_i^\sigma = 1 \wedge \beta_{i+1}^\sigma = j$ for all $i \in [n], j,k \in \{0,1\}$. Let $e := (b_1, g_1)$ and define $m := \max\{i \colon \beta_i^\sigma = 1\}$ and $u := \min\{i \colon \beta_i^\sigma = 0\}$. Then σe is a well-behaved phase-5-strategy for $\mathfrak{b}$ with $\mu^{\sigma e} = u$, $\sigma e \in \rho(\sigma_0)$ and*

$$I_{\sigma e} = (I_\sigma \setminus \{e\}) \cup \bigcup_{\substack{i'=u+1\\ \beta_i^\sigma = 0}}^{m-1} \{(d_{i,1-\beta_{i+1}^\sigma,0}, F_{i,1-\beta_{i+1}^\sigma}), (d_{i,1-\beta_{i+1}^\sigma,1}, F_{i,1-\beta_{i+1}^\sigma})\}.$$

Proof. As usual, the choice of e implies $\beta^{\sigma} = \beta^{\sigma e} = \mathfrak{b} + 1 =: \beta$.

We begin by proving $\mu^{\sigma e} = u$. Since σ is a phase-3-strategy for $\mathfrak{b}$, it has Property (EV1)$_i$ and Property (EV2)$_i$ for all $i > 1$. This implies $i \notin \mathcal{I}^{\sigma}$ and thus $i \notin \mathcal{I}^{\sigma e}$ for all $i > 1$. Since $\sigma e(b_1) = g_1$, it suffices to show $\overline{\sigma e}(g_i) = \overline{\sigma e}(b_{i+1})$, implying $\mathcal{I}^{\sigma e} = \emptyset$ and $\mu^{\sigma e} = u$. This however follows directly as σ has Property (CC2).

We now show that σe is a phase-5-strategy for $\mathfrak{b}$. Since σ is a phase-3-strategy for $\mathfrak{b}$ and $e = (b_1, b_2)$, it suffices to show that the σ has Properties (EV1)$_1$, (EV2)$_1$, (EV3)$_1$ and (CC2). Note that $\nu = 1$ implies $S_3 \neq \emptyset$, implying that σe does not have Property (ESC1). By definition, $1 = \beta_1 = \overline{\sigma e}(d_{1,\beta_2})$. Thus, since $\sigma e(b_1) = g_1$, σe has Property (EV1)$_1$. It also has Property (EV2)$_1$ and Property (CC2) since σ has Property (CC2). Since $\neg\overline{\sigma}(d_{1,1-\beta_2})$ by assumption, also $\neg\overline{\sigma e}(d_{1,1-\beta_2})$, hence σe has Property (EV3)$_1$. Thus, σe is a phase-5-strategy for $\mathfrak{b}$.

We prove that σe is well-behaved. Since $e = (b_1, g_1), \mu^{\sigma} = 1$ and $\mu^{\sigma e} = u > 1$, it suffices to investigate the following properties.

(S2) Let $i < \mu^{\sigma e}$. Then, since $\mu^{\sigma e} = u$, we have $\sigma e(b_i) = g_i$. Consequently, by Property (EV1)$_i$, Property (EV2)$_i$ and Property (USV1)$_i$, $\overline{\sigma e}(d_i)$ and $\overline{\sigma e}(s_i)$.

(B1) As $\mu^{\sigma e} = u$, the premise can never be correct as $i < \mu^{\sigma e} - 1$ implies $\sigma e(b_i) = g_i$.

(B2) This again holds since $\mu^{\sigma e} = u$.

(BR1) Let $i := \mu^{\sigma e} - 1$. Then $\sigma e(b_i) = g_i$ and $\sigma e(b_{i+1}) = b_{i+2}$. Thus, by Properties (EV1)$_i$ and (EV1)$_{i+1}$ as well as Property (EV2)$_i$, we have $\sigma e(g_i) = F_{i,0}$. For $i < \mu^{\sigma e} - 1$, we have $\sigma e(g_i) = F_{i,1}$ as we then have $\sigma e(b_{i+1}) = g_{i+1}$.

(BR2) Since $i < \mu^{\sigma e}$ implies $\sigma e(b_i) = g_i$, Property (EV1)$_i$ implies $\overline{\sigma e}(d_i)$ and thus $\neg\overline{\sigma e}(eg_i)$.

(D2) This follows by the same argument used for Property (BR2).

(EG5) By Property (USV1)$_i$, $\overline{\sigma e}(s_{i,j})$ implies $\overline{\sigma e}(b_{i+1}) = j$.

(EB*) Any pair of indices $i \in [n], j \in \{0,1\}$ either fulfills Property (ESC5)$_{i,j}$, Property (ESC3)$_{i,j}$ or $\overline{\sigma e}(d_{i,j})$. Hence, there are no indices such that $\overline{\sigma e}(eg_{i,j}) \wedge \neg\overline{\sigma e}(eb_{i,j})$, so the premise of any of the Properties (EB*) is always incorrect.

(EBG4) Since $\mu^{\sigma e} > 1$ implies $\sigma e(b_1) = g_1$, it is impossible that both $\sigma e(g_1) = F_{1,0}$ and $\sigma e(b_2) = g_2$.

(EBG5) Since $\mu^{\sigma e} > 1$ implies $\sigma e(b_1) = g_1$, it is impossible that both $\sigma e(g_1) = F_{1,1}$ and $\sigma e(b_2) = b_3$.

It remains to show that

$$I_{\sigma e} = (I_{\sigma} \setminus \{e\}) \cup \bigcup_{\substack{i=u+1 \\ \beta_i = 0}}^{m-1} \{(d_{i,1-\beta_{i+1},0}, F_{i,1-\beta_{i+1}}), (d_{i,1-\beta_{i+1},1}, F_{i,1-\beta_{i+1}})\}.$$

Let $i \in \{u+1, \ldots, m-1\}, \beta_i = 0, j = 1 - \beta_{i+1}$ and $k \in \{0,1\}$. We prove $(d_{i,j,k}, F_{i,j}) \in I_{\sigma e}$. By Property (ESC3)$_{i,j}$, it holds that $\overline{\sigma e}(eg_{i,j}) \wedge \neg\overline{\sigma e}(eb_{i,j})$. In addition, $\sigma(d_{i,j,k}) \neq F_{i,j}$. It thus suffices to show $\mathrm{Val}^*_{\sigma e}(e_{i,j,k}) \prec \mathrm{Val}^*_{\sigma e}(F_{i,j})$.

Consider the case $G_n = S_n$ first. Since $\sigma e(d_{i,j,k}) = e_{i,j,k}$, Property (ESC3)$_{i,j}$ implies $\mathrm{Val}^{\mathrm{S}}_{\sigma e}(e_{i,j,k}) = \{e_{i,j,k}\} \cup \mathrm{Val}^{\mathrm{S}}_{\sigma e}(g_1)$. Now, by Lemma 6.1.16, we obtain

$$\mathrm{Val}^{\mathrm{S}}_{\sigma e}(F_{i,j}) = \{F_{i,j}, d_{i,j,k'}, e_{i,j,k'}\} \cup \mathrm{Val}^{\mathrm{S}}_{\sigma e}(g_1)$$

for some $k' \in \{0,1\}$ as $\mu^{\sigma e} \neq 1$. This however implies the statement for $G_n = S_n$ as the priority of $F_{i,j}$ is even and larger than the priorities of both $d_{i,j,k'}, e_{i,j,k'}$.

Consider the case $G_n = M_n$. Then $\mathrm{Val}^{\mathrm{M}}_{\sigma e}(F_{i,j}) = (1-\varepsilon)\,\mathrm{Val}^{\mathrm{M}}_{\sigma e}(g_1) + \varepsilon\,\mathrm{Val}^{\mathrm{M}}_{\sigma e}(s_{i,j})$, it therefore suffices to prove $\mathrm{Val}^{\mathrm{M}}_{\sigma e}(s_{i,j}) > \mathrm{Val}^{\mathrm{M}}_{\sigma e}(g_1)$. This however follows directly as Property (USV1)$_i$, the choice of j and $\sigma e(b_1) = g_1$ imply $\mathrm{Val}^{\mathrm{M}}_{\sigma e}(s_{i,j}) = \langle s_{i,j}\rangle + \mathrm{Val}^{\mathrm{M}}_{\sigma e}(g_1)$.

It remains to show that no other improving switch is created and that switches that are improving for σ are improving for σe (with the exception of e). By applying $e = (b_1, g_1)$, the valuation of b_1 increases. The only vertices that have an edge towards b_1 are the vertices $s_{i,j}, i \in [n], j \in \{0,1\}$. To show that no other improving switch is created and that switches that are improving for σ are also improving for σe, it suffices to show that one of the following holds for all $i \in [n], j \in \{0,1\}$ not considered earlier:

1. $\sigma e(s_{i,j}) = h_{i,j}$ implies $\mathrm{Val}^*_{\sigma e}(h_{i,j}) \succ \mathrm{Val}^*_{\sigma e}(b_1)$.
2. $\sigma e(s_{i,j}) = b_1$ and $(i,j) \notin S_4$ implies $(d_{i,j,k}, F_{i,j}) \in I_\sigma \Leftrightarrow (d_{i,j,k}, F_{i,j}) \in I_{\sigma e}$.
3. If $G_n = S_n$, then $\sigma e(s_{i,j}) = b_1$ implies either $\tau^\sigma(F_{i,j}) = \tau^{\sigma e}(F_{i,j}) \neq s_{i,j}$ or $\sigma e(g_i) = F_{i,j}$ and $\sigma e(b_i) = b_{i+1} \wedge (b_i, g_i) \notin I_{\sigma e}$ as well as $\sigma e(s_{i-1,1}) = b_1 \wedge (s_{i-1,1}, h_{i-1,1}) \notin I_{\sigma e}$.
4. If $G_n = M_n$, then $\sigma e(s_{i,j}) = b_1$ and $\overline{\sigma e}(g_i) = 1-j$ implies $\mathrm{Val}^{\mathrm{M}}_{\sigma e}(F_{i,1-j}) > \mathrm{Val}^{\mathrm{M}}_{\sigma e}(F_{i,j})$.
5. If $G_n = M_n$, then $\sigma e(s_{i,j}) = b_1$ and $\overline{\sigma e}(g_i) = j$ implies $\mathrm{Val}^{\mathrm{M}}_{\sigma e}(g_i) - \mathrm{Val}^{\mathrm{M}}_{\sigma}(g_i) \in (0,1)$.

We now prove these statements one after another.

1. Fix indices i,j with $\sigma e(s_{i,j}) = h_{i,j}$. Then, by Property (USV1)$_i$, $j = \beta_{i+1}$. Consequently, by Property (EV1)$_{i+1}$, $\mathrm{rVal}^*_{\sigma e}(h_{i,j}) = [\![h_{i,j}]\!] \oplus \mathrm{rVal}^*_{\sigma e}(b_{i+1})$. Since σe has Property (EV1)$_{i'}$ and Property (EV2)$_{i'}$ for all $i' < \mu^{\sigma e}$, there is no $i' < \mu^{\sigma e}$ with $\neg\overline{\sigma e}(d_{i'})$. Consequently, by Lemma 6.1.12, $\mathrm{rVal}^*_{\sigma e}(g_1) = R^*_1$ in any case. Now, since $[\![h_{i,j}]\!] \succ \bigoplus_{i' \leq i} W^*_{i'}$, this implies $\mathrm{rVal}^*_{\sigma e}(h_{i,j}) \succ \mathrm{rVal}^*_{\sigma e}(g_1)$ for both possible cases $\mathrm{rVal}^*_{\sigma e}(b_{i+1}) = L^*_{i+1}$ and $\mathrm{rVal}^*_{\sigma e}(b_{i+1}) = R^*_{i+1}$.
2. Consider some edge $(d_{i,j,k}, F_{i,j})$ for which we did not prove that $(d_{i,j,k}, F_{i,j}) \in I_{\sigma e}$, i.e., assume $(i,j) \notin S_4$. We show that $(d_{i,j,k}, F_{i,j}) \in I_\sigma \Leftrightarrow (d_{i,j,k}, F_{i,j}) \in I_{\sigma e}$. Let $(d_{i,j,k}, F_{i,j}) \in I_\sigma$. Then, by our assumptions on I_σ, $\sigma(e_{i,j,k}) = b_2$. This implies that $\bar\sigma(eg_{i,j}) \wedge \bar\sigma(eb_{i,j})$ has to hold and, due to $(d_{i,j,k}, F_{i,j}) \in I_\sigma$ and $\bar\sigma(g_1) = \bar\sigma(b_2)$,

$$\mathrm{rVal}^{\mathrm{M}}_\sigma(F_{i,j}) = \frac{1}{2}\,\mathrm{rVal}^{\mathrm{M}}_\sigma(g_1) + \frac{1}{2}\,\mathrm{rVal}^{\mathrm{M}}_\sigma(b_2) > \mathrm{rVal}^{\mathrm{M}}_\sigma(b_2) = \mathrm{rVal}^{\mathrm{M}}_\sigma(e_{i,j,k}),$$
$$\mathrm{Val}^{\mathrm{S}}_\sigma(F_{i,j}) = \{F_{i,j}, d_{i,j,k'}, e_{i,j,k'}\} \cup \mathrm{Val}^{\mathrm{S}}_\sigma(b_2) \rhd \{e_{i,j,k'}\} \cup \mathrm{Val}^{\mathrm{S}}_\sigma(b_2) = \mathrm{Val}^{\mathrm{S}}_\sigma(e_{i,j,k}).$$

   But then, $\mathrm{rVal}^*_\sigma(e_{i,j,k}) = \mathrm{rVal}^*_{\sigma e}(e_{i,j,k})$ and $\mathrm{rVal}^*_{\sigma e}(g_1) \succeq \mathrm{rVal}^*_\sigma(g_1)$, the same estimation holds for σe, implying $(d_{i,j,k}, F_{i,j}) \in I_{\sigma e}$. Now let $(d_{i,j,k}, F_{i,j}) \notin I_\sigma$, implying $\sigma(e_{i,j,k}) = \sigma e(e_{i,j,k}) = g_1$. If $\sigma(d_{i,j,k}) = F_{i,j}$, then there is nothing to show hence assume $\sigma(d_{i,j,k}) = e_{i,j,k}$. This implies $\bar\sigma(eg_{i,j})$ and $\overline{\sigma e}(eg_{i,j})$. Assume $\bar\sigma(eb_{i,j})$. Then, using the same estimations used for the case $(d_{i,j,k}, F_{i,j}) \in I_\sigma$, we can show $(d_{i,j,k}, F_{i,j}) \notin I_{\sigma e}$. Thus assume $\neg\bar\sigma(eb_{i,j})$. Then, $\bar\sigma(eg_{i,j}) \wedge \neg\bar\sigma(eb_{i,j})$, implying $(i,j) \in S_4$. This however contradicts our choice of i and j, proving the statement.
3. Let $G_n = S_n$ and $\sigma(s_{i,j}) = \sigma e(s_{i,j}) = b_1$. Then, by Property (USV1)$_i$, $j = 1 - \beta_{i+1}$. By our assumptions, $F_{i,j}$ is thus either mixed or g_1-open. Consider the case that it is mixed first. Then, by Property (EV2)$_1$, $\overline{\sigma e}(g_1) = \overline{\sigma e}(b_2)$, implying $\bar\sigma(g_1) = \bar\sigma(b_2)$

by the choice of e. Consequently, by Lemma 6.1.16, $\mathrm{rVal}_{\sigma}^{\mathrm{S}}(F_{i,j}) = \mathrm{rVal}_{\sigma}^{\mathrm{S}}(b_2)$ and $\mathrm{rVal}_{\sigma e}^{\mathrm{S}}(F_{i,j}) = \mathrm{rVal}_{\sigma e}^{\mathrm{S}}(b_2)$. Thus, $\tau^{\sigma}(F_{i,j}), \tau^{\sigma e}(F_{i,j}) \neq s_{i,j}$.

Now assume that $F_{i,j}$ is g_1-open. Then, by our assumptions on the cycle centers, $(i,j) \in S_4$, implying $i \in \{u+1,\dots,m-1\}$ with $\beta_i = 0$ and $j = 1-\beta_{i+1}$. We prove $\mathrm{Val}_{\sigma}^{\mathrm{S}}(F_{i,j}) \rhd \mathrm{Val}_{\sigma}^{\mathrm{S}}(F_{i,1-j})$, implying $\sigma(g_i) = \sigma e(g_i) = F_{i,j}$ by our assumptions on I_{σ}. We then prove that $\sigma e(b_i) = b_{i+1}$ and $(b_i, g_i) \notin I_{\sigma e}$ as well as $\sigma e(s_{i-1,1}) = b_1$ and $(s_{i-1,1}, h_{i-1,1}) \notin I_{\sigma e}$ (if $i > 1$), implying that the valuation of no further vertex changes, proving the statement.

Since $1 - j = \beta_{i+1}$, we have $(i, 1-j) \in S_3$ by assumption. Consequently, by Property (ESC5)$_{i,j}$, it holds that $\bar{\sigma}(eg_{i,1-j}) \wedge \bar{\sigma}(eb_{i,1-j})$. As pointed out earlier, this implies $\mathrm{rVal}_{\sigma}^{\mathrm{S}}(F_{i,1-j}) = \mathrm{rVal}_{\sigma}^{\mathrm{S}}(b_2)$. Since $\mu^{\sigma} = 1$, Lemma 6.1.16 yields $\mathrm{rVal}_{\sigma}^{\mathrm{S}}(F_{i,j}) = \{s_{i,j}\} \cup \mathrm{rVal}_{\sigma e}^{\mathrm{S}}(b_2)$. This implies $\mathrm{rVal}_{\sigma}^{\mathrm{S}}(F_{i,j}) \rhd \mathrm{rVal}_{\sigma e}^{\mathrm{S}}(F_{i,1-j})$. We hence need to have $\sigma(g_i) = \sigma e(g_i) = F_{i,j}$ by our assumptions on I_{σ}.

By $\beta_i = 0$, Property (EV1)$_i$ implies $\sigma e(b_i) = b_{i+1}$. Since $\nu = 1$ implies $\beta_1 = 1$, we have $i > 1$. Thus, Property (USV1)$_{i-1}$ implies $\sigma e(s_{i-1,1}) = b_1$. Consequently, $\mathrm{rVal}_{\sigma e}^{\mathrm{S}}(b_{i+1}) = L_{i+1}^{\mathrm{S}}$. Since $\overline{\sigma e}(eg_{i,j}) \wedge \neg\overline{\sigma e}(eb_{i,j}) \wedge \mu^{\sigma e} > 1$, Lemma 6.1.16 implies $\mathrm{rVal}_{\sigma e}^{\mathrm{S}}(F_{i,j}) = \mathrm{rVal}_{\sigma e}^{\mathrm{S}}(g_1) = R_1^{\mathrm{S}}$. Consequently, since $i \geq \mu^{\sigma e}$ and $\sigma e(b_i) = b_{i+1}$ imply $R_i^{\mathrm{S}} = L_i^{\mathrm{S}} = L_{i+1}^{\mathrm{S}}$,

$$\begin{aligned}\mathrm{rVal}_{\sigma e}^{\mathrm{S}}(g_i) &= \langle g_i \rangle + R_1^{\mathrm{S}} = \langle g_i \rangle + R_{1,i-1}^{\mathrm{S}} + R_i^{\mathrm{S}} \\ &= \langle g_i \rangle + R_{1,i-1}^{\mathrm{S}} + L_{i+1}^{\mathrm{S}} \lhd L_{i+1}^{\mathrm{S}} = \mathrm{rVal}_{\sigma e}^{\mathrm{S}}(b_{i+1}).\end{aligned}$$

As $\mathrm{rVal}_{\sigma e}^{\mathrm{S}}(b_1) \rhd \mathrm{rVal}_{\sigma e}^{\mathrm{S}}(b_{i+1})$ and $\mathrm{rVal}_{\sigma e}^{\mathrm{S}}(h_{i-1,1}) = \langle h_{i-1,1} \rangle \cup \mathrm{rVal}_{\sigma e}^{\mathrm{S}}(g_i)$, a similar estimation yields $\mathrm{rVal}_{\sigma e}^{\mathrm{S}}(h_{i-1,1}) \lhd \mathrm{rVal}_{\sigma e}^{\mathrm{S}}(b_1)$. Consequently, $(b_i, g_i) \notin I_{\sigma e}$ as well as $(s_{i-1,1}, h_{i-1,1}) \notin I_{\sigma e}$.

4. Let $G_n = M_n$ and $\sigma e(s_{i,j}) = b_1$ and $\overline{\sigma e}(g_i) = 1-j$. Then, by Property (USV1)$_i$, $j = 1-\beta_{i+1}$ and $\sigma e(s_{i,1-j}) = h_{i,1-j}$. Assume that $F_{i,1-j}$ is closed. Then, by Property (EV1)$_{i+1}$, we have $\mathrm{rVal}_{\sigma e}^{\mathrm{M}}(F_{i,1-j}) = \langle s_{i,1-j}, h_{i,1-j} \rangle + \mathrm{rVal}_{\sigma e}^{\mathrm{M}}(b_{i+1})$. Since it is not possible that $F_{i,j}$ is closed, either

$$\mathrm{rVal}_{\sigma e}^{\mathrm{M}}(F_{i,j}) = \mathrm{rVal}_{\sigma e}^{\mathrm{M}}(g_1) \quad \text{or} \quad \mathrm{rVal}_{\sigma e}^{\mathrm{M}}(F_{i,j}) = \frac{1}{2}\mathrm{rVal}_{\sigma e}^{\mathrm{M}}(g_1) + \frac{1}{2}\mathrm{rVal}_{\sigma e}^{\mathrm{M}}(b_2).$$

As $\mathrm{rVal}_{\sigma e}^{\mathrm{M}}(b_2) < \mathrm{rVal}_{\sigma e}^{\mathrm{M}}(g_1)$ and $\sum_{\ell \in [i]} W_{\ell}^{\mathrm{M}} < \langle s_{i,1-j}, h_{i,1-j} \rangle$, this implies that we have $\mathrm{rVal}_{\sigma e}^{\mathrm{M}}(F_{i,1-j}) > \mathrm{rVal}_{\sigma e}^{\mathrm{M}}(F_{i,j})$ in any case.

Thus assume that $F_{i,1-j}$ is not closed. Since $1-j = \beta_{i+1}$, it then follows that $(i, 1-j) \in S_3$. If $(i,j) \in S_3$, then both cycle centers are in the same state. Since σe has Property (USV1)$_i$, Property (EV1)$_{i+1}$ and since $i \geq \nu = 1$, the statement thus follows from Lemma 6.2.1.

Thus assume $(i,j) \in S_4$. Then, by Property (ESC5)$_{i,1-j}$ and Property (ESC4)$_{i,j}$, we have $\mathrm{rVal}_{\sigma}^{\mathrm{M}}(F_{i,1-j}) = \frac{1}{2}\mathrm{rVal}_{\sigma}(g_1) + \frac{1}{2}\mathrm{rVal}_{\sigma}(b_2)$ and $\mathrm{rVal}_{\sigma}^{\mathrm{M}}(F_{i,j}) = \mathrm{rVal}_{\sigma}(g_1)$. But then, $\mathrm{rVal}_{\sigma}^{\mathrm{M}}(F_{i,j}) > \mathrm{rVal}_{\sigma}^{\mathrm{M}}(F_{i,1-j})$, implying $(g_i, F_{i,1-j}) \in I_{\sigma}$, contradicting our assumption on I_{σ}.

5. Let $G_n = M_n$ and $\sigma e(s_{i,j}) = b_1$ and $\overline{\sigma e}(g_i) = j$. Then, Property (USV1)$_i$ implies $j = 1 - \beta_{i+1}$. In particular, $F_{i,1-j}$ is not closed. Thus, either $\bar{\sigma}(eg_{i,j}) \wedge \neg\bar{\sigma}(eb_{i,j})$ or $\bar{\sigma}(eg_{i,j}) \wedge \bar{\sigma}(eb_{i,j})$. In the first case, $\mathrm{Val}^{\mathrm{M}}_{\sigma}(F_{i,j}) = (1-\varepsilon)\,\mathrm{Val}^{\mathrm{M}}_{\sigma}(g_1) + \varepsilon\,\mathrm{Val}^{\mathrm{M}}_{\sigma}(s_{i,j})$. But then, since $\mathrm{Val}^{\mathrm{M}}_{\sigma}(g_1) = \mathrm{Val}^{\mathrm{M}}_{\sigma e}(b_1) = R^{\mathrm{M}}_1$, this implies $\mathrm{Val}^{\mathrm{M}}_{\sigma e}(F_{i,j}) - \mathrm{Val}^{\mathrm{M}}_{\sigma}(F_{i,j}) \in (0,1)$. If $\bar{\sigma}(eg_{i,j}) \wedge \bar{\sigma}(eb_{i,j})$, the statement follows analogously since $\mathrm{Val}^{\mathrm{M}}_{\sigma}(b_2) = \mathrm{Val}^{\mathrm{M}}_{\sigma e}(b_2)$. □

Lemma 6.2.31. *Let $G_n = S_n$. Let $\sigma \in \rho(\sigma_0)$ be a well-behaved phase-4-strategy for $\mathfrak{b} \in \mathfrak{B}_n$ with $\nu > 1$. Assume that there is an index $i < \nu$ such that $e := (s_{i,j}, b_1) \in I_\sigma$ where $j := 1 - \beta^{\sigma}_{i+1}$. Further assume the following:*

1. *σ has Property (USV1)$_{i'}$ for all $i' > i$.*
2. *For all i', j', k', it holds that $\sigma(d_{i',j',k'}) = F_{i',j'}$ if and only if $\beta^{\sigma}_{i'} = 1 \wedge \beta^{\sigma}_{i'+1} = j'$.*
3. *$i' < \nu$ implies $\bar{\sigma}(g_{i'}) = 1 - \beta^{\sigma}_{i'+1}$.*
4. *$i' < i$ implies $\sigma(s_{i',*}) = h_{i',*}$.*

If there is an index $i' < i$ such that $(s_{i',1-\beta^{\sigma}_{i'+1}}, b_1) \in I_\sigma$, then σe is a well-behaved phase-4-strategy for $\mathfrak{b}$. Otherwise, it is a well-behaved phase-5-strategy for $\mathfrak{b}$. In either case, it holds that $I_{\sigma e} = (I_\sigma \setminus \{e\}) \cup \{(d_{i,j,0}, F_{i,j}), (d_{i,j,1}, F_{i,j})\}$.

Proof. We first note that $\mu^{\sigma} = \mu^{\sigma e} = 1$ since σ is a phase-4-strategy for $\mathfrak{b}$ and by the choice of e. Furthermore, by the choice of e, $\beta^{\sigma} = \beta^{\sigma e} = \mathfrak{b} + 1 =: \beta$. We first prove that σe is well-behaved. By the choice of e, we need to reevaluate the following properties:

(S1) $\sigma e(b_i) = g_i$ implies $\beta_i = 1$ by Property (EV1)$_i$, hence $i \geq \nu$.

(MNS1) By Lemma 6.1.6, $m^{\sigma e}_b \leq \overline{m}^{\sigma e}_s, \overline{m}^{\sigma e}_g$ implies $m^{\sigma e}_b = \nu = 2$. But then, by assumption 3., $\overline{\sigma e}(g_1) = F_{1,0}$, implying $\overline{m}^{\sigma e}_g = 1$ and thus contradicting the premise.

(MNS2) By assumption 3, $\overline{m}^{\sigma e}_g = \nu - 1$. By the choice of i and j, we have $\overline{m}^{\sigma e}_s \leq \nu - 1$. Thus $\overline{m}^{\sigma e}_s \leq \overline{m}^{\sigma e}_g$ and the premise of this property cannot be fulfilled.

(MNS4) The conclusion is always true since $i' < \nu$ implies $(i', 1 - \beta_{i'+1}) \in S_1$, implying that $\neg\overline{\sigma e}(eb_{i'}) \wedge \neg\overline{\sigma e}(eg_{i'})$.

(MNS6) See Property (MNS4).

(EG3) For every pair of indices i, j, either $\bar{\sigma}(d_{i,j})$ or $\bar{\sigma}(eb_{i,j}) \wedge \neg\bar{\sigma}(eg_{i,j})$ or $\bar{\sigma}(eb_{i,j}) \wedge \bar{\sigma}(eg_{i,j})$ by either $\beta = \mathfrak{b} + 1$ or Property (ESC4)$_{i,j}$ resp. (ESC5)$_{i,j}$. Consequently, the premise is incorrect for σe,

(EBG2) By assumption 3 and Property (EV1)$_2$, $\overline{\sigma e}(g_1) = \bar{\sigma}(g_1) \neq \bar{\sigma}(b_2) = \overline{\sigma e}(b_2)$, hence the premise is incorrect.

We next prove that σ is a phase-4-strategy if there is an index $i' < i$ such that $(s_{i',1-\beta^{\sigma}_{i'+1}}, b_1) \in I_\sigma$ and a phase-5-strategy otherwise. By the definition of the phases, it suffices to prove that σe has Property (USV1)$_\ell$ for all $\ell \in [n]$ if there is no such index. Hence assume that no such index exists and let $i' < i$ as there is nothing to prove if $i' \geq i$ and let $j' := 1 - \beta_{i'+1}$. Then, since $(s_{i',j'}, b_1) \notin I_\sigma$, assumption 4 implies $\sigma(s_{i',j'}) = h_{i',j'}$ and $\mathrm{Val}^{\mathrm{S}}_{\sigma}(h_{i',j'}) \rhd \mathrm{Val}^{\mathrm{S}}_{\sigma}(b_1)$. It now suffices to prove that this cannot happen, hence we prove that $\sigma(s_{i',j'}) = h_{i',j'}$ implies $\mathrm{Val}^{\mathrm{S}}_{\sigma}(h_{i',j'}) \lhd \mathrm{Val}^{\mathrm{S}}_{\sigma}(b_1)$.

Since $i' < i < \nu$, we have $i' \leq \nu + 2$, implying $j' = 1 - \beta_{i'+1} = 1$. Consequently, $\mathrm{rVal}^{\mathrm{S}}_{\sigma}(h_{i',j'}) = \{h_{i',j'}\} \cup \mathrm{rVal}^{\mathrm{S}}_{\sigma}(g_{i'+1})$. Since σ has Property (USV1)$_{i'+1}$ by assumption 1 and $\bar{\sigma}(g_{i'+1}) = 1 - \beta_{i'+1}$ by assumption 3, we have $\sigma(s_{i'+1,\bar{\sigma}(g_{i'+1})}) = b_1$. This implies $\lambda^{\mathrm{S}}_{i'+1} = i'+1$. Since $i'+1 < \nu$, the cycle center $F_{i'+1,\bar{\sigma}(g_{i'+1})}$ cannot be closed by assumption 2. As also $\neg\bar{\sigma}(s_{i'+1})$ and $\neg\bar{\sigma}(b_{i'+1})$ by Property (EV1)$_{i'+1}$, Corollary 6.1.18, implies that either

$$\begin{aligned}
\mathrm{rVal}^{\mathrm{S}}_{\sigma}(g_{i'+1}) &= \{g_{i'+1}\} \cup \mathrm{rVal}^{\mathrm{S}}_{\sigma}(g_1) \quad \text{or}\\
\mathrm{rVal}^{\mathrm{S}}_{\sigma}(g_{i'+1}) &= \{g_{i'+1}\} \cup \mathrm{rVal}^{\mathrm{S}}_{\sigma}(b_2) \quad \text{or}\\
\mathrm{rVal}^{\mathrm{S}}_{\sigma}(g_{i'+1}) &= \{g_{i'+1}, s_{i'+1,\bar{\sigma}(g_{i'+1})}\} \cup \mathrm{rVal}^{\mathrm{S}}_{\sigma}(b_1).
\end{aligned}$$

As $\mu^{\text{œ}} = 1$ implies $\text{œ}(b_1) = b_2$, the statement directly follows in the last two cases. In the first case, it follows from $\mathrm{rVal}^{\mathrm{S}}_{\text{œ}}(g_1) \lhd \mathrm{rVal}^{\mathrm{S}}_{\text{œ}}(b_2)$ which can be shown by using Lemmas 6.1.6 and 6.2.18 and assumption 3.

We now show that $(d_{i,j,0}, F_{i,j}), (d_{i,j,1}, F_{i,j})$ are improving for œ. Let $k \in \{0,1\}$. It suffices to show $\mathrm{Val}^{\mathrm{S}}_{\text{œ}}(F_{i,j}) \rhd \mathrm{Val}^{\mathrm{S}}_{\text{œ}}(e_{i,j,k})$ since $\text{œ}(d_{i,j,k}) = e_{i,j,k}$ by assumption. By Property (ESC4)$_{i,j}$, we have $\bar{\text{œ}}(eb_{i,j}) \wedge \neg\bar{\text{œ}}(eg_{i,j})$. Since $\text{œ}(d_{i,j,k}) = e_{i,j,k}$, this implies $\text{œ}(e_{i,j,k}) = b_2$. Hence, by Lemma 6.1.16,

$$\mathrm{Val}^{\mathrm{S}}_{\text{œ}}(F_{i,j}) = \{F_{i,j}, d_{i,j,k}, e_{i,j,k}\} \cup \mathrm{Val}^{\mathrm{S}}_{\text{œ}}(b_2) \rhd \{e_{i,j,k}\} \cup \mathrm{Val}^{\mathrm{S}}_{\text{œ}}(b_2) = \mathrm{Val}^{\mathrm{S}}_{\text{œ}}(e_{i,j,k}).$$

We now explain how we prove $I_{\text{œ}} = (I_\sigma \setminus \{e\}) \cup \{(d_{i,j,0}, F_{i,j}), (d_{i,j,1}, F_{i,j})\}$. Applying the switch e increases the valuation of $F_{i,j}$. By the choice of j and assumption 3, the valuation of g_i increases as well. We thus begin by showing $\text{œ}(b_i) = b_{i+1}$ and $(b_i, g_i) \notin I_\sigma, I_{\text{œ}}$. However, applying the switch e also increases the valuation of several vertices contained in levels below level i. To be precise, since $\sigma(g_\ell) = F_{\ell,1}, \tau^\sigma(F_{\ell,1}) = s_{\ell,1}$ and $\sigma(s_{\ell,1}) = h_{\ell,1}$ for all $\ell < i$, the valuation of all of these vertices g_ℓ and $F_{\ell,1}$ increases. We thus show that the following statements hold:

1. $\text{œ}(b_\ell) = b_{\ell+1}$ and $(b_\ell, g_\ell) \notin I_\sigma, I_{\text{œ}}$.
2. The edges $(d_{\ell,1,0}, F_{\ell,1})$ and $(d_{\ell,1,1}, F_{\ell,1})$ are not improving for œ.

Since also the valuation of g_1 increases, we also prove that $\sigma(e_{i',j',k'}) = b_2$ implies $(e_{i',j',k'}, g_1) \notin I_\sigma, I_{\text{œ}}$ for any indices i', j', k'. This then proves the statement as $\text{œ}(b_1) = b_2$ due to $\mu^{\text{œ}} = 1$, implying that the valuation of no further vertex can change.

First, since $i < \nu$, Property (EV1)$_i$ implies $\text{œ}(b_i) = b_{i+1}$. By the choice of i and j, Property (ESC4)$_{i,j}$ implies $\bar{\sigma}(eb_{i,j}) \wedge \neg\bar{\sigma}(eg_{i,j})$. As $\mu^{\text{œ}} = 1$ and $\text{œ}(s_{i,j}) = b_1$ by the choice of e, Corollary 6.1.18 implies

$$\mathrm{rVal}^{\mathrm{S}}_{\text{œ}}(g_i) = \{g_i, s_{i,\bar{\sigma}(g_i)}\} \cup \mathrm{rVal}^{\mathrm{S}}_{\text{œ}}(b_2) < \mathrm{rVal}^{\mathrm{S}}_{\text{œ}}(b_2) = \mathrm{rVal}^{\mathrm{S}}_{\text{œ}}(b_{i+1})$$

as $i+1 \leq \nu$. Since $\mathrm{rVal}^{\mathrm{S}}_{\sigma}(b_{i+1}) = \mathrm{rVal}^{\mathrm{S}}_{\text{œ}}(b_{i+1}) = L^{\mathrm{S}}_{i+1}$ and $\mathrm{rVal}^{\mathrm{S}}_{\sigma}(g_i) \leq \mathrm{rVal}^{\mathrm{S}}_{\text{œ}}(g_i)$, this implies $(b_i, g_i) \notin I_\sigma, I_{\text{œ}}$. Now, for any $\ell < i$, $\text{œ}(b_\ell) = b_{\ell+1}$ follows also by Property (EV1)$_\ell$ and $(b_\ell, g_\ell) \notin I_\sigma, I_{\text{œ}}$ follows as

$$\mathrm{rVal}^{\mathrm{S}}_{\text{œ}}(g_\ell) = \bigcup_{i'=\ell}^{i-1} W^{\mathrm{S}}_{\ell} \cup \{g_i, s_{i,\bar{\sigma}(g_i)}\} \cup \mathrm{rVal}^{\mathrm{S}}_{\text{œ}}(b_2) \lhd \mathrm{rVal}^{\mathrm{S}}_{\text{œ}}(b_2).$$

Similarly, as $\text{rVal}^{\text{S}}_{\sigma e}(F_{\ell,1}) = \text{rVal}_{\sigma e}(g_\ell) \setminus \{g_\ell\}$ and $\text{rVal}^{\text{S}}_{\sigma e}(e_{\ell,1,k}) = \text{rVal}^{\text{S}}_{\sigma e}(b_2)$ by Property (ESC4)$_{\ell,1}$, the same estimation yields $(d_{\ell,1,0}, F_{\ell,1}), (d_{\ell,1,1}, F_{\ell,1}) \notin I_\sigma, I_{\sigma e}$. Finally, if $\sigma e(e_{i',j',k'}) = b_2$, then the same estimation implies $(e_{i',j',k'}, g_1) \notin I_\sigma, I_{\sigma e}$. Consequently, $I_{\sigma e} = (I_\sigma \setminus \{e\} \cup \{(d_{i,j,0}, F_{i,j}), (d_{i,j,1}, F_{i,j})\}$. □

Lemma 6.2.32. *Let $\sigma \in \rho(\sigma_0)$ be a well-behaved phase-5-strategy for $\mathfrak{b} \in \mathfrak{B}_n$. Let $i \in [n]$ and $j, k \in \{0,1\}$ with $e := (e_{i,j,k}, t^{\rightarrow}) \in I_\sigma$ and $\bar{\sigma}(eb_{i,j}) \wedge \bar{\sigma}(eg_{i,j})$. Furthermore assume that $G_n = S_n$ implies*

$$j = 1 \wedge \nu > 1 \implies \neg\bar{\sigma}(eg_{i,1-j}) \quad \textit{and} \quad j = 1 \wedge \nu = 1 \implies \neg\bar{\sigma}(eb_{i,1-j}).$$

Similarly, assume that $G_n = M_n$ implies

$$j = 1 - \beta^\sigma_{i+1} \wedge \nu > 1 \implies \neg\bar{\sigma}(eg_{i,1-j}) \quad \textit{and} \quad j = 1 - \beta^\sigma_{i+1} \wedge \nu = 1 \implies \neg\bar{\sigma}(eb_{i,1-j}).$$

Moreover, assume that $\nu = 2$ implies $\sigma(g_1) = F_{1,0}$ if $G_n = S_n$. Then the following hold.

1. *If there are indices $(i', j', k') \neq (i, j, k)$ with $(e_{i',j',k'}, t^{\rightarrow}) \in I_\sigma$ or if there is an index i' such that σ does not have Property (SVG)$_{i'}$/(SVM)$_{i'}$, then σe is a phase-5-strategy for $\mathfrak{b}$.*
2. *The strategy σe is well-behaved.*
3. *If there are no indies $(i', j', k') \neq (i, j, k)$ with $(e_{i',j',k'}, t^{\rightarrow}) \in I_\sigma$ and if σ has Property (SVG)$_{i'}$/(SVM)$_{i'}$ for all $i'[n]$, then σe is a phase-1-strategy for $\mathfrak{b} + 1$.*
4. *If $G_n = S_n$, then*

$$(g_i, F_{i,j}) \in I_{\sigma e} \iff \beta^\sigma_i = 0 \wedge \overline{\sigma e}(g_i) = 1 \wedge j = 0 \wedge \begin{cases} \bar{\sigma}(eb_{i,1-j}), & \nu > 1 \\ \bar{\sigma}(eg_{i,1-j}), & \nu = 1 \end{cases}.$$

If $G_n = M_n$, then

$$(g_i, F_{i,j}) \in I_{\sigma e} \iff \beta^\sigma_i = 0 \wedge \overline{\sigma e}(g_i) = 1 - \beta^\sigma_{i+1} \wedge j = \beta^\sigma_{i+1} \wedge \begin{cases} \bar{\sigma}(eb_{i,1-j}), & \nu > 1 \\ \bar{\sigma}(eg_{i,1-j}), & \nu = 1 \end{cases}.$$

If the corresponding conditions are fulfilled, then

$$I_{\sigma e} = (I_\sigma \setminus \{e\}) \cup \{(d_{i,j,1-k}, F_{i,j}), (g_i, F_{i,j}))\}.$$

Otherwise, $I_{\sigma e} = (I_\sigma \setminus \{e\}) \cup \{(d_{i,j,1-k}, F_{i,j})\}$.

Proof. As usual, $\beta^\sigma = \beta^{\sigma e} =: \beta$ by the choice of e. Since σ is a phase-5-strategy, it has Property (REL1). Thus, $\mu^\sigma = \min\{i' : \sigma(b_{i'}) = b_{i'+1}\}$. Also, by the choice of e, $\mu^{\sigma e} = \mu^\sigma$. We first discuss some statements that will be used several times during this proof.

Since σ is well-behaved, $\nu > 1$ implies that there is no cycle center $F_{i',j'}$ with $\bar{\sigma}(eg_{i',j'}) \wedge \neg\bar{\sigma}(eb_{i',j'})$. More precisely, for the sake of a contradiction, assume there was such a cycle center. By Properties (EG2) and (EG3), this implies $\bar{\sigma}(d_1)$ and $\bar{\sigma}(s_1)$. Thus, by Property (USV1)$_1$, the cycle center F_{1,β^σ_2} is closed. This however contradicts $\nu > 1$.

Similarly, it is easy to verify that $\nu = 1$ implies that there is no cycle center $F_{i',j'}$ with $\bar{\sigma}(eb_{i',j'}) \wedge \neg\bar{\sigma}(eg_{i',j'})$. As we assume $\bar{\sigma}(eb_{i,j}) \wedge \bar{\sigma}(eg_{i,j})$, the choice of e implies

$$\mathrm{Val}^{\mathrm{S}}_{\sigma e}(F_{i,j}) = \{F_{i,j}, d_{i,j,*}, e_{i,j,*}\} \cup \mathrm{Val}^{\mathrm{S}}_{\sigma e}(t^{\rightarrow})$$

and

$$\mathrm{Val}^{\mathrm{M}}_{\sigma e}(F_{i,j}) = (1-\varepsilon)\,\mathrm{Val}^{\mathrm{M}}_{\sigma e}(t^{\rightarrow}) + \varepsilon \cdot \mathrm{Val}^{\mathrm{M}}_{\sigma e}(s_{i,j}).$$

Using these observations, we now prove the statements of the lemma.

1. If there are indices $i', j', k', (i,j,k) \neq (i',j',k')$ such that $(e_{i',j',k'}, t^{\rightarrow}) \in I_\sigma$, then σe cannot have Property (ESC1). This implies that σe is a phase-5-strategy for $\mathfrak{b}$. If there is an index i' such that σ does not have Property (SVG)$_{i'}$ resp. (SVM)$_{i'}$, then σe can also not have the corresponding property. Consequently, due to the special condition of phase 5, σe is a phase-5-strategy for $\mathfrak{b}$.
2. We next show that the strategy σe is well-behaved. Depending on ν, we thus need to investigate the following properties:
   - (MNS1) Assume that the premise of this property is correct. Then, by Lemma 6.1.6, $m_b^{\sigma e} = 2$ and consequently $\sigma e(g_1) = F_{1,1} \wedge \sigma e(s_{1,1}) = h_{1,1}$. But this contradicts that $\nu = 2$ implies $\sigma e(g_1) = \sigma(g_1) = F_{1,0}$ if $G_n = S_n$.
   - (MNS2) Assume that the premise of this property is correct. Then $\sigma e(g_1) = F_{1,1}$ and $\sigma e(s_{1,1}) = h_{1,1}$. But then, by Property (USV1)$_1$, $\sigma e(b_2) = g_2$, implying $m_b^{\sigma e} = 2$, contradicting the assumption.
   - (MNS3) This follows by the same arguments used for Property (MNS2).
   - (MNS4) We only need to investigate this property if $i = \overline{m}_s^{\sigma e} = \overline{m}_s^{\sigma}$ and if the premise is true for σe. But then, the premise was already true for σ, implying $\bar{\sigma}(eb_{\overline{m}_s^{\sigma}}) \wedge \neg\bar{\sigma}(eg_{\overline{m}_s^{\sigma}})$. Since $\mu^\sigma = 1$ implies $\nu > 1$, we apply an improving switch $(e_{i,j,k}, b_2)$. But then, also $\overline{\sigma e}(eb_{\overline{m}_s^{\sigma e}}) \wedge \neg\overline{\sigma e}(eg_{\overline{m}_s^{\sigma e}})$.
   - (MNS5) This follows by the same arguments used for Property (MNS2).
   - (MNS6) This follows by the same arguments used for Property (MNS4).
   - (EG1) By assumption, $\bar{\sigma}(eb_{i,j}) \wedge \bar{\sigma}(eg_{i,j})$. In order to have $\overline{\sigma e}(eg_{i,j}) \wedge \neg\overline{\sigma e}(eb_{i,j}) \wedge \mu^{\sigma e} = 1$ we thus need to have applied a switch $(e_{i,j,*}, g_1)$. This however implies $\mu^{\sigma e} \neq 1$, contradicting the premise.
   - (EG2) Follows by the same arguments.
   - (EG4) Follows by the same arguments.
   - (EG3) Assume the premise was true. Since $\bar{\sigma}(eg_{i,j}) \wedge \bar{\sigma}(eb_{i,j})$, we need to have $\mu^\sigma \neq 1$. But then, $\bar{\sigma}(b_1) = 1$ by Property (EV1)$_1$, implying $\bar{\sigma}(s_1)$ by Property (EV2)$_1$ and Property (USV1)$_1$.
   - (EG5) If the premise is true, then $\bar{\sigma}(s_{i,j})$. Hence, $j = \beta_{i+1} = \bar{\sigma}(b_{i+1})$ by Property (USV1)$_i$ and Property (EV1)$_{i+1}$. Since $\bar{\sigma}(s_{i,j}) = \overline{\sigma e}(s_{i,j})$ and $\bar{\sigma}(b_{i+1}) = \overline{\sigma e}(b_{i+1})$ by the choice of e, the statement follows.
   - (EB*) Assume $\overline{\sigma e}(eb_{i,j}) \wedge \neg\overline{\sigma e}(eg_{i,j})$. Since $\bar{\sigma}(eb_{i,j}) \wedge \bar{\sigma}(eg_{i,j})$, we need to have applied a switch $(e_{i,j,*}, b_2)$. But this implies $\mu^\sigma = 1$ and thus $\sigma(b_1) = b_2$, contradicting all of the premises.

(EBG*) Since $\bar{\sigma}(eb_{i,j}) \wedge \bar{\sigma}(eg_{i,j})$ by assumption, it is impossible to have $\overline{\sigma e}(eb_{i,j}) \wedge \overline{\sigma e}(eg_{i,j})$ after applying e. Hence the premise of any of these properties is incorrect.

3. Assume that there are no indices i', j', k' such that $(e_{i',j',k'}, t^{\rightarrow}) \in I_\sigma$ and that σ has Property (SVG)$_{i'}$/(SVM)$_{i'}$ for all $i' \in [n]$ We prove that σe is a phase-1-strategy for $\mathfrak{b}+1$. Since σ is a phase-5-strategy for $\mathfrak{b}$ and by the choice of e, we have $\beta^\sigma = \mathfrak{b}+1 = \beta^{\sigma e}$. Also, by our assumptions and the definition of phase-1-strategies and phase-5-strategies, it suffices to show that σe has Property (ESC1).

   Consider the case $\nu > 1$, implying $\mu^{\sigma e} = 1$. Then, by assumption, there are no further indices i', j', k' such that $(e_{i',j',k'}, b_2) \in I_\sigma$. Thus, for all these indices, it either holds that $\sigma(e_{i',j',k'}) = b_2$ or $\sigma(e_{i',j',k'}) = g_1 \wedge \mathrm{Val}^*_\sigma(g_1) \succeq \mathrm{Val}^*_\sigma(b_2)$. It suffices to prove that the second case cannot occur. We do so by proving

$$\nu > 1 \implies \mathrm{Val}^*_\sigma(g_1) \prec \mathrm{Val}^*_\sigma(b_2) \tag{A.12}$$

   and showing that the arguments also apply to σe. Consider the different cases listed in Lemma 6.1.14. If the conditions of either the first or the "otherwise" case are fulfilled, then the statement follows. It thus suffices to prove that the conditions of the second and third case cannot be fulfilled.

   Assume that the conditions of the second case were fulfilled. Since $\mu^\sigma = 1$ implies $\sigma(b_1) = b_2$ and thus $m_b^\sigma \geq 2$, we then have $\sigma(g_1) = F_{1,1}$ and $\sigma(s_{1,1}) = h_{1,1}$. But then Property (USV1)$_1$ yields $\nu = 2$ which is a contradiction if $G_n = S_n$. We thus need to have $G_n = M_n$ and $\bar{\sigma}(d_1)$. But then, the cycle center F_{1,β_2} is closed, implying $\beta_1 = 1$ by definition. This however contradicts $\nu > 1$. Thus the conditions of the second case cannot be fulfilled.

   Assume that the conditions of the third case were fulfilled. Let $\overline{m}_g^\sigma > 1$, implying $m_b^\sigma > 2$. Then $\sigma(g_1) = F_{1,1}$ and $\sigma(s_{1,1}) = h_{1,1}$. Then Property (USV1)$_1$ implies $m_b^\sigma = 2$ which is a contradiction. Thus let $\overline{m}_g^\sigma = 1$. Then $\sigma(g_1) = F_{1,0}$ and $\sigma(s_{1,0}) = h_{1,0}$. Consequently, by Property (USV1)$_1$, $\beta_2 = 0$, so in particular $\neg\bar{\sigma}(b_{\overline{m}_g^\sigma+1})$. For the sake of a contradiction, assume $\neg\bar{\sigma}(eb_1)$. Since $\nu > 1$ and $\beta_2 = 0$ imply that $F_{1,0}$ cannot be closed, it thus needs to hold that $\bar{\sigma}(eg_{1,0}) \wedge \neg\bar{\sigma}(eb_{1,0})$. But then, Property (EG1) implies $\sigma(s_{1,0}) = b_1$, contradicting $\sigma(s_{1,0}) = h_{1,0}$. Thus the conditions of the third case of Lemma 6.1.14 cannot be fulfilled, implying Equation (A.12). Note that these arguments can also be applied to σe, hence the same statement holds for σe.

   Now consider the case $\nu = 1$, implying $\mu^\sigma \neq 1$ and $\sigma(b_1) = g_1$. Then, by assumption, there are no further indices i', j', k' such that $(e_{i',j',k'}, g_1) \in I_\sigma$. Thus, either $\sigma(e_{i',j',k'}) = g_1$ or $\sigma(e_{i',j',k'}) = b_2$ and $\mathrm{Val}^*_\sigma(b_2) \succeq \mathrm{Val}^*_\sigma(g_1)$ by the choice of e. We now show that the second case cannot occur by proving

$$\nu = 1 \implies \mathrm{Val}^*_\sigma(b_2) \prec \mathrm{Val}^*_\sigma(g_1). \tag{A.13}$$

   It holds that $\mathrm{rVal}^*_\sigma(g_1) = R_1^*$ as $i' < \mu^\sigma$ implies $\bar{\sigma}(d_{i'})$ by Corollary 6.1.5. Consider the case $\sigma(b_2) = g_2$ first. Then, $\mu^\sigma > 2$, implying $\mathrm{rVal}^*_\sigma(b_2) = R_2^*$. Hence $\mathrm{rVal}^*_\sigma(b_2) = R_2^* \prec R_1^* = \mathrm{rVal}^*_\sigma(g_1)$.

Now let $\sigma(b_2) = b_3$. Then, $\mu^\sigma = 2$ and $\mathrm{rVal}_\sigma^*(b_2) = L_2^*$, so Lemma 6.1.10 implies $\mathrm{rVal}_{\sigma e}^*(b_2) \prec R_1^* = \mathrm{rVal}_{\sigma e}^*(g_1)$. Therefore, $\mathrm{rVal}_\sigma^*(b_2) \prec \mathrm{rVal}_\sigma^*(g_1)$ in any case, contradicting $\mathrm{Val}_\sigma^*(b_2) \succeq \mathrm{Val}_\sigma^*(g_1)$. Note again that the same arguments apply to σe.

* Before we prove the remaining aspects, we prove that $(d_{i,j,1-k}, F_{i,j}) \in I_{\sigma e}$ in any case. As we assume $\bar{\sigma}(eb_{i,j}) \wedge \bar{\sigma}(eg_{i,j})$, it holds that $\sigma e(d_{i,j,1-k}) = e_{i,j,1-k}$. It thus suffices to show $\mathrm{Val}_{\sigma e}^*(F_{i,j}) \succ \mathrm{Val}_{\sigma e}^*(e_{i,j,1-k})$. By assumption $\sigma(e_{i,j,1-k}) = \sigma e(e_{i,j,1-k}) = t^{\rightarrow}$. If $G_n = S_n$, the statement thus follows from

$$\mathrm{Val}_{\sigma e}^{\mathrm{S}}(F_{i,j}) = \{F_{i,j}, d_{i,j,*}, e_{i,j,*}\} \cup \mathrm{Val}_{\sigma e}^{\mathrm{S}}(t^{\rightarrow}) \rhd \{e_{i,j,1-k}\} \cup \mathrm{Val}_{\sigma e}^{\mathrm{S}}(t^{\rightarrow}) = \mathrm{Val}_{\sigma e}^{\mathrm{S}}(e_{i,j,1-k}).$$

If $G_n = M_n$, we then have $\mathrm{Val}_{\sigma e}^{\mathrm{M}}(F_{i,j}) = (1-\varepsilon)\,\mathrm{Val}_{\sigma e}^{\mathrm{M}}(t^{\rightarrow}) + \varepsilon\,\mathrm{Val}_{\sigma e}^{\mathrm{M}}(s_{i,j})$. To prove the statement, it thus suffices to prove $\mathrm{Val}_{\sigma e}^{\mathrm{M}}(s_{i,j}) > \mathrm{Val}_{\sigma e}^{\mathrm{M}}(t^{\rightarrow})$.

We only consider the case $\nu > 1$, the case $\nu = 1$ follows analogously. In this case, $t^{\rightarrow} = b_2$. If $j = \beta_{i+1}$, then $\mathrm{rVal}_{\sigma e}^{\mathrm{M}}(s_{i,j}) = \langle s_{i,j}, h_{i,j}\rangle + \mathrm{rVal}_{\sigma e}^{\mathrm{M}}(b_{i+1})$ by Property (EV1)$_i$. The statement thus follows since $\langle h_{i,j}\rangle > \sum_{\ell \in [i]} W_\ell^{\mathrm{M}}$. If $j \neq \beta_{i+1}$, then $\mu^{\sigma e} = 1$ implies $\sigma e(b_1) = b_2$ and thus $\mathrm{rVal}_{\sigma e}^{\mathrm{M}}(s_{i,j}) = \langle s_{i,j}\rangle + \mathrm{rVal}_{\sigma e}^{\mathrm{M}}(b_1) = \langle s_{i,j}\rangle + \mathrm{rVal}_{\sigma e}^{\mathrm{M}}(b_2)$.

4. We prove that $(g_i, F_{i,j})$ is improving for σe if and only if the corresponding conditions are fulfilled. Assume that the corresponding conditions are fulfilled. We distinguish the following cases.
   a) The cycle center $F_{i,1-j}$ cannot be closed as either $\bar{\sigma}(eb_{i,1-j})$ or $\bar{\sigma}(eg_{i,1-j})$.
   b) Let $F_{i,1-j}$ be $t^{\rightarrow}$-open. Then, if $G_n = S_n$, the statement follows since $j = 0$ and

$$\begin{aligned}\mathrm{Val}_{\sigma e}^{\mathrm{S}}(F_{i,0}) &= \{F_{i,0}, d_{i,j,*}, e_{i,j,*}\} \cup \mathrm{Val}_{\sigma e}^{\mathrm{S}}(t^{\rightarrow}) \\ &\rhd \{F_{i,1}, d_{i,1,*}, e_{i,1,*}\} \cup \mathrm{Val}_{\sigma e}^{\mathrm{S}}(t^{\rightarrow}) = \mathrm{Val}_{\sigma e}^{\mathrm{S}}(F_{i,1}).\end{aligned}$$

   If $G_n = M_n$, then the statement follows by Lemma 6.2.1 since $j = \beta_{i+1}$ and $F_{i,j}$ is also $t^{\rightarrow}$-open.
   c) Let $F_{i,1-j}$ be $t^{\rightarrow}$-halfopen. If $G_n = S_n$, then the statement follows analogously to the last case. If $G_n = M_n$, then the statement follows by an easy but tedious calculation.
   d) Let $F_{i,1-j}$ be mixed. Then, by either Equation (A.12) or Equation (A.13), it holds that $\mathrm{rVal}_{\sigma e}^*(F_{i,1-j}) \preceq \mathrm{rVal}_{\sigma e}^*(t^{\rightarrow})$. We can thus use the same arguments used in one of the last two cases to prove the statement.

As we proved that it is not possible that any cycle center escapes only to $t^{\leftarrow}$ at the beginning of this proof, these are all cases that need to be covered. Hence, if all the stated conditions are fulfilled, then the edge $(g_i, F_{i,j})$ is an improving switch for σe. We now prove that $(g_i, F_{i,j})$ is not improving for σe if any of these conditions is not fulfilled, proving the claimed equivalence. We consider the different conditions one after another.

   a) Let $\beta_i = 1$. Then, since σe is a phase-5-strategy for $\mathfrak{b}$, it holds that $\sigma e(b_i) = g_i$ and $\sigma e(g_i) = F_{i,\beta_{i+1}}$. Furthermore, this cycle center is then closed. Since $\bar{\sigma}(eb_{i,j}) \wedge \bar{\sigma}(eg_{i,j})$ by assumption, we then need to have $j = 1 - \beta_{i+1}$ and consequently $\overline{\sigma e}(g_i) = 1 - j$. By Properties (USV1)$_i$ and (EV1)$_{i+1}$, this thus yields

$\mathrm{rVal}^*_{\sigma e}(F_{i,1-j}) = [\![s_{i,1-j}, h_{i,1-j}]\!] \oplus \mathrm{rVal}^*_{\sigma e}(b_{i+1})$. Since $\mathrm{rVal}^*_{\sigma e}(F_{i,j}) = \mathrm{rVal}^*_{\sigma e}(t^{\rightarrow})$, the statement thus follows from $[\![s_{i,j}, h_{i,j}]\!] \succ \bigoplus_{\ell \leq i} W_\ell$.

b) Let $\bar{\sigma}(g_i) = 0$ resp. $\bar{\sigma}(g_i) = \beta_{i+1}$ depending whether we consider $G_n = S_n$ or $G_n = M_n$. Due to the first case, we may assume $\beta_i = 0$. We furthermore may assume $j = 1$ resp. $j = 1 - \beta_{i+1}$ if $G_n = S_n$ resp. $G_n = M_n$ as we otherwise already have $(g_i, F_{i,j}) \notin I_{\sigma e}$ by definition. We prove $\mathrm{Val}^*_{\sigma e}(F_{i,1-j}) \preceq \mathrm{Val}^*_{\sigma e}(F_{i,j})$ by distinguishing the different possible states of $F_{i,1-j}$.

   i. Let $F_{i,1-j}$ be closed. Since $\beta_i = 0$, this implies $1 - j = 1 - \beta_{i+1}$. Thus, by Property (USV1)$_i$,

$$\mathrm{rVal}^*_{\sigma e}(F_{i,1-j}) = [\![s_{i,1-j}]\!] \oplus \mathrm{rVal}^*_{\sigma e}(b_1) \succ \mathrm{rVal}^*_{\sigma e}(b_1) = \mathrm{rVal}^*_{\sigma e}(F_{i,j}).$$

   ii. Let $F_{i,1-j}$ be $t^{\rightarrow}$-open or $t^{\rightarrow}$-halfopen. Since $1-j = 0$ resp. $1-j = \beta_{i+1}$, the same arguments used when proving that $(g_i, F_{i,j}) \in I_{\sigma e}$ can be applied if the corresponding conditions are fulfilled to obtain $\mathrm{Val}^*_{\sigma e}(F_{i,1-j}) \succeq \mathrm{Val}^*_{\sigma e}(F_{i,j})$ in either case.

   iii. Let $F_{i,1-j}$ be mixed. Then, $\overline{\sigma e}(eb_{i,1-j}) \wedge \overline{\sigma e}(eg_{i,1-j})$ by the choice of e. However, in any context and for any ν, this contradicts the assumptions of the lemma.

   By the observations made at the beginning of this proof, these are all cases that can occur.

c) Let $j = 1$ resp. $j = 1 - \beta_{i+1}$. Due to the first two cases, we may assume $\beta_{i+1} = 0$ and $\overline{\sigma e}(g_i) = 1$ resp. $\overline{\sigma e}(g_i) = 1 - \beta_{i+1}$. But this implies $(g_i, F_{i,j}) \notin I_{\sigma e}$ by the definition of an improving switch.

d) We only discuss the last condition for $\nu > 1$ as the statement follows for $\nu = 1$ analogously. Hence let $\neg\bar{\sigma}(eb_{i,1})$ resp. $\neg\overline{\sigma e}(eb_{i,1-\beta_{i+1}})$. Due to the last cases, we may assume $\beta_i = 0$, $\bar{\sigma}(g_i) = 1$ resp. $\bar{\sigma}(g_i) = 1 - \beta_{i+1}$ and $j = 0$ resp. $j = \beta_{i+1}$. By the observations made at the beginning of the proof, we cannot have $\overline{\sigma e}(eg_{i,1})$ resp. $\overline{\sigma e}(eg_{i,1-\beta_{i+1}})$. This implies that we need to have $\overline{\sigma e}(d_{i,1})$ resp. $\overline{\sigma e}(d_{i,1-\beta_{i+1}})$, implying the statement in either case. More precisely, we then either have

$$\mathrm{rVal}^*_{\sigma e}(F_{i,1-j}) = [\![s_{i,1-j}]\!] \oplus \mathrm{rVal}^*_{\sigma e}(b_1) = [\![s_{i,1-j}]\!] \oplus \mathrm{rVal}^*_{\sigma e}(b_2) \succ \mathrm{rVal}^*_{\sigma e}(F_{i,j})$$

   or

$$\mathrm{rVal}^*_{\sigma e}(F_{i,1-j}) = [\![s_{i,1-j}, h_{i,1-j}]\!] \oplus \mathrm{rVal}^*_{\sigma e}(b_{i+1}) \succ \mathrm{rVal}^*_{\sigma e}(b_2) = \mathrm{rVal}^*_{\sigma e}(F_{i,j}).$$

Thus, if any of the given conditions is not fulfilled, then $(g_i, F_{i,j}) \notin I_{\sigma e}$. Consequently, $(g_i, F_{i,j}) \in I_{\sigma e}$ if and only if the stated conditions are fulfilled.

It remains to show that no other improving switches are created in any case. By assumption, $\bar{\sigma}(eb_{i,j}) \wedge \bar{\sigma}(eg_{i,j})$, implying $\sigma e(d_{i,j,*}) = e_{i,j,*}$. Since the application of e increases the valuation of $F_{i,j}$, we begin by proving $(d_{i,j,k}, F_{i,j}) \in I_\sigma, I_{\sigma e}$. This however follows easily by Equations (A.12) and (A.13) since

$$\begin{aligned}
\mathrm{rVal}^*_\sigma(d_{i,j,k}) &= \mathrm{rVal}^*_\sigma(t^{\leftarrow}) \prec \mathrm{rVal}^*_\sigma(t^{\rightarrow}) = \mathrm{rVal}^*_\sigma(F_{i,j}),\\
\mathrm{Val}^{\mathrm{S}}_{\sigma e}(d_{i,j,k}) &= \{e_{i,j,k}\} \cup \mathrm{Val}^{\mathrm{S}}_{\sigma e}(t^{\rightarrow}) \lhd \{F_{i,j}, e_{i,j,k^*}, d_{i,j,k^*}\} \cup \mathrm{Val}^{\mathrm{S}}_{\sigma e}(t^{\rightarrow}) = \mathrm{Val}^{\mathrm{S}}_{\sigma e}(F_{i,j})\\
, \mathrm{Val}^{\mathrm{M}}_{\sigma e}(d_{i,j,k}) &= \mathrm{Val}^{\mathrm{M}}_{\sigma e}(t^{\rightarrow}) < (1-\varepsilon)\,\mathrm{Val}^{\mathrm{M}}_{\sigma e}(t^{\rightarrow}) + \varepsilon\,\mathrm{Val}^{\mathrm{M}}_{\sigma e}(s_{i,j}) = \mathrm{Val}^{\mathrm{M}}_{\sigma e}(F_{i,j})
\end{aligned}$$

for some $k^* \in \{0,1\}$ since $\mathrm{Val}^{\mathrm{M}}_{\sigma e}(s_{i,j}) > \mathrm{Val}^{\mathrm{M}}_{\sigma e}(b_1)$. Moire precisely, if $j = 1 - \beta_{i+1}$, then this follows directly since we have $\mathrm{Val}_{\sigma e}(s_{i,j}) = \langle s_{i,j}\rangle + \mathrm{Val}^{\mathrm{M}}_{\sigma e}(b_1)$ in that case. If $j = \beta_{i+1}$ then this follows as $\mathrm{Val}^{\mathrm{M}}_{\sigma e}(s_{i,j}) = \langle s_{i,j}, h_{i,j}\rangle + \mathrm{Val}^{\mathrm{M}}_{\sigma e}(b_{i+1})$ in that case.

We now consider the possible change of the valuation of g_i. If $\sigma e(g_i) \neq j$, then the edge $(g_i, F_{i,j})$ is the only edge (besides the edges $(d_{i,j,*}, F_{i,j})$ that we already considered) that might become improving for σe. We however already completely described the conditions under which this edge becomes improving. Hence consider the case $\sigma e(g_i) = j$. We first observe that we cannot have $i = 1$ since $\bar{\sigma}(eg_{i,j})$ would then contradict the fact that G_n is a sink game resp. weakly unichain. We prove that we then have $\sigma e(b_i) = b_{i+1}$ and $\mathrm{Val}^*_{\sigma e}(b_{i+1}) \succeq \mathrm{rVal}^*_{\sigma e}(g_i)$ as well as $\sigma e(s_{i-1,1}) = b_1$ and $\mathrm{Val}^*_{\sigma e}(b_1) \succeq \mathrm{Val}^*_{\sigma e}(h_{i-1,1})$. By assumption and the choice of e, the cycle center $F_{i,j}$ is not closed for σe. If $j = \beta_{i+1}$, this implies $\sigma e(b_i) = b_{i+1}$ and $\sigma e(s_{i-1,1}) = b_1$ by Property (EV1)$_i$ resp. Property (USV1)$_{i-1}$. If $j = 1 - \beta_{i+1}$, then we need to have $\beta_i = 0$ since Property (EV1)$_i$ and Property (EV2)$_i$ would imply $\sigma e(g_i) = \beta_{i+1} = 1 - j$ otherwise. Thus $\sigma e(b_i) = b_{i+1}$ and $\sigma e(s_{i-1,1}) = b_1$ in any case. We now prove $\mathrm{Val}^*_{\sigma e}(b_{i+1}) \succeq \mathrm{Val}^*_{\sigma e}(g_i)$. Since $\sigma e(b_i) = b_{i+1}$ implies $i \geq \mu^{\sigma e}$ by Property (REL1), we have $\mathrm{rVal}^*_{\sigma e}(b_{i+1}) = L^*_{i+1}$. If $\nu = 1$, then $\overline{\sigma e}(eg_{i,j}) \wedge \neg\overline{\sigma e}(eb_{i,j}) \wedge \mu^{\sigma e} \neq 1$. Then, by Lemma 6.1.15 resp. Lemma 6.1.16, $\mathrm{rVal}^*_{\sigma e}(F_{i,j}) = \mathrm{rVal}^*_{\sigma e}(g_1) = R^*_1$. This in particular implies $\mathrm{rVal}^*_{\sigma e}(g_i) = [\![g_i]\!] \oplus R^*_1$. The statement thus follows directly since $[\![g_i]\!] \prec \bigoplus_{\ell=1}^{i-1} W^*_\ell$. If $\nu > 1$, then $\overline{\sigma e}(eb_{i,j}) \wedge \neg\overline{\sigma e}(eg_{i,j}) \wedge \mu^{\sigma e} = 1$. Since Properties (EV1)$_{i+1}$ and (USV1)$_i$ imply that either $\sigma e(s_{i,j}) = b_1$ or $\sigma e(b_{i+1}) = j$, Lemma 6.1.15 resp. Lemma 6.1.16 thus imply $\mathrm{rVal}^*_{\sigma e}(g_i) = [\![g_i]\!] \oplus \mathrm{rVal}^*_{\sigma e}(b_2)$. Then, the statement again follows since $[\![g_i]\!] \prec \bigoplus_{\ell=1}^{i-1} W^*_\ell$. Therefore $\mathrm{rVal}^*_{\sigma e}(b_{i+1}) \prec \mathrm{rVal}^*_{\sigma e}(g_i)$ in any case. Since $\mu^{\sigma e} = 1 \Leftrightarrow \sigma e(b_1) = b_2 \Leftrightarrow \nu > 1$, the same arguments imply

$$\mathrm{rVal}^*_{\sigma e}(h_{i-1,1}) = [\![h_{i-1,1}, g_i]\!] \oplus \mathrm{rVal}^*_{\sigma e}(b_1) \prec \mathrm{rVal}^*_{\sigma e}(b_1).$$

□

Lemma 6.2.33 (First row of Table 6.7). *Let $\sigma \in \rho(\sigma_0)$ be a well-behaved phase-5-strategy for $\mathfrak{b} \in \mathfrak{B}_n$. Let $i \in [n], j = 1 - \beta^\sigma_{i+1}, k \in \{0,1\}$ with $e := (d_{i,j,k}, F_{i,j}) \in I_\sigma$ and assume $\sigma(b_i) = b_{i+1}, \bar{\sigma}(g_i) = 1 - \beta^\sigma_{i+1}$ and $i \neq 1$. Then σe is a well-behaved Phase-5-strategy for $\mathfrak{b}$ with $\sigma e \in \rho(\sigma_0)$ and $I_{\sigma e} = I_\sigma \setminus \{e\}$.*

Proof. As in the last proofs, we have $\beta^\sigma = \beta^{\sigma e} =: \beta$ by the choice of e. Let $j := 1 - \beta_{i+1}$. We begin by showing that σe is a phase-5-strategy for $\mathfrak{b}$. If $\sigma(d_{i,j,1-k}) \neq F_{i,j}$, then the same cycle centers are closed with respect to σ and σe. In this case, σ being a phase-5-strategy immediately implies that σe is a phase-5-strategy. Thus assume $\sigma(d_{i,j,1-k}) = F_{i,j}$. Then, the cycle center $F_{i,j} = F_{i,1-\beta_{i+1}}$ is closed with respect to σ but not with respect to σe. It thus sufficient to investigate Property (EV3)$_i$ and Property (CC1)$_i$. Since, by assumption, $\sigma e(b_i) = \sigma(b_i) = b_{i+1}$, Property (EV3)$_i$ remains valid. As σ has Property (REL1), σe also has Property (REL1). This implies $\mu^{\sigma e} = \min\{i' : \sigma e(b_{i'}) = b_{i'+1}\}$. Thus, $i \geq \mu^{\sigma e}$, so σe has Property (CC1)$_i$. Therefore, σe is a phase-5-strategy for $\mathfrak{b}$.

We now show that σe is well-behaved. If $\sigma(d_{i,j,1-k}) \neq F_{i,j}$, this follows immediately since σ is well-behaved. Hence assume $\sigma(d_{i,j,1-k}) = F_{i,j}$ and note that we have $i \geq \mu^{\sigma e}$ as argued earlier. Since we close a cycle center $F_{i,\bar{\sigma}(g_i)}$ with $i \geq \mu^{\sigma e}$, we investigate the following properties.

(MNS4) Since σ is well-behaved, this property only needs to be reevaluated if $i = \overline{m}_s^{\sigma e}$. Since the case $i = 1$ cannot occur by assumption, assume $i > 1$. This implies $1 < \overline{m}_s^{\sigma e} \leq \overline{m}_g^{\sigma e} < m_b^{\sigma e}$. Thus, in particular $\sigma e(b_1) = b_2, \sigma e(g_1) = F_{1,1}$ and $\sigma e(s_{1,1}) = h_{1,1}$. By Property (USV1)$_1$ and Property (EV1)$_2$, $\sigma e(b_2) = g_2$, implying $m_b^{\sigma e} = 2$. But this contradicts the premise since $1 < \overline{m}_s^{\sigma e} < m_b^{\sigma e}$ implies $m_b^{\sigma e} \geq 3$.

(MNS6) Since σ is well-behaved, this only needs to be reevaluated if $i = \overline{m}_s^{\sigma e}$. Since $i \neq 1$ by assumption, assume $i > 1$. Then, $1 < \overline{m}_s^{\sigma e} \leq \overline{m}_g^{\sigma e} < m_b^{\sigma e}$, implying the same contradiction as in the last case.

(DN*) Since there is no cycle center $F_{n,1}$ by construction, we cannot have $i = n$.

Consequently, σe is well-behaved.

It remains to show that $I_{\sigma e} = I_\sigma \setminus \{e\}$. We distinguish three different cases.

- The cycle center $F_{i,j}$ is closed with respect to σe. Then, since $j = 1 - \beta_{i+1}$, we have $\sigma e(s_{i,j}) = b_1$ by Property (USV1)$_i$, implying $\mathrm{rVal}^*_{\sigma e}(F_{i,j}) = [\![s_{i,j}]\!] \oplus \mathrm{rVal}^*_{\sigma e}(b_1)$. The only vertices that have an edge towards $F_{i,j}$ are $d_{i,j,0}, d_{i,j,1}$ and g_i. Since $F_{i,j}$ is closed for σe and $\bar{\sigma}(g_i) = j$ by assumption, the valuation of these vertices might change when applying e. However, $\sigma e(d_{i,j,k}) = F_{i,j}$ implies that $(d_{i,j,0}, F_{i,j}), (d_{i,j,1}, F_{i,j}) \notin I_{\sigma e}$. Since no player 0 vertex has an edge to $d_{i,j,*}$, consider the vertex g_i. The only vertices having an edge towards g_i are b_i and $h_{i-1,1}$. Since $\sigma e(b_i) = b_{i+1}$ by assumption and $\sigma e(s_{i-1,1}) = b_1$ by Property (USV1)$_i$, it suffices to show $(b_i, g_i), (s_{i-1,1}, h_{i-1,1}) \notin I_{\sigma e}$. We begin by showing $(b_i, g_i) \notin I_{\sigma e}$. It suffices to show $\mathrm{rVal}^*_{\sigma e}(b_{i+1}) \rhd \mathrm{rVal}^*_{\sigma e}(g_i)$. As mentioned before, we have $\mathrm{rVal}^*_{\sigma e}(g_i) = [\![g_i, s_{i,j}]\!] \oplus \mathrm{rVal}^*_{\sigma e}(b_1)$. If $\mu^{\sigma e} = 1$, the statement follows since

  $$\mathrm{rVal}^*_{\sigma e}(g_i) = [\![g_i, s_{i,j}]\!] \oplus \mathrm{rVal}^*_{\sigma e}(b_1) = [\![g_i, s_{i,j}]\!] \oplus L_1^* \prec L_{i+1}^* = \mathrm{rVal}^*_{\sigma e}(b_{i+1}).$$

  Now consider the case $\mu^{\sigma e} \neq 1$, implying $\mathrm{rVal}^*_{\sigma e}(b_1) = R_1^*$ since $i' < \mu^{\sigma e}$ implies $\bar{\sigma}(d_{i'})$ by Corollary 6.1.5 and Property (REL1). Since $\sigma e(b_i) = b_{i+1}$ implies $i + 1 > \mu^{\sigma e}$, we have $\mathrm{rVal}^*_{\sigma e}(b_{i+1}) = L_{i+1}^*$. Consequently,

  $$\mathrm{rVal}^*_{\sigma e}(g_i) = [\![g_i, s_{i,j}]\!] \oplus R_1^* \prec \bigoplus_{i' \geq i+1} \{W_{i'}^* : \sigma e(b_{i'}) = g_{i'}\} = L_{i+1}^* = \mathrm{rVal}^*_{\sigma e}(b_{i+1}).$$

  It remains to prove $(s_{i-1,1}, h_{i-1,1}) \notin I_{\sigma e}$ by showing $\mathrm{Val}^*_{\sigma}(b_1) \succ \mathrm{Val}^*_{\sigma e}(h_{i-1,1})$. This however follows by $\mathrm{rVal}^*_{\sigma e}(h_{i-1,1}) = [\![h_{i-1,1}, g_i, s_{i,j}]\!] \oplus \mathrm{rVal}^*_{\sigma e}(b_1) \prec \mathrm{rVal}^*_{\sigma e}(b_1)$.

- The cycle center $F_{i,j}$ is not closed and $\overline{\sigma e}(eb_{i,j})$. Then, since $\sigma e(d_{i,j,k}) = F_{i,j}$, we have $\overline{\sigma e}(eb_{i,j}) \wedge \neg\overline{\sigma e}(eg_{i,j})$. Consider the case that $G_n = S_n$. Then, by Lemma 6.1.16, either $\mathrm{rVal}^{\mathrm{S}}_{\sigma e}(F_{i,j}) = \mathrm{rVal}^{\mathrm{S}}_{\sigma e}(s_{i,j})$ or $\mathrm{rVal}^{\mathrm{S}}_{\sigma e}(F_{i,j}) = \mathrm{rVal}^{\mathrm{S}}_{\sigma e}(b_2)$. In the first case we can use the same arguments as before to prove $(b_i, g_i), (s_{i-1,1}, h_{i-1,1}) \notin I_{\sigma e}$. Hence consider the second case. Then, by Lemma 6.1.16, we need to have $\mu^{\sigma e} = 1$. But then, it follows that $\mathrm{rVal}^{\mathrm{S}}_{\sigma e}(g_i) = \{g_i\} \cup \mathrm{rVal}^{\mathrm{S}}_{\sigma e}(b_1) = \{g_i\} \cup L^{\mathrm{S}}_1 \lhd L^{\mathrm{S}}_{i+1} = \mathrm{rVal}^{\mathrm{S}}_{\sigma e}(b_{i+1})$ and $\mathrm{rVal}^{\mathrm{S}}_{\sigma e}(h_{i-1,1}) = \{h_{i-1,1}, g_1\} \cup \mathrm{rVal}^{\mathrm{S}}_{\sigma e}(b_1) \lhd \mathrm{rVal}^{\mathrm{S}}_{\sigma e}(b_1)$, so $(b_i, g_i), (s_{i-1,1}, h_{i-1,1}) \notin I_{\sigma e}$. Now consider the case $G_n = M_n$. Since $F_{i,j}$ is b_2-halfopen, $\mathrm{rVal}^{\mathrm{M}}_{\sigma e}(F_{i,j}) = \mathrm{rVal}^{\mathrm{M}}_{\sigma e}(b_2)$. It suffices to prove $\mu^{\sigma e} = 1$ as we can then apply the same arguments used in the case $G_n = S_n$. But this follows from Property (EB5) as $\overline{\sigma e}(eb_{i,j}) \wedge \neg\overline{\sigma e}(eg_{i,j}) \wedge \mu^{\sigma e} \neq 1$ implies $\sigma e(b_{\mu^{\sigma e}}) = g_{\mu^{\sigma e}}$, contradicting Property (REL1).
- The cycle center $F_{i,j}$ is not closed and $\overline{\sigma e}(eg_{i,j})$. Similar to the last case, we then have $\overline{\sigma e}(eg_{i,j}) \wedge \neg\overline{\sigma e}(eg_{i,j})$. Consider the case $G_n = S_n$. Then, by Lemma 6.1.16, either $\mathrm{rVal}^{\mathrm{S}}_{\sigma e}(F_{i,j}) = \mathrm{rVal}^{\mathrm{S}}_{\sigma e}(s_{i,j})$ or $\mathrm{rVal}^{\mathrm{S}}_{\sigma e}(F_{i,j}) = \mathrm{rVal}^{\mathrm{S}}_{\sigma e}(g_1)$. Since the second case implies $\mu^{\sigma e} \neq 1$, similar arguments as the ones used previously can be used to show $(b_i, g_i), (s_{i-1,1}, h_{i-1,1}) \notin I_{\sigma e}$ in both cases. If $G_n = M_n$, then $\mathrm{rVal}^{\mathrm{M}}_{\sigma e}(F_{i,j}) = \mathrm{rVal}^{\mathrm{M}}_{\sigma e}(g_1)$ and it again suffices to prove $\mu^{\sigma e} \neq 1$. This follows from Property (EG1) and Property (EG2) since these properties would imply that the cycle center F_{1,β_2} was closed. Then, Property $(\mathrm{EV1})_1$ would imply $\sigma e(b_1) = g_1$, contradicting $\mu^{\sigma e} = 1$. □

Lemma 6.2.34 (Second row of of Table 6.7). *Let $\sigma \in \rho(\sigma_0)$ be a well-behaved phase-5-strategy for $\mathfrak{b} \in \mathfrak{B}_n$. Let $i \in [n], j \in \{0,1\}$ with $e := (g_i, F_{i,j}) \in I_\sigma$ and $\beta^\sigma_i = 0$. Assume that $\nu = 1$ implies $\bar\sigma(eg_{i,j}) \wedge \neg\bar\sigma(eb_{i,j})$ and that $\nu > 1$ implies $\bar\sigma(eb_{i,j}) \wedge \neg\bar\sigma(eg_{i,j})$. Further assume that $\mu^\sigma = 1$ implies that for any $i' \geq i$ and $j' \in \{0,1\}$, either $\bar\sigma(d_{i',j'})$ or $\bar\sigma(eb_{i',j'}) \wedge \neg\bar\sigma(eg_{i',j'})$. If $\sigma(e_{i',j',k'}) = t^{\rightarrow}$ for all $i' \in [n], j', k' \in \{0,1\}$ and if σe has Property $(\mathrm{SVG})_{i'}/(\mathrm{SVM})_{i'}$ for all $i' \in [n]$, then σe is a phase-1-strategy for $\mathfrak{b}+1$. Otherwise it is a phase-5-strategy for $\mathfrak{b}$. In either case, σe is well-behaved and $I_{\sigma e} = I_\sigma \setminus \{e\}$.*

Proof. We first show that σe is a phase-1-strategy for $\mathfrak{b}+1$ resp. a phase-5-strategy for $\mathfrak{b}$. We observe that we have $\beta^\sigma = \beta^{\sigma e} =: \beta$ as the status of no cycle center or entry vertex is changed. Since we change the target of a selector vertex with $\beta_i = 0$, it suffices to check Properties (REL1), (CC2), $(\mathrm{EV2})_i$ and $(\mathrm{SVG})_i/(\mathrm{SVM})_i$.

It is immediate that σe has Property (REL1) as $\beta_i = 0$. To prove that it has Property (CC2) assume $i = \nu$. But this implies $\beta = 1$, contradicting again the assumption. By definition, σe has Property (ESC1) if and only if there are no indies i', j', k' with $\sigma(e_{i',j',k'}) \neq t^{\rightarrow}$. Thus, if there are no such indices and if σe has Property $(\mathrm{SVG})_{i'}/(\mathrm{SVM})_{i'}$ for all $i' \in [n]$, then σe is a phase-1-strategy for $\mathfrak{b}+1$. Otherwise, it is a phase-5-strategy for $\mathfrak{b}$. In particular, $\mu^\sigma = \mu^{\sigma e} = \min\{i' \colon \sigma e(b_{i'}) = b_{i'+1}\}$.

We next show that σe is well-behaved. Since we change the target of a selector vertex and $i \neq n$, we need to investigate the following assumptions:

(S2) Since $\beta_i = 0$ implies $i \geq \mu^{\sigma e}$, it cannot hold that $i < \mu^{\sigma e}$.

(D2) This follows by the same argument.

(B3) Since σe has Property $(\mathrm{USV1})_i$ and Property $(\mathrm{EV1})_{i+1}$, the premise of this assumption is always incorrect.

(BR1) Since $\beta_i = 0$ implies $i \geq \mu^{\sigma e}$, it cannot hold that $i < \mu^{\sigma e}$.

(MNS1) If the premise is correct for both σ and σe, then σe has this property as σ has it. The implication is also fulfilled if the premise is incorrect for σe. Hence assume that the premise is correct for σe but incorrect for σ. Since σe is well-behaved, Lemma 6.1.6 implies $m_b^{\sigma e} = 2$. Thus, $\sigma e(b_2) = g_2$, hence $\sigma e(s_{1,1}) = h_{1,1}$ and $\sigma e(s_{1,0}) = b_1$. As we assume that the premise is incorrect for σ, the choice of e implies $\overline{m}_g^{\sigma} = 1$ and thus $e = (g_1, F_{1,1})$. We thus need to have $\sigma(g_1) = F_{1,0}$ and $\mathrm{Val}_{\sigma}^{\mathrm{S}}(F_{1,1}) \rhd \mathrm{Val}_{\sigma}^{\mathrm{S}}(F_{1,0})$. We show that this cannot be true.

Since we have $\mu^{\sigma e} = \mu^{\sigma} = 1$ and $\sigma e(b_2) = g_2$, the cycle center $F_{1,1}$ cannot be closed. Consequently, by assumption, $\overline{\sigma}(eb_{1,1}) \wedge \neg\overline{\sigma}(eg_{1,1})$. Thus,

$$\mathrm{Val}_{\sigma}^{\mathrm{S}}(F_{1,1}) = \{F_{1,1}, d_{1,1,k^*}, e_{1,1,k^*}\} \cup \mathrm{Val}_{\sigma}^{\mathrm{S}}(b_2)$$

for some $k^* \in \{0,1\}$ by Lemma 6.1.16. If also $\overline{\sigma}(eb_{1,0}) \wedge \neg\overline{\sigma e}(eg_{1,0})$, then this also yields $\mathrm{Val}_{\sigma}^{\mathrm{S}}(F_{1,0}) = \{F_{1,0}, d_{1,0,k^*}, e_{1,0,k^*}\}$ for some $k^* \in \{0,1\}$. The statement then follows since $\Omega(F_{1,0}) > \Omega(F_{1,1})$ and since the priority of $F_{1,0}$ is even. If this is not the case, then $F_{1,0}$ is closed by assumption. But this implies

$$\mathrm{rVal}_{\sigma}^{\mathrm{S}}(F_{1,0}) = \{s_{1,0}\} \cup \mathrm{rVal}_{\sigma}^{\mathrm{S}}(b_2) > \mathrm{rVal}_{\sigma}^{\mathrm{S}}(b_2) = \mathrm{rVal}_{\sigma}^{\mathrm{S}}(F_{1,1}).$$

(MNS2) Assume $\mu^{\sigma e} = 1$, let $i' < \overline{m}_g^{\sigma e} < \overline{m}_s^{\sigma e}, m_b^{\sigma e}$ and let $G_n = S_n$ imply $\neg\overline{\sigma e}(b_{\overline{m}_g^{\sigma e}+1})$. Then $\sigma e(b_2) = b_3$ since $\sigma e(b_2) = g_2$ implies $m_b^{\sigma e} = 2$, contradicting the premise. Consequently, $\beta_2 = 0$. However, since $1 < \overline{m}_g^{\sigma e} < \overline{m}_s^{\sigma e}$, it holds that $\sigma e(g_1) = F_{1,1}$ and $\sigma e(s_{1,1}) = h_{1,1}$. But this implies $\beta_2 = 1$ by Property (USV1)$_1$ and Property (EV1)$_2$ which is a contradiction.

(MNS3) If the premise is true, then $\beta_2 = 0$ since we need to have $\sigma e(b_2) = b_3$. But, since $1 < \overline{m}_s^{\sigma e} \leq \overline{m}_g^{\sigma e}$ implies $\sigma e(g_1) = F_{1,1}$ and $\sigma e(s_{1,1}) = h_{1,1}$, we also have $\beta_2 = 1$ which is a contradiction.

(MNS4) Let $\mu^{\sigma e} = 1$ and $\overline{m}_s^{\sigma e} \leq \overline{m}_g^{\sigma e} < m_b^{\sigma e}$. If $\overline{m}_s^{\sigma e} > 1$, then the same arguments used for proving that σe has Property (MNS3) can be used to prove that σe has Property (MNS4) as follows. Thus assume $1 = \overline{m}_s^{\sigma e}$, implying that we have $\sigma e(s_{1,\overline{\sigma e}(g_1)}) = b_1$. In particular, by Property (USV1)$_1$ and Property (EV2)$_2$, it holds that $\overline{\sigma e}(g_1) \neq \overline{\sigma e}(b_2) = \beta_2$. If $\overline{m}_s^{\sigma} = \overline{m}_s^{\sigma e}$ and $\overline{m}_g^{\sigma} = \overline{m}_g^{\sigma e}$, then the statement follows by applying Property (MNS4) to σ. Thus assume $\overline{m}_s^{\sigma e} \neq \overline{m}_s^{\sigma}$. Then $\sigma e(s_{1,\overline{\sigma}(g_1)}) = h_{1,\overline{\sigma}(g_1)}$. But this implies $e = (g_i, F_{i,j}) = (g_1, F_{1,1-\beta_{i+1}^{\sigma e}})$ and thus, by assumption, $\overline{\sigma e}(eb_{1,j}) \wedge \neg\overline{\sigma e}(eg_{i,j})$. Hence assume $\overline{m}_s^{\sigma e} = \overline{m}_s^{\sigma}$ and $\overline{m}_g^{\sigma e} \neq \overline{m}_g^{\sigma}$. If $\overline{m}_g^{\sigma} < m_b^{\sigma}$, then the statement follows since we can again apply Property (MNS4) to σ. Thus assume $\overline{m}_g^{\sigma} \geq m_b^{\sigma}$. But then, $\overline{m}_s^{\sigma} < m_b^{\sigma} \leq \overline{m}_g^{\sigma}$, hence $\overline{m}_s^{\sigma e} = \overline{m}_s^{\sigma}$ and applying Property (MNS6) to σ imply $\overline{\sigma e}(eb_{\overline{m}_s^{\sigma e}}) \wedge \overline{\sigma e}(eg_{\overline{m}_g^{\sigma e}})$.

(MNS5) If the premise is true, then $1 < \overline{m}_s^{\sigma e} < m_b^{\sigma e} \leq \overline{m}_g^{\sigma e}$. In particular, $\sigma e(g_1) = F_{1,1}$ and $\sigma e(s_{1,1}) = h_{1,1}$. By Property (EV1)$_2$, this implies $\sigma e(b_2) = g_2$ and thus $m_b^{\sigma e} = 2$. This however is a contradiction since the premise implies $m_b^{\sigma e} \geq 3$.

(MNS6) If $\overline{m}_s^{\sigma e} > 1$, then the same arguments used for Property (MNS5) can be used to prove that the premise cannot be correct. Hence assume $\overline{m}_s^{\sigma e} = 1$, implying

$\sigma e(g_1) = F_{1,1}$ and $\sigma e(s_{1,1}) = b_1$. This in particular implies $1 = \bar{\sigma e}(g_1) \neq \beta_2^{\sigma e} = 0$ and thus $\sigma(b_2) = \sigma e(b_2) = b_3$ by Property (USV1)$_1$ and Property (EV1)$_2$. If $\bar{m}_s^\sigma = \bar{m}_s^{\sigma e}$ and $\bar{m}_g^\sigma = \bar{m}_g^{\sigma e}$, then the statement follows by applying Property (MNS6) to σ. Assume $\bar{m}_s^\sigma \neq \bar{m}_s^{\sigma e}$. Then $\sigma(s_{1,\bar{\sigma}(g_1)}) = h_{1,\bar{\sigma}(g_1)}$ and thus $\bar{\sigma}(g_1) = \beta_2^{\sigma e}$. But then, $e = (g_i, F_{i,j}) = (g_1, F_{1,1-\beta_2^{\sigma e}})$, so $\bar{\sigma e}(eb_{i,j}) \wedge \neg\bar{\sigma e}(eg_{i,j})$ by assumption. Thus assume $\bar{m}_s^\sigma = \bar{m}_s^{\sigma e}$ and $\bar{m}_g^\sigma \neq \bar{m}_g^{\sigma e}$. Since the statement follows by applying Property (MNS6) to σ if $\bar{m}_s^\sigma < m_b^\sigma \leq \bar{m}_g^\sigma$, assume $\bar{m}_g^\sigma < m_b^\sigma$. But then, $1 = \bar{m}_s^\sigma \leq \bar{m}_g^\sigma < m_b^\sigma$. Since applying an improving switch in level 1 implies $\bar{m}_s^\sigma \neq \bar{m}_s^{\sigma e}$, we have $\bar{\sigma}(g_1) = \bar{\sigma e}(g_1)$. But then the statement follows by applying Property (MNS4) to σ.

(EG*) Since $\mu^{\sigma e} = 1$ implies $\nu > 1$, we have $\bar{\sigma e}(eb_{i,j}) \wedge \neg\bar{\sigma e}(eg_{i,j})$. Hence the premise of any of the Properties (EG1) to (EG4) is incorrect. Note that we do not need to validate Property (EG5).

(EBG*) By assumption, we cannot have $\bar{\sigma e}(eb_{i,j}) \wedge \bar{\sigma e}(eg_{i,j})$, hence the premise of any of these assumptions is incorrect.

Hence σe is a well-behaved strategy.

It remains to show that $I_{\sigma e} = I_\sigma \setminus \{e\}$. Since we apply the improving switch $e = (g_i, F_{i,j})$, the valuation of g_i increases. If $i \neq 1$, then there are only two vertices that have an edge to g_i, namely b_i and $h_{i-1,1}$. However, if $i = 1$, then also the valuation of escape vertices and hence cycle centers might be influenced. We prove that $\sigma e(b_i) = b_{i+1} \wedge (b_i, g_i) \notin I_{\sigma e}$ for all $i \in [n]$ and $\sigma e(s_{i-1,1}) = b_1 \wedge (s_{i-1,1}, h_{i-1,1}) \notin I_{\sigma e}$ if $i > 1$. We then discuss the case $i = 1$ at the end of this proof.

Thus let $i \in [n]$. Since $\beta_i = 0$ and by Property (EV1)$_i$, it holds that $\sigma e(b_i) = b_{i+1}$. It thus suffices to prove $\mathrm{Val}^*_{\sigma e}(b_{i+1}) \succ \mathrm{Val}^*_{\sigma e}(g_i)$. We distinguish the following cases.

1. Let $\mu^{\sigma e} = 1$. Then $\mathrm{rVal}^*_{\sigma e}(b_{i+1}) = L^*_{i+1}$. By assumption, $\bar{\sigma e}(eb_{i,j}) \wedge \neg\bar{\sigma e}(eg_{i,j})$. Thus, depending on whether $G_n = S_n$ or $G_n = M_n$, Lemma 6.1.16 and Property (USV1)$_i$ respectively Lemma 6.1.15 imply $\mathrm{rVal}^*_{\sigma e}(F_{i,j}) = \mathrm{rVal}^*_{\sigma e}(b_2)$. Consequently,

$$\mathrm{rVal}^*_{\sigma e}(g_i) = [\![g_i]\!] \oplus \mathrm{rVal}^*_{\sigma e}(b_2) = [\![g_i]\!] \oplus L^*_2 = [\![g_i]\!] \oplus L^*_{2,i-1} \oplus L^*_{i+1} \prec L^*_{i+1}.$$

2. Let $\mu^{\sigma e} \neq 1$. Since $\beta_i = 0$, it cannot hold that $\mathrm{rVal}^*_{\sigma e}(b_{i+1}) = R^*_{i+1}$. This implies that $\mathrm{rVal}^*_{\sigma e}(b_{i+1}) = L^*_{i-1}$ and $i \geq \mu^{\sigma e}$. By assumption, $\bar{\sigma e}(eg_{i,j}) \wedge \neg\bar{\sigma e}(eb_{i,j})$. Thus, by Lemma 6.1.15 resp. 6.1.16, $\mathrm{rVal}^*_{\sigma e}(g_i) = [\![g_i]\!] \oplus \mathrm{rVal}^*_{\sigma e}(g_1)$. Note that $\mathrm{rVal}^*_{\sigma e}(b_1) = R^*_1$ in any case by Corollary 6.1.5. Thus, by Property (USV1)$_i$ and since $i \geq \mu^{\sigma e}$,

$$\begin{aligned}\mathrm{rVal}^*_{\sigma e}(g_i) &= [\![g_i, s_{i,j}]\!] \oplus \mathrm{rVal}^*_{\sigma e}(b_1) = [\![g_i, s_{i,j}]\!] \oplus R^*_1\\ &= [\![g_i, s_{i,j}]\!] \oplus \bigoplus_{i'=1}^{\mu^{\sigma e}-1} W^*_{i'} \oplus L^*_{\mu^{\sigma e}+1,i-1} \oplus L^*_{i+1} \prec L^*_{i+1} = \mathrm{rVal}^*_{\sigma e}(b_{i+1}).\end{aligned}$$

Thus $\mathrm{rVal}^*_{\sigma e}(g_i) \prec \mathrm{rVal}^*_{\sigma e}(b_{i+1})$ in any case, implying $(b_i, g_i) \notin I_{\sigma e}$.

We prove that $i \neq 1$ implies $\sigma e(s_{i-1,1}) \neq h_{i-1,1}$ and $(s_{i-1,1}, h_{i-1,1}) \notin I_{\sigma e}$. The first statement follows since $\beta_i = 0$ and Property (USV1)$_{i-1}$ imply $\sigma e(s_{i-1,1}) = b_1$. It thus remains to prove $\mathrm{Val}^*_{\sigma e}(b_1) \succ \mathrm{Val}^*_{\sigma e}(h_{i-1,1})$. We again distinguish the following cases.

1. Assume $\mu^{\sigma e} = 1$ first. Then $\mathrm{rVal}^*_{\sigma e}(b_1) = L^*_1$. By assumption, $\sigma e(eb_{i,j}) \wedge \neg\sigma e(eg_{i,j})$. By Property (USV1)$_i$ and since $\mu^{\sigma e} = 1$ implies $\sigma e(b_1) = b_2$, we then have

$$\mathrm{rVal}^*_{\sigma e}(h_{i-1,1}) = \{h_{i-1,1}, g_i\} \oplus \mathrm{rVal}^*_{\sigma e}(b_2) \prec \mathrm{rVal}^*_{\sigma e}(b_2) = \mathrm{rVal}^*_{\sigma e}(b_1).$$

2. Let $\mu^{\sigma e} \neq 1$. Then $\sigma e(b_1) = g_1$, implying $\mathrm{rVal}^*_{\sigma e}(b_1) = \mathrm{rVal}^*_{\sigma e}(g_1) = R^*_1$ by Corollary 6.1.5. By assumption, $\overline{\sigma e}(eg_{i,j}) \wedge \neg\overline{\sigma e}(eb_{i,j})$. Thus, by Lemma 6.1.15 resp. Lemma 6.1.16, $\mathrm{rVal}^*_{\sigma e}(h_{i-1,1}) = [\![h_{i-1,1}, g_i]\!] \oplus \mathrm{rVal}^*_{\sigma e}(g_1) \prec \mathrm{rVal}^*_{\sigma e}(b_1)$.

It remains to discuss the case $i = 1$. Since $\beta_i = 0$ by assumption, we then have $\mu^{\sigma e} = 1$. In particular, for any cycle center $F_{i',j'}$, either $\overline{\sigma e}(d_{i',j'})$ or $\overline{\sigma e}(eb_{i',j'}) \wedge \neg\overline{\sigma e}(eg_{i',j'})$ by assumption. Thus, the valuation of no cycle center is increased as this could only happen if $\overline{\sigma e}(eg_{i',j'})$. Moreover, there is $d_{i',j',k'}$ with $\sigma e(d_{i',j',k'}) = e_{i',j',k'}$ and $\sigma e(e_{i',j',k'}) = g_1$. It thus suffices to prove that $\overline{\sigma e}(d_{i',j'}) \wedge \sigma e(e_{i',j',k'}) = g_1$ implies $(d_{i',j',k'}, e_{i',j',k'}) \notin I_\sigma, I_{\sigma e}$ and that $\sigma e(e_{i',j',k'}) = b_2$ implies $(e_{i',j',k'}, g_1) \notin I_\sigma, I_{\sigma e}$.

Consider the second statement first. It suffices to prove $\mathrm{rVal}^*_{\sigma e}(b_2) \succ \mathrm{rVal}^*_{\sigma e}(g_1)$. If $\overline{\sigma e}(eb_1) \wedge \neg\overline{\sigma e}(eg_1)$, then this follows since $\mathrm{rVal}^*_{\sigma e}(g_1) = [\![g_1]\!] \oplus \mathrm{rVal}^*_{\sigma e}(b_2)$ in that case. If $\overline{\sigma e}(d_1)$, then we need to have $\overline{\sigma e}(g_i) = 1 - \beta_2$ due to $\mu^{\sigma e} = 1$ and $\nu > 1$. But then, the statement follows since $\mathrm{rVal}^*_{\sigma e}(g_1) = [\![g_1, s_{1,\beta_2}]\!] \oplus \mathrm{rVal}^*_{\sigma e}(b_2)$. Since the same arguments hold for σ, the statement follows. Thus consider some cycle center $F_{i',j'}$ closed with respect to σe. It suffices to show $\mathrm{rVal}_{\sigma e}(F_{i',j'}) \succ \mathrm{rVal}^*_{\sigma e}(g_1)$. If $j' = 1 - \beta_{i'+1}$, then the statement follows since $\mathrm{rVal}^*_{\sigma e}(F_{i',j'}) = [\![s_{i',j'}]\!] \oplus \mathrm{rVal}^*_{\sigma e}(b_2)$ in this case and since $\mathrm{rVal}^*_{\sigma e}(b_2) \succ \mathrm{rVal}^*_{\sigma e}(g_1)$ as proven before. If $j' = \beta_{i'+1}$, then $\mathrm{rVal}^*_{\sigma e}(F_{i',j'}) = [\![s_{i',j'}, h_{i',j'}]\!] \oplus \mathrm{rVal}^*_{\sigma e}(b_{i'+1})$ and the statement follows since $\mathrm{rVal}^*_{\sigma e}(F_{i',j'}) \succ \mathrm{rVal}^*_{\sigma e}(b_2)$. □

Omitted proofs of Section 6.3

The following statements are claims that are used within proofs of the statements in Section 6.3. Each claim thus refers to the notation used in the corresponding proof, and this notation is not restated here.

Claim 1. If an edge $(g_i, F_{i,j'})$ with $i \in [n]$ and $j' \neq \mathfrak{b}_{\nu+1}$ becomes improving during the application of improving switches contained in $I^{<\mathrm{m}}$, then it is applied immediately. Its application is described by row 4 of Table 6.4.

Proof. Consider the first phase-1-strategy σ such that after applying an improving switch $e = (d_{i,j',k}, F_{i,j'})$ to σ, the edge $(g_i, F_{i,j'})$ becomes improving for σe. Then, $\mathfrak{A}^{\sigma e}_{\sigma_\mathfrak{b}} \subseteq \mathbb{D}^1$. Furthermore, σe is a phase-1-strategy for $\mathfrak{b}$ by Lemma 6.2.11 and $I_{\sigma e} = \mathfrak{D}^{\sigma e} \cup \{(g_i, F_{i,j'})\}$. Moreover, $F_{i,j'}$ is closed for σe and $\phi^{\sigma_\mathfrak{b}}(g_i, F_{i,j'}) \leq \phi^{\sigma_\mathfrak{b}}(d_{i,j',k}, F_{i,j'})$ by Table 5.6. Since $(d_{i,j',k}, F_{i,j'})$ minimized the occurrence record for σ, the switch $(g_i, F_{i,j'})$ minimizes the occurrence record for σe. By the tie-breaking rule, this switch is thus applied next. Since $e \in I^{<\mathrm{m}}_{\sigma_\mathfrak{b}}$, Lemma 6.2.5 implies that it cannot happen the cycle center $F_{1,1-\mathfrak{b}_2}$ was closed by applying e, so $i \neq 1$. It is easy to verify that the other conditions of row 4 of Table 6.4 hold as well, since $(g_i, F_{i,j'})$ would not have become an improving switch otherwise. Thus, by row 4 of Table 6.4, the strategy σ' obtained by applying $(g_i, F_{i,j'})$ to σe is a well-behaved phase-1-strategy for $\mathfrak{b}$ with $\sigma' \in \rho(\sigma_0)$ and $I_{\sigma'} = \mathfrak{D}^{\sigma'}$. This proves that the first switch of the type $(g_*, F_{*,*})$ is applied immediately when it becomes an improving switch. The same arguments can however also be applied for any edge $(g_*, F_{*,*})$ that becomes improving. □

Claim 2. Let $\nu > 1$ and let σ denote the strategy obtained after applying all improving switches contained in $I_\sigma^{<\mathfrak{m}}$. For all suitable indices $i \in [n], j' \in \{0,1\}$ it holds that $\sigma(d_{i,j',1}) = F_{i,j'}$, implying that no cycle center is open for σ.

Proof. Assume there were indices $i \in [n], j' \in \{0,1\}$ with $\sigma(d_{i,j',1}) \neq F_{i,j'}$ and let $e := (d_{i,j',1}, F_{i,j'})$. Then $e \in I_\sigma$, so $\phi^\sigma(e) = \mathfrak{m}$. Since $\sigma(d_{i,j',1}) \neq F_{i,j'}$ implies that $F_{i,j'}$ is not closed, $\mathfrak{b}_i = 0 \vee \mathfrak{b}_{i+1} \neq j'$ as σ is a phase-1-strategy for $\mathfrak{b}$. As e was not applied during $\sigma_\mathfrak{b} \to \sigma$, this yields

$$\phi^{\sigma_\mathfrak{b}}(e') = \phi^\sigma(e') = \min\left(\left\lfloor \frac{\mathfrak{b}+1-k}{2} \right\rfloor, \ell^\mathfrak{b}(i,j',k) + t_\mathfrak{b}\right)$$

for a feasible $t_\mathfrak{b}$ for $\mathfrak{b}$. In particular, $\phi^\sigma(e) \leq \lfloor (\mathfrak{b}+1-1)/2 \rfloor = \lfloor \mathfrak{b}/2 \rfloor$. But this is a contradiction, since $\phi^\sigma(e) = \mathfrak{m} > \lfloor \mathfrak{b}/2 \rfloor$ since $\mathfrak{b}$ is odd. □

Claim 3. Let $i \in [n], j,k \in \{0,1\}$ such that $(d_{i,j,k}, F_{i,j}) \in \mathfrak{A}_{\sigma_\mathfrak{b}}^{\sigma^{(3)}}$. The occurrence records of $(d_{i,j,k}, F_{i,j})$ with respect to $\sigma^{(3)}$ is specified by Table 5.6 when interpreted for $\mathfrak{b}+1$.

Proof. Consider some fixed indices $i \in [n], j,k \in \{0,1\}$ such that $(d_{i,j,k}, F_{i,j}) \in \mathfrak{A}_{\sigma_\mathfrak{b}}^{\sigma^{(3)}}$. As argued previously, this edge is contained in a cycle center $F_{i,j}$ which is open for $\sigma_\mathfrak{b}$. If the occurrence record of one of the cycle edges of $F_{i,j}$ is $\mathfrak{m}-1$, then the application of $(d_{i,j,k}, F_{i,j})$ is described by Lemma 6.2.11 and we do not need to consider it here. Also, due to the tie-breaking rule, we do not apply an improving switch contained in halfopen cycle centers (with the exception of $F_{\nu,\mathfrak{b}_{\nu+1}}$) as we only consider switches contained in $I_{\sigma_\mathfrak{b}}^{\mathfrak{m}}$. We may thus assume that $F_{i,j}$ is open with respect to $\sigma_\mathfrak{b}$ and that both cycle edges have an occurrence record of $\mathfrak{m}$.

We now distinguish between several possible indices. Consider the case $i \neq \nu$ or $i = \nu \wedge j \neq \mathfrak{b}_{i+1}$ first. By the tie-breaking rule, we then need to have $k=0$ as the edge $e := (d_{i,j,0}, F_{i,j})$ is then applied as improving switch. Let σ denote the strategy in which e is applied. Since $\mathfrak{b}$ is even, $\phi^{\sigma e}(e) = \mathfrak{m}+1 = \lfloor (\mathfrak{b}+2)/2 \rfloor$. It thus suffices to show that there is a parameter $t_{\mathfrak{b}+1}$ feasible for $\mathfrak{b}+1$ such that

$$\left\lfloor \frac{\mathfrak{b}}{2} \right\rfloor + 1 \leq \ell^{\mathfrak{b}+1}(i,j,k) + t_{\mathfrak{b}+1}. \tag{$\star$}$$

By the choice of i and j, Lemma 6.2.4 implies $\ell^\mathfrak{b}(i,j,k) + 1 = \ell^{\mathfrak{b}+1}(i,j,k)$. Therefore, $\phi^\sigma(e) + 1 \leq \ell^\mathfrak{b}(i,j,k) + t_\mathfrak{b} + 1 \leq \ell^{\mathfrak{b}+1}(i,j,k) + t_\mathfrak{b}$ for some $t_\mathfrak{b}$ feasible for $\mathfrak{b}$. Since $\mathfrak{b}$ is even, Property (OR4)$_{i,j,0}$ implies $\phi^{\sigma_\mathfrak{b}}(e) \neq \ell^\mathfrak{b}(i,j,k) - 1$. In addition, by Property (OR2)$_{i,j,0}$, it holds that $t_\mathfrak{b} \neq 1$ as this would imply $\sigma_\mathfrak{b}(d_{i,j,0}) = F_{i,j}$, contradicting our assumption. Consequently, $t_\mathfrak{b} = 0$, implying that $t_{\mathfrak{b}+1} = 0$ is a feasible parameter that yields ($\star$).

Consider the case $i = \nu$ and $j = \mathfrak{b}_{\nu+1}$ next. Then, both switches $(d_{i,j,*}, F_{i,j})$ are applied. Using Lemma 6.2.3, it is easy to verify that $\phi^{\sigma_\mathfrak{b}}(d_{i,j,k}, F_{i,j}) = \lfloor (\mathfrak{b}+1-k)/2 \rfloor$ for both

$k \in \{0,1\}$. Also, by the tie-breaking rule, $F_{i,j}$ is closed once there are no more open cycle centers. In particular, both cycle edges of $F_{i,j}$ are then applied and their application is described by row 1 resp. 5 of Table 6.4. Let σ denote the strategy obtained after closing $F_{i,j}$. Then, by definition and the choice of i and j, it holds that $\mathfrak{b}+1 = \operatorname{lfn}(\mathfrak{b}+1, \nu, \{(\nu+1, j)\})$. Since

$$\left\lceil \frac{(\mathfrak{b}+1)+1-k}{2} \right\rceil = \left\lfloor \frac{\mathfrak{b}+1+2-k}{2} \right\rfloor = \left\lfloor \frac{\mathfrak{b}+1-k}{2} \right\rfloor + 1,$$

this then implies

$$\phi^{\sigma}(d_{i,j,k}, F_{i,j}) = \left\lceil \frac{\operatorname{lfn}(\mathfrak{b}+1, \nu, \{(\nu+1, j)\}) + 1 - k}{2} \right\rceil$$

as required. □

Claim 4. Let $\nu = 2$ and consider the phase-2-strategy σ obtained after the application of (b_ν, g_ν). Then, the edge $(s_{1,1}, h_{1,1})$ is applied next, and the obtained strategy is a well-behaved phase-3-strategy for $\mathfrak{b}$ described by the respective rows of Tables 5.8 and 5.9.

Proof. As a reminder, the current strategy is denoted by $\sigma\!e$ and the set of improving switches for $\sigma\!e$ is given by $I_{\sigma\!e} = \mathfrak{D}^{\sigma\!e} \cup \{(b_1, b_2), (s_{1,1}, h_{1,1})\} \cup \{(e_{*,*,*}, b_2)\}$. By Table 5.6, $\phi^{\sigma\!e}(e_{*,*,*}, b_2) = \lfloor \mathfrak{b}/2 \rfloor, \phi^{\sigma\!e}(b_1, b_2) = \operatorname{fl}(\mathfrak{b}, 1) - 1$ and $\phi^{\sigma\!e}(s_{1,1}, h_{1,1}) = \operatorname{fl}(\mathfrak{b}, 2)$. Since $\mathfrak{b}$ is odd, Lemma 6.2.6 implies $\operatorname{fl}(\mathfrak{b}, 1) - 1 = \lfloor \frac{\mathfrak{b}}{2} \rfloor$. Consequently, $\phi^{\sigma\!e}(b_1, b_2) = \phi^{\sigma\!e}(e_{*,*,*}, b_2)$. If $\mathfrak{b} = 1$, then $\operatorname{fl}(\mathfrak{b}, 2) = \lfloor \frac{\mathfrak{b}+2}{4} \rfloor = 0 = \lfloor \frac{\mathfrak{b}}{2} \rfloor$ and $\phi^{\sigma\!e}(s_{1,1}, h_{1,1}) = \phi^{\sigma\!e}(b_1, b_2)$. In this situation $(s_{1,1}, h_{1,1})$ is applied next as this is exactly the exception described in which the tie-breaking rule behaves differently, see Definition 5.3.5. If $\mathfrak{b} > 1$, then $\nu = 2$ implies $\mathfrak{b} \geq 5$. But this implies $\phi^{\sigma\!e}(s_{1,1}, h_{1,1}) < \phi^{\sigma\!e}(b_1, b_2)$, so $(s_{1,1}, h_{1,1})$ is applied next. Consequently, $e' := (s_{1,1}, h_{1,1})$ is applied next in any case.

We now prove that the requirements of row 2 Table 6.5 are fulfilled. Since $\mu^{\sigma\!e} = \nu = 2$, we show the following statements:

1. $\overline{\sigma\!e}(d_1)$ **:** No switch of the type $(d_{*,*,*}, e_{*,*,*})$ was applied during $\sigma_{\mathfrak{b}} \to \sigma^{(2)}$ by Lemma 6.3.1. Also, no such switch or switch of the type $(g_*, F_{*,*,})$ was applied during $\sigma^{(2)} \to \sigma\!e$. Thus, by Lemma 6.3.1, $\overline{\sigma\!e}(d_1)$.
2. $\sigma\!e$ **has Property (USV3)$_1$:** Since $\nu = 2$, we have $\beta_2^{\sigma\!e} = 1$. Since we did not apply any improving switch of the type $(s_{*,*}, *)$ during $\sigma_{\mathfrak{b}} \to \sigma\!e$, the statement then follows by applying Property (USV1)$_1$ to $\sigma_{\mathfrak{b}}$.
3. $\sigma\!e$ **has Property (EV2)$_2$ and Property (CC2):** We already argued that $\sigma\!e$ has these properties when applying the statement described by row 1 of Table 6.5.

Thus, all requirements of row 2 of Table 6.5 are met.

To simplify notation ,we denote the strategy obtained by applying $e' = (s_{1,1}, h_{1,1})$ to $\sigma\!e$ again by σ. Then, σ is a phase-3-strategy for $\mathfrak{b}$ with $\sigma \in \rho(\sigma_0)$ and

$$I_\sigma = I_{\sigma\!e} \setminus \{e'\} = \mathfrak{D}^{\sigma} \cup \{(b_1, b_2)\} \cup \{(e_{*,*,*}, b_2)\}.$$

Since $\mathfrak{b}$ is odd,

$$\phi^{\sigma}(e') = \operatorname{fl}(\mathfrak{b}, 2) + 1 = \left\lfloor \frac{\mathfrak{b}+2}{4} \right\rfloor + 1 = \left\lfloor \frac{(\mathfrak{b}+1)+2}{4} \right\rfloor = \operatorname{fl}(\mathfrak{b}+1, 2)$$

and Table 5.6 describes the occurrence record of $(s_{1,1}, h_{1,1})$ with respect to $\mathfrak{b}+1$. Since we did not apply any improving switch $(g_*, F_{*,*})$ or $(d_{*,*,*}, e_{*,*,*})$, the conditions on cycle centers in levels below ν hold for $\sigma^{(3)}$ as they held for $\sigma^{(2)}$. Therefore, σ is a strategy as described by the respective rows of Tables 5.8 and 5.9. □

Claim 5. After the application of $(b_{\nu-1}, b_\nu)$ in the case $\nu > 2$, the switch $e = (s_{\nu-1,1}, h_{\nu-1,1})$ is applied next. Its application can be described by row 2 of Table 6.5 and Table 5.6 specifies its occurrence record after the application correctly when interpreted for $\mathfrak{b}+1$.

Proof. By the definition of ν, there is a number $k \in \mathbb{N}$ such that $\mathfrak{b} = k \cdot 2^{\nu-1} - 1$. By Table 5.6, Lemma 6.2.6 and using $\nu > 2$, we obtain the following:

$$\phi^\sigma(s_{\nu-1,1}, h_{\nu-1,1}) = \mathrm{fl}(\mathfrak{b}, \nu) = \left\lfloor \frac{k \cdot 2^{\nu-1} - 1 + 2^{\nu-1}}{2^\nu} \right\rfloor = \left\lfloor \frac{k}{2} \right\rfloor$$

$$\phi^\sigma(s_{\nu-2,0}, h_{\nu-2,0}) = \mathrm{fl}(\mathfrak{b}, \nu-1) - 1 = k \cdot 2^0 - 1 = k - 1$$

$$\phi^\sigma(s_{\nu-3,0}, h_{\nu-3,0}) = \phi^\sigma(b_{\nu-2}, b_{\nu-1}) = \mathrm{fl}(\mathfrak{b}, \nu-2) - 1 = k \cdot 2^{2-1} - 1 = 2k - 1$$

$$\phi^\sigma(e_{*,*,*}, b_2) = \left\lfloor \frac{\mathfrak{b}}{2} \right\rfloor = \left\lfloor \frac{k \cdot 2^{\nu-1} - 1}{2} \right\rfloor = \left\lfloor k \cdot 2^{\nu-2} - \frac{1}{2} \right\rfloor = 2^{\nu-2}k - 1.$$

If $k > 2$, then $(s_{\nu-1,1}, h_{\nu-1,1})$ is the unique improving switch minimizing the occurrence records. If $k \leq 2$, then the occurrence records of $(s_{\nu-1,1}, h_{\nu-,1})$ and $(s_{\nu-2,0}, h_{\nu-2,0})$ are identical and lower than the occurrence record of every other improving switch. Since the tie-breaking rule applies improving switches at selection vertices contained in higher levels first, $(s_{\nu-1,1}, h_{\nu-1,1})$ is also applied first then. Consequently, $e := (s_{\nu-1,1}, h_{\nu-1,1})$ is applied next in any case.

We prove that σ fulfills the conditions of row 2 of Table 6.5. By our previous arguments, it suffices to prove that σ has Property (USV3)$_{\nu-1}$. As $\beta_\nu = 1$, this however follows since $(s_{\nu-1,1}, h_{\nu-1,1}) \in I_\sigma$ and since σ has Property (USV2)$_{\nu-1,0}$ by the definition of a phase-2-strategy. By our previous arguments and row 2 of Table 6.5, σe then has Properties (USV2)$_{\nu-1,1}$, (CC2), (EV1)$_\nu$ and (USV3)$_{i,1-\beta_{i+1}}$ for all $i < \nu - 1$. Furthermore,

$$I_{\sigma e} = \mathfrak{D}^{\sigma e} \cup \{(s_{\nu-2,0}, h_{\nu-2,0}), (b_{\nu-2}, b_{\nu-1}), (s_{\nu-3,0}, h_{\nu-3,0})\}$$

if $\nu - 1 > 2$ and $\nu > 2$ implies

$$I_{\sigma e} = \mathfrak{D}^{\sigma e} \cup \{(e_{*,*,*}, b_2)\} \cup \{(b_1, b_2), (s_{1,0}, h_{1,0})\}.$$

Also note that $\nu > 2$ implies $\phi^{\sigma e}(s_{\nu-1,1}, h_{\nu-1,1}) = \mathrm{fl}(\mathfrak{b}, \nu) + 1 = \mathrm{fl}(\mathfrak{b}+1, \nu)$ by Lemma 6.2.6, so Table 5.6 specifies its occurrence record with respect to $\mathfrak{b}+1$. □

Lemma 6.3.10. *Let $\sigma \in \rho(\sigma^{(3)})$ be a well-behaved phase-3-strategy for $\mathfrak{b}$ obtained through the application of a sequence $\mathfrak{A}^\sigma_{\sigma^{(3)}} \subseteq \mathbb{E}^1 \cup \mathbb{D}^0$ of improving switches. Assume that the conditions of row 1 of Table 6.6 were fulfilled for each intermediate strategy σ' of the transition $\sigma^{(3)} \to \sigma$. Let $t^\to := b_2$ if $\nu > 1$ and $t^\to := g_1$ if $\nu = 1$. Let $i \in [n], j, k \in \{0, 1\}$ such that*

$e := (d_{i,j,k}, e_{i,j,k}) \in I_\sigma$ is applied next and assume $\sigma(e_{i,j,k}) = t^{\rightarrow}, \beta_i^\sigma = 0 \vee \beta_{i+1}^\sigma \neq j$ and $I_\sigma \cap \mathbb{D}^0 = \{e\}$. Further assume that either $i \geq \nu$ or that we consider the case $G_n = S_n$. Then σe is a phase-3-strategy for $\mathfrak{b}$ with $I_{\sigma e} = (I_\sigma \setminus \{e\})$.

Proof. As usual, we define $t^{\leftarrow} := \{g_1, b_2\} \setminus \{t^{\rightarrow}\}$. We consider the case $G_n = S_n$ and $j = \beta_{i+1}$ first. Then, since σ is a phase-3-strategy for $\mathfrak{b}$, it cannot happen that $\bar{\sigma}(d_{i,j})$ as this would imply $\beta_i = 1$. In particular, $\bar{\sigma}(d_{i,j})$ thus implies $\beta_{i+1} \neq j$. But then, by Lemma 6.2.21, σe is a well-behaved phase-3-strategy for $\mathfrak{b}$ with $I_{\sigma e} = I_\sigma \setminus \{e\}$.

Hence consider the case $G_n = M_n$ and $\beta_i = 1$. Then, by assumption, $\beta_{i+1} = 1 - j$, so the statement follows by Lemma 6.2.22. It thus suffices to consider the case $\beta_i^\sigma = 0$, implying $i > \nu$ by assumption. We remind here that $\mu^\sigma = \nu$ by Property (REL2) and distinguish two cases.

1 : Let $\beta_{i+1} = j$. We prove that the application of e is then described by Lemma 6.2.23 or Lemma 6.2.24. We begin by proving that $F_{i,j}$ is $t^{\leftarrow}$-halfopen and then discuss the possible states of $F_{i,1-j}$.

Since $\beta_i = 0 \wedge \beta_{i+1} = j$, the cycle center $F_{i,j}$ was not closed for $\sigma^{(3)}$. In particular, as the choice of e implies $\sigma(d_{i,j,k}) = \sigma^{(3)}(d_{i,j,k}) = F_{i,j}$, Corollary 6.3.7 and $\mathfrak{A}_{\sigma^{(3)}}^\sigma \cap \mathbb{D}^1 = \emptyset$ imply $\sigma(d_{i,j,1-k}) = e_{i,j,1-k}$. As improving switches were applied according to Zadeh's pivot rule and our tie-breaking rule, this implies $(e_{i,j,1-k}, t^{\rightarrow}) \notin \mathfrak{A}_{\sigma^{(3)}}^\sigma$. Consequently, $\sigma(d_{i,j,1-k}) = e_{i,j,1-k} \wedge \sigma(e_{i,j,1-k}) = t^{\leftarrow}$, so $F_{i,j}$ is $t^{\leftarrow}$-halfopen with respect to σ.

We now discuss the possible states of $F_{i,1-j}$. First, $F_{i,1-j}$ cannot be $t^{\leftarrow}$-open for σ as this would imply that it is also $t^{\leftarrow}$-open for $\sigma^{(3)}$, contradicting Corollary 6.3.7.

Also, $F_{i,1-j}$ cannot be closed as it would then be the unique closed cycle center in level i. Then, the tie-breaking rule would have applied some switch $(e_{i,1-j,*}, t^{\rightarrow})$. But this would have made the corresponding edge $(d_{i,1-j,*}, e_{i,1-j,*})$ improving by Lemma 6.2.19. Furthermore, this switch would then already have been applied, contradicting the assumption that $F_{i,1-j}$ was closed.

Let, for the sake of contradiction, $F_{i,1-j}$ be mixed. Then, $\sigma(d_{i,1-j,*}) = e_{i,1-j,*}$ as well as $\sigma(e_{i,1-j,k'}) = t^{\rightarrow}$ and $\sigma(e_{i,1-j,1-k'}) = t^{\leftarrow}$ for some $k' \in \{0,1\}$. This implies that $F_{i,1-j}$ was $t^{\leftarrow}$-halfopen with respect to $\sigma^{(3)}$ and that $(e_{i,1-j,k'}, t^{\rightarrow}) \in \mathfrak{A}_{\sigma^{(3)}}^\sigma$. Hence, this switch was ranked higher by the tie-breaking rule. But this is a contradiction as the tie-breaking rule ranks switches contained in $F_{i,\beta_{i+1}^\sigma} = F_{i,j}$ higher if both cycle centers are $t^{\leftarrow}$-halfopen.

It is also immediate that $F_{i,1-j}$ cannot be $t^{\rightarrow}$-halfopen as the tie-breaking rule would then choose the edge $(e_{i,1-j,k'}, t^{\rightarrow})$ with $\sigma(e_{i,1-j,k'}) = t^{\leftarrow}$ as next improving switch.

Now assume that $F_{i,1-j}$ is $t^{\rightarrow}$-open. We show that this implies that $F_{i,1-j}$ was closed at the end of phase 1. As the cycle center $F_{i,1-j}$ is $t^{\rightarrow}$-open and $\sigma^{(3)}(e_{i,1-j,*}) = t^{\leftarrow}$, this implies $(e_{i,1-j,0}, t^{\rightarrow}), (e_{i,1-j,1}, t^{\rightarrow}) \in \mathfrak{A}_{\sigma^{(3)}}^\sigma$. As all improving switches $(e_{*,*,*}, t^{\rightarrow})$ have the same occurrence records, this implies that the tie-breaking rule ranked $(e_{i,1-j,0}, t^{\rightarrow})$ and $(e_{i,1-j,1}, t^{\rightarrow})$ higher than $(e_{i,j,k}, t^{\rightarrow})$. However, since $j = \beta_{i+1}$, this can only happen if $F_{i,1-j}$ was closed with respect to $\sigma^{(3)}$. If $F_{i,1-j}$ was not closed for $\sigma_{\mathfrak{b}}$, then Corollary 6.3.2 and $\mathfrak{A}_{\sigma_3}^\sigma \subseteq \mathbb{D}^0 \cup \mathbb{E}^1$ imply $\sigma^{(3)}(g_i) = \sigma(g_i) = F_{i,1-j}$. If it was closed for $\sigma_{\mathfrak{b}}$, then $\sigma_{\mathfrak{b}}(g_i) = F_{i,1-j}$ by Definition 5.2.1. Moreover, by the choice of j and i and Corollary 6.3.2, it is not possible that the cycle center $F_{i,j}$ was closed during phase 1. Consequently, also $\sigma(g_i) = F_{i,1-j}$. Thus, the statement follows by Lemma 6.2.23.

Finally, assume that $F_{i,1-j}$ is $t^{\leftarrow}$-halfopen. Then, since $\mathfrak{A}^{\sigma}_{\sigma^{(3)}} \subseteq \mathbb{E}^1 \cup \mathbb{D}^0$, this implies that $F_{i,1-j}$ was $t^{\leftarrow}$-halfopen for $\sigma^{(3)}$. In particular, this implies that no cycle center of level i was closed during phase 1. But this implies $\sigma(g_i) = \sigma^{(3)}(g_i) = \sigma_{\mathfrak{b}}(g_i) = F_{i,\beta_{i+1}} = F_{i,j}$ by Corollary 6.3.2. Since $i \geq \nu + 1 = \mu^{\sigma} + 1$ by assumption, Lemma 6.2.24 implies the statement. This concludes the case $j = \beta_{i+1}$.

2 : Let $1 - \beta_{i+1} = j$. We investigate $F_{i,j}$ first. As $j = 1 - \beta_{i+1}$, it is possible that $F_{i,j}$ was closed with respect to $\sigma^{(3)}$. Depending on whether or not improving switches corresponding to $F_{i,j}$ were applied during $\sigma^{(3)} \to \sigma$, the cycle center is either (a) closed, (b) $t^{\rightarrow}$-halfopen or (c) $t^{\leftarrow}$-halfopen for σ. Consider the cycle center $F_{i,1-j}$. It cannot be closed as $1 - j = \beta_{i+1}$ and $\beta_i = 0$. If $F_{i,1-j}$ was $t^{\leftarrow}$-open with respect to σ, then the assumption $\mathfrak{A}^{\sigma}_{\sigma^{(3)}} \subseteq \mathbb{E}^1 \cup \mathbb{D}^0$ implies that it was $t^{\leftarrow}$-open with respect to $\sigma^{(3)}$, contradicting Corollary 6.3.7. If $F_{i,1-j}$ was $t^{\rightarrow}$-open, then $(e_{i,1-j,k'}, t^{\rightarrow}) \in \mathfrak{A}^{\sigma}_{\sigma^{(3)}}$ for both $k' \in \{0,1\}$. This implies that $\sigma^{(3)}(d_{i,1-j,k'}) = F_{i,1-j}$, hence $F_{i,1-j}$ was closed with respect to $\sigma^{(3)}$. But this is not possible as $1 - j = \beta_{i+1}$ and $i > \nu$ then imply $\beta_i = 1$, contradicting that $F_{i,1-j}$ is $t^{\rightarrow}$-open. By the same argument, $F_{i,1-j}$ cannot be $t^{\rightarrow}$-halfopen for σ.

Now assume that $F_{i,1-j}$ is mixed. Then, $(e_{i,1-j,k'}, b_2), (d_{i,1-j,k'}, e_{i,1-j,k'}) \in \mathfrak{A}^{\sigma}_{\sigma_3}$ for some $k' \in \{0,1\}$. This implies that $(e_{i,1-j,k'}, t^{\rightarrow})$ precedes $(e_{i,j,k}, t^{\rightarrow})$ within the tie-breaking rule. Consequently, $F_{i,j}$ cannot be closed or $t^{\rightarrow}$-halfopen and is hence $t^{\leftarrow}$-halfopen. Furthermore, this implies that $F_{i,1-j} = F_{i,\beta_{i+1}}$ was also $t^{\leftarrow}$-halfopen for $\sigma^{(3)}$. Therefore, no cycle center of level i was closed at the end of phase 1. Thus, by Corollary 6.3.2, $\sigma(g_i) = \sigma^{(3)}(g_i) = \sigma_{\mathfrak{b}}(g_i) = F_{i,1-j}$. The statement thus follows by Lemma 6.2.24.

Next, assume that $F_{i,1-j}$ is g_1-halfopen. Then $F_{i,j}$ cannot be g_1-halfopen since the tie-breaking rule would then choose to apply an improving switch involving $F_{i,1-j}$ as $1 - j = \beta_{i+1}$. Thus consider the case that $F_{i,j}$ is closed. We show that we can apply Lemma 6.2.25. By assumption, $I_{\sigma} \cap \mathbb{D}^0 = \{e\}$, hence there is no other improving switch $(d_{*,*,*}, e_{*,*,*})$. As $\beta_i = 0$ and since $F_{i,1-j}$ is $t^{\leftarrow}$-halfopen, we also need to prove $\sigma(g_i) = F_{i,j}$. This however follows by Corollary 6.3.2 if $F_{i,j}$ is closed during phase 1 resp. Definition 5.2.1 if it was already closed with respect to $\sigma_{\mathfrak{b}}$. Since σ is a phase-3-strategy, $i' > i > \nu$ implies $\sigma(s_{i,1-\beta_{i'+1}}) = b_1$ by Property (USV1)$_i$. Now, let $i' > i$ and $\beta_{i'} = 0$. Then, due to the tie-breaking rule, all improving switches $(e_{i',*,*}, b_2) \in \mathbb{E}^1$ have already been applied. Since $\beta_{i'} = 0$, the cycle center $F_{i',\beta_{i'+1}}$ cannot have been closed with respect to $\sigma^{(3)}$. If both cycle centers of level i' were $t^{\leftarrow}$-halfopen for $\sigma^{(3)}$, then they are mixed for σ, and, in addition, $\sigma(g_{i'}) = \sigma^{(3)}(g_{i'}) = \sigma_{\mathfrak{b}}(g_{i'}) = F_{i',\beta_{i'+1}}$. If the cycle center $F_{i',1-\beta_{i'+1}}$ is closed for $\sigma^{(3)}$, then $F_{i',\beta_{i'+1}}$ can only be $t^{\leftarrow}$-halfopen for $\sigma^{(3)}$. Consequently, by Corollary 6.3.2 resp. Definition 5.2.1 and our previous arguments, this implies $\bar{\sigma}(g_{i'}) = 1 - \beta_{i'+1}$. Furthermore, $F_{i',1-\beta_{i'+1}}$ is then $t^{\rightarrow}$-open and $F_{i',\beta_{i'+1}}$ is $t^{\rightarrow}$-halfopen (for σ). Similarly, if $i' > i$ and $\beta_{i'+1} = 1$, then $F_{i',1-\beta_{i'+1}}$ is $t^{\rightarrow}$-open if it was closed for $\sigma^{(3)}$ and mixed if it was $t^{\leftarrow}$-halfopen. Hence, all requirements of Lemma 6.2.25 are met and the statement follows since $i > \nu$.

Finally, assume that $F_{i,j}$ is $t^{\rightarrow}$-halfopen. If we can prove that $\sigma(g_i) = F_{i,j}$, then the statement follows by Lemma 6.2.27. This however follows immediately since $F_{i,j}$ can only be $t^{\rightarrow}$-halfopen if it was closed with respect to $\sigma^{(3)}$, implying $\sigma(g_i) = F_{i,j}$ by the same statements used several times before. □

Claim 6. Let σ denote the phase-3-strategy in which the improving switch $(b_1, t^{\rightarrow})$ should be applied next. If $\nu > 1$, then $\bar{\sigma}(eb_{i,j}) \wedge \neg\bar{\sigma}(eg_{i,j})$ for all $(i,j) \in S_1$ and, in addition, $\bar{\sigma}(eb_{i,j}) \wedge \bar{\sigma}(eg_{i,j})$ for all $(i,j) \in S_2$. If $\nu = 1$, then $\bar{\sigma}(eg_{i,j}) \wedge \neg\bar{\sigma}(eb_{i,j})$ for all $(i,j) \in S_4$ and $\bar{\sigma}(eb_{i,j}) \wedge \bar{\sigma}(eg_{i,j})$ for all $(i,j) \in S_3$.

Proof. The definition of the sets S_1 to S_4 implies that $\beta_i = 0 \vee \beta_{i+1} \neq j$ for all of the relevant indices. We thus begin by considering some fixed but arbitrary indices $i \in [n], j, k \in \{0,1\}$ with $\beta_i = 0 \vee \beta_{i+1} \neq j$. Then, due to the previous application of the improving switches during phase 3, it holds that $(e_{i,j,k}, t^{\rightarrow}) \in \mathfrak{A}^{\sigma}_{\sigma^{(3)}}$ if and only if $\sigma^{(3)}(d_{i,j,k}) = F_{i,j}$. Thus, $(e_{i,j,k}, t^{\rightarrow}) \notin \mathfrak{A}^{\sigma}_{\sigma^{(3)}}$ if and only if $\sigma^{(3)}(d_{i,j,k}) = e_{i,j,k}$. Since $e_{i,j,k}$ has an outdegree of 2 by construction, this implies that $\sigma(e_{i,j,k}) = t^{\leftarrow}$ if and only if $\sigma^{(3)}(d_{i,j,k}) = e_{i,j,k}$. In particular, due to $\beta_i = 0 \vee \beta_{i+1} \neq j$, the switch $(d_{i,j,k}, e_{i,j,k})$ was then also applied. Hence, if there is a $k' \in \{0,1\}$ with $\sigma^{(3)}(d_{i,j,k'}) = e_{i,j,k'}$, then $\bar{\sigma}(eg_{i,j})$ if $\nu > 1$ resp. $\bar{\sigma}(eb_{i,j})$ if $\nu = 1$.

Now, consider some fixed indices $i \in [n], j \in \{0,1\}$ and the corresponding cycle center $F_{i,j}$. Since every cycle center is closed or escapes to $t^{\rightarrow}$ with respect to σ, either $\bar{\sigma}(eb_{i,j}) \wedge \bar{\sigma}(eg_{i,j})$ or $\bar{\sigma}(eb_{i,j}) \wedge \neg\bar{\sigma}(eg_{i,j})$ or $\bar{\sigma}(d_{i,j})$ if $\nu > 1$. Similarly, if $\nu = 1$, either $\bar{\sigma}(eg_{i,j}) \wedge \bar{\sigma}(eb_{i,j})$ or $\bar{\sigma}(eg_{i,j}) \wedge \neg\bar{\sigma}(eb_{i,j})$ or $\bar{\sigma}(d_{i,j})$. Consequently, $\bar{\sigma}(eb_{i,j}) \wedge \bar{\sigma}(eg_{i,j})$ holds if and only if there is a$k \in \{0,1\}$ such that that $\sigma_{\mathfrak{b}}(d_{i,j,k}) \neq F_{i,j}$ and $(d_{i,j,k}, F_{i,j})$ was *not* applied during phase 1. By Lemma 6.3.6, all improving switches of the type $(d_{*,*,*}, F_{*,*})$ not applied in phase 1 had $\phi^{\sigma_{\mathfrak{b}}}(d_{*,*,*}, F_{*,*}) = \mathfrak{m}$. By Corollary 6.3.7, it thus suffices to prove that there is a $k \in \{0,1\}$ with $\phi^{\sigma_{\mathfrak{b}}}(d_{i,j,k}, F_{i,j}) = \mathfrak{m}$ to prove $\bar{\sigma}(eb_{i,j}) \wedge \bar{\sigma}(eg_{i,j})$. Analogously, to prove $\bar{\sigma}(eb_{i,j}) \wedge \neg\bar{\sigma}(eg_{i,j})$ resp. $\bar{\sigma}(eg_{i,j}) \wedge \neg\bar{\sigma}(eb_{i,j})$, it suffices to show that $F_{i,j}$ was closed at the end of phase 1.

Let $\nu > 1$. Let $m = \max\{i\colon \sigma(b_i) = g_i\}$ and $u = \min\{i\colon \sigma(b_i) = b_{i+1}\}$.

1. We prove that $\phi^{\sigma_{\mathfrak{b}}}(d_{i,j,0}, F_{i,j}) = \mathfrak{m}$ for all $(i,j) \in S_2$.
   - Let $i \leq \nu - 1, j = \beta_{i+1}$ and $k \in \{0,1\}$. Then, $\mathfrak{b}_{i+1} \neq (\mathfrak{b}+1)_{i+1} = \beta_{i+1}$ by the choice of i. In particular, $j \neq \mathfrak{b}_{i+1}$. Thus, there is a feasible $t_{\mathfrak{b}}$ for $\mathfrak{b}$ with
   $$\phi^{\sigma_{\mathfrak{b}}}(d_{i,j,k}, F_{i,j}) = \min\left(\left\lfloor \frac{\mathfrak{b}+1-k}{2} \right\rfloor, \ell^{\mathfrak{b}}(i,j,k) + t_{\mathfrak{b}}\right).$$
   However, the choice of i implies $\mathfrak{b}_i = 1$ and thus $t_{\mathfrak{b}} = 0$ is the only feasible parameter. It thus suffices to show $\ell^{\mathfrak{b}}(i,j,0) \geq \mathfrak{m}$. Since $\mathfrak{b}_i = 1$ and $j \neq \mathfrak{b}_{i+1}$, this follows from Lemma 6.2.3.
   - Let $i \in \{\nu+1, \dots, m\}, \beta_i = 1$ and $j = 1 - \beta_{i+1}$. Since $i > \nu$ implies $\beta_i = \mathfrak{b}_i$ and $\beta_{i+1} = \mathfrak{b}_{i+1}$, we can deduce $\ell^{\mathfrak{b}}(i,j,0) \geq \mathfrak{m}$ as in the previous case.
   - Let $i \in \{\nu, \dots, m-1\} \wedge \beta_i = 0$ and $j = \beta_{i+1}$. Since $i + 1 > \nu$ implies that we have $\beta_{i+1} = \mathfrak{b}_{i+1}, \mathfrak{b}_{\nu-1} = 1$ and $\nu \geq 2$, we obtain $\ell^{\mathfrak{b}}(i,j,0) > \mathfrak{m}+1$ as
   $$\begin{aligned}\ell^{\mathfrak{b}}(i,j,0) &= \left\lceil \frac{\mathfrak{b} + 2^{i-1} + \sum(\mathfrak{b}, i) + 1}{2} \right\rceil \geq \left\lceil \frac{\mathfrak{b} + 2^{i-1} + 2^{\nu-2} + 1}{2} \right\rceil \\ &\geq \left\lceil \frac{\mathfrak{b} + 2^{\nu-1} + 2^{\nu-2} + 1}{2} \right\rceil \geq \left\lceil \frac{\mathfrak{b}+4}{2} \right\rceil = \left\lfloor \frac{\mathfrak{b}+5}{2} \right\rfloor > \left\lfloor \frac{\mathfrak{b}+1}{2} \right\rfloor + 1.\end{aligned}$$
   Thus, $\ell^{\mathfrak{b}}(i,j,0) + t_{\mathfrak{b}} > \mathfrak{m}$ for every $t_{\mathfrak{b}}$ feasible for $\mathfrak{b}$, implying the statement.

- Let $i > m$ and $j \in \{0,1\}$. Then, $\operatorname{lfn}(\mathfrak{b}, i+1) = \operatorname{lufn}(\mathfrak{b}, i+1) = 0$ since $\mathfrak{b}'_i = 0$ for all $\mathfrak{b}' \leq \mathfrak{b}$. Hence, by Lemma 6.2.3, $\ell^{\mathfrak{b}}(i,j,k) \geq \mathfrak{b}$. Consequently, $\phi^{\sigma_{\mathfrak{b}}}(d_{i,j,0}, F_{i,j}) = \mathfrak{m}$
- Let $\mathfrak{b}+1 = 2^l$ for some $l \in \mathbb{N}$. Then $\nu = l+1$ and $\mathfrak{b}_\nu = 0$. This implies $\operatorname{lfn}(\mathfrak{b}, \nu, \{(\nu,1)\}) = \operatorname{lfn}(\mathfrak{b}, \nu+1) = \operatorname{lufn}(\mathfrak{b}, \nu+1) = 0$ and consequently $\phi^{\sigma_{\mathfrak{b}}}(d_{i,j,0}, F_{i,j}) = \mathfrak{m}$.

2. We prove that either $\overline{\sigma_{\mathfrak{b}}}(d_{i,j})$ or $\phi^{\sigma_{\mathfrak{b}}}(d_{i,j,k}, F_{i,j}) < \mathfrak{m}$ for both $k \in \{0,1\}$ holds for all $(i,j) \in S_1$.
   - Let $i \leq \nu - 1$ and $j = 1 - \beta_{i+1}$. Then $\mathfrak{b}_i = 1$ and $j = 1 - \beta_{i+1} = \mathfrak{b}_{i+1}$. Hence $F_{i,j}$ was closed with respect to $\sigma_{\mathfrak{b}}$.
   - Let $i \in \{\nu, \dots, m-1\}, \beta_i = 0, j = 1 - \beta_{i+1}$ and $k \in \{0,1\}$. Then $\mathfrak{b}_i = \beta_i = 0, \beta_{i+1} = \mathfrak{b}_{i+1}$ and $\beta_i = 0$ implies $i \neq \nu$. In particular, $\nu \leq i-1$ and $\mathfrak{b}_\nu = 0$. Using Lemma 6.2.3, this implies $\ell^{\mathfrak{b}}(i,j,k) \leq \lfloor (\mathfrak{b}+1-k)/2 \rfloor - 1$. Rearranging this inequality implies $\phi^{\sigma_{\mathfrak{b}}}(d_{i,j,k}, F_{i,j}) \leq \ell^{\mathfrak{b}}(i,j,1) + 1$. If this inequality is strict, the statement follows. If the inequality is tight, then $\sigma_{\mathfrak{b}}(d_{i,j,k}) = F_{i,j}$ by Property $(\text{OR2})_{i,j,k}$ and thus $\phi^{\sigma_{\mathfrak{b}}}(d_{i,j,k}, F_{i,j}) < \mathfrak{m}$ by Property $(\text{OR1})_{i,j,k}$.
   - Assume that there is no $l \in \mathbb{N}$ with $\mathfrak{b}+1 = 2^l$ and let $i = \nu$ and $j = 1 - \mathfrak{b}_{\nu+1}$. Since $\mathfrak{b}$ is odd, Property $(\text{OR3})_{i,j,0}$ implies $\phi^{\sigma_{\mathfrak{b}}}(d_{i,j,0}, F_{i,j}) < \mathfrak{m}$. For $k=1$, $\mathfrak{b}$ being odd implies $\phi^{\sigma_{\mathfrak{b}}}(d_{i,j,1}, F_{i,j}) \leq \lfloor (\mathfrak{b}+1-1)/2 \rfloor < \mathfrak{m}$.

We now consider the case $\nu = 1$, implying $\mathfrak{b}_i = (\mathfrak{b}+1)_i$ for all $i > 1$.

1. We prove that $\phi^{\sigma_{\mathfrak{b}}}(d_{i,j,0}, F_{i,j}) = \mathfrak{m}$ for all $(i,j) \in S_3$.
   - Let $i \in [u]$ and $j = 1 - \beta_{i+1}$. By the definition of u, it holds that $\beta_i = \mathfrak{b}_i = 1$ if $i < u \wedge i \neq 1$ and $\mathfrak{b}_i = 0$ if $i = u \vee i = 1$. In either case, $j = 1 - \mathfrak{b}_{i+1}$. Hence, in the first case, Lemma 6.2.3 implies

$$\ell^{\mathfrak{b}}(i,j,0) = \left\lceil \frac{\mathfrak{b} + \sum(\mathfrak{b}, i) + 1}{2} \right\rceil \geq \left\lceil \frac{\mathfrak{b}+1}{2} \right\rceil \geq \left\lfloor \frac{\mathfrak{b}+1}{2} \right\rfloor.$$

     This implies $\phi^{\sigma_{\mathfrak{b}}}(d_{i,j,0}, F_{i,j}) = \mathfrak{m}$ since -1 is not a feasible parameter as $\mathfrak{b}$ is even. Consider the second case, implying

$$\ell^{\mathfrak{b}}(i,j,0) = \left\lceil \frac{\mathfrak{b} - 2^{i-1} + \sum(\mathfrak{b}, i) + 1}{2} \right\rceil.$$

     If $i = 1$, then $\ell^{\mathfrak{b}}(i,j,0) = \lceil \mathfrak{b}/2 \rceil = \mathfrak{m}$. If $i = u$, then $\mathfrak{b}_{i'} = 1$ for all indices $i' \in \{2, \dots, u-1\}$ and $\mathfrak{b}_1 = 0$. This implies $\ell^{\mathfrak{b}}(i,j,0) = \mathfrak{m}$, and hence the statement since $\mathfrak{b}$ is even.
   - Let $i \in \{u+1, \dots, m\}, \beta_i = 1$ and $j = 1 - \beta_{i+1}$. Then $i \geq 2, \mathfrak{b}_i = 1$ and $j = 1 - \mathfrak{b}_{i+1}$. Thus, $\phi^{\sigma_{\mathfrak{b}}}(d_{i,j,0}, F_{i,j}) = \mathfrak{m}$ follows by the same arguments used in the last case.
   - Let $i \in \{u+1, \dots, m-1\}, \beta_i = 0$ and $j = \beta_{i+1}$. Then $i \geq 2$ as well as $\mathfrak{b}_i = 0$ and $j = \mathfrak{b}_{i+1}$ and $\phi^{\sigma_{\mathfrak{b}}}(d_{i,j,0}, F_{i,j}) = \mathfrak{m}$ follows from Lemma 6.2.3 and

$$\ell^{\mathfrak{b}}(i,j,0) = \left\lceil \frac{\mathfrak{b} + 2^{i-1} + \sum(\mathfrak{b}, i) + 1}{2} \right\rceil \geq \left\lceil \frac{\mathfrak{b}+3}{2} \right\rceil \geq \left\lfloor \frac{\mathfrak{b}+1}{2} \right\rfloor + 1.$$

- Let $i > m$ and $j \in \{0, 1\}$. Then, $\mathbf{1}_{j=0}\text{lfn}(\mathfrak{b}, i+1) + \mathbf{1}_{j=1}\text{lufn}(\mathfrak{b}, i+1) = 0$ by the definition of m. Hence, by Lemma 6.2.3, $\ell^{\mathfrak{b}}(i, j, k) \geq \mathfrak{b}$. This implies $\phi^{\sigma_{\mathfrak{b}}}(d_{i,j,0}, F_{i,j}) = \mathfrak{m}$.
- Finally consider the pair $(u, \beta^{\sigma}_{u+1})$. Then, by definition, $\beta_u = 0$ and $\beta_{u+1} = \mathfrak{b}_{u+1}$. If $u > 1$, the statement follows as in the third case. The case $u = 1$ is not possible since $\nu = 1$.

2. We prove that $\phi^{\sigma_{\mathfrak{b}}}(d_{i,j,k}, F_{i,j}) < \mathfrak{m}$ for both $k \in \{0, 1\}$ for all $(i, j) \in S_4$. First, $(i, j) \in S_4$ implies $i \in \{u+1, \dots, m-1\}, \beta_i = 0$ and $j = 1 - \beta_{i+1}$. Since $i > u$ implies $i > 1$, we have $\mathfrak{b}_i = 0$ and $j = 1 - \mathfrak{b}_{i+1}$. Consequently, by Lemma 6.2.3,

$$\ell^{\mathfrak{b}}(i, j, k) = \left\lceil \frac{\mathfrak{b} - 2^{i-1} + \sum(\mathfrak{b}, i) + 1 - k}{2} \right\rceil = \left\lceil \frac{\mathfrak{b} - 2^{i-1} + \sum_{l=2}^{u-1} 2^{l-1} + \sum_{l=u+1}^{i-1} \mathfrak{b}_l 2^{l-1} + 1 - k}{2} \right\rceil.$$

We prove that this implies $\ell^{\mathfrak{b}}(i, j, k) < \mathfrak{m}$, implying the statement as we then either have $\phi^{\sigma_{\mathfrak{b}}}(d_{i,j,k}, F_{i,j}) < \mathfrak{m}$ or $\phi^{\sigma_{\mathfrak{b}}}(d_{i,j,k}, F_{i,j}) = \ell^{\mathfrak{b}}(i, j, k) + 1$. If $u = 2$, then

$$\begin{aligned}
\ell^{\mathfrak{b}}(i, j, k) &= \left\lceil \frac{\mathfrak{b} - 2^{i-1} + \sum_{l=3}^{i-1} \mathfrak{b}_l 2^{l-1} + 1 - k}{2} \right\rceil \leq \left\lceil \frac{\mathfrak{b} - 2^{i-1} + 2^{i-1} - 4 + 1 - k}{2} \right\rceil \\
&= \left\lceil \frac{\mathfrak{b} - 3 - k}{2} \right\rceil = \left\lceil \frac{\mathfrak{b} - 1 - k}{2} \right\rceil - 1 \leq \left\lfloor \frac{\mathfrak{b} + 1 - k}{2} \right\rfloor - 1 \\
&\leq \left\lfloor \frac{\mathfrak{b} + 1}{2} \right\rfloor - 1 < \left\lfloor \frac{\mathfrak{b} + 1}{2} \right\rfloor.
\end{aligned}$$

If $u > 2$, then

$$\begin{aligned}
\ell^{\mathfrak{b}}(i, j, k) &\leq \left\lceil \frac{\mathfrak{b} - 2^{i-1} + \sum_{l=2}^{u-1} 2^{l-1} + \sum_{l=u+1}^{i-1} 2^{l-1} + 1 - k}{2} \right\rceil \\
&= \left\lceil \frac{\mathfrak{b} - 2^{i-1} + 2^{u-1} - 2 + 2^{i-1} - 2^u + 1 - k}{2} \right\rceil = \left\lceil \frac{\mathfrak{b} - 2^{u-1} + 1 - k}{2} \right\rceil \\
&\leq \left\lceil \frac{\mathfrak{b} - 4 + 1 - k}{2} \right\rceil < \left\lfloor \frac{\mathfrak{b} + 1}{2} \right\rfloor.
\end{aligned}$$

□

Claim 7. The strategy $\sigma_{\mathfrak{e}}$ meets the five requirements of Lemma 6.2.25 and the lemma thus describes the application of the improving switch $(d_{i,j,k}, e_{i,j,k})$.

Proof. As a reminder, we have $i = \nu - 1, j = 1 - \mathfrak{b}_{i+1}$ and $k \in \{0, 1\}$. There are no other indices i', j', k' with $(d_{i',j',k'}, e_{i',j',k'}) \in I_{\mathfrak{e}}$. Also, since no such switch was applied previously in any level below level i, the cycle center $F_{i,j}$ is closed for $\sigma_{\mathfrak{e}}$ as it was closed for $\sigma^{(3)}$ by Lemma 6.3.6. As $i < \nu$ and $\mathfrak{b}_i = 1 \wedge \mathfrak{b}_{i+1} \neq \beta^{\mathfrak{e}}_{i+1}$, Definition 5.2.1 implies that $\sigma_{\mathfrak{b}}(g_i) = F_{i,j}$. By the same arguments used when discussing the case $G_n = S_n$ resp. Claim 6, it can be proven that $F_{i,1-j}$ was not closed during phase 1 as $(i, 1-j) \in S_2$. Consequently, $\sigma e(g_i) = F_{i,j}$ follows from Corollary 6.3.2. By the tie-breaking rule, no improving switch involving $F_{i,1-j}$ was applied yet. Therefore, $\sigma e(e_{i,1-j,*}) = \sigma^{(3)}(e_{i,1-j,*}) = g_1$ as well as $\sigma e(d_{i,1-j,*}) = \sigma^{(3)}(d_{i,1-j,*})$. By Corollary 6.3.7, $F_{i,1-j}$ cannot be open for $\sigma^{(3)}$, so it is not open for σe. Therefore, as $\beta_i = 0$ and $1 - j = \beta_{i+1}$, it is g_1-halfopen. Thus, the first requirement of Lemma 6.2.25 is met.

By Lemma 6.3.6 and since $(s_{i',*}, h_{i',*}) \notin \mathfrak{A}^{\sigma e}_{\sigma(3)}$ for all $i' < \nu$, it follows that we have $\sigma e(s_{i',*}) = \sigma^{(3)}(s_{i',*}) = h_{i',*}$ for all $i' < \nu$. Furthermore, $i' < \nu$ implies $\mathfrak{b}_{i'} = 1$ and no improving switch $(d_{*,*,*}, e_{*,*,*})$ below level ν was applied yet. Consequently, $\overline{\sigma e}(d_{i'})$ for all $i' < \nu$. Now consider some cycle center $F_{i',j'}$ where $i' < i$ and $j' = 1 - \overline{\sigma e}(g_{i'})$. We prove that $F_{i',j'}$ is g_1-halfopen. The cycle center $F_{i',\beta_{i'+1}}$ is not closed while $F_{i',1-\beta_{i'+1}}$ is closed due to $1 - \beta_{i'} = \mathfrak{b}_{i'}$. Thus, by Corollary 6.3.2 and the same arguments used before, $\overline{\sigma e}(g_{i'}) = \overline{\sigma_{\mathfrak{b}}}(g_{i'}) = 1 - \beta_{i'+1}$ and, in particular, $j' = \beta_{i'+1}$. However, by Corollary 6.3.7 and the tie-breaking rule, this implies that $F_{i',j'}$ is g_1-halfopen as before. Thus, the second requirement is met.

The third requirement is met as $i' > i = \nu - 1$ and since σe has Property (USV1)$_{i'}$.

Consider the fourth requirement. Let $i' > i$ and $\beta_{i'} = 0$. Then, due to the tie-breaking rule, all improving switches $(e_{i',j',k'}, b_2)$ with $\sigma^{(3)}(d_{i',j',k'}) = F_{i',j'}$ have already been applied. Since $\beta_{i'} = 0$, $F_{i',\beta_{i'+1}}$ cannot have been closed for $\sigma^{(3)}$. If both cycle centers of level i' were g_1-halfopen for $\sigma^{(3)}$, then they are mixed for σ, and $\sigma(g_{i'}) = \sigma^{(3)}(g_{i'}) = \sigma_{\mathfrak{b}}(g_{i'}) = F_{i',\beta_{i'+1}}$. If $F_{i',1-\beta_{i'+1}}$ is closed for $\sigma^{(3)}$, then $F_{i',\beta_{i'+1}}$ can only be g_1-halfopen for $\sigma^{(3)}$. Consequently, by Corollary 6.3.2 resp. Definition 5.2.1, $\bar{\sigma}(g_{i'}) = 1 - \beta_{i'+1}$. Furthermore, $F_{i',1-\beta_{i'+1}}$ is then b_2-open and $F_{i',\beta_{i'+1}}$ is b_2-halfopen (for σ). Thus, the fourth requirement is met.

By the same argument, if $i' > i$ and $\beta_{i'+1} = 1$, then $F_{i',1-\beta_{i'+1}}$ is b_2-open if it was closed for $\sigma^{(3)}$ and mixed if it was g_1-halfopen. Thus, the fifth and final requirement is met. □

Claim 8. Let σ denote the first phase-4-strategy in S_n for $\nu > 1$. Then, the switch $(s_{\nu-1,0}, b_1)$ is applied next and the application of this switch is described by Lemma 6.2.31.

Proof. We first consider the case that $\mathfrak{b}+1$ is a power of two, implying $\mathfrak{b} = 2^{\nu-1} - 1$. We distinguish four kinds of improving switches.

1. Let $e = (s_{\nu-1,0}, b_1)$. Then, $\phi^\sigma(e) = 0$ follows from
$$\phi^\sigma(e) = \mathrm{fl}(\mathfrak{b}, \nu) = \left\lfloor \frac{\mathfrak{b} + 2^{\nu-1}}{2^\nu} \right\rfloor < \left\lfloor \frac{2^{\nu-1} + 2^{\nu-1}}{2^\nu} \right\rfloor.$$

2. Let $e = (s_{i,1}, b_1)$ for $i \leq \nu - 2$. Then, $\phi^\sigma(e) = \mathrm{fl}(\mathfrak{b}, i+1) - j \cdot \mathfrak{b}_{i+1} = \lfloor (\mathfrak{b} + 2^i)/2^{i+1} \rfloor - 1$. If $i = \nu - 2$, then $\nu \geq 3$ and
$$\phi^\sigma(e) = \left\lfloor \frac{2^{\nu-1} - 1 + 2^{\nu-2}}{2^{\nu-1}} \right\rfloor - 1 = \left\lfloor 1 + \frac{2^{\nu-2} - 1}{2^{\nu-1}} \right\rfloor - 1 = 0.$$
If $i \leq \nu - 3$, then $\nu \geq 4$ and
$$\phi^\sigma(e) \geq \left\lfloor \frac{2^{\nu-1} - 1 + 2^{\nu-3}}{2^{\nu-2}} \right\rfloor - 1 = \left\lfloor 2 + \frac{2^{\nu-3} - 1}{2^{\nu-2}} \right\rfloor - 1 = 1.$$

3. Let $e = (d_{i,j,k}, F_{i,j})$ for some indices $i \in [n], j, k \in \{0,1\}$ with $\sigma(e_{i,j,k}) = g_1$. Then $\phi^\sigma(e) = \mathfrak{m} \geq 1$.
4. Let $e = (e_{i,j,k}, b_2)$ for some indices $i \in [n], j, k \in \{0,1\}$ with $\sigma(e_{i,j,k}) = g_1$. Then, $\phi^\sigma(e) = \lfloor \mathfrak{b}/2 \rfloor \geq 1$ if $\mathfrak{b} > 1$ and $\phi^\sigma(e) = 0$ if $\mathfrak{b} = 1$.

Thus, $(s_{\nu-1,0}, b_1)$ and $(s_{\nu-2,1}, b_1)$ both minimize the occurrence record if $\mathfrak{b} > 1$. If $\mathfrak{b} = 1$, then all switches $(e_{i,j,k}, b_2)$ with $i \in [n], j, k \in \{0,1\}$ and $\sigma(e_{i,j,k}) = g_1$ also minimize the occurrence record. Due to the tie-breaking rule, $(s_{\nu-1,0}, b_1)$ is thus applied next in either case.

Now consider the case that $\mathfrak{b}+1$ is not a power of two. Then $\mathfrak{b} \geq 2^\nu + 2^{\nu-1} - 1$ and $\mathfrak{b} \geq 6$, implying $\lfloor (\mathfrak{b}+2)/4 \rfloor < \mathfrak{m}$ and $\lfloor (\mathfrak{b}+2)/4 \rfloor < \lfloor \mathfrak{b}/2 \rfloor$. We prove that $(s_{\nu-1,0}, b_1)$ minimizes the occurrence record.

1. Let $e = (s_{\nu-1,0}, b_1)$. Then, $\phi^\sigma(e) = \mathrm{fl}(\mathfrak{b}, \nu) = \lfloor (\mathfrak{b} + 2^{\nu-1})/2^\nu \rfloor \leq \lfloor (\mathfrak{b}+2)/4 \rfloor$ as $\mathfrak{b} \geq 6$ implies $\nu \geq 2$.
2. Let $e = (d_{i,j,k}, F_{i,j})$ with $i \in [n], j, k \in \{0,1\}$ and $\sigma(e_{i,j,k}) = g_1$. Then $\phi^\sigma(e) = \mathfrak{m}$, implying that $\phi^\sigma(e) > \phi^\sigma(s_{\nu-1,0}, b_1)$.
3. Let $e = (e_{i,j,k}, b_2)$ with $i \in [n], j, k \in \{0,1\}$ and $\sigma(e_{i,j,k}) = g_1$. Then $\phi^\sigma(e) = \lfloor \mathfrak{b}/2 \rfloor$ by Table 5.6, implying that $\phi^\sigma(e) > \phi^\sigma(s_{\nu-1,0}, b_1)$.
4. Let $e = (s_{i,1}, b_1)$ with $i \leq \nu - 2$. Then, $\phi^\sigma(e) = \mathrm{fl}(\mathfrak{b}, i+1) - \mathfrak{b}_{i+1} = \mathrm{fl}(\mathfrak{b}, i+1) - 1$ by Table 5.6. Hence, $\phi^\sigma(e) = \mathrm{fl}(\mathfrak{b}, i+1) - 1 > \mathrm{fl}(\mathfrak{b}, \nu) - 1 > \phi^\sigma(s_{\nu-1,0}, b_1) - 1$ by Lemma 6.2.6. Thus, by integrality, $\phi^\sigma(e) \geq \phi^\sigma(s_{\nu-1,0}, b_1)$.
5. Let $e = (d_{i,j,k}, F_{i,j})$, with $i = \nu, j = 1 - \beta_{i+1}$ and $k \in \{0,1\}$. By the definition of a canonical strategy, $\sigma_\mathfrak{b}(d_{i,j,k}) \neq F_{i,j}$. Hence $\phi^{\sigma_\mathfrak{b}}(e) = \min(\lfloor \frac{\mathfrak{b}+1-k}{2} \rfloor, \ell^\mathfrak{b}(i,j,k) + t_\mathfrak{b})$, where $t_\mathfrak{b}$ is feasible for $\mathfrak{b}$. Since $\mathfrak{b}_i = \mathfrak{b}_\nu = 0$ and $\beta_{i+1} = \mathfrak{b}_{i+1}$, Lemma 6.2.3 then implies
$$\ell^\mathfrak{b}(i,j,k) = \left\lceil \frac{\mathfrak{b} - 2^{i-1} + \sum(\mathfrak{b}, i) + 1 - k}{2} \right\rceil$$
$$= \left\lceil \frac{\mathfrak{b} - 2^{i-1} + 2^{i-1} - 1 + 1 - k}{2} \right\rceil = \left\lfloor \frac{\mathfrak{b}+1-k}{2} \right\rfloor.$$
Hence, by Property $(\text{OR3})_{i,j,k}$, $\phi^{\sigma_\mathfrak{b}}(e) = \ell^\mathfrak{b}(i,j,k) - 1 = \mathfrak{m} - 1$ for $k = 0$ and $\phi^{\sigma_\mathfrak{b}}(e) = \ell^\mathfrak{b}(i,j,k) = \mathfrak{m} - 1$ for $k = 1$. If $k = 1$, Corollary 6.3.3 implies that the edge e was applied during phase 1. Consequently, $\phi^\sigma(e) = \mathfrak{m} > \phi^\sigma(s_{\nu-1,0}, b_1)$.
6. Let $e = (d_{i,j,k}, F_{i,j})$ with $i \in \{\nu+1, \dots, m-1\}, \beta_i = 0, j = 1 - \beta_{i+1}$ and $k \in \{0,1\}$. By the choice of i, it then follows that $\mathfrak{b}_i = 0$ and $j = 1 - \mathfrak{b}_{i+1}$. If we have $\phi^{\sigma_\mathfrak{b}}(e) = \lfloor (\mathfrak{b}+1-k)/2 \rfloor$, then it either holds that $\phi^{\sigma_\mathfrak{b}}(e) = \mathfrak{m} - 1$ or $\phi^{\sigma_\mathfrak{b}}(e) = \mathfrak{m}$. In both cases, $\mathfrak{b} \geq 6$ implies that $\phi^\sigma(s_{\nu-1,0}, b_1) \leq \phi^{\sigma_\mathfrak{b}}(e) \leq \phi^\sigma(e)$. Thus assume $\phi^{\sigma_\mathfrak{b}}(e) = \ell^\mathfrak{b}(i,j,k) + t_\mathfrak{b}$ for some $t_\mathfrak{b}$ feasible for $\mathfrak{b}$ and $\phi^{\sigma_\mathfrak{b}}(e) \neq \lfloor (\mathfrak{b}+1-k)/2 \rfloor$. By Lemma 6.2.3,
$$\ell^\mathfrak{b}(i,j,k) = \left\lceil \frac{\mathfrak{b} - 2^{i-1} + \sum(\mathfrak{b}, i) + 1 - k}{2} \right\rceil \geq \left\lceil \frac{\mathfrak{b} - 2^{i-1} + 2^{\nu-1} - k}{2} \right\rceil$$
$$= \left\lfloor \frac{\mathfrak{b} - 2^{i-1} + 2^{\nu-1} + 1 - k}{2} \right\rfloor.$$

We prove that $\ell^{\mathfrak{b}}(i,j,k) > \lfloor(\mathfrak{b}+2)/4\rfloor$, implying $\phi^\sigma(e) \geq \phi^\sigma(s_{\nu-1,0}, b_1)$. We begin by observing

$$\ell^{\mathfrak{b}}(i,j,k) = \left\lfloor \frac{\mathfrak{b} - 2^{i-1} + 2^{\nu-1} + 1 - k}{2} \right\rfloor = \left\lfloor \frac{2\mathfrak{b} - 2^i + 2^\nu + 2 - 2k}{4} \right\rfloor \geq \left\lfloor \frac{2\mathfrak{b} - 2^i + 2^\nu}{4} \right\rfloor.$$

By the choice of i and j, the cycle center $F_{i,j}$ was closed at least once during some previous transition. But, since $\mathfrak{b}_i = 0$, the cycle center was also opened again later. This implies $\mathfrak{b} \geq 2^{i-1} + 2^{i-1} + 2^{\nu-1} - 1 = 2^i + 2^{\nu-1} - 1$. Thus,

$$\begin{aligned} 2\mathfrak{b} - 2^i + 2^\nu - [\mathfrak{b}+2] &= \mathfrak{b} - 2^i + 2^\nu - 2 \\ &\geq 2^i + 2^{\nu-1} - 1 - 2^i + 2^\nu - 2 \\ &= 2^\nu + 2^{\nu-1} - 3 \geq 4 + 1 - 3 = 2. \end{aligned}$$

Since $2\mathfrak{b} - 2^i + 2^\nu$ is even and larger than 0 and since $\mathfrak{b}$ being odd implies $\mathfrak{b}+2$ being odd, this difference is at least 3. It is easy to show that, in general, x being even and larger than 0, y being odd and $x - y \geq 3$ implies $\lfloor x/4\rfloor > \lfloor y/4\rfloor$. This yields $\ell^{\mathfrak{b}}(i,j,k) > \lfloor(\mathfrak{b}+2)/4\rfloor$.

It remains to prove that Lemma 6.2.31 describes the application of e. Since σ is a phase-4-strategy and since $i' > i = \nu - 1$ implies $i' \geq \nu$, σ has Property (USV1)$_{i'}$ for all $i' > i$. By Lemma 6.3.9, it follows that σ also meets the other requirements of Lemma 6.2.31. □

Claim 9. For all $e \in I_\sigma$, it holds that $\phi^\sigma(e) \leq \mathfrak{m}$. Let $e \in I_\sigma$ with $\phi^\sigma(e) < \mathfrak{m}$. Then, $e = (d_{i,j,k}, F_{i,j})$ with $i \in \{u+1, \dots, m-1\}, j = 1 - \beta^{i+1}, k \in \{0,1\}$ and $\sigma_{\mathfrak{b}}(d_{i,j,k}) = F_{i,j}$.

Proof. By Lemma 6.3.12, the set of improving switches can be partitioned as follows:

1. Let $e = (d_{i,j,k}, F_{i,j})$ resp. $e = (e_{i,j,k}, g_1)$ with $i \in [n], j,k \in \{0,1\}$ and $\sigma(e_{i,j,k}) = b_2$. Then $\phi^\sigma(e) = \mathfrak{m}$ resp. $\phi^\sigma(e) = \phi^{\sigma_{\mathfrak{b}}}(e) = \lceil \mathfrak{b}/2\rceil = \mathfrak{m}$ by Lemma 6.3.12.
2. Let $e = (d_{i,j,k}, F_{i,j})$ with $\beta_i = 0, i \in \{u+1, \dots, m-1\}, j = 1 - \beta_{i+1}$ and $k \in \{0,1\}$. Then, $\mathfrak{b}_i = 0$ and $j = 1 - \mathfrak{b}_{i+1}$ since $i \geq u+1 > 1$ and $\nu = 1$. In addition, $\mathfrak{b}_1 = 0$ and, due to $i > u$, there is at least one $l \in \{2, \dots, i-1\}$ with $(\mathfrak{b}+1)_l = \mathfrak{b}_l = 0$. Consequently, Lemma 6.2.3 yields

$$\ell^{\mathfrak{b}}(i,j,k) = \left\lceil \frac{\mathfrak{b} - 2^{i-1} + \sum(\mathfrak{b}, i) + 1 - k}{2} \right\rceil \leq \left\lceil \frac{\mathfrak{b} - 3 - k}{2} \right\rceil = \left\lfloor \frac{\mathfrak{b} - k}{2} \right\rfloor - 1.$$

Since there is a $t_{\mathfrak{b}}$ feasible for $\mathfrak{b}$, $\phi^{\sigma_{\mathfrak{b}}}(e) = \min(\lfloor(\mathfrak{b}+1-k)/2\rfloor, \ell^{\mathfrak{b}}(i,j,k) + t_{\mathfrak{b}})$. We thus distinguish the following cases.

a) Let $\phi^{\sigma_{\mathfrak{b}}}(e) = \ell^{\mathfrak{b}}(i,j,k) + 1$. Then, by Property (OR2)$_{i,j,k}$, $\sigma_{\mathfrak{b}}(d_{i,j,k}) = F_{i,j}$ and e was not applied during $\sigma_{\mathfrak{b}} \to \sigma$ as switches of this type were only applied during phase 1 so far. Consequently, $\phi^\sigma(e) = \phi^{\sigma_{\mathfrak{b}}}(e) < \mathfrak{m}$ by Property (OR1)$_{i,j,k}$.

b) Let $\phi^{\sigma_\mathfrak{b}}(e) = \ell^\mathfrak{b}(i,j,k)$. Then, $\phi^{\sigma_\mathfrak{b}}(e) \leq \phi^\sigma(e) \leq \lfloor(\mathfrak{b}-k)/2\rfloor - 1 < \mathfrak{m}$ as well as $\sigma_\mathfrak{b}(d_{i,j,k}) \neq F_{i,j}$ by Property $(\text{OR2})_{i,j,k}$. Using Property $(\text{OR4})_{i,j,k}$, this implies $\phi^{\sigma_\mathfrak{b}}(e) = \mathfrak{m} - 1$. Hence, by Corollary 6.3.3, e was applied during phase 1. Consequently, $\phi^\sigma(e) = \phi^{\sigma_\mathfrak{b}}(e) + 1 = \mathfrak{m}$.

c) The case $\phi^{\sigma_\mathfrak{b}}(e) = \ell^\mathfrak{b}(i,j,k) - 1$ cannot occur since the parameter $t_\mathfrak{b} = -1$ is not feasible as $\mathfrak{b}$ is even.

d) Let $\phi^{\sigma_\mathfrak{b}}(e) = \lfloor(\mathfrak{b}+1-k)/2\rfloor$ but $\lfloor(\mathfrak{b}+1-k)/2\rfloor \neq \ell^\mathfrak{b}(i,j,k), \ell^\mathfrak{b}(i,j,k)+1$. This implies that we need to have $\lfloor(\mathfrak{b}+1-k)/2\rfloor < \ell^\mathfrak{b}(i,j,k)$ since it holds that $\phi^{\sigma_\mathfrak{b}}(e) = \min(\lfloor(\mathfrak{b}+1-k)/2\rfloor, \ell^\mathfrak{b}(i,j,k) + t_\mathfrak{b})$. But this is a contradiction since $\ell^\mathfrak{b}(i,j,k) \leq \lfloor(\mathfrak{b}-k)/2\rfloor - 1$.

Since the only improving switches with an occurrence record lower than $\mathfrak{m}$ are the switches described in case 2.a), the second part of the statement follows. □

Claim 10. Let σ denote the phase-5-strategy at the beginning of phase 5 for $\nu = 1$. Let $i, \in [n], j, k \in \{0,1\}$ such that $e = (d_{i,j,k}, F_{i,j}) \in I_\sigma$ and $\phi^\sigma(e) < \mathfrak{m}$. Row 1 of Table 6.7 can be applied to describe the application of e.

Proof. We currently consider the first phase-5-strategy σ as described by Lemma 6.3.12 and an improving switch $e = (d_{i,j,k}, F_{i,j})$ with $i \in \{u+1, \dots, m-1\}, \beta_i = 0, j = 1 - \beta_{i+1}$ and $k \in \{0,1\}$ as well as $\sigma_\mathfrak{b}(d_{i,j,k}) = F_{i,j}$. We have to prove $\sigma(b_i) = b_{i+1}, j = 1 - \beta_{i+1}, \bar{\sigma}(g_i) = 1 - \beta_{i+1}$ and $i \neq 1$. The first two statements follow directly since σ is a phase-5-strategy and $\beta_i = 0$ as well as by the choice of j. Also, $i \neq 1$ follows from $i \geq u+1 > 1$. It thus suffices to show $\bar{\sigma}(g_i) = 1 - \beta_{i+1}$.

For the sake of a contradiction, let $\sigma(g_i) = F_{i,\beta_{i+1}}$. Since $\beta_i = 0$ and $\nu = 1$, it holds that $\mathfrak{b}_i = (\mathfrak{b}+1)_i = 0$, implying $i \neq \nu$. By Corollary 6.3.2, the only improving switch from a selector vertex towards the active cycle center of a level that can be performed during phase 1 is $(g_\nu, F_{\nu,\mathfrak{b}_{\nu+1}})$. This implies $(g_i, F_{i,\beta_{i+1}}) \notin \mathfrak{A}_{\sigma_\mathfrak{b}}^\sigma$, hence $\sigma_\mathfrak{b}(g_i) = F_{i,\beta_{i+1}}$. If $\sigma_\mathfrak{b}(d_{i,j,1-k}) = F_{i,j}$, then $F_{i,j} = F_{i,1-\beta_{i+1}}$ was closed at the beginning of phase 1 as $\sigma_\mathfrak{b}(d_{i,j,k}) = F_{i,j}$. But this implies $\sigma_\mathfrak{b}(g_i) = F_{i,1-\beta_{i+1}}$ by the definition of a canonical strategy which is a contradiction. Thus let $\sigma_\mathfrak{b}(d_{i,j,1-k}) \neq F_{i,j}$, implying that we have $\phi^{\sigma_\mathfrak{b}}(d_{i,j,1-k}, F_{i,j}) \neq \ell^\mathfrak{b}(i,j,1-k)+1$. Then, by the same arguments used when proving Claim 9, it follows that $\ell^\mathfrak{b}(i,j,1-k) \leq \lfloor(\mathfrak{b}-(1-k))/2\rfloor - 1$. Also, by these arguments, it cannot happen that $\phi^{\sigma_\mathfrak{b}}(d_{i,j,1-k}) = \lfloor(\mathfrak{b}+1-(1-k))/2\rfloor \neq \ell^\mathfrak{b}(i,j,1-k)$. Since the parameter $t_\mathfrak{b} = -1$ is not feasible, we thus have

$$\phi^{\sigma_\mathfrak{b}}(d_{i,j,1-k}, F_{i,j}) = \ell^\mathfrak{b}(i,j,1-k) \leq \left\lfloor \frac{\mathfrak{b}-(1-k)}{2} \right\rfloor - 1 < \left\lfloor \frac{\mathfrak{b}+1}{2} \right\rfloor.$$

But this implies that $(d_{i,j,1-k}, F_{i,j})$ was applied in phase 1 by Corollary 6.3.3. Hence, $F_{i,j}$ was closed in phase 1. But then, by Corollary 6.3.2, $(g_i, F_{i,j})$ became improving during phase 1 and was thus applied. This implies $\sigma(g_i) = F_{i,j} = F_{i,1-\beta_{i+1}}$, contradicting the assumption. Consequently, $\sigma(g_i) = F_{i,1-\beta_{i+1}}$. □

Claim 11. Let $\nu = 1$ and let σ denote the strategy obtained after applying all improving switches with an occurrence record less than $\mathfrak{m}$ during phase 5. Then, Lemma 6.2.32 can be applied to describe the application of $e = (e_{i,j,k}, g_1)$.

Proof. First, we show that $F_{i,j}$ is mixed. Since $e = (e_{i,j,k}, g_1) \in I_\sigma$ implies $(d_{i,j,k}, F_{i,j}) \in I_\sigma$ by Equation (6.3), we have $\bar{\sigma}(eb_{i,j})$. In particular, $F_{i,j}$ is not closed, so $\beta_i = 0 \vee \beta_{i+1} \neq j$. Consequently, $(i,j) \in S_3$ or $(i,j) \in S_4$. By Lemma 6.3.12, $\bar{\sigma}(eb_{i,j})$, and as no improving switch $(e_{*,*,*}, b_2)$ was applied during $\sigma^{(5)} \to \sigma$, we need to have $(i,j) \in S_3$, implying the statement. We now prove that $j = 1$ implies $\neg\bar{\sigma}(eb_{i,1-j})$ if $G_n = S_n$. Since $j = 1$, we need to prove $\neg\bar{\sigma}(eb_{i,0})$. If $F_{i,0}$ is closed, then the statement follows. If $F_{i,0}$ is not closed, then $\beta_i = 0 \vee \beta_{i+1} \neq j$ as $F_{i,1}$ cannot be closed by the choice of e. Consequently, $(i,0) \in S_3$ or $(i,0) \in S_4$. In the second case, $\neg\bar{\sigma}^{(5)}(eb_{i,0})$ by Lemma 6.3.12 and the statement follows as no improving switch $(e_{*,*,*}, b_2)$ was applied during $\sigma^{(5)} \to \sigma$. Consider the case $(i,0) \in S_3$. Then, by Lemma 6.3.12, $F_{i,0}$ and $F_{i,1}$ are mixed with respect to $\sigma^{(5)}$. Thus, as we consider the case $G_n = S_n$, the tie-breaking rule must have applied the improving switches $(e_{i,0,*}, g_1)$ prior to $(e_{i,1,k}, g_1)$, implying the statement. Note that the statement "$j = 1 - \beta_{i+1} \implies \neg\bar{\sigma}(eb_{i,1-j})$ if $G_n = M_n$" follows by the same arguments and since the tie-breaking rule applies improving switches $(e_{i,\beta_{i+1},*}, g_1)$ first. □

Corollary 6.3.15. *Let $\nu = 1$ and $i \in [n], j,k \in \{0,1\}$. If the edge $\tilde{e} = (d_{i,j,1-k}, F_{i,j})$ becomes improving during phase 5 due to the application of $(e_{i,j,k}, g_1)$, then the corresponding strategy has Property (OR4)$_{i,j,1-k}$.*

Proof. Let σ denote the strategy before the application of the switch $(e_{i,j,k}, g_1)$ and let σe denote the strategy obtained after the application of this switch. By the same arguments used in the proof of Claim 11, it follows that $F_{i,j}$ is mixed with respect to σ. By the characterization of I_σ given in Equation (6.3), it holds that $(d_{i,j,k}, F_{i,j}) \in I_\sigma$, implying $\sigma(d_{i,j,k}) \neq F_{i,j}$ and $\sigma e(d_{i,j,k}) \neq F_{i,j}$. Since σ is a phase-5-strategy for $\mathfrak{b}$, this furthermore implies $\beta_i = 0 \vee \beta_{i+1} \neq j$.

Assume that $\tilde{e}$ was applied previously in this transition. It is not possible that $\tilde{e}$ was applied during phase 5 as this would imply $\sigma(d_{i,j,1-k}) = F_{i,j}$, contradicting that $F_{i,j}$ is mixed with respect to σ. Consequently, $\tilde{e}$ was applied during phase 1. We thus need to have $\sigma_{\mathfrak{b}}(d_{i,j,1-k}) \neq F_{i,j}$ and $\phi^{\sigma_{\mathfrak{b}}}(\tilde{e}) \in \{\mathfrak{m}-1, \mathfrak{m}\}$ by Property (OR4)$_{i,j,1-k}$. This in particular implies $\phi^{\sigma e}(\tilde{e}) \in \{\mathfrak{m}, \mathfrak{m}+1\}$ and hence the statement.

Now assume that $\tilde{e}$ was not applied previously in this transition, implying $\phi^{\sigma e}(\tilde{e}) = \phi^{\sigma_{\mathfrak{b}}}(\tilde{e})$. Let $\sigma_{\mathfrak{b}}(d_{i,j,1-k}) \neq F_{i,j}$. Then, by Property (OR4)$_{i,j,1-k}$ and Corollary 6.3.3, it follows that $\phi^{\sigma_{\mathfrak{b}}}(\tilde{e}) = \mathfrak{m}$. Thus let $\sigma_{\mathfrak{b}}(d_{i,j,1-k}) = F_{i,j}$. Then, by Properties (OR1)$_{i,j,1-k}$ and (OR2)$_{i,j,1-k}$, it holds that $\phi^{\sigma_{\mathfrak{b}}}(\tilde{e}) = \ell^{\mathfrak{b}}(i,j,1-k) + 1 < \mathfrak{m}$. We now prove that this yields a contradiction.

1. Let, for the sake of a contradiction, $\beta_i = 1 \wedge \beta_{i+1} \neq j$. Assume $\mathfrak{b}_i = 0 \wedge \mathfrak{b}_{i+1} \neq j$. Since $\nu = 1$, this implies $i = \nu = 1$. Thus, as $\mathfrak{b}$ is even,

$$\ell^{\mathfrak{b}}(i,j,1-k) = \left\lceil \frac{\mathfrak{b} - 2^0 + \sum(\mathfrak{b}, i) + 1 - (1-k)}{2} \right\rceil = \left\lfloor \frac{\mathfrak{b}+1}{2} \right\rfloor = \mathfrak{m}$$

by Lemma 6.2.3. But then, $\phi^{\sigma_{\mathfrak{b}}}(\tilde{e}) = \ell^{\mathfrak{b}}(i,j,1-k)+1 = \mathfrak{m}+1$ which is a contradiction. Assuming $\mathfrak{b}_i = 1 \wedge \mathfrak{b}_{i+1} \neq j$ also results in a contradiction since

$$\ell^{\mathfrak{b}}(i,j,1-k) = \left\lceil \frac{\mathfrak{b} + \sum(\mathfrak{b}, i) + 1 - (1-k)}{2} \right\rceil \geq \left\lceil \frac{\mathfrak{b}+k}{2} \right\rceil \geq \left\lfloor \frac{\mathfrak{b}+1}{2} \right\rfloor.$$

2. Let, for the sake of a contradiction, $\beta_i = 0 \wedge \beta_{i+1} = j$. Then also $\mathfrak{b}_i = 0 \wedge \mathfrak{b}_{i+1} = j$ and $i \geq 2$ since $\nu = 1$. Then, Lemma 6.2.3 implies
$$\ell^{\mathfrak{b}}(i,j,1-k) \geq \left\lceil \frac{\mathfrak{b}+3-(1-k)}{2} \right\rceil \geq \left\lfloor \frac{\mathfrak{b}+1}{2} \right\rfloor + 1.$$
This yields a contradiction as before.
3. Let, for the sake of a contradiction, $\beta_i = 0 \wedge \beta_{i+1} \neq j$. This implies that $i \geq u$. If $i > m$, then Lemma 6.2.3 implies $\ell^{\mathfrak{b}}(i,j,1-k) \geq \mathfrak{b}$, contradicting that we have $\phi^{\sigma_{\mathfrak{b}}}(\tilde{e}) = \ell^{\mathfrak{b}}(i,j,1-k)+1 < \mathfrak{m}$. We hence may assume $i \in \{u,\ldots,m-1\}$. If $i \neq u$, then Lemma 6.3.12 implies $(d_{i,j,1-k}, F_{i,j}) \in I_{\sigma^{(5)}}$. But then, the switch was applied during $\sigma_{\mathfrak{b}} \to \sigma e$, contradicting the assumption. Hence let $i = u$. Then, $\beta_{i'} = 1$ for all $i' < u = i$. Consequently,
$$\begin{aligned}\ell^{\mathfrak{b}}(i,j,1-k) &= \left\lceil \frac{\mathfrak{b}-2^{i-1}+\sum(\mathfrak{b},i)+1-(1-k)}{2} \right\rceil \\ &= \left\lceil \frac{\mathfrak{b}-2^{i-1}+2^{i-1}-1-1+1-1+k}{2} \right\rceil = \left\lceil \frac{\mathfrak{b}-2+k}{2} \right\rceil \\ &= \left\lfloor \frac{\mathfrak{b}-1+k}{2} \right\rfloor.\end{aligned}$$
But this implies $\phi^{\sigma_{\mathfrak{b}}}(d_{i,j,1-k}) = \ell^{\mathfrak{b}}(i,j,1-k)+1 = \mathfrak{m}$ and thus contradicts Property $(\text{OR1})_{i,j,1-k}$. □

Claim 12. If $\nu = 1$, then the strategy σ obtained before the application of the switch $e := (e_{1,1-\beta_2,k}, g_1)$ has Property $(\text{SVG})_i/(\text{SVM})_i$ for all $i \in [n]$.

Proof. Consider some arbitrary but fixed index $i \in [n]$. If $\beta_i = 1$, then the statements follow from the definition of a phase-5-strategy. If $\beta_i = 0$ and $(g_i, F_{i,j}) \in \mathfrak{A}^{\sigma}_{\sigma^{(5)}}$, then this follows from Corollary 6.3.16. Thus, let $\beta_i = 0$ and $(g_i, F_{i,j}) \notin \mathfrak{A}^{\sigma}_{\sigma^{(5)}}$, implying $i \neq 1$ since $\nu = 1$. We now prove the following statement. If $\bar{\sigma}(g_i) = 1$ resp. $\bar{\sigma}(g_i) = 1-\beta_{i+1}$ and $\neg\bar{\sigma}(d_{i,1})$ resp. $\neg\bar{\sigma}(d_{i,1-\beta_{i+1}})$ then $(g_i, F_{i,0}) \in I_\sigma$ resp. $(g_i, F_{i,\beta_{i+1}}) \in I_\sigma$. This is sufficient to prove the statement as $I_\sigma \cap \mathbb{G} = \emptyset$.

Thus, let $j := 0$ (if $G_n = S_n$) resp. $j := \beta_{i+1}$ (if $G_n = M_n$) and assume $\neg\bar{\sigma}(d_{i,1-j})$. It suffices to prove $\mathrm{Val}^*_\sigma(F_{i,j}) \succ \mathrm{Val}^*_\sigma(F_{i,1-j})$. Since $\sigma(e_{i',j',k'}) = g_1$ for all $(i',j',k') \neq (1,\beta_2,k), i \neq 1$ and $\mu^\sigma = u \neq 1$, the two cycle centers $F_{i,*}$ are either closed or escape only to g_1.

Consider the case $G_n = S_n$. If both cycle centers escape towards g_1, then the statement follows from
$$\begin{aligned}\mathrm{Val}^{\mathrm{S}}_\sigma(F_{i,0}) &= \{F_{i,0}, d_{i,0,*}, e_{i,0,*}, b_1\} \cup \mathrm{Val}^{\mathrm{S}}_\sigma(g_1) \\ &\rhd \{F_{i,1}, d_{i,1,*}, e_{i,1,*}, b_1\} \cup \mathrm{Val}^{\mathrm{S}}_\sigma(g_1) = \mathrm{Val}^{\mathrm{S}}_\sigma(F_{i,1}).\end{aligned}$$
Since we currently consider the case $\beta_i = 0$, only $F_{i,1-\beta_{i+1}}$ can be closed, so assume this is the case. Assume that $0 = 1-\beta_{i+1}$, so $j = 1-\beta_{i+1}$. Then, by Property $(\text{USV1})_i$ and

$\sigma(b_1) = g_1$, the statement follows since $\mathrm{Val}^{\mathrm{S}}_{\sigma}(F_{i,0}) = \{s_{i,0}, b_1\} \cup \mathrm{Val}^{\mathrm{S}}_{\sigma}(g_1)$ and $\mathrm{Val}^{\mathrm{S}}_{\sigma}(F_{i,1}) = \{F_{i,1}, d_{i,1,k}, e_{i,1,k}, b_1\} \cup \mathrm{rVal}^{\mathrm{S}}_{\sigma}(g_1)$ for some $k \in \{0,1\}$. Thus assume that $0 = \beta_{i+1}$, so $j = \beta_{i+1}$. Then, the cycle center $F_{i,1-j} = F_{i,1}$ is closed, contradicting the assumption developed at the beginning of the proof.

Consider the case $G_n = M_n$ next and note that we thus have $j = \beta_{i+1}$ from now on. If both cycle centers of level i are g_1-open or g_1-halfopen, then the statement follows by Lemma 6.2.1 since $i > \nu = 1$. Thus consider the case that $F_{i,j}$ is g_1-open and that $F_{i,1-j}$ is g_1-halfopen. By assumption, $j = \beta_{i+1}$, implying $\sigma(s_{i,j}) = h_{i,j}$ and $\sigma(s_{i,1-j}) = b_1$ by Property (USV1)$_i$. Thus, by Property (EV1)$_{i+1}$,

$$\mathrm{Val}^{\mathrm{M}}_{\sigma}(F_{i,j}) = (1-\varepsilon)\,\mathrm{Val}^{\mathrm{M}}_{\sigma}(g_1) + \varepsilon\left[\langle s_{i,j}, h_{i,j}\rangle + \mathrm{Val}^{\mathrm{M}}_{\sigma}(b_{i+1})\right]$$

and

$$\mathrm{Val}^{\mathrm{M}}_{\sigma}(F_{i,1-j}) = \mathrm{Val}^{M}_{\sigma}(g_1) + \frac{2\varepsilon}{1+\varepsilon}\langle s_{i,1-j}\rangle\,.$$

To prove $\mathrm{Val}^{\mathrm{M}}_{\sigma}(F_{i,j}) > \mathrm{Val}^{\mathrm{M}}_{\sigma}(F_{i,1-j})$, it thus suffices to prove

$$\langle s_{i,j}, h_{i,j}\rangle + \mathrm{Val}^{\mathrm{M}}_{\sigma}(b_{i+1}) - \mathrm{Val}^{\mathrm{M}}_{\sigma}(g_1) - \frac{2}{1+\varepsilon}\langle s_{i,1-j}\rangle > 0.$$

This can be shown by an easy but tedious calculation using $\mathrm{Val}^{\mathrm{M}}_{\sigma}(g_1) = R^{\mathrm{M}}_1$, $\beta_i = 0$, $i+1 > \mu^{\sigma}$, and $\mathrm{Val}^{\mathrm{M}}_{\sigma}(b_{i+1}) = L^{\mathrm{M}}_{i+1}$. Now let $F_{i,j}$ be g_1-halfopen and $F_{i,1-j}$ be g_1-open. Then, by the same arguments used before,

$$\mathrm{Val}^{\mathrm{M}}_{\sigma}(F_{i,j}) = \frac{1-\varepsilon}{1+\varepsilon}\mathrm{Val}^{\mathrm{M}}_{\sigma}(g_1) + \frac{2\varepsilon}{1+\varepsilon}[\langle s_{i,j}, h_{i,j}\rangle + \mathrm{Val}^{\mathrm{M}}_{\sigma}(b_{i+1})]$$

and

$$\mathrm{Val}^{\mathrm{M}}_{\sigma}(F_{i,1-j}) = \mathrm{Val}^{\mathrm{M}}_{\sigma}(g_1) + \varepsilon\langle s_{i,1-j}\rangle\,.$$

It thus suffices to prove

$$\langle s_{i,j}, h_{i,j}\rangle + \mathrm{Val}^{\mathrm{M}}_{\sigma}(b_{i+1}) - \mathrm{Val}^{\mathrm{M}}_{\sigma}(g_1) - \frac{1+\varepsilon}{2}\langle s_{i,1-j}\rangle > 0$$

which follows analogously. Since only $F_{i,1-\beta_{i+1}} = F_{i,1-j}$ can be closed in level i, the statement then follows by the same argument used for the case $G_n = S_n$. □

Claim 13. It holds that $I_{\sigma_{\mathfrak{b}+1}} = \{(d_{i,j,k}, F_{i,j})\colon \sigma_{\mathfrak{b}+1}(d_{i,j,k}) \neq F_{i,j}\}$.

Proof. To simplify notation, let $\sigma := \sigma_{\mathfrak{b}+1}$. Consider the strategy $\sigma^{(5)}$. Using the characterization of the strategy that was obtained after having applied all switches $(d_{i,j,k}, F_{i,j})$ with an occurrence record smaller than $\mathfrak{m}$ (see Equation (6.3)), we obtain

$$\begin{aligned} I_{\sigma} &= \{(d_{i,j,*}, F_{i,j})\colon \sigma^{(5)}(e_{i,j,*}) = b_2\} \\ &\cup \bigcup_{\substack{i=u+1\\ \beta^{\sigma}_i=0}}^{m-1}\left\{e = (d_{i,1-\beta^{\sigma}_{i+1},*}, F_{i,1-\beta^{\sigma}_{i+1}})\colon \phi^{\sigma^{(5)}}(e) = \left\lfloor\frac{\mathfrak{b}+1}{2}\right\rfloor\right\}. \end{aligned}$$

In particular, $I_\sigma \subseteq \{(d_{i,j,k}, F_{i,j})\colon \sigma(d_{i,j,k}) \neq F_{i,j}\}$ and every improving switch has an occurrence record of at least $\mathfrak{m}$. To prove $\{(d_{i,j,k}, F_{i,j})\colon \sigma(d_{i,j,k}) \neq F_{i,j}\} \subseteq I_\sigma$, let $e := (d_{i,j,k}, F_{i,j})$ with $\sigma(d_{i,j,k}) \neq F_{i,j}$. It suffices to show $\mathrm{Val}^*_\sigma(F_{i,j}) \succ \mathrm{Val}^*_\sigma(e_{i,j,k})$. Property (ESC1) and $\nu = 1$ imply $\bar{\sigma}(eg_{i,j}) \wedge \neg\bar{\sigma}(eb_{i,j})$. Furthermore, Property (REL1) yields $\mu^\sigma = \min\{i'\colon \sigma(b_{i'}) = b_{i'+1}\} \neq 1$. This implies $\mathrm{Val}^{\mathrm{S}}_\sigma(F_{i,j}) = \{F_{i,j}\} \cup \mathrm{Val}^{\mathrm{S}}_\sigma(e_{i,j,k})$, implying the statement if $G_n = S_n$. If $G_n = M_n$, it suffices to prove $\mathrm{Val}^{\mathrm{M}}_\sigma(s_{i,j}) > \mathrm{Val}^{\mathrm{M}}_\sigma(g_1)$ as this implies $\mathrm{Val}^{\mathrm{M}}_\sigma(F_{i,j}) > \mathrm{Val}^{\mathrm{M}}_\sigma(g_1)$. Since $\sigma(d_{i,j,k}) \neq F_{i,j}$, either $\beta_i = 0$ or $\beta_{i+1} \neq j$. In the second case, Property (USV1)$_i$ implies $\sigma(s_{i,j}) = b_1$ and the statement follows since $\mathrm{Val}^{\mathrm{M}}_\sigma(s_{i,j}) = \langle s_{i,j}\rangle + \mathrm{Val}^{\mathrm{M}}_\sigma(g_1)$ due to $\sigma(b_1) = g_1$. Thus let $\beta_i = 0 \wedge \beta_{i+1} = j$. Then, the statement follows since $\mathrm{Val}^{\mathrm{M}}_\sigma(s_{i,j}) = \langle s_{i,j}, h_{i,j}\rangle + \mathrm{Val}^{\mathrm{M}}_\sigma(b_{i+1})$ by Property (EV1)$_{i+1}$ and $\langle s_{i,j}, h_{i,j}\rangle > \sum_{\ell<i} \langle g_\ell, s_{\ell,\bar{\sigma}(g_\ell)}, h_{\ell,\bar{\sigma}(g_\ell)}\rangle$. □

Claim 14. Let $\nu > 1$. The occurrence records of the improving switches with respect to the phase-5-strategy σ described by Lemma 6.3.12 is described correctly by Table 6.8.

Proof. We consider each cell of the table individually. We also observe that $\sigma(e_{i,j,k}) = g_1$ implies $(e_{i,j,k}, b_2) \in I_\sigma$, it holds that $\mathrm{Val}^*_\sigma(g_1) \prec \mathrm{Val}^*_\sigma(b_2)$.

1. Let $e = (d_{i,j,k}, F_{i,j})$ with $\sigma(e_{i,j,k}) = g_1$. Then, $\phi^\sigma(e) = \phi^{\sigma_\mathfrak{b}}(e) = \mathfrak{m}$ by Lemma 6.3.12.
2. Let $e = (e_{i,j,k}, b_2)$ with $\sigma(e_{i,j,k}) = g_1$. Then, $\phi^\sigma(e) = \phi^{\sigma_\mathfrak{b}}(e) = \mathfrak{m} - 1$ by Table 5.6.
3. Let $e = (d_{\nu,j,k}, F_{\nu,j})$ with $j := 1 - \beta_{\nu+1}$ for some $k \in \{0,1\}$. This edge is only an improving switch if $\mathfrak{b}+1$ is not a power of two. Note that this in particular implies $\mathbf{1}_{j=0}\mathrm{lfn}(\mathfrak{b}, \nu+1) + \mathbf{1}_{j=1}\mathrm{lufn}(\mathfrak{b}, \nu+1) \neq 0$. Since $\mathfrak{b}_\nu = 0 \wedge \mathfrak{b}_{\nu+1} \neq j$, Lemma 6.2.3 thus implies
$$\begin{aligned}\ell^\mathfrak{b}(\nu, j, k) &= \left\lceil \frac{\mathfrak{b} - 2^{\nu-1} + \sum(\mathfrak{b}, \nu) + 1 - k}{2} \right\rceil = \left\lceil \frac{\mathfrak{b} - 2^{\nu-1} + 2^{\nu-1} - 1 + 1 - k}{2} \right\rceil \\ &= \left\lceil \frac{\mathfrak{b} - k}{2} \right\rceil = \left\lfloor \frac{\mathfrak{b} + 1 - k}{2} \right\rfloor = \left\lfloor \frac{\mathfrak{b} + 1}{2} \right\rfloor - k.\end{aligned}$$
Since $\mathfrak{b}+1$ is not a power of two, the parameter $t_\mathfrak{b} = -1$ is not feasible by Property (OR3)$_{i,j,k}$. Hence $\phi^{\sigma_\mathfrak{b}}(d_{\nu,j,k}, F_{\nu,j}) = \mathfrak{m} - k$. Then, Corollary 6.3.3 implies that $(d_{\nu,j,1}, F_{\nu,j})$ was applied during $\sigma_\mathfrak{b} \to \sigma$. Consequently, for both $k \in \{0,1\}$, it holds that $\phi^\sigma(d_{\nu,j,k}, F_{\nu,j}) = \mathfrak{m}$.
4. Let $e = (d_{i,j,k}, F_{i,j})$ with $i \in \{\nu+1, \dots, m-1\}, \beta_i = 0, j := 1 - \beta_{i+1}$ and $k \in \{0,1\}$. This edge is only an improving switch if $\mathfrak{b}+1$ is not a power of two. Since $i > \nu$, $\beta_i = 0$ implies $\mathfrak{b}_i = 0 \wedge \mathfrak{b}_{i+1} \neq j$. Also, $i < m$ implies $\mathbf{1}_{j=0}\mathrm{lfn}(\mathfrak{b}, i+1) + \mathbf{1}_{j=1}\mathrm{lufn}(\mathfrak{b}, i+1) \neq 0$ since $j = 1 - \beta_{i+1}$ and $\mathfrak{b} \geq 1$ by the choice of i. Since $\mathfrak{b}_\nu = 0$, this yields
$$\begin{aligned}\ell^\mathfrak{b}(i, j, k) &= \left\lceil \frac{\mathfrak{b} - 2^{i-1} + \sum(\mathfrak{b}, i) + 1 - k}{2} \right\rceil \leq \left\lceil \frac{\mathfrak{b} - 2^{i-1} + 2^{i-1} - 1 - 2^{\nu-1} + 1 - k}{2} \right\rceil \\ &= \left\lceil \frac{\mathfrak{b} - 2^{\nu-1} - k}{2} \right\rceil \leq \left\lceil \frac{\mathfrak{b} - 2 - k}{2} \right\rceil = \left\lfloor \frac{\mathfrak{b} - 1 - k}{2} \right\rfloor \leq \left\lfloor \frac{\mathfrak{b} - 1}{2} \right\rfloor \leq \mathfrak{m} - 1.\end{aligned}$$
There are two cases. If $\sigma_\mathfrak{b}(d_{i,j,k}) = F_{i,j}$, then $\phi^{\sigma_\mathfrak{b}}(d_{i,j,k}, F_{i,j}) = \ell^\mathfrak{b}(i,j,k) + 1 \leq \mathfrak{m} - 1$ by Property (OR1)$_{i,j,k}$. If $\sigma_\mathfrak{b}(d_{i,j,k}) \neq F_{i,j}$, then $\phi^{\sigma_\mathfrak{b}}(d_{i,j,k}, F_{i,j}) = \ell^\mathfrak{b}(i,j,k) \leq \mathfrak{m} - 1$.

In the first case, e was not applied during phase 1 and $\phi^{\sigma_{\mathfrak{b}}}(e) = \phi^{\sigma}(e) \leq \mathfrak{m} - 1$. In the second case, $\phi^{\sigma_{\mathfrak{b}}}(e) = \mathfrak{m} - 1$ by Property (OR4)$_{i,j,k}$. Then, e was applied during phase 1, implying $\phi^{\sigma}(d_{i,j,k}, F_{i,j}) = \mathfrak{m}$.

5. Let $e = (d_{i,j,k}, F_{i,j})$ with $i \leq \nu - 1$ and $j := 1 - \beta_{i+1}$. Then, bit i and bit $i+1$ switched during $\sigma_{\mathfrak{b}} \to \sigma^{(5)}$. In particular, $F_{i,j}$ was closed with respect to $\sigma_{\mathfrak{b}}$ and consequently $(d_{i,j,k}, F_{i,j}) \notin \mathfrak{A}^{\sigma}_{\sigma_{\mathfrak{b}}}$. Hence, by Table 5.6,

$$\begin{aligned}\phi^{\sigma}(e) = \phi^{\sigma_{\mathfrak{b}}}(e) &= \left\lceil \frac{\operatorname{lfn}(\mathfrak{b}, i, \{(i+1, j)\}) + 1 - k}{2} \right\rceil = \left\lceil \frac{\mathfrak{b} - \sum(\mathfrak{b}, i) + 1 - k}{2} \right\rceil \\ &= \left\lceil \frac{\mathfrak{b} - 2^{i-1} + 1 + 1 - k}{2} \right\rceil = \left\lfloor \frac{\mathfrak{b} - 2^{i-1} + 3 - k}{2} \right\rfloor.\end{aligned}$$

We now distinguish several cases.

- For $i = 1$, $\phi^{\sigma}(e) = \lfloor (\mathfrak{b} + 2 - k)/2 \rfloor = \mathfrak{m}$ independent of k.
- For $i = 2$, $\phi^{\sigma}(e) = \lfloor (\mathfrak{b} + 1 - k)/2 \rfloor$, so $\phi^{\sigma}(e) = \mathfrak{m} - k$.
- For $i = 3$, $\phi^{\sigma}(e) = \lfloor (\mathfrak{b} - 1 - k)/2 \rfloor$, so $\phi^{\sigma}(e) = \mathfrak{m} - 1$ if $k = 0$ and $\phi^{\sigma}(e) = \mathfrak{m} - 2$ if $k = 1$.
- For $i > 3$, it is easy to see that the occurrence record is always strictly smaller than $\mathfrak{m} - 1$.

□

Claim 15. Let $\nu > 1$ and consider the first phase-5-strategy. The application of type 3 switches is described by row 1 of Table 6.7.

Proof. As a reminder, $e = (d_{i,j,k}, F_{i,j})$ being a type 3 switch implies that we either have $i < \nu - 1, j = 1 - \beta_{i+1}$ or $i \in \{\nu + 1, \ldots, m - 1\}, \beta_i = 0, j := 1 - \beta_{i+1}$. In the second case, $\sigma_{\mathfrak{b}}(d_{i,j,k}) = F_{i,j}$ holds as well. Since it is easy to verify that $i \neq 1$ and $\sigma(b_i) = b_{i+1}$ (for example by the arguments used in the proof of Claim 14), we only show $\bar{\sigma}(g_i) = 1 - \beta_{i+1}$. By Lemma 6.3.12, this holds for all $i \leq \nu - 1$. It thus suffices to prove this for $i \in \{\nu + 1, \ldots, m - 1\} \wedge \beta_i = 0$. We show the statement by proving that $\bar{\sigma}(g_i) = \beta_{i+1}$ implies $(g_i, F_{i,1-\beta_{i+1}}) \in I_{\sigma}$, contradicting the characterization of I_{σ} given in Equation (6.4).

Since $j = 1 - \beta_{i+1}$, it suffices to prove $\operatorname{Val}^*_{\sigma}(F_{i,j}) \succ \operatorname{Val}^*_{\sigma}(F_{i,1-j})$. We have $(i,j) \in S_1$ and $(i, 1-j) \in S_2$. Thus, by Lemma 6.3.12, $\bar{\sigma}(eb_{i,j}) \wedge \neg\bar{\sigma}(eg_{i,j})$ as well as $\bar{\sigma}(eb_{i,1-j}) \wedge \bar{\sigma}(eg_{i,1-j})$. Also, by the choice of j and Property (USV1)$_i$, $\sigma(s_{i,j}) = b_1$. Thus, by Lemmas 6.1.15 and 6.1.16, $\operatorname{rVal}^*_{\sigma}(F_{i,j}) = \operatorname{rVal}^*_{\sigma}(b_2)$ regardless of whether $G_n = S_n$ or $G_n = M_n$. Also, since $\nu \geq 2$, $\bar{\sigma}(g_1) = 1 - \beta_2 \neq \bar{\sigma}(b_2)$ by Lemma 6.3.12. Thus, if $G_n = S_n$, then Lemma 6.1.16 implies $\operatorname{rVal}^{\mathrm{S}}_{\sigma}(F_{i,1-j}) = \operatorname{rVal}^{\mathrm{S}}_{\sigma}(g_1) \lhd \operatorname{rVal}^{\mathrm{S}}_{\sigma}(b_2) = \operatorname{rVal}^{\mathrm{S}}_{\sigma}(F_{i,j})$ as player 1 minimizes the valuation. If $G_n = M_n$, then $\operatorname{rVal}^{\mathrm{M}}_{\sigma}(F_{i,1-j}) = \frac{1}{2}\operatorname{rVal}^{\mathrm{M}}_{\sigma}(g_1) + \frac{1}{2}\operatorname{rVal}^{\mathrm{M}}_{\sigma}(b_2)$, hence the statement follows since $\operatorname{rVal}^{\mathrm{M}}_{\sigma}(g_1) < \operatorname{rVal}^{\mathrm{M}}_{\sigma}(b_2)$. □

Claim 16. Let $\nu > 1$ and let σ denote the strategy obtained after the application of all improving switches of type 3 during phase 5. The application of type 2 switches of the form $(e_{*,*,*}, b_2)$ is described by row 1 of Lemma 6.2.32.

Proof. Let $i \in [n], j, k \in \{0,1\}$ and let $e := (e_{i,j,k}, b_2)$ be an improving switch. We begin by proving that the cycle center $F_{i,j}$ is mixed. Since only improving switches of type 3 were applied so far during phase 5, $\sigma(e_{i,j,k}) = g_1$ implies $\sigma(d_{i,j,k}) = e_{i,j,k}$. Consequently, we have $\bar{\sigma}(eg_{i,j})$. In particular, $F_{i,j}$ is not closed, so $\beta_i = 0 \vee \beta_{i+1} = j$. Thus, either $(i,j) \in S_1$ or $(i,j) \in S_2$. By Lemma 6.3.12, $\bar{\sigma}(eg_{i,j})$ and as no switch $(e_{*,*,*}, g_1)$ was applied during $\sigma^{(5)} \to \sigma$, we need to have $(i,j) \in S_2$, implying that $F_{i,j}$ is mixed.

We next prove that $j = 1$ resp. $j = 1 - \beta_{i+1}$ (depending on whether $G_n = S_n$ or $G_n = M_n$) implies $\neg\bar{\sigma}(eg_{i,1-j})$. Consider the case $G_n = S_n$ and thus $j = 1$ first. We prove $\neg\bar{\sigma}(eg_{i,0})$. If $F_{i,0}$ is closed, then the statement follows. If it is not closed, then $\beta_i = 0 \vee \beta_{i+1} \neq 0$. Consequently, either $(i,0) \in S_1$ or $(i,0) \in S_2$. In the first case, $\neg\bar{\sigma}(eg_{i,0})$ follows from Lemma 6.3.12 as no improving switch $(e_{*,*,*}, b_2)$ was applied during $\sigma^{(5)} \to \sigma$, so assume $(i,0) \in S_2$. Then, by the same lemma, both cycle centers $F_{i,0}, F_{i,1}$ were mixed for $\sigma^{(5)}$. Thus, as we consider the case $G_n = S_n$, the tie-breaking rule must have applied the improving switches $(e_{i,0,*}, b_2)$ prior to $(e_{i,j,k}, b_2)$, implying $\neg\bar{\sigma}(eg_{i,0})$. If $G_n = M_n$, then $\neg\bar{\sigma}(eg_{i,1-j})$ follows by the same arguments as the tie-breaking rule applied the improving switches $(e_{i,\beta_{i+1},*}, b_2)$ first. Finally, as no improving switch $(g_*, F_{*,*})$ was applied during $\sigma^{(5)} \to \sigma$, $\nu = 2$ implies $\sigma(g_1) = F_{1,0}$ if $G_n = S_n$ by Lemma 6.3.12. Thus, all requirements of Lemma 6.2.32 are met. □

Corollary 6.3.19. *Let $i \in [n], j, k \in \{0,1\}$ and let σ denote the strategy obtained after the application of an improving switch $(e_{i,j,k}, b_2)$ during phase 5. If $(d_{i,j,1-k}, F_{i,j}) \in I_\sigma$, then$\sigma$ has Property (OR4)$_{i,j,1-k}$ and it holds that* $\min_{k' \in \{0,1\}} \phi^{\sigma_\mathfrak{b}}(d_{i,j,k'}, F_{i,j}) \leq \mathfrak{m} - 1$.

Proof. To simplify the notation, let σ denote the strategy obtained after the application of $(e_{i,j,k}, b_2)$ and let $e := (d_{i,j,1-k}, F_{i,j}) \in I_\sigma$. We prove that σ has Property (OR4)$_{i,j,1-k}$ by proving

$$\phi^{\sigma}(d_{i,j,1-k}, F_{i,j}) \in \left\{ \left\lfloor \frac{\mathfrak{b}+1+1}{2} \right\rfloor - 1, \left\lfloor \frac{\mathfrak{b}+1+1}{2} \right\rfloor \right\}. \tag{A.14}$$

The second statement is shown along the way.

Consider the case $e \in \mathfrak{A}^{\sigma}_{\sigma_\mathfrak{b}}$. Since $e \in \mathfrak{A}^{\sigma}_{\sigma^{(5)}}$ would imply $\sigma(d_{i,j,1-k}) = F_{i,j}$, we need to have $e \in \mathfrak{A}^{\sigma^{(5)}}_{\sigma_\mathfrak{b}}$. This implies that the switch was applied during phase 1 as well as $\sigma_\mathfrak{b}(d_{i,j,1-k}) \neq F_{i,j}$ and $\phi^{\sigma_\mathfrak{b}}(e) \in \{\mathfrak{m}-1, \mathfrak{m}\}$. The only improving switches of type $(d_{*,*,*}, F_{*,*})$ with an occurrence record of $\mathfrak{m}$ applied in phase 1 are the cycle edges of $F_{\nu,\beta_{\nu+1}}$. Consequently, $\phi^{\sigma_\mathfrak{b}}(e) = \mathfrak{m} - 1$ as $F_{\nu,\beta_{\nu+1}}$ is closed and its cycle edges cannot become improving switches. Hence

$$\phi^{\sigma}(e) = \phi^{\sigma_\mathfrak{b}}(e) + 1 \in \left\{ \left\lfloor \frac{\mathfrak{b}+1+1}{2} \right\rfloor - 1, \left\lfloor \frac{\mathfrak{b}+1+1}{2} \right\rfloor \right\},$$

proving both parts of the statement.

Consider the case $e \notin \mathfrak{A}^{\sigma}_{\sigma_\mathfrak{b}}$ next. Since the switch was not applied, this then implies $\phi^{\sigma_\mathfrak{b}}(e) = \phi^{\sigma}(e)$. We distinguish two cases.

1. Consider the case $\sigma_\mathfrak{b}(d_{i,j,1-k}) = F_{i,j}$ first. Then, Property (OR1)$_{i,j,1-k}$ implies $\phi^{\sigma_\mathfrak{b}}(e) = \ell^\mathfrak{b}(i,j,1-k) + 1 \leq \mathfrak{m} - 1$. Assume $\phi^{\sigma_\mathfrak{b}}(d_{i,j,1-k}, F_{i,j}) < \mathfrak{m} - 1$. This implies $\ell^\mathfrak{b}(i,j,1-k) \leq \mathfrak{m} - 3$ by integrality. Since $\ell^\mathfrak{b}(i,j,0)$ and $\ell^\mathfrak{b}(i,j,1)$ differ by at most 1,

it follows that $\ell^{\mathfrak{b}}(i,j,k) \leq \mathfrak{m}-2$. This implies that $\phi^{\sigma_{\mathfrak{b}}}(d_{i,j,k}, F_{i,j}) \leq \mathfrak{m}-1$. But this is a contradiction since $(e_{i,j,k}, b_2) \in I_{\sigma^{(5)}}$ implies $\phi^{\sigma_{\mathfrak{b}}}(d_{i,j,k}, F_{i,j}) = \mathfrak{m}$ by Lemma 6.3.12. Hence the statement follows from $\phi^{\sigma_{\mathfrak{b}}}(e) = \phi^{\sigma}(e) = \mathfrak{m} - 1$.

2. Let $\sigma_{\mathfrak{b}}(d_{i,j,1-k}) \neq F_{i,j}$. Then, $\phi^{\sigma}(e) = \phi^{\sigma_{\mathfrak{b}}}(e) = \mathfrak{m}$ by Property (OR4)$_{i,j,1-k}$ and Corollary 6.3.3. This already implies the first of the two statements. In addition, either $\sigma_{\mathfrak{b}}(d_{i,j,k}) = F_{i,j}$ and $\phi^{\sigma_{\mathfrak{b}}}(d_{i,j,k}, F_{i,j}) \leq \mathfrak{m} - 1$ or $\sigma_{\mathfrak{b}}(d_{i,j,k}) \neq F_{i,j}$ and $\phi^{\sigma_{\mathfrak{b}}}(d_{i,j,k}, F_{i,j}) = \mathfrak{m} - 1$. If none of these were true, then $F_{i,j}$ would be open at the end of phase 1, contradicting Corollary 6.3.4. This proves the second part of the statement. □

Claim 17. Let $i \in [n], j,k \in \{0,1\}$ and let σ denote the strategy obtained after the application of an improving switch $(e_{i,j,k}, b_2)$ during phase 5. If $(g_i, F_{i,j}) \in I_\sigma$, then $(g_i, F_{i,j}) \notin \mathfrak{A}^{\sigma}_{\sigma_{\mathfrak{b}}}$.

Proof. Let $e = (g_i, F_{i,j})$. By Lemma 6.2.32, $e \in I_\sigma$ if and only if $\beta_i = 0, \bar{\sigma}(eb_{i,1-j})$ and $[j = 0 \wedge \bar{\sigma}(g_i) = 1]$ if $G_n = S_n$ resp. $[j = \beta_{i+1} \wedge \bar{\sigma}(g_i) = 1 - \beta_{i+1}]$ if $G_n = M_n$. Let, for the sake of contradiction, $(g_i, F_{i,j}) \in \mathfrak{A}^{\sigma}_{\sigma_{\mathfrak{b}}}$. The conditions on j and $\bar{\sigma}(g_i)$ imply $(g_i, F_{i,j}) \notin \mathfrak{A}^{\sigma}_{\sigma^{(5)}}$. Since $\beta_i = 0$ implies $i \neq \nu$, also $(g_i, F_{i,j}) \neq (g_\nu, F_{\nu,*})$. Thus, by Lemma 6.3.12, $\mathfrak{b}_i = 0 \wedge \mathfrak{b}_{i+1} \neq j$. Consequently, $0 = \mathfrak{b}_i = \beta_{i+1} = (\mathfrak{b}+1)_{i+1}$ and $j = 1 - \mathfrak{b}_{i+1}$. Since all bits below level ν have $\mathfrak{b}_i = 1 \wedge (\mathfrak{b}+1)_i = 0$, this implies $i > \nu$. Therefore, $\mathfrak{b}_{i+1} = (\mathfrak{b}+1)_{i+1} = 1 - j$ and in particular $j = 1 - \beta_{i+1}$ This is a contradiction if $G_n = M_n$ as $j = \beta_{i+1}$. Hence consider the case $G_n = S_n$. Then, $j = 1 - \beta_{i+1} = 0$, implying $\beta_{i+1} = 1$. Thus, $i \in \{\nu+1, \ldots, m-1\}, \beta_i = 0$ and $j = 1 - \beta_{i+1}$, implying $(i,j) \in S_1$. Therefore, $\bar{\sigma}^5(eb_{i,j}) \wedge \neg\bar{\sigma}^5(eg_{i,j})$, contradicting $(e_{i,j,k}, b_2), (d_{i,j,k}, e_{i,j,k}) \in I_{\sigma^{(5)}}$. Thus, $(g_i, F_{i,j}) \notin \mathfrak{A}^{\sigma^{(5)}}_{\sigma_{\mathfrak{b}}}$, implying the statement. □

Claim 18. Let $\nu > 1$. The strategy σ obtained after the application of the final improving switch of phase 5 has Property (SV*)$_i$ for all $i \in [n]$.

Proof. As a reminder, σ is the strategy obtained after the application of the switch $(e_{1,\beta_2,k}, b_2)$ resp. after the application of the switch (g_1, F_{1,β_2}) if it becomes improving. First consider some $i \geq 2$. If $\beta_i = 1$, then σ has Property (SV*)$_i$ as it is a phase-5-strategy. If $\beta_i = 0$ and $(g_i, F_{i,j}) \in \mathfrak{A}^{\sigma}_{\sigma^{(5)}}$, then this follows from Corollary 6.3.16. Thus, let $\beta_i = 0$ and $(g_i, F_{i,j}) \notin \mathfrak{A}^{\sigma}_{\sigma^{(5)}}$. For the sake of a contradiction, assume that σ does not have Property (SV*)$_1$. Then, $\bar{\sigma}(g_i) = 1$ resp. $\bar{\sigma}(g_i) = 1 - \beta$ (depending on whether $G_n = S_n$ or $G_n = M_n$) and $\neg\bar{\sigma}(d_{i,1})$ resp. $\neg\bar{\sigma}(d_{i,1-\beta_{i+1}})$. To simplify notation, let $j := 0$ resp. $j := \beta_{i+1}$. We show that we then have $\operatorname{Val}^*_\sigma(F_{i,j}) \succ \operatorname{Val}^*_\sigma(F_{i,1-j})$, implying $(g_i, F_{i,j}) \in I_\sigma$. But this is a contradiction as any improving switch of this kind is applied immediately and $i \geq 2$ implies that the application of $(e_{1,\beta_2^\sigma,k}, b_2)$ cannot have unlocked this switch.

As the last improving switch of the type $(e_{*,*,*}, b_2)$ was just applied, any cycle center is either closed or escapes to b_2. We first consider the case $G_n = S_n$. Since σ is a phase-5-strategy for $\mathfrak{b}$, it has Property (USV1)$_i$ and Property (EV1)$_{i+1}$. Consequently, either $\bar{\sigma}(b_{i+1}) = j$ or $\neg\bar{\sigma}(s_{i,j})$ If both cycle centers of level i escape towards b_2, then the statement follows since

$$\operatorname{Val}^P_\sigma(F_{i,j}) = \{F_{i,0}, e_{i,0,*}, d_{i,0,*}\} \cup \operatorname{Val}^P_\sigma(b_2) \rhd \{F_{i,1}, e_{i,1,*}, d_{i,1,*}\} \cup \operatorname{Val}^P_\sigma(b_2) = \operatorname{Val}^P_\sigma(F_{i,1-j})$$

by Lemma 6.1.16. Since $\beta_i = 0$, only $F_{i,1-\beta_{i+1}}$ can be closed in level i. Let this cycle center be closed. If $j = 1 - \beta_{i+1} = 0$, then Property (USV1)$_i$ and $\sigma(b_1) = b_2$ implies $\mathrm{rVal}_\sigma^P(F_{i,0}) = \{s_{i,0}\} \cup \mathrm{rVal}_\sigma^P(b_2)$ and the statement follows from $\mathrm{rVal}_\sigma^{\mathrm{S}}(F_{i,1}) = \mathrm{rVal}_\sigma^{\mathrm{S}}(b_2)$. If $j = \beta_{i+1} = 0$, then $F_{i,1-j} = F_{i,1}$ is closed, contradicting the assumption $\neg\bar{\sigma}(d_{i,1})$.

Consider the case $G_n = M_n$. If both cycle centers are b_2-open or b_2-halfopen, then the statement follows by Lemma 6.2.1 since σ has Property (REL1). If $F_{i,j}$ is b_2-open and $F_{i,1-j}$ is b_2-halfopen, then the statement follows by an easy but tedious calculation. Thus consider the case that $F_{i,j}$ is b_2-halfopen and that $F_{i,1-j}$ is b_2-open. Then, by the choice of j and Property (USV1)$_i$,

$$\begin{aligned}
\mathrm{Val}_\sigma^{\mathrm{M}}(F_{i,j}) &= \frac{1-\varepsilon}{1+\varepsilon}\mathrm{Val}_\sigma^{\mathrm{M}}(b_2) + \frac{2\varepsilon}{1+\varepsilon}\mathrm{Val}_\sigma^{\mathrm{M}}(s_{i,j}) \\
&= \frac{1-\varepsilon}{1+\varepsilon}\mathrm{Val}_\sigma^{\mathrm{M}}(b_2) + \frac{2\varepsilon}{1+\varepsilon}[\langle s_{i,j}, h_{i,j}\rangle + \mathrm{Val}_\sigma^{\mathrm{M}}(b_{i+1})] \\
\mathrm{Val}_\sigma^{\mathrm{M}}(F_{i,1-j}) &= (1-\varepsilon)\mathrm{Val}_\sigma^{\mathrm{M}}(b_2) + \varepsilon\,\mathrm{Val}_\sigma^{\mathrm{M}}(s_{i,1-j}) = \mathrm{Val}_\sigma^{\mathrm{M}}(b_2) + \varepsilon\,\langle s_{i,1-j}\rangle, \\
\mathrm{Val}_\sigma^{\mathrm{M}}(F_{i,j}) - \mathrm{Val}_\sigma^{\mathrm{M}}(F_{i,1-j}) &= \frac{2\varepsilon}{1-\varepsilon}(\langle s_{i,j}, h_{i,j}\rangle + \mathrm{Val}_\sigma^{\mathrm{M}}(b_{i+1})) - \frac{2\varepsilon}{1+\varepsilon}\mathrm{Val}_\sigma^{\mathrm{M}}(b_2) - \varepsilon\,\langle s_{i,j}\rangle \\
&= \varepsilon\left[\frac{2}{1+\varepsilon}(\langle s_{i,j}, h_{i,j}\rangle + \mathrm{Val}_\sigma^{\mathrm{M}}(b_{i+1}) - \mathrm{Val}_\sigma^{\mathrm{M}}(b_2)) - \langle s_{i,1-j}\rangle\right]
\end{aligned}$$

It thus suffices to show that the last term is larger than zero which follows easily from $\beta_i = 0$.

In level i, only $F_{i,1-\beta_{i+1}} = F_{i,1-j}$ can be closed. Then, the statement follows by the same argument used for the case $G_n = S_n$.

We now consider Property (SV*)$_1$. Assume that (g_1, F_{1,β_2}) does not become improving when applying $(e_{1,\beta_2,k}, b_2)$. Then, by Lemma 6.2.32, we need to have $\bar{\sigma}(g_i) = \beta_{i+1}$ if $G_n = M_n$. Consider the case $G_n = S_n$. If $\beta_2 = 0$, then Lemma 6.2.32 implies that we need to have $\bar{\sigma}(g_1) = 0$. If $\beta_2 = 1$, then $\nu = 2$. But this implies $\sigma_{\mathfrak{b}}(g_1) = F_{1,0}$ since the cycle center $F_{1,0}$ was then closed with respect to $\sigma_{\mathfrak{b}}$. For this reason, the switch $(g_1, F_{1,1})$ was not applied during phase 1. Since a switch involving a selection vertex gi can only be applied during phase 5 if $\bar{\sigma}(g_i) = 1$ by Lemma 6.2.32, the switch cannot have been applied during phase 5. Consequently, $\sigma(g_1) = \sigma_{\mathfrak{b}}(g_1) = F_{1,0}$ Thus, σ has Property (SV*)$_1$. If the edge (g_1, F_{1,β_2}) becomes an improving switch, then the strategy obtained after applying it has Property (SV*)$_i$ by Corollary 6.3.16. □

Claim 19. Let σ denote the strategy obtained after applying the final improving switch of phase 5 for $\nu > 1$. Then $I_\sigma = \{(d_{i,j,k}, F_{i,j}) \colon \sigma(d_{i,j,k}) \neq F_{i,j}\}$.

Proof. Let $\sigma^{(5)}$ denote the phase-5-strategy of Lemma 6.3.12 with $\sigma \in \rho(\sigma^{(5)})$. We first observe that $\beta^\sigma = \beta^{\sigma^{(5)}}$, so the upper index can be omitted. It is easy to verify that I_σ can be partitioned as

$$\begin{aligned}
I_\sigma = &\{(d_{i,j,*}, F_{i,j}) \colon \sigma^{(5)}(e_{i,j,*}) = g_1\} \cup \{(d_{\nu,1-\beta_{\nu+1},*}, F_{\nu,1-\beta_{\nu+1}})\} \\
&\cup \{e = (d_{i,1-\beta_{i+1},*}, F_{i,1-\beta_{i+1}}) \colon i \in \{\nu+1, \dots, m-1\}, \beta_i = 0, \phi^{\sigma_5}(e) = \mathfrak{m} - 1\} \\
&\cup \left\{e = (d_{i,1-\beta_{i+1},*}, F_{i,1-\beta_{i+1}}) \colon i < \nu, \phi^{\sigma^{(5)}}(e) \geq \mathfrak{m} - 1\right\},
\end{aligned}$$

if $\mathfrak{b}+1$ is not a power of two. A similar partition can be derived if $\mathfrak{b}+1$ is a power of two. In particular, $I_\sigma \subseteq \{(d_{i,j,k}, F_{i,j}) : \sigma(d_{i,j,k}) \neq F_{i,j}\}$. We prove that $e = (d_{i,j,k}, F_{i,j})$ implies $e \in I_\sigma$ if $\sigma(d_{i,j,k}) \neq F_{i,j}$.

If $\sigma^{(5)}(e_{i,j,k'}) = g_1$ for some $k' \in \{0,1\}$, then $e \in I_\sigma$ as one of the cycle edges of $F_{i,j}$ is improving for $\sigma^{(5)}$ while the other becomes improving after applying $(e_{i,j,k'}, b_2)$. Thus let $\sigma^{(5)}(e_{i,j,*}) = b_2$, implying $\neg\bar{\sigma}^{(5)}(eg_{i,j})$. Then, by Lemma 6.3.12, either $\bar{\sigma}^{(5)}(d_{i,j})$ or $\bar{\sigma}^{(5)}(eb_{i,j}) \wedge \neg\bar{\sigma}^{(5)}(eg_{i,j})$. In the first case, $\beta_i = 1 \wedge \beta_{i+1} = j$ by Lemma 6.3.12. But this implies $\bar{\sigma}(d_{i,j})$ since σ is a phase-5-strategy for $\mathfrak{b}$ and thus has Property (EV1)$_i$. This however contradicts $\sigma(d_{i,j,k}) \neq F_{i,j}$. Hence, assume that $\bar{\sigma}^{(5)}(eb_{i,j}) \wedge \neg\bar{\sigma}^{(5)}(eg_{i,j})$. Then, by Lemma 6.3.12, $(i,j) \in S_1$. We distinguish three cases.

1. Let $(i,j) \in \{(i, 1-\beta_{i+1}) : i \leq \nu - 1\}$. If $\phi^{\sigma^{(5)}}(e) < \mathfrak{m} - 1$, then e was an improving switch of type 3 for $\sigma^{(5)}$ and thus applied during phase 5. But this contradicts $\sigma(d_{i,j,k}) \neq F_{i,j}$ since no switch $(d_{*,*,*}, e_{*,*,*})$ is applied during phase 5. This implies $(i,j) \in \{(i, 1-\beta_{i+1}) : i \leq \nu - 1, \phi^{\sigma^{(5)}}(e) \geq \mathfrak{m} - 1\}$, hence $e \in I_\sigma$.
2. Let $(i,j) \in \{(i, 1-\beta_{i+1}) : i \in \{\nu+1, \dots, m-1\}, \beta_i = 0\}$ which can only occur if $\mathfrak{b}+1$ is not a power of 2. As proved when discussing $I_{\sigma^{(5)}}$, we then either have $\sigma_{\mathfrak{b}}(d_{i,j,k}) = F_{i,j}$, implying $\phi^{\sigma_{\mathfrak{b}}}(d_{i,j,k}, F_{i,j}) \leq \mathfrak{m} - 1$ or $\sigma_{\mathfrak{b}}(d_{i,j,k}) \neq F_{i,j}$ and $\phi^{\sigma_{\mathfrak{b}}}(d_{i,j,k}, F_{i,j}) = \mathfrak{m} - 1$. Consider the first case. If the inequality is strict, the switch was applied previously during phase 5, yielding a contradiction. Otherwise, $(d_{i,j,k}, F_{i,j}) \in I_\sigma$. In the second case, the switch was applied during phase 1, hence it was a switch of type 1 during phase 5, also implying $(d_{i,j,k}, F_{i,j}) \in I_\sigma$.
3. Finally, let $i = \nu \wedge j = 1 - \beta_{\nu+1}$ which only needs to be considered if $\mathfrak{b}+1$ is not a power of 2. In this case we however have $e \in I_{\sigma^{(5)}}$, implying $e \in I_\sigma$.

Thus, $e \in I_\sigma$ in all case, proving the statement. □

Claim 20. Let σ denote the strategy obtained after applying the final improving switch of phase 5 for $\nu > 1$. Then σ is a canonical strategy for $\mathfrak{b}+1$.

Proof. Consider Definitions 5.1.2 and 5.2.1. As σ is a phase-5-strategy for $\mathfrak{b}$, it holds that $\beta = \mathfrak{b}+1$. Thus, condition 1 follows since $\sigma(e_{*,*,*}) = b_2$ and $\nu > 1$. This also implies that conditions 2(a), 2(c), 3(a) and 3(b) are fulfilled as σ has Property (EV1)$_*$ and Property (EV2)$_*$.

Consider condition 2(b) and let $i \in [n]$. Since $(\mathfrak{b}+1)_i = 1$ implies that $F_{i,(\mathfrak{b}+1)_{i+1}}$ is closed, we prove that $F_{i,1-(\mathfrak{b}+1)_{i+1}}$ is not closed. Let $j := 1 - (\mathfrak{b}+1)_{i+1}$. Then, by Lemma 6.3.12, $\sigma^{(5)}(d_{i,j,*}) = e_{i,j,*}$ and it suffices to prove $(d_{i,j,0}, F_{i,j}) \notin \mathfrak{A}^{\sigma}_{\sigma^{(5)}}$. As such a switch is applied during $\sigma^{(5)} \to \sigma$ if and only if it is of type 3 by Corollary 6.3.18, we prove

$$\phi^{\sigma^{(5)}}(d_{i,j,0}, F_{i,j}) \geq \left\lfloor \frac{\mathfrak{b}+1}{2} \right\rfloor - 1. \tag{A.15}$$

This follows if $\mathbf{1}_{j=0}\text{lfn}(\mathfrak{b}, i+1) + \mathbf{1}_{j=1}\text{lufn}(\mathfrak{b}, i+1) = 0$ since this implies $\ell^{\mathfrak{b}}(i,j,k) \geq \mathfrak{b}$. Thus suppose that this term is not 0. Then, since $\mathfrak{b}_1 = 1$ and by the choice of i and j,

$$\ell^{\mathfrak{b}}(i,j,k) = \left\lceil \frac{\mathfrak{b} + \sum(\mathfrak{b}, i) + 1 - k}{2} \right\rceil \geq \left\lceil \frac{\mathfrak{b} + 2 - k}{2} \right\rceil = \left\lfloor \frac{\mathfrak{b} + 1 - k}{2} \right\rfloor + 1.$$

But this implies Inequality (A.15) since $\ell^{\mathfrak{b}}(i,j,0) \geq \lfloor(\mathfrak{b}+1)/2\rfloor + 1$.

Consider condition 3(c) and let $i \in [n]$ and $j := 1-(\mathfrak{b}+1)_{i+1}$. It is easy to prove that σ has condition 3(c) since $(\mathfrak{b}+1)_i = 0$ and $F_{i,j}$ being closed imply $\mathrm{Val}^*_\sigma(F_{i,j}) \succ \mathrm{Val}^*_\sigma(F_{i,1-j})$. Since $F_{i,j}$ is closed, $\mathrm{rVal}^*_\sigma(F_{i,j}) = [\![s_{i,j}]\!] \oplus \mathrm{rVal}^*_\sigma(b_2)$ by Property $(\mathrm{USV1})_i$, Lemma 6.1.16 and $\sigma(b_1) = b_2$. Since $F_{i,1-j}$ cannot be closed due to the choice of j and $(\mathfrak{b}+1)_i = 0$, we have $\bar{\sigma}(eb_{i,1-j}) \wedge \neg\bar{\sigma}(eg_{i,1-j})$. Consequently, $\mathrm{rVal}^*_\sigma(F_{i,1-j}) = \mathrm{rVal}^*_\sigma(b_2)$ since $\sigma(b_1) = b_2$. But this implies $\mathrm{rVal}^*_\sigma(F_{i,1-j}) \prec \mathrm{rVal}^*_\sigma(F_{i,j})$. Next, consider condition 3(d) and consider a level i with $(\mathfrak{b}+1)_i = 0$. Let $j := 0$ resp. $j := \beta_{i+1}$ depending on whether $G_n = S_n$ or $G_n = M_n$. We prove that $\mathrm{Val}^*_\sigma(F_{i,j}) \succ \mathrm{Val}^*_\sigma(F_{i,1-j})$ if none of the two cycle centers is closed. If $G_n = M_n$, this either follows from Lemma 6.2.1 since σ has Property (REL1) or by an easy but tedious calculation. If $G_n = S_n$, this follows since $\Omega(F_{i,0}) > \Omega(F_{i,1})$ and as these priorities are even.

Property (USV1) implies that σ fulfills conditions 4 and 5 for all indices. Finally, consider condition 6 and let $i = \ell(\mathfrak{b}+2), j = \beta_{\ell(\mathfrak{b}+2)+1}$. By the same argument used for condition 3(c), it suffices to prove $\phi^\sigma(d_{i,j,k}, F_{i,j}) \geq \lfloor(\mathfrak{b}+1)/2\rfloor - 1$ for both $k \in \{0,1\}$. This however follows from $\ell(\mathfrak{b}+1) = 1, \beta_2 = 1-\mathfrak{b}_2$ and $\mathfrak{b}_1 = 1$ by

$$\ell^{\mathfrak{b}}(1, 1-\mathfrak{b}_2, k) = \left\lceil \frac{\mathfrak{b} + \sum(\mathfrak{b},1) + 1 - k}{2} \right\rceil = \left\lceil \frac{\mathfrak{b}+1-k}{2} \right\rceil = \left\lfloor \frac{\mathfrak{b}+2-k}{2} \right\rfloor = \left\lfloor \frac{\mathfrak{b}+1}{2} \right\rfloor,$$

implying the statement. Hence, σ is a canonical strategy for $\mathfrak{b}+1$. □

Lemma 6.3.21. *Let $\mathfrak{b} \in \mathfrak{B}_n$ be even, $i := \ell(\mathfrak{b}+2)$ and $j := 1-(\mathfrak{b}+2)_{i+1}$. If $\mathfrak{b}+2$ is a power of 2, then $\phi^{\sigma_\mathfrak{b}}(d_{i,j,*}, F_{i,j}) = \mathfrak{m}$. Otherwise, $\phi^{\sigma_\mathfrak{b}}(d_{i,j,0}, F_{i,j}) = \lfloor(\mathfrak{b}+1)/2\rfloor$ and $\phi^{\sigma_\mathfrak{b}}(d_{i,j,1}, F_{i,j}) = \mathfrak{m}-1$. In any case, $\sigma_\mathfrak{b}(d_{i,j,k}) \neq F_{i,j}$ for both $k \in \{0,1\}$.*

Proof. Assume $\mathfrak{b}+2 = 2^l$ for some $l \in \mathbb{N}$. Then, the choice of i and j implies $\mathfrak{b}+2 = 2^{i-1}$ and $j = 1$. In particular $\mathrm{lufn}(\mathfrak{b}, i+1) = 0$, implying $\ell^{\mathfrak{b}}(i,j,k) \geq \mathfrak{b}$ by Lemma 6.2.3. Consequently, $\phi^{\sigma_\mathfrak{b}}(d_{i,j,k}, F_{i,j}) = \lfloor(\mathfrak{b}+1-k)/2\rfloor = \mathfrak{m}$ since $\mathfrak{b}+1$ is odd. In addition, $\phi^{\sigma_\mathfrak{b}}(d_{i,j,k}, F_{i,j}) \neq \ell^{\mathfrak{b}}(i,j,k)+1$, hence $\sigma_\mathfrak{b}(d_{i,j,k}) \neq F_{i,j}$ as $\sigma_\mathfrak{b}$ has the canonical properties.

Thus assume that $\mathfrak{b}+2$ is not a power of 2. Since $\mathfrak{b}$ is even and by the choice of i, it holds that $\mathfrak{b}_1 = 0$ and $\mathfrak{b}_2 = \cdots = \mathfrak{b}_{i-1} = 1$. In particular $\mathbf{1}_{j=0}\mathrm{lfn}(\mathfrak{b}, i+1) + \mathbf{1}_{j=1}\mathrm{lufn}(\mathfrak{b}, i+1) \neq 0$. Hence, by Lemma 6.2.3,

$$\begin{aligned}\ell^{\mathfrak{b}}(i,j,k) &= \left\lceil \frac{\mathfrak{b} - 2^{i-1} + \sum(\mathfrak{b},i) + 1 - k}{2} \right\rceil = \left\lceil \frac{\mathfrak{b} - 2^{i-1} + 2^{i-1} - 2 + 1 - k}{2} \right\rceil \\ &= \left\lceil \frac{\mathfrak{b}-1-k}{2} \right\rceil = \left\lfloor \frac{\mathfrak{b}-k}{2} \right\rfloor.\end{aligned}$$

Furthermore, $\mathfrak{b}$ being even and Property $(\mathrm{OR4})_{i,j,0}$ implies $\phi^{\sigma_\mathfrak{b}}(d_{i,j,k}, F_{i,j}) \neq \ell^{\mathfrak{b}}(i,j,k) - 1$. Hence $\phi^{\sigma_\mathfrak{b}}(d_{i,j,0}, F_{i,j}) = \lfloor(\mathfrak{b}+1)/2\rfloor$ and $\ell^{\mathfrak{b}}(i,j,1) = \lfloor(\mathfrak{b}-1)/2\rfloor = \mathfrak{m}-1$. Also, $\sigma_\mathfrak{b}(d_{i,j,k}) \neq F_{i,j}$ for both $k \in \{0,1\}$ since $\phi^{\sigma_\mathfrak{b}}(d_{i,j,k}, F_{i,j}) = \ell^{\mathfrak{b}}(i,j,k)+1 < \mathfrak{m}$ otherwise, contradicting the previous arguments. □

Claim 21. Let $i \in [n], j,k \in \{0,1\}$ and consider the two equations

$$\phi^\sigma(e) \neq \ell^{\mathfrak{b}+1}(i,j,k) - 1, \tag{6.8}$$

$$\phi^{\sigma}(e) = \left\lfloor \frac{\mathfrak{b}+1+1-2}{2} \right\rfloor. \tag{6.9}$$

1. If $j = (\mathfrak{b}+2)_{i+1}$, then either Equation (6.8) or Equation (6.9) holds.
2. If $i \neq \ell(\mathfrak{b}+2)$ and $j \neq (\mathfrak{b}+2)_{i+1}$, then either Equation (6.8) or Equation (6.9) holds.
3. If $\mathfrak{b}+1$ is even, $i = \ell(\mathfrak{b}+2)$ and $j \neq (\mathfrak{b}+2)_{i+1}$, then Equation (6.9) holds.
4. If $\mathfrak{b}+1$ is odd, $i = \ell(\mathfrak{b}+2), j = 1-(\mathfrak{b}+2)_{i+1}, k \in \{0,1\}$ and $\mathfrak{b}+2$ is a power of two, then Equation (6.9) holds.
5. If $\mathfrak{b}$ is even, $i = \ell(\mathfrak{b}+2), j \neq (\mathfrak{b}+2)_{i+1}, k = 1$ and $\mathfrak{b}+2$ is not a power of two, then Equation (6.9) holds.

Proof. As a remainder, we currently consider a canonical strategy σ for $\mathfrak{b}+1$ with $I_\sigma = \mathfrak{D}^\sigma$. We prove the statements one after another.

1. We distinguish several cases.
   a) Let $\mathfrak{b}_i = 1 \wedge \mathfrak{b}_{i+1} = j$. This implies $i \neq 1$ since $i = 1$ contradicts the choice of j. Also $\mathfrak{b} \geq 4$ for the same reason. Let

   $$\mathbf{1}_{j=0}\operatorname{lfn}(\mathfrak{b}+1, i+1) + \mathbf{1}_{j=1}\operatorname{lufn}(\mathfrak{b}+1, i+1) = 0.$$

   Then $\ell^{\mathfrak{b}+1}(i,j,k) - 1 \geq \mathfrak{b}$ by Lemma 6.2.3. Since $\phi^{\sigma}(e) \leq \phi^{\sigma_{\mathfrak{b}}}(e)+1$, we then have $\phi^{\sigma}(e) \leq \mathfrak{m}+1$. Since $\mathfrak{b} \geq 4$, this implies $\phi^{\sigma}(e) < \ell^{\mathfrak{b}+1}(i,j,k)-1$, implying the statement. Let $\mathbf{1}_{j=0}\operatorname{lfn}(\mathfrak{b}+1, i+1) + \mathbf{1}_{j=1}\operatorname{lufn}(\mathfrak{b}+1, i+1) \neq 0$ and observe

   $$\phi^{\sigma_{\mathfrak{b}}}(e) = \left\lceil \frac{\operatorname{lfn}(\mathfrak{b}, i, \{(i+1,j)\}) + 1 - k}{2} \right\rceil = \left\lceil \frac{\mathfrak{b} - \sum(\mathfrak{b}, i) + 1 - k}{2} \right\rceil.$$

   We distinguish two more cases.
   i. Let $(\mathfrak{b}+1)_i = 1$, implying $(\mathfrak{b}+1)_{i+1} = j$. Then $e \notin \mathfrak{A}^{\sigma}_{\sigma_{\mathfrak{b}}}$, and consequently $\phi^{\sigma_{\mathfrak{b}}}(e) = \phi^{\sigma}(e)$. It is easy to verify that

   $$\mathbf{1}_{j=0}\operatorname{lfn}(\mathfrak{b}+1, i+1) + \mathbf{1}_{j=1}\operatorname{lufn}(\mathfrak{b}+1, i+1) = \mathfrak{b}+1-\sum(\mathfrak{b}+1, i) - 2^{i-1} - 2^{i-1}$$

   in this case. Hence, by the definition of $\ell^{\mathfrak{b}+1}(i,j,k)$, we have

   $$\begin{aligned}\ell^{\mathfrak{b}+1}(i,j,k) &= \left\lceil \frac{\operatorname{lfn}(\mathfrak{b}+1, i, \{(i+1,j)\}) + 1 - k}{2} \right\rceil - \sum(\mathfrak{b}+1, i) + 2^i \\ &= \phi^{\sigma_{\mathfrak{b}+1}}(e) + \sum(\mathfrak{b}+1, i) + 2^i \geq \phi^{\sigma_{\mathfrak{b}+1}}(e) + 4,\end{aligned}$$

   so $\phi^{\sigma_{\mathfrak{b}+1}}(e) \leq \ell^{\mathfrak{b}+1}(i,j,k) - 4$, implying Equation (6.8).
   ii. Assume $(\mathfrak{b}+1)_i = 0$, implying $i < \nu$ and thus $(\mathfrak{b}+1)_{i+1} \neq \mathfrak{b}_i = j$. Then $\mathfrak{b}_1 = \cdots = \mathfrak{b}_i = 1$ and $(\mathfrak{b}+1)_1 = \cdots = (\mathfrak{b}+1)_i = 0$. Hence, by Lemma 6.2.6,

   $$\ell^{\mathfrak{b}+1}(i,j,k) = \left\lceil \frac{\mathfrak{b}+1-2^{i-1}+\sum(\mathfrak{b}+1, i) + 1 - k}{2} \right\rceil$$

$$= \left\lceil \frac{\mathfrak{b} - (2^{i-1} - 1) + 1 - k}{2} \right\rceil$$
$$= \left\lceil \frac{\mathfrak{b} - \sum(\mathfrak{b}, i) + 1 - k}{2} \right\rceil = \phi^{\sigma_\mathfrak{b}}(e).$$

This implies $\phi^\sigma(e) \geq \phi^{\sigma_\mathfrak{b}}(e) = \ell^{\mathfrak{b}+1}(i,j,k)$, and thus Equation (6.8).

b) Let $\mathfrak{b}_i = 1 \wedge \mathfrak{b}_{i+1} \neq j$. Since $j = (\mathfrak{b}+2)_{i+1}$, this implies $(\mathfrak{b}+2)_{i+1} \neq \mathfrak{b}_{i+1}$. Hence, bit $i+1$ was switched when transitioning from $\sigma_\mathfrak{b}$ to $\sigma_{\mathfrak{b}+2}$. In one of the two transitions, the first bit switched from 0 to 1 and this bit was the only bit that was switched in this transition. Thus, either $[i < \ell(\mathfrak{b}+1)$ and $\ell(\mathfrak{b}+2) = 1]$ or $[\ell(\mathfrak{b}+1) = 1$ and $i < \ell(\mathfrak{b}+2)]$. Consider $[i < \ell(\mathfrak{b}+1)$ and $\ell(\mathfrak{b}+2) = 1]$ first. Since $\mathfrak{b}_i = 1$ and $\mathfrak{b}_{i+1} \neq j$, Lemma 6.2.3 implies that it holds that $\ell^\mathfrak{b}(i,j,k) = \lceil(\mathfrak{b} + \sum(\mathfrak{b}, i) + 1 - k)/2\rceil$. Now, since $i < \ell(\mathfrak{b}+1)$, we have $\mathfrak{b}_l = 1$ for all $l < i$, implying $\ell^\mathfrak{b}(i,j,k) = \lceil(\mathfrak{b} + 2^{i-1} - k)/2\rceil$. If $i = 1$, then

$$\ell^\mathfrak{b}(i,j,k) = \left\lceil \frac{\mathfrak{b}+1-k}{2} \right\rceil \geq \left\lfloor \frac{\mathfrak{b}+1-k}{2} \right\rfloor.$$

This implies Equation (6.9) since the only feasible tolerance for $i = 1$ is 0. If $i > 1$, then Equation (6.9) follows from

$$\ell^\mathfrak{b}(i,j,k) \geq \left\lceil \frac{\mathfrak{b}+2-k}{2} \right\rceil = \left\lfloor \frac{\mathfrak{b}+3}{2} \right\rfloor = \left\lfloor \frac{\mathfrak{b}+1-k}{2} \right\rfloor + 1.$$

Thus, $\phi^{\sigma_\mathfrak{b}}(e) = \lfloor(\mathfrak{b}+1-k)/2\rfloor \neq \ell^\mathfrak{b}(i,j,k) + 1$. Consequently, by Property (OR1)$_{i,j,k}$, $\sigma_\mathfrak{b}(d_{i,j,k}) \neq F_{i,j}$ for both $k \in \{0,1\}$. Since $\ell(\mathfrak{b}+1) > i \geq 1$ implies that $\mathfrak{b}$ is odd, this yields $\phi^{\sigma_\mathfrak{b}}(d_{i,j,1}, F_{i,j}) < \phi^{\sigma_\mathfrak{b}}(d_{i,j,0}, F_{i,j})$. Combining these implies that $(d_{i,j,1}, F_{i,j})$ is applied during phase 1 of $\sigma_\mathfrak{b} \to \sigma_{\mathfrak{b}+1}$, so

$$\phi^\sigma(d_{i,j,1}, F_{i,j}) = \left\lfloor \frac{\mathfrak{b}}{2} \right\rfloor + 1 = \left\lfloor \frac{\mathfrak{b}+1}{2} \right\rfloor = \left\lfloor \frac{\mathfrak{b}+1+1-1}{2} \right\rfloor \quad \text{and}$$
$$\phi^\sigma(d_{i,j,0}, F_{i,j}) = \left\lfloor \frac{\mathfrak{b}+1}{2} \right\rfloor = \left\lfloor \frac{\mathfrak{b}+1+1-0}{2} \right\rfloor.$$

Next let $\ell(\mathfrak{b}+1) = 1$ and $i < \ell(\mathfrak{b}+2)$. Since $\ell(\mathfrak{b}+1) = 1$ implies $\mathfrak{b}_1 = 0$, $\mathfrak{b}_i = 1$ implies $i > 1$. In addition, $i < \ell(\mathfrak{b}+2)$ implies $\mathfrak{b}_{i'} = 1$ for all $i' \in \{2, \dots, i-1\}$. Consequently, as in the last case,

$$\ell^\mathfrak{b}(i,j,k) = \left\lceil \frac{\mathfrak{b} + \sum(\mathfrak{b}, i) + 1 - k}{2} \right\rceil = \left\lceil \frac{\mathfrak{b} + 2^{i-1} - 2 + 1 + k}{2} \right\rceil$$
$$\geq \left\lceil \frac{\mathfrak{b}+1-k}{2} \right\rceil \geq \left\lfloor \frac{\mathfrak{b}+1-k}{2} \right\rfloor.$$

Since $\mathfrak{b}$ is even, $t_\mathfrak{b} = -1$ is not a feasible parameter for $\mathfrak{b}$. This implies that $\phi^{\sigma_\mathfrak{b}}(e) = \lfloor(\mathfrak{b}+1-k)/2\rfloor$ and in particular $\phi^{\sigma_\mathfrak{b}}(e) \neq \ell^\mathfrak{b}(i,j,k) + 1$. Thus,

Property $(\text{OR1})_{i,j,k}$ implies $\sigma_{\mathfrak{b}}(d_{i,j,k}) \neq F_{i,j}$ for both $k \in \{0,1\}$. Since $\mathfrak{b}$ is even, $\phi^{\sigma_{\mathfrak{b}}}(d_{i,j,0}, F_{i,j}) = \phi^{\sigma_{\mathfrak{b}}}(d_{i,j,1}, F_{i,j})$. Hence, as discussed previously, the switch $(d_{i,j,0}, F_{i,j})$ is applied during $\sigma_{\mathfrak{b}} \to \sigma$. Thus, Equation (6.9) follows from

$$\phi^{\sigma_{\mathfrak{b}+1}}(d_{i,j,0}, F_{i,j}) = \left\lfloor \frac{\mathfrak{b}+1}{2} \right\rfloor + 1 = \left\lfloor \frac{\mathfrak{b}+1+1-0}{2} \right\rfloor \quad \text{and}$$

$$\phi^{\sigma_{\mathfrak{b}+1}}(d_{i,j,1}, F_{i,j}) = \left\lfloor \frac{\mathfrak{b}}{2} \right\rfloor = \left\lfloor \frac{\mathfrak{b}+1}{2} \right\rfloor = \left\lfloor \frac{\mathfrak{b}+1+1-1}{2} \right\rfloor.$$

c) Let $\mathfrak{b}_i = 0$ and $(\mathfrak{b}+1)_i = 1 \wedge j = (\mathfrak{b}+1)_{i+1}$, implying $i = \ell(\mathfrak{b}+1)$. Hence, as the occurrence record of e with respect to σ is described by Table 5.6,

$$\phi^{\sigma}(e) = \left\lceil \frac{\text{lfn}(\mathfrak{b}+1, i, \{(i+1,j)\}) + 1 - k}{2} \right\rceil = \left\lceil \frac{\mathfrak{b}+1+1-k}{2} \right\rceil = \left\lfloor \frac{\mathfrak{b}+1-k}{2} \right\rfloor + 1.$$

Since $\mathbf{1}_{j=0}\text{lfn}(\mathfrak{b}+1, i+1) + \mathbf{1}_{j=1}\text{lufn}(\mathfrak{b}+1, i+1) < \mathfrak{b}+1-2^i$, this implies $\ell^{\mathfrak{b}+1}(i,j,k) > \phi^{\sigma}(e) + 2$ and consequently Equation (6.8).

d) Let $\mathfrak{b}_i = 0$ and $(\mathfrak{b}+1)_i = 1 \wedge j \neq (\mathfrak{b}+1)_{i+1}$. Then, $i = \ell(\mathfrak{b}+1)$. Since $j = (\mathfrak{b}+2)_{i+1}$ by assumption, the bit with index $i+1$ has switched when transitioning from $\mathfrak{b}+1$ to $\mathfrak{b}+2$. This is however only possible if $i = \ell(\mathfrak{b}+1) = 1$. As this also implies $\mathfrak{b}_{i+1} = (\mathfrak{b}+1)_{i+1} \neq j$, this implies

$$\ell^{\mathfrak{b}}(i,j,k) = \left\lceil \frac{\mathfrak{b} - 2^{i-1} + \sum(\mathfrak{b}, i) + 1 - k}{2} \right\rceil = \left\lceil \frac{\mathfrak{b}-1+1-k}{2} \right\rceil = \left\lfloor \frac{\mathfrak{b}+1-k}{2} \right\rfloor$$

Property $(\text{OR3})_{i,j,k}$ applied to $\sigma_{\mathfrak{b}}$ thus implies $\phi^{\sigma_{\mathfrak{b}}}(e) = \lfloor (\mathfrak{b}+1-k)/2 \rfloor$. By the same arguments used in the earlier cases, we devise $(d_{i,j,0}, F_{i,j}) \in \mathfrak{A}^{\sigma}_{\sigma_{\mathfrak{b}}}$. But, similar to the previous cases, this implies Equation (6.9).

e) Let $\mathfrak{b}_i = 0$ and $(\mathfrak{b}+1)_i = 0$. Then, $i > 1$ and $\mathfrak{b}_{i+1} = (\mathfrak{b}+1)_{i+1} = (\mathfrak{b}+2)_{i+1}$, hence $j = \mathfrak{b}_{i+1}$. Thus, by Lemma 6.2.3,

$$\ell^{\mathfrak{b}}(i,j,k) = \left\lceil \frac{\mathfrak{b} + 2^{i-1} + \sum(\mathfrak{b}, i) + 1 - k}{2} \right\rceil \geq \left\lceil \frac{\mathfrak{b} + 2^{i-1} + 1 - k}{2} \right\rceil \geq \left\lceil \frac{\mathfrak{b}+2+1-k}{2} \right\rceil = \left\lceil \frac{\mathfrak{b}+1-k}{2} \right\rceil + 1 \geq \left\lfloor \frac{\mathfrak{b}+1-k}{2} \right\rfloor + 1.$$

Therefore $\lfloor (\mathfrak{b}+1-k)/2 \rfloor \leq \ell^{\mathfrak{b}}(i,j,k) - 1$, implying $\phi^{\sigma_{\mathfrak{b}}}(e) = \lfloor (\mathfrak{b}+1-k)/2 \rfloor$. This implies Equation (6.9) by using the same arguments as in the last cases.

2. Since $i \neq \ell(\mathfrak{b}+2)$, it is not possible that $(\mathfrak{b}+1)_{i+1} = 0 \wedge (\mathfrak{b}+2)_{i+1} = 1$. It thus suffices to investigate the following cases.

a) Let $\mathfrak{b}_i = 0 \land (\mathfrak{b}+1)_i = 1$, i.e., $i = \ell(\mathfrak{b}+1) = \nu$. Then $\mathfrak{b}_{i+1} = (\mathfrak{b}+1)_{i+1}$ and $(\mathfrak{b}+1)_{i+1} = (\mathfrak{b}+2)_{i+1}$ if and only if $i \neq 1$. Consider the case $i \neq 1$ first. Then, $j \neq \mathfrak{b}_{i+1}$, hence $j = 1 - \mathfrak{b}_{i+1}$. Since $i = \ell(\mathfrak{b}+1)$, Lemma 6.2.3 then implies

$$\begin{aligned}\ell^{\mathfrak{b}}(i,j,k) &= \left\lceil \frac{\mathfrak{b} - 2^{i-1} + \sum(\mathfrak{b}, i) + 1 - k}{2} \right\rceil = \left\lceil \frac{\mathfrak{b} - 2^{\nu-1} + 2^{\nu-1} - 1 + 1 - k}{2} \right\rceil \\ &= \left\lfloor \frac{\mathfrak{b} + 1 - k}{2} \right\rfloor .\end{aligned}$$

Now, $\phi^{\sigma_{\mathfrak{b}}}(e) = \lfloor(\mathfrak{b}+1-k)/2\rfloor$ or $\phi^{\sigma_{\mathfrak{b}}}(e) = \ell^{\mathfrak{b}}(i,j,k) - 1 \neq \lfloor(\mathfrak{b}+1-k)/2\rfloor$. Consider the first case. Then $\sigma_{\mathfrak{b}}(d_{i,j,k}) \neq F_{i,j}$ by Property (OR2)$_{i,j,k}$. Using the same arguments used when proving the first statement, this implies $\phi^{\sigma}(d_{i,j,k}, F_{i,j}) = \lfloor(\mathfrak{b}+1+1-k)/2\rfloor$. Consider the second case. By our previous calculation and by Property (OR3)$_{i,j,k}$, $\phi^{\sigma_{\mathfrak{b}}}(d_{i,j,k}, F_{i,j}) = \mathfrak{m} - 1$ and $k = 0$. Also, by Property (OR2)$_{i,j,k}$, $\sigma_{\mathfrak{b}}(d_{i,j,0}) \neq F_{i,j}$. Thus, by Corollary 6.3.3, $e \in \mathfrak{A}_{\sigma_{\mathfrak{b}}}^{\sigma}$. Hence $\phi^{\sigma}(e) = \lfloor(\mathfrak{b}+1)/2\rfloor = \lfloor(\mathfrak{b}+2)/2\rfloor$ since $\mathfrak{b}$ is odd. Consequently, $\phi^{\sigma}(e) = \lfloor(\mathfrak{b}+1+1-0)/2\rfloor$, so Equation (6.9) holds.

Now consider the case $i = 1$, i.e., $(\mathfrak{b}+1)_{i+1} \neq (\mathfrak{b}+2)_{i+1}$. Then $j = \mathfrak{b}_{i+1}$, hence

$$\ell^{\mathfrak{b}}(i,j,k) = \left\lceil \frac{\mathfrak{b} + 2^{i-1} + \sum(\mathfrak{b}, i) + 1 - k}{2} \right\rceil = \left\lceil \frac{\mathfrak{b} + 2 - k}{2} \right\rceil = \left\lfloor \frac{\mathfrak{b} + 1 - k}{2} \right\rfloor + 1.$$

Thus, $\ell^{\mathfrak{b}}(i,j,k) - 1 = \lfloor(\mathfrak{b}+1-k)/2\rfloor$, implying $\phi^{\sigma_{\mathfrak{b}}}(e) = \lfloor(\mathfrak{b}+1-k)/2\rfloor$. Since $F_{1,j}$ is the cycle center that is closed during the transition from $\sigma_{\mathfrak{b}}$ to $\sigma_{\mathfrak{b}+1}$, the switch $(d_{1,j,k}, F_{1,j})$ is applied for both k. Since $\mathfrak{b}$ is even, Equation (6.9) follows from $\phi^{\sigma}(e) = \lfloor(\mathfrak{b}+1-k)/2\rfloor + 1 = \lfloor(\mathfrak{b}+1+1-k)/2\rfloor$.

b) Let $\mathfrak{b}_i = 0 \land (\mathfrak{b}+1)_i = 0$. Since $i \neq \ell(\mathfrak{b}+2)$, we have $(\mathfrak{b}+2)_i = 0$. This implies $\mathfrak{b}_{i+1} = (\mathfrak{b}+1)_{i+1} = (\mathfrak{b}+2)_{i+1}$, so $j = 1 - \mathfrak{b}_{i+1}$ and $i > 2$. Thus, by Lemma 6.2.3,

$$\ell^{\mathfrak{b}}(i,j,k) = \left\lceil \frac{\mathfrak{b} - 2^{i-1} + \sum(\mathfrak{b}, i) + 1 - k}{2} \right\rceil .$$

Since $(\mathfrak{b}+1)_i = 0$ implies $i \neq \ell(\mathfrak{b}+1)$, Property (OR3)$_{i,j,k}$ implies that either $\phi^{\sigma_{\mathfrak{b}}}(d_{i,j,k}, F_{i,j}) \neq \ell^{\mathfrak{b}}(i,j,k) - 1$ or $\phi^{\sigma_{\mathfrak{b}}}(d_{i,j,k}, F_{i,j}) = \lfloor(\mathfrak{b}+1-k)/2\rfloor$. Assume $\phi^{\sigma_{\mathfrak{b}}}(d_{i,j,k}, F_{i,j}) = \lfloor(\mathfrak{b}+1-k)/2\rfloor$ and let $\ell^{\mathfrak{b}}(i,j,k) + 1 = \lfloor(\mathfrak{b}+1-k)/2\rfloor$ first. Then, since $\ell^{\mathfrak{b}}(i,j,k) + 1 = \ell^{\mathfrak{b}+1}(i,j,k)$ by Lemma 6.2.6, we obtain

$$\phi^{\sigma}(e) \geq \phi^{\sigma_{\mathfrak{b}}}(d_{i,j,k}, F_{i,j}) = \ell^{\mathfrak{b}}(i,j,k) + 1 = \ell^{\mathfrak{b}+1}(i,j,k).$$

Hence $\phi^{\sigma}(d_{i,j,k}, F_{i,j}) \neq \ell^{\mathfrak{b}+1}(i,j,k) - 1$. Now let $\phi^{\sigma_{\mathfrak{b}}}(d_{i,j,k}, F_{i,j}) \neq \ell^{\mathfrak{b}}(i,j,k) + 1$. Then, by Property (OR2)$_{i,j,k}$, $\sigma_{\mathfrak{b}}(d_{i,j,k}) \neq F_{i,j}$. Since $\phi^{\sigma_{\mathfrak{b}}}(e) = \lfloor(\mathfrak{b}+1-k)/2\rfloor$, we can apply the same arguments used when discussing previous cases to obtain Equation (6.9).

Let $\phi^{\sigma_{\mathfrak{b}}}(d_{i,j,k}, F_{i,j}) \neq \ell^{\mathfrak{b}}(i,j,k) - 1 \neq \lfloor(\mathfrak{b}+1-k)/2\rfloor$ as we could apply the same arguments used before otherwise. Thus, either $\phi^{\sigma_{\mathfrak{b}}}(d_{i,j,k}, F_{i,j}) = \ell^{\mathfrak{b}}(i,j,k)$

or $\phi^{\sigma_{\mathfrak{b}}}(d_{i,j,k}, F_{i,j}) = \ell^{\mathfrak{b}}(i,j,k) + 1$. Since $\ell^{\mathfrak{b}+1}(i,j,k) = \ell^{\mathfrak{b}}(i,j,k) + 1$, the statement follows directly if $\phi^{\sigma_{\mathfrak{b}}}(d_{i,j,k}, F_{i,j}) = \ell^{\mathfrak{b}}(i,j,k) + 1$. Hence assume $\phi^{\sigma_{\mathfrak{b}}}(d_{i,j,k}, F_{i,j}) = \ell^{\mathfrak{b}}(i,j,k)$. Then, Property (OR2)$_{i,j,k}$ implies $\sigma_{\mathfrak{b}}(d_{i,j,k}) \neq F_{i,j}$. As $\phi^{\sigma_{\mathfrak{b}}}(d_{i,j,k}, F_{i,j}) < \lfloor(\mathfrak{b}+1-k)/2\rfloor$, e is applied when transitioning from $\sigma_{\mathfrak{b}}$ to σ. Thus $\phi^{\sigma}(e) = \varphi^{\sigma_{\mathfrak{b}}}(e) + 1 = \ell^{\mathfrak{b}+1}(i,j,k)$, implying Equation (6.8).

c) Let $\mathfrak{b}_i = 1 \wedge j = 1 - \mathfrak{b}_{i+1}$. Then, Lemma 6.2.3 implies

$$\ell^{\mathfrak{b}}(i,j,k) = \left\lceil \frac{\mathfrak{b} + \sum(\mathfrak{b}, i) + 1 - k}{2} \right\rceil \geq \left\lfloor \frac{\mathfrak{b}+1-k}{2} \right\rfloor.$$

Since $i \neq \ell(\mathfrak{b}+1)$ as $\mathfrak{b}_i = 1$, this yields $\phi^{\sigma_{\mathfrak{b}}}(e) = \lfloor(\mathfrak{b}+1-k)/2\rfloor$ by Property (OR3)$_{i,j,k}$. In particular, $\sigma_{\mathfrak{b}}(d_{i,j,k}) \neq F_{i,j}$, so the same arguments used previously yield Equation (6.9).

d) Let $\mathfrak{b}_i = 1 \wedge j = \mathfrak{b}_{i+1}$, implying $\phi^{\sigma_{\mathfrak{b}}}(e) = \lceil(\operatorname{lfn}(\mathfrak{b}, i, \{(i+1, j)\}) + 1 - k)/2\rceil$. Since $j \neq (\mathfrak{b}+2)_{i+1}$, bit $i+1$ switched. As $i \neq \ell(\mathfrak{b}+2)$ and $\mathfrak{b}_i = 1$ yields $i \neq 1$, this implies $i \leq \nu - 1$. In particular, $\phi^{\sigma_{\mathfrak{b}}}(e) = \lceil(\mathfrak{b} - 2^{i-1} + 2 - k)/2\rceil$. Furthermore, we then have $(\mathfrak{b}+1)_{i+1} = 0 \wedge (\mathfrak{b}+1)_{i+1} \neq j$. Hence, by Lemma 6.2.3,

$$\begin{aligned}\ell^{\mathfrak{b}+1}(i,j,k) &= \left\lceil \frac{\mathfrak{b} + 1 - 2^{i-1} + \sum(\mathfrak{b}+1) + 1 - k}{2} \right\rceil \\ &= \left\lceil \frac{\mathfrak{b} - 2^{i-1} + 2 - k}{2} \right\rceil = \phi^{\sigma_{\mathfrak{b}}}(e).\end{aligned}$$

Thus, $\phi^{\sigma}(e) \geq \phi^{\sigma_{\mathfrak{b}}}(e) = \ell^{\mathfrak{b}+1}(i,j,k)$, so $\phi^{\sigma}(e) \neq \ell^{\mathfrak{b}+1}(i,j,k) - 1$.

3. Since $i = \ell(\mathfrak{b}+2)$, we have $(\mathfrak{b}+1)_i = 0$. This further implies $(\mathfrak{b}+1)_{i+1} = (\mathfrak{b}+2)_{i+1}$, so $j \neq (\mathfrak{b}+1)_{i+1}$. Thus, by Lemma 6.2.3,

$$\begin{aligned}\ell^{\mathfrak{b}+1}(i,j,k) &= \left\lceil \frac{\mathfrak{b} + 1 - 2^{i-1} + \sum(\mathfrak{b}+1, i) + 1 - k}{2} \right\rceil \\ &= \left\lceil \frac{\mathfrak{b} + 1 - 2^{\nu-1} + 2^{\nu-1} - 1 + 1 - k}{2} \right\rceil \\ &= \left\lceil \frac{\mathfrak{b}+1-k}{2} \right\rceil = \left\lfloor \frac{\mathfrak{b}+1+1-k}{2} \right\rfloor.\end{aligned}$$

Since $\mathfrak{b}+1$ is even, the parameter $t_{\mathfrak{b}+1} = -1$ is not feasible. This implies Equation (6.9).

4. By the choice of i, there is no number $\mathfrak{b}' \leq \mathfrak{b}+1$ with $i = \ell(\mathfrak{b}')$. Consequently, it holds that $\operatorname{lfn}(\mathfrak{b}+1, i+1) = \operatorname{lufn}(\mathfrak{b}+1, i+1) = 0$. Thus, by Lemma 6.2.3, $\ell^{\mathfrak{b}+1}(i,j,k) \geq \mathfrak{b}+1 > \lfloor(\mathfrak{b}+1+1-k)/2\rfloor$, implying Equation (6.9).

5. Since $k = 1$, it suffices to show $\phi^{\sigma}(d_{i,j,1}, F_{i,j}) = \mathfrak{m}$. Since $\mathfrak{b}$ is even, we have $\ell(\mathfrak{b}+1) = 1$, hence $\mathfrak{b}_i = 0$. As shown in the proof of Lemma 6.3.21, this implies $\ell^{\mathfrak{b}}(i,j,k) = \lfloor(\mathfrak{b}-1)/2\rfloor = \mathfrak{m} - 1$. Consequently, by Lemma 6.3.21, it holds that $\phi^{\sigma_{\mathfrak{b}}}(d_{i,j,1}, F_{i,j}) = \ell^{\mathfrak{b}}(i,j,k) < \phi^{\sigma_{\mathfrak{b}}}(d_{i,j,0}, F_{i,j})$. Since Property (OR2)$_{i,j,1}$ now implies

$\sigma_{\mathfrak{b}}(d_{i,j,1}) \neq F_{i,j}$, this implies $(d_{i,j,1}, F_{i,j}) \in \mathfrak{A}^{\sigma}_{\sigma_{\mathfrak{b}}}$. But then, the statement follows since we then have $\phi^{\sigma}(d_{i,j,1}, F_{i,j}) = \mathfrak{m}$.

□

Claim 22. Let $\nu = 1$. The occurrence records of edges of the type $(d_{*,*,*}, F_{*,*})$ not applied during $\sigma_{\mathfrak{b}} \to \sigma$ is described correctly by Table 5.6.

Proof. Let $i \in [n], j, k \in \{0, 1\}$. We distinguish four cases.

1. Let $\mathfrak{b}_i = 1 \wedge \mathfrak{b}_{i+1} = j$. Since $\nu = 1$, this implies $(\mathfrak{b}+1)_i = 1 \wedge (\mathfrak{b}+1)_{i+1} = j$. Hence, $F_{i,j}$ is closed for both $\sigma_{\mathfrak{b}}$ and σ and the switch was not applied during $\sigma_{\mathfrak{b}} \to \sigma$, implying $i \neq 1$. Consequently, $\operatorname{lfn}(\mathfrak{b}, i, \{(i+1, j)\}) = \operatorname{lfn}(\mathfrak{b}+1, i, \{(i+1, j)\})$, implying the statement.

2. Let $\mathfrak{b}_i = 0 \wedge \mathfrak{b}_{i+1} \neq j$. Consider the case $(\mathfrak{b}+1)_i = 0 \wedge (\mathfrak{b}+1)_{i+1} \neq j$, implying $i \neq 1$. Let $\mathbf{1}_{j=0}\operatorname{lfn}(\mathfrak{b}, i+1) + \mathbf{1}_{j=1}\operatorname{lufn}(\mathfrak{b}, i+1) = 0$, implying $\ell^{\mathfrak{b}}(i, j, k) \geq \mathfrak{b}$ by Lemma 6.2.3. Since $\phi^{\sigma_{\mathfrak{b}}}(d_{i,j,k}, F_{i,j}) \leq \mathfrak{m}$, this implies $\phi^{\sigma}(e) = \phi^{\sigma_{\mathfrak{b}}}(e) = \mathfrak{m}$ independent of k since $\mathfrak{b}+1$ is odd. Note that this implies $\sigma_{\mathfrak{b}}(d_{i,j,*}) \neq F_{i,j}$ by Property $(\text{OR1})_{i,j,*}$. Consequently, both $(d_{i,j,0}, F_{i,j})$ and $(d_{i,j,1}, F_{i,j})$ could have been applied during phase 1. However, due to the tie-breaking rule, only $(d_{i,j,0}, F_{i,j})$ was applied during phase 1. It thus suffices to investigate $e := (d_{i,j,1}, F_{i,j})$. Since $\ell^{\mathfrak{b}+1}(i, j, 1) \geq \mathfrak{b}$ by Lemma 6.2.3 and $\phi^{\sigma}(e) = \phi^{\sigma_{\mathfrak{b}}}(e)$, we thus obtain

$$\phi^{\sigma}(e) = \left\lfloor \frac{\mathfrak{b}+1}{2} \right\rfloor = \left\lfloor \frac{(\mathfrak{b}+1)+1-k}{2} \right\rfloor = \min\left(\left\lfloor \frac{(\mathfrak{b}+1)+1-k}{2} \right\rfloor, \ell^{\mathfrak{b}+1}(i, j, k) \right),$$

hence choosing $t_{\mathfrak{b}+1} = 0$ yields the correct description of the occurrence record.

Let $\mathbf{1}_{j=0}\operatorname{lfn}(\mathfrak{b}, i+1) + \mathbf{1}_{j=1}\operatorname{lufn}(\mathfrak{b}, i+1) \neq 0$. Then, by Lemma 6.2.3 and since $\mathfrak{b}_1 = 0$,

$$\ell^{\mathfrak{b}}(i, j, k) \leq \left\lceil \frac{\mathfrak{b} - 2^{i-1} + 2^{i-1} - 1 - 1 + 1 - k}{2} \right\rceil = \mathfrak{m} - k. \tag{A.16}$$

If $\sigma_{\mathfrak{b}}(d_{i,j,k}) \neq F_{i,j}$ and $\phi^{\sigma_{\mathfrak{b}}}(e) < \mathfrak{m}$, then $(d_{i,j,k}, F_{i,j})$ was applied in phase 1 by Corollary 6.3.3. By Property $(\text{OR1})_{i,j,k}$, $\sigma_{\mathfrak{b}}(d_{i,j,k}) = F_{i,j}$ and $\phi^{\sigma_{\mathfrak{b}}}(e) = \mathfrak{m}$ is not possible. Consider the case $\sigma_{\mathfrak{b}}(d_{i,j,k}) = F_{i,j}$ and $\phi^{\sigma_{\mathfrak{b}}}(d_{i,j,k}, F_{i,j}) < \mathfrak{m}$. We show that e was then applied during phase 5. By Corollary 6.3.17, we need to show $i > u$ and $i < m$. If $i \neq u$, the first statement follows as $\mathfrak{b}_i = 0$ and $(\mathfrak{b}+1)_i = 0$. If $i = u$, Inequality (A.16) is tight, contradicting $\sigma_{\mathfrak{b}}(d_{i,j,k}) = F_{i,j}$ since we then had $\phi^{\sigma_{\mathfrak{b}}}(d_{i,j,k}, F_{i,j}) = \ell^{\mathfrak{b}}(i, j, k) + 1 \geq \mathfrak{m}$. Assume, for the sake of contradiction, $i > m$. Then $\mathfrak{b} < 2^{i-1}$, hence $\operatorname{lfn}(\mathfrak{b}, i+1) = \operatorname{lufn}(\mathfrak{b}, i+1) = 0$. Consequently, by Lemma 6.2.3, $\ell^{\mathfrak{b}}(i, j, k) \geq \mathfrak{b}$, contradicting Properties $(\text{OR1})_{i,j,k}$ and $(\text{OR2})_{i,j,k}$. Therefore, e was applied during phase 5 and we do not consider it here. It thus suffices to consider the case $\sigma_{\mathfrak{b}}(d_{i,j,k}) \neq F_{i,j}$ and $\phi^{\sigma_{\mathfrak{b}}}(e) = \mathfrak{m}$. It then holds that $\phi^{\sigma_{\mathfrak{b}}}(e) = \ell^{\mathfrak{b}}(i, j, k) + t_{\mathfrak{b}}$ for some feasible $t_{\mathfrak{b}}$ due to Inequality (A.16). This implies $\phi^{\sigma}(e) = \ell^{\mathfrak{b}}(i, j, 1) + 1$ if $k = 1$, contradicting $\sigma_{\mathfrak{b}}(d_{i,j,1}) \neq F_{i,j}$. Thus $k = 0$ and $\phi^{\sigma_{\mathfrak{b}}}(e) = \ell^{\mathfrak{b}}(i, j, 0)$. But this implies that Inequality (A.16) is an equality. Consequently, $i = \ell(\mathfrak{b}+2)$. Also, $\mathbf{1}_{j=0}\operatorname{lfn}(\mathfrak{b}, i+1) + \mathbf{1}_{j=1}\operatorname{lufn}(\mathfrak{b}, i+1) \neq 0$ by assumption, hence $\mathfrak{b}+2$ is not a power of two. Hence, by Property (OR3)$_{i,j,0}$, the parameter $t_{\mathfrak{b}+1} = -1$ is feasible. Since $\ell^{\mathfrak{b}+1}(i,j,0) = \ell^{\mathfrak{b}}(i,j,0) + 1$ by Lemma 6.2.4 and $\mathfrak{m} = \lfloor(\mathfrak{b}+1+1-0)/2\rfloor - 1$, this parameter describes the occurrence record with respect to σ. Hence,

$$\phi^{\sigma}(e) = \min\left(\left\lfloor\frac{\mathfrak{b}+1+1-0}{2}\right\rfloor, \ell^{\mathfrak{b}+1}(i,j,0) + t_{\mathfrak{b}+1}\right) < \left\lfloor\frac{\mathfrak{b}+1+1-0}{2}\right\rfloor$$

for $t_{\mathfrak{b}+1} = -1$, so the occurrence record is correctly described by Table 5.6. This concludes the case $(\mathfrak{b}+1)_i = 0 \wedge (\mathfrak{b}+1)_{i+1} \neq j$.

Consider the case $(\mathfrak{b}+1)_i = 1$ and $(\mathfrak{b}+1)_{i+1} \neq j$ next, implying $i = \nu = 1$. If $\mathbf{1}_{j=0}\text{lfn}(\mathfrak{b},2) + \mathbf{1}_{j=1}\text{lufn}(\mathfrak{b},2) = 0$, we can use the same arguments used for the case $(\mathfrak{b}+1)_i = 0 \wedge (\mathfrak{b}+1)_{i+1} \neq j$. Hence let $\mathbf{1}_{j=0}\text{lfn}(\mathfrak{b},2) + \mathbf{1}_{j=1}\text{lufn}(\mathfrak{b},2) \neq 0$. Then by Lemma 6.2.3, $\ell^{\mathfrak{b}}(1,j,k) = \mathfrak{m}$ for both choices of $k \in \{0,1\}$. Since the parameter $t_{\mathfrak{b}} = -1$ is not feasible as $\mathfrak{b}$ is even and choosing $t_{\mathfrak{b}} = 1$ violates Lemma 6.2.5, it thus holds that $\phi^{\sigma}(e) = \ell^{\mathfrak{b}}(i,j,k) = \lfloor(\mathfrak{b}+1)/2\rfloor$ for both $k \in \{0,1\}$. In particular, $\sigma_{\mathfrak{b}}(d_{i,j,k}) \neq F_{i,j}$ for both $k \in \{0,1\}$. Hence, by the tie-breaking rule, $(d_{i,j,0}, F_{i,j})$ is applied during phase 1. Consequently, $(d_{i,j,1}, F_{i,j})$ is not applied during phase 1 and the same arguments used previously can be used to show that choosing $t_{\mathfrak{b}+1} = 0$ is feasible and implies the desired characterization. This concludes the case $(\mathfrak{b}+1)_i = 1 \wedge (\mathfrak{b}+1)_{i+1} \neq j$. Since only the first bit switches during $\sigma_{\mathfrak{b}} \to \sigma$, this also concludes the case $\mathfrak{b}_i = 0 \wedge \mathfrak{b}_{i+1} \neq j$.

3. Let $\mathfrak{b}_i = 0 \wedge \mathfrak{b}_{i+1} = j$. Since only the first bit switches, it suffices to consider $i \neq 1$ and $(\mathfrak{b}+1)_i = 0 \wedge (\mathfrak{b}+1)_{i+1} = j$. As before, if $\mathbf{1}_{j=0}\text{lfn}(\mathfrak{b},i+1) + \mathbf{1}_{j=1}\text{lufn}(\mathfrak{b},i+1) = 0$, then the statement follows directly. Hence assume $\mathbf{1}_{j=0}\text{lfn}(\mathfrak{b},i+1) + \mathbf{1}_{j=1}\text{lufn}(\mathfrak{b},i+1) \neq 0$. Then, $\ell^{\mathfrak{b}}(i,j,k) \geq \lceil(\mathfrak{b}+2+1-k)/2\rceil \geq \mathfrak{m}+1$ by Lemma 6.2.3. Since the parameter $t_{\mathfrak{b}} = -1$ is not feasible, this implies $\phi^{\sigma_{\mathfrak{b}}}(d_{i,j,*}, F_{i,j}) = \mathfrak{m}$ and $\sigma_{\mathfrak{b}}(d_{i,j,*}) \neq F_{i,j}$. By the same arguments used before, $(d_{i,j,1}, F_{i,j})$ is not applied and its occurrence record with respect to σ is described by Table 5.6 when interpreted for $\mathfrak{b}+1$.
4. Finally, consider the case $\mathfrak{b}_i = 1 \wedge \mathfrak{b}_{i+1} \neq j$. Since only the first bit switches, this implies $i \neq 1$ and $(\mathfrak{b}+1)_i = 1$ and $(\mathfrak{b}+1)_{i+1} \neq j$. It is easy to see that this enables us to use the same arguments used previously. □

Claim 23. Equation (6.10), $e = (g_i, F_{i,j}) \in \mathfrak{A}_{\sigma_{\mathfrak{b}-1}}^{\sigma_{\mathfrak{b}}}$ and $\mathfrak{b}_i = 0$ either imply Inequality (6.11) directly or that exactly one of the cycle edges of $F_{i,j}$ is switched during $\sigma_{\mathfrak{b}-1} \to \sigma_{\mathfrak{b}}$.

Proof. By Equation (6.10), at most one of the two edges of the cycle center $F_{i,j}$ is switched. We distinguish the following cases.

1. Let $F_{i,j}$ be open for $\sigma_{\mathfrak{b}-1}$. Then, one of the two cycle edges is applied during phase 1 of $\sigma_{\mathfrak{b}-1} \to \sigma_{\mathfrak{b}}$ since no cycle center is open at the end of phase 1 by Corollary 6.3.4 resp. 6.3.7.
2. Let $F_{i,j}$ be closed for $\sigma_{\mathfrak{b}-1}$. Then, since $\mathfrak{b}_i = 0$, either $(\mathfrak{b}-1)_i = 1 \wedge (\mathfrak{b}-1)_{i+1} = j$ or $(\mathfrak{b}-1)_i = 0 \wedge (\mathfrak{b}-1)_{i+1} \neq j$. Consider the first case. This case can only happen if $i < \ell(\mathfrak{b})$, additionally implying $j \neq \mathfrak{b}_{i+1}$. In addition, we then have

$$\phi^{\sigma_{\mathfrak{b}-1}}(d_{i,j,k}, F_{i,j}) = \left\lfloor\frac{\text{lfn}(\mathfrak{b}-1, i, \{(i+1,j)\}) - k}{2}\right\rfloor + 1$$

$$= \left\lfloor \frac{\mathfrak{b} - 1 - 2^{i-1} + 1 - k}{2} \right\rfloor + 1$$
$$= \left\lfloor \frac{\mathfrak{b} - 2^{i-1} - k}{2} \right\rfloor + 1 = \left\lfloor \frac{\mathfrak{b} - 2^{i-1} + 2 - k}{2} \right\rfloor.$$

for $k \in \{0, 1\}$. For $i = 2$, this implies

$$\phi^{\sigma_{\mathfrak{b}-1}}(d_{i,j,0}, F_{i,j}) = \left\lfloor \frac{\mathfrak{b} - 2 + 2 - 0}{2} \right\rfloor = \left\lfloor \frac{\mathfrak{b}}{2} \right\rfloor$$
$$\phi^{\sigma_{\mathfrak{b}-1}}(d_{i,j,1}, F_{i,j}) = \left\lfloor \frac{\mathfrak{b} - 2 + 2 - 1}{2} \right\rfloor = \left\lfloor \frac{\mathfrak{b} - 1}{2} \right\rfloor \geq \left\lfloor \frac{\mathfrak{b}}{2} \right\rfloor - 1$$

and thus the statement. For $i \geq 3$ it is easy to verify that this implies that the occurrence record of at least one of the cycle edges is so low that the corresponding edge is applied as improving switch during phase 5 of $\sigma_{\mathfrak{b}-1} \to \sigma_{\mathfrak{b}}$. This concludes the first case, hence assume $(\mathfrak{b}-1)_i = 0 \wedge (\mathfrak{b}-1)_{i+1} \neq j$. Then, $F_{i,j}$ being closed implies $\phi^{\sigma_{\mathfrak{b}-1}}(d_{i,j,k}, F_{i,j}) = \ell^{\mathfrak{b}-1}(i,j,k) + 1 \leq \lfloor \mathfrak{b}/2 \rfloor - 1$ and $\sigma_{\mathfrak{b}-1}(d_{i,j,k}) = F_{i,j}$ for both $k \in \{0,1\}$. If this inequality is met with equality for both k, then the statement follows as the occurrence record of the edges is sufficiently high. If the inequality is strict for at least one k, then the corresponding switch is applied during phase 5 of the transition $\sigma_{\mathfrak{b}-1} \to \sigma_{\mathfrak{b}}$.

3. Let $F_{i,j}$ be halfopen for $\sigma_{\mathfrak{b}-1}$. Then, for some $k \in \{0,1\}$, $\sigma_{\mathfrak{b}-1}(d_{i,j,k}) = F_{i,j}$ as well as $\phi^{\sigma_{\mathfrak{b}-1}}(d_{i,j,k}, F_{i,j}) = \ell^{\mathfrak{b}-1}(i,j,k) + 1 \leq \lfloor \mathfrak{b}/2 \rfloor - 1$. Furthermore, $\sigma_{\mathfrak{b}-1}(d_{i,j,1-k}) = F_{i,j}$ and $\phi^{\sigma_{\mathfrak{b}-1}}(d_{i,j,k}, F_{i,j}) \in \{\lfloor \mathfrak{b}/2 \rfloor - 1, \lfloor \mathfrak{b}/2 \rfloor\}$. If $\phi^{\sigma_{\mathfrak{b}-1}}(d_{i,j,k}, F_{i,j}) = \lfloor \mathfrak{b}/2 \rfloor - 1$, then the edge is applied as an improving switch and the statement follows. Hence assume $\phi^{\sigma_{\mathfrak{b}-1}}(d_{i,j,k}, F_{i,j}) = \lfloor \mathfrak{b}/2 \rfloor$. Then, due to $\sigma_{\mathfrak{b}-1}(d_{i,j,k}) \neq F_{i,j}$, this implies $\ell^{\mathfrak{b}-1}(i,j,1-k) + 1 \neq \lfloor \mathfrak{b}/2 \rfloor$. However, $\ell^{\mathfrak{b}-1}(i,j,k) \leq \lfloor \mathfrak{b}/2 \rfloor - 2$ since it then holds that $\phi^{\sigma_{\mathfrak{b}-1}}(d_{i,j,k}, F_{i,j}) = \ell^{\mathfrak{b}-1}(i,j,k) + 1 \leq \lfloor \mathfrak{b}/2 \rfloor - 1$. But, since $\ell^{\mathfrak{b}-1}(i,j,k)$ and $\ell^{\mathfrak{b}-1}(i,j,1-k)$ differ by at most one, this implies $\ell^{\mathfrak{b}-1}(i,j,1-k) \leq \lfloor \mathfrak{b}/2 \rfloor - 1$. But this is a contradiction to $\phi^{\sigma_{\mathfrak{b}-1}}(d_{i,j,1-k}, F_{i,j}) = \lfloor \mathfrak{b}/2 \rfloor$. □

Claim 24. Assume that Equation (6.10), $e = (g_i, F_{i,j}) \in \mathfrak{A}_{\sigma_{\mathfrak{b}-1}}^{\sigma_{\mathfrak{b}}}$, $\mathfrak{b}_i = 0$ hold and that exactly one of the two cycle edges $(d_{i,j,0}, F_{i,j}), (d_{i,j,1}, F_{i,j})$ is applied during $\sigma_{\mathfrak{b}-1} \to \sigma_{\mathfrak{b}}$. Then $(\mathfrak{b}-1)_i = 0$.

Proof. Let, for the sake of contradiction, $(\mathfrak{b}-1)_i = 1$. Then, as $\mathfrak{b}_i = 0$, we have $i < \ell(\mathfrak{b})$ and consequently $(\mathfrak{b}-1)_{i+1} \neq \mathfrak{b}_{i+1}$. It further implies that $\mathfrak{b}$ is even. Then, since $e \in \mathfrak{A}_{\sigma_{\mathfrak{b}-1}}^{\sigma_{\mathfrak{b}}}$ by assumption, the switch $e = (g_i, F_{i,j})$ was applied during phase 5 of $\sigma_{\mathfrak{b}-1} \to \sigma_{\mathfrak{b}}$. This implies $j = 0$ if $G_n = S_n$ resp. $j = \mathfrak{b}_{i+1} = 1 - (\mathfrak{b}-1)_{i+1}$ if $G_n = M_n$. Consider the case $G_n = M_n$ first. Then, since $(\mathfrak{b}-1)_i = 1 \wedge j = 1 - (\mathfrak{b}-1)_{i+1}$ imply $\ell^{\mathfrak{b}-1}(i,j,k) \geq \lfloor (\mathfrak{b}-k)/2 \rfloor + 1$ for $k \in \{0,1\}$ by Lemma 6.2.3, we obtain $\phi^{\sigma_{\mathfrak{b}-1}}(d_{i,j,k}, F_{i,j}) = \lfloor (\mathfrak{b}-k)/2 \rfloor$.

In addition, since $\phi^{\sigma_{\mathfrak{b}-1}}(d_{i,j,k}, F_{i,j}) \neq \ell^{\mathfrak{b}-1}(i,j,k)+1$ for both $k \in \{0,1\}$, $F_{i,j}$ is then open with respect to $\sigma_{\mathfrak{b}-1}$ by Property (OR2)$_{i,j,*}$. This implies that $(d_{i,j,1}, F_{i,1})$ is applied during phase 1 of $\sigma_{\mathfrak{b}-1} \to \sigma_{\mathfrak{b}}$. But then $\min_{k \in \{0,1\}} \phi^{\sigma_{\mathfrak{b}-1}}(d_{i,j,k}, F_{i,j}) < \min_{k \in \{0,1\}} \phi^{\sigma_{\mathfrak{b}}}(d_{i,j,k}, F_{i,j})$, contradicting Equation (6.10). Now consider the case $G_n = S_n$. If $j = 0 = 1 - (\mathfrak{b}-1)_{i+1}$, then the statement follows by the same arguments. This is the case if and only if $i < \ell(\mathfrak{b})-1$, so let $i = \ell(\mathfrak{b}) - 1$. By Definition 5.1.2, this implies $\sigma_{\mathfrak{b}-1}(g_i) = F_{i,0}$. Since $F_{i,0}$ is then closed during phase 1 of the transition $\sigma_{\mathfrak{b}-1} \to \sigma_{\mathfrak{b}}$ and since $(g_i, F_{i,1})$ cannot be applied during phase 5 in S_n, this is a contradiction. □

Claim 25. Assume that Equation (6.10), $e = (g_i, F_{i,j}) \in \mathfrak{A}_{\sigma_{\mathfrak{b}-1}}^{\sigma_{\mathfrak{b}}}, \mathfrak{b}_i = 0$ hold and that exactly one of the two cycle edges $(d_{i,j,0}, F_{i,j}), (d_{i,j,1}, F_{i,j})$ is applied during $\sigma_{\mathfrak{b}-1} \to \sigma_{\mathfrak{b}}$. If $(g_i, F_{i,j})$ is applied during phase 1 of $\sigma_{\mathfrak{b}-1} \to \sigma_{\mathfrak{b}}$, then

1. $\mathfrak{b}$ is even and $i \neq 2$,
2. $\sum(\mathfrak{b}, i) = 2^{i-1} - 2$ and
3. if $(g_i, F_{i,j}) \in \mathfrak{A}_{\sigma_{\mathfrak{b}-2}}^{\sigma_{\mathfrak{b}-1}}$, then $(g_i, F_{i,j}) \notin \mathfrak{A}_{\sigma_{\mathfrak{b}-3}}^{\sigma_{\mathfrak{b}-2}}$.

Proof. By Corollary 6.3.2, there is some index $k \in \{0,1\}$ such that $\sigma_{\mathfrak{b}-1}(d_{i,j,k}) = F_{i,j}$ as well as $\phi^{\sigma_{\mathfrak{b}-1}}(d_{i,j,k}, F_{i,j}) = \ell^{\mathfrak{b}-1}(i,j,k) + 1 \leq \lfloor \mathfrak{b}/2 \rfloor - 1$. Moreover, it holds that $\phi^{\sigma_{\mathfrak{b}-1}}(d_{i,j,1-k}, F_{i,j}) = \lfloor \mathfrak{b}/2 \rfloor - 1$ and $(d_{i,j,1-k}, F_{i,j})$ is applied during phase 1 of $\sigma_{\mathfrak{b}-1} \to \sigma_{\mathfrak{b}}$. By Equation (6.10), $(d_{i,j,k}, F_{i,j})$ is not applied as improving switch during $\sigma_{\mathfrak{b}-1} \to \sigma_{\mathfrak{b}}$. It is easy to verify that this implies that we need to have $\phi^{\sigma_{\mathfrak{b}-1}}(d_{i,j,k}, F_{i,j}) = \lfloor \mathfrak{b}/2 \rfloor - 1$ as well as $\ell(\mathfrak{b}) > 1$. This implies that $\mathfrak{b}$ is even and, since $(\mathfrak{b}-1)_i = 0$ and that we cannot have $i = 2$. In particular, the first statement holds and we have $\ell^{\mathfrak{b}-1}(i,j,k) = \lfloor \mathfrak{b}/2 \rfloor - 2$. Since $\ell^{\mathfrak{b}-1}(i,j,1) \leq \ell^{\mathfrak{b}-1}(i,j,0)$, this implies $k = 1$ as $\phi^{\sigma_{\mathfrak{b}-1}}(d_{i,j,1-k}, F_{i,j}) \leq \lfloor \mathfrak{b}/2 \rfloor - 2$ held otherwise. This implies $\lfloor (\mathfrak{b}-2)/2 \rfloor = \lfloor (\mathfrak{b} - 2^{i-1} + \sum(\mathfrak{b}-1, i) + 1)/2 \rfloor$ as

$$\left\lfloor \frac{\mathfrak{b}-2}{2} \right\rfloor = \left\lfloor \frac{\mathfrak{b}}{2} \right\rfloor - 1 = \ell^{\mathfrak{b}-1}(i,j,1) + 1 = \left\lceil \frac{\mathfrak{b} - 1 - 2^{i-1} + \sum(\mathfrak{b}-1, i)}{2} \right\rceil + 1$$
$$= \left\lfloor \frac{\mathfrak{b} - 2^{i-1} + \sum(\mathfrak{b}-1, i) + 2}{2} \right\rfloor = \left\lfloor \frac{\mathfrak{b} - 2^{i-1} + \sum(\mathfrak{b}-1, i) + 1}{2} \right\rfloor.$$

where the last equality follows since $i \geq 3$ and since $\sum(\mathfrak{b}-1, i)$ is odd as $\mathfrak{b}$ is even. Since $\mathfrak{b}$ is even and $i \neq 1$, the nominators on both sides are then even. But this implies that the nominators have to be equal. Since $\sum(\mathfrak{b}, i) = \sum(\mathfrak{b}-1, i) + 1$, this implies $\sum(\mathfrak{b}, i) = 2^{i-1} - 2$. More precisely,

$$\sum(\mathfrak{b}, i) = \sum(\mathfrak{b}-1, i) + 1 = \mathfrak{b} - 2 - \mathfrak{b} + 2^{i-1} - 1 + 1 = 2^{i-1} - 2,$$

implying the second statement.

We now prove that it is not possible that $(g_i, F_{i,j})$ was applied during both $\sigma_{\mathfrak{b}-2} \to \sigma_{\mathfrak{b}-1}$ and $\sigma_{\mathfrak{b}-3} \to \sigma_{\mathfrak{b}-2}$, implying (c). First note that $i \geq 3$ implies $\mathfrak{b} - 3 \geq \bar{\mathfrak{b}}$, i.e., $\mathfrak{b} - 3$ is indeed "contributing" to $\mathfrak{N}(\bar{\mathfrak{b}}, \mathfrak{b}-1)$. Since the statement follows if $(g_i, F_{i,j}) \notin \mathfrak{A}_{\sigma_{\mathfrak{b}-2}}^{\sigma_{\mathfrak{b}-1}}$, assume that this was the case. Since we assume that $(g_i, F_{i,j})$ is applied during phase 1 of $\sigma_{\mathfrak{b}-1} \to \sigma_{\mathfrak{b}}$ and since $i \neq \ell(\mathfrak{b})$ and $i \neq \ell(\mathfrak{b}-1)$, it is not possible that $(g_i, F_{i,j})$ was applied during phase 5 of $\sigma_{\mathfrak{b}-2} \to \sigma_{\mathfrak{b}-1}$. It was thus applied during phase 1 of $\sigma_{\mathfrak{b}-2} \to \sigma_{\mathfrak{b}-1}$. Since $\mathfrak{b}$ is even, this implies $\phi^{\sigma_{\mathfrak{b}-2}}(d_{i,j,k}, F_{i,j}) \leq \lfloor (\mathfrak{b}-2+1)/2 \rfloor - 1 = \lfloor \mathfrak{b}/2 \rfloor - 2$ for both $k \in \{0,1\}$. Now, for the sake of contradiction, assume $(g_i, F_{i,j}) \in \mathfrak{A}_{\sigma_{\mathfrak{b}-3}}^{\sigma_{\mathfrak{b}-2}}$. By the same argument used previously, it is not possible that this switch was applied during phase 5 of that transition. It thus needs to be applied during phase 1. This is an immediate contradiction if $(\mathfrak{b}-3)_i = 1$. Thus assume $(\mathfrak{b}-3)_i = 0$. Then, since $(g_i, F_{i,j})$ was applied during phase 1 of $\sigma_{\mathfrak{b}-3} \to \sigma_{\mathfrak{b}-2}$, there is some $k \in \{0,1\}$ such that $\sigma_{\mathfrak{b}-3}(d_{i,j,1-k}) \neq F_{i,j}$ as well as $\phi^{\sigma_{\mathfrak{b}-3}}(d_{i,j,1-k}, F_{i,j}) = \lfloor(\mathfrak{b}-3+1)/2\rfloor - 1 = \lfloor\mathfrak{b}/2\rfloor - 2$. Since this switch is then applied during phase 1, this implies $\phi^{\sigma_{\mathfrak{b}-2}}(d_{i,j,1-k}, F_{i,j}) = \lfloor\mathfrak{b}/2\rfloor - 1$ which is a contradiction. □

Claim 26. Assume that Equation (6.10), $e = (g_i, F_{i,j}) \in \mathfrak{A}_{\sigma_{\mathfrak{b}-1}}^{\sigma_{\mathfrak{b}}}, \mathfrak{b}_i = 0$ hold and that exactly one of the two cycle edges $(d_{i,j,0}, F_{i,j}), (d_{i,j,1}, F_{i,j})$ is applied during $\sigma_{\mathfrak{b}-1} \to \sigma_{\mathfrak{b}}$. If $(g_i, F_{i,j})$ is applied during phase 5 of $\sigma_{\mathfrak{b}-1} \to \sigma_{\mathfrak{b}}$ and $\bar{\sigma}_{\mathfrak{b}-1}(g_i) = 1-j$, then

1. $i \neq 2$,
2. $\sum(\mathfrak{b}, i) = 2^{i-1} - 2$ and
3. if $(g_i, F_{i,j}) \in \mathfrak{A}_{\sigma_{\mathfrak{b}-2}}^{\sigma_{\mathfrak{b}-1}}$, then$(g_i, F_{i,j}) \notin \mathfrak{A}_{\sigma_{\mathfrak{b}-3}}^{\sigma_{\mathfrak{b}-2}}$.

Proof. We remind here that we have $\mathfrak{b}_i = 0$ and $(\mathfrak{b}-1)_i = 0$, implying $\mathfrak{b}_{i+1} = (\mathfrak{b}-1)_{i+1}$. By the conditions describing under which circumstances an improving switch $(g_i, F_{i,j})$ becomes improving in phase 5 (see Lemma 6.2.32) and the assumption, we have $j = 0$ if $G_n = S_n$ resp. $j = \beta_{i+1}^{\sigma} = \mathfrak{b}_{i+1} = (\mathfrak{b}-1)_{i+1}$ if $G_n = M_n$. By Definition 5.1.2 resp. 5.2.1, $\sigma_{\mathfrak{b}-1}(g_i) = F_{i,1-j}$ then implies that $F_{i,1-j}$ has to be closed with respect to $\sigma_{\mathfrak{b}-1}$. As $(\mathfrak{b}-1)_i = 0$, it also implies that $1-j = 1-(\mathfrak{b}-1)_{i+1}$ needs to hold for both M_n and S_n But this implies that

$$\phi^{\sigma_{\mathfrak{b}-1}}(d_{i,1-j,k}, F_{i,1-j}) = \ell^{\mathfrak{b}-1}(i, 1-j, k) + 1 \leq \left\lfloor \frac{\mathfrak{b}}{2} \right\rfloor - 1$$

for both $k \in \{0,1\}$. We show this implies $i \neq 2$ as assuming $\phi^{\sigma_{\mathfrak{b}-1}}(d_{2,1-j,0}, F_{2,1-j}) = \ell^{\mathfrak{b}-1}(2, 1-j, 0) + 1$ contradicts $\phi^{\sigma_{\mathfrak{b}-1}}(d_{2,1-j,0}, F_{2,1-j}) \leq \lfloor\mathfrak{b}/2\rfloor - 1$.

Assume $i = 2$. Then, since $(\mathfrak{b}-1)_2 = \mathfrak{b}_2 = 0$, we need to have $\mathfrak{b}_1 = 1$, so $\mathfrak{b}$ is odd. Hence, $\ell^{\mathfrak{b}-1}(2, 1-j, 0) + 1 = \lfloor(\mathfrak{b}-2+1)/2\rfloor = \lfloor\mathfrak{b}/2\rfloor$. Also, $\lfloor(\mathfrak{b}-2)/2\rfloor = \lfloor\mathfrak{b}/2\rfloor - 1$ due to the parity of $\mathfrak{b}$ and thus $\lfloor(\mathfrak{b}-2)/2\rfloor \geq \lfloor(\mathfrak{b}-2^{i-1}+\sum(\mathfrak{b}-1, i)+1)/2\rfloor$, which contradicts the previously given inequality. Using the same arguments used when proving Claim 25, this implies $\sum(\mathfrak{b}, i) \leq 2^{i-1} - 2$.

It remains to prove that $(g_i, F_{i,j}) \in \mathfrak{A}_{\sigma_{\mathfrak{b}-2}}^{\sigma_{\mathfrak{b}-1}}$ implies $(g_i, F_{i,j}) \notin \mathfrak{A}_{\sigma_{\mathfrak{b}-3}}^{\sigma_{\mathfrak{b}-2}}$. As we have $\bar{\sigma}_{\mathfrak{b}-1}(g_i) = 1-j$, the switch $(g_i, F_{i,j})$ cannot have been applied during phase 5 of $\sigma_{\mathfrak{b}-2} \to \sigma_{\mathfrak{b}-1}$. It was thus applied during phase 1 of $\sigma_{\mathfrak{b}-2} \to \sigma_{\mathfrak{b}-1}$. Assume $\mathfrak{b}-2 = \bar{\mathfrak{b}}$ which can happen if $\mathfrak{b}$ is odd. But then, $(g_i, F_{i,j})$ was not applied during phase 1 of $\sigma_{\mathfrak{b}-2} \to \sigma_{\mathfrak{b}-1}$ as we then have $(\mathfrak{b}-2)_i = 1$. Thus assume $\bar{\mathfrak{b}} \leq \mathfrak{b}-3$ and that $(g_i, F_{i,j})$ was applied during phase 1 of $\sigma_{\mathfrak{b}-2} \to \sigma_{\mathfrak{b}-1}$. Towards a contradiction, assume that $(g_i, F_{i,j})$ is applied during $\sigma_{\mathfrak{b}-3} \to \sigma_{\mathfrak{b}-2}$. Since we apply the same switch during phase 1 of $\sigma_{\mathfrak{b}-2} \to \sigma_{\mathfrak{b}-1}$ and $i \neq \ell(\mathfrak{b}-1)$ due to $(\mathfrak{b}-1)_i = 0$, the switch must have been applied during phase 1 of $\sigma_{\mathfrak{b}-3} \to \sigma_{\mathfrak{b}-2}$. If this was not the case, i.e., if it was applied during phase 5, we had $\sigma_{\mathfrak{b}-2}(g_i) = F_{i,j}$ and it would not be possible to apply $(g_i, F_{i,j})$. Note that this implies $(\mathfrak{b}-3)_i = 0$ and in particular $(\mathfrak{b}-3)_{i+1} = j$. Then, since $(g_i, F_{i,j})$ is applied during both $\sigma_{\mathfrak{b}-3} \to \sigma_{\mathfrak{b}-2}$ and $\sigma_{\mathfrak{b}-2} \to \sigma_{\mathfrak{b}-1}$, the improving switch $(g_i, F_{i,1-j})$ has to be applied in between. This switch can only be applied during phase 1 of $\sigma_{\mathfrak{b}-2} \to \sigma_{\mathfrak{b}-1}$ since $j = (\mathfrak{b}-3)_{i+1} = \mathfrak{b}_{i+1}$ and since $(g_i, F_{i,j})$ was applied during phase 5 of $\sigma_{\mathfrak{b}-1} \to \sigma_{\mathfrak{b}}$. But this is a contradiction as we apply $(g_i, F_{i,j})$ during phase 1 of that transition and $i \neq \ell(\mathfrak{b}-2)$. □

Index

Curriculum Vitae

Alexander Vincent Hopp

10/2020	Promotion am Fachbereich Mathematik Technische Universität Darmstadt
11/2019 – 10/2020	Wissenschaftlicher Mitarbeiter am Fachbereich Mathematik Technische Universität Darmstadt
11/2016 – 10/2019	Stipendiat an der Graduiertenschule "Computational Engineering" Technische Universität Darmstadt
09/2016	M.Sc. in Mathematik Technische Universität Berlin
10/2014	B.Sc. in Mathematik mit Nebenfach Philosophie Technische Universität Berlin
10/2011 – 09/2016	Studium der Mathematik Technische Universität Berlin